JOURNAL OF CHROMATOGRAPHY LIBRARY — volume 69A

chromatography
6th edition

fundamentals and applications of chromatography and related differential migration methods

part A: fundamentals and techniques

JOURNAL OF CHROMATOGRAPHY LIBRARY — volume 69A

chromatography
6th edition

fundamentals and applications of chromatography and related differential migration methods

part A: fundamentals and techniques

edited by

E. Heftmann

2004

ELSEVIER

**Amsterdam – Boston – Heidelberg – London – New York – Oxford – Paris
San Diego – San Francisco – Singapore – Sydney – Tokyo**

ELSEVIER B.V. ELSEVIER Inc. ELSEVIER Ltd ELSEVIER Ltd
Sara Burgerhartstraat 25 525 B Street, Suite 1900 The Boulevard, Langford Lane 84 Theobalds Road
P.O. Box 211, 1000 AE Amsterdam San Diego, CA 92101-4495 Kidlington, Oxford OX5 1GB London WC1X 8RR
The Netherlands USA UK UK

First edition 2004

Library of Congress Cataloging in Publication Data
A catalog record is available from the Library of Congress.

British Library Cataloguing in Publication Data
A catalogue record is available from the British Library.

ISBN: 0-444-51107-5
Set ISBN: 0-444-51106-7
ISSN: 0-301-4770

♾ The paper used in this publication meets the requirements of ANSI/NISO Z39.48-1992 (Permanence of Paper). Printed in The Netherlands.

TO BIBI

CONTENTS

List of Authors XI

List of Abbreviations XIII

List of Italic Symbols XXIX

List of Greek Symbols XXXIII

Foreword (S. Hjertén) XXXV

Preface (E. Heftmann) XXXIX

1 Theory of chromatography (V.L. McGuffin) 1
    1.1 Introduction 2
    1.2 Resolution 2
    1.3 Principles of solute-zone dispersion 9
    1.4 Principles of solute-zone separation 38
    1.5 Optimization of chromatographic separations 72
    1.6 Summary 88
    References 88

2 Column liquid chromatography (R.M. Smith) 95
    2.1 Introduction 96
    2.2 Equipment for liquid-phase separations 98
    2.3 Stationary-phase materials 104
    2.4 Detection 115
    2.5 Separation methods 121
    2.6 Liquid chromatography in sample preparation 128
    2.7 High-temperature separations 129
    2.8 Miniaturization 129
    2.9 Large-scale separations 130
    References 131

3 Affinity chromatography (F.B. Anspach) 139
    3.1 Introduction 139
    3.2 Chromatographic operations 141

    3.3   Affinity interactions at solid interfaces 142
    3.4   Affinity ligands 153
    3.5   Summary 160
    References 161

4   Ion chromatography (C.A. Lucy and P. Hatsis) 171
    4.1   Introduction 172
    4.2   Instrumentation 173
    4.3   Selectivity in ion chromatography 186
    4.4   Sample preparation 203
    4.5   Future directions 207
    Acknowledgments 208
    References 209

5   Size-exclusion chromatography (J. Silberring, M. Kowalczuk, J. Bergquist,
   A. Kraj, P. Suder, T. Dylag, M. Smoluch, J.-P. Chervet, and J. Ekman). . . . 213
    5.1   Introduction 213
    5.2   Theory 215
    5.3   Columns 217
    5.4   Mobile phase 219
    5.5   Detectors 222
    5.6   Calibration 228
    5.7   Applications 230
    Acknowledgments 241
    References 241

6   Planar chromatography (Sz. Nyiredy) 253
    6.1   Introduction 254
    6.2   Classification of planar chromatographic techniques 255
    6.3   Principles of planar chromatographic methods 257
    6.4   Principal factors in planar chromatography 258
    6.5   Instrumentation 273
    6.6   Qualitative and quantitative analysis 277
    6.7   Preparative planar chromatography 282
    6.8   Special planar chromatographic techniques 285
    6.9   Comparison of various planar chromatographic techniques 287
    6.10 Trends in planar chromatography 287
    References 291

7   Electrokinetic chromatography (E. Kenndler and A. Rizzi) 297
    7.1   Introduction 298
    7.2   Electro-osmotic flow in open and packed capillaries 301

7.3   Electrochromatography with stationary phases   304

7.4   Electrokinetic chromatography with pseudo-stationary phases . . .   310

7.5   Electrically driven *vs.* pressure-driven chromatography   315

7.6   Conclusions   316

References   317

8   Gas chromatography (P.J. Marriott)   319

8.1   Introduction   320

8.2   Basic operating variables   321

8.3   Enhanced and fast separations   332

8.4   Sample introduction   340

8.5   Detection   348

References   364

9   Capillary zone electrophoresis (P.G. Righetti, A. Bossi, L. Castelletti, and B. Verzola)   369

9.1   Introduction   369

9.2   The instrument   370

9.3   The capillary   374

9.4   How to modulate the EOF   377

9.5   The buffers   382

9.6   Modes of operation   384

9.7   Micellar electrokinetic chromatography   390

9.8   Biosensors   394

9.9   Conclusions   394

Acknowledgments   395

References   395

10   Combined techniques (W.M.A. Niessen)   403

10.1   Introduction   404

10.2   Coupled columns   405

10.3   Chromatography–spectrometry   413

10.4   Liquid chromatography–mass spectrometry   418

References   425

11   Microfabricated analytical devices (A. Guttman and J. Khandurina)   431

11.1   Introduction   431

11.2   Capillary electrophoresis on microchips   432

11.3   Applications   437

11.4   System integration   451

11.5   Modeling by computational fluid dynamics   461

Acknowledgment 462
References 462

12   Instrumentation (R. Stevenson) 469
     12.1   High-performance liquid chromatography 470
     12.2   Gas chromatography 507
     12.3   Thin-layer chromatography 511
     12.4   Supercritical-fluid chromatography 512
     12.5   Flash chromatography 512
     12.6   Electrophoresis 513
     12.7   Electrochromatography 513
     12.8   Future developments 515
     Acknowledgments 515
     References 516

Subject Index A1

List of Authors

Friedrich Birger Anspach, Department of Applied Natural Sciences, Hamburg University of Applied Sciences, Lohbrügger Kirchstraße 65, D-21033 Hamburg, GERMANY, Fax: (49-040)428-912-785, e-mail: birger.anspach@rzbd.haw-hamburg.de

Jonas Bergquist, Department of Analytical Chemistry, Institute of Chemistry, Uppsala University, P. O. Box 599, SE-751 21 Uppsala, SWEDEN, Fax: (46-18)471-3692, e-mail: jonas.bergquist@kemi.uu.se

Alessandra Bossi, Department of Agricultural & Industrial Biotechnologies, Università di Verona, Strada le Grazie 15, I-37134 Verona, ITALY, Fax: (39-458)027-901, e-mail: bossi@sci@unvr.it

Laura Castelletti, Department of Agricultural & Industrial Biotechnologies, Università di Verona, Strada le Grazie 15, I-37134 Verona, ITALY, Fax: (39-458)027-901, e-mail: castelletti@sci.univr.it

J. P. Chervet, LC Packings (Dionex), Aberdaan 114, NL-1046 AA Amsterdam, THE NETHERLANDS, Fax: (31-20)685-3452, e-mail: jp.chervet@lcpackings.nl

Tomasz Dylag, Faculty of Chemistry, Jagellonian University, Ingardena 3, PL-30-060 Kraków, POLAND, Fax: (48-12) 292-7949, e-mail: Dylag@chemia.uj.edu.pl

Rolf Ekman, Neurochemical Unit, Institute of Clinical Neuroscience, Sahlgrenska University Hospital, SE-431-80 Mölndal, SWEDEN, Fax: (46-31)343-2421, e-mail: Rolf@neuro.gu.se

András Guttman, DIVERSA Co., 4955 Directors Place, B-1-131, San Diego, CA 92121, UNITED STATES, Fax: (1-858) 526-5921, e-mail: aguttman@diversa.com

Panos Hatsis, Institute for Marine Biosciences, National Research Council, Halifax, Nova Scotia B3H 3Z1, CANADA, Fax: (1-902)426-9413, e-mail: panos.hatsis@nrc-cnrc.gc.ca

Erich Heftmann, P. O. Box 928, Orinda, CA 94563, UNITED STATES, Fax: (1-925)254-7754, e-mail: chromatography@comcast.net

Ernst Kenndler, Institut für Analytische Chemie, Universität Wien, Währingerstraße 38, A-1090 Wien, AUSTRIA, Fax: (43-1)4277-9523, e-mail: ernst.kenndler@univie.ac.at

Julia Khandurina, Torrey Mesa Research Institute, 3115 Merryfield Row, San Diego, CA 92121, UNITED STATES, Fax: (1-858)812-1097, e-mail: julia.khandurina@syngeta.com

Marek Kowalczuk, Centre for Polymer Chemistry, Polish Academy of Sciences, Sklodowskiej Curie Str. 34, 41-800 Zabrze, POLAND, Fax: (48-32)271-2969, e-mail: cchpmk@bachus.ck.gliwice.pl

Agnieszka Kraj, Faculty of Chemistry, Jagellonian University, Ingardena 3, P-30-060 Kraków, POLAND, Fax: (48-12) 292-7949, e-mail: sciubisz@chemia.uj.edu.pl

Charles A. Lucy, Department of Chemistry, University of Alberta, Edmonton, Alta. T6G 2G2, CANADA, Fax: (1-780)492-8231, e-mail: charles.lucy@ualberta.ca

Philip J. Marriott, Australian Centre for Research on Separation Science, Department of Applied Chemistry, Royal Melbourne Institute of Technology, GPO Box 2476V, Melbourne, Vic 3001, AUSTRALIA, Fax: (61-3)9639-1321, e-mail: philip.marriott@rmit.edu.au

Victoria L. McGuffin, Department of Chemistry, Michigan State University, East Lansing, MI 48824-1322, UNITED STATES, Fax: (1-517)353-1793, e-mail: jgshabus@aol.com

Wilfried M. A. Niessen, hyphen MassSpec, De Wetstraat 8, NL-2332 XT Leiden, THE NETHERLANDS, Fax: (31-71)528-9330, e-mail: mail@hyphenms.nl

Szabolcs Nyiredy, Research Institute for Medicinal Plants, P.O. Box 11, H-2011 Budakalász, HUNGARY, Fax: (36-26)340-426, e-mail: rimp@axelero.hu

Pier Giorgio Righetti, Department of Agricultural & Industrial Biotechnologies, Università di Verona, Strada le Grazie, Cà Vignal, I-37134 Verona, ITALY, Fax: (39-458)027-901, e-mail: piergiorgio.righetti@univr.it

Andreas M. Rizzi, Institute of Analytical Chemistry, University of Vienna, Währingerstraße 38, A-1090 Wien, AUSTRIA, Fax: (43-1)4277-9523, e-mail: Andreas.Rizzi@univie.ac.at

Jerzy Silberring, Faculty of Chemistry, Jagellonian University, Ingardena 3, P-30-060 Kraków, POLAND, Fax: (48-12) 292-7949, e-mail: silber@chemia.uj.edu.pl

Roger M. Smith, Department of Chemistry, Loughborough University, Loughborough, Leics. LE11 3TU, GREAT BRITAIN, Fax: (44-1509)223-925, e-mail: R.M.Smith@lboro.ac.uk

Marek Smoluch, LC Packings (Dionex), Abberdaan 114, NL-1046 AA Amsterdam, THE NETHERLANDS, Fax: (31-20)685-3452, e-mail: marek.smoluch@lcpackings.nl

Robert Stevenson, The Abacus Group, 333 Carlyle Terrace, Lafayette, CA 94549-5202, UNITED STATES, Fax: (1-925)283-5621, e-mail: rlisteven@comcast.net

Piotr Suder, Neurobiochemistry Unit, Faculty of Chemistry, Jagellonian University, Ingardia Street 3, PL-30-060 Kraków, POLAND, Fax: (48-12) 292-7949, e-mail: Suder@chemia.uj.edu.pl

Barbara Verzola, Department of Agricultural & Industrial Biotechnologies, Università di Verona, Strada le Grazie 15, I-37134 Verona, ITALY, Fax: (39-458)027-901, e-mail: verzola@sci.univr.it

List of Abbreviations

A

A	ampere, adenine, amphetamine
Å	ångstrom = 10^{-8} cm
AA	amino acids, arachidonic acid
AAEE	acryloylaminoethoxyethanol
AAS	atomic absorption spectrometry
ABEE	4-aminobenzoic acid ethyl ester
AC	alternating current, acetylcodeine, affinity chromatography
Ac	acetyl
AcP	acyl phosphatase
ADCC	antibody-dependent cellular cytotoxicity
ADME	absorption, distribution, metabolism, and excretion
AED	atomic-emission detector
AEDA	aroma extract-dilution analysis
AEO	alcohol ethoxylates
AEOC	2-(9-anthryl)ethyl chloroformate
AES	atomic emission spectrometry
AFS	atomic fluorescence spectrometry
ag	attogram = 10^{-18} g
AGP	acid glycoprotein
AIA	Analytical Instrument Association
Ala	alanine
AMAC	2-aminoacridine
AMD	automated multiple development
amol	attomol = 10^{-18} mol
AMS	accelerator mass spectrometer
amu	atomic mass units
ANTS	aminonaphthalene-1,3,6-trisulfonic acid
ANDI	analytical-data interchange
AP	alkylphenols
APCI	atmospheric-pressure chemical ionization
APD	avalanche photodiode
APEC	alkylphenoxy carboxylates
APEO	alkylphenol ethoxylates
API	atmospheric-pressure ionization
APOC	1-(9-anthryl)-2-propyl chloroformate
aq.	aqueous
AQC	6-aminoquinolyl-*N*-hydroxy-succinimidyl carbamate
ARC	acridone-*N*-acetyl

Arg	arginine
ASB	amidosufobetaine
ASE	accelerated solvent extraction
Asn	asparagine
Asp	aspartate
atm	atmosphere = 1 bar = 760 torr = *ca.* 14.7 psi = 10^5 Pa
AUC	area under the curve

B

bar	atmosphere = *ca.* 14.7 psi
BBP	butylbenzyl phthalate
BCIP	5-bromo-4-chloro-3-indolyl phosphate
BDB	benzoxodioxoazolylbutanamine
BDE	brominated diphenyl ethers
BGE	background electrolyte
BHT	2,6-di-*t*-butyl-*p*-cresol (butylated hydroxytoluene)
BN-chamber	Brenner-Niederwieser chamber
BP	buprenorphine
bp	base pair
BPA	bisphenol A
BrNP	brominated nonylphenol
BrNPEC	brominated nonylphenol carboxylates
BrNPEO	brominated nonylphenol ethoxylates
BSA	bovine serum albumin
BTEX	benzene, toluene, ethylbenzene and xylenes
Bu	butyl
BZE	benzoylecgonine
BZITC	benzyl isothiocyanate

C

C	centigrade, celsius, cytosine, codeine
CA	carrier ampholyte
CAD	computer-assisted design
CAE	capillary array electrophoresis
CAGE	capillary affinity gel electrophoresis
cap	capillary
CAPEC	dicarboxylated alkylphenol ethoxylate
CB	chlorinated biphenyls
CBD	cannabidiol
CBN	cannabinol
CBQ	3-(*p*-carboxybenzoyl)quinoline 2-carboxaldehyde
CBQCA	3-(4-carboxybenzoyl)-2-quinoline carboxyaldehyde
CCD	chemical composition distribution
CCD	charge-coupled device
CCLC	column/column liquid chromatography
CD	cyclodextrin
CD	circular dichroism
CD	continuous development
CDEA	coconut diethanolamide

CDICT	(1R,2R)-N-[(2-isothiocyanato)cyclo-hexyl]-6-methoxy-4-quinolinylamide
CE	capillary electrophoresis, cholesterol esters
CEC	capillary electro(kinetic)chromatography
CFD	computational fluid dynamics
CFLSI-MS	continuous-flow-liquid secondary-ion mass spectrometry
CGE	capillary gel electrophoresis
CHARM	combined hedonic and response measurement
CHO	Chinese hamster ovarian (cells)
CI	chemical ionization
CID	collision-induced dissociation
cIEF	capillary isoelectric focusing
CINP	chlorinated nonylphenol
CINPEC	chlorinated nonylphenol carboxylates
CINPEO	chlorinated nonylphenol ethoxylates
CIS	coordinated ion spray
CLC	column liquid chromatography, conjoint liquid chromatography
CLEC	chiral ligand exchange chromatography
CLND	chemiluminescent nitrogen detector
CM	carboxymethyl
cm	centimeter $= 10^{-2}$ m
CMA	carbazole-N-(2-methyl)acetyl
CMC	critical micelle concentration
COC	cocaine
conc.	concentrated
CRA	carbazole-9-acetyl
CRF	charge-remote fragmentation
CRMV	collagenase-released matrix vesicle
CRP	carbazole-9-propionyl
CSF	cerebrospinal fluid
CSGE	conformation-sensitive gel electrophoresis
CSP	chiral stationary phase
CTAB	cetyltrimethylammonium bromide
CTAC	cetyltrimethylammonium chloride
Cys	cysteine
CZE	capillary zone electrophoresis

D

2-D	2-dimensional
Da	dalton
dabsyl	4-dimethylaminoazobenzene 4'-sulfonyl chloride
DABTH	4,4-N,N-dimethylaminoazobenzene 4'-isothiocyanate
DAD	diode-array detector
DAG	diacylglycerols
DAM	diacetylmorphine
DANI	1,3-diacetoxy-1-(4-nitrophenyl)-2-propyl isothiocyanate
DAR	digital autoradiography
DAT	diacetyl-L-tartaric anhydride
DATS	dialkyltetralinesulfonate
DBD	degree-of-branching distribution

DBP	dibutylphthalate
DBT	dibenzoyl-L-tartaric anhydride
DC	direct current
DCCC	droplet counter-current chromatography
DCM	dichloromethane
DCP	direct-current plasma
DDT	dichlorodiphenyltrichloroethane
DEAE	diethylaminoethyl
DEHP	di(2-ethylhexyl)phthalate
DEP	diethylphthalate
DEP	di-electrophoresis
des	desamido
DGGE	denaturing gradient gel electrophoresis
DHA	docosahexaenoic acid
DHB	2,5-dihydroxybenzoic acid
DHC	dihydrocodeine
DHET	dihydroxyeicosatrienoic acid
DHM	dihydromorphine
DHPLC	denaturing HPLC
disc	discontinuous
DLS	dynamic light scattering
DMA	dimethylaniline
DMALS	depolarization multi-angle light scattering
DMOX	2-alkenyl-4,4-dimethyloxazolines
DMP	dimethylphthalate
DMSO	dimethylsulfoxide
DMT	dimethoxytrityl
DNA	deoxyribonucleic acid
DNB	dinitrobenzoyl
DnOP	di-*n*-octylphthalate
DNP	dinitrophenyl
DNPH	dinitrophenylhydrazine
DNPU	dinitrophenylurethane
DNS	5-dimethylaminonaphthalene-1-sulfonyl (dansyl)
DNT	dinitrotoluene
dNTP	deoxynucleoside triphosphate
DOC	dissolved organic carbon
DP	degree of polymerization
dpm	disintegrations per minute
DPPP	diphenyl-1-pyrenylphosphine
DRIFT	diffuse-reflectance Fourier transform
dsDNA	double-stranded DNA
DTAB	dodecyltrimethylammonium bromide
DTDP	dithiodipyridine
DTDP	3-(4,6-dichloro-1,3,5-triazinylamino)-7-dimethylamino-2-methylphenazine
DTE	dithioerythrol
DTPA	diethylenetriaminepentaacetic acid
DTT	dithiothreitol
DVB	divinylbenzene

E

EAD	electro-antennographic detection
EAG	electro-antennograph
EC	electrokinetic chromatography
ECD	electron-capture detector
ECG	ecgonine
ECN	equivalent carbon number
EDC	endocrine-disrupting compounds
EDDP	2-ethylidene-1,5-dimethyl-3,3-diphenylpyrrolidine
EDTA	ethylenediaminetetraacetic acid
ee	enantiomeric excess
EGDN	ethyleneglycol dinitrate
EI	electron ionization, electron impact
EIC	electrostatic ion chromatography
EKC	electrokinetic capillary chromatography
ekd	electrokinetically driven
ELISA	enzyme-linked immunosorbent assay
ELSD	evaporative light-scattering detector
em	emission
EMDP	2-ethyl-5-methyl-3,3-diphenyl-1-pyrrolidine
EME	electrostatic/magnetic/electrostatic geometry
EME	ecgonine methyl ester
EOF	electro-osmotic flow
EPA	eicosapentaenoic acid
EPA	Environmental Protection Agency
EPC	electro planar chromatography
EPF	zero electro-osmotic flow
EPO	erythropoietin
EpPUFA	epoxypolyunsaturated fatty acids
Eqn.	Equation
ERIN	electrochemically regenerated ion neutralizer
ESI	electrospray ionization
ESR	electron-spin resonance
Et	ethyl
ET	electrothermal
ET	energy transfer
EU	endotoxin unit
eV	electron volt
EVB	ethylvinylbenzene
ex	excitation

F

FVIII	blood coagulation Factor VIII
FA	fatty acids
FAAS	flame atomic absorption spectrometry
FAB	fast atom bombardment
FACE	fluorophore-assisted carbohydrate electrophoresis
FACS	fluorescence-activated cell sorting

FAD	full adsorption/desorption
FAME	fatty acid methyl esters
FCSE	fully concurrent solvent evaporation
FD	field desorption
FDA	Food & Drug Administration
FDAA	1-fluoro-2,4-dinitrophenyl-5-L-alanine amide (Marfey's reagent)
FFA	free fatty acids
FFPC	forced-flow planar chromatography
fg	femtogram $= 10^{-15}$ g
FIA	flow injection analysis
FID	flame-ionization detector
Fig.	Figure
FLD	fluorescence detection
FLEC	1-(9-fluorenyl)-ethyl chloroformate
FMOC	9-fluorenylmetyl chloroformate
fmol	femtomol $= 10^{-15}$ mol
FPD	flame-photometric detector
FPLC	Fast Protein Liquid Chromatography
FRES	forward recoil spectrometry
FS	fused silica
FT	Fourier transform
ft.	foot $= 30.48$ cm
FTH	fluorescein isothiocyanate
FT-ICR-MS	Fourier-transform ion-cyclotron resonance mass spectrometry
FTID	flame-thermionic ionization detector

G

g	gram
G	guanine
GABA	gamma-aminobutyric acid
GBL	gamma-butyrolactone
GC	gas chromatography
GC × GC	two-dimensional gas chromatography
GD	glow discharge
GH	growth hormone
GHB	gamma-hydroxybutyrate
GITC	2,3,4,6-tetra-O-acetyl-1-thio-β-D-glucopyranosyl isothiocyanate
GLC	gas/liquid chromatography
Gln	glutamine
Glu	glutamate
Gly	glycine
GM	glycidylmethacrylate
GPC	gel permeation chromatography
GPEC	gradient polymer elution chromatography
GPL	glycerophospholipid

H

h	hour(s)
HA	heteroduplex analysis
Hb	hemoglobin
HDC	hydrodynamic chromatography
HDL	high-density lipoproteins
HEC	hydroxyethylcellulose
HEDTC	*bis*(2-hydroxyethyl)dithiocarbamate
HEPES	*N*-2-hydroxyethylpiperazine-*N'*-2-ethanesulfonic acid
HETP	height equivalent to a theoretical plate
HFB	hexafluorobenzoyl
HFBA	heptafluorobutyric acid
HFBAA	heptafluorobutyric anhydride
hGH	human growth hormone
HGP	human genome project
HIBA	α-hydroxyisobutyric acid
HIC	hydrophobic-interaction chromatography
HILIC	hydrophilic-interaction chromatography
His	histidine
HMA	hexamethonium
HPAEC	high-performance anion-exchange chromatography
HPC	hydroxypropylcellulose
HPCE	high-performance capillary electrophoresis
HPLC	high-performance liquid chromatography
HPMC	hydroxypropylmethylcellulose
HPODE	hydroperoxyoctadecadienoic acid
HPSEC	high-performance size-exclusion chromatography
HPTLC	high-performance thin-layer chromatography
HQS	8-hydroxyquinolinesulfonate
HRGC	high-resolution gas chromatography
HRMS	high-resolution mass spectrometry
HS	headspace
HSA	human serum albumin
HSGC	high-speed gas chromatography
HTA	hexdecyltrimethylammonium
HTS	high-throughput screening
Hyp	hydroxyproline
Hz	hertz

I

IC	ion chromatography
ICAT	isotope-coded affinity tag
ICP	inductively coupled plasma
ICR	ion cyclotron resonance
ID	internal diameter
IDA	iminodiacetic acid
IDA	1-methoxycarbonylindolizine 3,5-dicarbaldehyde
IEC	ion-exchange chromatography

IEF	isoelectric focusing
IgG	immunoglobulin G
Ile	isoleucine
IMAC	immobilized-metal-ion affinity chromatography
IMP	ion-moderated partitioning
in.	inch = 2.54 cm
InsP	inositol phosphate
IP	ion pair
IPC	isopycnic centrifugation
IPG	immobilized pH gradient
IR	infrared, isotope ratio
ISEC	inverse size-exclusion chromatography
ITMS	ion-trap mass spectrometry
ITP	isotachophoresis

K

K	kelvin
kbp	kilobase pair = 10^3 base pairs
kDa	kilodalton = 10^3 daltons
KDO	3-deoxy-D-*manno*-octulosonic acid
kPa	kilopascal = 10^3 pascal

L

L	liter, lambert
LALLS	low-angle laser-light scattering
LAS	linear alklbenzenesulfonate
LC	liquid chromatography
LC-CAP	liquid chromatography at the critical adsorption point
LC-CC	liquid chromatography at the critical condition
LCM	laser capture microdissection
LCR	ligase chain reaction
LDL	low-density lipoproteins
LEC	ligand-exchange chromatography
Leu	leucine
LIF	laser-induced fluorescence
LIMS	laboratory information management system
LINAC	linear accelerating high-pressure collision cell
L/L	liquid/liquid
LLC	liquid/liquid chromatography
LLE	liquid/liquid extraction
LOD	limit of detection
LOQ	limit of quantification
LOSI	limit of spectroscopic identification
LPA	linear polyacrylamide
LPA	lysophosphatidic acid
LRMS	low-resolution mass spectrometry
LSC	liquid/solid chromatography
LSD	light-scattering detector

LSIMS	liquid secondary-ion mass spectrometry
Lys	lysine

M

M	molar
m	meter
μ	micro
μLC	micro liquid chromatography
MA	methamphetamine
mA	milliampere $= 10^{-3}$ A
μA	microampere $= 10^{-6}$ A
MACS	magnetic-activated cell sorting
MADGE	microplate-array diagonal gel electrophoresis
MAE	microwave-assisted extraction
MAG	monoacylglycerol
MALDI	matrix-assisted laser desorption ionization
MALS	multi-angle light scattering
MAM	monoacetylmorphine
MBBr	monobromobimane
MBDB	*N*-methylbenzoxodialolylbutanamine
Mbp	megabase pair $= 10^6$ bp
M-chamber	micro-chamber
MC	methylcellulose
MCA	metal chelate affinity
MCIC	metal chelate interaction chromatography
MCT	mercury-cadmium telluride
MD	multi-dimensional
MDA	methylenedioxyamphetamine
MDEA	methylenedioxyethylamphetamine
MDGC	multi-dimensional gas chromatography
MDMA	methylenedioxymethamphetamine
MDMAES	mono(dimethylaminoethyl)succinyl
Me	methyl
MEC	micellar electrochromatography
MEEKC	micro-emulsion electrokinetic chromatography
MEKC	micellar electrokinetic capillary chromatography
MEMS	micro-electromechanical system
MEP	4-mercaptoethylpyridine
MES	morpholinoethane sulfonate
Met	methionine
meq	milliequivalent $= 10^{-3}$ equivalent
μeq	microequivalent $= 10^{-6}$ equivalent
MG	morphine glucuronide
MIMS	membrane-introduction mass spectrometry
min	minutes
MIP	microwave-induced plasma, molecularly imprinted polymer
MISPE	molecular imprint solid-phase extraction
mg	milligram $= 10^{-3}$ g
mL	milliliter $= 10^{-3}$ L

XXII

µL	microliter = 10^{-6} L
mm	millimeter 10^{-3} M
µm	micrometer = 10^{-6} m
mM	millimolar 10^{-3} M
µmol	micromol = 10^{-6} mol
MMS	micromembrane suppressor
MOPS	3-(N-morpholino)propanesulfonic acid
MP	medium pressure
MPa	megapascal = 10^{6} Pa
MP	methyl prednisolone
MPS	methyl prednisolone hemisuccinate
MRA	mass-rate attenuator
MRM	multiple-reaction monitoring
mRNA	messenger RNA
MS	mass spectrometry
MS^n	multiple mass spectrometry
MSA	methanesulfonic acid
msec	millisecond = 10^{-3} sec
MS/MS	tandem mass spectrometry
MS^n	multiple mass spectrometry
MSPD	matrix solid-phase dispersion
µTAS	micro total analysis system
MTBA	methyl tetrabutyl ether
MUX	multiplex
mV	millivolt = 10^{-3} V
mW	milliwatt = 10^{-3} W
mw	molecular weight
MWD	molecular mass distribution
mu	mass units

N

N	normal
nA	nanoampere = 10^{-9} ampere
NADP	nicotinamide adenine dinucleotide phosphate
NBP	norbuprenorphine
NC	norcodeine
NCA	N-carboxyanhydride
N-chamber	normal chamber
NCI	negative chemical ionization
NDA	naphthalene 2,3-dicarboxaldehyde
NEFA	nonesterified fatty acids
NG	nitroglycerin
ng	nanogram = 10^{-9} g
NICI	negative-ion chemical ionization
nL	nanoliter = 10^{-9} L
nm	nanometer = 10^{-9} m
NM	normorphine
NMIFA	non-methylene-interrupted fatty acids
nmol	nanomol = 10^{-9} mol

NMR	nuclear magnetic resonance
NOM	natural organic matter
NP	nonylphenol
NP	normal-phase
NPC	normal-phase chromatography
NPEC	nonylphenol carboxylates
NPEO	nonylphenol ethoxylates
NPD	nitrogen/phosphorus detector
NPLC	normal-phase liquid chromatography
NSAID	non-steroidal anti-inflammatory drugs
NSIC	non-suppressed ion chromatography
NTA	nitrilotriacetic acid

O

O	olfactometry
oaTOF-MS	orthogonal-acceleration time-of-flight mass spectrometry
OCEC	open-channel electrochromatography
OD	outside diameter
OP	octylphenol
OPA	o-phthaldialdehyde
OPEC	octylphenol carboxylates
OPEO	octylphenol ethoxylates
OPLC	overpressured-layer chromatography
OQ	operational qualification
ORM	overlapping resolution mapping
OT	open-tubular
OVM	ovomucoid

P

Pa	pascal $= 10^{-5}$ bar
pA	picoampere $= 10^{-12}$ A
PAD	pulsed-amperometric detector
PAE	phthalate esters
PAF	platelet-activating factor
PAGE	polyacrylamide gel electrophoresis
PAS	photoacoustic spectrometry
PAH	polycyclic aromatic hydrocarbons
PAN	4-(2-pyridylazo)naphthol
PANI	polyaniline
PAR	4-(2-pyridylazo)resorcinol
PATRIC	position- and time-resolved ion counting
PB	particle beam
PBA	phenylboronic acid
PBD	poly(butadiene)
PBDE	polybrominated diphenyl ethers
PBS	phosphate-buffered saline
PC	planar chromatography, personal computer, poly(bisphenol A carbonate)
PCA	principal component analysis
PCB	poly(chlorinated biphenyls)

PCDD	poly(chlorinated dibenzo-p-dioxins)
PCDF	poly(chlorinated dibenzofurans)
PCI	positive chemical ionization
PCR	post-column reagent, polymerase chain reaction
PCS	photon correlation spectroscopy
PCSE	partially concurrent solvent evaporation
PD	polydispersity
pd	pressure-driven
PDA	photodiode array
PDD	pulse-discharge detector
PDDAC	poly(diallyldimethylammonium) chloride
PDECD	pulsed-discharge electron-capture detector
PDMA	poly(N,N-dimethylacrylamide)
PDMS	poly(dimethylsiloxane)
PDPID	pulsed-discharge photoionization detector
PEC	pressurized capillary electrochromatography
PED	pulsed-electrochemical detector
PEEK	poly(ether ethyl ketone)
PEG	poly(ethylene glycol)
PEI	poly(ethylene imine)
PEMA	poly(ethylmethacrylate)
PEO	poly(ethylene oxide)
PETN	pentaerythritol tetranitrate
PFB	pentafluorobenzoyl
PFE	pressurized-fluid extraction
PFPD	pulsed flame-photometric detector
pg	picogram $= 10^{-12}$ g
PGC	porous graphitic carbon
Ph	phenyl
PHB	poly(3-hydroxybutyrate)
PHBV	poly(3-hydroxybutyrate-co-3-hydroxyvalerate)
Phe	phenylalanine
PI	polyisoprene
pI	isoelectric point
PIBM	poly(isobutylmethacrylate)
PICES	passive in $situ$ concentration/extraction sampler
PICI	positive-ion chemical ionization
PID	photoionization detector
pL	picoliter $= 10^{-12}$ L
PLE	pressurized-liquid extraction
PLOT	porous-layer open-tubular
PMA	poly(methacrylate)
PMD	programmed multiple development
PMMA	poly(methylmethacrylate)
pmol	picomol $= 10^{-12}$ mol
PMP	1-phenyl-3-methyl-5-pyrazolone
PMT	photomultiplier tube
PNB	p-nitrobenzylhydroxylamine
POP	persistent organic pollutants

ppb	parts per billion $= 10^{-9}$ parts
PPCP	pharmaceuticals and personal-care products
ppm	parts per million $= 10^{-6}$ parts
ppt	parts per trillion $= 10^{-12}$ parts
PPO	poly(propylene oxide)
ppq	parts per quadrillion $= 10^{-15}$ parts
Pr	propyl
Pro	proline
PrP	prion protein
PS	polystyrene
Ps	phosphatides
psi	pounds per square inch $= 51.77$ torr
PtdCho	phosphatidylcholine
PtdIns	phosphatidylinositols
PTFE	poly(tetrafluoroethylene)
PTH	phenylisothiocyanate
PTV	programmed-temperature vaporizer
PUFA	polyunsaturated fatty acids
PVA	poly(vinyl alcohol)
PVAc	poly(vinyl acetate)
PVC	poly(vinyl chloride)
PVDF	poly(vinylidene fluoride)
PVP	poly(vinylpyrrolidone)

Q

QAP	quaternary ammonium phosphates
QqQ	triple quadrupole
QSAR	quantitative structure/activity relationships
QTOF-MS	quadrupole time-of-flight mass spectrometry

R

r	recombinant
RAM	restricted access medium
Ref.	Reference
RF	radiofrequency
RFLP	restriction-fragment-length polymorphism
RI	refractive index
RIA	radio-immuno assay
RID	refractive-index detector
RNA	ribonucleic acid
ROMP	ring-opening metathesis polymerization
RP	reversed-phase
RPC	reversed-phase chromatography, rotation planar chromatography
rpm	rotations per minute
RRA	radio-receptor assay
RRF	relative response factor
RSD	relative standard deviation
RT	retention time
RT-PCR	real-time polymerase chain reaction

S

S	siemens
SAMBI	α-methylbenzyl isothiocyanate
satd.	saturated
SAX	strong-anion exchange
SB	short-bed
SB	sulfobetaine
SBSE	stir-bar sorptive extraction
S-chamber	sandwich chamber
SCD	sulfur chemiluminescence detector
SDE	simultaneous steam distillation/solvent extraction
SCAN	sample concentrator and neutralizer
SCOT	support-coated open-tubular
SCX	strong-cation exchange
SD	standard deviation
SDA	strand displacement amplification
SDM	stoichiometric displacement model
SDS	sodium dodecyl sulfate
sec	seconds
SEC	size-exclusion chromatography
SEEC	size-exclusion electrochromatography
SELDI	surface-enhanced laser-desorption ionization
Ser	serine
SERS	surface-enhanced Raman spectroscopy
SFC	supercritical-fluid chromatography
SFE	supercritical-fluid extraction
SGC	solvating gas chromatography
SIC	self-interaction chromatography
SIC	suppressed-ion chromatography
SID	surface-ionization detection
SIM	single-ion monitoring, selected-ion monitoring
SIMS	secondary-ion mass spectrometry
SLAB	inter-laboratory standard deviation
SLE	solid/liquid extraction
SLM	supported-liquid membrane
SM	sphingomyelin
SMA	steric mass action
S/N ratio	signal-to-noise ratio
SNEIT	1-(1-naphtyl)-ethyl isothiocyanate
SNP	single-nucleotide polymorphism
SPC	sulfophenylcarboxylate
SPE	solid-phase extraction
SPMD	semi-permeable membrane device
SPME	solid-phase micro-extraction
SPR	surface plasmon resonance
sq.	square
SRM	selective reaction monitoring
SRS	self-regenerating suppressor
SSCP	single-strand conformation polymorphism

SSDNA	single-stranded DNA
SSO	sequence-specific oligonucleotide
STP	sewage treatment plant
STR	short tandem repeats

T

T	thymine
TAG	triacylglycerol
TAPS	3-*tris*[(hydroxymethyl)methylamino] 1-propanesulfate
TAS	total analysis system
TBA	tetrabutylammonium
TBMB	(*S*)-(+)-2-*tert*-butyl-2-methyl-1,3-benzodioxole
TBP	tributyl phosphine
TBQCA	3-(4-tetrazolbenzoyl) 2-quinolinecarboxyaldehyde
TBE	Tris/borate/EDTA
t-BOC	*N-tert*-butyloxycarbonyl
tert-BOOH	*tert*-butyl hydroperoxide
TCA	trichloroacetic acid
TCD	thermal conductivity detector
TCDD	tetrachlorodibenzo-*p*-dioxin
TEA	thermal-energy analyzer
TEAA	triethylammonium acetate
TEAF	triethylammonium formate
TEAP	triethylammonium phosphate
TED	triscarboxymethyl ethylenediamine
TEMED	N,N,N',N'-tetramethylethylenediamine
temp.	temperature
TEPA	tetraethylenepentamine
TFA	trifluoroacetic acid
TFC	turbulent-flow chromatography
TFE	trifluoroethanol
TGGE	temperature-gradient gel electrophoresis
TGIC	temperature-gradient interaction chromatography
THC	tetrahydrocannabinol
THCA	tetrahydrocannabinolic acid
THCCOOH	11-nor-Δ^9-tetrahydrocannabinol 9-carboxylic acid
THF	tetrahydrofuran
Thr	threonine
TIC	total-ion chromatogram
TIE	toxicity identification evaluation
TID	thermionic ionization detector
TLC	thin-layer chromatography
TMAE	trimethylaminoethyl
TMAOH	trimethylammonium hydroxide
TMS	trimethylsilyl
TMSO	trimethylsiloxy
TNP	trinitrophenyl
TNT	trinitrotoluene
TOF-MS	time-of-flight mass spectrometry

TPA	tetrapentylammonium 3-[*tris*(hydroxymethyl) methylamino] 1-propanesulfate
Tris	tris(hydroxymetyl)aminomethane
TRITC	tetramethylrhodamine isothiocyanate
Trp	tryptophan
TSI	thermospray ionization
TTA	tetradecyltrimethylammonium
Tyr	tyrosine

U

U	uracil
UCC	universal calibration curve
U-chamber	ultra-micro chamber
UV	ultraviolet

V

V	volt
Val	valine
VBC	vinylbenzyl chloride
vis.	visible range
VNTR	variable number of tandem repeats
vol.	volume
2VP	2-vinylpyridine
v/v	volume-by-volume
vWF	Van Willebrand Factor

W

W	watt
WCOT	wall-coated open-tubular
w/w	weight-by-weight
WWCOT	whiskered-wall-coated open-tubular

X

XNP	halogenated nonylphenols

List of Italic Symbols

(Subscripts and superscripts are not always listed)

A

a	activity
A_s	surface area of the stationary phase
$A^{(k)}$	surface area of phase k

B

B_0	chromatographic permeability
b	ion-binding coefficient

C

C	concentration
C_i	counter-ion concentration
C_m, C_M	solute concentration in the mobile phase
C_s, C_S	solute concentration in the stationary phase
COF	chromatographic optimization function
CRF	chromatographic response function
CRS	chromatographic resolution statistic

D

D	diffusion coefficient, diffusivity, displacing ion
D_{hb}	energy of the hydrogen bond
D_M	diffusion coefficient in the mobile (gas) phase
d	dimension
d_c	diameter of the channel
d_{col}	column diameter
d_{conn}	connector diameter
d_{det}	flow-cell diameter
d_f	stationary-phase film thickness
d_p	particle diameter

E

E	electric field strength
e	charge on an electron

F

F	Faraday constant, net flow-rate
f/f_0	frictional ratio

G

G	Gibbs free energy
ΔG	change in free energy
g	surface tension, activity coefficient
$g(r)$	pair distribution function

H

H	column efficiency, plate height, enthalpy

H_{min}	minimum plate height
h	reduced plate height
$\boldsymbol{I}$	
I	ionic strength, ionization energy
I_0	zero-order modified Bessel function
ISR	inverse sum of resolutions
$\boldsymbol{J}$	
J	mass flux
$\boldsymbol{K}$	
K	equilibrium constant
K_A	affinity constant
K_B	binding constant
K_D	distribution coefficient, dissociation constant
K_s	salting-out constant
k^l	capacity factor (obsolete)
k	retention factor, Debye screening parameter
k_B	Boltzmann constant
k_d	rate constant for desorption
k_i	distribution coefficient
k_{sm}	rate constant for transfer from stationary to mobile phase
$\boldsymbol{L}$	
L	length, distance
L_{col}	column length
L_{conn}	connector length
L_{det}	flow-cell length
$\boldsymbol{M}$	
M, m	molecular mass
$M(r)$	molecular mass
M_η	viscosity-average molecular weight
M_n	number-average molecular weight
M_r	molecular radius
M_w	weight-average molecular weight
M_z	z-average molecular weight
m/z	mass-to-charge ratio
m_s	molality of the salt
m(subscript)	mobile phase
$\boldsymbol{N}$	
N	plate number, Avogadro's number
n	number, mole number, peak capacity, molecular area
NPR	normalized product of resolutions, non-porus resin
nR_S	number of peaks resolved
$\boldsymbol{P}$	
P, p	pressure
P^l	solvent polarity parameter
p_c	critical pressure
pH_f	eluting pH
$\boldsymbol{Q}$	
Q	ion-exchange capacity, polydispersity
Q_l	group adsorption energy
q	volume ratio
q_{chr}	net charge of a protein

R

R	radius, gas constant, response, coefficient of correlation
$R\eta$	viscosity-based Stokes radius
R_F	relative rate of migration
Rg	radius of gyration
R_S	resolution, Stokes radius (frictional coefficient)
Re	Reynolds number
RI	retention index
r	radius, radial coordinate, mass, separation distance
r(subscript)	resin phase
r_g	radius of gyration
r_{hb}	separation distance of the hydrogen bond
r_p	particle radius

S

S	sensitivity, solubility, entropy
s, S(subscript)	stationary phase
Sig	signal
S_G	saturation grade
s	molal surface tension
s_i	solvent strength
s_v	selectivity value

T

T	absolute temperature
T_C	compensation temperature, critical temperature
t	time
t_0	time zero, elution time of nonretained solute
t_{mc}	time due to micellar electrokinetic chromatography
t_R	retention time

U

U	molecular inhomogeneity, voltage, internal or potential energy
$U(r)$	sum of the internal energies
U_d	internal energy of dispersion forces
U_{hb}	internal energy of hydrogen-bonding forces
U_i	internal energy of dipole induction forces
U_{II}	internal energy of ion–ion forces
U_{Ii}	internal energy of ion-induced dipole forces
U_{Io}	internal energy of ion–dipole orientation forces
U_o	internal energy of dipole orientation forces
u	solvent velocity
$\bar{u}$	carrier flow-rate, linear velocity
u_{opt}	optimum flow-rate
$u(r)$	radial velocity

V

V	molar volume, molal volume
V_{det}	detection volume
V_{inj}	injection volume
V_m, V_M	volume of the mobile phase
V_o	interstitial volume, void volume
V_P	pore volume
V_R	retention volume
V_s, V_S	volume of the stationary phase

V_t	total volume
v	velocity, molar volume
v_{eo}	velocity due to electro-osmosis
v_{ep}	velocity due to electrode potential
v_i	total velocity
v_{mc}	velocity due to micellar electrokinetic chromatography
W	
w	weight, water solubility, zone width
w_b	width at the base
$w_{1/2}$	width at half-height
X	
X	mole fraction
x	distance
Z	
Z, z	charge, charged species
Z_e	electronic charge
Z_f	solvent front travel

List of Greek Symbols

(Subscripts and superscripts are not always listed)

α	selectivity factor, degree of dissociation, polarizability
β	phase ratio, buffering power
Δ	difference
δ	Hildebrand solubility parameter
δ_i	overall solubility polarity
δ_M	mobile-phase strength
δ_S	stationary-phase solubility parameter
ε	permittivity
ε_0	permittivity of vacuum
ε_i	intra-particle porosity, interstitial porosity
ε_r	relative permittivity, dielectric constant
ε_t	total porosity
ε_u	inter-particle porosity
ε^0	solvent strength
Φ	phase ratio
ϕ	volume fraction
θ^2	fraction of extra-column variance
Γ	preferential-interaction parameter
γ	activity coefficient, tortuosity factor
η	viscosity
φ	fraction of modifier in the mobile phase
κ	solvent velocity, Debye length
κ_o	conductance of an open capillary
κ_p	conductance of a packed capillary
κ^{-1}	Debye length
Λ	salting-in coefficient, binding capacity
λ	ionic equivalent conductance, mean free path, packing-structure constant
μ	mobility, chemical potential, dipole moment, ionic strength
μ_0	standard-state chemical potential
μ_{ep}	electrophoretic mobility
ρ	density
ρ_c	critical density
ρ_r	reduced density
σ	surface charge density, surface-tension increment
σ^2	variance
σ_i	steric factor, standard deviation
σ_P	charge density on the protein surface
σ_v	standard deviation in volume units
$\Omega\sigma$	salting-out coefficient

τ	exponential time constant
Ψ	shape factor
Ψ_B	solvent association factor
ζ	zeta potential

Foreword

A prerequisite for a meaningful study of the chemical and biological properties of a given molecule or particle is that the sample is homogeneous. If it is not, one can only determine the properties of the *mixture*, which may be of little interest. Therefore, the importance of high-resolution separation methods for both analytical and (micro)preparative purposes is obvious and, accordingly, books like this have a place in life science, pharmaceutical, medical and many other types of laboratories.

In general, forewords are focused on the importance of the particular subject the books deal with. However, chromatography is a very well-known and established separation technique and, therefore, it might be superfluous to give examples of its enormous potential. Instead, I will take another approach; namely, to emphasize the *analogies* between chromatography and other separation methods, such as electrophoresis and centrifugation, rather than to discuss the dissimilarities. A common feature of these three methods is that the separations are based on differences in the migration velocity, v, of the sample constituents. Therefore, one can understand intuitively that there exists an equation (or more) which is valid for all of these methods, provided that the migration velocity is chosen as the independent variable. One of these equations is:

$$c_j^\alpha \cdot v_j^\alpha - c_j^\beta \cdot v_j^\beta = v^{\alpha\beta}(c_j^\alpha - c_j^\beta)$$

where α and β are two phases separated by a moving boundary; c_j^α and c_j^β are the concentrations of the ion j in the α and β phases, respectively; v_j^α and v_j^β are the velocities of the ion j in the α and β phases, respectively; and $v^{\alpha\beta}$ is the velocity of the moving boundary.

From the mere existence of an equation which is universal for chromatography, electrophoresis and centrifugation one can conclude that any characteristic feature of one of these methods (say electrophoresis) has a counterpart in the other methods (chromatography, centrifugation). This finding is of fundamental importance, because as soon as a new technique has been developed in one of these differential velocity-based separation methods, one should start to develop the corresponding technique for the other analogous separation methods. The more methods that are available, the easier and faster one can purify, say, a protein, which is very important in proteomics studies. Similarly, any new phenomenon observed in one of these separation methods has its analog in the other methods.

Ignorance of the analogy between different separation techniques has, unfortunately, delayed the progress of separation science and, thereby, also of life sciences and related areas. For instance, isoelectric focusing and indirect detection in electrophoresis were introduced as early as 1961 and 1967, respectively, but the chromatographic counterparts did not appear until 1977 and 1979.

Another example of lack of thinking about the analogy is the unfortunate exchange of the term displacement electrophoresis for isotachophoresis, which made Professor A.J.P. Martin, Nobel Prize Laureate and one of the pioneers in this field, dismayed. Perhaps commercial interests played also a role in this case, as they did when companies exchanged the term "continuous beds" for "monoliths". A representative of a company admitted that this new notation is linguistically not correct, but "sells" better. Another reason for the misuse of terms is the lack of theoretical knowledge. Nor is it uncommon that a new term is introduced for an already existing technique, which may mislead the reader of the paper to believe that the author is the inventor. For instance, the well-known zone sharpening in electrophoresis (enrichment of a sample by transferring it into a dilute buffer) is often erroneously called "stacking" (enrichment by displacement electrophoresis). I believe that this approach is seldom used consciously by the author to become recognized – rather it is likely that he was not aware that the method had been described earlier. However, it is common that this mode of action is consciously employed in commercial advertisements – ethically dubious tactics.

Electrochromatography has gained much in popularity since the previous edition of this series was published. One reason might be that it is often wrongly stated that the electroosmosis in a packed bed generates a plug flow and, therefore, that the resolution is considerably higher than that in conventional chromatography. A higher resolution can be obtained if the pores in the beads are large enough to create an electroosmotic flow which transports the solutes through the beads faster than does diffusion. However, the resolution cannot be higher than that obtained in beds packed with nonporous beads. Homogeneous gels yield a perfect plug flow and are from that point of view the ideal electrochromatographic medium. Electrochromatography in such gels might be the only chromatographic method that can give a resolution comparable to that obtained in electrophoresis and should, therefore, be used more frequently. However, for project-scale separations chromatography is far superior to electrophoresis.

Much effort is being devoted to improving the resolution of chromatographic media and, in some cases, one has succeeded in approaching the theoretical limit, i.e., the method cannot be improved much more. In such cases one should intensify the research on selective chromatographic techniques, which have the advantage to permit very fast purification of both low- and high-molecular-weight substances, as well as viruses and cells, such as bacteria, also for complex samples. The recently introduced "artificial antibodies" in the form of highly selective gel granules will, no doubt, play an important role for rapid one-step isolation of a given substance or bioparticle (by affinity chromatography), as well as for clinical diagnosis, for instance for detection of biomarkers for different diseases and as the sensing element in biosensors.

With the introduction of the continuous beds (monoliths) and the homogeneous gels the prerequisite for high resolving chromatographic and electrochromatographic experiments in microchips was fulfilled. There are, accordingly, no restrictions on the choice of appropriate chromatographic, electrochromatographic or electrophoretic media. The crucial point is rather the design and the cost of the microchip. It will be interesting to see whether the hybrid microdevice will become an attractive alternative.

This year it is one hundred years since Mikhail Semenovich Tswett presented at the Meeting of Warsaw Society of Natural Scientists his famous lecture "On a New Category of Adsorption Phenomena and Its Application in Biochemical Analysis". During the intervening years the chromatographic technique has continuously been refined and the resolution in analytical experiments is, in some modes, close to what is theoretically possible (see above), *i.e.*, the performance of the column cannot be improved significantly. The situation resembles that in electrophoresis. Therefore, to attain an ultra-high resolution one can expect a trend toward appropriate combinations of these two methods or – as discussed above – toward the use of selective adsorption.

STELLAN HJERTÉN
Uppsala

November 2003

Preface

Although the fifth edition of this textbook, published in 1992, is still being bought, it is high time to bring out a completely revised new edition. Techniques, such as paper chromatography, countercurrent chromatography, and field-flow fractionation, have been eclipsed by more convenient, faster, more specific and sensitive methods of differential migration, and electronic technology has revolutionized every aspect of chromatography and electrophoresis. This will be the last edition of CHROMATOGRAPHY, not only because of my age, but because hard-copy books cannot compete with their electronic counterparts, which can be continuously updated and rapidly accessed.

Again, I am proud to present a star-studded cast of authors, offering their extensive knowledge of specialized areas in analytical chemistry and ample leads for further reading. I have modified their contributions to avoid overlap as much as possible and to create a more uniform style. I have deleted such expressions as "see also, *e.g.*, the reviews XY and references therein", because most readers will realize that the authors have cited only a selection of references, some of which are review articles. As an editor of the *Journal of Chromatography* I have long fought a losing battle against lab slang, *e.g.*, "hyphenated methods", but I have not yielded to widely practiced abuses of grammar, such as changing transitive verbs to intransitive forms and *vice versa* (*e.g.*, "to elute" and "to adsorb") and confusing misnomers, such as "support", or fuzzy terms, such as "phase". Combined methods are not "hyphenated"; ion-exchange chromatography and size-exclusion chromatography are hyphenated.

Each chapter has been refereed by another author of this book, and authors have checked the edited versions of their chapters, but I am ultimately responsible for the accuracy of the contents of the entire book. This includes lists of abbreviations and symbols and the subject index for the entire book. In the acronyms for the combination of methods ("hyphenation") I have used slashes instead of hyphens. All chapters were up to date in 2003, and some of the references will undoubtedly soon be outdated. Each chapter is either new or completely revised. Only five authors of the fifth edition have survived in the sixth edition. The chapters on Electrochromatography, Combined Techniques, Microfluidics, Instrumentation, Phytochemical Analysis, Forensic Applications, and Computer Resources are entirely new. Proteins are now treated in two separate chapters, one on chromatography and one on electrophoresis.

As before, this book is divided into two parts: Part A deals with fundamentals and techniques of chromatography and electrophoresis, and Part B with their applications. The lists of abbreviation and symbols and the subject index are the same in both parts. I hope this textbook will serve the novice as an introduction to the vast literature in this branch of analytical chemistry and give the experienced research scientist a perspective of activities outside his/her own laboratory. The last chapter in each part will probably be the first to become obsolete, but these chapters are, in my opinion, the most useful introductions to the practice of chromatography and electrophoresis.

ERICH HEFTMANN
Orinda, California

Erich Heftmann (Editor)
Chromatography, 6th edition
Journal of Chromatography Library, Vol. 69A

1

Chapter 1

Theory of chromatography

VICTORIA L. McGUFFIN

CONTENTS

1.1 Introduction 2
1.2 Resolution 2
    1.2.1 Effect of plate number on resolution 7
    1.2.2 Effect of retention factor on resolution 7
    1.2.3 Effect of selectivity factor on resolution 7
    1.2.4 Implications 9
1.3 Principles of solute-zone dispersion 9
    1.3.1 Diffusion phenomena 10
    1.3.2 Flow phenomena 14
    1.3.3 Column contributions to dispersion 17
        1.3.3.1 Multiple paths 17
        1.3.3.2 Longitudinal diffusion 18
        1.3.3.3 Resistance to mass transfer in the mobile phase 19
        1.3.3.4 Resistance to mass transfer in the stationary phase 20
        1.3.3.5 Sum of column contributions 20
        1.3.3.6 Implications 23
    1.3.4 Extra-column contributions to dispersion 33
    1.3.5 Implications 38
1.4 Principles of solute-zone separation 38
    1.4.1 Molecular interactions 39
    1.4.2 Thermodynamics 43
    1.4.3 Partition 47
    1.4.4 Adsorption 54
    1.4.5 Complexation 60
    1.4.6 Ion exchange, ion exclusion, and ion pairing 64
    1.4.7 Hydrodynamic and size exclusion 67
    1.4.8 Implications 71
1.5 Optimization of chromatographic separations 72
    1.5.1 Performance criteria 72
    1.5.2 Parameter space 74

  1.5.3 Optimization methods 75
      1.5.3.1 Simultaneous methods 75
      1.5.3.2 Sequential methods 75
      1.5.3.3 Regression methods 77
      1.5.3.4 Theoretical methods 79
  1.5.4 Univariate optimization 79
      1.5.4.1 Temperature 79
      1.5.4.2 Stationary-phase composition 80
      1.5.4.3 Mobile-phase composition 82
      1.5.4.4 Other parameters 83
  1.5.5 Multivariate optimization 84
      1.5.5.1 Temperature and pressure 84
      1.5.5.2 Temperature and flow-rate 84
      1.5.5.3 Temperature and stationary-phase composition 85
      1.5.5.4 Temperature and mobile-phase composition 86
      1.5.5.5 Mobile-phase and stationary-phase compositions 87
  1.5.6 Implications 88
1.6 Summary 88
References 88

1.1 INTRODUCTION

In this chapter, the theoretical and fundamental background as broadly applicable to gas, supercritical-fluid, and liquid chromatography will be presented. A comprehensive treatment of the theory of chromatography would require several volumes to cover thoroughly all of the important topics. Hence, the present treatment must be limited in both scope and depth. The scope of the present treatment will be focused on the concepts of resolution, solute-zone dispersion or broadening, solute-zone retention and selectivity, and the consequent methods for optimization. For more detail on these or other theoretical and fundamental topics, the reader is referred to the many excellent texts [1–4].

1.2 RESOLUTION

Ultimately, the goal of all chromatographic methods is to separate the individual components or solutes in the sample. The resolution, R_s, is used to express the extent of separation between two adjacent solutes, *1* and *2*

$$R_s = \frac{2(t_{R2} - t_{R1})}{(w_1 + w_2)} \tag{1.1}$$

where t_R is the retention time and w is the width of each solute zone. The retention time is used for qualitative analysis, *i.e.* to determine the identity of the solute by comparison with known standards. The peak height, h, or area is used for quantitative analysis, i.e. to determine the concentration of the solute by comparison with known standards. When the separation is incomplete, however, the retention time and height of the overlapped solute zones may be different from those of the individual zones. Table 1.1 shows a series of solute zones with the extent of separation ranging from severely overlapped ($R_s = 0.5$) to resolved at the baseline ($R_s = 1.5$). The errors in t_R and h are summarized for each of the solute zones. It is important to note that the error in t_R is always positive for Solute *1* and negative for Solute *2*, because the overlap of the zones adds to the trailing edge of Solute *1* and the leading edge of Solute *2*. However, the error in h is always positive for both solutes. The errors in t_R and h are greater for Solute *1* than for Solute *2* when the zones are of equal height. This trend is observed because the zone for Solute *1* is slightly narrower than that for Solute *2*, which provides less overlap and error. In general, a resolution of 1.0 is usually sufficient for qualitative analysis and a resolution of 1.5 or greater is needed for accurate quantitative analysis. Table 1.2 shows solute zones with the same resolution ($R_s = 1.0$) but varying height ratios from 1:1 to 16:1. It is apparent that the errors in both t_R and h are more significant for the smaller zone. The errors become greater for the smaller zone (Solute *2*) but become lesser for the larger zone (Solute *1*) as the height ratio increases. For large height ratios, resolution in excess of 1.5 may be needed for accurate qualitative and quantitative analysis of the smaller zone [5].

The peak width in Eqn. 1.1 is associated with the extent of solute-zone dispersion, which will be discussed in more detail in Sec. 1.3. The peak width is related to the plate number, N

$$N = \frac{t_R^2}{\sigma^2} \qquad (1.2)$$

where σ is the standard deviation of the solute zone. For a symmetric Gaussian peak, as shown in Fig. 1.1, the standard deviation can be determined from the peak width at the baseline ($w = 4\sigma$) or the peak width at half-height ($w_{1/2} = 2.354\sigma$). Accordingly, the plate height can be written as

$$N = 16\left(\frac{t_R^2}{w^2}\right) = 5.545\left(\frac{t_R^2}{w_{1/2}^2}\right) \qquad (1.3)$$

In general, the width at half-height can be measured with greater accuracy than the width at the baseline, especially if there is significant overlap of the solute zones.

The retention time in Eqn. 1.1 is associated with the extent of solute-zone separation, which will be discussed in more detail in Sec. 1.4. The retention time is related to the retention factor, k, by

$$k = \frac{t_R - t_0}{t_0} \qquad (1.4)$$

where t_0 is the elution time of a nonretained solute. The retention factor is a normalized measure of solute retention that is independent of the column radius, length, and flow-rate. As such, it is more reliable and reproducible than retention time alone. It is also useful

TABLE 1.1

EFFECT OF RESOLUTION ON THE ERROR IN RETENTION TIME (t_R) AND PEAK HEIGHT
(h) FOR CHROMATOGRAPHIC PEAKS OF EQUAL HEIGHT

	Resolution	Peak height h_1:h_2	Error t_{R1} (%)	Error t_{R2} (%)	Error h_1 (%)	Error h_2 (%)
	0.50	1:1	3.22	−1.41	22.7	13.3
	0.75	1:1	0.241	−0.105	2.03	0.581
	1.00	1:1	0.013	−0.002	0.112	0.014
	1.25	1:1	0.000	0.000	0.002	0.000
	1.50	1:1	0.000	0.000	0.000	0.000

TABLE 1.2

EFFECT OF RESOLUTION ON THE ERROR IN RETENTION TIME (t_R) AND PEAK HEIGHT (h) FOR CHROMATOGRAPHIC PEAKS OF VARYING HEIGHT

	Resolution	Peak height $h_1:h_2$	Error t_{R1} (%)	Error t_{R2} (%)	Error h_1 (%)	Error h_2 (%)
	1.00	1:1	0.013	−0.002	0.112	0.014
	1.00	2:1	0.005	−0.005	0.056	0.027
	1.00	4:1	0.003	−0.009	0.028	0.054
	1.00	8:1	0.003	−0.019	0.014	0.109
	1.00	16:1	0.003	−0.037	0.007	0.222

to express the ratio of the retention factors for two solutes by means of the selectivity factor (α)

$$\alpha = \frac{k_2}{k_1} \tag{1.5}$$

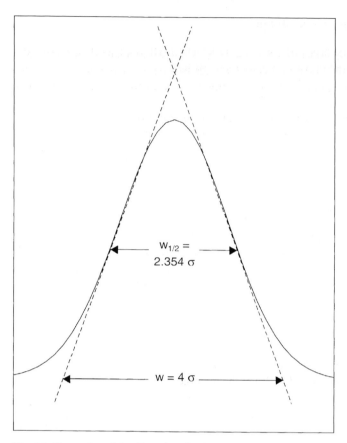

Fig. 1.1. Properties of the Gaussian distribution. w is determined by drawing tangents at the inflection point to measure width at the base; $w_{1/2}$ is determined by measuring peak width at half of maximum peak height.

where $k_2 > k_1$. The selectivity factor is a measure of the relative affinity of the mobile and stationary phases for two adjacent solutes.

By substitution of Eqns. 1.3 to 1.5 into Eqn. 1.1, we obtain a useful expression for the resolution known as the Purnell Equation [1]

$$R_s = \frac{N^{1/2}}{4}\left(\frac{\alpha - 1}{\alpha}\right)\left(\frac{k}{1 + k}\right) \tag{1.6}$$

Although there are other valid expressions for the resolution derived with different assumptions [6], the Purnell Equation is relatively common and will be used throughout this chapter. The Purnell Equation clearly shows the relationship between the resolution and the retention factor, selectivity factor, and plate number. In Secs. 1.2.1 to 1.2.3, we will explore the effect of each of these parameters on the resolution and the quality of the separation.

1.2.1 Effect of plate number on resolution

In Fig. 1.2A, a separation is shown for two solutes under standard conditions ($k = 3.0$, $\alpha = 1.10$, $N = 3500$, $R_s = 1.00$). When the plate number is increased to 7000 with all other parameters remaining constant, the resolution increases to 1.42. This increase in resolution is shown in Fig. 1.2B.

In order to examine the effect of plate number factor on resolution in a more quantitative manner, it is helpful to calculate the partial derivative of Eqn. 1.6

$$\frac{\partial R_s}{\partial N} = \frac{1}{8N^{1/2}} \left(\frac{\alpha - 1}{\alpha} \right) \left(\frac{k}{1 + k} \right) \tag{1.7}$$

When this partial derivative is evaluated for the values of the parameters given above for Fig. 1.2A, the resulting magnitude is 1.44×10^{-4}. In other words, a unit change in plate number is expected to cause the resolution to change by 0.0144%. In practice, the plate number is generally controlled by varying the particle diameter in packed columns, the diameter in open-tubular columns, and the column length. The effect of these parameters will be discussed in more detail in Sec. 1.3.

1.2.2 Effect of retention factor on resolution

In Fig. 1.2A, a separation is shown for two solutes under standard conditions ($k = 3.0$, $\alpha = 1.10$, $N = 3500$, $R_s = 1.00$). When the average retention factor is increased to 6.0 with all other parameters remaining constant, the resolution increases to 1.16. This increase in resolution is shown in Fig. 1.2C.

In order to examine the effect of retention factor on resolution in a more quantitative manner, it is helpful to calculate the partial derivative of Eqn. 1.6

$$\frac{\partial R_s}{\partial k} = \frac{N^{1/2}}{4} \left(\frac{\alpha - 1}{\alpha} \right) \left(\frac{1}{(1 + k)^2} \right) \tag{1.8}$$

When this partial derivative is evaluated for the values of the parameters given above for Fig. 1.2A, the resulting magnitude is 8.4×10^{-2}. In other words, a unit change in retention factor is expected to cause the resolution to change by 8.4%. In practice, the retention factor is generally controlled by varying the temperature in gas chromatography and by varying the mobile-phase composition or the temperature in liquid chromatography. These parameters will be discussed in more detail in Sec. 1.4.

1.2.3 Effect of selectivity factor on resolution

In Fig. 1.2A, a separation is shown for two solutes under standard conditions ($k = 3.0$, $\alpha = 1.10$, $N = 3500$, $R_s = 1.00$). When the selectivity factor is increased to 1.20 with all other parameters remaining constant, the resolution increases to 1.83. This increase in resolution is shown in Fig. 1.2D.

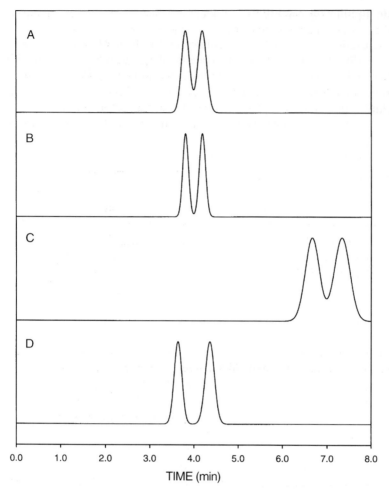

Fig. 1.2. Effect of retention factor, selectivity factor, and plate number on resolution. (A) $k = 3.0$, $\alpha = 1.10$, $N = 3500$, $R_s = 1.00$, (B) $k = 3.0$, $\alpha = 1.10$, $N = 7000$, $R_s = 1.42$, (C) $k = 6.0$, $\alpha = 1.10$, $N = 3500$, $R_s = 1.16$, (D) $k = 3.0$, $\alpha = 1.20$, $N = 3500$, $R_s = 1.83$.

In order to examine the effect of selectivity factor on resolution in a more quantitative manner, it is helpful to calculate the partial derivative of Eqn. 1.6

$$\frac{\partial R_s}{\partial \alpha} = \frac{N^{1/2}}{4}\left(\frac{1}{\alpha^2}\right)\left(\frac{k}{1+k}\right) \tag{1.9}$$

When this partial derivative is evaluated for the values of the parameters given above for Fig. 1.2A, the resulting magnitude is 9.17. In other words, a unit change in selectivity factor is expected to cause the resolution to change by 917%. In practice, the selectivity factor is generally controlled by varying the type of stationary phase in gas chromatography and by varying the type of mobile and stationary phases in liquid chromatography. These parameters will be discussed in more detail in Sec. 1.4.

1.2.4 Implications

From the results shown above, it is apparent that resolution is influenced most by the selectivity factor (Eqn. 1.9), followed by the retention factor (Eqn. 1.8), and then by the plate number (Eqn. 1.7). This conclusion suggests the most effective order for optimizing a separation: First the selectivity is adjusted by choosing appropriate mobile and stationary phases for the solutes of interest. Next, the retention factors of the solutes are adjusted within a reasonable range $(0.5 < k < 20)$ by means of the temperature or mobile-phase composition. Finally, the plate number is adjusted to provide the required resolution for the separation.

It is important to note that the parameters that influence resolution are under our direct control in an experimental system. If we develop an understanding of the physical and chemical processes occurring on the column, the separation of solutes becomes a systematic science rather than a trial-and-error process or art. In the foreword to *Dynamics of Chromatography*, Giddings observed [2]:

> Theory (when correct) and experiment (if carefully executed) describe the same truths. The science of chromatography requires both approaches if it is to grow in proportion to the demands made on it… If some attempt is not made by the field's active workers to correlate the two, the study of chromatography will be in danger of becoming, on one side, an unrelated array of tens of thousands of separate facts and observations and, on the other side, a meaningless set of mathematics.

In order to facilitate this correlation between theory and experiment, we will discuss in detail the principles of solute-zone dispersion and separation. The principles of solute-zone dispersion, which govern the plate number, are discussed in Sec. 1.3. The principles of solute-zone separation, which govern the retention factor and selectivity factor, are discussed in Sec. 1.4. With this detailed understanding, we can undertake the systematic optimization of separations, as established in Sec. 1.5.

1.3 PRINCIPLES OF SOLUTE-ZONE DISPERSION

In order to understand the concept of resolution, it is important to consider the dispersion or broadening of solute zones along the column. The classical description of dispersion was based on the "plate theory", originally developed by Martin and Synge [7] and described in many texts [2,8]. Although this model is simple and easily understood, it does not offer sufficient physical insight about the effects of column dimensions, flow-rate etc. For this reason, we will rely on the "rate theory", which provides a more realistic and accurate model of dispersion processes [2]. In chromatographic systems, all molecular motion arises from two sources: diffusion and flow phenomena. We will consider the fundamental aspects of these processes separately before describing their influence on solute-zone dispersion.

1.3.1 Diffusion phenomena

Because of their kinetic energy, all molecules undergo random Brownian motion. This random motion can lead to a net migration or flux of molecules in a system that has a difference in concentration, temperature, pressure, etc. For a gradient in concentration, C, with respect to distance, x, the mass flux of solute per unit area, J, is given by Fick's first law [9]

$$J = -D\frac{dC}{dx} \tag{1.10}$$

This equation shows that the mass flux is proportional to the concentration gradient, but in the opposite direction. In other words, the solute molecules will migrate from a region of higher concentration to lower concentration. The proportionality constant is called the diffusion coefficient, D, which is characteristic of the solute and the medium in which it is diffusing. Fick's second law is given by

$$\frac{dC}{dt} = D\frac{d^2C}{dx^2} \tag{1.11}$$

where t is the time and all other variables are as previously defined. This equation states that the temporal rate of change in concentration is proportional to the spatial rate of change in the concentration gradient. Again, the constant of proportionality is the diffusion coefficient. For a discrete source, as we might have in elution chromatography, a solution to Fick's second law is

$$C = \frac{1}{2(\pi Dt)^{1/2}} \, exp\left(\frac{-x^2}{4Dt}\right) \tag{1.12}$$

This equation can be compared with a normalized Gaussian function

$$C = \frac{1}{(2\pi\sigma^2)^{1/2}} \, exp\left(\frac{-x^2}{2\sigma^2}\right) \tag{1.13}$$

It is apparent that diffusion from a discrete source leads to a Gaussian distribution, where the variance is given by the Einstein Equation [10]

$$\sigma^2 = 2Dt \tag{1.14}$$

This relationship is shown in Fig. 1.3, where the dispersion increases with the diffusion coefficient or with time. For a continuous source, as we might have in frontal chromatography, a solution to Fick's second law is

$$C = \frac{1}{2}\left(1 - erf\left(\frac{x}{2(Dt)^{1/2}}\right)\right) \tag{1.15}$$

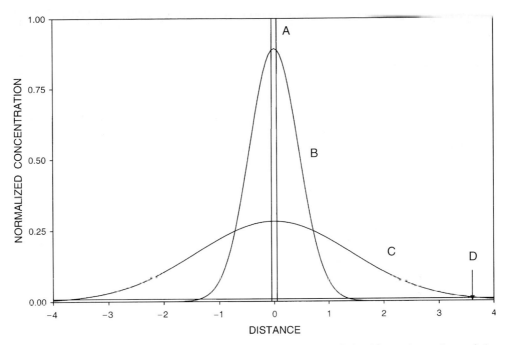

Fig. 1.3. Diffusion from a discrete source (elution chromatography) with varying values of the product of diffusion coefficient, D, and time, t. (A) $Dt = 0.00$, (B) $Dt = 0.01$, (C) $Dt = 1.00$, (D) $Dt = \infty$.

Thus, diffusion from a continuous source leads to a sigmoidal distribution. This relationship is shown in Fig. 1.4, where the dispersion increases with the diffusion coefficient or with time.

Diffusion coefficients have been measured for a large number of solutes and solvents [11,12]. However, as many solutes of interest have not been measured, it is necessary to have a reliable means to estimate diffusion coefficients. Both theoretical and empirical relationships have been used for this purpose. In the gas phase, a number of theoretical relationships have been derived. The Hirschfelder–Bird–Spotz Equation [13,14] is representative of those derived from the kinetic theory of gases

$$D_{AB} = \frac{1.86 \times 10^{-3} T^{3/2}}{p \delta_{AB}^2 \Omega_{AB}} \left(\frac{1}{M_A} + \frac{1}{M_B} \right)^{1/2} \tag{1.16}$$

where p is the pressure, T is the absolute temperature, Species A is the solute (infinite dilution), Species B is the solvent (bulk), M is the molecular weight, δ_{AB} is the minimum distance between A and B when they collide with zero kinetic energy, and Ω_{AB} is a function of the temperature and the intermolecular potential field between A and B. This equation has a firm theoretical foundation and provides an accurate estimation of the diffusion coefficient. However, the molecular parameters δ_{AB} and Ω_{AB} are difficult to determine, and the temperature dependence is known to vary from the power of

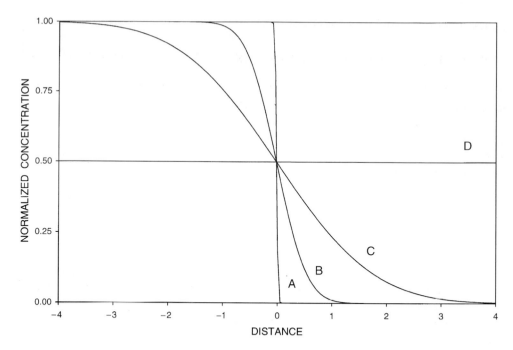

Fig. 1.4. Diffusion from a continuous source (frontal chromatography) with varying values of the product of diffusion coefficient, D, and time, t. (A) $Dt = 0.00$, (B) $Dt = 0.01$, (C) $Dt = 1.00$, (D) $Dt = \infty$.

1.5 to 2.0 [14]. A more practical approach is the empirical equation developed by Fuller *et al.* [15]

$$D_{AB} = \frac{1.00 \times 10^{-3} T^{1.75}}{p(\sum v_A^{1/3} + \sum v_B^{1/3})^2} \left(\frac{1}{M_A} + \frac{1}{M_B}\right)^{1/2} \tag{1.17}$$

where the empirical atomic diffusion volumes (v) are summed for Solute A and Solvent B. The volumes are given for carbon (16.5), hydrogen (1.98), oxygen (5.48), nitrogen (5.69), sulfur (17.0), and chlorine (19.5), as well as an adjustment for an aromatic ring (-20.2). The Fuller–Schettler–Giddings Equation is generally able to estimate diffusion coefficients within ± 5–10%.

Some typical values of the diffusion coefficient in the gas phase are summarized in Table 1.3 [1]. In general, diffusion coefficients are in the range of 0.01–1.0 cm^2/sec in the gas phase. A number of important trends are evident in these data. First, as the molecular weight of the solute increases, the diffusion coefficient decreases (*e.g.*, *n*-heptane and *n*-octane in hydrogen). For solutes with the same molecular weight, a smaller molecular volume results in a larger diffusion coefficient (*e.g.*, *n*-octane and 2,2,4-trimethylpentane in hydrogen). For the same solute, an increase in the molecular weight of the solvent results in a decrease in the diffusion coefficient (*e.g.*, *n*-octane in hydrogen, deuterium, and nitrogen). Because the molecular weight of the solvent is much smaller than the solute,

TABLE 1.3

TRENDS AND TYPICAL VALUES FOR GAS-PHASE DIFFUSION COEFFICIENTS $(D_{AB})^*$

Solute (A)	Solvent (B)	D_{AB} (cm^2/sec)
n-Heptane	Hydrogen	0.283
n-Octane	Hydrogen	0.277
2,2,4-Trimethylpentane	Hydrogen	0.292
n-Octane	Deuterium	0.208
n-Octane	Nitrogen	0.0726

* Diffusion coefficients measured at 30 °C and 1 atm [1].

it has a greater effect on the diffusion coefficient. All of these trends are evident from Eqn. 1.17. It is also evident that diffusion coefficients will increase with temperature and decrease with pressure.

In the liquid phase, one of the most common theoretical models is the Stokes–Einstein Equation [12,14]

$$D_{AB} = \frac{RT}{6\pi r_A \eta_B N} \tag{1.18}$$

where r_A is the solute radius, η_B is the solvent viscosity, and N is Avogadro's number. Because this model assumes that the molecules are non-interacting hard spheres, deviations from ideal behavior can lead to errors on the order of ±50–200%. More accurate predictions can be obtained from semi-empirical models, such as the Wilke–Chang Equation [16]

$$D_{AB} = \frac{7.4 \times 10^{-8}(\Psi_B M_B)^{0.5} T}{\eta_B V_A^{0.6}} \tag{1.19}$$

where V_A is the solute molal volume, which can be determined according to the method of LeBas [17]. The volume increments are summed for carbon (14.8), hydrogen (3.7), oxygen (7.4 for ethers and alcohols, 9.1–11 for esters, 12 for acids), and nitrogen (10.5–15.6 for primary to tertiary amines), with an adjustment for an aromatic ring (− 15). The solvent association factor, Ψ_B, is given as 1.0 for nonpolar solvents such as hexane, benzene, and ether, 1.5 for ethanol, 1.9 for methanol, and 2.6 for water. Other empirical equations have been reviewed and compared by Li and Carr [18,19].

Some typical values of the diffusion coefficient in the liquid phase are summarized in Table 1.4 [1]. In general, diffusion coefficients are on the order of $0.5–5.0 \times 10^{-5}$ cm^2/sec in the liquid phase. Many of the same general trends arise for molecular weight and volume as in the gas phase. However, there are also some interesting effects that arise from solvation. For example, the diffusion coefficients for benzene, ethyl acetate, and acetic acid systematically decrease because of the ability to form hydrogen bonds with the

TABLE 1.4

TRENDS AND TYPICAL VALUES FOR LIQUID-PHASE DIFFUSION COEFFICIENTS $(D_{AB})^*$

Solute (A)	Solvent (B)	D_{AB} (cm^2/sec)
Benzene	Methanol	2.12×10^{-5}
Ethyl acetate	Methanol	2.10×10^{-5}
Acetic acid	Methanol	1.54×10^{-5}
Acetic acid	Benzene	1.92×10^{-5}

* Diffusion coefficients measured at 15 °C and 1 atm [1].

solvent methanol. When hydrogen bonding with the solvent is not possible, the diffusion coefficient increases (*i.e.* acetic acid in methanol and benzene).

In the supercritical phase, diffusion coefficients can be estimated by the same theoretical and semi-empirical equations as for liquids [20]. In general, diffusion coefficients are on the order of $10^{-3} - 10^{-4}$ cm^2/sec in the supercritical phase.

1.3.2 Flow phenomena

In general, the mobile phases in gas, supercritical-fluid, and liquid chromatography behave as Newtonian fluids. For such fluids, an applied stress, such as pressure, causes a linearly proportional change in the velocity or position. Most theoretical models of flow in chromatographic systems presume Newtonian behavior. However, all fluids have an initial time in which non-Newtonian behavior is observed. During this time, the applied stress serves to compress the molecules until they reach the appropriate density at the given temperature and pressure. This behavior may lead to deviations from the theoretical models described herein. However, such deviations are generally small once the system has reached steady state [21].

There are three general regimes of flow in chromatographic systems: molecular, laminar, and turbulent, given in order of increasing velocity [3,14]. To distinguish between these different flow regimes, it is helpful to define the Reynolds number, *Re*,

$$Re = \frac{\rho u d}{\eta} \tag{1.20}$$

where ρ and η are the density and absolute viscosity of the fluid, u is the linear velocity, and d is the characteristic dimension of the system, either the column diameter for an open-tubular column or the particle diameter for a packed column. The Reynolds number is a ratio of the inertial forces (ρu^2) to the viscous forces ($\eta u/d$) in the fluid. The inertial forces tend to preserve the fluid in motion while the viscous forces tend to restrain motion. The balance of these forces influences the nature of flow within the system. Molecular or Knudsen flow occurs when the mean free path of the molecules is large with respect to the characteristic dimension d. The mean free path, λ, represents the average distance traveled between molecular collisions.

$$\lambda = \frac{1.256\eta}{p}\left(\frac{RT}{M}\right)^{1/2}$$

(1.21)

For the gases nitrogen, hydrogen, and helium, the mean free paths are 60, 120, and 190 nm, respectively, at 20 °C and 1 atm. When the mean free path is larger than the characteristic dimension, the molecules are more likely to collide with the walls of the system than with other molecules. This influences the net flow-rate, which is given by

$$F = \frac{-4}{3}\left(\frac{2\pi RT}{M}\right)^{1/2}\left(\frac{d_c^3}{8L}\right)\left(\frac{\Delta p}{p}\right)$$

(1.22)

where d_c is the diameter and L is the length of the flow channel, and Δp is the difference between the outlet and inlet pressures. Molecular flow is important in macropores (>25 nm) and mesopores ($1-25$ nm) for gases and in micropores (<1 nm) for liquids.

Laminar or Poiseuille flow occurs when viscous forces prevail, so that the Reynolds number is less than $2000-2500$ for gases and liquids flowing through an open tube. This is the most common type of flow in chromatography, since Reynolds numbers are typically in the range of $0.1-10$ [22]. Because the viscous forces predominate, there is friction along the wall that results in well-defined streamlines with a parabolic radial velocity profile

$$u = 2\bar{u}\left(1 - \frac{r^2}{R^2}\right)$$

(1.23)

where r is the radial coordinate and R is the radius of the open tube. The average linear velocity, $\bar{u}$, is given by the Hagen–Poiseuille Equation [14,23,24]

$$\bar{u} = \frac{-R^2\Delta p}{8\eta L}$$

(1.24)

Although the streamlines are not as well defined in a packed bed, the average linear velocity is given by a similar relationship

$$\bar{u} = \frac{-B_0\Delta p}{\eta L}$$

(1.25)

where the chromatographic permeability, B_0, is given by the Kozeny–Carman Equation [25] as

$$B_0 = \frac{d_p^2\varepsilon_u^3}{180\Psi^2(1 - \varepsilon_u)^2\varepsilon_t}$$

(1.26)

In this equation, d_p is the particle diameter and Ψ^2 is a shape factor that ranges from 1.0 for solid, spherical particles to 1.7 for porous, irregular particles. The total porosity, ε_t, of the packed bed is the volume occupied by the mobile phase divided by the total column volume. The total porosity consists of two distinct fractions, the interparticle and intraparticle porosities. The interparticle porosity, ε_u, is the volume between particles divided by the total column volume, which has values of 0.26 for closest packed spherical particles, 0.38 for randomly packed spherical particles, and 0.42 for randomly packed irregular particles. For typical chromatographic columns, the interparticle porosity is in the range of 0.40–0.45. The intraparticle porosity, ε_i, is the volume within the particle pores divided by the total column volume, which has a typical value of 0.5 for many chromatographic packing materials [26]. The volumetric flow-rate, F, is given by

$$F = \pi R^2 \varepsilon_t u = \frac{\pi R^2 \varepsilon_t L}{t_0} \tag{1.27}$$

Thus, the total porosity can be determined by measuring the flow-rate and the elution time of a nonretained solute on a column of known radius and length. The interparticle porosity can be determined by a similar measurement, using a nonretained solute that is too large to enter the pores of the packing material. The intraparticle porosity can then be determined by difference.

Eqn. 1.24 for open-tubular columns and Eqns. 1.25 and 1.26 for packed columns indicate several important trends in the laminar flow regime. First, the linear velocity increases with the square of the radius for an open tube or the square of the particle diameter for a packed column. For a column of fixed dimensions, the linear velocity increases with the applied pressure. The viscosity has an inverse effect, as the increased frictional forces reduce the velocity. It is interesting to note that an increase in temperature causes an increase in viscosity for gases and a concomitant decrease in velocity, whereas it causes a decrease in viscosity for liquids and a concomitant increase in velocity.

Turbulent flow occurs when inertial forces prevail, so that the Reynolds number is greater than 2000–2500 for gases and liquids. Some discrepancy exists in the literature regarding the point of onset for turbulence, because it is a gradual rather than sudden occurrence. In fully developed turbulent flow, a laminar sub-layer exists near the wall where frictional forces serve to maintain the parabolic velocity profile. In the central core, where there is little friction and great momentum, turbulent eddies arise that facilitate radial mass transport and cause a flat velocity profile. The transitional region between the laminar sub-layer and the turbulent core is known as the buffer layer. Theoretically, turbulent flow has two important advantages over laminar flow: a decrease in zone dispersion and a decrease in analysis time [2]. To achieve turbulent flow and the requisite high Reynolds numbers (Eqn. 1.20), the linear velocity and column or particle diameter should be increased and the kinematic viscosity, η/ρ, should be as small as possible. Accordingly, turbulent flow has been achieved using both open-tubular and packed columns in gas chromatography because of the favorable properties of the mobile phase [27,28]. However, there are technological problems in generating sufficient pressure to attain high linear velocities in liquid chromatography, such that turbulent flow has only been achieved for nonretained solutes in open tubes [28,29]. As a consequence, other methods have been investigated to enhance radial mass transport in chromatographic columns [30]. Golay has examined a variety of column cross-sections, including round, elliptical, square, and rectangular [31–33]. Others have investigated the effect of distortion of the column cross-section [34], tight coiling [35,36], and liquid flow segmented by gas bubbles [37]. Although these approaches have merit, they have not yielded a substantive improvement in performance over conventional open-tubular or packed columns. An approach that does appear to be promising is electro-osmotic flow, which uses an applied voltage rather than applied pressure to induce flow [38]. In a glass or silica capillary, the silanol groups at the surface are weakly acidic and carry a net negative charge at pH values greater than 7. A double layer, predominantly consisting of H^+ but also containing Na^+, K^+, and other cations, is formed to counterbalance this charge. Because the layer of cations is not covalently bound to the surface, it can migrate in an applied electric field. Thus, a flow of cations originates at the wall that carries along the solvent and other ions in the bulk liquid. The radial velocity profile, given by the Rice–Whitehead Equation [39], is nearly flat

$$u = \bar{u}\left(1 - \frac{I_0(\kappa r)}{I_0(\kappa R)}\right) \tag{1.28}$$

where κ^{-1} is the Debye length and I_0 is the zero-order modified Bessel function of the first kind. The maximum velocity is given by the Helmholtz–Smoluchowski Equation [40,41] as

$$\bar{u} = \frac{-\varepsilon\varepsilon_0\zeta\Delta V}{\eta L} \tag{1.29}$$

where ε is the permittivity of the mobile phase, ε_0 is the permittivity of vacuum, ΔV is the difference between the outlet and inlet voltages, and ζ is the zeta potential, which is on the order of $100-150$ mV for glass or silica surfaces in contact with an aqueous buffer system [42]. The flat radial velocity profile of electro osmotic flow can reduce zone dispersion compared with traditional laminar flow in open-tubular and packed columns, as will be discussed in Sec. 1.3.3.3.

1.3.3 Column contributions to dispersion

The rate theory presumes that there are several sources of zone dispersion, each of which acts independently in the chromatographic system. Because these sources are independent, their variances, σ_i^2, are additive

$$\sigma = \sum \sigma_i^2 \tag{1.30}$$

The contributions to dispersion from the chromatographic column include multiple paths (A), longitudinal diffusion (B), and resistance to mass transfer (C) in the mobile and stationary phases. There are several different mathematical approaches to describe these contributions, all of which lead to rather similar conclusions [2,31,43]. The general results are summarized in Secs. 1.3.3.1 to 1.3.3.6.

1.3.3.1 Multiple paths

In an open-tubular column, there is a single path through the column. In a packed column, however, molecules can take many different paths around the particles. Each of these paths will have a different distance or time traveled, leading to broadening of the solute zone. In the older literature, this process is frequently called "eddy diffusion", which is misleading because it suggests that turbulent eddies and mixing are responsible for the broadening. A more accurate and descriptive term is "multiple paths", which will be used in this chapter. The paths most likely to be taken are those that offer the least impedance to mobile-phase flow, *i.e.* those with the largest diameters and shortest lengths. For this reason, packing uniformity is very important, because void spaces or large channels through the packing will destroy column efficiency. For a uniformly packed column, the variance contribution due to multiple paths is given by

$$\sigma_A^2 = 2\lambda d_p L \tag{1.31}$$

where λ is a constant of approximate value 0.5–1.0 that reflects the packing structure. Hence, to a first approximation, the extent of broadening reflects only the particle diameter and the physical structure of the packed bed, but is independent of the linear velocity.

1.3.3.2 Longitudinal diffusion

The random motion of molecules also leads to broadening of the solute zone. As discussed in Sec. 1.3.1, this broadening results in a symmetric Gaussian zone profile with a variance given by Eqn. 1.14. The extent of diffusion depends upon the diffusion coefficient and the time over which diffusion occurs. For the mobile phase, the variance contribution due to diffusion in the mobile phase, $\sigma_{B,M}^2$, is given by

$$\sigma_{B,M}^2 = \frac{2\gamma_M D_M L}{u} \tag{1.32}$$

where D_M is the diffusion coefficient in the mobile phase and γ_M is a tortuosity factor. The tortuosity factor has a value of 1.00 for an open-tubular column, 0.73 for a packed column with nonporous, spherical particles, and 0.63 for a packed column with porous, irregular particles [44,45]. For the stationary phase, the variance contribution due to diffusion in the stationary phase, $\sigma_{B,S}^2$, is given by

$$\sigma_{B,S}^2 = \frac{2\gamma_S D_S k L}{u} \tag{1.33}$$

where D_S is the diffusion coefficient in the stationary phase and γ_S is a tortuosity factor. The tortuosity factor has a value of 1.00 for a uniform film but will decrease if the stationary phase is not uniform in thickness or forms droplets. The tortuosity factor will also vary if the stationary phase is not chemically homogeneous, as may occur in bonded phases for liquid chromatography. It is noteworthy that the variance in Eqn. 1.33 increases with the retention factor, since this increases the time spent diffusing in the stationary phase.

In gas chromatography, the diffusion coefficient in the mobile phase is typically 10^{-1} cm^2/sec, whereas that in the stationary phase is typically 10^{-6} cm^2/sec. Hence, the variance contribution in the mobile phase is approximately 5 orders of magnitude greater than that in the stationary phase, which can usually be neglected. In liquid chromatography, however, the diffusion coefficient in the mobile phase is typically

10^{-5} cm^2/sec and is more comparable to that in the stationary phase. Thus, both mobile and stationary phase contributions must usually be considered in liquid chromatography, but their magnitude is much smaller than those in gas chromatography.

1.3.3.3 Resistance to mass transfer in the mobile phase

Resistance to mass transfer refers to the tendency of a molecule to remain in its present location because of inertial forces. Molecules in the mobile phase will tend to stay in the same flow-stream and, unless something is done to enhance radial transport, only diffusion will allow them to change flow-streams. As each flow-stream has a different velocity, this leads to broadening of the solute zone. For laminar flow in an open tube, the parabolic radial velocity profile given by Eqn. 1.23 leads to a variance $\sigma_{C,M}^2$ [31]

$$\sigma_{C,M}^2 = \frac{(1 + 6k + 11k^2)d_c^2 Lu}{96(1 + k)^2 D_M} \tag{1.34}$$

For laminar flow in a packed column, a similar expression applies [43]

$$\sigma_{C,M}^2 = \frac{k^2 d_p^2 Lu}{100(1 + k)^2 D_M} \tag{1.35}$$

In each case, the extent of broadening depends upon the diffusion coefficient as well as the column or particle diameter, which represents the distance over which diffusion must occur in order to average the velocities.

For electro-osmotic flow, the nearly flat radial velocity profile, given by Eqn. 1.28, leads to a variance [46]

$$\sigma_{C,M}^2 = \frac{C_1 R^2 Lu}{D_M} \tag{1.36}$$

where

$$C_1 = \left(\frac{1}{4(1 - C_2)^2} - \frac{6}{(\kappa R)^2(1 - C_2)} + \frac{C_2^2}{2(1 - C_2)^2} \right)$$
$$- \left(\frac{1}{4(1 - C_2)} - \frac{2}{(\kappa R)^2} \right)\left(\frac{2}{1 + k} \right) + \frac{1}{4(1 + k)^2} \tag{1.37}$$

$$C_2 = \left(\frac{2}{\kappa R} \right)\left(\frac{I_1(\kappa R)}{I_0(\kappa R)} \right) \tag{1.38}$$

and all other symbols are as previously defined. Other contributions from resistance to mass transfer in the mobile phase have been described in detail by Giddings [2]. In general, all of these contributions are proportional to the linear velocity.

1.3.3.4 Resistance to mass transfer in the stationary phase

Similarly, once a molecule is in the stationary phase, inertial forces tend to make it remain there. Broadening occurs because some molecules reside longer than others in the stationary phase. For a partition mechanism (Sec. 1.4.3), the variance ($\sigma_{C,S}^2$) in an open-tubular or packed column is given by [31]

$$\sigma_{C,S}^2 = \frac{2kd_f^2 Lu}{3(1+k)^2 D_S} \tag{1.39}$$

Hence, the extent of broadening depends upon the diffusion coefficient as well as the thickness of the stationary phase film, d_f, which represents the distance over which diffusion must occur. If there is interfacial resistance to transport between the mobile and stationary phases, perhaps owing to high surface tension or specific configurational or orientational effects, an additional variance contribution must be considered [2]

$$\sigma_{C,S}^2 = \frac{2kLu}{(1+k)^2 k_{sm}} \tag{1.40}$$

where k_{sm} is the rate constant for transfer from stationary to mobile phase. For an adsorption mechanism (Sec. 1.4.4), the variance is given by a similar expression

$$\sigma_{C,S}^2 = \frac{2kLu}{(1+k)^2 k_d} \tag{1.41}$$

where k_d is the rate constant for desorption. Other contributions from resistance to mass transfer in the stationary phase have been described in detail by Giddings [2]. In general, all of these contributions are proportional to the linear velocity.

1.3.3.5 Sum of column contributions

As noted in Eqn. 1.30, the variances from all of the processes described above are additive in determining the total variance. The total variance is related to the plate height, H, and plate number, N, in the following manner

$$H = \frac{L}{N} = \frac{\sigma^2}{L} \tag{1.42}$$

Thus, for an open-tubular column, the plate height is given by summation of Eqns. 1.32, 1.33, 1.34, and 1.39 as

$$H = \frac{2D_M}{u} + \frac{2\gamma_S D_S k}{u} + \frac{(1+6k+11k^2)d_c^2 u}{96(1+k)^2 D_M} + \frac{2kd_f^2 u}{3(1+k)^2 D_S} \tag{1.43}$$

This is the well-known Golay Equation [31]. Similarly, for a packed column, the plate height is given by summation of Eqns. 1.31, 1.32, 1.33, 1.35, and 1.39

$$H = 2\lambda d_p + \frac{2\gamma_M D_M}{u} + \frac{2\gamma_S D_S k}{u} + \frac{k^2 d_p^2 u}{100(1+k)^2 D_M} + \frac{2kd_f^2 u}{3(1+k)^2 D_S} \tag{1.44}$$

Eqns. 1.43 and 1.44 can be written in an abbreviated form as

$$H = A + \frac{B}{u} + Cu \tag{1.45}$$

where A represents the term for multiple paths, B represents longitudinal diffusion in the mobile and stationary phases, and C represents resistance to mass transfer in the mobile and stationary phases.

The total plate height and the individual contributions are shown in Fig. 1.5 for a typical packed column for gas chromatography. It is apparent that the multiple path term is constant, the longitudinal diffusion term decreases, and the mass transfer term increases with linear velocity. Accordingly, the total plate height goes through a minimum value at a certain value of the linear velocity. In general, it is best to operate the chromatographic column at or near this optimum velocity in order to achieve the least broadening and the highest resolution. The optimum velocity can be determined by taking the first derivative of the plate height with respect to linear velocity in Eqn. 1.45.

$$\frac{\partial H}{\partial u} = \frac{-B}{u^2} + C = 0 \tag{1.46}$$

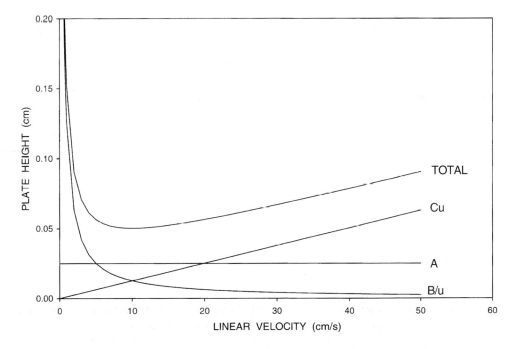

Fig. 1.5. Plate height *vs.* linear velocity for a typical packed column for gas chromatography. $d_p = 125\ \mu m$, $d_f = 1.0\ \mu m$, $D_M = 10^{-1}\ cm^2/sec$, $D_S = 10^{-6}\ cm^2/sec$, $k = 3.0$.

This derivative is set equal to zero to calculate the optimum velocity, u_{opt},

$$u_{opt} = \left(\frac{B}{C}\right)^{1/2}$$ (1.47)

The minimum plate height, H_{min}, can then be calculated by substituting this expression into Eqn. 1.45

$$H_{min} = A + \frac{B}{(B/C)^{1/2}} + C(B/C)^{1/2} = A + 2(BC)^{1/2}$$ (1.48)

For example, the typical packed column for gas chromatography shown in Fig. 1.5 is dominated by longitudinal diffusion in the mobile phase and resistance to mass transfer in the stationary phase. Accordingly, the optimum velocity is given by

$$u_{opt} = \left(\frac{3\gamma_M(1+k)^2 D_M D_S}{k d_f^2}\right)^{1/2} = 10 \text{ cm/sec}$$ (1.49)

and the minimum plate height is given by

$$H_{min} = 2\lambda d_p + \left(\frac{16\gamma_M k D_M d_f^2}{3(1+k)^2 D_S}\right)^{1/2} = 0.050 \text{ cm}$$ (1.50)

In many cases, the first term in this equation dominates and the minimum plate height for a packed column is approximately equal to two times the particle diameter.

The plate height *vs.* linear velocity graph is shown for a typical open-tubular column for gas chromatography in Fig. 1.6. The general shape of this graph is similar to that for the packed column in Fig. 1.5. However the plate heights are smaller because there is no contribution from multiple paths. The open-tubular column is dominated by longitudinal diffusion and resistance to mass transfer in the mobile phase. Accordingly, the optimum velocity is given by

$$u_{opt} = \left(\frac{192(1+k)^2 D_M^2}{(1 + 6k + 11k^2) d_c^2}\right)^{1/2} = 25 \text{ cm/sec}$$ (1.51)

and the minimum plate height is given by

$$H_{min} = \left(\frac{(1 + 6k + 11k^2) d_c^2}{12(1+k)^2}\right)^{1/2} = 0.016 \text{ cm}$$ (1.52)

As a general rule, the minimum plate height for an open-tubular column is approximately equal to 0.8 times the column diameter.

Finally, the plate height *vs.* linear velocity graph is shown for a typical packed column for liquid chromatography in Fig. 1.7. In liquid chromatography, the contributions to diffusion and mass transfer from the mobile and stationary phases are approximately equal. Consequently, none of the terms in Eqn. 1.44 can be neglected in calculating the optimum velocity and the minimum plate height. The optimum velocity (0.35 cm/sec) is substantially smaller than that for gas chromatography (10 cm/sec, Eqn. 1.49). This is

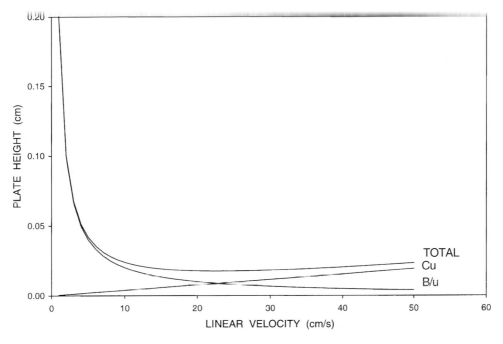

Fig. 1.6. Plate height *vs.* linear velocity for a typical open-tubular column for gas chromatography. $d_c = 200$ µm, $d_f = 0.25$ µm, $D_M = 10^{-1}$ cm^2/sec, $D_S = 10^{-6}$ cm^2/sec, $k = 3.0$.

because the optimum velocity is directly related to the diffusion coefficient in the mobile phase, which is four orders of magnitude smaller in liquids than in gases. The minimum plate height (0.0011 cm) is also substantially smaller than that for gas chromatography (0.050 cm, Eqn. 1.50), owing to the smaller particle diameter.

1.3.3.6 Implications

Upon comparison of Figs. 1.5 and 1.6, it is apparent that a typical open-tubular column has a smaller plate height and a higher optimum velocity than a typical packed column in gas chromatography. Consequently, open-tubular columns generally provide higher resolution and/or shorter analysis time than packed columns. This is demonstrated for the separation of a complex sample, calmus oil, in Fig. 1.8 [47]. Open-tubular columns offer other practical advantages as well. Because of their higher permeability (Eqn. 1.24), open-tubular columns of long length can be operated at modest pressures. Typical open-tubular columns are 10–100 m in length, whereas typical packed columns are 1–5 m. In addition, the surface of open-tubular columns is fused silica, which is more inert and has fewer contaminants than most packing materials. As a result, there are fewer problems with irreversible adsorption, and stationary phases can be deposited with smaller film thickness.

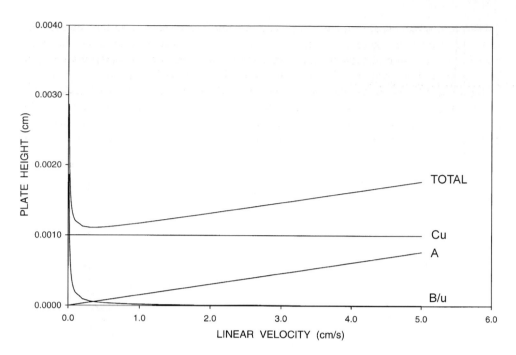

Fig. 1.7. Plate height *vs.* linear velocity for a typical packed column for liquid chromatography. $d_p = 5\ \mu m$, $d_f = 0.1\ \mu m$, $D_M = 10^{-5}\ cm^2/sec$, $D_S = 10^{-6}\ cm^2/sec$, $k = 3.0$.

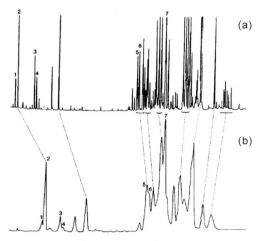

Fig. 1.8. Separation of calmus oil by gas chromatography using open-tubular and packed columns. (a) Open-tubular column, $d_c = 300\ \mu m$, $L = 50\ m$, OV-1 stationary phase. Temperature program: 60 to 200 °C at 4 °C/min. (b) Packed column, $d_c = 3.0$ mm, $d_p = 177–250\ \mu m$ (60/80 mesh) Gas Chrom Q with 5% OV-1 stationary phase, $L = 4$ m. Temperature program: 80 to 220 °C at 5 °C/min. Mobile phase: hydrogen. Adapted from Ref. 47.

All of these factors lead to higher plate numbers and higher resolution for open-tubular columns. However, there are some disadvantages of open-tubular columns, most notably that injection, detection, and connection volumes must be substantially reduced. These extra-column sources of dispersion will be discussed in Sec. 1.3.4.

For packed columns, the plate height is strongly dependent upon the particle diameter. As shown in Eqns. 1.31 and 1.35, the multiple-path term is linearly dependent and the resistance-to-mass-transfer term in the mobile phase is dependent upon the square of the particle diameter. The effect of particle diameter on plate height is shown in Fig. 1.9 for

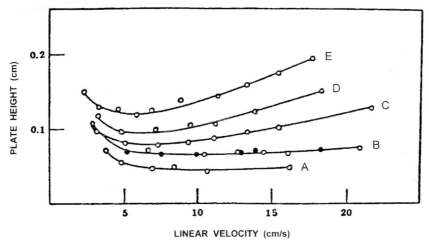

Fig. 1.9. Effect of particle diameter on plate height *vs.* linear velocity for gas chromatography. Column: d_p = (A) 100–149 μm (100/150 mesh), (B) 250–297 μm (50/60 mesh), (C) 297–420 μm (40/50 mesh), (D) 420–590 μm (30/40 mesh), (E) 590–840 μm (20/30 mesh) Silocel with 20% polyethylene glycol stationary phase. Mobile phase: nitrogen. Temperature: 47 °C. Solute: acetone. Adapted from Ref. 48.

gas chromatography [48]. Similar effects are, of course, observed for liquid chromatography as well. It is evident that the minimum plate height is decreased (Eqn. 1.48) and the optimum velocity is increased (Eqn. 1.47) as the particle diameter is decreased. Thus, columns with smaller particles can be used at higher linear velocity without sacrificing efficiency for analysis time. This suggests that the particle diameter should be decreased insofar as possible to obtain the highest resolution. However, there are technological problems in producing uniform particles of small diameter, which usually are limited to 37 μm (400 mesh) for gas chromatography and 1–3 μm for liquid chromatography. It is also important that the particles have a narrow distribution or range of diameters. Particles of differing sizes do not pack very efficiently, as shown in Fig. 1.10 [48], leading to plate heights that are higher than expected. In general, the range should be no more than 10–20% of the average particle diameter. Finally, there are other practical limitations to reducing the particle diameter. The pressure drop increases with the square

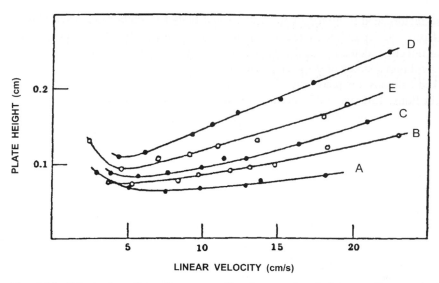

Fig. 1.10. Effect of particle diameter uniformity on plate height *vs.* linear velocity for gas chromatography. Column: (A) 0%, (B) 25%, (C) 50%, (D) 75%, (E) 100% of 420–590 μm (30/40 mesh) Silocel mixed with 250–297 μm (50/60 mesh) Silocel with 20% polyethylene glycol stationary phase. Mobile phase: nitrogen. Temperature: 47 °C. Solute: benzene. Adapted from Ref. 48.

of the particle diameter (Eqns. 1.25 and 1.26), which ultimately limits the column length that can be used. Moreover, the compressibility of gases, supercritical fluids, and even liquids must be considered at higher pressure. Finally, columns packed with small particles can easily become fouled, so samples must be filtered of all particulates (>0.45 μm).

For open-tubular columns, the plate height is strongly dependent upon the column diameter. As shown in Eqn. 1.34, the resistance-to-mass-transfer term in the mobile phase is dependent upon the square of the column diameter. Hence, the minimum plate height is decreased (Eqn. 1.52) and the optimum velocity is increased (Eqn. 1.51) as the column diameter is decreased. The effect of column diameter on the plate height *vs.* linear velocity is similar to that shown for particle diameter in Fig. 1.9. Hence, the column diameter should also be decreased insofar as possible to obtain the highest resolution. Open-tubular columns are commercially available with diameters from 180–530 μm for gas chromatography. However, columns developed for supercritical-fluid chromatography with diameters as small as 50 μm can also be used. For best results in liquid chromatography, open-tubular columns must have diameters on the order of 5–10 μm. Columns of these dimensions have been prepared in research laboratories [49,50], but are not commercially available.

For both packed and open-tubular columns, the plate height is dependent upon the thickness of the stationary phase. As shown in Eqn. 1.39, the resistance-to-mass-transfer term in the stationary phase is dependent upon the square of the film thickness. The effect of film thickness on plate height is shown in Fig. 1.11 for gas chromatography [51]. Similar effects are, of course, observed for liquid chromatography as well. It is evident that the minimum plate height is decreased (Eqn. 1.50) and the optimum velocity is increased

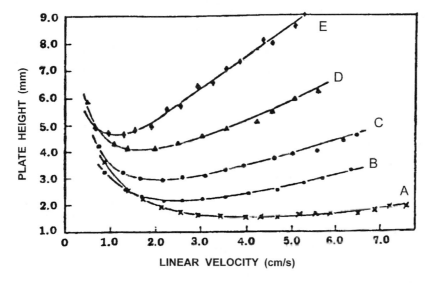

Fig. 1.11. Effect of stationary-phase film thickness on plate height *vs.* linear velocity for gas chromatography. Column: d_p = 250–590 μm (30/60 mesh) Kromat FB with (A) 18%, (B) 25%, (C) 34%, (D) 43%, (E) 50% polyethylene glycol stationary phase. Mobile phase: helium. Temperature: 25 °C. Solute: 2,3-dimethylbutane. Adapted from Ref. 51.

(Eqn. 1.49) as the film thickness is decreased. Thus, columns with smaller film thickness can be used at higher linear velocity without sacrificing efficiency for analysis time. This suggests that the film thickness should be decreased insofar as possible to obtain the highest resolution. However, there are technological problems in producing uniform, thin films of the stationary phase, which usually are limited to 3–5% (w/w) for packed columns and 0.1 μm for open-tubular columns for gas chromatography. In addition, there are other practical limitations to reducing the film thickness. Columns with very thin films can easily become overloaded, so sample volume and mass must be reduced.

Moreover, changes in film thickness and column or particle diameter have another important influence on separation. The phase ratio, β, which is the ratio of the volumes of the mobile and stationary phases, varies concurrently with these parameters. For an open-tubular column,

$$\beta = \frac{V_M}{V_S} = \frac{(d_c - 2d_f)^2}{d_c^2 - (d_c - 2d_f)^2} \tag{1.53}$$

This relationship is shown graphically in Fig. 1.12 [52]. A practical example of this effect is illustrated in the chromatogram of gasoline in Fig. 1.13 [47]. The film thickness and the corresponding phase ratio for each chromatogram are summarized in Table 1.5. It is

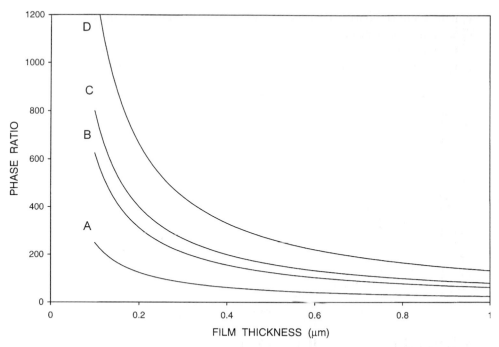

Fig. 1.12. Phase ratio for open-tubular columns with different diameter and stationary-phase film thickness for gas chromatography according to Eqn. 1.53. Column: $d_c =$ (A) 100 μm, (B) 250 μm, (C) 320 μm, (D) 530 μm.

apparent that the retention factors for all solutes increase with decreasing phase ratio, as will be discussed in Sec. 1.4.2 (Eqn. 1.81). The retention factor for toluene, which increases from 0.4 to 16.2, is cited as an example in Table 1.5. Accordingly, the chromatogram with the smallest film thickness (0.05 μm) shows very little retention of all solutes, but excellent plate height and resolution. As the film thickness increases to the maximum value of 2.0 μm, the solutes are more retained but also show much greater broadening. It is clear that some compromise is necessary to achieve the best separation. Film thickness and column diameter or particle diameter cannot be considered alone, but must be decreased together in order to maintain a reasonable phase ratio. In general, the phase ratio should be maintained within the range of 500–1000 for gas chromatography and 1–10 for liquid chromatography. This provides for a reasonable range of retention factors as well as a reasonable plate height. The retention factor has a direct influence on the plate height. As shown in Eqn. 1.33, diffusion in the stationary phase is linearly dependent upon the retention factor. In addition, the resistance-to-mass-transfer terms in the mobile and stationary phases have a complex dependence on retention factor, as illustrated for open-tubular columns in Eqns. 1.34 and 1.39 as well as Fig. 1.14. If the mobile-phase term predominates, the plate height will increase consistently with retention factor. If the stationary phase term predominates, the plate height will increase significantly with retention factors up to $k = 1.0$, and decrease thereafter.

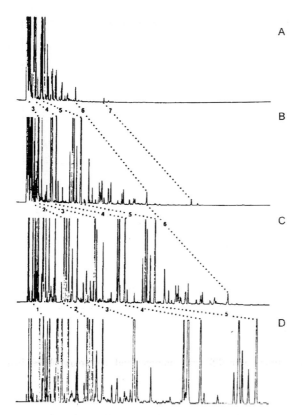

Fig. 1.13. Separation of gasoline by gas chromatography using open-tubular columns with varying film thickness. Column: $d_c = 300$ μm, $L = 15$ m, SE-52 stationary phase with $d_f =$ (A) 0.05, (B) 0.15, (C) 0.80, and (D) 2.00 μm. Temperature program: 20 to 80 °C at 2 °C/min. Solutes: (1) *n*-pentane, (2) benzene, (3) toluene, (4) *o*-xylene, (5) 1,2,4-trimethylbenzene, (6) naphthalene, (7) 2-methylnaphthalene. Adapted from Ref. 47.

TABLE 1.5

FILM THICKNESS, PHASE RATIO, AND RETENTION FACTOR FOR TOLUENE FROM THE CHROMATOGRAMS IN FIG. 1.13 [47]

Film thickness (μm)	Phase ratio	Retention factor
0.05	3000	0.4
0.15	1000	1.3
0.80	190	6.6
2.00	75	16.2

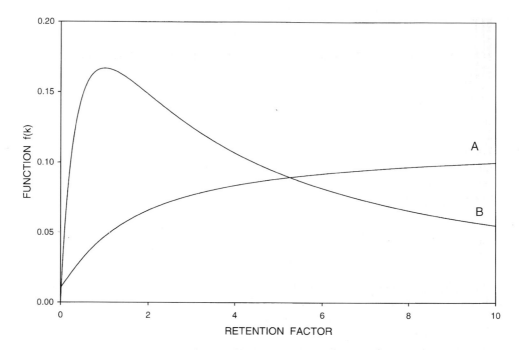

Fig. 1.14. Function of the retention factor for (A) mobile-phase and (B) stationary-phase resistance to mass transfer, according to Eqns. 1.34 and 1.39.

Depending upon the relative magnitude of the mobile- and stationary-phase terms, very complex relationships may be observed. A practical example of the effect of retention factor is illustrated in the chromatogram of *n*-alkanes in Fig. 1.15 [52]. The data from this chromatogram, summarized in Table 1.6, show that the plate height increases consistently with retention factor. This indicates that resistance to mass transfer in the mobile phase

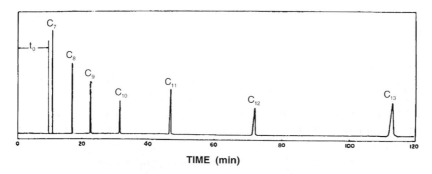

Fig. 1.15. Effect of the retention factor on plate height in gas chromatography. Column: $d_c = 250\ \mu m$, $L = 100$ m, SE-30 stationary phase. Mobile phase: nitrogen. Temperature: 130 °C. Solutes: (1) *n*-heptane, (2) *n*-octane, (3) *n*-nonane, (4) *n*-decane, (5) *n*-undecane, (6) *n*-dodecane, (7) *n*-tridecane. Adapted from Ref. 52.

TABLE 1.6

RETENTION FACTOR, PLATE HEIGHT, AND PLATE NUMBER FOR THE *n*-ALKANES IN FIG. 1.15 [52]

Solute	Retention factor	Plate height (cm)	Plate number
n-Heptane	0.14	0.0263	380,000
n-Octane	0.8	0.0296	338,000
n-Nonane	1.4	0.0344	291,000
n-Decane	2.3	0.0369	271,000
n-Undecane	3.9	0.0425	235,000
n-Dodecane	6.5	0.0469	213,000
n-Tridecane	10.8	0.0508	197,000

predominates in this open-tubular column. When mobile-phase effects are dominant, the minimum plate height is increased (Eqn. 1.52) and the optimum velocity is decreased (Eqn. 1.51) as the retention factor increases. Effects similar to those described above are observed for packed columns.

In addition to the retention factor, other solute properties also influence the plate height. As shown in Eqns. 1.32 and 1.33, diffusional broadening is linearly dependent upon the solute diffusion coefficient in the mobile and stationary phases. In addition, the resistance-to-mass-transfer terms are inversely dependent upon the solute diffusion coefficient in the mobile and stationary phases, as shown in Eqns. 1.34, 1.35, and 1.39. As a consequence, the plate height is strongly influenced by the diffusion coefficient and the solute properties on which it depends, such as molecular weight and volume (Eqns. 1.16 to 1.19). In addition, the mobile-phase and stationary-phase properties of molecular weight, density, and viscosity also affect the diffusion coefficient. For example, the effect of mobile-phase selection on the plate height *vs.* velocity graph is shown in Fig. 1.16 and Table 1.7 [53,54]. At velocities less than the optimum, the plate height is dominated by longitudinal diffusion. In this region, the plate height is lower for mobile phases with higher molecular weight and viscosity, such as nitrogen. At velocities greater than the optimum, the plate height is dominated by resistance to mass transfer. In this region, the plate height is lower for mobile phases with lower molecular weight and viscosity, such as helium and hydrogen. It is interesting that the minimum plate height is not dependent upon the diffusion coefficient (Eqn. 1.52), even though diffusion is implicated in the individual terms. This occurs because the longitudinal diffusion term, which is directly dependent upon D_M, and the resistance-to-mass-transfer term, which is inversely related to D_M, are exactly equal at the optimum velocity. Although the minimum plate height is unaffected, the optimum velocity increases with D_M, as indicated in Eqn. 1.51. Thus, the mobile phases with lower molecular weight and viscosity can be operated at higher velocities without sacrificing efficiency. This benefit is of particular importance for high-speed separations.

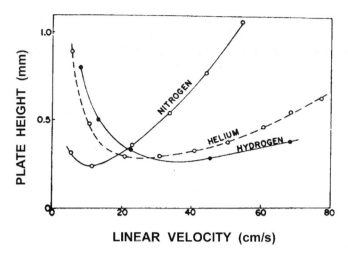

Fig. 1.16. Effect of mobile phase selection on plate height *vs.* linear velocity for gas chromatography. Column: $d_c = 250$ μm, $L = 25$ m, OV-101 stationary phase. Mobile phase: hydrogen, helium, nitrogen. Temperature: 175 °C. Solute: *n*-heptadecane. Adapted from Ref. 53.

Finally, the physical variables of temperature and pressure influence the plate height. Although not directly implicated in the plate height equations, these variables have an indirect effect through the retention factor and the diffusion coefficient. The effect of temperature on the retention factor will be discussed in Sec. 1.4.2 (Eqn. 1.82). An increase in temperature will cause a decrease in the retention factor, which subsequently influences the resistance-to-mass-transfer terms in Eqns. 1.34, 1.35, and 1.39. If the mobile-phase term predominates, the plate height will decrease consistently as temperature increases. If the stationary-phase term predominates, however, the plate height may decrease ($k < 1.0$)

TABLE 1.7

MOBILE-PHASE PROPERTIES, OPTIMUM VELOCITY, AND MINIMUM PLATE HEIGHT FROM FIG. 1.16 [53]

Mobile phase	Molecular weight	Viscosity (μP)[*]	D_M (cm^2/sec)[**]	Optimum velocity (cm/sec)	Minimum plate height (cm)
Nitrogen	28.0	242	0.093	13	0.022
Helium	4.0	264	0.306	21	0.028
Hydrogen	2.0	118	0.384	37	0.028

[*] Viscosity at 175 °C [54].
[**] Diffusion coefficient calculated from Eqn. 1.17 for *n*-heptadecane at 175 °C.

or increase ($k > 1.0$) as temperature increases. The effect of temperature and pressure on the diffusion coefficient is given by Eqns. 1.16 to 1.19. In general, the diffusion coefficient will increase with temperature and decrease with pressure, which subsequently influences the diffusional broadening in Eqns. 1.32 and 1.33 and the mass-transfer broadening in Eqns. 1.34, 1.35, and 1.39. At velocities less than the optimum, the plate height is dominated by longitudinal diffusion. In this region, the plate height will increase with increasing diffusion coefficient and, thus, with increasing temperature and decreasing pressure. At velocities greater than the optimum, the plate height is dominated by resistance to mass transfer. In this region, the plate height will decrease with increasing temperature and decreasing pressure. The minimum plate height is unaffected, but the optimum velocity will increase with increasing temperature and decreasing pressure.

1.3.4 Extra-column contributions to dispersion

In order to obtain the highest resolution possible from the chromatographic system, it is necessary to reduce all sources of dispersion. In the previous sections, we have discussed how to reduce column contributions to dispersion by operating at the optimum velocity, by reducing column diameter, particle diameter, stationary-phase film thickness, etc. Now, we will discuss the extra-column sources of dispersion and how they affect the chromatographic efficiency of the system [55,56].

As noted in Eqn. 1.30, the rate theory is based on the premise that all sources of variance in the chromatographic system are independent and, therefore, additive

$$\sigma^2_{total} = \sigma^2_{col} + \sigma^2_{conn} + \sigma^2_{inj} + \sigma^2_{det} + \sigma^2_{elec} \tag{1.54}$$

where σ^2_{col} is given by the plate-height equations in Eqns. 1.43 and 1.44. The other components of the system, such as connecting tubing and unions (*conn*), injectors (*inj*), detectors (*det*), and signal processing electronics (*elec*), also add variance and dead-volume. This dispersion is detrimental, because it reduces the overall plate number and resolution. We might ask the following questions: How much extra-column dispersion is acceptable in a chromatographic system? What sources of dispersion are most important to consider and how can they be minimized? The answer to these questions is dependent upon the specific requirements of the application. If the chromatographic system is dedicated to the analysis of a single type of sample, then the column variance and the permissible extra-column variance will depend only on the solutes of interest. In general, however, the chromatographic system should be designed to be sufficiently flexible for a wide range of applications. In this case, the extra-column variance should be no more than a small fraction (θ^2) of the variance of the most efficient column to be used in the chromatographic system.

$$\sigma^2_{total} = (1 + \theta^2)\sigma^2_{col} \tag{1.55}$$

For example, the extra-column variance might be limited to 10% ($\theta^2 = 0.10$) of the smallest column variance. When expressed in units of volume, the column variance is given by

$$\sigma_{col}^2 = \frac{(\pi d_{col}^2 \varepsilon_t)^2 H_{col} L_{col}}{16} \tag{1.56}$$

where d_{col} and L_{col} are the diameter and length, respectively, of the column. The plate height, H_{col}, may be estimated from the minimum value for an open-tubular column ($H_{min} \approx d_{col}$) or a packed column ($H_{min} \approx 2d_p$), if the column is operated at the optimum velocity. The fraction θ^2 may be expressed as a function of the individual sources of extra-column dispersion

$$\theta^2 = \theta_{conn}^2 + \theta_{inj}^2 + \theta_{det}^2 + \theta_{elec}^2 \tag{1.57}$$

Some of these sources of dispersion are entirely detrimental; that is, they provide no benefits in the chromatographic system. For example, connecting tubes and unions only serve to adjoin different components of the system and their dispersion should be reduced as far as possible. With other sources, however, some benefits are to be gained by increasing the dispersion. For example, if we increase the injection volume, we will increase the amount of sample and, hence, the concentration limit of detection. If we increase the volume of certain types of detectors, we will also increase the concentration limit of detection. For these sources of dispersion, a compromise is necessary between the added variance and the added sensitivity. In general, we want these sources of dispersion to be the predominant contributions to the extra-column variance. Thus, we might allow injection and detection variance to be 5% of the column variance ($\theta_{inj}^2 = \theta_{det}^2 = 0.05$) and connection and electronic variance to be 1% of the column variance ($\theta_{conn}^2 = \theta_{elec}^2 = 0.01$).

A connecting tube is essentially like an open-tubular column without a stationary phase. Hence, its variance can be developed using the equations given in Sec. 1.3.3. When expressed in units of volume, the variance is given by

$$\sigma_{conn}^2 = \theta_{conn}^2 \sigma_{col}^2 = \frac{\pi^2 d_{conn}^4 L_{conn} d_{col}^2 \varepsilon_t u_{col}}{1536 D_M} \tag{1.58}$$

where d_{conn} and L_{conn} are the diameter and length, respectively, of the connecting tube. The maximum permissible values of d_{conn} and L_{conn} can be determined by substitution of the column variance from Eqn. 1.56 and subsequent rearrangement to yield

$$d_{conn}^4 L_{conn} = \frac{96 \theta_{conn}^2 D_M d_{col}^2 \varepsilon_t H_{col} L_{col}}{u_{col}} \tag{1.59}$$

It is apparent from this equation that the diameter of the connecting tube has a much greater effect on the variance than its length. This would suggest, at first glance, that the diameter should be reduced as far as possible. However, these equations were derived by assuming that the only contributions to broadening arise from laminar flow in the connecting tube. In fact, if there are changes in diameter between the connecting tube and the column fittings or other unions, there may be additional broadening due to unswept volume at the junction. For this reason, changes in diameter should be judiciously avoided. The connecting tube should be no smaller than the inner diameter of the adjacent fittings and unions (which can be replaced, if necessary, with ones of smaller diameter). Any remaining reduction in variance should be accomplished by reducing the length of the connecting tube. Typical values for the maximum permissible connection dimensions ($\theta^2_{conn} = 0.01$) are summarized for gas and liquid chromatography in Tables 1.8 and 1.9, respectively.

The variance for an injector is given by

$$\sigma^2_{inj} = \theta^2_{inj}\sigma^2_{col} = \frac{V^2_{inj}}{K^2} \tag{1.60}$$

TABLE 1.8

MAXIMUM PERMISSIBLE EXTRA-COLUMN SOURCES OF VARIANCE IN GAS CHRO-MATOGRAPHY SYSTEMS

Parameter	Column type			
	Packed	Open tubular		
Column				
d_{col}	4.0 mm	500 μm	320 μm	200 μm
L_{col}	2.0 m	10 m	10 m	10 m
d_p	125 μm			
$H_{col}{}^*$	250 μm	500 μm	320 μm	200 μm
N	8,000	20,000	31,000	50,000
u_{col}	10 cm/sec	10 cm/sec	15 cm/sec	25 cm/sec
F	64 mL/min	1.2 mL/min	0.72 mL/min	0.47 mL/min
σ^2_{col}	5.7×10^{-2} cm^6	1.9×10^{-4} cm^6	2.1×10^{-5} cm^6	2.0×10^{-6} cm^6
Connection**				
σ^2_{conn}	5.7×10^{-4} cm^6	1.9×10^{-6} cm^6	2.1×10^{-7} cm^6	2.0×10^{-8} cm^6
d_{conn}	1.0 mm	500 μm	320 μm	200 μm
L_{conn}	65 cm	190 cm	200 cm	190 cm
Injection***				
σ^2_{inj}	2.8×10^{-3} cm^6	9.6×10^{-6} cm^6	1.0×10^{-6} cm^6	9.9×10^{-8} cm^6
V_{inj}	110 μL	6.2 μL	2.0 μL	0.6 μL
Detection***				
σ^2_{det}	2.8×10^{-3} cm^6	9.6×10^{-6} cm^6	1.0×10^{-6} cm^6	9.9×10^{-8} cm^6
V_{det}	110 μL	6.2 μL	2.0 μL	0.6 μL
Electronic**				
σ^2_{elec}	5.7×10^{-4} cm^6	1.9×10^{-6} cm^6	2.1×10^{-7} cm^6	2.0×10^{-8} cm^6
τ	0.022 sec	0.071 sec	0.038 sec	0.018 sec

* $H_{min} \approx d_{col}$ for open-tubular column and $H_{min} \approx 2d_p$ for packed column.
** $\theta^2_{conn} = \theta^2_{elec} = 0.01$.
*** $\theta^2_{inj} = \theta^2_{det} = 0.05$, $K^2 = 4$.

TABLE 1.9

MAXIMUM PERMISSIBLE EXTRA-COLUMN SOURCES OF VARIANCE IN LIQUID
CHROMATOGRAPHY SYSTEMS

Parameter	Column type			
	Preparative	Analytical	Microbore	Capillary
Column				
d_{col}	10 mm	4.6 mm	1.0 mm	200 μm
L_{col}	25 cm	25 cm	50 cm	100 cm
d_p	5 μm	5 μm	5 μm	5 μm
H_{col} *	10 μm	10 μm	10 μm	10 μm
N	25,000	25,000	50,000	100,000
u_{col}	0.1 cm/sec	0.1 cm/sec	0.1 cm/sec	0.1 cm/sec
F	4.0 mL/min	0.85 mL/min	40 μL/min	1.6 μL/min
σ^2_{col}	1.1×10^{-2} cm^6	5.0×10^{-4} cm^6	2.2×10^{-6} cm^6	7.1×10^{-9} cm^6
Connection**				
σ^2_{conn}	1.1×10^{-4} cm^6	5.0×10^{-6} cm^6	2.2×10^{-8} cm^6	7.1×10^{-11} cm^6
d_{conn}	250 μm (0.010 in)	180 μm (0.007 in)	100 μm	50 μm
L_{conn}	5.2 cm	4.6 cm	4.1 cm	5.2 cm
Injection***				
σ^2_{inj}	5.6×10^{-4} cm^6	2.5×10^{-5} cm^6	1.1×10^{-7} cm^6	3.5×10^{-10} cm^6
V_{inj}	47 μL	10 μL	0.67 μL	38 nL
Detection***				
σ^2_{det}	5.6×10^{-4} cm^6	2.5×10^{-5} cm^6	1.1×10^{-7} cm^6	3.5×10^{-10} cm^6
V_{det}	47 μL	10 μL	0.67 μL	38 nL
Electronic**				
σ^2_{elec}	1.1×10^{-4} cm^6	5.0×10^{-6} cm^6	2.2×10^{-8} cm^6	7.1×10^{-11} cm^6
τ	0.16 sec	0.16 sec	0.22 sec	0.22 sec

* $H_{min} \approx 2\, d_p$ for packed column.
** $\theta^2_{conn} = \theta^2_{elec} = 0.01$.
*** $\theta^2_{inj} = \theta^2_{det} = 0.05$, $K^2 = 4$.

where V_{inj} is the injection volume and K^2 is a constant that is characteristic of the injection profile. This constant has a value of 36 for a Gaussian profile, 12 for a rectangular profile, and 1 for an exponential profile. For a well-behaved chromatographic injector, with a combination of Gaussian and exponential functions, the constant typically has a value of 4. Valve injectors generally give higher values for the constant K^2 than syringe injectors. The maximum permissible injection volume can be determined by substitution of the column variance from Eqn. 1.56 and subsequent rearrangement to yield

$$V_{inj}^2 = \frac{\pi^2 K^2 \theta_{inj}^2 d_{col}^4 \varepsilon_t^2 H_{col} L_{col}}{16} \tag{1.61}$$

Typical values for the maximum permissible injection volume ($\theta_{inj}^2 = 0.05$) are summarized for gas and liquid chromatography in Tables 1.8 and 1.9, respectively.

The variance for a detector is given by a similar equation

$$\sigma_{det}^2 = \theta_{det}^2 \sigma_{col}^2 = \frac{V_{det}^2}{K^2} \tag{1.62}$$

where V_{det} is the detection volume and K^2 is a constant that is characteristic of the detection profile, which has the same values as given above. The maximum permissible detection volume can be determined by substitution of the column variance from Eqn. 1.56 and subsequent rearrangement to yield

$$V_{det}^2 = \frac{\pi^2 K^2 \theta_{det}^2 d_{col}^4 \varepsilon_t^2 H_{col} L_{col}}{16} \tag{1.63}$$

Typical values for the maximum permissible detection volume ($\theta_{det}^2 = 0.05$) are summarized for gas and liquid chromatography in Tables 1.8 and 1.9, respectively. For the specific case of a capillary flow-cell that is transversely illuminated for absorbance or fluorescence detection, the variance is given by

$$\sigma_{det}^2 = \theta_{det}^2 \sigma_{col}^2 = \frac{V_{det}^2}{12} = \frac{(\pi d_{det}^2 L_{det})^2}{192} \tag{1.64}$$

where d_{det} and L_{det} are the diameter and length, respectively, of the capillary flow-cell. The maximum permissible values of d_{det} and L_{det} can be determined by substitution of the column variance from Eqn. 1.56 and subsequent rearrangement to yield

$$d_{det}^2 L_{det} = (12 \theta_{det}^2 H_{col} L_{col})^{1/2} d_{col}^2 \varepsilon_t \tag{1.65}$$

These equations are helpful to establish the permissible dimensions of detectors for capillary liquid chromatography and capillary electrophoresis.

All of the previous sources of extra-column variance are volumetric in origin. Temporal variance can also arise from electronic components with an exponential time response due to resistance and capacitance, such as detector amplifiers. The variance from electronic sources is given by

$$\sigma_{elec}^2 = \theta_{elec}^2 \sigma_{col}^2 = \frac{(\pi d_{col}^2 \varepsilon_t u_{col} \tau)^2}{16} \tag{1.66}$$

where τ is the exponential time constant. The maximum permissible time constant can be determined by substitution of the column variance from Eqn. 1.56 and subsequent rearrangement to yield

$$\tau^2 = \frac{\theta_{elec}^2 H_{col} L_{col}}{u_{col}^2} \tag{1.67}$$

Typical values for the maximum permissible time constant ($\theta_{elec}^2 = 0.01$) are summarized for gas and liquid chromatography in Tables 1.8 and 1.9, respectively.

Upon examination of the permissible extra-column sources of variance for gas chromatography in Table 1.8, it is evident that the requirements for connection can be easily met. The requirements for injection can be easily met by syringe injection for packed columns and by syringe injection with splitting for open-tubular columns. The requirements for detection can be met by nearly all detectors for packed columns and by flame-based detectors for open-tubular columns. The most rigorous requirements are presented by the electronic contributions to variance, which are not met by many commercial detectors and data acquisition systems. This remains the most critical aspect of gas chromatography systems.

Upon examination of the permissible extra-column sources of variance for liquid chromatography in Table 1.9, it is evident that the requirements for injection can be met by valve injection for preparative and analytical columns and by valve injection with splitting for microbore and capillary columns. The requirements for detection can be met by most detectors for preparative and analytical columns and by spectroscopic detectors with capillary flow-cells for microbore and capillary columns. The requirement for electronic sources can be met by judicious choice of commercial detectors and data acquisition systems. However, the most rigorous requirements are presented by the connection tubing and unions, which are not met by most commercial systems. This remains the most critical aspect of liquid chromatography systems.

1.3.5 Implications

In Sec. 1.3, the origin and extent of solute-zone dispersion from both column and extra-column sources have been presented. The influence of such variables as linear velocity, column diameter, column length, particle size, stationary-phase film thickness, retention factor, etc., have been discussed from a theoretical and practical perspective. An awareness and understanding of the many variables that influence dispersion is necessary in order to enable judicious control of plate height and plate number. This, in turn, will permit the resolution to be achieved according to the fundamental resolution equation (Eqn. 1.6).

1.4 PRINCIPLES OF SOLUTE-ZONE SEPARATION

In order to understand the concept of resolution, it is also important to consider the separation of solute zones along the column. The separation mechanism influences the retention factor and the selectivity factor, both of which are implicated directly in Eqn. 1.6. By understanding the physical and chemical aspects of the separation mechanism in greater detail, methods can be developed that are most suitable for the solutes of interest. In this section, we will review the fundamental principles of molecular interactions and thermodynamics. These principles will then be applied to the most common separation mechanisms, including partition, adsorption, complexation, ion exchange, and size exclusion.

1.4.1 Molecular interactions

In each separation mechanism, the solute molecules interact with molecules of the mobile and stationary phases. These cohesive or attractive interactions may be of several types, each of which depends on different properties of the solute, mobile phase, and stationary phase. We will discuss these interactions in order of increasing strength [3,57,58].

Dispersion forces, also known as London forces, result from the interaction of a temporary dipole in one atom or molecule with an induced dipole in another atom or molecule. The internal or potential energy of dispersion forces, U_d, is given by

$$U_d = \frac{-3}{2}\left(\frac{\alpha_i\alpha_j}{r_{ij}^6}\right)\left(\frac{I_iI_j}{I_i+I_j}\right) \tag{1.68}$$

where α and I are the polarizability and ionization energy of each atom or molecule, and r is the separation distance between them. This equation suggests that the interaction energy becomes stronger (more negative) with an increase in polarizability. The polarizability, typical values of which are summarized in Table 1.10 [59], represents the ease with which electrons can be redistributed around the atoms. It is apparent that many single bonds (*e.g.*, C–C, C–H, and N–H) have low polarizability, with most of the electron mobility occurring parallel to the bond axis. As represented by the sequence of C–C bonds, the

TABLE 1.10

POLARIZABILITY VOLUMES ($\alpha/4\pi\varepsilon_0$) OF INDIVIDUAL BONDS CALCULATED PARALLEL ($\alpha_\parallel$) AND PERPENDICULAR ($\alpha_\perp$) TO THE BOND AXIS. THE TOTAL POLARIZABILITY VOLUME AVERAGED OVER ALL ORIENTATIONS IS $\alpha = 1/3$ ($\alpha_\parallel + 2\alpha_\perp$) [59]

Bond	$\alpha_\parallel \times 10^{25}$ (cm^3)	$\alpha_\perp \times 10^{25}$ (cm^3)	$\alpha \times 10^{25}$ (cm^3)
C–C (aliphatic)	18.8	0.2	6.4
C–C (aromatic)	22.5	4.8	10.7
C=C	28.6	10.6	16.6
C≡C	35.4	12.7	20.3
C–H	7.9	5.8	6.5
C–Cl	36.7	20.8	26.1
C–Br	50.4	28.8	36.0
C=O (carbonyl)	19.9	7.5	11.6
C=O (CO$_2$)	20.5	9.6	13.2
C=S (CS$_2$)	75.7	27.7	43.7
C≡N	31	14	19.7
N–H	5.8	8.4	7.5
S–H	23.0	17.2	19.1

polarizability both parallel and perpendicular to the bond axis increases significantly with bond order. Homonuclear bonds generally have higher polarizability than heteronuclear bonds for atoms within the same row of the periodic table (*e.g.*, C=C > C=O, C≡C > C≡N). For other heteronuclear bonds, the polarizability increases as the size of the atom increases (*e.g.*, C–Cl < C–Br, C=O < C=S). All of these trends indicate the types of functional groups that will have high polarizability and, hence, greater interaction. Eqn. 1.68 also suggests that the interaction energy becomes stronger with an increase in the ionization energy. The ionization energy, typical values of which are summarized in Table 1.11 [54], represents the ease of removing an electron from an atom or molecule.

TABLE 1.11

IONIZATION ENERGY OF THE ELEMENTS [54]

Element	Ionization energy (eV)	Element	Ionization energy (eV)
H	13.598	Si	8.151
He	24.587	P	10.486
C	11.260	S	10.360
N	14.534	Na^+	47.286
O	13.618	K^+	31.625
F	17.422	Cu^+	20.292
Cl	12.967	Ag^+	21.49
Br	11.814	Fe^{2+}	30.651
I	10.451	Fe^{3+}	54.8

Within a row of the periodic table, the ionization energy increases as the electronegativity increases (*e.g.*, C < N ≈ O < F). Within a group or column of the periodic table, the ionization energy decreases with increasing atomic size (*e.g.*, O > S, F > Cl > Br > I). In general, the ionization energy is rather similar for the atoms in organic molecules (*e.g.*, C, O, N, etc.), but may vary substantially for metals. Finally, Eqn. 1.68 suggests that the interaction energy decreases dramatically with increasing separation distance (r^{-6}). Generally, dispersion forces are rather weak but universal, so that they increase with the number of atoms in a molecule. As a consequence, they may account for a major part of the total interaction energy for simple organic molecules.

Dipole induction forces, also known as Debye forces, result from the interaction of a permanent dipole in one molecule with a temporary, induced dipole in another molecule. The internal energy of dipole induction forces, U_i, is given by

$$U_i = \frac{-1}{4\pi\varepsilon_0} \left(\frac{\alpha_i \mu_j^2 + \alpha_j \mu_i^2}{r_{ij}^6} \right) \tag{1.69}$$

where μ is the dipole moment and ε_0 is the permittivity of vacuum. This equation suggests that the interaction energy becomes stronger (more negative) with an increase in polarizability and dipole moment. The dipole moment, typical values of which are summarized in Table 1.12 [60], represents the extent of electronic charge separation. For homonuclear bonds, there is no permanent dipole moment as each atom has equal electronegativity. For heteronuclear bonds, the dipole moment increases with the difference in electronegativity of the two atoms (*e.g.*, C–C < C–N < C–O < C–F). As the bond order for heteronuclear bonds increases, the dipole moment increases (*e.g.*, C–N < C=N < C≡N, C–O < C=O). In general, there is little effect of size on the dipole moment (*e.g.*, C–F ≈ C–Cl ≈ C–Br ≈ C–I). All of these trends indicate the types of functional groups that will have high dipole moments and, hence, greater interaction. Generally, dipole induction forces represent a relatively small fraction of the total interaction energy, but increase with the number of permanent dipoles.

Dipole orientation forces, also known as Keesom forces, result from the interaction of a permanent dipole in one molecule with a permanent dipole in another molecule.

TABLE 1.12

DIPOLE MOMENT OF INDIVIDUAL BONDS. VALUES ARE LISTED FOR BOTH ASSUMED DIRECTIONS OF THE DIPOLE MOMENT OF THE C–H BOND (0.4 DEBYE). THE ATOM TO THE LEFT OF THE BOND IS THE POSITIVE END OF THE MOMENT AND, UNLESS OTHERWISE INDICATED, ALL MULTIVALENT ATOMS ARE sp^3 HYBRIDS [60]

Bond	Dipole moment (debye)	
	$C \rightarrow H$	$C \leftarrow H$
C–C (sp^3–sp^2)	0.69	
C–C (sp^3–sp)	1.48	
C–C (sp^2–sp)	1.15	
C–N	1.26	0.45
C=N		1.4
C≡N	3.94	3.1
C–O (alcohol)	1.9	0.7
C–O (ether)	1.5	0.7
C=O	3.2	2.4
C–F	2.19	1.39
C–Cl	2.27	1.47
C–Br	2.22	1.42
C–I	2.05	1.25
C–S	1.6	0.9
H–O		1.51
H–N		1.31
H–S		0.7

The internal energy of dipole orientation forces, U_o, is given by

$$U_o = \frac{-1}{4\pi\varepsilon_0} \left(\frac{2\mu_i^2 \mu_j^2 \cos\theta}{3k_B T r_{ij}^6} \right)$$ (1.70)

where k_B is the Boltzmann constant and T is the absolute temperature. This equation suggests that the interaction energy becomes stronger (more negative) with an increase in the dipole moments, but is also dependent upon the cosine of the angle between the dipoles. This angular dependence accounts for the greatest cohesive attraction when the dipoles are aligned ($0°$), no attraction when the dipoles are at $90°$, and repulsion when the dipoles are at $180°$. The temperature dependence arises because kinetic energy increases the rate of translational and rotational motion, thereby decreasing the interaction between the dipoles. As with dispersion and dipole induction forces, dipole orientation forces decrease dramatically with increasing separation distance (r^{-6}).

In addition to the weak cohesive forces discussed above, collectively called van der Waals forces, stronger and more selective forces can arise when one of the molecules involved is an ion. For example, an ion can induce a temporary dipole in another molecule. The internal energy of ion–induced dipole forces, U_{Ii}, is given by

$$U_{Ii} = \frac{-1}{4\pi\varepsilon_0} \left(\frac{z_i^2 e^2 \alpha_j}{2r_{ij}^4} \right)$$ (1.71)

where z is the ion charge, and e is the charge of an electron. This equation suggests that the interaction energy becomes stronger with an increase in ion charge and an increase in the polarizability. Ion–induced dipole forces also decrease with increasing separation distance (r^{-4}), but at a slower rate than the van der Waals forces (r^{-6}).

An ion can also interact with another molecule having a permanent dipole. The internal energy of ion–dipole orientation forces, U_{Io}, is given by

$$U_{Io} = \frac{-1}{4\pi\varepsilon_0} \left(\frac{z_i e \mu_j \cos\theta}{r_{ij}^2} \right)$$ (1.72)

This equation suggests that the interaction energy becomes stronger with an increase in the ion charge and the dipole moment, but is also dependent upon the cosine of the angle between them. As with dipole orientation forces, the greatest cohesive attraction occurs when the ion and dipole are aligned ($0°$) and repulsion occurs when they are at $180°$. Ion-dipole forces decrease with increasing separation distance (r^{-2}).

An ion can also interact with another ion, which is known as Coulombic attraction. The internal energy of ion–ion forces, U_{II}, is given by

$$U_{II} = \frac{-1}{4\pi\varepsilon_0} \left(\frac{z_i z_j e^2}{\varepsilon r_{ij}} \right)$$ (1.73)

where ε is the relative permittivity of the medium, that is the mobile or stationary phase. This equation suggests that the interaction energy becomes stronger with an increase in the ion charge. Ion–ion forces exhibit the smallest decrease with increasing separation distance (r^{-1}).

Finally we can consider hydrogen bonding, which is a combination of electrostatic and covalent interactions. According to the Bronsted definition, an acid is a proton donor and a base is a proton acceptor. Although there are several ways to calculate the energy of hydrogen bonds, we will use an empirical force-field approach similar to those used in molecular modeling programs [61]. The internal energy of hydrogen-bonding forces, U_{hb}, is given by

$$U_{hb} = D_{hb}\left(5\left(\frac{r_{hb}}{r_{ij}}\right)^{12} - 6\left(\frac{r_{hb}}{r_{ij}}\right)^{10}\right)\cos^4\theta \tag{1.74}$$

where D_{hb} and r_{hb} are the defined energy and separation distance of the hydrogen bond. The energy is given as 4.0 kcal/mol by experiment, 7.0 kcal/mol by the Gasteiger model, and 9.0 kcal/mol by the Dreiding model. The separation distance is given as 2.75 Å by experiment and by these models.

In order to calculate the overall interaction energy between two molecules, the individual energies from Eqns. 1.68 to 1.74 are summed and integrated over all of space

$$U = \frac{\pi N}{V}\int U(r)g(r)r^2 dr \tag{1.75}$$

The molecular energies are then converted to molar quantities by means of Avogadro's number (N) and the molar volume (V). In this equation, $U(r)$ is the sum of the individual energies and $g(r)$ is the pair distribution function, which represents the probability of finding a molecule at distance, r, relative to that in a random distribution at constant temperature and pressure. It is helpful to establish the average distance between molecules in order to understand the relative strength of interactions in gas, supercritical-fluid, and liquid chromatography. For gases, the density is typically on the order of 10^{-3} g/cm^3, which corresponds to an average distance of 20–50 Å. Because of the long distance between molecules, intermolecular interactions are relatively weak and only the strongest forces (*e.g.*, ion–ion) can influence the behavior of the molecules. For supercritical fluids, the density is typically 0.3–0.8 g/cm^3, which corresponds to an average distance of 5–10 Å. For liquids, the density is typically 1.0 g/cm^3, which corresponds to an average distance of 2–5 Å. Because of the shorter distances, intermolecular interactions are very strong and even van der Waals forces have an important influence on the behavior.

1.4.2 Thermodynamics

In chromatography, we are interested in the interaction energy of the solute, i, with the mobile (M) and stationary (S) phases

$$\Delta U = U_{iS} - U_{iM} \tag{1.76}$$

where the individual energies are calculated from Eqn. 1.75. The term ΔU represents the change in molar internal or potential energy as the solute is transferred from the mobile to stationary phase. This change in internal energy is related to other important thermodynamic quantities [57,62], such as the change in molar enthalpy (ΔH)

$$\Delta H = \Delta U + p\Delta V \tag{1.77}$$

where p is the pressure and ΔV is the change in molar volume. For one mole of an ideal gas, this equation can be written as

$$\Delta H = \Delta U + RT \tag{1.78}$$

where R is the gas constant and T is the absolute temperature. The change in molar Gibbs free energy is given by

$$\Delta G = \Delta H - T\Delta S \tag{1.79}$$

The molar Gibbs free energy, thus, has two contributions: the molar enthalpy, which is related to the molecular interactions, and the molar entropy (ΔS), which is related to the disorder of the system. The Gibbs free energy reflects the tendency of a reaction, in this case the transfer of solute from mobile to stationary phase, to occur spontaneously. It is related to the thermodynamic equilibrium constant for solute transfer by

$$ln\ K = \frac{-\Delta G}{RT} = \frac{-\Delta H}{RT} + \frac{\Delta S}{R} \tag{1.80}$$

This equilibrium constant is related to the retention factor, which was introduced previously in Eqn. 1.4

$$k = \frac{K}{\beta} = \left(\frac{a_S}{a_M}\right)\left(\frac{V_S}{V_M}\right) \tag{1.81}$$

The equilibrium constant is the ratio of solute activity, a, in the mobile and stationary phases. The phase ratio, β, is the ratio of the volume, V, of the mobile and stationary phases. Thus, the retention factor represents the ratio of solute molecules (or moles) in the two phases. A retention factor less than unity means that the solute is present in greater proportion in the mobile phase, whereas a retention factor greater than unity means that the solute is present in greater proportion in the stationary phase.

By substitution of Eqn. 1.81 into Eqn. 1.80, we obtain

$$ln\ k = \frac{-\Delta H}{RT} + \frac{\Delta S}{R} - ln\ \beta \tag{1.82}$$

This equation, known as the van't Hoff Equation, illustrates several important trends that are observed in chromatography. First, the molecular interactions in Eqns. 1.68 to 1.74 generally increase with the number of atoms in the molecule, thereby increasing the molar enthalpy. Hence, the logarithm of the retention factor generally increases in a linear manner with carbon number, functional group number, etc. In addition, if the molar enthalpy is constant, then the logarithm of the retention factor is inversely related to temperature. Hence, retention decreases logarithmically as temperature is increased. This relationship is especially important in gas and supercritical-fluid chromatography, where temperature is often used to control retention behavior. From Eqn. 1.81, the equilibrium constant is defined in terms of the solute activity in the mobile and stationary phases. The activity is related to the solute concentration, C, in each phase by means of the activity coefficient, γ

$$K = \frac{a_S}{a_M} = \frac{\gamma_S C_S}{\gamma_M C_M} \tag{1.83}$$

In order to determine the activity coefficient, it is necessary to define the standard state. There are two common definitions of the standard state that are used in chromatography. The first definition uses the pure material as the reference (Raoult's Law), whereas the second uses a solution at infinite dilution as the reference (Henry's Law). Although there are merits to each approach, we will use the former definition in this chapter. When the pure material is chosen to define the standard state, we can draw conclusions about chromatographic retention and selectivity from the physical properties of the pure material, which are readily available in the literature.

For example, consider the phase diagram of a hypothetical pure solute, shown in Fig. 1.17. This diagram defines the regions of temperature and pressure where the solute

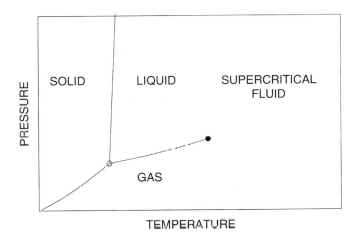

Fig. 1.17. Phase diagram for a hypothetical pure solute, showing gas, supercritical-fluid, liquid, and solid phases; (o) triple point, (●) critical point.

exists as a gas, liquid, and solid. Above the critical point, the solute no longer exists as separate gas and liquid phases, but as a single supercritical phase. The lines separating the regions represent the locus of temperatures and pressures at which a phase transition occurs, such as vaporization (liquid to gas), sublimation (solid to gas), and liquefaction (solid to liquid). The slopes of these lines are given by the Clapeyron Equation [57]

$$\frac{dp}{dT} = \frac{\Delta H}{T \Delta V} \tag{1.84}$$

where the changes in molar enthalpy and volume are those accompanying the phase transition. We will consider the physical meaning of this equation in gas chromatography, where vaporization is the most important process governing retention. Inasmuch as the volume of one mole of a gas, V_G, is much greater than that of a liquid, V_L, we can write

$$\Delta V = V_G - V_L \approx V_G \tag{1.85}$$

For an ideal gas,

$$V_G = \frac{nRT}{p} \tag{1.86}$$

By substitution in Eqn. 1.84, we obtain the Clausius–Clapeyron Equation

$$\frac{dp}{p(dT)} = \frac{d(\ln p)}{dT} = \frac{\Delta H}{RT^2} \tag{1.87}$$

and by integration

$$\ln p = \frac{-\Delta H}{RT} + C \tag{1.88}$$

where C is a constant of integration. This equation illustrates two important points: At constant temperature, the logarithm of the standard vapor pressure is linearly related to the molar enthalpy of vaporization. At constant pressure, the boiling temperature is linearly related to the molar enthalpy of vaporization. Since the enthalpy directly reflects the extent of solute retention (Eqn. 1.82), either the standard vapor pressure or the boiling point can be used as an estimate. Thus, to a first approximation, these data can be used to predict retention and selectivity in gas chromatography. For example, we may expect the retention order of phenol, benzyl alcohol, 2-nitrophenol, and benzoic acid on the basis of their boiling points of 182, 205, 216, and 249 °C, respectively, at 1 atm. The prediction of retention order is usually quite accurate for solutes such as these with widely varying boiling points, as verified in EPA Method 8270. However, because this prediction uses the pure solute to define the standard state, it considers only interactions of the solute with other molecules of its kind but does not consider interactions with the stationary phase. These interactions can alter the retention order significantly, especially when there are strong and selective interactions between the solute and stationary phase.

This brings to our attention one additional point: The equilibrium constant is defined as the ratio of activities or concentrations in Eqn. 1.83. A graph of the solute activity (or concentration) in the stationary phase as a function of the solute activity (or concentration) in the mobile phase at fixed temperature is called an "isotherm". An isotherm will be linear, as shown in Fig. 1.18A, if the interactions of the solute with the mobile and stationary phases do not vary with concentration. However, an isotherm may be nonlinear if the stationary phase has a limited number of sites or has multiple types of sites with different interaction energy. Two common examples of nonlinear isotherms are shown in Figs. 1.18B and 1.18C. The origin of these effects will be discussed, where relevant, in the following sections that describe the most common retention mechanisms in chromatography.

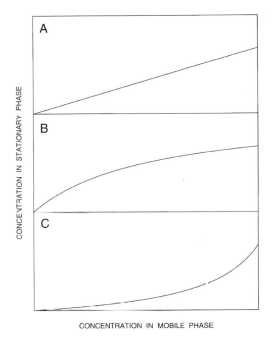

Fig. 1.18. Common shapes of isotherms in chromatography. (A) Linear isotherm, (B) convex, also known as Langmuir or BET Type 1 isotherm, (C) concave, also known as BET Type III isotherm.

1.4.3 Partition

In the partition (absorption) mechanism, the mobile phase may be a gas, supercritical fluid, or liquid. The stationary phase is a permeable liquid or polymer, such that the solute is completely surrounded and solvated by the molecules of the stationary phase. In this mechanism, retention is based entirely on the solubility of the solute in the mobile and stationary phases. In order to understand this mechanism in greater depth and detail, we will use a model based on regular solution theory, which is an extension of standard gas-phase thermodynamics to the liquid phase. There are several assumptions of this model:

(a) The interaction forces between two molecules i and j are assumed to be additive. The presence of other molecules does not influence the interaction energy.

(b) The molecule is considered to be a point mass and charge at the center of a volume V. There is no change in net volume upon mixing molecules i and j.

(c) The solution of molecules i and j is randomly mixed, so that there is no preferential order or orientation.

This model was originally developed by Hildebrand and Scott for nonpolar molecules with dispersion forces only [63]. It was later extended to include other forces, such as dipole induction, orientation, etc. [64]. However some of the assumptions may not be valid, so restraint must be exercised in extrapolating the results of this model to polar molecules. The basis for this model is the Hildebrand solubility parameter, δ, which is defined by

$$\delta^2 = \frac{-U}{V} \tag{1.89}$$

The solubility parameter represents the cohesive energy density, or internal energy per unit volume, of a pure material (solute, mobile phase, or stationary phase). As such, it is a direct numerical measure of polarity; the greater the solubility parameter, the greater the polarity. The overall solubility parameter has been divided into three components

$$\delta^2 = \delta_d^2 + \delta_p^2 + \delta_h^2 \tag{1.90}$$

where δ_d^2 represents the dispersion forces, δ_p^2 represents the polar interactions due to dipole induction and orientation forces, and δ_h^2 represents the hydrogen-bonding forces. Barton [64] has tabulated values of the overall solubility parameter as well as the individual components for many organic molecules, a representative selection of which is given in Table 1.13. For the normal alkanes, dispersion is the only important cohesive force, and the solubility parameter increases slightly with carbon number. The interaction energy for *n*-hexane is similar to that for cyclohexane, but the molar volume for cyclohexane is smaller, so that the solubility parameter increases according to Eqn. 1.89. For benzene, the dispersion contribution is slightly higher than for *n*-hexane or cyclohexane because of the smaller molar volume. The polar contribution is negligible for benzene, as there is no permanent dipole moment. However, there is a noteworthy contribution from hydrogen bonding, since benzene can act as a base (proton acceptor or π-electron donor). The addition of an alkyl group, with its smaller interaction energy, leads to a decrease in the solubility parameter for toluene and ethyl benzene. However, an extension of the aromatic structure, as for naphthalene, increases the individual components as well as the overall solubility parameter. The substitution of halogens has an interesting effect. Methylene chloride is much more polar than the alkanes, with larger contributions to the solubility parameter from dispersion, polar, and hydrogen-bonding components. However, the further addition of chlorine decreases the polarity; the solubility parameter for chloroform is slightly less and for carbon tetrachloride it is substantially less than for methylene chloride. Chlorobenzene is only slightly more polar than benzene. For oxygen-containing compounds, the ether group has the smallest effect on polarity. Relative to the alkanes, there is a small increase in the polar contribution and a larger increase in the hydrogen bonding contribution, since the ether group can serve as a base (proton acceptor or electron pair donor). The interaction energy for tetrahydrofuran is similar to that for diethyl ether, but the solubility parameter is greater, owing to the smaller molar volume. The ketone, ester, and carboxylic acid groups have progressively greater effects, as indicated by acetone, ethyl acetate, and acetic acid. The hydroxyl groups have the greatest effect, with a very substantial polar and hydrogen-bonding contribution. As indicated by the series of methanol, ethanol, and propanol, the addition of alkyl substituents to the polar hydroxyl group serves to decrease the overall polarity of the molecule. For nitrogen-containing compounds, the nitrile group has the smallest effect on polarity, followed by nitro and amide groups. Finally, water has the highest solubility parameter with very substantial contributions from hydrogen bonding.

TABLE 1.13

TRENDS AND TYPICAL VALUES FOR THE MOLAR VOLUME, V, HILDEBRAND
SOLUBILITY PARAMETER, δ, AND ITS DISPERSION, δ_d, POLAR, δ_p, AND HYDROGEN-
BONDING, δ_h, COMPONENTS [64]

Solute	V (cm^3/mol)	δ (cal/cm^3)$^{1/2}$	δ_d (cal/cm^3)$^{1/2}$	δ_p (cal/cm^3)$^{1/2}$	δ_h (cal/cm^3)$^{1/2}$
n-Pentane	116.2	7.0	7.1	0.0	0.0
n-Hexane	131.6	7.3	7.3	0.0	0.0
n-Heptane	147.4	7.4	7.5	0.0	0.0
n-Octane	163.5	7.6	7.6	0.0	0.0
n-Dodecane	228.6	7.9	7.8	0.0	0.0
Cyclohexane	108.7	8.2	8.2	0.0	0.1
Benzene	89.4	9.2	9.0	0.0	1.0
Toluene	106.6	8.9	8.8	0.7	1.0
Ethyl benzene	123.1	8.8	8.7	0.3	0.7
Naphthalene	111.5	9.9	9.4	1.0	2.9
Dichloromethane	63.9	9.7	8.9	3.1	3.0
Chloroform	80.7	9.3	8.7	1.5	2.8
Carbon tetrachloride	97.1	8.6	8.7	0.0	0.3
Chlorobenzene	102.1	9.5	9.3	2.1	1.0
Tetrahydrofuran	81.7	9.1	8.2	2.8	3.9
Diethyl ether	104.8	7.4	7.1	1.4	2.5
Acetone	74.0	9.9	7.6	5.1	3.4
Ethyl acetate	98.5	9.1	7.7	2.6	3.5
Methanol	40.7	14.5	7.4	6.0	10.9
Ethanol	58.5	12.7	7.7	4.3	9.5
1-Propanol	75.2	11.9	7.8	3.3	8.5
Acetic acid	57.1	10.1	7.1	3.9	6.6
Acetonitrile	52.6	11.9	7.5	8.8	3.0
Nitromethane	54.3	12.7	7.7	9.2	2.5
Formamide	39.8	19.2	8.4	12.8	9.3
Water	18.0	23.4	7.6	7.8	20.7

The solubility parameter can be related to the equilibrium constant for solute transfer
between the mobile and stationary phases in the partition mechanism [65,66]

$$\ln K = \frac{V_i}{RT}\left((\delta_i - \delta_M)^2 - (\delta_i - \delta_S)^2\right) \tag{1.91}$$

where δ_i, δ_M, and δ_S are the solubility parameters of the solute, mobile phase, and
stationary phase, respectively. This equation provides a great deal of insight into the nature
of the partition mechanism. There are two parenthetical terms, the first is concerned with
the difference in polarity between the solute and the mobile phase $(\delta_i - \delta_M)$, the second

with the difference in polarity between the solute and the stationary phase $(\delta_i - \delta_S)$. It is this balance of forces that controls the extent of retention. As the solute polarity approaches that of the mobile phase $(\delta_i = \delta_M)$, then $\ln K$ reaches a minimum value, determined by the solubility in the stationary phase $(\delta_i - \delta_S)^2$. As the solute polarity approaches that of the stationary phase $(\delta_i = \delta_S)$, then $\ln K$ reaches a maximum value determined by the solubility in the mobile phase $(\delta_i - \delta_M)^2$. If the solute polarity is midway between that of the mobile and stationary phases $(\delta_i = (\delta_M + \delta_S)/2)$, then $(\delta_i - \delta_M)^2 = (\delta_i - \delta_S)^2$ and $K = 1$. Under these conditions, the solute activity (or concentration) is equal in the mobile and stationary phases.

Eqn. 1.91 can also be used to derive an expression for the selectivity factor of two solutes with equal molar volume $(V_i = V_j = V)$

$$\ln \alpha_{ij} = \ln \frac{K_j}{K_i} = \frac{2V}{RT}(\delta_i - \delta_j)(\delta_M - \delta_S) \qquad (1.92)$$

This equation shows that the extent of separation between two solutes is influenced by the difference in polarity between the two solutes $(\delta_i - \delta_j)$. Solutes with greatly differing polarity should be easily separated. However, the separation becomes more difficult as the polarities of the solutes approach one another. For the case of $\delta_i = \delta_j$, the selectivity reaches a minimum value $(\alpha = 1)$ and no separation is possible. The extent of the separation also depends on the difference in polarity between the mobile and stationary phases $(\delta_M - \delta_S)$. The greater the difference in polarity between the two phases, the greater the ability of the chromatographic system to separate solutes of similar polarity.

Let us now consider the application of this model to gas chromatography, as illustrated schematically in Fig. 1.19A. The molar volume of the mobile phase is very large (22.4 L for an ideal gas), so that $\delta_M \approx 0\,(\mathrm{cal/cm^3})^{1/2}$. Because the mobile phase is essentially inert, δ_S must be used to control retention and selectivity. As a consequence, there are more than 100 different stationary phases available for gas chromatography [67], with solubility parameters ranging from ~ 7 $(\mathrm{cal/cm^3})^{1/2}$ for nonpolar phases such as polydimethylsiloxane polymers to ~ 15 $(\mathrm{cal/cm^3})^{1/2}$ for the most polar phases. This range of solubility parameters for the mobile and stationary phases suggests that solutes ranging from 3.5 to 7.5 $(\mathrm{cal/cm^3})^{1/2}$ can be separated with $K = 1$, as shown in Fig. 1.19A. Thus, gas chromatography is limited to relatively nonpolar solutes. The solutes are eluted in order of increasing polarity according to Eqn. 1.91. As δ_M and δ_S are fixed in this equation, temperature is the only variable that is available to adjust retention during the separation. As temperature is increased, solute retention decreases logarithmically. Hence, isothermal and gradient temperature programs are commonly employed for complex samples in gas chromatography.

Next, let us consider the application of this model to liquid chromatography. We can distinguish two distinct cases, normal and reversed phase, as illustrated schematically in Figs. 1.19B and 1.20C, respectively. In normal-phase liquid chromatography, the mobile phase is less polar than the stationary phase $(\delta_M < \delta_S)$. Common stationary phases are silica, modified with aminopropyl or cyanopropyl groups, for which we may estimate a solubility parameter of ~ 15 $(\mathrm{cal/cm^3})^{1/2}$. Common mobile phases consist of hexane, mixed with polar modifiers such as chloroform, tetrahydrofuran, and acetonitrile. These modifiers are chosen on the basis of their differences in selective interactions; specifically,

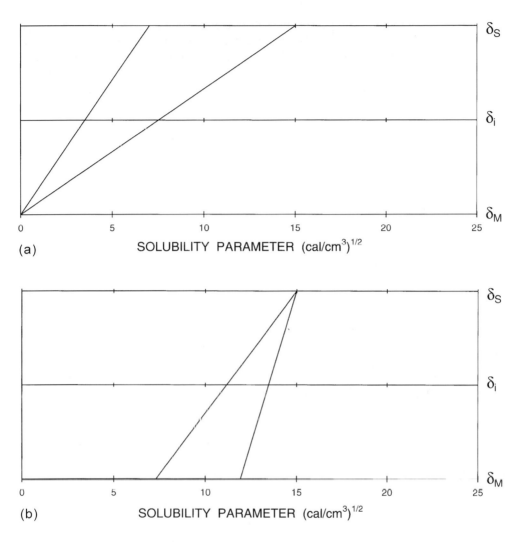

Fig. 1.19. Schematic representation of gas chromatography (a), normal-phase liquid chromatography (b), reversed-phase liquid chromatography (c), and supercritical-fluid chromatography (d) according to the regular solution model, with typical values for the solubility parameters of the mobile phase, δ_M, stationary phase, δ_S, and solute δ_i.

chloroform has strong acidic hydrogen bonding, tetrahydrofuran has strong basic hydrogen bonding, and acetonitrile has strong dipole interactions (Table 1.13). Hence, the solubility parameter of the mobile phase may range from 7.3 $(\text{cal/cm}^3)^{1/2}$ for pure hexane to 11.9 $(\text{cal/cm}^3)^{1/2}$ for pure acetonitrile. This range of solubility parameters for the mobile and stationary phases suggests that solutes ranging from 11.1 to 13.5 $(\text{cal/cm}^3)^{1/2}$ can be separated with $K = 1$, as shown in Fig. 1.19B. Thus, normal-phase liquid

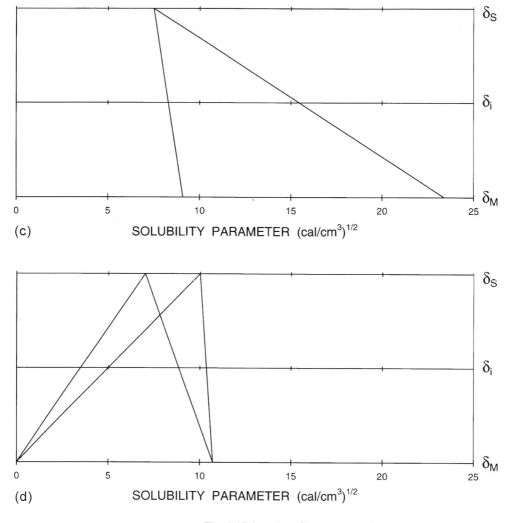

δ_S

δ_i

δ_M

| 0 | 5 | 10 | 15 | 20 | 25 |

(c)   SOLUBILITY PARAMETER $(cal/cm^3)^{1/2}$

δ_S

δ_i

δ_M

| 0 | 5 | 10 | 15 | 20 | 25 |

(d)   SOLUBILITY PARAMETER $(cal/cm^3)^{1/2}$

Fig. 1.19 (*continued*)

chromatography is useful for the separation of a relatively narrow range of polar molecules. The solutes are eluted in order of increasing polarity according to Eqn. 1.91. Although δ_S is fixed in this equation, either δ_M or temperature can be used to adjust retention during the separation. Of these options, mobile-phase composition is most commonly employed in liquid chromatography. For a mixed mobile phase, the solubility parameter can be calculated as

$$\delta = \sum_{i=1}^{n} \phi_i \delta_i \qquad (1.93)$$

where ϕ is the volume fraction and δ the solubility parameter of each constituent. As the solubility parameter of the mobile phase is increased, solute retention decreases logarithmically. Hence, isocratic and gradient mobile-phase compositions are commonly employed for complex samples in normal-phase liquid chromatography.

In reversed-phase liquid chromatography, the mobile phase is more polar than the stationary phase ($\delta_M > \delta_S$). Common stationary phases are silica, modified with octyl or octadecyl groups, for which we may estimate a solubility parameter of ~ 7.5 (cal/cm^3)$^{1/2}$. Common mobile phases consist of water mixed with organic modifiers, such as methanol, tetrahydrofuran, and acetonitrile. These modifiers are chosen on the basis of their differences in selective interactions. Specifically, methanol has strong acidic hydrogen bonding, tetrahydrofuran has strong basic hydrogen bonding, and acetonitrile has strong dipole interactions (Table 1.13). Hence, the solubility parameter of the mobile phase may range from 9.1 (cal/cm^3)$^{1/2}$ for pure tetrahydrofuran to 23.4 (cal/cm^3)$^{1/2}$ for pure water. This range of solubility parameters for the mobile and stationary phases suggests that solutes ranging from 8.3 to 15.5 (cal/cm^3)$^{1/2}$ can be separated with $K = 1$, as shown in Fig. 1.19C. Thus, reversed-phase liquid chromatography is useful for the separation of a very wide range of nonpolar and polar molecules. The solutes are eluted in order of decreasing polarity according to Eqn. 1.91. For solutes with similar polarity, such as the polycyclic aromatic hydrocarbons, the solutes are eluted in order of increasing molar volume according to Eqn. 1.91. Although δ_S is fixed in this equation, either δ_M or temperature can be used to adjust retention during the separation. Of these options, isocratic and gradient mobile-phase compositions are most commonly employed for the separation of complex samples. Reversed-phase separations are used for a vast majority of all applications in liquid chromatography because of the broad range of solutes that can be separated [68].

It is helpful to examine a hypothetical example in more detail. Suppose that we have a solute with a solubility parameter of 12.5 (cal/cm^3)$^{1/2}$, which is amenable to separation by either normal- or reversed-phase liquid chromatography. For a reversed-phase separation, the solubility parameter of the octadecylsilica stationary phase is approximately 7.5 (cal/cm^3)$^{1/2}$. Using Eqn. 1.91, we calculate that the solubility parameter of the mobile phase must be 17.5 (cal/cm^3)$^{1/2}$ to elute the hypothetical solute with $K = 1$. This mobile phase may be achieved in several different ways according to Eqn. 1.93: 66% methanol/34% water, 41% tetrahydrofuran/59% water, or 51% acetonitrile/49% water. Each of these mobile phases provides the same overall polarity and, hence, the same elution strength (*isoeluotropic*). However, each will have slightly different selective interactions with the solutes of interest. If the solutes differ in hydrogen-bonding character, we may choose methanol (acidic) or tetrahydrofuran (basic) solvents as the modifiers. However, if the solutes differ in dipole moment, we may choose acetonitrile as the modifier. Because of the great breadth of mobile-phase strength and selectivity that can be achieved, only a few stationary phases are needed for liquid chromatography.

Finally, let us consider the application of this model to supercritical-fluid chromatography, as illustrated schematically in Fig. 1.19D. The solubility parameter of a supercritical fluid has been given by Giddings *et al.* [69] as

$$\delta_M = 1.25 p_c \left(\frac{\rho_r}{\rho_r (liq)} \right) \qquad (1.94)$$

where p_c is the critical pressure, ρ_c is the critical density, $\rho_r = \rho/\rho_c$ is the reduced density, and $\rho_r(liq)$ is the reduced density of the liquid, which is approximately 2.66 for many fluids. One of the most common mobile phases is carbon dioxide, whose critical temperature is 31 °C, critical pressure is 73 atm, and critical density is 0.464 g/cm^3. The solubility parameter can range from the gas phase ($\delta_M \approx 0$ (cal/cm^3)$^{1/2}$) to the liquid phase ($\delta_M = 10.7$ (cal/cm^3)$^{1/2}$) by suitable adjustment of temperature and pressure. Common stationary phases are polydimethylsiloxane (~ 7 (cal/cm^3)$^{1/2}$) and polyethylene oxide (~ 10 (cal/cm^3)$^{1/2}$), which are also widely used for gas chromatography. This range of solubility parameters for the mobile and stationary phases suggests that solutes ranging from 3.5 to 10.3 (cal/cm^3)$^{1/2}$ can be separated with $K = 1$, as shown in Fig. 1.19D. Thus, supercritical-fluid chromatography is useful for the separation of a wide range of nonpolar and moderately polar molecules. Retention can be adjusted by means of the temperature, pressure, or composition of the mobile phase through the addition of a polar modifier, such as methanol or acetonitrile. Because of the broad range of mobile-phase polarity that can be achieved, supercritical-fluid chromatography effectively bridges the gap between gas and liquid chromatography.

In summary, partition is one of the most common separation mechanisms in gas, supercritical-fluid, and liquid chromatography. Retention and selectivity are based on the relative solubility of the solute in the mobile and stationary phases. In general, the isotherms are linear (Fig. 1.18A) and the equilibrium constants do not vary significantly with solute concentration. The regular solution model, which is based on the Hildebrand solubility parameter, provides detailed insight into the nature of this separation mechanism.

1.4.4 Adsorption

There are several features that distinguish the adsorption mechanism from the partition mechanism [70]. First, the stationary phase is an impermeable, solid surface. One of the most common stationary phases is silica, with the nominal chemical formula SiO_2. At the surface, some of the sites remain as siloxane groups (Si–O–Si) while others are hydrolyzed to silanol groups (Si–OH). The siloxane groups are weakly basic, whereas the silanol groups are weakly acidic. Another common stationary phase is alumina, with the nominal formula Al_2O_3. Alumina is not hydrolyzed to the same extent as silica, so the surface remains weakly basic. Carbonaceous materials are widely used as adsorbents, including charcoal, carbon black, graphite, etc. These materials consist primarily of aromatic sheets, with relatively weak dispersion, dipole-induction, and dipole-orientation interactions. However, at edges and other defect sites, oxidation can occur to form hydroxyl, carbonyl, and carboxyl groups. These groups have much stronger interactions, including both acidic and basic hydrogen bonding. These stronger interactions are usually detrimental to the performance of the material. Hence, porous graphitic carbon, which has the most stable surface and is relatively free of surface oxides, is the most suitable carbonaceous material for chromatographic separations. There are a variety of other, less common adsorbents, such as magnesia (magnesium oxide/hydroxide), zirconia (zirconium oxide/hydroxide), zeolite (sodium or calcium aluminosilicates), and

Florisil (magnesium silicate). Finally, there are a number of rigid polymers such as polystyrene, polystyrene/divinylbenzene, acrylonitrile/divinylbenzene, etc.

Most of these adsorbents are heterogeneous, having surface sites with different functional groups. These functional groups generally involve strong selective interactions, such as dipole orientation and hydrogen bonding. Owing to these strong interactions, there is often a fixed stoichiometry between the solute and the stationary phase, most frequently 1:1, but occasionally 1:2 or higher. However, because the stationary phase is a solid with a fixed surface area, there are a limited number of surface sites. Consequently, the solute molecules are in competition for the surface sites and may be displaced by more strongly interacting species. These features of the adsorption mechanism make it distinctly different from partition.

The extent of retention is described by the equilibrium constant, which is related to the free energy of adsorption (Eqn. 1.80). Snyder [70,71] has defined the group interaction energy as the free energy of adsorption for a specific functional group minus the free energy for an equivalent volume of *n*-pentane. Selected values of the group interaction energy are summarized in Table 1.14 for silica and alumina [71]. The adsorption energy for methyl and methylene groups is nearly zero, whereas those for aromatic and higher bond orders are slightly greater. In general, the adsorption energy for specific functional groups is greater in aliphatic solutes than in aromatic solutes. This trend is observed because the electron density of the functional group remains localized in aliphatic molecules, but is dispersed throughout the aromatic ring, thereby decreasing its availability for interaction with the adsorbent. Accordingly, aliphatic halogen compounds have very high adsorption energy relative to their aromatic counterparts. This results from the high polarizability and dipole moment of the halogens (Tables 1.10 and 1.12), leading to greater interactions from dipole induction and orientation. For acidic groups (proton donors), such as thiol, hydroxyl, and carboxyl, the adsorption energy is greater on alumina than on silica. For basic groups (electron-pair donors), such as carbonyl, amine, and nitrile, the adsorption energy is greater on silica than on alumina. These selective interactions are a consequence of the weakly acidic and weakly basic character of silica and alumina, respectively.

If the functional groups of the solute act independently then, to a first approximation, the group interaction energies are additive. Hence, we can estimate the overall interaction energy by summation of the individual values for each functional group. For example, the interaction energy of 1,4-dinitrocyclohexane on silica is given by $6(-0.05) + 2(5.71) = 11.1$ kcal/mol. However, if the functional groups of the solute do not act independently, then secondary structural effects can arise [70]. Some secondary effects are concerted or synergistic in nature. For example, the functional groups in 1,2-dinitrocyclohexane are in close proximity, such that they can both be adsorbed on the surface at the same time. Their interaction is subsequently greater than would be expected for two isolated functional groups, as in 1,4-dinitrocyclohexane. Hence, synergistic effects lead to an increase in solute retention. Other secondary effects are antagonistic in nature. For example, the functional groups in 1,2-aminohydroxycyclohexane are in close proximity, such that hydrogen bonding can occur between the acidic hydroxyl group and the basic amine group. Their interaction with the surface is subsequently less than would be expected for two isolated functional groups, as in 1,4-aminohydroxycyclohexane. Hence, antagonistic

TABLE 1.14

TRENDS AND TYPICAL VALUES FOR THE GROUP ADSORPTION ENERGY, Q_i, IN ALIPHATIC AND AROMATIC SOLUTES ON ALUMINA AND SILICA [71]

Group	Group adsorption energy (kcal/mol)					
	Aliphatic[*]		Aliphatic–aromatic[**]		Aromatic[***]	
	Al_2O_3	SiO_2	Al_2O_3	SiO_2	Al_2O_3	SiO_2
CH_3 (sp^3)	-0.03	0.07	–	–	0.06	0.11
CH_2 (sp^3)	0.02	-0.05	0.07	0.01	0.12	0.07
C (sp^2)	0.31	0.25	0.31	0.25	0.31	0.25
F	1.64	1.54	–	–	0.11	-0.15
Cl	1.82	1.74	–	–	0.20	-0.20
Br	2.00	1.94	–	–	0.33	-0.17
I	2.00	1.94	–	–	0.51	-0.15
S	2.65	2.94	1.32	1.29	0.76	0.48
SH	2.80	1.70	–	–	8.70	0.67
O	3.50	3.61	1.77	1.83	1.04	0.87
OH	6.50	5.60	–	–	7.40	4.20
H–C=O	4.73	4.97	–	–	3.35	3.48
C=O	5.00	5.27	3.74	4.69	3.30	2.68
O–C=O	5.00	5.27	3.40	3.45	2.67	2.68
HO–C=O	21	7.60	–	–	19	6.10
H_2N–C=O	8.90	9.60	–	–	6.20	6.60
NH_2	6.24	8.00	–	–	4.41	5.10
NO_2	5.40	5.71	–	–	2.75	2.77
C≡N	5.00	5.27	–	–	3.25	3.33

[*] R–i, R–i–R, or R–i–R$_2$.
[**] Ar–i–R, or Ar–i–R$_2$.
[***] Ar–i, Ar–i–Ar, or Ar–i–Ar$_2$.

effects lead to a decrease in solute retention. Another type of secondary effect arises from intramolecular electronic induction. For example, consider the aromatic molecules *p*-methoxyaniline and *p*-nitroaniline. Of the functional groups on these molecules, the amine group has the strongest interaction energy (Table 1.14). The methoxy group donates electrons to the aromatic ring, thereby making the amine group more basic and increasing retention. Any electron-donating substituent (NH_2 > OH > OCH_3 > C_6H_5 > CH_3) will have this effect. The nitro group withdraws electrons from the aromatic ring, thereby making the amine group less basic and decreasing retention. Any electron-withdrawing substituent (NH_3^+ > NO_2 > C≡N > SO_3H > HO–C=O > H–C=O > R–C=O) will have this effect. The inductive effects in the *ortho-*, *meta-*, and *para*-positions in aromatic molecules correlate very well with Hammett acidity functions [72]. The final type of secondary effect arises from steric constraints. For example, consider *ortho-* and *para-t-* butylaniline. The bulky *t*-butyl group in the *ortho*-position limits access of the amine group to the surface. The interaction is subsequently less than would be expected for groups in the *para*-position. Hence, steric hindrance leads to a decrease in solute retention. Consider also naphthalene and biphenyl. The aromatic rings in naphthalene lie within the same plane, such that they can both be adsorbed on the surface at the same time. The interaction is subsequently greater than would be expected for aromatic rings that are not coplanar, as in biphenyl. Hence, a planar structure can lead to an increase in solute retention. Finally, consider tetracene and pyrene. The linear aromatic rings in tetracene can interact very effectively with multiple sites, especially on alumina, which has a very regular surface structure. Hence, a linear structure can lead to an increase in solute retention. All of these steric effects provide excellent selectivity for isomeric solutes in the adsorption mechanism.

The role of the mobile phase in adsorption is distinctly different from that in the partition mechanism. In gas chromatography, the mobile phase is inert and has little or no effect on solute retention. In liquid and supercritical-fluid chromatography, the mobile phase has distinct chemical properties and acts as a displacing agent for solutes at the surface sites. Thus, it is the strength of mobile-phase interactions with the adsorbent, rather than with the solute, that is important. The solvent-strength parameter, ε°, was developed as a numerical scale of the relative elution strength on alumina. Selected values are summarized in Table 1.15 [70]. The solvent strength on other adsorbents is usually proportional but slightly smaller than that on alumina, for example

$$\varepsilon^{\circ}(\text{silica}) = 0.77\varepsilon^{\circ}(\text{alumina}) \tag{1.95}$$

$$\varepsilon^{\circ}(\text{magnesia}) = 0.58\varepsilon^{\circ}(\text{alumina}) \tag{1.96}$$

$$\varepsilon^{\circ}(\text{Florisil}) = 0.52\varepsilon^{\circ}(\text{alumina}) \tag{1.97}$$

The mobile phase is prepared by selecting a bulk constituent with a relatively low solvent-strength parameter, such as pentane, hexane, carbon tetrachloride, or chloroform. Polar modifiers are then added to act as displacing agents. The modifier is selected for its miscibility with the bulk constituent and its solvent strength, rather than selectivity. The strength of the mixed mobile phase, ε°_{AB}, is given by

$$\varepsilon^{\circ}_{AB} = \varepsilon^{\circ}_{B} + \frac{log(X_A 10^{\alpha n_A(\varepsilon^{\circ}_A - \varepsilon^{\circ}_B)} + 1 - X_A)}{\alpha n_A} \tag{1.98}$$

where X_A and n_A are the mole fraction and molecular area (Table 1.15), respectively, of the modifier, and ε°_A and ε°_B are the solvent strength parameters of the modifier and bulk constituent. The strength of mixed mobile phases, when pentane is used as the bulk constituent with several different modifiers, is shown in Fig. 1.20. For weak modifiers, such as carbon tetrachloride, the overall solvent strength changes smoothly and continuously with the mole fraction. As the difference in strength between the modifier and the bulk constituent increases, the overall solvent strength changes more significantly and over smaller ranges of the mole fraction. For the strongest modifiers, such as ethanol, the most significant changes in solvent strength occur at mole fractions less than 0.01. In general, the

TABLE 1.15

TRENDS AND TYPICAL VALUES FOR THE SOLVENT STRENGTH PARAMETER, ε°, AND MOLECULAR AREA, *n*, ON ALUMINA [70]

Solvent	Strength parameter	Molecular area
n-Pentane	0.00	5.9
n-Hexane	0.01	6.8
n-Decane	0.04	10.3
Cyclohexane	0.04	6.0
Benzene	0.32	6.0
Toluene	0.29	6.8
Dichloromethane	0.42	4.1
Chloroform	0.40	5.0
Carbon tetrachloride	0.18	5.0
Chlorobenzene	0.30	6.8
Tetrahydrofuran	0.45	5.0
Diethyl ether	0.38	4.5
Acetone	0.56	4.2
Ethyl acetate	0.58	5.7
Methanol	0.95	8.0
Ethanol	0.88	8.0
Acetic acid	Large	8.0
Acetonitrile	0.65	10.0
Nitromethane	0.64	3.8
Water	Large	

mobile-phase composition should be adjusted within the regions where solvent strength is linearly related to mole fraction, so that retention can be accurately controlled.

As noted in Fig. 1.20, modifiers with very high solvent strength have significant effects even at low concentrations. As a consequence, trace concentrations of water can have a dramatic effect on retention. The primary problem is variability in water concentration in solvents and in adsorbents, which can lead to fluctuations in retention and poor reproducibility. As a consequence, the water content must be carefully controlled. For solvents, the most reliable approach is to prepare them at 50% of the water-saturation level. Two batches of solvent are required, one completely dry and the other fully saturated with water, at a fixed temperature and pressure. Each day, these two batches of solvent are mixed together in equal volumes to prepare reproducible mobile phases. For the adsorbent, the water content is usually specified by the Brock activity scale. The adsorbent is initially dried at 800 °C to remove all physisorbed and chemisorbed water from the surface (Grade I). Then, water is selectively added to achieve Grade II (3%), Grade III (6%), Grade IV (10%), and Grade V (15%) activity. An increase in the water content decreases the activity of the adsorbent and decreases retention.

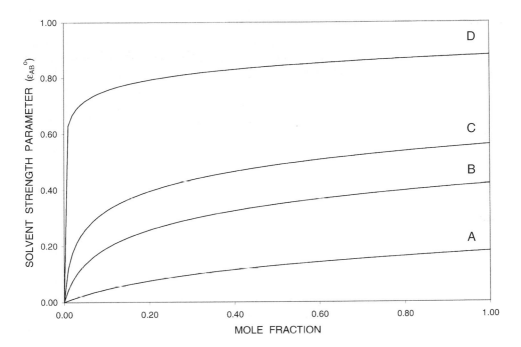

Fig. 1.20. Solvent strength of mixed mobile phases for adsorption. (A) pentane ($\varepsilon_B^o = 0.00$) and carbon tetrachloride ($\varepsilon_A^o = 0.18$), (B) pentane and dichloromethane ($\varepsilon_A^o = 0.42$), (C) pentane and acetone ($\varepsilon_A^o = 0.56$), (D) pentane and ethanol ($\varepsilon_A^o = 0.88$).

Finally, it is important to consider the shape of the isotherm in the adsorption mechanism [73,74]. At very low solute concentrations, the isotherm may be linear, as shown in Fig. 1.18A. A linear isotherm leads to a symmetric, Gaussian zone profile with a constant value of the equilibrium constant. As the concentration increases, however, deviations from linearity may be observed. There are several common shapes of the isotherm. If the surface is homogeneous with a limited number of surface sites, then the isotherm may be convex, as shown in Fig. 1.18B. In this case, the equilibrium constant has a constant value until the solute concentration exceeds the number of surface sites, whereafter it approaches zero. This isotherm can be described by the empirical equation of Freundlich [75]

$$C_S = k_1 C_M^{1/k_2} \tag{1.99}$$

where k_1 and k_2 are adjustable parameters. Alternatively, this isotherm can be described by the kinetic equation of Langmuir [76]

$$C_S = \frac{k_1 C_M}{1 + k_2 C_M} \tag{1.100}$$

The convex isotherm leads to an asymmetric (tailing) zone profile.

If the surface is homogeneous but has very little interaction with the solute, then the isotherm may be concave as shown in Fig. 1.18C. In this case, the equilibrium constant has a small constant value initially but becomes larger as solute molecules begin to interact with one another rather than with the surface. This isotherm can also be described by the empirical equation of Freundlich. The concave isotherm leads to an asymmetric (fronting) zone profile. More complex shapes of the isotherm may be observed if the surface is heterogeneous, if there is multilayer rather than monolayer coverage, or if there are pores. All of these isotherms, both simple and complex, can be described by the comprehensive equation of Brunauer, Emmett, and Teller [77,78]. The practical outcome is that solute concentrations should be maintained within the linear region of the isotherm to provide reproducible retention time, symmetrical peak shape, and good resolution.

1.4.5 Complexation

Adsorption is a special case of the more general mechanism of complexation. Complexation may take place in either the mobile or stationary phase. If the complexing agent is in the mobile phase, an increase in the equilibrium constant will generally lead to a decrease in the retention factor. Conversely, if the complexing agent is in the stationary phase, an increase in the equilibrium constant will lead to an increase in the retention factor. As in adsorption, there are a limited number of interaction sites in the mobile or stationary phase for complexation. Hence, similar problems arise with regard to nonlinear isotherms, and any of the models described for adsorption are also applicable to complexation.

Complexation involves strong selective forces, usually arising from electron donor/acceptor interactions [79]. There are several types of electron donors: nonbonding electron pairs (n-donors) such as amines, sulfides, ketones, and ethers, bonding σ-orbitals (σ-donors) such as alkanes and alkylpolysiloxanes, and bonding π-orbitals (π-donors) such as alkenes, alkynes, and aromatics. There are also several types of electron acceptors: vacant orbitals (n-acceptors) such as Ag^+, Cu^{2+}, and other metal cations, antibonding σ^*-orbitals (σ^*-acceptors) such as halogenated alkanes, and antibonding π^*-orbitals (π^*-acceptors) such as alkenes, alkynes, and aromatics with electron-withdrawing subsituents. From among these possibilities for electron donor/acceptor interactions, some combinations are more common than others. The most common are $n-n$, $n-\sigma^*$, $n-\pi^*$, $\pi-n$, and $\pi-\pi^*$.

Nearly any reversible complexation reaction that is sufficiently rapid for analytical applications such as titration, spectroscopy, etc. can be adapted for separations by gas, supercritical-fluid, or liquid chromatography. Accordingly, there are hundreds of reactions that have already been used and thousands more that are potentially useful. In these reactions, there may be a single site or multiple sites for interaction. The equilibrium constants for multiple sites are generally larger than would be expected simply on the basis of the strength of the individual interaction sites. As shown in Table 1.16, the equilibrium constants for cadmium increase significantly with the number of amine groups in the complexing agent [80]. This increase is due, in part, to the enthalpic contributions, which increase proportionately for each additional amine group. More important, however, is the increase due to the entropic contributions. When ammonia or methylamine bind to the

TABLE 1.16

ILLUSTRATION OF THE CHELATE EFFECT FOR COMPLEXES OF CADMIUM WITH AMINE LIGANDS[*]

Complex	Log K	ΔG (kcal/mol)	ΔH (kcal/mol)	ΔS (cal/mol K)
$Cd(NH_3)_4^{2+}$	7.44	-10.15	-12.70	-8.52
$Cd(NH_2CH_3)_4^{2+}$	6.55	-8.94	-13.70	-16.00
$Cd(NH_2CH_2CH_2NH_2)_2^{2+}$	10.62	-14.50	-13.50	3.29

[*] Equilibrium constant, K, and associated thermodynamic parameters ($\Delta G, \Delta H, \Delta S$) measured in water at 25 °C [80].

metal, one solvent molecule is released; when ethylenediamine binds, two solvent molecules are released; and so on. This increase in the number of solvent molecules increases the net entropy of the system. Although the entropy increase in Table 1.16 appears to be small, it is multiplied by the temperature to determine the net Gibbs free-energy and equilibrium constant in Eqn. 1.82. Accordingly, the entropic contributions are generally the most important for complexing agents with multiple binding sites. This general trend is known as the "chelate effect", and is commonly encountered with polyamines, polyethers, polyalcohols, polycarboxylic acids, etc. We may similarly compare complexing agents with the same number of binding sites, arranged in a linear or cyclic structure. As shown in Table 1.17, the cyclic structure has a significantly larger equilibrium constant than the corresponding linear structure of triethylenetetramine [81].

TABLE 1.17

ILLUSTRATION OF THE MACROCYCLIC EFFECT FOR COMPLEXES OF COPPER WITH AMINE LIGANDS[*]

Complex	Log K	ΔG (kcal/mol)	ΔH (kcal/mol)	ΔS (cal/mol K)
$Cu(NH_2CH_2CH_2NHCH_2$-$CH_2NHCH_2CH_2NH_2)^{2+}$	20.2	-27.6	-21.6	19.5
$Cu(cyclo\text{-}NH_2CH_2CH_2NHCH_2CH_2$-$NHCH_2CH_2NH_2CH_2CH_2)^{2+}$	24.8	-33.8	-22.7	36.2

[*] Equilibrium constant, K, and associated thermodynamic parameters ($\Delta G, \Delta H, \Delta S$) measured in water at 25 °C [81].

For the linear structure, a dynamic equilibrium exists in which metal ions may be partially bound to a different number of amine groups. In contrast, the cyclic tetramine has only two states, fully free and fully bound by metal. Accordingly, the enthalpic contributions are greater for the cyclic structure. However, it is the entropic contributions that are the most important. This general trend is known as the "macrocyclic effect". It is most commonly encountered with cyclic polyethers (*e.g.*, crown ethers) and cyclic polysaccharides (*e.g.*, cyclodextrins). The data in Tables 1.16 and 1.17 illustrate another important advantage of complexing agents with multiple sites and cyclic structures: they can impart a selectivity based on size and shape. The distance between complexing groups and the size of the cyclic structure can be adjusted to optimize interactions with specific solutes. As a consequence, complexation is especially beneficial for the separation of positional isomers, *cis/trans-* or *E/Z*-isomers, and optical isomers. Some examples of these separations will be described below.

One of the earliest and most successful examples of complexation is known as argentation chromatography. In this method, alkenes and aromatic molecules donate π-electrons to the vacant 5s orbital of Ag^+. In general, isolated double bonds have greater interaction than conjugated double bonds because of the higher and more localized electron density. An increase in the degree of substitution on the double bond or aromatic ring reduces interactions with Ag^+ because of steric constraints. These features are illustrated in the separation of the 1,5,9-cyclododecatriene isomers by Vonach and Schomburg [82]. When separated by reversed-phase liquid chromatography (partition) with octadecylsilica stationary phase and 75% methanol/25% water mobile phase, there is little discrimination of these isomers. When 3×10^{-3} M $AgClO_4$ is added to the mobile phase, the *E,E,E*-isomer has the least accessibility whereas the *Z,Z,Z*-isomer has the greatest accessibility and the greatest interaction. Accordingly, the retention of the *E,E,E*-isomer decreases only slightly, the *Z,E,E*- and *Z,Z,E*-isomers decrease more significantly, and the *Z,Z,Z*-isomer is nearly nonretained. Argentation has been very successful for the separation of unsaturated fatty acids, prostaglandins, terpenes, and other isomeric alkene and aromatic solutes.

Other transition metals such as Cr^{2+}, Mn^{2+}, Fe^{2+}, Co^{2+}, Ni^{2+}, Cu^{2+}, Zn^{2+}, Pd^{2+}, Pt^{2+}, Cd^{2+}, and Hg^{2+} have also been used for complexation [79,83]. An example is the separation of the enantiomers of amino acids by ternary complexation with Cu^{2+} by Lam *et al.* [84,85]. L-Histidine is added to the mobile phase, which forms a complex with Cu^{2+} by bonding in the plane through the amine group and the ring nitrogen and axially through the carboxylate group. The amino acids interact at the same time by bonding in the plane through their amine and carboxylate groups. The L-isomers form a less stable complex than the D-isomers, because the functional group of the amino acid interferes with the axial binding of the carboxylate group of L-histidine [86]. The complexes, which are neutral, are then separated by reversed-phase liquid chromatography (partition) with octadecylsilica as the stationary phase and 20% acetonitrile/80% water with an ammonium acetate buffer at pH 7.0. There is no separation of D,L-alanine because the methyl group provides no significant difference in stability of the complexes. As the size of the functional group increases, however, the difference in stability and the extent of separation increases. The propyl and isopropyl groups of D,L-norvaline and D,L-valine provide selectivities of 1.20 and 1.16, respectively. The butyl and *iso*-butyl groups of D,L-norleucine and D,L-leucine provide selectivities of 1.45 and 1.41, respectively. This complexation enables excellent resolution of the enantiomers of the amino acids.

Chiral separations have been achieved by many other approaches, most of which are derived from complexation reactions. A seminal study by Armstrong and DeMond [87] demonstrated that cyclic polysaccharides are useful for chiral separations in liquid chromatography. Based on the size of the solutes, inclusion complexes may be formed preferentially with α-, β-, or γ-cyclodextrins, which have six, seven, and eight units of D-glucose, respectively. For example, β-cyclodextrin provides excellent resolution of the enantiomers of D,L-mephobarbital, but also provides resolution of a variety of other structural isomers of barbiturates. Since this seminal study, cyclodextrins in their native and derivatized forms have been broadly applied in gas, supercritical-fluid, and liquid chromatography as well as capillary electrophoresis. Because of the large number of chiral sites within the cyclodextrin rings, many chiral solutes can be well resolved. However, it is difficult to know *a priori* whether these stationary phases will provide a successful separation for any specific solutes. A more predicable approach has been proposed by Pirkle [88] based on the "three-point interaction model" of Cram [89]. In this approach, chiral stationary phases are specifically designed to have three sites of interaction that are appropriate for the functional groups surrounding the chiral center of the solute. These interaction sites may include any of the types discussed above, most commonly n-donors and acceptors as well as π-donors and π^*-acceptors. For example, the chiral stationary phase N-(3,5-dinitrobenzoyl)-D-phenylglycine has n-donor sites at the carbonyl groups and a π^*-acceptor site at the dinitrobenzoyl group. This phase interacts with solutes having donor and acceptor sites at complementary positions. Although separations are more predictable with chiral stationary phases of this type, the number of solutes that can be separated with each phase is more limited. Accordingly, several dozen stationary phases have already been synthesized, and a few are now commercially available.

In addition to the intentional use of complexation reactions discussed above, complexation may also arise inadvertently or by accident within a chromatographic system. In such cases, the interactions are more likely to be detrimental, leading to irreproducible retention times, poor peak shape, and poor resolution. One source of such interactions is trace metal impurities in the packing materials and solvents. Silica, the most common support for partition and adsorption mechanisms, may contain part-per-million concentrations of metals such as iron, aluminum, magnesium, copper, etc. [90]. If not removed by acid washing during the column manufacturing process, these metals can exhibit strong interactions with solutes containing ketone and amine functional groups. To test commercial columns, Verzele [91] has advocated the use of acetylacetone as a probe of trace-metal activity. In the absence of metals, this solute shows good peak height and peak symmetry and reproducible retention time. If metals are present, the solute exhibits extreme tailing and substantial reduction in peak height. These changes arise because of the strong interactions and the limited number of metal sites, which lead to nonlinear isotherms (*vide supra*). Although it may be possible to use displacing agents, such as citric acid, to improve chromatographic performance, it is best to avoid columns with metal impurities for solutes containing ketone and amine groups.

1.4.6 Ion exchange, ion exclusion, and ion pairing

Consider an organic solute with acidic functional groups, such as aliphatic or aromatic carboxylic acids

$$RCOOH \Leftrightarrow RCOO^- + H^+ \qquad K_a = \frac{(a_{RCOO^-})(a_{H^+})}{a_{RCOOH}} \tag{1.101}$$

where the extent of ionization is reflected in the acid dissociation constant, K_a. In general, weakly acidic solutes are predominantly neutral at pH values less than the pK_a and predominantly anionic at pH values greater than the pK_a. Consider also an organic solute with basic functional groups, such as aliphatic or aromatic amines

$$RNH_3^+ \Leftrightarrow RNH_2 + H^+ \qquad K_a = \frac{(a_{RNH_2})(a_{H^+})}{a_{RNH_3^+}} \tag{1.102}$$

In general, weakly basic solutes are predominantly cationic at pH values less than the pK_a and predominantly neutral at pH values greater than the pK_a. For such solutes, the separation may be performed in several distinct ways. The solutes may be converted to the ionic form and separated by ion exchange or ion chromatography. Alternatively, the solutes may be converted to the neutral form and separated by ion exclusion, ion pairing, or reversed-phase liquid chromatography. Each of these separation methods will be described in detail below.

Ion exchange or ion chromatography [Chap. 4] is useful for the separation of ions, whether they are weak electrolytes that partially dissociate, as in Eqns. 1.101 and 1.102, or strong electrolytes that completely dissociate [68,92–94]. Ion exchange can also be used for the separation of complexes, as discussed in Sec. 1.4.5, if they are ionic. To separate ions by ion exchange, a stationary phase of opposite charge is chosen. For acidic solutes and others with negative charge, an anion-exchange polymer with quaternary amine groups is commonly used. For basic solutes and others with positive charge, a cation-exchange polymer with sulfate or sulfonate groups is commonly used. The mechanism of ion exchange is similar to that of adsorption, with competition for the limited number of surface sites and displacement as the means of elution. As in adsorption, the mobile phase is chosen on the basis of its strength as a displacing agent. Hence, aqueous solutions of strong or weak acids are used for anion exchange and aqueous solutions of strong or weak bases are used for cation exchange. The pH of the mobile phase must also be controlled to ensure that weakly acidic or basic solutes are in the fully ionized form. Although ion–induced dipole and ion–dipole forces are important, ion–ion interactions (Eqn. 1.73) are generally predominant. Consequently, the equilibrium constant and, hence, the extent of retention is determined primarily by the charge and size of the ionic solute.

In their neutral form, weakly acidic and basic solutes can be separated by ion-exclusion chromatography. In this approach, the separation is performed on an ion-exchange resin with the same, rather than opposite, charge as the ionic form of the solutes. Ions with the same charge as the stationary phase are repelled and not retained, whereas nonionic and weakly ionic solutes are retained and separated by partitioning into the pores and interstitial volume. The separation is based, in large part, on the extent of dissociation, such that solutes are generally eluted in order of their pK_a values. It is also dependent upon the extent of solute interactions with the stationary phase, primarily from ion–induced dipole and ion–dipole forces.

Weakly acidic or basic solutes can also be separated by the partition mechanism in reversed-phase liquid chromatography. According to Eqn. 1.91, the ionic form is not retained because of its high polarity and high solubility parameter, whereas the neutral form has a finite retention. The retention factor can be controlled between these two limits by means of the pH. For acidic solutes, the predominant form is neutral and retained at pH values less than their pK_a, but ionic and nonretained at pH values greater than their pK_a. Conversely for basic solutes, the predominant form is neutral and retained at pH values greater than their pK_a, but ionic and not retained at pH values less than their pK_a. The greatest changes in retention occur when the pH is in the vicinity of the pK_a. An example of the effect of pH is shown in Fig. 1.21 [95]. The neutral solute, phenacetin, is largely unaffected by pH. The acidic solutes, salicylic acid and phenobarbital, have finite retention at low pH and become nearly nonretained as he pH increases. The basic solutes, nicotine and methylamphetamine, have a finite retention at high pH and become nearly nonretained at low pH. For both groups of solutes, an inflection point is visible when pH $\approx pK_a$. Separation of these solutes could be achieved at several pH values, such as 5.0 or 7.0–8.0, where there is little overlap and a reasonable range of retention factors.

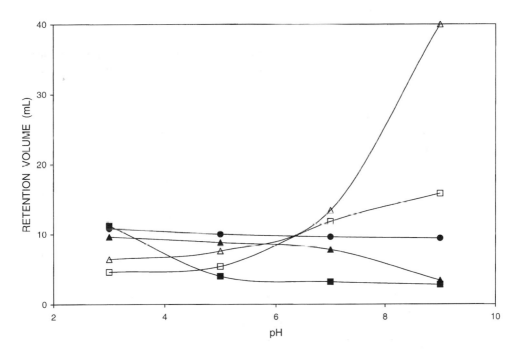

Fig. 1.21. Effect of mobile phase pH on solute retention in reversed-phase liquid chromatography. Column: $d_c = 0.40$ cm, $L = 30$ cm, $d_p = 10$ μm octadecylsilica. Mobile phase: 40% methanol/60% water with 0.025 M NaH_2PO_4/Na_2HPO_4 buffer, 2 mL/min. Solutes: (●) phenacetin, (■) salicylic acid ($pK_a = 2.98$), (▲) phenobarbital ($pK_a = 7.3$), (□) nicotine ($pK_a = 6.16$), (△) methylamphetamine ($pK_a = 10.1$). Adapted from Ref. 95.

In addition to the hydrogen and hydroxide ions used for pH control, other ions can be used to impart neutrality to weakly acidic and basic solutes. In the ion-pairing technique, an alkylamine or alkylsulfonate is added to the mobile phase, and the separation is performed by reversed-phase liquid chromatography [96]. The alkyl chain of the ion-pair agent partitions into the nonpolar stationary phase, thereby creating a charged surface. Solute ions with the same charge as the ion-pair agent are repelled and not retained. Solute ions with opposite charge form ion pairs at the interface and then partition as neutral species into the nonpolar stationary phase. These trends are illustrated by using octylamine hydrochloride as the ion-pair agent in Fig. 1.22 [97]. Neutral solutes, such as toluene, are unaffected by the concentration of the ion-pair agent. Anionic solutes, such as benzenesulfonic acid and chromatropic acid, are attracted to the positively charged amine, so that their retention increases with the concentration of the ion-pair agent. Conversely, cationic solutes, such as aniline and benzylamine, are repelled, so that their retention decreases with the concentration of the ion-pair agent. Hence, solute retention can be controlled by means of the concentration, charge, and alkyl chain length of the ion-pair agent as well as the pH and composition of the mobile phase.

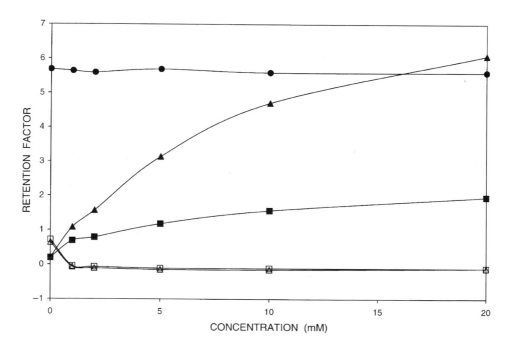

Fig. 1.22. Effect of octylamine hydrochloride concentration on solute retention in ion-pairing reversed-phase liquid chromatography. Column: $d_c = 0.40$ cm, $L = 30$ cm, $d_p = 10$ µm octa-decylsilica. Mobile phase: 32.5% methanol/67.5% water, pH 3.0, containing the indicated concentration of octylamine hydrochloride. Solutes: (●) toluene, (■) benzenesulfonic acid, (▲) chromatropic acid, (□) aniline, (△) benzylamine. Adapted from Ref. 97.

Of the alternatives presented herein, only ion exchange can be used for the separation of permanent ions, such as strong acids, bases, and electrolytes. Because of the strong ion–ion forces, the resulting slow kinetics and nonlinear isotherms can often lead to broad and asymmetric solute zones. Hence, pH control and ion pairing with reversed-phase liquid chromatography are preferable alternatives for the separation of weak acids, bases, and electrolytes. Because of the weaker forces involved, peak shapes are generally symmetric and provide much higher plate number and resolution.

1.4.7 Hydrodynamic and size exclusion

As discussed in the previous sections, small molecules can be separated on the basis of the number and type of functional groups. However, many larger molecules have repeating structural units with the same functional groups. Examples include industrial polymers, such as polystyrenes, polyesters, polyacrylic acids, polyamides, etc. as well as biopolymers, such as peptides, proteins, nucleic acids, and polysaccharides. For such molecules, separation cannot usually be achieved by partition, adsorption, complexation, etc. because each repeating unit has the same interaction energy per unit volume. As discussed in Sec. 1.4.2, this interaction energy contributes to the molar enthalpy in Eqn. 1.82. Hence, it is more beneficial to minimize the enthalpic contributions and attempt to maximize the entropic contributions. Changes in the molar entropy may arise when a large molecule displaces many smaller solvent molecules. There are several separation mechanisms that exploit changes in entropy, including hydrodynamic and size-exclusion chromatography.

In hydrodynamic chromatography, the separation is based solely on flow phenomena in open and packed columns [98]. Flow phenomena are discussed in detail in Sec. 1.3.2. In an open-tubular column, the velocity is zero at the wall because of frictional forces and reaches a maximum at the center of the tube. Small molecules can reside in all of the flow-streams and are eluted with an average velocity of

$$\bar{u}_M = \frac{2}{R^2} \int_0^R u(r) r \, dr \tag{1.103}$$

where $u(r)$ is the radial velocity profile given in Eqn. 1.23. In contrast, large molecules and particles can only reside in flow streams that are more than one particle radius, r_p, from the wall and have an average velocity of

$$\bar{u}_R = \frac{2}{(R - r_p)^2} \int_0^{R-r_p} u(r) r \, dr \tag{1.104}$$

From these equations, the equilibrium constant, K_{HYD}, is given by

$$K_{HYD} = \frac{\bar{u}_M}{\bar{u}_R} - 1 = \left(1 + 2\left(\frac{r_p}{R}\right) - \left(\frac{r_p}{R}\right)^2 \right)^{-1} - 1 \tag{1.105}$$

The equilibrium constant is always negative, indicating that all of the solutes are eluted before a small, nonretained solute. The solutes are eluted in the order of decreasing size;

large solutes are eluted first because they cannot travel in the slowest flow-streams near the wall and have the highest velocity. The largest solutes can travel no faster than $2\ \bar{u}_M$ (Eqn. 1.23), so the minimum value of K_{HYD} is -0.5. The maximum value is zero for a small, nonretained solute. Hence, the range of equilibrium constants is finite, leading to a limited dynamic range. As the column radius is increased, the solute size or molecular weight that can be separated increases as well [99]. However, as shown in Eqn. 1.105, hydrodynamic flow in open-tubular columns is relatively insensitive to changes in size or molecular weight.

Hydrodynamic chromatography can also be performed in columns packed with nonporous particles. There is greater sensitivity to changes in size or molecular weight because two different flow phenomena are important in packed columns. For inter-channel flow phenomena, the smallest channels between particles have the slowest average velocity. Large molecules cannot enter into the smallest channels, so they have a higher velocity than small molecules that can enter all channels. For intra-channel flow phenomena, the velocity within each channel varies with the distance from the particle surface (as in open-tubular columns, discussed above). Large molecules cannot travel in the slowest flow-streams near the particle surface, so they have a higher velocity than small molecules that can travel in all flow-streams. Both of these flow phenomena accentuate the differential velocities based on size or molecular weight. Consequently, nonporous packed columns are more effective and have a greater dynamic range than open-tubular columns. However, there is not a clear theoretical relationship between the solute size or molecular weight and the retention time or volume.

In size-exclusion chromatography [Chap. 5], the separation is based on flow and diffusion phenomena in columns packed with porous particles [100]. Flow phenomena are discussed in Sec. 1.3.2, whereas diffusion phenomena are discussed in Sec. 1.3.1. Size exclusion combines the flow-based separation of hydrodynamic chromatography, as discussed above, with diffusion into pores of controlled size. Molecules that are larger than the pore diameter cannot enter the pores and, hence, they are eluted first at the interstitial volume. Molecules of intermediate size can enter some fraction of the pores. The smallest molecules can enter all of the pores and, hence, they are eluted last at the void volume. The retention volume, V_R, is given by

$$V_R = K_{SEC}V_P + V_0 \tag{1.106}$$

As discussed in Sec. 1.3.2, the interstitial volume, V_0, is equal to the interparticle porosity, ε_u, multiplied by the column volume; the pore volume, V_P, is equal to the intraparticle porosity, ε_i, multiplied by the column volume; and the void volume is equal to the total porosity $(\varepsilon_t = \varepsilon_u + \varepsilon_i)$ multiplied by the column volume. The equilibrium constant, K_{SEC}, reflects the fraction of the pore volume that is accessible to the solute.

There have been many theoretical models that enable the estimation of the equilibrium constant from the size and shape of the solute and pores [101]. The solutes have been modeled as hard spheres, rods, random coils, etc. The pores have been modeled as cylinders, spheres, squares, rectangles, parallel planes, random planes, etc. For example, if the solute is a hard sphere of radius r_p and the pore is a cylinder of radius a, the equilibrium constant is given by

$$K_{SEC} = \left(1 - \frac{r_p}{a}\right)^2 \tag{1.107}$$

Thus, K_{SEC} varies from unity for small molecules to zero for molecules that are equal to or larger than the pore size. The relationship between the hard-sphere radius, the radius of gyration, and the molecular weight of the solute can be determined experimentally from measurements of the intrinsic viscosity. Some representative results for polystyrene standards are summarized in Table 1.18 [102]. Accordingly, size-exclusion chromatography can provide a means of separation but it can also provide direct information about the molecular-weight distribution. In practice, this is usually achieved by using standards of known molecular weight to generate a calibration curve of retention volume *vs.* molecular weight. An example for polystyrene standards is shown in Fig. 1.23 [100]. The molecular-weight distribution of unknown samples is then evaluated on the basis of their retention-volume distribution.

The equilibrium constant, K_{SEC}, is dependent upon the pore size, according to Eqn. 1.107. Accordingly, the range of molecular weights that can be separated is increased by means of the pore size, as shown in Fig. 1.23 for pore sizes from 125 to 1000 Å. By connecting these columns in series, an extended range of molecular weights can be separated. However, an improved linear range can be obtained by combining the particles

TABLE 1.18

ESTIMATED HARD-SPHERE RADIUS AND RADIUS OF GYRATION
FOR POLYSTYRENE STANDARDS [102]

Molecular weight	Radius of gyration (Å)	Hard-sphere radius (Å)
1.80×10^6	660	585
8.60×10^5	428	379
4.11×10^5	277	246
1.60×10^5	159	141
9.72×10^4	119	105
5.10×10^4	81	72
2.04×10^4	47	42
1.03×10^4	32	28
4.00×10^3	18	16
2.25×10^3	13	12
1.25×10^3	9	9

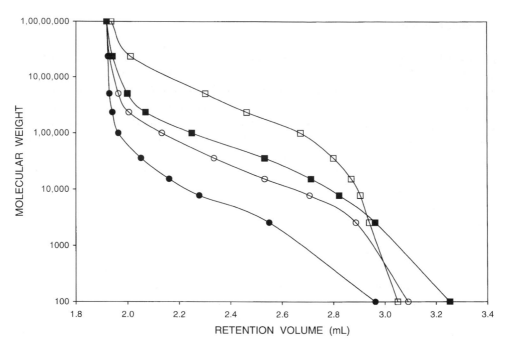

Fig. 1.23. Calibration curves for polystyrene in size-exclusion chromatography. Column: $d_c =$ 0.40 cm, $L = 30$ cm, μ-Bondagel E with pore sizes of (●) 125 Å, (○) 300 Å, (■) 500 Å, (□) 1000 Å. Mobile phase: toluene. Adapted from Ref. 100.

in the correct proportions within a single column. This column provides excellent molecular-weight discrimination over the range from 6.0×10^2 to 2.1×10^6, as demonstrated by Yau *et al.* [100].

There are two common modes for size-exclusion chromatography. Gel filtration chromatography uses porous particles of hydrophilic polymer or silica for separations in aqueous solvents. Gel permeation chromatography uses porous particles of hydrophobic polymer for separations in organic solvents. The mode is chosen to minimize enthalpic interactions of the solute with the stationary phase, as discussed above, so that the entropic interactions can control the separation. Under these conditions, the separation is based wholly on size-exclusion phenomena. However, it is also possible to combine the size-exclusion mechanism with partition or adsorption in a beneficial way. The goal is to create a "critical condition" where the entropic contributions of size exclusion are exactly compensated by the enthalpic contributions of partition or adsorption. When this compensation occurs, all solutes with the same repeating structural feature will be eluted together at the void volume. Solutes with minor variations in structural features will have greater or lesser enthalpic interactions and can be readily separated. This approach is particularly useful for the characterization of mixtures of polymers and co-polymers [103].

1.4.8 Implications

Now that we have described the most common retention mechanisms, it is desirable to elucidate which of these mechanisms is most likely to be successful for a specific application. In order to do so, we must collect detailed information about the sample and the solutes. For the sample, it is necessary to know the identity of the solutes of interest, potential interferences, and the solvent or matrix in which they are dissolved. For the solutes, it is desirable to know the molecular structure, molecular weight, boiling point or vapor pressure, dissociation constants (K_a), charge, and solubility in common solvents such as hexane, chloroform, acetone, methanol or acetonitrile, and water. It is also helpful to identify differences in the structure and properties of the solutes, such as differences in size, polarity, functional groups, geometric or positional isomers, optical isomers, etc. On the basis of this collected information, we can select the most promising retention mechanism.

A scheme for selecting the retention mechanism is shown in Fig. 1.24. To begin, we will examine the information on solute volatility. Note that boiling points at various pressures or vapor pressures at various temperatures can be easily interconverted by using standard nomographs [104]. Gas chromatography is the best alternative for separation if the highest boiling point is less than 500 °C or 1.5 times the maximum column temperature. If gas chromatography is not feasible, then liquid chromatography is a viable alternative if the lowest boiling point is greater than 100 °C or the boiling point of the solvent. Supercritical-fluid chromatography should be considered if there is a wide range of boiling points that span the range between gas and liquid chromatography.

If gas chromatography is selected, the most common and versatile mechanism is partition. The stationary phase polarity is usually used to control the selectivity, whereas the temperature is usually used to adjust the retention factor. If the solutes do not vary substantially in boiling point or polarity, then adsorption or size exclusion may be successful for the separation.

If liquid chromatography is selected, we will next examine the information on molecular weight. If the lowest molecular weight is greater than 2000 and the difference between solutes is at least 15–20% of the average molecular weight, then size exclusion may be successful. If the solutes are soluble in organic solvents, such as hexane, toluene, or chloroform, then gel permeation chromatography is advisable. If the solutes are soluble in aqueous solvents, then gel filtration chromatography is advisable. In either case, the pore size is used to control the selectivity.

If the molecular weight is less than 2000, then we will next examine the information on solubility to distinguish the proper mode of separation. If the solutes are soluble in organic solvents, then normal-phase separations are appropriate. Adsorption is most promising if the solutes vary substantially in the number and type of functional groups and their adsorption energy, Q_i. The stationary-phase type (silica, alumina, carbon, etc.) is usually used to control the selectivity, whereas the mobile-phase strength (ε^0) or temperature is used to adjust the retention factor. Partition is most promising if the solutes vary in overall polarity, δ_i. The stationary-phase type (aminopropylsilica, cyanopropylsilica, etc.) is usually used to control the selectivity, whereas the mobile-phase strength, δ_M, is used to adjust the retention factor.

If the solutes are soluble in aqueous or mixtures of organic and aqueous solvents, then we will next examine whether the solute is ionizable. If the solutes are nonionic under all conditions, then reversed-phase separations by the partition mechanism are appropriate. The stationary-phase type (octadecylsilica, octylsilica, phenylsilica, etc.) and the mobile-phase type (methanol, acetonitrile, or tetrahydrofuran in water) are usually used to control the selectivity, whereas the mobile-phase strength (δ_M) is used to adjust the retention factor. If the solutes are ionizable, then we examine the equilibrium constant for acid–base or other reaction. For solutes with pK_a values between 2 and 9, pH control may be used to adjust the extent of ionization with the reversed-phase partition mechanism. For solutes with pK_a values less than 2 or greater than 9, traditional silica-based packings cannot be used with pH control. Polymeric reversed-phase materials, such as polystyrene/divinylbenzene polymers, or zirconia-based packings can be used over a wider range of pH. Alternatively, an ion-pairing agent can be used to alter the charge state of the solute with the reversed-phase partition mechanism. In addition to the variables mentioned previously, the charge of the ion-pair agent is used to control selectivity and the chain-length and concentration of the ion-pair agent are used to adjust the retention factor. For strong acids, bases, and electrolytes, ion exchange is the preferred retention mechanism. The charge of the stationary phase is used to control selectivity, and the strength of the displacing agent in the mobile phase is used to adjust the retention factor. This scheme should not be considered definitive, but merely illustrative of the manner in which an appropriate separation mechanism may be chosen.

1.5 OPTIMIZATION OF CHROMATOGRAPHIC SEPARATIONS

Now that we have a detailed understanding of the principles of solute-zone separation and dispersion, we can apply this knowledge to optimization. Optimization refers to the process of identifying the experimental conditions required to achieve the most desirable separation in the most time- and cost-effective manner. Because of the many experimental parameters that directly and indirectly influence retention and dispersion, optimization is not a trivial process. It has been the subject of numerous theoretical and practical studies [105–109]. In order to determine the most appropriate experimental conditions, a systematic approach must be adopted in the optimization process. This process consists of three distinct steps: selection of the performance criterion, selection of the experimental parameters, and selection of the method to identify the optimal value or level of each parameter. These steps are discussed sequentially below.

1.5.1 Performance criteria

The first step involves the formulation of a criterion that is an objective and quantitative measure of the quality of the separation. Because this criterion is used to guide the decision-making process, it must inherently define and rank the goals of the optimization. One of the most important goals in chromatography is to maximize the resolution between adjacent solutes, which may be calculated from Eqn. 1.1 or 1.6. Several performance criteria have been developed based on resolution alone. These range from simple functions, such as the limiting resolution of the least-resolved solute pair [110–112], to more complex functions, such as the sum or product of the resolutions for all adjacent solute pairs. The simple sum of resolutions is not very useful, because the smallest resolutions, which represent the greatest problem in optimization, contribute the least to the sum. However, the inverse sum of resolutions, *ISR*, does not suffer from this problem

$$ISR = \sum_{i=1}^{n-1} \frac{1}{R_{s,i}} \tag{1.108}$$

where $R_{s,i}$ is the resolution between solutes i and $i + 1$ and n is the total number of solutes. From this expression, it is evident that the smallest resolutions contribute most greatly to the inverse sum, which should be minimized in order to optimize the separation. The normalized product of resolutions, *NPR*, was introduced by Schoenmakers *et al.* [106,113,114]

$$NPR = \prod_{i=1}^{n-1} \frac{R_{s,i}}{\bar{R}_s} \tag{1.109}$$

where

$$\bar{R}_s = \frac{1}{n-1} \sum_{i=1}^{n-1} R_{s,i} \tag{1.110}$$

This function should be maximized in order to optimize the separation. Although the sum and product of resolutions are useful criteria to maximize the resolution, they may lead to excessively long analysis times.

In addition to resolution, other goals have been incorporated in the performance criterion in either a simultaneous or sequential manner [105]. Among these more complex criteria are the chromatographic response function, *CRF*, proposed by Watson and Carr [115] and the chromatographic optimization function, *COF*, proposed by Glajch *et al.* [116], both of which optimize resolution and analysis time. For example,

$$COF = \sum_{i=1}^{n-1} w_i \ln\left(\frac{R_{s,i}}{R_{s,opt}}\right) + w(t_{opt} - t_n) \tag{1.111}$$

where w_i and w are user-defined weighting functions, $R_{s,opt}$ is the desired resolution, t_{opt} is the desired maximum analysis time, and t_n is the retention time of the last (n^{th}) component. This function approaches zero as the separation is optimized. However, the logarithmic term will approach negative infinity whenever any pair of adjacent solutes is completely overlapped ($R_s = 0$), which leads to problems when automated search methods (*vide infra*) are used to seek the optimum. Although terms have been added to compensate for this problem [117], this correction remains largely empirical in nature.

Other performance criteria attempt to optimize resolution, analysis time, and the number of solutes and to incorporate more sophisticated goals [118–120]. For example, the chromatographic resolution statistic, *CRS*, of Schlabach and Excoffier [120] is given by

$$CRS = \left(\sum_{i=1}^{n-1} \frac{(R_{s,i} - R_{s,opt})^2}{(R_{s,i} - R_{s,min})^2} \left(\frac{1}{R_{s,i}} \right) + \sum_{i=1}^{n-1} \frac{R_{s,i}^2}{(n-1)\bar{R}_s^2} \right) \frac{t_n}{n} \tag{1.112}$$

In this equation, the first term evaluates the resolution of each adjacent solute pair with respect to the desired optimum resolution, $R_{s,opt}$, and the minimum acceptable resolution, $R_{s,min}$. The second term evaluates the uniformity of spacing by comparing the individual resolution values to the average resolution, which is given by Eqn. 1.110. The last term, which serves as a multiplier to the other terms, evaluates the analysis time relative to the number of solutes. The *CRS* function reaches a minimum value as the optimum conditions are approached. Each of these performance criteria, as well as many others reported in the literature, have inherent advantages and limitations that should be considered in view of the specific application and its goals [105,106,117, 121–124].

1.5.2 Parameter space

Next, the experimental parameters that have the greatest influence on the performance criterion are identified. These parameters may have a limited number of discrete or finite values, or they may be continuously variable within a specified range. Among the chromatographic parameters that are discrete in nature are the type of mobile and stationary phases. In addition, physical dimensions such as the particle diameter, column diameter, and column length may be discrete if restricted to columns that are commercially available. Parameters that are continuously variable in nature include the composition of the mobile and stationary phases, as well as the temperature, pressure, and flow-rate. The proper selection of these parameters is crucial, since the success of any optimization strategy is dependent exclusively upon them. It is, therefore, assumed that all experimental factors not explicitly chosen as parameters will be held constant during the optimization.

Depending upon the number and nature of the parameters, the relationship with the performance criterion can be represented by individual point(s), by line(s) in two-dimensional space, by plane(s) in three-dimensional space, or by higher-order surface(s) in N-dimensional space. For simplicity in the subsequent discussion, the relationship between the parameter(s) and the performance criterion will be referred to as the response surface, regardless of its continuity and dimensionality.

1.5.3 Optimization methods

Once the performance criterion and the experimental parameters are chosen, a logical procedure must be used to locate the optimum value of the criterion on the response surface. Because the response surface may be highly complex, with many local minima and maxima, this procedure must be sufficiently robust and reliable to locate the primary or global optimum rather than secondary or local optima. The procedures that have been adopted to locate the optimum can be classified into four general categories: simultaneous, sequential, regression, and theoretical [108]. Each of these procedures combines a specific method or experimental design for data acquisition together with a method for data analysis.

1.5.3.1 Simultaneous methods

In the simultaneous methods, all of the parameters to be varied, as well as their levels, are chosen *a priori*. Experimental data are acquired at each level for each of the parameters chosen. The data are then analyzed, either by visual inspection or by a computer-assisted sorting routine, to determine which of these conditions yields the best value of the performance criterion. The simultaneous method is also referred to as a grid search or mapping method, since the parameter space can be likened to a grid or framework of experimental conditions, generally spaced in uniform increments. The number of experiments to be performed depends on the number of the parameters and their levels [106]; for n parameters at q levels, the number of experiments is $q \times n$. In order to describe the response surface completely and accurately, each parameter must be examined over a wide range in small increments of the level. However, this may require a prohibitively large number of experiments. In order to reduce the number of experiments, constrained parameters may be chosen to reduce the degrees of freedom of the parameter space. A constrained parameter is one in which the sum of the levels is constrained to a constant value; for example, the sum of the volume fractions of the mobile-phase solvents is unity. Another approach to reduce the number of experiments is through experimental design, common examples of which are the factorial and star-design methods [125,126].

Because of the inherent nature of simultaneous methods, they are most suitable for parameters that are discrete and independent. An advantage of these methods is that they are derived directly from the experimental data, and do not require or rely upon a theoretical or mathematical model to locate the optimum. These methods can be used to elucidate the complete response surface, if the parameter levels are varied over a wide range in small increments. Thus, it is possible, in principle, to identify the global optimum in the presence of other local optima. The most significant disadvantage is that the number of experiments increases rapidly with the number of parameters and their levels.

1.5.3.2 Sequential methods

In the sequential methods, a few initial experiments are performed at selected levels of the parameter(s). These data are analyzed, either by visual inspection or by computer, and the results are used to direct and guide subsequent experiments. The success of this method is highly dependent upon the type of performance criterion and the type of search routine used to locate the optimum. A variety of search routines have been developed [126,127], the most common of which is the simplex routine [128–133]. A simplex is a geometric figure that consists of $N + 1$ vertices in N-dimensional space; for example, a two-dimensional simplex is a triangle. Thus, three initial experiments are required to optimize two parameters. To illustrate the search procedure, consider the response surface shown as a contour diagram in Fig. 1.24 [108]. The Vertices *1, 2,* and *3* define the initial simplex and

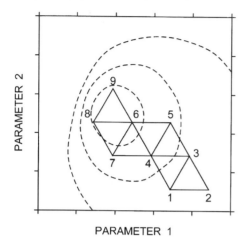

PARAMETER 1

Fig. 1.24. Sequential optimization for two parameters by the simplex approach. The initial simplex is comprised of Vertices *1, 2,* and *3*. The vertex with the worst response is identified, then replaced by projection through the midpoint of the remaining vertices. The simplex eventually converges at the optimum, which is located in the vicinity of Vertex *6* in this example. Adapted from Ref. 108.

represent the initial levels of the parameters at which an experiment is performed. The results of these experiments are then compared by means of the performance criterion; in this example, Vertex *2* has the worst response. Consequently, the simplex routine replaces this vertex with one obtained by reflection through the midpoint of the line connecting Vertices *1* and *3*. The simplex is now described by Vertices *1, 3,* and *4*. An experiment is performed at the new Vertex *4*, and the performance criterion is compared to that obtained previously at Vertices *1* and *3*. Vertex *1* now has the worst response, and reflection in the opposite direction leads to Vertex *5*. The objective of this iterative procedure is to direct successive experiments toward more favorable regions of the response surface. This procedure is continued within the defined boundary conditions of the parameters until the optimum is obtained in the vicinity of Vertex *6*. In the sequential simplex method, originally proposed by Spendley *et al.* [128], the vertices are equally spaced to create a simplex that is an equilateral polyhedron. Such equilateral simplexes may not readily converge, in some cases circling around the optimum or becoming stranded on a surface ridge. These limitations have been largely overcome by the modified procedure of Nelder and Mead [129], which allows for expansion of the simplex in a favorable direction and contraction in an unfavorable direction during the search procedure.

The simplex and related sequential methods are most suitable for parameters that are continuous, since an experiment must be performed at each new vertex that is calculated. An important advantage is that these methods are able to optimize several parameters simultaneously with no prior knowledge about their interaction. Additionally, these methods do not rely on any theoretical or mathematical model to identify the optimum conditions. However, sequential procedures have some disadvantages, the foremost being that the number of experiments required to locate the optimum is indeterminate. Moreover, for response surfaces consisting of many local maxima and minima, a local optimum may be erroneously identified as the global optimum. This problem may be circumvented by repeating the optimization procedure with the initial simplex of various sizes and in various locations on the response surface. Additionally, because the performance criterion is only evaluated at selected points, a limited insight is provided about the response surface. Despite these potential drawbacks, sequential optimization procedures, such as the simplex method, are versatile with wide applicability in chromatography [130–134].

1.5.3.3 Regression methods

Regression methods, also known as interpretive methods, rely on the use of a semi-empirical mathematical model to describe the performance criterion as a function of the experimental parameters. In these techniques, a chromatographic property, such as the retention factor, plate number, etc. that is related to the performance criterion is measured at certain levels of the parameters. The number of experiments depends on the mathematical model; a linear model requires a minimum of two experiments, a quadratic model requires three experiments, etc. However, it is generally desirable to acquire data at additional levels of the parameter in excess of the minimum requirements in order to allow for greater statistical confidence in the results of the regression. The resulting data are analyzed by the least-squares method in order to determine the best value for the coefficient(s) in the model equation.

Once the regression coefficients have been determined, the equation may be used to calculate the chromatographic property and, hence, the performance criterion at any intermediate level of the parameter. Numerical or graphical methods may then be used to represent the response surface and to determine the optimal value of the parameter. A well-known method of representation for a single parameter is the window diagram approach, originally developed by Laub and Purnell [135–137]. As shown in Fig. 1.25 [108], a performance criterion such as the resolution is graphed as a function of the parameter for each solute pair. The regions of the graph beneath the limiting resolution for the least-resolved solute pair are shaded to create "windows". The optimum value of the parameter may then be identified from the highest point in the highest window, which represents the maximum value of the performance criterion. This approach may be extended to two parameters by using the overlapping resolution mapping (ORM) technique, described by Glajch and Kirkland [138]. In this case, the resolution of each solute pair is represented by a surface, rather than a line, and the regions of intersection must be determined in order to create the three-dimensional "windows".

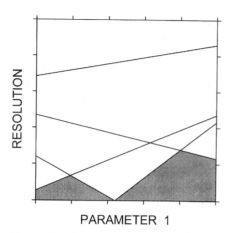

Fig. 1.25. The regression method of optimization for one parameter by the window-diagram approach. The resolution or other performance criterion is calculated by regression for each solute pair. The shaded window regions represent the limiting resolution of the least-resolved solute pair. The optimum is located at the highest point in the highest window, which is ~0.8 for Parameter *1* in this example. Adapted from Ref. 108.

A simpler graphical approach, illustrated in Fig. 1.26 [108], is to shade those regions with resolution greater than a selected threshold value (*e.g.*, $R_{s,i} = 1.5$) for each solute pair and then overlap these maps to identify the optimum value of the two parameters. The extension of these graphical techniques to three or more parameters remains a challenging problem [139].

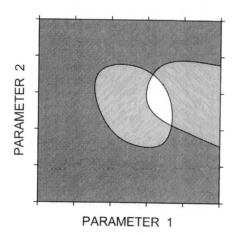

Fig. 1.26. The regression method of optimization for two parameters by the overlapping resolution mapping approach. The resolution or other performance criterion is calculated by regression for each solute pair. The regions with resolution greater than a threshold value (e.g., $R_{s,i} = 1.5$) are shaded. The optimum is located by overlapping the maps for all solute pairs, which is ~0.7 for Parameter *1* and ~0.6 for Parameter *2* in this example. Adapted from Ref. 108.

When compared with the simultaneous and sequential methods, the regression techniques require fewer experiments to describe the complete response surface. However, the successful application of this method is greatly reliant on the accuracy of the mathematical model. For this reason, regression methods are most appropriate for continuous parameters with a well-defined and verifiable relationship to the chromatographic property or performance criterion, such as mobile-phase composition, temperature, and pressure.

1.5.3.4 Theoretical methods

Theoretical methods, like regression methods, use a mathematical equation to describe the effect of the parameter on a chromatographic property, such as the retention factor, plate number, etc. The important distinction is that the equation is theoretically derived and requires no preliminary experimental data for implementation. This approach is most appropriate for discrete or continuous parameters that have a well-known and theoretically predictable effect, such as flow-rate, column diameter, and column length. The optimal value of the parameter may be determined by any of the numerical or graphical methods discussed previously.

1.5.4 Univariate optimization

Univariate optimization involves the optimization of a single parameter or of multiple parameters in a wholly independent manner. This is achieved by varying a given parameter to investigate its influence on the performance criterion, while all other parameters are held constant. When an optimum value for that parameter has been established, its value is held constant while the next parameter is varied to determine its own optimum value. This classical approach is relatively simple and is suitable for parameters that are independent. While it can be applied over restricted ranges for parameters that influence or interact with one another [140,141], the resulting error can become problematic. Univariate optimization can be accomplished by any of the simultaneous, sequential, regression, or theoretical methods described previously. In this section, the use of these methods for the univariate optimization of important chromatographic parameters will be reviewed.

1.5.4.1 Temperature

Temperature is, by far, the most common parameter used for optimization in gas chromatography. It is used less frequently in liquid chromatography. The practical temperature range for most common solvents is somewhat limited, being proscribed by the boiling point, freezing point, and viscosity. Nevertheless, temperature changes within this range have been reported to have a significant influence on solute retention and selectivity in liquid chromatography [142–144].

In both gas and liquid chromatography, temperature optimization is often performed by the simultaneous method. By comparing the performance criterion at several different temperatures, the optimum temperature may be readily identified [145]. A more systematic approach has been adopted by using a regression method based on the classical thermodynamic relationship of van't Hoff, which is given in Eqn. 1.82 [146–148]. If the molar enthalpy and molar entropy are invariant with temperature, then

$$ln \, k = A_0 + \frac{A_1}{T} \tag{1.113}$$

By measuring the retention factor at two temperatures, the regression coefficients, A_i, can be determined by linear regression. The resulting equation can be used to predict the retention factor and, hence, the performance criterion at any intermediate temperature. The optimum temperature may be identified by the window diagram approach [149] or by other graphical or numerical methods.

In addition to isothermal optimization, temperature programming is also commonly employed in gas chromatography. The optimization is more complicated in this case, as the retention of the solute zone changes continuously during the temperature program. The net retention must be determined by applying a numerical method to calculate the retention, according to Eqn. 1.113, in small increments of time or length [150]. This approach can be used to optimize simple linear temperature programs as well as more complex programs with multiple stepwise or linear changes in temperature. For example, Suliman [151] used a factorial design to locate the optimum conditions for initial temperature, rate for the first linear ramp, and rate for the second linear ramp for the separation of 2,4-dinitrophenylhydrazone derivatives of eight carbonyl compounds. Snijders *et al.* [150,152] have developed a theoretical method, based on Kovats retention indices, to predict optimum conditions for both isothermal and temperature-programmed separations. This method has the advantage that it uses published values of the retention index and, in principle, requires no experimental measurements for implementation. Retention times can generally be predicted with average relative errors of ±5.1% for isothermal and ±2.7% for linear temperature-programmed separations.

Temperature programming has also been implemented in liquid chromatography. This is often accomplished by using microbore or capillary columns in order to minimize the effect of radial temperature gradients [153]. Many workers have demonstrated that, under well-controlled experimental conditions, temperature programming may be comparable or superior to solvent programming [154–156]. Although several applications of temperature programming in liquid chromatography have been reported, theoretical models that enable optimization of the separation are rare [156]. The optimization is particularly complicated, because temperature influences both solute retention and band-broadening during the separation, often in a manner that is not as accurately predictable as for gas chromatography.

1.5.4.2 Stationary-phase composition

In many cases, a single stationary phase is sufficient to achieve the desired chromatographic separation. Because it is a discrete parameter, the type of stationary phase is most often optimized by the simultaneous method in both gas and liquid chromatography [105,106]. This approach is clearly illustrated in the work of Glajch *et al.* [157], who examined octyl-, cyano-, and phenethyl-modified silica stationary phases for the separation of phenylthiohydantoin derivatives of twenty common amino acids by liquid chromatography. They applied the mixture-design statistical method to optimize the mobile-phase composition separately on each column. By comparison of the three overlapping resolution maps, the cyano-modified silica was identified as the optimal stationary phase.

When a single stationary phase is not sufficient, mixed or multimodal stationary phases can be used to combine dissimilar but compatible retention mechanisms. These stationary phases may be combined in several ways: by incorporating or chemically bonding the stationary phases on the same particles or in the same column, or by incorporating the phases on different particles or in different columns that are connected in series [158]. For gas chromatography, many nonpolar and polar stationary phases have been combined in this manner [135,136]. In general, when the phases are mixed within the same column, the solute retention is accurately predictable from the equilibrium constant, K, and the fraction, ϕ, of the individual stationary phases

$$K = \sum_{i=1}^{n} \phi_i K_i \qquad (1.114)$$

Similarly, when the phases are combined in separate columns, the retention time is accurately predictable from the dimensions of the individual columns

$$t_R = \sum_{i=1}^{n} \frac{\pi d_{col,i}^2 \varepsilon_{t,i} L_i (1 + k_i)}{4F} \qquad (1.115)$$

This latter approach, sometimes called tandem gas chromatography, can be readily optimized by the regression method. The solute retention factor is measured on the individual stationary phases, from which the retention time and, hence, the resolution can then be calculated for any column length by means of Eqn. 1.115. The window diagram approach is usually used to identify the optimum column length for each phase [159].

For liquid chromatography, there are greater constraints because the stationary phases must have compatible mobile phases. Mixed phases that have been successfully used include reversed-phase with size-exclusion material [160–163], cation- or anion-exchange with size-exclusion material [163], reversed-phase with cation- or anion-exchange material [164–168], and cation- with anion-exchange material [167–169]. Isaaq *et al.* [162] have investigated solute retention as a function of stationary-phase composition for columns containing β-cyclodextrin- and octadecyl-modified silica. They observed that retention on chemically or physically mixed supports may not be predictable as a linear function of the stationary-phase composition. This nonlinear additivity of solute retention may arise as a consequence of interactions between the two stationary phases. On the other hand, retention on serially coupled columns is generally additive and predictable, just as in gas chromatography, because the stationary phases are physically separated. Similar results were reported by Glajch *et al.* [157] for columns containing cyano- and phenethyl-modified silica.

These investigations suggest that the regression technique may be used to optimize stationary-phase composition for serially coupled columns in liquid chromatography.

1.5.4.3 Mobile-phase composition

The mobile-phase composition is, by far, the most common parameter used for optimization in liquid chromatography. As such, it has already been the subject of several excellent books [105–107] and comprehensive review papers [114,170–172]. Mobile-phase optimization will generally include the type of solvent or other modifier, which influences the selectivity factor, as well as its concentration, which influences the retention factor. Because of the discrete nature of solvent or modifier type, optimization is most often performed by the simultaneous method. This approach has been utilized to select from among several organic solvents, buffers, ion-pair agents, etc. [171,173,174]. However, the simultaneous approach is time-consuming and ineffective for optimizing solvent or modifier concentration, which is most often performed by regression methods.

For normal- and reversed-phase chromatography, regression models have been developed to describe solute retention as a function of mobile-phase composition. By extension of classical thermodynamics to binary solvent mixtures (Sec. 1.4.3, Eqns. 1.91 and 1.93), the retention factor can be described as a quadratic function of the volume fraction, ϕ, of each component

$$ln\ k = A_0 + A_1\phi_1 + A_{11}\phi_1^2 + A_2\phi_2 + A_{22}\phi_2^2 + A_{12}\phi_1\phi_2 \tag{1.116}$$

If the volume fraction is a constrained parameter (*e.g.*, $\phi_2 = 1 - \phi_1$), this expression can be simplified as follows

$$ln\ k = A_0 + A_1\phi_1 + A_{11}\phi_1^2 \tag{1.117}$$

In order to reduce the number of experiments required for optimization, this quadratic model is frequently approximated by a linear equation [171]

$$ln\ k = A_0 + A_1\phi_1 \tag{1.118}$$

With this simplification, retention measurements at two solvent compositions are sufficient to describe the complete response surface. In general, the linear approximation is accurate over a more limited range of solvent compositions than the quadratic model. Caution must be exercised when extrapolating either the linear or quadratic model beyond the range of available experimental data. Significant deviations may arise as a consequence of non-ideal solute/solvent and solvent/solvent interactions, especially in polar solvents.

By extension of classical thermodynamics to ternary solvent mixtures (Sec. 1.4.3, Eqns. 1.91 and 1.93), the retention factor can be described as a quadratic function of the volume fraction of each component

$$ln\ k = A_0 + A_1\phi_1 + A_{11}\phi_1^2 + A_2\phi_2 + A_{22}\phi_2^2 + A_3\phi_3 + A_{33}\phi_3^2$$
$$+ A_{12}\phi_1\phi_2 + A_{13}\phi_1\phi_3 + A_{23}\phi_2\phi_3 \tag{1.119}$$

If the volume fraction is a constrained parameter (*e.g.*, $\phi_3 = 1 - \phi_1 - \phi_2$), this expression can be simplified as follows

$$\ln k = A_0 + A_1\phi_1 + A_{11}\phi_1^2 + A_2\phi_2 + A_{22}\phi_2^2 + A_{12}\phi_1\phi_2 \tag{1.120}$$

Alternatively, the semi-empirical expression proposed by Weyland *et al.* [175] can be obtained by neglecting the second-order terms in Eqn. 1.119 and including a term for first-order ternary solvent interactions

$$\ln k = A_1\phi_1 + A_2\phi_2 + A_3\phi_3 + A_{12}\phi_1\phi_2 + A_{13}\phi_1\phi_3 + A_{23}\phi_2\phi_3 + A_{123}\phi_1\phi_2\phi_3 \tag{1.121}$$

This latter polynomial expression requires retention measurements at seven solvent compositions to describe the complete response surface.

The equations cited above have been widely and successfully applied for the univariate optimization of mobile-phase composition in normal- and reversed-phase partition mechanisms, as well as in ion-exchange, ion-pair, and other chromatographic methods [105–107,114,170–172]. Although there have been many noteworthy accomplishments, the most exemplary are those by Snyder, Dolan, *et al.* [171,176–180] for binary mobile phases and by Glajch, Kirkland, *et al.* [116,138] for ternary and quaternary mobile phases. In the latter work, a mixture-design statistical method is applied, which combines the simultaneous method with regression techniques to optimize both selectivity and strength of the mobile phase. The mobile phases are chosen such that they exhibit the greatest possible difference in chemical interactions, including proton-donor (methanol), proton-acceptor (tetrahydrofuran), and dipole (acetonitrile) interactions. A minimum of seven preliminary experiments are performed, and the performance criterion is evaluated by using the overlapping resolution mapping technique in order to locate the optimum ternary or quaternary mobile phase.

In addition to isocratic optimization, gradient mobile phases have also been optimized by a variety of methods. These methods involve optimization of the solvent selectivity, as discussed above, the initial and final solvent concentration, as well as the shape and duration of the gradient. The sequential simplex approach has been used by Watson and Carr [115], whereas regression techniques have been favored by Jandera, Churacek, *et al.* [181–183] and by Snyder, Dolan, *et al.* [172,177–180,184]. Kirkland and Glajch [185] extended the mixture-design statistical method described above to gradient elution. Experimental data were acquired, using a series of seven gradients with iso-eluotropic mobile phases of aqueous methanol, tetrahydrofuran, and acetonitrile. From these results, the performance criterion was evaluated by using the overlapping resolution mapping technique.

1.5.4.4 Other parameters

In addition to the parameters discussed above, optimization techniques have also been developed for physical parameters, such as particle diameter, column diameter, column length, flow-rate, etc. Snyder, Dolan, *et al.* [171,172,176–180] have applied regression methods for particle diameter, and theoretical methods for column dimensions and flow-rate. These methods are especially successful when combined with univariate optimization of the mobile-phase composition in the DryLab I (isocratic) and DryLab G (gradient) programs from LC Resources, Inc.

1.5.5 Multivariate optimization

If the optimum of one parameter depends on and varies with the value of another, those parameters are mutually interacting. The optimization of interacting parameters is best achieved by concurrent variation, rather than independent variation as in univariate optimization. All of the optimization procedures described previously, individually or in combination, may be used to implement multivariate optimization. However, the most widely used and successful methods for univariate optimization, simultaneous and regression methods, can become cumbersome and ineffective, because the number of characterizing experiments increases rapidly with the number of parameters. In this section, the extension of these methods to multivariate optimization of important chromatographic parameters will be reviewed.

1.5.5.1 Temperature and pressure

Temperature and pressure are important and interacting variables of state in gas chromatography. These parameters have been optimized in a multivariate manner by Morris *et al.* [186], using a central-composite experimental design with multiple linear regression (15 experiments). This approach enabled the selection of the optimum values for inlet pressure, initial temperature, and temperature-programming rate for 17 substituted phenols. Li *et al.* [187] have applied a simplex approach to optimize inlet pressure, initial temperature and time, temperature-programming rate, and final temperature and time for the separation of polychlorinated biphenyls.

1.5.5.2 Temperature and flow-rate

Temperature and flow-rate are also interacting parameters in gas chromatography. In order to avoid the potential complications, many workers choose to maintain conditions of constant flow-rate for optimization. However, Guillaume *et al.* [188] have developed an approach to optimizing the isothermal temperature and flow-rate concurrently by using a central-composite experimental design (9 experiments). The data were fitted by regression to a simple empirical equation, with an average relative error in retention time of $\pm 2.4\%$. This method was demonstrated for the separation of eight esters of *p*-hydroxybenzoic acid.

The optimization of temperature and flow-rate, as an alternative to mobile-phase composition in liquid chromatography, has been accomplished by Scott and Lawrence [189], using the simultaneous method. Using methyl palmitate and dinonylphthalate as models from among 11 aliphatic and aromatic solutes, the resolution was examined as a function of linear velocity at six temperatures (26 experiments). The optimum conditions were selected from among those studied, and the results were compared under constant conditions (isothermal/isotachic) as well as with gradient programming in both temperature and flow-rate.

Jinno and Yamagami [190] used the regression method to optimize temperature and flow-rate concurrently under gradient-programming conditions in liquid chromatography. In preliminary studies, a linear equation was used to relate the retention factor of phenylthiohydantoin derivatives of five amino acids with a constant parameter that is characteristic of the physico-chemical properties, shape, and structure. This equation enabled the estimation of the retention factor for other amino acids in the absence of experimental data, with an average relative error of $\pm 2.5\%$. Since the retention factor is independent of flow-rate, this equation also allowed the prediction of retention time or volume under conditions of constant and gradient flow-rate. In order to consider the effects of both flow-rate and temperature programming simultaneously, a quadratic expression was included to relate the logarithm of the retention factor with temperature

$$ ln\ k = A_0 + \frac{A_1}{T} + \frac{A_2}{T^2} \tag{1.122} $$

This approach was validated by optimizing the separation of the phenylthiohydantoin derivatives of 20 amino acids. The results showed good agreement between the predicted and observed chromatograms, with an average relative error in predicted retention time of $\pm 1.7\%$.

1.5.5.3 Temperature and stationary-phase composition

Whenever mixed stationary phases are needed to achieve a separation in gas chromatography, it is usually necessary to optimize temperature as well. If the stationary phases are mixed within a single column, then the optimization entails the determination of the best stationary-phase composition and the best isothermal or gradient-temperature program. Such optimizations are usually performed by regression methods. Several columns are prepared with different stationary-phase compositions, and retention data are acquired at various temperatures. The data are fitted independently to Eqns. 1.113 and 1.114, and the optimum conditions are determined by the window-diagram approach.

If the stationary phases are contained in serially coupled columns, then the optimization entails the determination of the best column length and temperature. As an alternative to column length, pressure or flow-rate can be used to adjust the retention time in the individual columns. The temperature can be controlled in two ways: all columns can be maintained in a single oven with a common isothermal or gradient-temperature program, or each column can be maintained in a separate oven with its own, usually isothermal temperature. Bakeas and Siskos [191] have used the simplex method to optimize the separation of simple hydrocarbons on serially coupled columns in a single oven. The parameters that were optimized included the initial temperature and time, the temperature-programming rate, and the pressure at the midpoint of the two columns. Baycan-Keller and Oehme [192] optimized the separation of 18 toxaphene isomers on serially coupled columns by using the window-diagram approach, described above. Under isothermal conditions, the optimum separation was achieved with a 19-m column containing derivatized β-cyclodextrin in OV-1701, and a 6-m column containing polymethylbiphenylsiloxane. An independent optimization of the temperature program allowed separation of the most important congeners without interference.

The optimization of stationary phase and temperature in liquid chromatography has been accomplished by Mao and Carr [193,194]. In their approach, two serially coupled columns, containing stationary phases with widely differing selectivity for the solutes of interest, were used with a common mobile phase. The stationary phases that have been successfully combined include reversed-phase partition materials, such as polybutadiene-coated zirconia or octadecylsilica, with a reversed-phase adsorption material, carbon-coated zirconia. Each of the columns was maintained in a separate, isothermally controlled oven. The variation of temperature served to adjust the relative retention time on each column. Retention measurements on each column at a few selected temperatures is sufficient for the optimization. The temperature dependence of the retention factors is given by linear regression to Eqn. 1.113, and the net retention time is given by Eqn. 1.115. The window-diagram approach is used to determine the optimal temperature for each column. This approach has been applied to the separation of substituted benzenes, barbiturates, derivatized amino acids, and others.

1.5.5.4 Temperature and mobile-phase composition

The optimization of temperature and mobile-phase composition has been accomplished by using a combination of the simultaneous and regression methods [156,195–197]. Diasio and Wilburn [195] measured solute retention as a function of mobile-phase composition and analyzed the results by linear regression. The optimum mobile-phase composition was then used to measure solute retention at various temperatures, and the temperature that yielded the best quality criterion was identified as the optimum.

In the fractional-factorial design of Yoo *et al.* [156], solute retention was measured as a function of mobile-phase composition at four different temperatures (20 experiments), and then analyzed by linear regression to Eqn. 1.118. By using the resolution of the least-resolved solute pair as the quality criterion, the optimum mobile-phase composition was identified by the window-diagram approach at each temperature. These results were then used to optimize the separation of 14 saturated and unsaturated fatty acids under temperature-programming conditions. The optimum separation conditions enabled baseline resolution ($R_{s,i} > 1.5$) of the least-resolved solute pair, with an average relative error in predicted retention time of $\pm 0.8\%$. A similar approach was demonstrated by Hancock *et al.* [196,197] for the optimization of gradients in both temperature and mobile-phase composition.

Because these simple approaches do not account for interaction between the parameters, others have adopted the use of regression methods to describe retention as a combined function of temperature and mobile-phase composition. Melander and Horváth [148,198] have proposed the following analytical expression for reversed-phase liquid chromatography

$$ln\ k = A_0 + \frac{A_1}{T} + A_2\phi\left(1 - \frac{T_c}{T}\right) \qquad\qquad (1.123)$$

where T_c is the compensation temperature, which was examined as a constant (625 K) and an adjustable coefficient. This expression was later extended by Melander *et al.* [199] in order to improve the accuracy and precision of regression results

$$\ln k = A_0 + \frac{A_1}{T} + A_2\phi\left(1 - \frac{T_c}{T}\right) + A_3\phi \tag{1.124}$$

Inherent in these equations is the assumption of linear relationships between the logarithm of the retention factor and mobile-phase composition, and between the logarithm of the retention factor and inverse temperature. As a consequence of these assumptions, the average relative error in the predicted retention factor was $\pm 7.8\%$ and $\pm 6\%$ for Eqns. 1.123 and 1.124, respectively, when 28 experimental measurements were used for each solute in the regression.

1.5.5.5 Mobile-phase and stationary-phase compositions

Mobile- and stationary-phase compositions have been optimized concurrently by Glajch *et al.* [157] for the separation of phenylthiohydantoin derivatives of 20 common amino acids. In an extension of the univariate study described previously, the solute retention factors were measured on three different columns, containing octyl-, cyano-, and phenethyl-modified silica stationary phases, using seven different ternary solvent compositions of methanol, acetonitrile, and tetrahydrofuran (21 experiments). By using the regression model, a response surface was generated as a function of mobile-phase composition for each solute on each column. In order to establish the optimum composition of the mobile and stationary phases, the solute retention on a mixed bed was assumed to be a linear combination of the retention on each individual stationary phase at the same mobile-phase composition. The quality criterion, either the limiting resolution of the least-resolved solute pair or the COF function (Eqn. 1.111), was then calculated for all possible combinations of stationary and mobile phases. From these results, the optimum stationary-phase composition was predicted to be 28% cyano/72% phenethyl, and the optimum mobile-phase composition was 9.2% methanol/10.6% acetonitrile/13% tetrahydrofuran/67.2% water with phosphoric acid at pH 2.1. This prediction was validated by using a single column, containing a physically mixed bed as well as two serially coupled columns. A good separation of the amino acids was achieved that was similar, but not identical, in the two systems. Unfortunately, it was not possible to assess the predictive accuracy of these methods from the results presented by Glajch *et al.* [157].

A different approach, called parametric modulation, was developed by Lukulay and McGuffin [108,200] for the concurrent optimization of mobile and stationary phases. The fundamental strategy of this approach is that retention can be accurately predicted if the solute undergoes independent interactions with each of the mobile and stationary phases. Hence, the mobile phases are introduced as separate zones of variable length (similar to a step gradient) and the stationary phases are contained in separate columns of variable length. The overall retention time can then be predicted by summation of the retention time in each individual environment, similar to Eqn. 1.115. This approach was demonstrated by the separation of isomeric four- and five-ring polycyclic aromatic hydrocarbons on octadecylsilica and β-cyclodextrin silica columns with aqueous methanol and acetonitrile stationary phases. Although there are 64 possible permutations of these mobile and stationary phases, only 8 experiments were necessary to fully optimize the separation by the parametric-modulation approach. Under the predicted optimum conditions, the average relative error in retention time was ± 3.5%.

1.5.6 Implications

In this section, we have reviewed the various methods that have been used for univariate and multivariate optimization in gas and liquid chromatography. These methods allow the systematic optimization of temperature, mobile- and stationary-phase composition, flow-rate, etc. in order to obtain the desired resolution or other performance criterion. Many of these methods are already available to practicing chromatographers in commercially available software. More recently, artificial neural networks and related chemometric methods have been applied to chromatographic separations [109]. Although still in the early phases of development, these methods show much promise for the future.

1.6 SUMMARY

In this chapter, we have identified resolution as the most important factor underlying the practical and theoretical applications of chromatographic separations. The extent of resolution determines the purity of the solute zones, which influences the accuracy of qualitative and quantitative measurements for practical applications. In addition, it influences the accuracy of thermodynamic and kinetic measurements for theoretical applications. We have shown how resolution arises from three distinct contributions: plate number, retention factor, and selectivity factor. The plate number is related to the extent of solute-zone dispersion or broadening, which is discussed in Sec. 1.3. These contributions may be properly controlled by judicious adjustment of the linear velocity, column diameter, column length, particle size, etc. The retention factor and selectivity factor are related to the extent of solute-zone separation, which is discussed in Sec. 1.4. These contributions may be properly controlled by selection of a suitable retention mechanism and judicious adjustment of the stationary-phase type, mobile-phase type and composition, temperature, etc. With a detailed and thorough understanding of the fundamental principles underlying solute-zone dispersion and separation, it is possible to optimize chromatographic separations in a systematic manner, as discussed in Sec. 1.5. In this manner, successful separation becomes an efficient and methodical scientific process rather than a time-consuming trial-and-error process.

REFERENCES

1   H. Purnell, *Gas Chromatography*, Wiley, New York, 1962.
2   J.C. Giddings, *Dynamics of Chromatography*, Marcel Dekker, New York, 1965.

3   B.L. Karger, L.R. Snyder and C. Horvath, *An Introduction to Separation Science*, Wiley, New York, 1973.

4   C.F. Poole and S.K. Poole, *Chromatography Today*, Elsevier, Amsterdam, 1991.

5   N. Dyson, *Chromatographic Integration Methods*, Springer, New York, 1998.

6   J.P. Foley, *Analyst*, 116 (1991) 1275.

7   A.J.P. Martin and R.L.M. Synge, *Biochem. J.*, 35 (1941) 1358.

8   A.B. Littlewood, *Gas Chromatography*, 2nd Edn. Academic Press, New York, 1970.

9   A. Fick, *Ann. Phys. Leipzig*, 170 (1855) 59.

10 A. Einstein, *Ann. Phys.*, 17 (1905) 549.

11 E.W. Washburn, *International Critical Tables of Numerical Data, Physics, Chemistry, Technology*, McGraw-Hill, New York, 1926–1930.

12 C.Y. Ho, *Transport Properties of Fluids: Thermal Conductivity, Viscosity, and Diffusion Coefficient*, Hemisphere, New York, 1988.

13 J.O. Hirschfelder, C.F. Curtis and R.B. Bird, *Molecular Theory of Gases and Liquids*, Wiley, New York, 1954.

14 R.B. Bird, W.E. Stewart and E.N. Lightfoot, *Transport Phenomena*, Wiley, New York, 2002.

15 E.N. Fuller, P.D. Schettler and J.C. Giddings, *Ind. Eng. Chem.*, 58 (1966) 19.

16 C.R. Wilke and P. Chang, *A.I.Ch.E. J.*, 1 (1955) 264.

17 R.C. Reid, J.M. Prausnitz and T.K. Sherwood, *The Properties of Gases and Liquids*, McGraw-Hill, New York, 1977.

18 J. Li and P.W. Carr, *Anal. Chem.*, 69 (1997) 2530.

19 J. Li and P.W. Carr, *Anal. Chem.*, 69 (1997) 2550.

20 S.V. Olesik and J.L. Woodruff, *Anal. Chem.*, 63 (1991) 670.

21 J.P. Foley, J.A. Crow, B.A. Thomas and M. Zamora, *J. Chromatogr.*, 478 (1989) 287.

22 C.A. Cramers, J.A. Rijks and C.P.M. Schutjes, *Chromatographia*, 14 (1981) 439.

23 G. Hagen, *Poggendorfs Ann.*, 46 (1839) 423.

24 J.L.M. Poiseuille, *Compt. Rend.*, 1840 (1840) 11.

25 J. Kozeny, *Sitzber. Akad. Wiss. Mn.*, 136 (1927) 271.

26 R. Ohmacht and I. Halasz, *Chromatographia*, 14 (1981) 155.

27 V. Pretorius and T.W. Smuts, *Anal. Chem.*, 38 (1966) 274.

28 J.H. Knox, *Anal. Chem.*, 38 (1966) 253.

29 K. Hofmann and I. Halasz, *J. Chromatogr.*, 173 (1979) 211.

30 S.R. Sumter and M.L. Lee, *J. Microcolumn Sep.*, 3 (1991) 91.

31 M.J.E. Golay, in D.A. Desty (Ed.), *Gas Chromatography 1958*, Academic Press, New York, 1958, p. 36.

32 M.J.E. Golay, *J. Chromatogr.*, 196 (1980) 349.

33 M.J.E. Golay, *J. Chromatogr.*, 216 (1981) 1.

34 I. Halasz and P. Walking, *Ber. Bunsenges. Phys. Chem.*, 74 (1970) 66.

35 J.C. Giddings, *J. Chromatogr.*, 3 (1960) 520.

36 R. Tijssen, *Sep. Sci. Technol.*, 13 (1978) 681.

37 L.R. Snyder and J.W. Dolan, *J. Chromatogr.*, 185 (1979) 43.

38 V. Pretorius, B.J. Hopkins and J.D. Schieke, *J. Chromatogr.*, 99 (1974) 23.

39 C.L. Rice and R. Whitehead, *J. Phys. Chem.*, 69 (1965) 4017.

40 M. von Smoluchowski, *Int. Acad. Sci. Cracovic*, Vol. 1903 (1903) 184.

41 E. Huckel, *Physik. Z.*, 25 (1924) 204.

42 M.F.M. Tavares and V.L. McGuffin, *Anal. Chem.*, 67 (1995) 3687.

43 J.J. van Deemter, F.J. Zuiderweg and A. Klinkenberg, *Chem. Eng. Sci.*, 5 (1956) 271.

44 P. Thumneum and S. Hawkes, *Chromatographia*, 14 (1981) 576.

45 J.H. Knox and L. McLaren, *Anal. Chem.*, 36 (1964) 1477.

46   J.P. McEldoon and R. Datta, *Anal. Chem.*, 64 (1992) 227.

47   K. Grob and G. Grob, *J. High Resolut. Chromatogr. Chromatogr. Commun.*, 4 (1979) 109.

48   J. Bohemen and J.H. Purnell, in D.H. Desty (Ed.), *Gas Chromatography 1958*, Academic Press, New York, 1958, pp. 6–22.

49   D.M. Dohmeier and J.S. Jorgenson, *J. Microcolumn Sep.*, 3 (1991) 311.

50   R. Swart, J.C. Kraak and H. Poppe, *Trends Anal. Chem.*, 16 (1997) 332.

51   J.J. Duffield and L.B. Rogers, *Anal. Chem.*, 32 (1960) 340.

52   W. Jennings, *Gas Chromatography with Glass Capillary Columns*, 2nd Edn. Academic Press, New York, 1980.

53   R.R. Freeman, *High Resolution Gas Chromatography*, Hewlett-Packard Co., Palo Alto, CA, 1981.

54   D.R. Lide, *Handbook of Chemistry and Physics*, CRC Press, Boca Raton, 1990.

55   J.C. Sternberg, *Adv. Chromatogr.*, 2 (1966) 205.

56   M. Martin, C. Eon and G. Guiochon, *J. Chromatogr.*, 108 (1975) 229.

57   P.W. Atkins, *Physical Chemistry*, 3rd Edn. W.H. Freeman & Co., New York, 1986.

58   C. Reichardt, *Solvents and Solvent Effects in Organic Chemistry*, 2nd Edn. VCH, New York, 1988.

59   K.G. Denbigh, *Trans. Faraday Soc.*, 36 (1940) 936.

60   V.I. Minkin, O.A. Osipov and Y.A. Zhadanov, *Dipole Moments in Organic Chemistry*, Plenum, New York, 1970.

61   L.M. Stephen, B.D. Olafson and W.A. Goddard, *J. Phys. Chem.*, 94 (1990) 8897.

62   J.R. Conder and C.L. Young, *Physicochemical Measurement by Gas Chromatography*, Wiley, New York, 1979.

63   J.H. Hildebrand and R.L. Scott, *Regular Solutions*, Prentice-Hall, Englewood Cliffs, 1962.

64   A.F.M. Barton, *Chem. Rev.*, 75 (1975) 731.

65   R. Tijssen, H.A.H. Billiet and P.J. Schoenmakers, *J. Chromatogr.*, 122 (1976) 185.

66   P.J. Schoenmakers, H.A.H. Billiet and L. de Galan, *Chromatographia*, 15 (1982) 205.

67   G.E. Baiulescu and V.A. Ilie, *Stationary Phases in Gas Chromatography*, Pergamon, New York, 1975.

68   L.R. Snyder and J.J. Kirkland, *Introduction to Modern Liquid Chromatography*, 2nd Edn. Wiley, New York, 1979.

69   J.C. Giddings, M.N. Myers, L. McLaren and R.A. Keller, *Science*, 162 (1968) 67.

70   L.R. Snyder, *Principles of Adsorption Chromatography*, Dekker, New York, 1968.

71   L.R. Snyder, in E. Heftmann (Ed.), *Chromatography*, Van Nostrand Reinhold, New York, 1975.

72   T.H. Lowry and K.S. Richardson, *Mechanism and Theory in Organic Chemistry*, Harper & Row, New York, 1987.

73   D.M. Young and A.D. Crowell, *Physical Adsorption of Gases*, Butterworths, London, 1962.

74   D.M. Ruthven, *Principles of Adsorption and Adsorption Processes*, Wiley, New York, 1984.

75   H. Freundlich, *Colloid and Capillary Chemistry*, Methuen, London, 1926.

76   I. Langmuir, *J. Am. Chem. Soc.*, 40 (1918) 1361.

77   S. Brunauer, P.H. Emmett and E. Teller, *J. Am. Chem. Soc.*, 60 (1938) 309.

78   S.J. Gregg and R. Stock, in D.A. Desty (Ed.), *Gas Chromatography 1958*, Academic Press, New York, 1958, p. 90.

79   D. Cagniant, *Complexation Chromatography*, Dekker, New York, 1992.

80   D.D. Perrin, *Stability Constants of Metal-Ion Complexes: Part B Organic Ligands*, Pergamon Press, New York, 1979.

81   A. Anichini, L. Fabbrizi, P. Paoletti and R.M. Clay, *J. Chem. Soc. Dalton*, 1978 (1978) 577.

82   B. Vonach and G. Schomburg, *J. Chromatogr.*, 149 (1978) 417.

83   O.K. Guha and J. Janak, *J. Chromatogr.*, 68 (1972) 325.
84   S. Lam, F. Chow and A. Karmen, *J. Chromatogr.*, 199 (1980) 295.
85   S. Lam and A. Karmen, *J. Chromatogr.*, 239 (1982) 451.
86   G. Brookes and L.D. Pettit, *J. Chem. Soc. Dalton*, 1977 (1977) 1918.
87   D.W. Armstrong and W. DeMond, *J. Chromatogr. Sci.*, 22 (1984) 411.
88   W.H. Pirkle and T.C. Pochapsky, *Chem. Rev.*, 89 (1989) 347.
89   R.C. Helgeson, J.M. Timko, P. Moreau, S.C. Peacock, J.M. Mayer and D.J. Cram, *J. Am. Chem. Soc.*, 96 (1974) 6762.
90   J. Nawrocki, D.L. Moir and W. Szczepaniak, *Chromatographia*, 28 (1989) 143.
91   M. Verzele, *LC Magazine*, 1 (1983) 217.
92   H.F. Walton and R.D. Rocklin, *Ion Exchange in Analytical Chemistry*, CRC Press, Boca Raton, 1990.
93   H. Small, *Ion Chromatography*, Plenum, New York, 1989.
94   J.S. Fritz and D.T. Gjerde, *Ion Chromatography*, Wiley, New York, 2000.
95   P.J. Twitchett and A.C. Moffat, *J. Chromatogr.*, 111 (1975) 149.
96   M.T.W. Hearn, *Ion-Pair Chromatography*, Dekker, New York, 1985.
97   B.A. Bidlingmeyer, S.N. Deming, W.P. Price, B. Sachok and M. Petrusek, *J. Chromatogr.*, 186 (1979) 419.
98   H. Small, *Adv. Chromatogr.*, 15 (1977) 113.
99   R. Tijssen, J. Box and M.E. Van Kreveld, *Anal. Chem.*, 58 (1986) 3036.
100 W.W. Yau, J.J. Kirkland and D.D. Bly, *Modern Size-Exclusion Liquid Chromatography*, Wiley, New York, 1979.
101 J.C. Giddings, E. Kucera, C.P. Russell and M.N. Myers, *J. Phys. Chem.*, 72 (1968) 4397.
102 M.E. Van Kreveld and N. Van Den Hoed, *J. Chromatogr.*, 83 (1973) 111.
103 M. Potschka and P.L. Dubin, *Strategies in Size-Exclusion Chromatography*, ACS Symposium Series, Vol. 635, American Chemical Society, Washington, DC, 1996.
104 A.J. Gordon and R.A. Ford, *The Chemist's Companion*, Wiley, New York, 1973, pp. 33–36.
105 J.C. Berridge, *Techniques for the Automated Optimization of HPLC Separation*, Wiley, New York, 1985.
106 P.J. Schoenmakers, *Optimization of Chromatographic Selectivity*, Elsevier, Amsterdam, 1986.
107 J.L. Glajch and L.R. Snyder, *Computer-Assisted Method Development for High-Performance Liquid Chromatography*, Elsevier, Amsterdam, 1990.
108 P.H. Lukulay and V.L. McGuffin, *J. Microcolumn Sep.*, 8 (1996) 211.
109 A.M. Siouffi and R. Phan-Tan-Luu, *J. Chromatogr. A*, 892 (2000) 75.
110 R.E. Kaiser, *Gas Chromatographie*, Portig, Leipzig, 1960, p. 33.
111 S.N. Deming and M.L.H. Turoff, *Anal. Chem.*, 50 (1978) 546.
112 P. Jones and C.A. Wellington, *J. Chromatogr.*, 213 (1981) 357.
113 P.J. Schoenmakers, A.C.J.H. Drouen, H.A.H. Billiet and L. de Galan, *Chromatographia*, 15 (1982) 48.
114 H.A.H. Billiet and L. de Galan, *J. Chromatogr.*, 485 (1989) 27.
115 M.W. Watson and P.W. Carr, *Anal. Chem.*, 51 (1979) 1835.
116 J.L. Glajch, J.J. Kirkland, K.M. Squire and J.M. Minor, *J. Chromatogr.*, 199 (1980) 57.
117 R. Cela, C.G. Barroso and J.A. Perez-Bustamante, *J. Chromatogr.*, 485 (1989) 477.
118 J. Vajda and L. Leisztner, in H. Kalasz (Ed.), *New Approaches in Liquid Chromatography*, Elsevier, Amsterdam, 1982, p. 103.
119 J.C. Berridge, *J. Chromatogr.*, 244 (1982) 1.
120 T.D. Schlabach and J.L. Excoffier, *J. Chromatogr.*, 439 (1988) 173.
121 H.J.G. Debets, B.L. Bajema and D.A. Doornbos, *Anal. Chim. Acta*, 151 (1983) 131.

122   J.W. Weyland, C.H.P. Bruins, H.J.G. Debets, B.L. Bajema and D.A. Doornbos, *Anal. Chim. Acta*, 153 (1983) 93.

123   A. Peeters, L. Buydens, D.L. Massart and P.J. Schoenmakers, *Chromatographia*, 26 (1988) 101.

124   E.J. Klein and S.L. Rivera, *J. Liq. Chromatogr. Rel. Technol.*, 23 (2000) 2097.

125   S.N. Deming and S.L. Morgan, *Experimental Design: A Chemometric Approach*, Elsevier, Amsterdam, 1987.

126   J.C. Miller and J.N. Miller, *Statistics for Analytical Chemists*, 3rd Edn. Ellis Horwood, Chichester, 1993, Ch. 7.

127   T.F. Edgar and D.M. Himmelblau, *Optimization of Chemical Processes*, McGraw-Hill, New York, 1988, Ch. 6.

128   W. Spendley, G.R. Hext and F.R. Hinsworth, *Technometrics*, 4 (1962) 44.

129   J.A. Nelder and R. Mead, *Comput. J.*, 7 (1965) 308.

130   M.L. Raney and W.C. Purdy, *Anal. Chim. Acta*, 93 (1977) 211.

131   V. Svoboda, *J. Chromatogr.*, 201 (1980) 241.

132   D.M. Fast, P.H. Culbreth and E.J. Sampson, *Clin. Chem.*, 28 (1982) 444.

133   J.C. Berridge, *J. Chromatogr.*, 244 (1982) 1.

134   J.A. Crow and J.P. Foley, *Anal. Chem.*, 62 (1990) 378.

135   R.J. Laub and J.H. Purnell, *J. Chromatogr.*, 112 (1975) 71.

136   R.J. Laub and J.H. Purnell, *Anal. Chem.*, 48 (1976) 1720.

137   R.J. Laub, in T. Kuwana (Ed.), *Physical Methods in Modern Chemical Analysis*, Vol. 3, Academic Press, New York, 1983, Ch. 5.

138   J.L. Glajch and J.J. Kirkland, *J. Chromatogr.*, 485 (1989) 51.

139   E.R. Tufte, *The Visual Display of Quantitative Information*, Graphics Press, Cheshire, CT, 1983.

140   J.W. Dolan, D.C. Lommen and L.R. Snyder, *J. Chromatogr.*, 535 (1990) 55.

141   L.R. Snyder, J.W. Dolan and D.C. Lommen, *J. Chromatogr.*, 535 (1990) 75.

142   L.R. Snyder, *J. Chromatogr.*, 179 (1979) 167.

143   H. Colin, J.C. Diez-Masa, G. Guiochon, T. Czajkowska and I. Miedziak, *J. Chromatogr.*, 167 (1978) 41.

144   J. Chmielowiec and H. Sawatzky, *J. Chromatogr.*, 17 (1979) 245.

145   L.C. Sander and S.A. Wise, *Anal. Chem.*, 61 (1989) 1749.

146   J.V. Hinshaw, *LC/GC Magazine*, 9 (1991) 94.

147   W.R. Melander, D.E. Campbell and C. Horvath, *J. Chromatogr.*, 158 (1978) 215.

148   W.R. Melander, B.K. Chen and C. Horvath, *J. Chromatogr.*, 185 (1979) 99.

149   R.J. Laub and J.H. Purnell, *J. Chromatogr.*, 161 (1978) 49.

150   H. Snijders, H.G. Janssen and C. Cramers, *J. Chromatogr. A*, 718 (1995) 339.

151   F.E.O. Suliman, *Talanta*, 56 (2002) 175.

152   H. Snijders, H.G. Janssen and C. Cramers, *J. Chromatogr. A*, 756 (1996) 175.

153   H.M. McNair and J. Bowermaster, *J. High Resolut. Chromatogr. Chromatogr. Commun.*, 10 (1987) 27.

154   K. Jinno, J.B. Phillips and D.P. Carne, *Anal. Chem.*, 57 (1985) 574.

155   G. Liu, N.M. Djordjevic and F. Erni, *J. Chromatogr.*, 592 (1992) 239.

156   J. Yoo, J.T. Watson and V.L. McGuffin, *J. Microcolumn Sep.*, 4 (1992) 349.

157   J.L. Glajch, J.C. Gluckman, J.G. Charikofsky, J.M. Minor and J.J. Kirkland, *J. Chromatogr.*, 318 (1985) 23.

158   H.J. Issaq and J. Gutierrez, *J. Liq. Chromatogr.*, 11 (1988) 2851.

159   D. Rood, *LC-GC Magazine*, 17 (1999) 914.

160   I.H. Hagestam and T.C. Pinkerton, *Anal. Chem.*, 57 (1985) 1757.

161 C.P. Desilets, M.A. Rounds and F.E. Regnier, *J. Chromatogr.*, 544 (1991) 25.

162 H.J. Issaq, D.W. Mellini and T.E. Beesley, *J. Liq. Chromatogr.*, 11 (1988) 333.

163 B. Feibush and C.T. Santasania, *J. Chromatogr.*, 544 (1991) 41.

164 J.B. Crowther and R.A. Hartwick, *Chromatographia*, 16 (1982) 349.

165 R. Bischoff and L.W. McLaughlin, *J. Chromatogr.*, 270 (1983) 117.

166 L.A. Kennedy, W. Kopaciewicz and F.E. Regnier, *J. Chromatogr.*, 359 (1986) 73.

167 Z. El Rassi and C. Horvath, *J. Chromatogr.*, 359 (1986) 255.

168 H.J. Issaq and J. Gutierrez, *J. Liq. Chromatogr.*, 11 (1988) 2851.

169 D.J. Pietrzyk, S.M. Senne and D.M. Brown, *J. Chromatogr.*, 546 (1991) 101.

170 S.N. Deming, J.G. Bower and K.D. Bower, *Adv. Chromatogr.*, 24 (1984) 35.

171 L.R. Snyder, J.W. Dolan and D.C. Lommen, *J. Chromatogr.*, 485 (1989) 65.

172 J.W. Dolan, D.C. Lommen and L.R. Snyder, *J. Chromatogr.*, 485 (1989) 91.

173 J.L. Glajch, J.C. Gluckman, J.G. Charikofsky, J.M. Minor and J.J. Kirkland, *J. Chromatogr.*, 318 (1986) 23.

174 M.T.W. Hearn, *Ion-Pair Chromatography*, Dekker, New York, 1985.

175 J.W. Weyland, C.H.P. Bruins and D.A. Doornbos, *J. Chromatogr. Sci.*, 22 (1984) 31.

176 M.A. Quarry, R.L. Grob, L.R. Snyder, J.W. Dolan and M.P. Rigney, *J. Chromatogr.*, 384 (1987) 163.

177 M.A. Quarry, R.L. Grob and L.R. Snyder, *J. Chromatogr.*, 285 (1984) 1.

178 M.A. Quarry, R.L. Grob and L.R. Snyder, *J. Chromatogr.*, 285 (1984) 19.

179 M.A. Quarry, R.L. Grob and L.R. Snyder, *Anal. Chem.*, 58 (1986) 907.

180 L.R. Snyder, M.A. Quarry and J.L. Glajch, *Chromatographia*, 24 (1987) 33.

181 P. Jandera and J. Churacek, *J. Chromatogr.*, 192 (1980) 1.

182 P. Jandera and J. Churacek, *J. Chromatogr.*, 192 (1980) 19.

183 P. Jandera and J. Churacek, *Gradient Elution in Column Liquid Chromatography – Theory and Practice*, Elsevier, Amsterdam, 1985.

184 J.W. Dolan, J.R. Gant and L.R. Snyder, *J. Chromatogr.*, 165 (1979) 31.

185 J.J. Kirkland and J.L. Glajch, *J. Chromatogr.*, 255 (1983) 27.

186 V.M. Morris, J.G. Hughes and P.J. Marriott, *J. Chromatogr. A*, 755 (1996) 235.

187 X.F. Li, W.R. Cullen and K.J. Reimer, in R. Clement and B. Burk, (Eds.), *Enviroanalysis 2002: Proceedings of the Biennial International Conference on Monitoring and Measurement of the Environment*, EnviroAnalysis 2002 Conference Secretariat, Ottawa, Canada, 2002.

188 Y. Guillaume, M. Thomassin and C. Guinchard, *J. Chromatogr. A*, 704 (1995) 437.

189 R.P.W. Scott and J.G. Lawrence, *J. Chromatogr. Sci.*, 7 (1969) 65.

190 K. Jinno and M. Yamagami, *Chromatographia*, 27 (1988) 417.

191 E.B. Bakeas and P.A. Siskos, *J. High Resolut. Chromatogr. Commun.*, 19 (1996) 277.

192 R. Baycan-Keller and M. Oehme, *J. Chromatogr. A*, 837 (1999) 201.

193 Y. Mao and P.W. Carr, *Anal. Chem.*, 72 (2000) 110.

194 Y. Mao and P.W. Carr, *Anal. Chem.*, 73 (2001) 1821.

195 R.B. Diasio and M.E. Wilburn, *J. Chromatogr. Sci.*, 17 (1979) 565.

196 W.S. Hancock, R.C. Chloupek, J.J. Kirkland and L.R. Snyder, *J. Chromatogr. A*, 686 (1994) 31.

197 R.C. Chloupek, W.S. Hancock, B.A. Marchylo, J.J. Kirkland, B.E. Boyes and L.R. Snyder, *J. Chromatogr. A*, 686 (1994) 45.

198 W.R. Melander and C. Horvath, *Chromatographia*, 18 (1984) 353.

199 W.R. Melander, B.K. Chen and C. Horvath, *J. Chromatogr.*, 318 (1985) 1.

200 P.H. Lukulay and V.L. McGuffin, *J. Chromatogr. A*, 691 (1995) 171.

Erich Heftmann (Editor)
Chromatography 6th edition
Journal of Chromatography Library, Vol. 69A
© 2004 Elsevier B.V. All rights reserved

Chapter 2

Column liquid chromatography

ROGER M. SMITH

CONTENTS

2.1 Introduction 96
    2.1.1 Modes of liquid chromatography 97
2.2 Equipment for liquid-phase separations 98
    2.2.1 Pumps, injectors, columns, and ovens 100
        2.2.1.1 Solvents and solvent delivery systems 100
        2.2.1.2 Sample injector systems 101
        2.2.1.3 Columns and ovens 103
2.3 Stationary-phase materials 104
    2.3.1 Silica-based stationary phases 104
        2.3.1.1 Silica-based stationary phases for normal-phase
                chromatography 104
        2.3.1.2 Silica-based stationary phases for reversed-phase
                chromatography 105
        2.3.1.3 Hybrid-silica stationary phases 107
        2.3.1.4 Other silica stationary phases 109
        2.3.1.5 Monolithic silica columns 109
    2.3.2 Stationary phases based on zirconia, titania, and alumina 109
    2.3.3 Polymeric stationary phases 110
    2.3.4 Stationary phases of porous graphitic carbon 110
    2.3.5 Chiral stationary phases 111
        2.3.5.1 Donor–acceptor stationary phases 112
        2.3.5.2 Cellulose-based stationary phases 112
        2.3.5.3 Inclusion-complex stationary phases 113
        2.3.5.4 Ligand chiral phases 113
        2.3.5.5 Protein chiral phases 113
        2.3.5.6 Macrocyclic chiral phases 114
    2.3.6 Molecularly imprinted polymeric stationary phases 114
    2.3.7 Evaluation and comparison of stationary-phase materials 115
2.4 Detection 115
    2.4.1 Spectroscopic detection 115

　　　　2.4.1.1　Ultraviolet spectroscopic detectors　. 116
　　　　2.4.1.2　Fluorescence detectors 116
　　2.4.2　Other common detectors　. 118
　　　　2.4.2.1　Refractive-index detectors　. 118
　　　　2.4.2.2　Light-scattering detectors　. 118
　　　　2.4.2.3　Electrochemical amperometric/coulometric detectors　. . . 118
　　　　2.4.2.4　Conductivity and suppressed-conductivity detectors　. . . . 119
　　　　2.4.2.5　Chiral detectors　. 119
　　　　2.4.2.6　Other detectors　. 119
　　2.4.3　Coupled detection methods 120
　　　　2.4.3.1　Liquid chromatography/mass spectrometry and liquid
　　　　　　　　chromatography/tandem mass spectrometry 120
　　　　2.4.3.2　Liquid chromatography/nuclear magnetic resonance
　　　　　　　　spectrometry　. 120
　　　　2.4.3.3　Liquid chromatography/infrared spectrometry 120
　　　　2.4.3.4　Liquid chromatography/atomic absorption spectrometry
　　　　　　　　and liquid chromatography/inductively coupled plasma
　　　　　　　　spectrometry　. 121
2.5　Separation methods　. 121
　　2.5.1　Normal-phase and supercritical-fluid separations　. 121
　　2.5.2　Reversed-phase separations　. 124
　　　　2.5.2.1　Ion-pair and buffered eluents 125
　　　　2.5.2.2　Relationship of retention to structure 127
　　2.5.3　Optimization methods　. 128
2.6　Liquid chromatography in sample preparation　. 128
2.7　High-temperature separations　. 129
2.8　Miniaturization　. 129
2.9　Large-scale separations　. 130
　　2.9.1　Simulated moving-bed separations　. 130
References　. 131

2.1 INTRODUCTION

Column liquid chromatography, in which a sample is applied to a column, containing a stationary-phase material, and a liquid mobile phase is passed through the column so that the components are eluted, is the oldest form of chromatography and was first described by Tswett over 100 years ago [1,2]. Originally carried out with a slurry-packed column and a liquid phase moving under gravity, it has developed into a modern instrumental method with a high resolving power and a sensitivity capable of providing qualitative and quantitative results with excellent reproducibility. The original open-column method is now rarely used in analytical chemistry, except during some sample preparation methods, and almost all modern instruments use closed columns, a mechanical pumping system for the eluent, and an on-line instrumental system to detect and often to identify the eluted analytes. Because of the higher performance of these systems compared to the older non-instrumental methods, these techniques have been known since the early 1980s as high-performance liquid chromatography (HPLC). HPLC is now the dominant analytical method worldwide, both in terms of the number of assays carried out and in the number of instruments employed. Its impact over the last 20 years, since it started to be widely adopted, has been dramatic, and there are very few application areas of analytical chemistry where it has not had a significant influence. This is mainly because essentially the same instrumentation is used for partition or adsorption separation methods, which are described in this chapter, as for ion (and ion-exchange) chromatography (Chap. 4) and size-exclusion separations (Chap. 5), albeit with different column packings and elution solvents. One instrumental system can therefore provide different separation techniques capable of handling a wide range of analytes and matrices. The method has also been scaled up to provide preparative separations for the synthetic or industrial chemist (Sec. 2.9).

Related methods in which the separation occurs on a planar system and the analyte is usually detected *in situ*, primarily thin layer-chromatography (TLC), are discussed in Chap. 6. Capillary electrochromatography (CEC), in which analytes are driven through a packed column by an electric field to give a mixed liquid/liquid phase and electrophoretic separation is covered in Chap. 7.

In HPLC, the stationary phase is usually bound to a static matrix, but true liquid–liquid chromatography is also known. However, its resolving power is usually much weaker and chromatographic runs are often longer than in HPLC. It is used primarily as a very mild preparative separation technique, in the form of counter-current chromatography [3–5] or droplet counter-current chromatography, typically for labile natural products [6].

The basic concepts and methods of HPLC have been extensively discussed in the scientific literature, and many major and numerous minor journals are largely devoted to this technique. Together with papers in general analytical chemistry journals, they provide an enormous database of the theory, practice, and potential applications of the method. Aspects of HPLC have also been covered in numerous research monographs on selected topics, review series and textbooks, including some recent typical general guides [7–16] and an electronic text [17]. Recommendations for the terminology of chromatography [18,19] and liquid-phase separations [20–22], including chiral chromatography [23], and supercritical-fluid chromatography [24] have been published by IUPAC. The historical development of HPLC has been covered in a number of articles and texts [25,26]. The wide range of applications of HPLC (covered in Part B of this book) derives from its versatility, as usually the only requirement for this method of analysis is that the analyte must be soluble in the mobile phase. HPLC can also handle a wide range of sample sizes, from preparative to trace levels, although this can vary with the detectability of the analyte (Sec. 2.4).

2.1.1 Modes of liquid chromatography

Chromatography is defined as "...a physical method of separation in which the components to be separated are distributed between two phases, one of which is stationary (the stationary phase), while the other (the mobile phase) moves in a definite direction" [18]. The dominance of liquid chromatography is due to its great versatility, as it encompasses a variety of separation modes, the principal modes being adsorption, partition, ion exchange and size exclusion. In adsorption chromatography, the analytes interact with a solid stationary surface and are displaced by competition with the eluent for active sites on the surface. The separation in partition chromatography results from a thermodynamic distribution between two liquid (or liquid-like) phases, whereas ion-exchange chromatography is primarily governed by ionic interactions between ionized analytes and an oppositely charged stationary-phase surface. In size exclusion chromatography the resistive force for separation is the physical size of the analytes, which determines their accessibility to differently sized pores in the stationary-phase material. Because it was the original chromatographic method, separations based on adsorption methods are usually referred to as *normal-phase* separations. In this mode the stationary phase is polar, usually silica or alumina, and the analytes are retained on the column according to their polarity by a polar/polar interaction. Thus, weakly polar analytes (see Table 2.1), such as alkanes and aromatic hydrocarbons, are readily eluted, whereas alcohols and amines can be highly retained. The strength of the eluent is determined by its interaction with the stationary phase and, hence, ability to compete with the analyte for the stationary phase. Thus, hexane is a weak eluent, dichloromethane and ethyl acetate are moderate eluents and methanol is a strong eluent.

In contrast, in *reversed-phase* (RP) chromatography, these parameters are effectively reversed. Normally, a non-polar stationary phase is employed, consisting of a bonded hydrocarbon layer on a solid support, with a polar mobile phase usually consisting of water and either methanol, acetonitrile, or tetrahydrofuran (THF). The analyte is partitioned between the mobile and stationary phases so that low-polarity analytes are more readily retained, because they dissolve in the hydrocarbon layer and are less soluble in the mobile phase, whereas polar analytes favor the mobile phase and are more rapidly eluted. Water is, thus, a weak eluent, because most analytes are water-insoluble and are partitioned into the stationary phase, while organic solvents are strong eluents. In addition, the size of the analyte is important, because larger analytes suffer a hydrophobic effect. They are less soluble in the mobile phase, as they have to create a larger cavity in the hydrogen-bonded matrix of the solution. Thus, homologs can be readily separated in RP systems, and this is often not possible in normal-phase separations, because in the latter a polar functional group can dominate the separation.

2.2 EQUIPMENT FOR LIQUID-PHASE SEPARATIONS

The operation of a liquid-chromatographic separation has not changed in principle since the technique was first developed. Originally, an open-topped column was packed with relatively large particles, usually of silica, and the sample was applied directly. For elution an organic solvent was used, which was flowing under gravity. The eluent from the column was collected in fractions, which were then assayed off-line, by spectroscopy or TLC. The first widely available instrumental developments came with the introduction of specialized ion-exchange separation systems for amino acid analyses, based on the work of Moore and Stein in the 1960s [27]. These employed a pumping system for the eluent,

TABLE 2.1

POLARITY AND pKa's OF TYPICAL FUNCTIONAL GROUPS

Polarity	Functional groups	Approximate pKa
Low polarity	R–H R–O–R R–Halogen	
Medium polarity		
High polarity		
	R—OH	14.2
Ionizable acidic groups	R—SO$_3$H	< 0.07
	R—CO$_2$H	4.75
	—CO$_2$H	4.19
	—OH	9.89
Ionizable basic groups[*]	—NH$_2$	4.63
	R–NH$_2$	10.77

[*] pKa of protonated bases.

a small closed column, gradient elution, post-column addition of a reagent, and on-line spectrophotometric detection of the analytes as derivatives. By the late 1960s a number of manufacturers were offering instrumental chromatographs [28], but these were often based on normal-phase separations with a stationary phase of silica, having a controlled particle size of 100–125 μm, packed in a long, thin steel tube, up to 12 ft long and of 0.093-in ID. A pump was needed to propel the solvent through the column against a moderate back-pressure, and an on-line spectrometer at 254 nm (based on a mercury lamp) with a flow-cell was generally used as a detector. The introduction of bonded phases with the correspondingly more viscous aqueous/organic eluents for reversed-phase separations and the use of smaller particles, first 10 μm and then 5 μm, packed in short columns, 250 to 100 mm in length, caused practical problems, as the back-pressures from the columns rose markedly. Although partly compensated for by the shorter column lengths, this required a change in injection methods and a considerable improvement in the technical quality of the pumps. Instruments have become simpler, as the earlier systems [29] often had complex degassing and equilibration systems, later found to be unnecessary.

2.2.1 Pumps, injectors, columns, and ovens

The basic HPLC system consists of a pumping system for the solvent, an injection device to introduce the analyte into the solvent flow, the column, and a detector.

2.2.1.1 Solvents and solvent delivery systems

Whereas in the early days of HPLC the purification of solvents was a major consideration, nowadays reasonably priced solvents with a consistent state of purity, specially prepared for LC, are readily available commercially [30]. The air, dissolved in liquids exposed to the atmosphere, can cause problems. As its solubility in reversed-phase eluents can vary with the composition, out-gassing can occur when mobile-phase components are mixed, causing bubble formation. The bubbles can interfere with check valve operation or cause spiking signals in spectrophotometric detectors. Many methods have been reported for degassing solvents but two have become routine, either sparging with helium gas, which has a very low solubility but flushes other gases from the eluent, or by using an on-line degasser system in which any dissolved gas diffuses through a membrane across a pressure drop. Together with improvements in pump design, degassing is now much less of an issue.

The basic design of the pumps used in HPLC has changed little over the years, although gradual improvements have made them easier to use and more reliable. The operational requirements are, typically, the ability to deliver a flow from 0.05 to 10 mL/min against a back-pressure of up to 400 bar. Typically, flows are 1–2 mL/min with a back-pressure of 150–300 bar. However, most routine analytical pumps can become less reliable below 0.1 mL/min, and specialized pumps are required for the low flow-rates used with narrow-bore columns. Large pump heads or dedicated pumps are needed for the high flows required in preparative systems.

Originally, all separations were carried out isocratically, meaning that the eluent composition and elution strength remained constant throughout the analysis. However, most pumps are now designed for gradient operation, in which the composition of the mobile phase is changed systematically during the separation. Although not practical in normal-phase separation because of the long equilibration time of the stationary phase surface, gradient elution is widely employed in reversed-phase chromatography. Increasing the proportion of the organic modifier (the stronger component of the eluent) during the separation can speed up the elution of more strongly retained analytes, thus enabling a wider range of analytes to be resolved during a single analysis. Gradient elution can be achieved by either high-pressure mixing (using two pumps running at variable speeds under computer control) or low-pressure mixing (using a pre-pump mixer and a single pump, with the proportion of the eluent taken from different reservoirs being altered by a computer). The latter method seems to be gaining preference as it requires only a single pump, running a constant speed, and causes fewer degassing problems. Normally, only two eluents are mixed, which can be either neat solvents, such as water and methanol, or, preferably, partially pre-mixed eluents, such as methanol/water in 20:80 and 80:20 ratios, corresponding to the start and end compositions of the gradient run. Although a number of pump designs have been used over the years, most modern pumps are based on a reciprocating piston design, in which the solvent is drawn into a chamber and then pumped out against a check valve (Fig. 2.1). Usually, either two pumps heads are used out

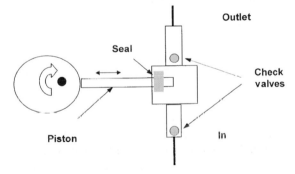

Fig. 2.1 Reciprocating pump.

of phase so that one is always delivering, or very frequently, a master/slave system is used, in which a second pump head, after the main pump head, takes up the flow, as the first pump refills the chamber (Fig. 2.2). One advantage of this latter system is that only one set of check valves is needed, as the second pump head is operating within the high-pressure part of the instrument.

2.2.1.2 Sample injector systems

Because of the small particle size of the stationary phase in the column, there is usually a significant back-pressure at the head of the column, from 100 to 400 bar, with normal flow-rates. Therefore, sample injection cannot be carried out directly by

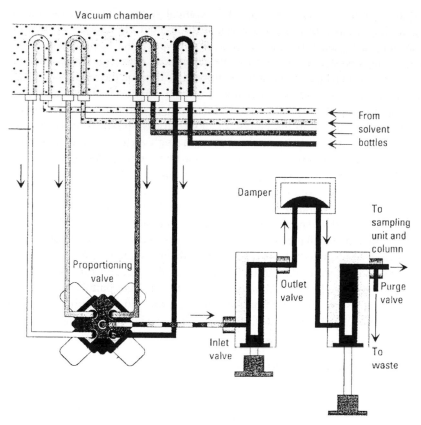

Fig. 2.2. Typical HPLC pumping system with mobile-phase vacuum degasser, solvent proportioning valves for gradient elution, and master/slave pumping system. (reproduced through the courtesy of Agilent.)

a syringe; a switching-valve system is needed. In most cases, a rotary valve is used in which the sample is loaded into a loop at atmospheric pressure and the loop is then switched into the high-pressure eluent flow, which carries the sample to the column. Two designs are used, a fixed-volume loop, which is filled with the sample (a 6-port valve) and, more commonly, a variable-volume valve, in which a sample volume defined by a syringe is injected into a much larger loop (a 7-port valve) (Fig. 12.1).

Nowadays the majority of HPLC systems include an autosampler. There are a number of different designs, but in most the sample is collected from the sample vial with a fixed needle/probe and injected into the loop of a rotary valve, which is then switched under computer control. The autosampler takes much of the operator variability out of the operation of the instrument, increases precision, and enables virtually 24-h operation of the equipment. Because the sample solution does not flow into the narrow-ID loop as a plug, the injected volume for partially filled loops should be no more than 50% of the total loop volume [31], whereas for filled-loop injection the loop must be flushed with 5–10 times the nominal value for accurate results.

It is important that the path of the sample through the instrument from injector to detector contains no dead-volumes, where the eluent flow could eddy, as they will cause band-broadening and loss of separation efficiency. This means that all the connecting tubing must have a relatively narrow bore [typically, 0.005 in. (0.127 mm) ID] and that the tubing must fit snugly into the injector, the ends of the column, and detector without gaps or spaces. Frequently, 1/16-in. stainless-steel tubing is used in the high-pressure parts of the instrument, and the ferrules, once fitted, cannot be removed. One problem is that the depths of the connections in columns and fittings from different equipment manufactures are not consistent. However, some of the problems can be overcome by using adjustable polyether ether ketone (PEEK) nuts and ferrules and PEEK tubing, although care must be taken, as this material is attacked by some solvents, particularly THF.

2.2.1.3 Columns and ovens

Although longer columns generate higher efficiencies (Chap. 1), the retention times are also longer. The selection of column dimensions is thus a balance between resolution and throughput. With 5-μm particles, columns longer than 30 cm generate an unacceptably high back-pressure at normal flow-rates, and smaller particles impose an even lower length limit. Typically, most HPLC columns are 250 mm (the "standard" length), 150 mm or 100 mm long, although 50-mm and 30-mm columns are used when very fast separations are required or when the detector can provide additional resolving power, as in LC/MS/MS (Chap. 10). The ID of most columns was originally 4.6 mm (based on the use of 1/4-in. OD stainless-steel pipe fittings and tubing), which, with flow-rates of 1 mL/min should give good efficiencies. More recently, dedicated HPLC columns have been designed with 4- or 5-mm-ID columns, as manufacturers try to improve the whole column package. Narrow columns can also be used with correspondingly lower eluent flow rates: 2-mm-ID columns are often termed "narrow-bore" and 1-mm and 0.5-mm are usually regarded as "microbore". Even narrower capillary columns are used in research or for special methods, such as capillary electrochromatography (Chap. 7), as the limited sample capacity makes analyte detection difficult [32]. Frits or meshes (typically 2 μm) in the column end-fittings hold the stationary phase in the columns. In recent years, there has been a trend away from stainless-steel fittings to PEEK or polymeric meshes, which can be more inert. The composition and nature of the stationary-phase material will be considered in Sec. 2.3. A second, very short (3- to 5-mm) column, packed with the same stationary phase as the main column, is often placed before (or in the inlet fitting) of the analytical column to act as a disposable guard column, intended to protect the main column from debris and other insoluble material.

One aspect of the operation of an HPLC that has often been ignored is the effect of temperature on retention times and selectivity. This is now recognized as an important factor in reproducibility [33], even though it had been known for many years that retention times vary as $1/T$ [34]. Consequently, columns (and the incoming eluent) should be thermostated to a constant temperature, typically above ambient conditions. Some control is provided by room temperature control but this can be unreliable and may be altered automatically at times when a laboratory is not in use. In recent years there has also been an interest in exploring temperature-induced effects in more detail and in examining separation at elevated temperatures (Sec. 2.7).

2.3 STATIONARY-PHASE MATERIALS

Columns can be divided according to the stationary phase or support materials and by the nature of the bonded phase. Most are based on silica, which shows good chromatographic and mechanical properties, but alternative phases are also available, including alumina, zirconia, and organic polymers. The background of the column materials and bonded phases is reported in books by Neue [35] and Scott [36].

2.3.1 Silica-based stationary phases

The properties of silica have been extensively reviewed [37–41]. One of the major limitations of silica as a stationary-phase material is that above pH 8–9 it dissolves, and below pH 3 any bonded groups can be hydrolyzed from the surface. Degradation is increased at higher temperature, and for normal use most silica-based phases are unstable above 60–70 °C. Newer materials (see later) can often extend these limits to high pH values and higher temperatures. Usually, columns are packed with 5-μm particles, but 3-μm particles are becoming more popular, even though they generate a higher back-pressure. Usually, there is a spread of size distribution around the nominal size. Although very accurately sized, mono-disperse silica particles have been produced [42], they seemed to offer little advantage in column efficiency. Porous silica is available with a range of pore diameters. Generally, analytical columns have pores measuring 80–120 Å and columns for natural and synthetic polymers *ca.* 300 Å. The surface areas are typically 400–500 m^2/g.

2.3.1.1 Silica-based stationary phases for normal-phase chromatography

Over the years the use of normal-phase chromatography, and with it the use of bare silica as a stationary phase, has declined compared to reversed-phase separations because of the much greater applicability and greater robustness of the latter. One of the main problems is that the high surface activity and acidity of a silica stationary phase makes it susceptible to traces of moisture, which deactivate the surface, or cause irreversible interactions with strongly basic analytes. The surface of silica can be partially protected by bonding cyanopropyl [43,44], aminopropyl, or hydroxyl (glyceryl) silanes to the surface [45] (Table 2.2). These provide a polar stationary phase while protecting the ionizable and water-susceptible silanol groups. The most widely used phases are probably cyano-bonded, but different formulation methods have resulted in nominally similar phases with different properties [46]. Polar-bonded materials have the added advantage of having faster equilibration times than silica on changing the mobile phase. Although not very popular as normal-phase columns, many are also used as reversed-phase stationary phases.

TABLE 2.2

TYPICAL SUBSTITUENT GROUPS IN POLAR BONDED STATIONARY PHASES
(R USUALLY METHYL)

Bonded phase	Substituent groups
Cyano	$-O-Si(R_2)-(CH_2)_3-CN$
Amino	$-O-Si(R_2)-(CH_2)_3-NH_2$
Hydroxyl	$-O-Si(R_2)-(CH_2)_3-O-CH_2-CH(OH)CH_2OH$
Nitro	$-O-Si(R_2)-(CH_2)_3-C_6H_4-NO_2$
Polyethylene glycol	$HO-(CH_2-CH_2-O)_n-H$

2.3.1.2 Silica-based stationary phases for reversed-phase chromatography

It was the introduction of alkyl-bonded phase materials and their ability to give reproducible results with polar analytes that lead to the rapid rise of HPLC in the early 1980s. The initial materials were based on irregularly shaped silica particles, which had been bonded with a range of alkyl-bonded silanes having side-chain lengths from 1–30 carbon atoms. The octadecylsilyl ($C_{18}H_{37}-$, ODS) alkyl chain, rapidly became the most popular. Although some of these early materials with shorter chain-length are still marketed, because they have become enshrined in established methods, they had many disadvantages. Chromatographically, they were often of limited efficiency and poor stability and could not be manufactured reproducibly. Similar alkyl-bonding reactions with spherical particles of silica were more successful. Two main silanization reactions were used with either monochlorodimethyl alkylsilanes or trichloroalkylsilanes (Fig. 2.3). However, these reagents do not react with all the silanol groups on the silica surface because of steric hindrance, and the unreacted acidic silanol groups can still ionize and cause poor peak shapes for basic analytes. The next development was the use of end-capping, where a smaller reagent (Me_3SiCl, trimethylsilyl chloride) is used to cap these groups. However, some acidic groups always remained and still caused problems. Although the pKa of the silanol groups on silica is about 8.0, many of these bonded materials appear to contain a few strongly acidic sites, associated with the presence of trace metal ions and causing particular interaction problems. This problem is greatly reduced with Type B silica. The manufacturers now offer a range of substituents (Table 2.3) Effectively, these primarily alter the depth of the stationary-phase layer and, hence, the overall retentive capacity of the column. The degree of bonding varies between brands of columns. This is reflected in the carbon loading (% of the column by mass that is carbon), which is a rough guide to the proportion of stationary phase and, hence, the overall retentivity of a column. Currently, most bonded phases on the market are C_{18} and, occasionally, C_8 phases.

The change from Type A silica, which is derived from inorganic silicates, to Type B silica, which is formed by the hydrolysis of highly purified organic silanes, has largely eliminated many of the problems due to metal-ion interactions. However, the same

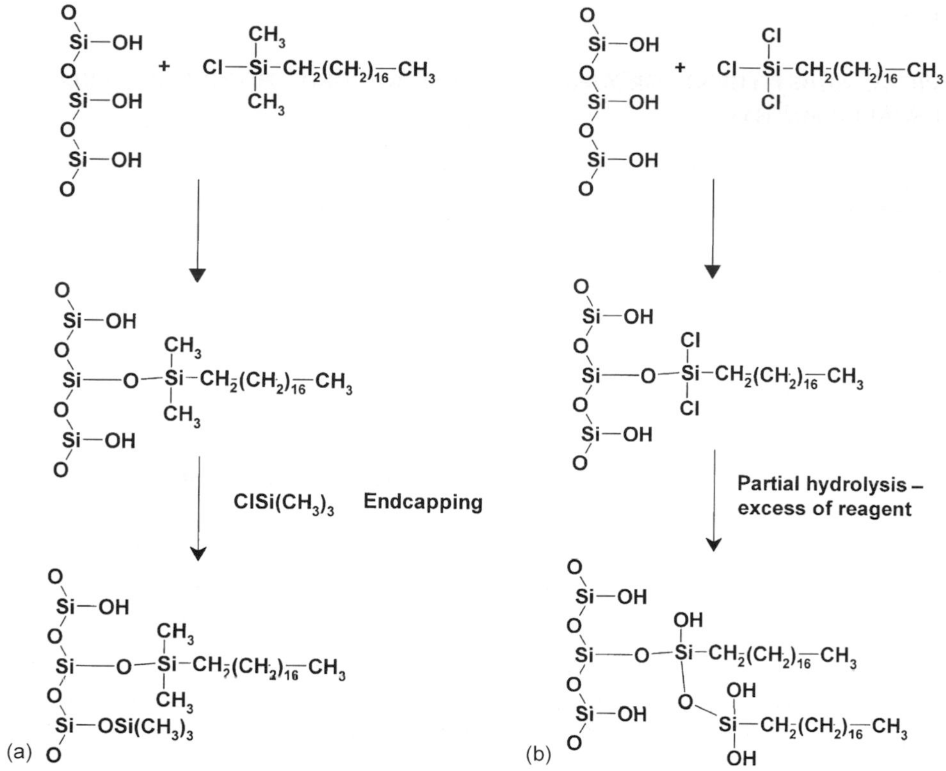

Fig. 2.3. Silanization reactions for introducing alkyl groups into the silica surface. (a) Reaction of monomeric reagents and end-capping with trimethylsilyl chloride. (b) Reaction of a polymeric reagent. In the second stage unreacted Si–Cl groups are hydrolyzed and additional reaction takes place.

description of ODS-bonded silica can now be applied to a wide range of materials, capped and uncapped, from two different silica types and from different bonding reactions. These differences can cause a variation in the carbon loading on the column. As a result, although nominally equivalent "ODS-silica" columns can be obtained, the replacement of one brand of column by another can often lead to major differences in the retention and selectivities of the separation. A further problem has been found with phases which have a low surface activity. If the mobile phase contains a high proportion of water (>95%), the eluent can be excluded from this very hydrophobic environment with the result that the bonded alkyl chains effectively collapse [47]. This creates a non-polar, impenetrable alkyl surface rather than a "liquid-like" phase, resulting in dramatic changes in retention.

Although end-capping reduces the surface activity, short alkyl chains are more susceptible to hydrolysis than longer chains, and there has been considerable interest in recent years in developing improved or alternative more stable bonding methods. For example, phases have been developed in which there is a multidentate bonding between the alkyl-chain and the silica surface [48]. These factors have also led to the

TABLE 2.3

TYPICAL SUBSTITUENT GROUPS ON NON-POLAR STATIONARY PHASES (R = USUALLY METHYL)

Bonded phase	Substituent groups
Octadecylsilyl (C_{18}) (ODS)	$-O-Si(CH_3)_2-(CH_2)_{17}CH_3$
	$-O-Si(C_4H_9)_2-(CH_2)_{17}CH_3$ (sterically hindered)
Octylsilyl (C_8) (and corresponding side chains for butyl and hexyl)	$-O-Si(CH_3)_2-(CH_2)_7CH_3$
Phenyl	$-O-Si(R_2)-(CH_3)_2-C_6H_5$ or
	$-O-Si(R_2)-(CH_2)_2-O-C_6H_5$
Amide phases	$-O-Si(CH_3)_2-(CH_2)_3-NHCO-C_{15}H_{31}$ or
	$-O-(CH_2)_3-O-CO-NH-C_8H_{17}$
Alkyl and aryl fluorinated phases	$-O-Si(R_2)-C_6F_{13}$
	$-O-Si(R_2)-C(CF_3)_2 \; C_3F_7$
	$-O-Si(R_2)-C_6F_5$

insertion of polar linking groups into the bonded chain near the silica surface. These protect the silanol groups and bonding sites from attack and, by creating a polar region, prevent the exclusion of aqueous mobile phases [49,50]. Some of the early linkages were generated with amines (but these tended to react with acidic groups) and even some with acidic functions but more recently amides (Table 2.3) appear to be favored, although not all the production methods have been revealed, as some of them are proprietary.

One further advantage of these newer materials is an improved stability to pH and temperature, columns being stable up to pH 11.5, and temperature stability up to 100 °C being claimed. Column technology is continually improving and changing, and there is currently a trend toward columns with bonded fluorinated alkyl or phenyl groups, claimed to offer a different selectivity (Table 2.3) [51,52]. The cyano- and amino-bonded columns (Table 2.2) used for normal-phase separations can also be employed in reversed-phase chromatography. The cyano-bonded columns have found application to a number of basic-drug assays [53]. They tend to behave primarily like short alkyl chains. The amino-bonded phases are used primarily for the separation of carbohydrates (Chap. 18). Recently, additional polar columns, based on a polyethylene glycol coating have been marketed (Chap. 12).

2.3.1.3 Hybrid-silica stationary phases

In an effort to overcome the stability and ionization limitations of silica, hybrid materials have been marketed [54], which are based on a polymerized methylsilicone, where the methyl groups are not just at the surface but are incorporated within the solid matrix. These materials are available as C_{18}, C_8, and phenyl-bonded phases (Fig. 2.4) and

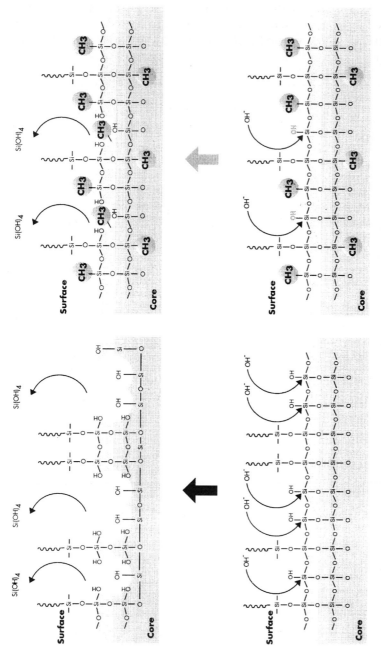

Fig. 2.4. Structure of XTerra column materials, showing hybrid structure with embedded methyl groups and resistance to partial dissolution of the surface (right) compared to conventional silica (left). (Reproduced through the courtesy of Waters.)

have enhanced thermal and pH stability (up to pH 12) with good efficiency and a selectivity similar to conventional C_{18}-bonded materials [55]. Further variations are being developed, using different starting materials.

2.3.1.4 Other silica stationary phases

Although unbonded silica columns are normally regarded as normal-phase materials, a number of investigators have reported their use in drug analysis under extreme pH conditions with non-aqueous eluents. For example, basic drugs can be separated on unbonded silica, using 90:10 methanol/buffer (pH 10.5) [56,57]. Even at this high pH, the column is stable for many months, primarily because silica is insoluble in the largely organic mobile phase. The mechanism of separation is not fully understood, but the material appears to behave as a form of ion exchanger [58]. Alternatively, silica columns can be used with eluents containing a strong acid, such as perchloric acid, in an eluent of high methanol content, also for drug screening [59].

2.3.1.5 Monolithic silica columns

An important recent advance has been the development and commercialization of monolithic columns, made from a porous silica rod, rather than individual particles [60–63]. The column is not filled with individual particles, but with a single lattice of C_{18}-bonded silica and the mobile phase flows through the pores. One advantage is that no end frits are required. The effect is a stable system with an excellent mass transfer (a small van Deemter C term) (Chap. 1). A low back-pressure enables separations to be carried out at high flow rates without pressure limitations. Monolithic polymeric materials have also been reported [64] (Chap. 12).

2.3.2 Stationary phases based on zirconia, titania, and alumina

Because of the problems often encountered with silica due to its acidic surface and limited pH range, other inorganic support materials have been proposed for normal- and reversed-phase chromatography. However, if the surface is less active, it is more difficult to form an alkyl-bonded stationary phase. Instead, polymeric films have been used to coat the surface. Earlier, alumina was often employed as a column material for open-column separations, because it is amphoteric and less reactive than silica, but it is rarely used in HPLC, even though it is better than silica for amines [65–67]. Use as the basis of a reversed-phase material has been attempted, but because it is difficult to form alkyl bonds [68], it may be coated with polybutadiene [69]. Titania has also been tested as a normal-phase material [70]. Use of C_{18}-coated titania has been reported [71] and attracted particular interest as a high-temperature material (up to 200 °C) [72], but few titania columns are available commercially.

Most interest has been in zirconia, as both an uncoated and alkyl-coated phase [73]. As a sorbent, zirconia has a surface with strong Lewis acid properties. Even with bonded phases, additives such as ammonium fluoride may be needed to reduce surface interactions. Although originally proposed some years ago [70], it has only come into prominence fairly recently with the work of Carr and co-workers [74]. They generated polybutadiene (PBD) phases [75,76], porous graphitic carbon-coated phases [77], and a PBD-coated phenyl phase. Subsequently, a C_{18}-bonded PBD-coated phase has become available. A particular area of application has been for high temperatures (up to 200 °C), as these lead to faster and better separations [78,79]. They have been used for the separation of steroids and pharmaceuticals with superheated water as the sole eluent [80]. In other studies, a combination of different columns at different temperatures have been employed to increase selectivity through the tuned-tandem column effect [81].

2.3.3 Polymeric stationary phases

Because silica materials are unstable outside a relatively limited pH range, there has been considerable interest in polymeric phases, principally based on polystyrene-divinylbenzene (PS-DVB). The macroporous structures of polymers provide a large surface area [82,83], unlimited pH stability from 1–14 [84,85], and high temperature stability up to 200 °C. Being totally organic stationary phases, they tend to have a much high retention capacity than C_{18}-bonded phases, but their main limitation is that column efficiency is dependent on the organic components of the mobile phase. Methanol/water gives poor separations; the preferred eluent is THF/water or, to a lesser extent, acetonitrile/water [86]. A wide range of buffers can be employed, so that even quite basic or acidic analytes can be analyzed in their neutral form. A variant on the PS-DVB material is polymeric material with bonded alkyl groups, which has similar pH stability, but is claimed to have a selectivity more like C_{18}-silica materials [87]. Polymeric beads also form the basis of many ion-exchange materials [88] (Chap. 4). Although other polymeric matrices, such as polymethacrylate, have been reported [89], few have found significant application.

2.3.4 Stationary phases of porous graphitic carbon

Another highly inert material is porous graphitic carbon (PGC), consisting of ribbons of graphite. It is very robust, stable over a wide pH range and at high temperatures. It was chiefly studied by Knox and co-workers [90,91], who recently reviewed its properties and applications in detail [92,93]. Recently, a number of investigators have examined the retention properties of PGC and concluded that it works by electron/electron surface interaction. Retention selectivity is very different from that of conventional partition columns, $\pi-\pi$ electronic interactions and ionized bonds having a much greater effect than hydrophobicity [94]. Steric and isomeric differences, which often alter the ability of an analyte to interact with a planar surface, can have a significant effect, often enabling discrimination of structural isomers [95] and chiral analytes [96].

2.3.5 Chiral stationary phases

Most conventional columns have no inherent chirality and, hence, two enantiomers will have identical retention properties and a racemic mixture will be eluted as a single fraction. In order to generate a chiral separation, three HPLC methods may be employed. Firstly, the enantiomeric analytes can be converted into diastereoisomers, either by derivatization with a single enantiomer of a chiral reagent or by the formation of transitory diastereoisomers with a chiral ion-pair reagent or additive in the mobile phase. This was a common method for some years but it requires either a separate reaction, which may be stereoselective (and hence give misleading results), or the continuous use of a chiral additive in the mobile phase, which is expensive. Secondly, in theory, a chiral mobile phase could be used but few are available which have a sufficiently low viscosity or cost to be practical. The third and now most popular methods is to use a chiral stationary phase (CSP) in which a chiral selector (a group with inherent chirality which interacts with the enantiomers) is part of the stationary phase.

The use of HPLC for this role has become increasingly important over the last 15 years and hundreds of different CPSs have been reported. These were divided into five main groups by Wainer [97,98] and have since expanded to six groups (Table 2.4). The different

TABLE 2.4

CLASSIFICATION OF CHIRAL STATIONARY PHASES

Type	Name	Properties
Type I	Donor–acceptor	π,π-interactions, dipole stacking and H-bonding Normal-phase eluents
Type II	Cellulose attractive	Attractive forces Normal-phase eluents
Type III	Inclusion interactions	Inclusion in chiral cavity steric and size factors Reversed-phase eluents
Type IV	Ligand interactions	Outer complexes with aminoacid groupings
Type V	Protein based	Multiple interaction sites low sample capacity
Type VI	Macrocyclic phases	Many interaction sites

materials and their applications have been reviewed in a number of texts [99–103] and articles [104–107]. This is one area where SFC has made a major impact. It can often generate a higher resolution than conventional mobile phases because the low mobile-phase polarity increases analyte/stationary phase interactions [108–110]. One difficulty in chiral chromatography is that, apart from a few very general rules, it is very difficult to predict which column will achieve the separation of a particular pair of enantiomers. This is because the energy differences in the interactions between the chiral phase and the two enantiomers are often small and are difficult to model. This means that column selection tends to be a process of trial and error, where success is most likely with a phase known to have separated a closely related analyte.

2.3.5.1 Donor–acceptor stationary phases

The first chiral phases to have a significant impact were developed by Pirkle [111,112]. The initial material was an ionic complex between an amino-bonded silica and the 3,5-dinitrobenzoyl derivative of an amino acid, such as phenylalanine. The ionic link was soon replaced by the more stable covalent amide bond. It provided a structure with positions for donor–acceptor π–π interactions, amide dipoles, and hydrogen-bonding groups with a clear steric selectivity. For good selectivity, a three-point interaction is required [113]

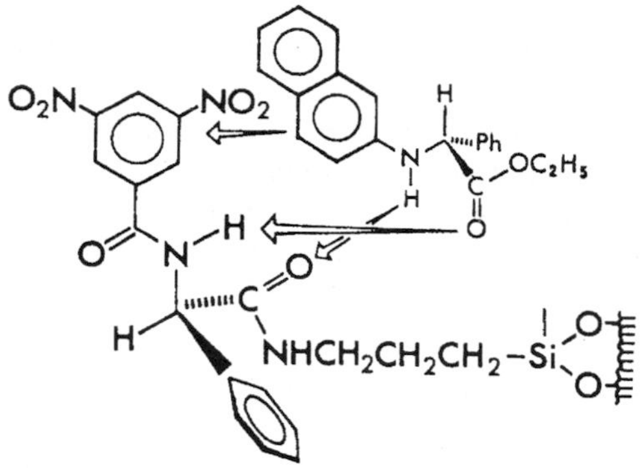

Fig. 2.5. Example of three-point interactions between dinitrobenzoylphenylglycine chiral sorbent and the most retained enantiomer of *N*-(2-naphthyl)phenylglycine methyl ester, showing π–π-electron interaction between aromatic rings, and hydrogen-bonding interaction with amide. (Reproduced from Ref. 115 with permission.)

(Fig. 2.5), and if this is present, a high chiral discrimination can be obtained. The initial concept has led to a wide range of related phases with either π-donor or π-acceptor properties and an amide or urea linkage [114,115].

Because selectivity depends on strong hydrogen-bonding interactions, these phases are restricted to normal-phase (or SFC) separations, as in an aqueous environment the polar mobile phase would preferentially interact with the active groups. The principal role for these phases is to generate a strong 3-point interaction when the analyte has sufficient functionality near the chiral center, and these phases show a very high selectivity.

2.3.5.2 Cellulose-based stationary phases

One of the most successful series of chiral stationary phases is made up of substituted cellulose phases developed by Okamoto [116–118]. Initially, they were produced as acetyl cellulose beads, but subsequently as silica-supported materials. A wide range of substituted acyl and carbamates have been developed, based on either cellulose or amylose substrates, including Chiracel OB (cellulose tribenzoate), Chiracel OD [cellulose tris(3,5-dimethylphenyl carbamate)], Chiracel OK (cellulose tricinnamate), and Chiralpak AD [amylose tris(3,5-dimethylphenyl carbamate)]. Chiral selectivity is generated by the inherent chirality of the cellulose backbone. A wide range of enantiomers can be resolved, the main requirements being the presence of a moderately polar functional group, such carbonyl, ester, amide, alcohol or amine, which will interact with the substituents on the cellulose backbone. The mobile phase is usually hexane, containing a low proportion of propanol. Recently, RP-compatible materials have become available (Chiralcel OD-R). The cellulose phases are also very successful as chiral phases in SFC [119,120].

2.3.5.3 Inclusion-complex stationary phases

Another group of widely available natural chiral materials is based on α- β-, and γ-cyclodextrins. They consist of a ring or toroid of 6, 7, or 8 glucose units with successively larger cavities [121]. The center of the each toroid forms a hydrophobic cavity, and the hydroxyl groups around the edges provide points of interaction for bonding the ring to a silica support or for interacting with the analyte [122]. To achieve a resolution, there needs to be a grouping on the analyte, such as a phenyl group, that can enter the cavity and bring the chiral center of the analyte close to the chiral edge of the cyclodextrin. One advantage of these chiral selectors is that they are compatible with reversed-phase eluents and so can potentially handle a wide range of polar analytes, such as drugs. Because of the steric requirements of the cavity, these phases can often also separate positional isomers of aromatic analytes [123]. Alternatively, free cyclodextrins can also be added to an eluent to form complexes in solution.

2.3.5.4 Ligand chiral phases

The interaction of amino acids and metals, such as copper, to form complexes has generated chiral columns [124] which can form an outer complex with amino acid-type groups ($R-C(NH_2)CO_2H$). Unfortunately, the range of application is limited to analytes, such as dopamine, although the selectivities obtained are high. The ligand complex can also be used as mobile-phase additive.

2.3.5.5 Protein chiral phases

A number of immobilized proteins, including α_1-acid glycoprotein [125], bovine serum albumin [126], and ovomucoid [127], have also been used as chiral selectors. The presence of numerous different potential interaction points on a complex protein means that these columns can separate a wide range of analytes. However, because in a complex protein any one site occurs infrequently, the sample capacity is limited and the protein phases are restricted in their use.

2.3.5.6 Macrocyclic chiral phases

The last major group of chiral phases to be developed were the macrocyclic phases based on the immobilized antibiotics vancomycin or teicoplanin [128]. These antibiotic-based materials offer a number of $\pi-\pi$ interactions, chiral hydrogen-bonding sites, ionizable groups, amide linkages, and cavity sites for inclusion interactions (Fig. 2.6). They provide both normal- and reversed-phase compatible systems with wide applicability.

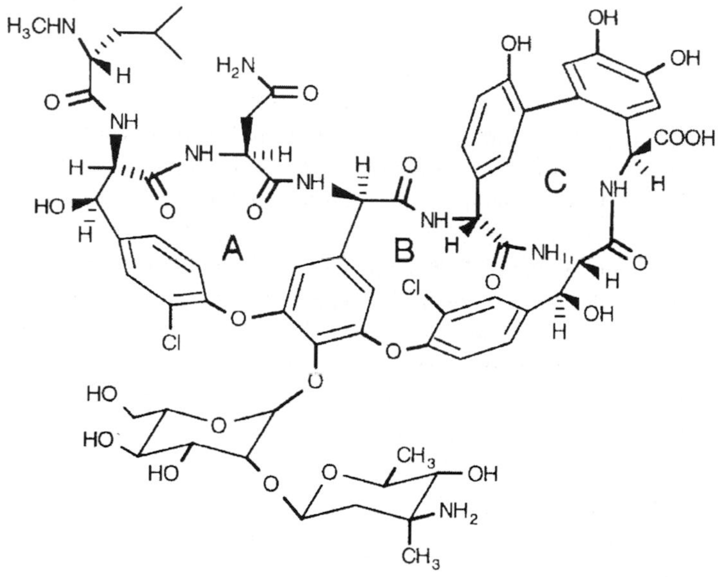

Fig. 2.6. Structure of vancomycin chiral phase (Chirobiotic V). A, B, and C are inclusion cavities. (Reprinted through the courtesy of Advanced Separations Technologies.)

2.3.6 Molecularly imprinted polymeric stationary phases

Whereas chiral columns employ the selectivity between analytes and the CSP to achieve resolution, a much more specific interaction can be seen in affinity chromatography, which uses the specificity of an antibody to selectively retain an analyte (Chap. 3). A number of attempts have been made to mimic this interaction by preparing synthetic receptor systems as molecularly imprinted polymers (MIP). A target analyte is mixed with a monomer containing a functional group, such an acrylic acid, which would be expected to generate a specific interaction with the target. The monomer is then polymerized, and when the target analyte is removed, it should leave a cavity in the polymeric bead, corresponding to the analyte. This material should then retain analytes containing the template grouping to a greater extent than an un-imprinted polymer [129]. The concept appears to work well in columns for sample preparation [130,131], where groups of related analytes can be selectively trapped, but attempts to generate a chiral phase specific for a single enantiomer have given poor chromatographic results.

2.3.7 Evaluation and comparison of stationary-phase materials

Numerous papers have reported methods for the evaluation and classification of columns, particular reversed-phase stationary phases, and have compiled comparisons of the different properties [132–140]. Typical measurements include hydrophobicity (based on the ratio of the retention factors of two homologous analytes – also called methylene selectivity), shape selectivity, silanol activity, and column efficiency. The shape selectivity tests compare the relative retention of either benz[*a*]pyrene, dibenzo[*g,p*]chrysene and phenanthro[3,4-*c*]phenanthrene [141], or the noncarcinogenic aromatic hydrocarbons, triphenylene and *o*-terphenyl [142]. More specialized tests include silanol activity, ion-exchange capacity at different pH values and metal-ion content [138–143]. The last of these can, however, be dependent on the usage of the column, as metals ions are introduced from the pump, tubing, and eluents. Most recent work has concentrated on the effect of different stationary phases on ionizable analytes [144–147]. However, there are no standardized tests or conditions, and most lists should be regarded as comparative studies. Also, a number of the tests have been found to be not very robust or transferable. Solid-state NMR spectroscopy has also been used to directly examine the nature of linkages, the types of free silanols and bonded groups, and the stability of stationary phases [148–150].

2.4 DETECTION

One of the distinguishing features of instrumental chromatography is the use of on-line detectors to continuously monitor the eluent from the column and to present an electrical analog or digital signal to a data system. This output is usually linked to a computer, which may also provide operational control of the detector as well as integration and storage of the data. Thus, it is important to understand the methods and decision-making associated with data collection [151–153]. The main advance in recent years have been a move from a single output, such as absorbance at a single wavelength, to an output with increased data content, such as continuous, full UV spectra, mass spectra or, more recently, NMR spectra, which can be used for both identification and confirmation of the analyte structure.

The principal detectors are: ultraviolet/visible spectrometers, fluorescence spectrometers, refractive-index detectors, evaporative light-scattering detectors, electrochemical detectors, conductivity detectors, mass spectrometers, and NMR spectrometers. Many other detection systems have been marketed, but their penetration of the market is small. The whole field of detectors for HPLC has been reviewed in monographs [154–156].

2.4.1 Spectroscopic detection

Spectrometers are the workhorses of chromatography laboratories and probably account for over 95% of the detectors on HPLC systems.

2.4.1.1 Ultraviolet spectroscopic detectors

These are the standard detectors used in HPLC and the presence of a detectable chromophore is often regarded as an essential feature of the sample to be assayed. Originally, cost was important, and simple, fixed-wavelength detectors were employed, based on the line emissions from the mercury (254 nm) or zinc (214 nm) lamps. Although now rarely used, they have left a legacy, as 254 nm often still regarded as the "standard" wavelength for detection. In practice, this is often a good starting point for most compounds. Most UV detectors now employ a deuterium lamp, which emits a continuum of radiation from 180 nm up to about 600 nm. A tungsten lamp could be used up to 850 nm for visible absorbance, but because colored analytes can usually also be detected at shorter wavelengths, a separate visible-light source is rarely used. Most current detectors are either variable-wavelength systems, based on a monochromator, which gives a single detection wavelength – although this may be programmable during the separation – or, commonly, photodiode-array detectors, in which reversed optics are used and the spectrum is spread across an array of photodiodes. These are interrogated by a computer to continuously provide the full spectrum of the eluate [157] or, simultaneously, the absorbance at a number of selected wavelengths. The cost of a diode-array system was initially high but has rapidly decreased, as computing and microchip technology has become more powerful and effectively cheaper.

All of these detectors are based on the use of the Beer–Lambert relationship (Absorbance = path length × molar concentration × molar absorptivity). Thus, for any particular analyte in the same detection cell, the signal is proportion to its concentration, and a calibration curve can be used to determine the concentration. However, different chromophores can produce very different responses and, thus, their sensitivities can differ markedly. For example, the addition of a hydroxyl group to an aromatic ring (*e.g.*, from benzene to phenol) will increase the absorbance 10-fold at *ca.* 250 nm (Table 2.5).

As a consequence, even large amounts of an analyte lacking a chromophore may not be detected, whereas highly absorbent trace impurities may give a large signal. Thus, the relative areas under the peaks in a chromatogram may represent very different masses of the analytes. It is also important to realize that because "stray light" gives a background signal from the photomultiplier in a UV detector, the linearity of the response decreases above an absorbance of 1.0; a higher absorbance should not be used for quantitative analysis. There is also a mode of indirect detection, in which the background response due to an ionized mobile-phase additive containing a chromophore is reduced by displacement with ionized analyte ions to maintain electrical neutrality [158].

2.4.1.2 Fluorescence detectors

The second-most common spectroscopic detection method is fluorescence spectroscopy, which is both highly selective and highly sensitive [159]. However, its applicability is limited by the restricted number of naturally fluorescent analytes. Since fluorescent analytes also contain a chromophore and can be detected by UV spectroscopy, fluorescence detection is only needed when either enhanced selectivity or sensitivity are

TABLE 2.5

EXAMPLES OF TYPICAL CHROMOPHORES

Chromophore	Wavelength maxima (nm)	Molar absorptivity (ε)
	170–180	7000–10,000
	214	21,000
	190 280	1900 13
	215–240	10,000–15,000
	245–270	15,000–20,000
	184 200 254 248	60,000 7400 204 14,800
	280	3000
	210 at high pH Ph–O⁻ 235	6200 9400
	230 at low pH NH₃ 254	8600 160
	254	500
	266	7800
	240	12,600
	230	10,000

important. Typical applications are for trace levels of polynuclear aromatic hydrocarbons (PAHs) in environmental analysis, but even here there is a problem, as each PAH has a different pair of optimum excitation and emission wavelengths. One application of fluorescence detection has been to increase the sensitivity for very small amounts of sample or in microflow systems when the inherent amount of the analytes is small. For some analytes, derivatization is used to introduce a fluorophore to enhance sensitivity, but although the literature contains a large number of potential reactions and reagents [160–163], in reality few are ever used, because the process of derivatization adds an extra step to sample preparation, which can introduce additional variability and reduce quantitative accuracy.

2.4.2 Other common detectors

2.4.2.1 Refractive-index detectors

Refractive-index detectors have been long regarded as the alternative to UV spectroscopy when an analyte lacks a chromophore. However, they are bulk-property detectors and operate by distinguishing the small change in the refractive index of the eluent when an analyte is present. Consequently, they are inherently insensitive. In the past, their main application has been for carbohydrates or lipids, but these are now often detected with a light-scattering detector.

2.4.2.2 Light-scattering detectors

The light-scattering (or mass evaporative) detector (LSD) operates by spraying the eluent into a drift tube and thermally evaporating the solvent [164–166]. Any residual involatile analyte particles are then detected as they fall through a light beam by measuring the scattered radiation. This detector has been developed steadily over the years, and it now often rivals or exceeds the sensitivity of the refractive-index detector. It is particularly advantageous with high-molecular-weight analytes [167,168], but improved design allows it to be used when the difference in volatility between the analyte and the mobile phase is relatively small. The main disadvantage is that the response can be non-linear, as the intensity of scattered light is a combination of Raleigh and Mie light-scattering mechanisms, which often depend on particle diameter rather than mass. The LSD is particularly useful in SFC, where the carbon dioxide mobile phase evaporates readily [169]. In recent years, an important role for the detector has been as part of a LC/MS system, as it enables the universal detection of all analytes, whereas the mass spectrometer can be selective, depending on the ionization process [170].

2.4.2.3 Electrochemical amperometric/coulometric detectors

The electrochemical group of detectors is based on the determination of the current that flows across a detector cell when an analyte is electrochemically oxidized or reduced by the application of a potential [171–173]. It works most effectively for the RP chromatography of easily oxidizable groups, such as phenols or arylamines at an applied potential of 0.8 to 1 V relative to the Ag/AgCl electrode. A number of different electrode designs have been marketed [174]. One difficulty is that the range of applications is limited and, for some samples, fouling of the electrode surfaces reduces its reliability. The method is very sensitive (up to 100 times greater than UV spectrometry) and very selective, as no response is generated by non-oxidizable analytes. Its principal application has been in the pharmaceutical field, where it is used, *e.g.*, for the determination of estrogens in urine, and clinically for the determination of catecholamines at natural levels.

Reducible analytes, such as nitroaromatic and azo compounds, can be detected by using a negative applied voltage, but the response is more difficult to achieve in practice, as oxygen is oxidized at a lower voltage and must therefore be excluded from the eluent and sample solutions. This mode has found application in the trace detection of explosives [175] and food dyes. If the applied potential is pulsed cyclically, the electrode surface can be cleaned between each measurement step, and this increases robustness and reduces the effects of contamination [176]. This method has been found to be suitable for the detection of carbohydrates in eluents of high pH [177], where its gives much better results than refractive-index or UV detection, and it can also be used for some antibiotics [178].

2.4.2.4 Conductivity and suppressed-conductivity detectors

The elution of ionized analytes (both cations and anions) can be detected by the change that they cause in the conductivity of the eluent. Direct measurement of the resistance of the eluent can be very successful, if the background conductivity is low. However, as the separation method for ionized analytes is often ion-exchange chromatography, a high buffer concentration is frequently used, which produces a potentially high background signal, making analyte detection harder. In these cases, suppressed conductivity detection can be employed, in which the buffer ionization is suppressed so that the analytes are more readily distinguished [179]. This technique forms the basis of the determination of small inorganic cations and anions by ion chromatography (Chap. 4).

2.4.2.5 Chiral detectors

The interest in the separation of enantiomers (Sec. 2.3.5) has also led to an interest in detectors selective for chiral analytes. Both polarimetry- [180,181] and circular dichroism-based [182] detectors have been reported, but neither has gained wide popularity because of limited sensitivity. In addition, it is difficult to determine the enantiomeric purity of an analyte from optical rotation alone, so that separation into the individual components is more powerful as an analytical tool.

2.4.2.6 Other detectors

Over the years, there have been reports of numerous other minor detection methods, including radiochemical detection for analytes containing radioactive isotopes [183] and flame-ionization detectors (where the mobile phase needs to be removed). Some satisfy niche applications, but none have gained widespread application.

2.4.3 Coupled detection methods

As well as a direct detection method for analytes, one of the areas of interest in recent years has been in the use of coupled detectors, where the output from the column is connected directly to a detector, which can provide additional structure or identification information on the analyte. These detectors are usually well established and widely used as stand-alone analytical instruments. The last 10 years has seen more interest in these information-rich detection methods, especially LC/MS, LS/MS/MS, LC/ICP, and LC/NMR as well as diode-array detection. Reduction in the cost of the computing power to handle the data and consequently increased sales have tended to reduce overall prices markedly. In some cases, these methods can also provide a high degree of selectivity by responding to the presence of selected ions or elements. Although usually used individually, one recent trend has been multiple coupling of detectors for NMR, IR, and MS, together with UV spectrometry (Fig. 12.2) [184,185]. A more detailed review of many of these detectors can be found in Chap. 10.

2.4.3.1 Liquid chromatography/mass spectrometry and liquid chromatography/tandem mass spectrometry

Mass spectrometry (MS) is a very sensitive detection method, which can provide mass and structural information. For many years, coupling LC to MS posed practical problems, but most of these are now overcome [186,187]. Application of LC/MS in areas, such as drug metabolism and for rapid identification in synthetic combinatorial chemistry, is now routine [188], but some analytes and matrices still cause problems and some analytes may go undetected [189,190]. The further stage of LC/MS/MS provides additional selectivity and, hence, sensitivity, which can be valuable in trace assays and metabolism studies.

2.4.3.2 Liquid chromatography/nuclear magnetic resonance spectrometry

LC/NMR [191,192] can provide a high level of structural information and is complementary to LC/MS, as it can often readily distinguish between isomers with identical mass spectra. However, it is a relatively expensive detector, and mobile-phase components can produce interfering signals. Its principal drawback is that the sensitivity is relatively poor, although improving with recent advances in design.

2.4.3.3 Liquid chromatography/infrared spectrometry

Although IR spectroscopy can yield information on the presence of functional groups, and the fingerprint region can provide a good characterization, LC/IR [193] has not been particularly popular because of severe interference from the mobile phases, which give numerous signals.

2.4.3.4 Liquid chromatography/atomic absorption spectrometry and liquid chromatography/inductively coupled plasma spectrometry

Interest in environmental samples and in organometallic drugs has led to the coupling of LC and atomic absorption spectrometry (AAS) [194] or, more recently, to inductively coupled plasma (ICP) spectrometry [195–197]. Whereas AAS effectively monitors a single element, ICP/MS can simultaneously detect a range of elements and isotopes [198]. Both have high sensitivity and can be very selective. As they normally are designed for a liquid inlet, interfacing poses few problems.

2.5 SEPARATION METHODS

The selection of a separation method depends on a number of factors, the purpose of the assay, the nature of the analyte of interest, any associated compounds from which it has to be separated, and the sample matrix, as well as the number of samples to be analyzed and the sensitivity required. In many cases, the best guide for a method is previously reported assays of the same or related analytes, but differences in stationary-phase materials mean that the conditions may need to be optimized or significantly altered to achieve the desired separation. An important factor in many assays is the need for robustness and reproducibility to satisfy the requirements of regulatory agencies.

The primary considerations are the nature and structure of the analyte. Its polarity, ionizability and size will largely determine the retention. Additionally, the presence or absence of a chromophore will govern detectability (Sec. 2.4). The primary choice for compounds with molecular weight <2000 is between normal- and reversed-phase separation methods and can be described by a decision tree (Fig. 2.7). For larger molecules, primarily biological and synthetic polymers, SEC should also be considered (Chap. 5).

The other components in a sample mixture and the sample solvent can also influence the method selection. Because normal-phase separations are primarily governed by polar–polar interactions, it is a suitable method for mixtures containing analytes with a range of polarities. But because water tends to deactivate silica surfaces, the sample solvent must be organic and relatively dry, making this an unsuitable method for many pharmaceutical samples. In contrast, in RPC an aqueous eluent is used, so that aqueous sample solvents cause no problems. However, organic sample solvents are strong eluents and can interfere with peak shapes. Reversed-phase methods separate on the basis of size and polarity (hydrophobicity), so they can usually resolve analytes with the same polar functionality but different alkyl groups, such as homologs, and compounds without functional groups. Much practical advice on the operation and practical problems in HPLC separations and systems can be found in recent monographs [199–201].

2.5.1 Normal-phase and supercritical-fluid separations

In normal-phase chromatography (NPC), the principal separation mechanism is a polar–polar interaction between the analytes and the stationary phase. As a consequence, very non-polar analytes, such as alkanes, aryl hydrocarbons, and ethers, which suffer very

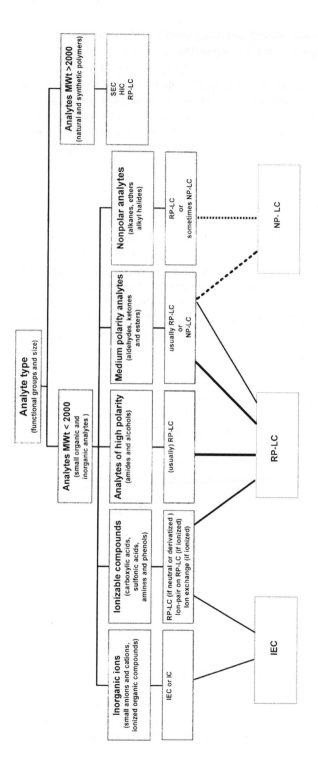

Fig. 2.7. Selection of a separation modes in HPLC according to the size, structure, and functional groups of an analyte.

little interaction with the stationary phase, are poorly retained and difficult to resolve. Very polar compounds, including amines, alcohols, amides, and ionizable analytes, are also difficult to separate as they interact strongly with silica columns and are difficult to elute. Amines are particularly problematic, as they can also suffer additional ion-exchange interactions with any acidic silanol groups. Normal-phase separations are therefore primarily useful for moderately polar analytes, such as esters, ketones, aldehydes, and some hydroxyl compounds.

As the main column material is silica, the principal variable in any method development is the mobile phase. Many potential eluents have been classified by their elution strength (Table 2.6) on a silica column [202] The selection of useful solvents is severely limited, because they must have no chromophore, as even a weak absorbance would interfere with the nearly universal UV spectroscopic detection. Normally, solvent UV absorbances of less than 1 are required (see Table 2.6, otherwise effectively all the light passing through the detector cell is absorbed and detection becomes difficult.

To increase selectivity, different combinations of solvents can be used to achieve the same overall elution strength. However, even small amounts of a strong component, such as an alcohol, can have a marked effect. On silica columns, the separations are usually carried out isocratically. The slow equilibration of the silica surface upon

TABLE 2.6

ELUTION STRENGTHS (SOLVENT POLARITY PARAMETERS, P') OF TYPICAL ORGANIC SOLVENTS USED IN THE MOBILE PHASE IN NORMAL-PHASE HPLC AND THEIR CUT-OFF WAVELENGTHS [202]

Solvent	P'	UV cut-off (nm)
Alkanes (hexane)	0.1	200
Toluene (UV – not used)	(2.4)	290
Isopropyl ethyl	2.4	220
Diethyl ether	2.8	218
Dichloromethane	3.1	235
Tetrahydrofuran	4.0	212
Chloroform	4.1	245
Ethyl acetate	4.4	256
Acetone	(5.1)	330
Ethanol	4.3	210
Dioxan	4.8	215
Methanol	5.1	205
Acetonitrile	5.8	190[*]
Water	10.2	<190

[*] High-purity grade.

changing the eluent means that gradient elution is not practical, because long re-equilibration periods would be needed after each run. Polar bonded phases, such as cyano-bonded silica, are better but are still best used isocratically. These restrictions have severely limited the application of NPC and they probably represent only a small fraction of the methods in current use. Many of the separations that are used are effectively adaptations of TLC methods or special cases, such as chiral separations in which non-polar solvents are needed to enhance the interaction between an analyte and the stationary phase (Sec. 2.3.5).

Supercritical-fluid chromatography, where the mobile phase, normally condensed carbon dioxide, is environmentally benign and has no background chromophore, can be regarded as a specialized form of normal-phase separation [203–205]. It seems particularly attractive, because the elution strength can be simply manipulated by altering the density by changing pressure and temperature. Although potentially very attractive, it has not gained the usage that was originally promised, largely because as a normal-phase method it is limited to relatively non-polar analytes and it is unsuitable for most pharmaceutical applications. Its principal application is now in chiral separations [108,206] and the fractionation of lipids and polymers. The ease of removing the solvent after chromatography has also made it attractive for preparative separations [207,208]. However, instrumental problems of working with a fluid under pressure meant that it was slow to be accepted. Apart from selected applications, it rarely offers advantages over RP methods. The main role of supercritical fluids in analytical chemistry is now in extraction, as part of sample preparation [209].

2.5.2 Reversed-phase separations

Reversed-phase chromatography (RPC) employs a non-polar bonded phase and usually an aqueous organic mobile phase. Separation is based on partition of the analyte between the mobile and stationary phases and is governed by polarity and hydrophobicity of the analytes. RPC can handle a wide range of polarities – from non-polar compounds, like PAHs, to polar pharmaceuticals. Equilibration is rapid and it is a very robust technique, ideal for quality control and regulated methods, and thus it has been adopted for the majority of HPLC separations.

The strength of the eluent is governed by the proportion of organic modifier, usually either methanol, acetonitrile, or THF (Table 2.7). Each solvent has advantages and disadvantages and all three are compatible with UV spectroscopic detection [210]. Because the interaction of each modifier with an analyte can be different, selectivity or relative retention changes for a mixture can often be obtained by comparing isoeluotropic mixtures (same elution strength combinations, for example, methanol/water 70:30, acetonitrile/water 50:50, or THF/water 35:65). Gradient elution can be used to speed up the elution of more highly retained analytes. With the bonded C_{18} alkylsilicone columns equilibration is rapid on changing the eluent. The normal preference for the organic component of the eluent is methanol for economy or acetonitrile for efficiency. THF is often avoided because of its high vapor pressure, which can cause degassing and vaporization problems. Other organic solvents (such as dioxan, ethanol, or propanol) have been tested and are occasionally employed. For most

TABLE 2.7

RELATIVE ELUTION STRENGTHS OF PRINCIPAL MOBILE-PHASE
COMPONENTS EMPLOYED IN REVERSED-PHASE CHROMATOGRAPHY
[210]

Solvent	Elution strength (S)	UV cut-off (nm)
H_2O	0.0	<200
MeOH	2.6	205
MeCN	3.1	230
		190 (far-UV grade)
THF	4.4	212

assays they do not offer significant selectivity differences and have higher viscosity and
back-pressure. The main reason for comparing the three principal solvents is that they
offer different retention selectivities, as they have different dipole, proton-accepting,
and proton-donating properties. Optimization methods (Sec. 1.5) usually interpolate
between the individual solvents to obtain the combination with the optimum selectivity
for a particular mixture.

Many different reversed phase column materials are available, but unless there is a
specific interaction, the selectivity differences between similar types of columns
(particularly with Type B silica columns) are usually less than the differences introduced
on changing the eluent solvent. The presence of basic analytes in a sample can cause
differences, because their retention will depend on the type of silica and degree of end-
capping. For non-polar molecules, the shape of PAHs can cause changes on different
phases according to the type of bonding and density of the C_{18} alkyl chains through a
shape-selectivity effect.

2.5.2.1 Ion-pair and buffered eluents

Conventional reversed-phase separations can be readily used for even highly polar
but neutral analytes. However, functional groups which can ionize, such as amines,
acids, or phenols play a major role in the pharmaceutical industry, because these
groups can increase water solubility and hence drug adsorption. Unfortunately, the
presence of an ionizable group causes problems, as the degree of ionization can change
during chromatography, altering the polarity of the analyte and causing poor peak
shapes and unreliable results. Two main approaches have been used to overcome these
problems.

Firstly, by controlling the pH of the eluent the analyte can be forced into either a fully
neutral (ion suppression) or ionized form. However, charged analytes are so water-soluble
that they are effectively unretained, even with neat water as the eluent, and cannot be

TABLE 2.8

BUFFERS USED IN HPLC TO CONTROL IONIZATION

Buffer	pKa	Effective range
Phosphate		
pKa1	2.1	1.1–3.1
pKa2	7.2	6.2–8.2
pKa3	12.3	11.2–13.3
Citrate		
pKa1	3.1	2.1–4.1
pKa2	4.7	3.7–5.7
pKa3	5.4	4.4–6.4
Formate		
pKa	3.8	2.8–4.8
Acetate		
pKa	4.8	3.8–5.8
Ammonia		
pKa	9.2	8.2–10.2

Typically used at concentrations of *ca.* 20–50 mM, buffers can control the pH accurately only within ±1 pH units of their pKa value. For LC/MS volatile buffers (usually ammonium salts) must be used.

retained or separated so that the neutral form is preferred. The buffers employed (Table 2.8) must be compatible with the type of detector used. The buffer must contain no chromophore absorbing in the UV range, and for LC/MS, must be volatile. However, the stability of silica-based stationary phases imposes the limitation of the pH 3–8 range for column stability, although some newer materials claim a greater stability. Alternatively, polymeric columns or PGC columns, which are more pH stable, can be employed. On most columns, ion suppression is therefore usually restricted to: carboxylic acids (tpically pKa = 4.75), phenols (pKa = 9.89), and aromatic amines (pKa = 4.63) (Table 2.1) but cannot be used to suppress the ionization of sulfonic acid groups (pKa = <0.7) or aliphatic amines (pKa = 10.77), remembering that the pH must be at least one unit above or below the pKa to achieve full control.

The second method is to employ ion-pair chromatography [211], in which a counter-ion of opposite charge to the ionized analyte is added to the mobile phase to create a neutral ion pair in solution (Table 2.9). This method can be applied to almost all ionizable analytes if the pH of the mobile phase is adjusted so that the analytes are fully charged. The retention of the ion pair depends only on the concentration of the ion-pair reagent and the degree of ionization (usually selected to be 100% for robustness) controlled by a buffer. A number of ion-pair reagents are available (Table 2.9). Typically, they contain an aliphatic chain, and reagents with longer chains give increased retention. Fluorinated alkyl chains are used to increase volatility for LC/MS [212].

TABLE 2.9

SELECTION OF COUNTER-IONS FOR ION-PAIR CHROMATOGRAPHY

Analyte	Counter-ion*	pH
Sulfonic acids (RSO_3^-)	NR_4^+	3–8
Carboxylic acids (RCO_2^-)	NR_4^+	5–8
Aliphatic and aromatic amines ($RNR'H^+$)	$RSO_3^-/RF-CO_2^-$	3–5
Quaternary ammonium salts (NR_4^+)	$RSO_3^-/RF-CO_2^-$	3–8

* Typical counter-ions—all aliphatic without chromophore: NR_4^+ Quaternary aliphatic amines; $(C_4H_9)_4N^+$ tetrabutyl ammonium salts TBA; $C_{16}H_{33}N^+(CH_3)_3$ cetrimide salts
RSO_3^- Aliphatic sulfonic acids: $CH_3SO_3^-$ to $C_{12}H_{25}SO_3^-$
$RF-CO_2^-$ Volatile fluorocarboxylic acids for LC–MS: $CF_3CO_2^-$ (trifluoroacetic acid, TFA), $C_3F_7CO_2^-$ (heptafluorobutanoic acid), and $C_6F_{13}CO_2^-$ (perfluoroheptanoic acid)

The ion-pair reagents must not contain a chromophore, so that when they are added to the mobile phase, they cause no interference with UV detection. The concentration of the ion-pair reagent is limited, because above a certain concentration (critical micelle concentration) they can form micelles having very different retention properties.

If neither ion pairing nor suppression can be used, the analyte can be derivatized to form a neutral species for separation (*e.g.*, by converting an amine into an amide or an acid into an ester) [160–163]. Although derivatization reactions were popular in the early days of HPLC, the technique is now rarely employed, because it requires an additional sample preparation stage, which can reduce quantitative accuracy. The only routine exceptions are cases where derivatization would also aid detection by the introduction of a chromophore, such as the conversion of amino acids to their *o*-phthaldialdehyde (OPA) derivatives. Inorganic ions whose ionization cannot be suppressed and will not form ion pairs are usually analyzed by ion-exchange chromatography [213] (Chap. 4).

2.5.2.2 Relationship of retention to structure

Because reversed-phase chromatography is a partition-based separation, it reflects other hydrophobicity-controlled distributions between polar and non-polar phases. The most important of these is the log P, octanol/water distribution, which is often used as a guide to the partition between body fluids and fatty tissues and hence the transport of drugs around the body and their activity. Extensive studies have related log P values to the chemical structure (size, functional groups and their positions) of drugs, using quantitative structure/activity relationships (QSAR) [214,215]. Similar relationships are found in RP-HPLC [216] and can be used as a guide to log P activity, being more convenient than the traditional "shake-flask" method [217,218].

This work is part of a larger series of investigations into the relationship between structure, functional groups, hydrophobicity, and retention [219–222]. Other studies, in particular by Abrahams and co-workers [223,224], have focused on the details of the effect of aromatic substituents and hydrogen bonding on solvatochromism. The hope is always that a full understanding of the factors influencing partitioning can be used to predict the retention of new compounds, thus facilitating method development.

2.5.3 Optimization methods

An important part of setting up a new HPLC method is to optimize the conditions that are being used, to achieve the best separation in a reasonable time. This is discussed in detail in Sec. 1.5. Here, it is necessary only to emphasize that one must also ensure that the method is robust and will not be altered by minor method changes [225]. The advent of computers has meant considerable progress in optimization methods [226,227], based on prediction, stepwise method improvement, and expert systems [228–230], or neural networks [231]. Many of these optimization methods are characterized by an empirical approach, which generally does not take into account the structure of the analytes as a parameter but alters the parameters of a separation to give the "best" resolution. These are usually based on three isoeluotropic starting points and interpolations between them, by using either a limited number of points or a mapping technique.

Optimization is based on the application of a separation function criterion to obtain maximum resolution (no overlaps) and maximum and minimum run times. The selection can also be based on interpolation between fixed points or by iterative methods [232–234]. This method can be quite successful and frequent selects suitable conditions after as few as seven trial separation conditions. It tends to favor binary or ternary eluent combinations. An alternative approach is to use computer interpolation between measured separations. This method (DryLab) tries to be more predictive [235] and includes the examination of the effects of solvent, pH, and temperature. Computer systems based on a widespread knowledge of analytes and functional group properties in different eluents and columns are the basis of other computational models, an example of which is ChromSword [236]. This approach has the advantage that it works in the same way as an experienced analyst, by taking the structure of the analytes and the effects of different conditions into account.

2.6 LIQUID CHROMATOGRAPHY IN SAMPLE PREPARATION

Although conventional gravity-fed columns are rarely used in analytical LC, the same partitioning principle is used in sample preparation in a number of different guises [237]. The most common is solid-phase extraction (SPE) [238–240], in which a small, disposable column is used to carry out the extraction of analytes from complex mixtures, such as biological or environmental samples, so that subsequent chromatographic separations are easier to perform. The column can be packed with either normal-phase, reversed-phase or ion-exchange materials so that they retain the analyte. The concentrated analyte is then eluted in a small volume of organic solvent. Alternatively, the column can allow the analyte to pass through and trap the rest of the sample. A variation on this method is solid-phase microextraction (SPME), in which the analyte is trapped on a coated fiber [241–244]. This has been used mainly for GC, but it can also be used for HPLC. The stir-bar extractor [245], in which the analyte is trapped on a coated magnetic stirrer, uses a similar concept. Again, the main application has been in GC, where the analyte can be removed by thermal desorption. The common feature of all these methods is an attempt to reduce the amount of organic solvent that is used and to enable the steps to be automated, from the initial sample to injection and beyond.

2.7 HIGH-TEMPERATURE SEPARATIONS

In the early years of HPLC, when separations were carried out with the column at ambient temperature, it was recognized that retention times were inversely dependent on temperature [246]. Ambient temperature variations have since been identified as a major source of retention variation. However, little use has been made of temperature as an operating variable. Increasing the temperature alters the thermodynamic distribution, and in almost all cases this results in a decrease of retention and, hence, faster separations. The increased temperature also reduces the mobile-phase viscosity and the lowered back-pressure enables higher flow-rates to be employed. A concurrent increase in diffusion rate is also claimed to improve the mass transfer within the column and should therefore also improve the efficiency of the separation. Confusingly there are also reports where an increase in temperature has resulted in a decrease in efficiency [247]. This is often attributed to a difference between the inlet temperature and that of the column. However, some studies have demonstrated that reducing the column inlet temperature can also actually increase the efficiency and that marked differences can occur between nominally similar instruments and columns [248].

There has been a recent upsurge of interest in high-temperature LC, in many cases with the eluent temperature above the boiling point of the solvent at atmospheric pressure [249,250]. Many of these separations, particularly on capillary or open-tubular columns, are particularly effective. Another approach has been to use water as the sole mobile phase. Its polarity decreases markedly as the temperature is raised, to the extent that it starts to mimic methanol/water mixtures in the range 150–200 °C [251]. This enables solvent-free separations to be carried out, and this has particular advantages in detection. LC/NMR can be carried out with D_2O as the eluent [252], or an aqueous eluent can be directed to a flame-ionizaton detector (FID) [253–255]. The main limitations are the stability of the stationary phases, although, PS-DVB, PGC, and zirconia-based columns can be used up to 200 °C.

2.8 MINIATURIZATION

Over the years, there has been an interest in the miniaturization of LC, partly by the use of narrower columns, as less solvent would be required, and there are claims of improved separations [256]. However, miniaturization causes other problems, as both the injection volume and detector volumes also have to be miniaturized to reduce band-broadening.

The reduction in the flow-cell dimensions causes sensitivity problem in UV detection as the path-length is reduced. In the 1980s there was some interest in microbore LC, but the practical instrumental problems were not balanced by savings. There is also some more recent interest, as LC/MS can provide sensitive detection with a limited flow-rate.

In recent years, the use of electrically driven separations, in which the mobile phase is moved by electro-osmosis with a planar front, has been developed as a result of interest in CE. This has enabled narrow, packed columns to be tested in capillary electrochromatography [257–259] (Chap. 7). This system also permits the use of much smaller particle sizes (down to 1 μm), as flow is no longer restrained by the back-pressure. These ideas are moving toward separation devices built on silicone or PS-DVB chips, potentially miniaturizing the HPLC process, but detection is still limiting (Chap. 11).

2.9 LARGE-SCALE SEPARATIONS

HPLC is gradually moving from being an analytical technique to also being a preparative method. Even though it is expensive to produce large columns and chromatographic systems, it is commercially viable for high-cost chemicals, such as some pharmaceuticals, flavors, perfume components, and natural products [260,261]. For much of the work, the process has been primarily one of scaling up the separations process, by using larger-diameter columns and higher flow-rates to maintain the same linear flow-rates [262–265]. Both normal- and reversed-phase separations have been carried out [266], and throughput could be increased by overloading the columns at the expense of efficiency. However, increased size brings problems of safety and solvent handling.

Most separations are carried out in the partition or distribution modes. With high sample loading, displacement chromatography can also be used, in which each component of the sample mixture is displaced by the next more highly retained component [267]. This can give very high throughputs, but the system is harder to set up and is not easily understood in terms of traditional separations methods. Peak-widths and separations between components depend on the amounts of each component rather than on retention distributions.

2.9.1 Simulated moving-bed separations

A major problem in scaling up is that chromatographic methods are a sequential or batch process and thus not compatible with continuous chemical and production processes. One approach that has been developed to overcome this problem is to alter the conventional column system, so that the stationary and mobile phase both effectively move through the system in opposite directions. Some analytes are then carried in one direction from the injection point (carried by the moving "stationary phase" and others in the opposite direction carried by the mobile phase). Continuous injection results in a splitting of the sample into two components.

Because it is experimentally very difficult to move stationary and mobile phases simultaneously, the simulated moving bed has been designed [268,269]. In this

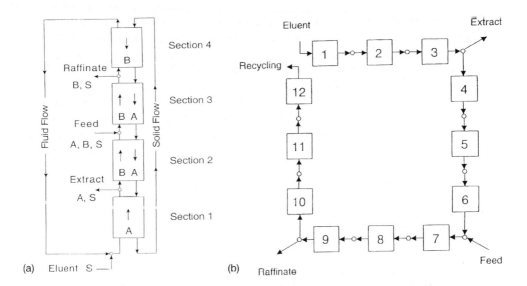

Fig. 2.8. Moving bed separations. (a) True moving bed separation in which the mobile and stationary phases move in opposite direction. (b) Simulated moving bed in which the eluent moves through a circle of columns and the inlet feed, product extraction and raffinate (unwanted components) points move in steps clockwise around the circle to mimic the movement of the stationary phase in the opposite direction to the mobile phase. (Reproduced from Ref. 272 with permission.)

system, a series of columns is arranged in a circle (Fig. 2.8). The stationary phase stays static but the mobile-phase input point, the sample input and extraction points move in steps around the circumference so that, effectively, the stationary phase is moved, relative to the injection point. Continuous operation can generate a high throughput, even from small column dimensions, and the system is ideal for simple mixtures, such as for the separation of a pair of enantiomers [270], and it can also be used in SFC [271,272].

REFERENCES

1   L.S. Ettre and K.I. Sakodynski, *Chromatographia*, 35 (1993) 223 and 329.
2   L.S. Ettre, *Chromatographia*, 42 (1996) 343.
3   Y. Ito and W.D. Conway (Eds.), *High-Speed Countercurrent Chromatography*, Wiley, New York, 1996.
4   A. Berthod and B. Billardello, *Adv. Chromatogr.*, 40 (2000) 503.
5   J.M. Menet and D. Thiebaut, *Countercurrent Chromatography*, Dekker, New York, 1999.
6   Gy. Kery, A. Turiak and P. Tetenyi, *J. Chromatogr.*, 446 (1988) 157.
7   A. Braithwaite and F.J. Smith, *Chromatographic Methods*, Blackie, London, 1996.

8   T. Hanai, *HPLC: A Practical Guide*, Royal Society of Chemistry, Cambridge, 1999.

9   E.D. Katz, *High Performance Liquid Chromatography: Principle and Methods in Biotechnology*, Wiley, Chichester, 1996.

10 V.R. Meyer, *Practical High-Performance Liquid Chromatography*, 3rd Edn. Wiley, Chichester, 1999.

11 I.W. Wainer, *High-performance Liquid Chromatography: Fundamental Principles and Practice*, Blackie, London, 1995.

12 R.P.W. Scott, *Techniques and Practice of Chromatography*, Dekker, New York, 1995.

13 M.C. MacMaster, *HPLC: A Practical User's Guide*, VCH, New York, 1994.

14 A. Weston and P.R. Brown, *HPLC and CE. Principles and Practice*, Academic, London, 1997.

15 E. Katz, R. Eksteen, R. Schoenmakers and N. Miller (Eds.), *Handbook of HPLC*, Dekker, New York, 1998.

16 T. Shibamoto, *Liquid Chromatography in Analysis*, Dekker, New York, 1994.

17 Y. Kazakevich and H.M. McNair, *Basic Liquid Chromatography*, (http://hplc.chem.shu.edu/new/hplc_book.html).

18 L.S. Ettre, *Pure Appl. Chem.*, 65 (1993) 819.

19 L.S. Ettre, *Chromatographia*, 38 (1994) 521.

20 J.A. García Domínguez, J.C. Díez-Masa and V.A. Davankov, *Pure Appl. Chem.*, 73 (2001) 969.

21 R.M. Smith and A. Marton, *Pure Appl. Chem.*, 66 (1994) 1739.

22 J.A. Jonsson, *Pure Appl. Chem.*, 68 (1996) 1591.

23 V.A. Davankov, *Pure Appl. Chem.*, 69 (1997) 1469.

24 R.M. Smith, *Pure Appl. Chem.*, 65 (1993) 2397.

25 L.S. Ettre, *J. Chromatogr.*, 535 (1990) 3.

26 C.W. Gehrke, R.L. Wixon and E. Bayer (Eds.), *Chromatography – a Century of Discovery 1900–2000. The Bridge to the Sciences/Technology*, Elsevier, Amsterdam, 2001.

27 S. Moore and W.H. Stein, *J. Biol. Chem.*, 211 (1954) 893.

28 K.J. Bombaugh, *Am. Lab.*, July (1969).

29 H. Felton, *J. Chromatogr. Sci.*, 7 (1969) 13.

30 P.C. Sadek, *The HPLC Solvent Guide*, 2nd Edn. Wiley, New York, 2002.

31 *Achieving Accuracy and Precision with Rheodyne Sample Injectors*, Technical Notes 5, Rheodyne, Cotati, 1983.

32 M. Szumski and B. Buszewski, *Crit. Rev. Anal. Chem.*, 32 (2002) 1.

33 J.W. Dolan, *J. Chromatogr. A*, 965 (2002) 195.

34 W.R. Melander, B.-K. Chen and Cs. Horvath, *J. Chromatogr.*, 185 (1979) 99.

35 U.D. Neue, *HPLC Columns, Theory, Technology, and Practice*, Wiley-VCH, New York, 1997.

36 R.P.W. Scott (Ed.), *Silica Gel and Bonded Phases, Their Production, Properties and Use in LC*, Wiley, Chichester, 1993.

37 J. Nawrocki and B. Buszewski, *J. Chromatogr.*, 449 (1988) 1.

38 G.B. Cox, *J. Chromatogr.*, 656 (1993) 353.

39 J. Nawrocki, *Chromatographia*, 31 (1991) 177.

40 J. Nawrocki, *Chromatographia*, 31 (1991) 193.

41 P. Dietrich, A. Kunath and B. Hoffmann, *Chromatographia*, 28 (1988) 428.

42 L. Jelinek, P. Dong, C. Rojaspazos, H. Taibi and E.S. Kovats, *Langmuir*, 8 (1992) 2152.

43 E.L. Weiser, A.W. Salotto, S.M. Flach and L.R. Snyder, *J. Chromatogr.*, 303 (1984) 1.

44 W.T. Cooper and P.L. Smith, *J. Chromatogr.*, 355 (1986) 57.

45 P.L. Smith and W.T. Cooper, *J. Chromatogr.*, 410 (1987) 249.

46 A. Houbenova, H.A. Claessens, J.W. de Haan, C.A. Cramers and K. Stulik, *J. Liq. Chromatogr.*, 17 (1994) 49.

47 R.D. Morrison and J.W. Dolan, *LC–GC Europe*, 13 (2000) 720.

48 J.J. Kirkland, J.L. Glajch and R.D. Farlee, *Anal. Chem.*, 61 (1989) 2.

49 U.D. Neue, Y.F. Cheng, Z. Lu, B.A. Alden, P.C. Iraneta, C.H. Phoebe and K. Van Tran, *Chromatographia*, 54 (2001) 169.

50 R.E. Majors and M. Przybyciel, *LC–GC North America*, 20 (2002) 584.

51 M. Turowski, T. Morimoto, K. Kimata, H. Monde, T. Ikegami, K. Hosoya and N. Tanaka, *J. Chromatogr. A*, 911 (2001) 177.

52 F.M. Yamamoto and S. Rokushika, *J. Chromatogr. A*, 898 (2000) 141.

53 M. de Smet and D.L. Massart, *Trends Anal. Chem.*, 8 (1989) 96.

54 Y.F. Cheng, T.H. Walter, Z.L. Lu, P. Iraneta, B.A. Alden, C. Gendreau, U.D. Neue, J.M. Grassi, J.L. Carmody, J.E. O'Gara and R.P. Fisk, *LC–GC North America*, 18 (2000) 1162.

55 H.K. Chepkwony, P. Dehouck, E. Roets and J. Hoogmartens, *Chromatographia*, 53 (2001) 159.

56 B. Law, R. Gill and A.C. Moffat, *J. Chromatogr.*, 301 (1984) 165.

57 R.M. Smith, T.G. Hurdley, R. Gill and M.D. Osselton, *J. Chromatogr.*, 398 (1987) 73.

58 R.J. Flanagan, E.J. Harvey and E.P. Spencer, *Forensic Sci. Int.*, 121 (2001) 97.

59 R.J. Flanagan, G.C.A. Storey, B.K. Bhrama and I. Jane, *J. Chromatogr.*, 247 (1982) 15.

60 N. Tanaka, H. Kobayashi, K. Nakanishi, H. Minakuchi and N. Ishizuka, *Anal. Chem.*, 73 (2001) 420A.

61 K. Cabrera, G. Wieland, D. Lubda, K. Nakanishi, N. Soga, H. Minakuchi and K.K. Unger, *Trends Anal. Chem.*, 17 (1998) 50.

62 K. Cabrera, D. Lubda, H.M. Eggenweiler, H. Minakuchi and K. Nakanishi, *J. High Res. Chromatogr.*, 23 (2000) 93.

63 I. Gusev, X. Huang and C. Horvath, *J. Chromatogr. A*, 855 (1999) 273.

64 M. Petro, F. Svec and J.M.J. Frechet, *J. Chromatogr. A*, 752 (1997) 59.

65 C.P. Jaroniec, M. Jaroniec and M. Kruk, *J. Chromatogr. A*, 797 (1998) 93.

66 J.E. Haky, S. Vemulapalli and L.F. Wieserman, *J. Chromatogr.*, 505 (1990) 307.

67 H. Billiet, C. Laurent and L. De Galan, *Trends Anal. Chem.*, 4 (1985) 100.

68 J.E. Haky and S. Vemulapelli, *J. Liq. Chromatogr.*, 13 (1990) 3111.

69 R.V. Arenas and J.P. Foley, *Analyst*, 119 (1994) 1303.

70 U. Trudinger, G. Muller and K.K. Unger, *J. Chromatogr.*, 535 (1990) 111.

71 A. Ellwanger, M.T. Matyska, K. Albert and J.J. Pesek, *Chromatographia*, 49 (1999) 424.

72 T.S. Kephart and P.K. Dasgupta, *Anal. Chim. Acta*, 414 (2000) 71.

73 C. Dunlap, C. McNeff, D. Stoll and P. Carr, *Anal. Chem.*, 73 (2001) 598A.

74 J. Nawrocki, M.P. Rigney, A. Mccormick and P.W. Carr, *J. Chromatogr. A*, 657 (1993) 229.

75 J.W. Li and P.W. Carr, *Anal. Chem.*, 68 (1996) 2857.

76 J.W. Li and P.W. Carr, *Anal. Chem.*, 69 (1997) 2193.

77 T.P. Weber, P.T. Jackson and P. Carr, *Anal. Chem.*, 67 (1995) 3042.

78 J.W. Li and P.W. Carr, *Anal. Chem.*, 69 (1997) 2202.

79 J.W. Li, Y. Hu and P.W. Carr, *Anal. Chem.*, 69 (1997) 3884.

80 S.M. Fields, C.Q. Ye, D.D. Zhang, B.R. Branch, X.J. Zhang and N. Okafo, *J. Chromatogr. A*, 913 (2001) 197.

81 Y. Mao and P.W. Carr, *Anal. Chem.*, 73 (2001) 1821.

82 L.L. Lloyd, *J. Chromatogr.*, 544 (1991) 210.

83 B. Gawdzik and J. Osypiuk, *Chromatographia*, 54 (2001) 595.

84 D.P. Lee, *J. Chromatogr., Sci.*, 20 (1982) 203.

85 J.V. Dawkins, L.L. Lloyd and F.P. Warner, *J. Chromatogr.*, 352 (1986) 157.

86   R.M. Smith and D.R. Garside, *J. Chromatogr.*, 407 (1988) 19.
87   N. Tanaka, T. Ebata, K. Hashizume, K. Hosoya and M. Araki, *J. Chromatogr.*, 475 (1989) 195.
88   J.R. Benson and D.J. Woo, *J. Chromatogr. Sci.*, 22 (1984) 386.
89   J.V. Dawkins, N.P. Gabbott, L.L. Lloyd, J.A. McConville and F.P. Warner, *J. Chromatogr.*, 452 (1988) 145.
90   M.T. Gilbert, J.H. Knox and B. Kaur, *Chromatographia*, 16 (1982) 138.
91   J.H. Knox, B. Kaur and G.R. Millward, *J. Chromatogr.*, 352 (1986).
92   J.H. Knox and P. Ross, *Adv. Chromatogr.*, 37 (1997) 73.
93   J.H. Knox and P. Ross, *Adv. Chromatogr.*, 37 (1997) 121.
94   J. Kriz, E. Adamcova, J.H. Knox and J. Hora, *J. Chromatogr. A*, 663 (1994) 151.
95   M.C. Hennion, V. Coquart, S. Gienu and C. Sella, *J. Chromatogr. A*, 712 (1995) 287.
96   A. Salvador, B. Herbreteau, M. Dreux, A. Karlsson and O. Gyllenhaal, *J. Chromatogr. A*, 929 (2001) 101.
97   I.W. Wainer, *Trends Anal. Chem.*, 6 (1987) 125.
98   I.W. Wainer, *LC–GC North America*, 7 (1989) 378.
99   A.M. Krstulovic (Ed.), *Chiral Separations by HPLC*, Ellis Horwood, Chichester, 1989.
100 S. Allenmark, *Chromatographic Enantioseparations*, Ellis Horwood, Chichester, 1991.
101 S. Ahuja, *Chiral Separations by Chromatography*, OUP-American Chemical Society, Washington, 2000.
102 T.E. Beesley and R.P.W. Scott, *Chiral Chromatography*, Wiley, Chichester, 1999.
103 B. Chankvetadze, *Chiral Separations*, Elsevier, Amsterdam, 2001.
104 D.W. Armstrong, *Anal. Chem.*, 59 (1987) 84A.
105 A.C. Mehta, *J. Chromatogr.*, 426 (1988) 1.
106 A.M. Krstulovic, *J. Pharm. Biomed. Anal.*, 6 (1988) 641.
107 G. Subramanian, *Chiral Separation Techniques. A Practical Approach*, 2nd Edn. Wiley-VCH, Weinheim, 2001.
108 R.M. Smith, in C.L. Berger and K. Anton (Eds.), *Supercritical Fluid Chromatography with Packed Columns. Techniques and Applications*, Dekker, New York, 1998, p. 223.
109 G. Terfloth, *J. Chromatogr. A*, 907 (2001) 310.
110 P. Macaudiere, M. Caude, R. Rosset and A. Tambute, *J. Chromatogr. Sci.*, 27 (1989) 583.
111 W.H. Pirkle, J.M. Finn, J.L. Schreiner and B.C. Hamper, *J. Am. Chem. Soc.*, 103 (1981) 3964.
112 W.H. Pirkle and T.C. Pochapsky, *Chem. Rev.*, 89 (1989) 347.
113 W.H. Pirkle, T.C. Pochapsky, G.S. Mahler and R.E. Field, *J. Chromatogr.*, 348 (1985) 89.
114 W.H. Pirkle, M.H. Hyun and B. Bank, *J. Chromatogr.*, 316 (1984) 585.
115 W.H. Pirkle, *ACS Symp. Ser.*, 297 (1986) 101.
116 Y. Okamoto, R. Aburatani, Y. Kaida, K. Hatada, N. Inotsume and M. Nakano, *Chirality*, 1 (1989) 239.
117 E. Yashima, C. Yamamoto and Y. Okamoto, *Synlett*, (1998) 344.
118 Y. Okamoto and E. Yashima, *Angew. Chem. Int. Ed.*, 37 (1998) 1021.
119 Y. Kaida and Y. Okamoto, *Bull. Chem. Soc. Japan*, 65 (1992) 2286.
120 A. Medvedovici, P. Sandra, L. Toribio and F. David, *J. Chromatogr. A*, 785 (1997) 159.
121 T. Cserhati and E. Forgacs, *Cyclodextrins in Chromatography*, Royal Society of Chemistry, Cambridge, 2003.
122 T. Hargitai and Y. Okamoto, *J. Liq. Chromatogr.*, 16 (1993) 843.
123 Y. Kawaguchi, M. Tanaka, M. Nakae, K. Funazo and T. Shono, *Anal. Chem.*, 55 (1983) 1852.
124 V.A. Davankov, *Chromatographia*, 27 (1989) 475.
125 J. Hermansson, *J. Chromatogr.*, 298 (1984) 67.
126 S. Allenmark, *J. Liq. Chromatogr.*, 9 (1986) 425.
127 K.M. Kirkland, K.L. Neilson and D.A. McCombs, *J. Chromatogr.*, 545 (1991) 43.

128 T.J. Ward and A.B. Farris, *J. Chromatogr. A*, 906 (2001) 73.

129 V.T. Remcho and Z.J. Tan, *Anal. Chem.*, (1999) 248A.

130 L.I. Andersson, *J. Chromatogr. B*, 739 (2000) 163.

131 F. Lanza and B. Sellergren, *Adv. Chromatogr.*, 41 (2001) 137.

132 H. Engelhardt and M. Jungheim, *Chromatographia*, 29 (1990) 59.

133 H. Engelhardt, H. Low and W. Gotzinger, *J. Chromatogr.*, 544 (1992) 371.

134 H.A. Claessens, M.A.van. Straten, C.A. Cramers, M. Jezierska and B. Buszewski, *J. Chromatogr. A*, 826 (1998) 135.

135 E. Cruz, M.R. Euerby, C.M. Johnson and C.A. Hackett, *Chromatographia*, 44 (1997) 151.

136 L.C. Sander, *J. Chromatogr. Sci.*, 26 (1988) 380.

137 M.R. Euerby, C.M. Johnson, I.D. Rushin and D.A.S. Sakunthala Teenekoon, *J. Chromatogr. A*, 705 (1995) 219.

138 M.R. Euerby, C.M. Johnson, I.D. Rushin and D.A.S. Sakunthala Teenekoon, *J. Chromatogr. A*, (1995) 229.

139 L. Rohrschneider, *J. Sep. Sci.*, 24 (2001) 3.

140 T. Ivanyi, Y. VanderHeyden, D. Visky, P. Baten, J. DeBeer, I. Lazar, D.L. Massart, E. Roets and J. Hoogmartens, *J. Chromatogr. A*, 954 (2002) 99.

141 S.A. Wise, W.J. Bonnett, F.R. Guenther and W.E. May, *J. Chromatogr. Sci.*, 19 (1981) 457.

142 K. Jinno, T. Nagoshi, N. Tanaka, M. Okamoto, J.C. Fetzer and W.R. Biggs, *J. Chromatogr.*, 392 (1987) 75.

143 H. Engelhardt and T. Lobert, *Anal. Chem.*, 71 (1999) 1885.

144 D.A. Barrett, V.A. Brwon, M.C. Davies and P.N. Shaw, *Anal. Chem.*, 68 (1996) 2170.

145 D.V. McCalley, *J. Chromatogr. A*, 769 (1997) 169.

146 U.D. Neue, E. Serowik, P. Iraneta, B.A. Alden and T.H. Walter, *J. Chromatogr. A*, 849 (1999) 87.

147 U.D. Neue, B.A. Alden and T.H. Walter, *J. Chromatogr. A*, 849 (1999) 101.

148 K. Albert and E. Bayer, *J. Chromatogr.*, 544 (1991) 345.

149 M.J.J. Hetem, *A Fundamental Study of Chemically Modified Silica Surfaces in Chromatography*, Hüthig, Heidelberg, 1993.

150 M.J.J. Hetem, J.W. de Haan, H.A. Claessens, L.J.M. van de Ven, C.A. Cramers and J.N. Kinkel, *Anal. Chem.*, 62 (1990) 2288.

151 N. Dyson, *Chromatographic Integration Methods*, 2nd Edn. Royal Society of Chemistry, Cambridge, 1998.

152 T.E. Beesley, B. Buglio and R.P.W. Scott, *Quantitative Chromatographic Analysis*, Dekker, New York, 2000.

153 A. Felinger, *Data Analysis and Signal Processing in Chromatography*, Elsevier, Amsterdam, 1998.

154 G. Patonay, *HPLC Detection. Newer Methods*, VCH, New York, 1992.

155 D. Parriott, *A Practical Guide to HPLC Detection*, Academic, San Diego, 1993.

156 R.P.W. Scott, *Chromatographic Detectors*, Dekker, New York, 1996.

157 L. Huber and S.A. George, *Diode Array Detection in HPLC*, Dekker, New York, 1993.

158 P.W. Carr, *J. Chromatogr.*, 316 (1984) 211.

159 M.B. Smalley and L.B. McGown, *Adv. Chromatogr.*, 37 (1997) 29.

160 H. Lingeman and W.J.M. Underberg, *Detection-Oriented Derivatization Techniques in Liquid Chromatography*, Dekker, New York, 1990.

161 K. Blau and J.M. Halket (Eds.), *Handbook of Derivatives for Chromatography*, 2nd Edn. Wiley, Chichester, 1993.

162 G. Lunn and L.C. Hellwig, *Handbook of Derivatisation Reactions for HPLC*, Wiley, New York, 1998.

163   T. Toyo'oka, *Modern Derivatization Methods for Separation Science*, Wiley, New York, 1998.

164   J.A. Koropchak, L.E. Magnusson, M. Heybroek, S. Sadain, X.H. Yang and M.P. Anisimov, *Adv. Chromatogr.*, 40 (2000) 275.

165   M. Dreux and M. Lafosse, *Analusis*, 20 (1992) 587.

166   C. Henry, *Anal. Chem.*, 69 (1997) 561A.

167   A. Criado, S. Cardenas, M. Gallego and M. Valcarcel, *Anal. Chim. Acta*, 435 (2001) 281.

168   A. Stolywo, H. Colin and G. Guiochon, *Anal. Chem.*, 57 (1985) 1345.

169   M. Dreux and M. Lafosse, *LC–GC Int.*, 10 (1997) 382.

170   K. Petritis, I. Gillaizeau, C. Elfakir, M. Dreux, A. Petit, N. Bongibault and W. Luijten, *J. Sep. Sci.*, 25 (2002) 593.

171   A.M. Krstulovic, M. Ante, H. Colin and G.A. Guiochon, *Adv. Chromatogr.*, 24 (1984) 83.

172   I.N. Acworth, M. Naoi, H. Parvez and S. Parvez (Eds.), *Coulometric Electrode Arrays Detection for HPLC*, VSP, Utrecht, 1997.

173   P.T. Kissinger, *J. Pharm. Biomed. Anal.*, 14 (1996) 871.

174   S.G. Weber and J.T. Long, *Anal. Chem.*, 60 (1988) 903A.

175   J.B.F. Lloyd, *J. Chromatogr.*, 330 (1985) 121.

176   W.R. LaCourse, *Pulsed Electrochemical Detection in High-Performance Liquid Chromatography*, Wiley, New York, 1997.

177   J.G. Chen, S.J. Woltman and S.G. Weber, *Adv. Chromatogr.*, 36 (1996) 273.

178   W.R. LaCourse and C.O. Dasenbrock, *Adv. Chromatogr.*, 38 (1998) 189.

179   H. Small, *J. Chromatogr.*, 546 (1991) 3.

180   W. Boehme, G. Wagner, U. Oehme and U. Priesnitz, *Anal. Chem.*, 54 (1982) 709.

181   D.R. Bobbitt and E.S. Yeung, *Anal. Chem.*, 56 (1984) 1577.

182   J. Zukowski, Y.B. Tang, A. Berthod and D.W. Armstrong, *Anal. Chim. Acta*, 258 (1992) 83.

183   A.R. Reich, H. Parvez, S. Lucas-Reich and S. Parvez (Eds.), *Flow Through Radioactive Detection in HPLC*, VSP, Utrecht, 1988.

184   I.D. Wilson, *J. Chromatogr. A*, 892 (2000) 315.

185   D. Louden, A. Handley, S. Taylor, I. Sinclair, E. Lenz and I.D. Wilson, *Analyst*, 126 (2001) 1625.

186   W.M.A. Niessen, *J. Chromatogr. A*, 794 (1998) 407.

187   W. Niessen, *Liquid Chromatography–Mass Spectrometry*, 2nd Edn. Dekker, New York, 1998.

188   C.K. Lim and G. Lord, *Biol. Pharmacol. Bull.*, 25 (2002) 547.

189   J. Smeraglia, S.F. Baldrey and D. Watson, *Chromatographia*, 55 (2002) S95.

190   M.D. Nelson and J.W. Dolan, *LC–GC Europe*, 15 (2002) 73.

191   J.C. Lindon, J.K. Nicholson and I.D. Wilson, *Adv. Chromatog.*, 36 (1996) 315.

192   K. Albert, *On-line LC-NMR and Related Techniques*, Wiley, New York, 2002.

193   G.W. Somsen, C. Gjooijer, N.H. Velthorst and U.A.Th. Brinkmann, *J. Chromatogr. A*, 811 (1998) 1.

194   L. Ebdon, S. Hill and R.W. Ward, *Analyst*, 112 (1987) 1.

195   M.W. Raynor, G.D. Dawson, M. Balcerzak, W.G. Pretorius and L. Ebdon, *J. Anal. Atom. Spec.*, 12 (1997) 1057.

196   G.K. Zoorob, J.W. McKiernan and J.A. Caruso, *Mikrochim. Acta*, 128 (1998) 145.

197   S.J. Hill, M.J. Bloxham and P.J. Worsfold, *J. Anal. Atom. Spec.*, 8 (1993) 499.

198   P.C. Uden, *J. Chromatogr. A*, 703 (1995) 393.

199   S. Kromidas, *Practical Problem Solving in HPLC*, Wiley-VCH, Weinheim, 2000.

200   P.C. Sadek, *Troubleshooting HPLC Systems: A Bench Manual*, Wiley-VCH, Weinheim, 1999.

201   V.R. Meyers, *Pitfalls and Errors of HPLC in Pictures*, Wiley-VCH, Weinheim, 1997.

202   L.R. Snyder and J.J. Kirkland, *Introduction to Modern Liquid Chromatography*, 2nd Edn. Wiley, New York, 1979.
203   R.M. Smith, *Supercritical Fluid Chromatography*, Royal Society of Chemistry, London, 1988.
204   R.M. Smith, *J. Chromatogr. A*, 856 (1999) 83.
205   K. Anton and C. Berger (Eds.), *Supercritical Fluid Chromatography with Packed Columns*, Dekker, New York, 1997.
206   G. Terfloth, *J. Chromatogr. A*, 907 (2001) 301.
207   C. Berger and M. Perrut, *J. Chromatogr.*, 505 (1990) 37.
208   K.D. Bartle, C.D. Bevan, A.A. Clifford, S.A. Jafar, N. Malak and M.S. Verrall, *J. Chromatogr. A*, 697 (1995) 579.
209   M.D. Luque de Castro, M. Valcarcel and M.T. Tena, *Analytical Supercritical Fluid Extraction*, Springer, Berlin, 1994.
210   J.L. Glajch and J.J. Kirkland, *Anal. Chem.*, 54 (1982) 2593.
211   M.T.V. Hearn, *Ion-Pair Chromatography*, Dekker, New York, 1985.
212   S.A. Gustavsson, J. Samskog, K.E. Markides and B. Langstrom, *J. Chromatogr. A*, 937 (2001) 41.
213   J.S. Fritz and D.T. Gjerde, *Ion Chromatography*, 3rd Edn. Wiley-VCH, New York, 2000.
214   C. Hansch and A. Leo, *Substituent Constants for Correlation Analysis in Chemistry and Biology*, Wiley, New York, 1987.
215   A. Leo, *J. Chem. Soc., Perkin Trans. II*, (1983) 825.
216   R. Kaliszan, *Quantitative Structure Retention Relationships*, Wiley, Chichester, 1987.
217   C.M. Du, K. Valko, C. Bevan, D. Reynolds and M.H. Abraham, *Anal. Chem.*, 70 (1998) 4228.
218   K. Valko, C. Bevan and D. Reynolds, *Anal. Chem.*, 69 (1997) 2022.
219   R.M. Smith, *J. Chromatogr. A*, 656 (1993) 381.
220   R.M. Smith (Ed.), *Retention and Selectivity in Liquid Chromatography, Prediction Standardisation and Phase Comparisons*, Elsevier, Amsterdam, 1995.
221   K. Jinno (Ed.), *Separations Based on Molecular Recognition*, Wiley, New York, 1997.
222   E. Forgacs and T. Cserhati, *Molecular Basis of Chromatographic Separation*, CRC Press, Boca Raton, 1997.
223   M.H. Abraham, *Chem. Soc. Rev.*, 22 (1993) 73.
224   C. Sadek, P.W. Carr, R.M. Doherty, M.J. Kamlet, R.W. Taft and M.H. Abraham, *Anal. Chem.*, 57 (1985) 2971.
225   L.R. Snyder, J.J. Kirkland and J.L. Glajch, *Practical HPLC Method Development*, 2nd Edn. Wiley, New York, 1997.
226   P.J. Schoenmakers, *Optimisation of Chromatographic Selectivity*, Elsevier, Amsterdam, 1986.
227   H.A.H. Billiet and G. Rippel, *Adv. Chromatogr.*, 39 (1998) 263.
228   T. Hamoir and D.L. Massart, *Adv. Chromatogr.*, 33 (1993) 97.
229   M. Peris, *Crit. Rev. Anal. Chem.*, 26 (1996) 219.
230   H.A.H. Billiet and G. Rippel, *Adv. Chromatogr.*, 39 (1998) 263.
231   J.R.M. Smits, W.J. Melssen, G.J. Daalmans and G. Kateman, *Comput. Chem.*, 18 (1994) 157.
232   A. Drouen, H. Billiet, P. Schoenmakers and L. De Galen, *Chromatographia*, 16 (1982) 48.
233   J. Strasters, in E. Katz, R. Eksteen, P. Schoenmakers and N. Miller (Eds.), *Handbook of HPLC*, Dekker, New York, 1998, p. 233.
234   S. Ahuja, *Selectivity and Detectability Optimizations in HPLC*, Wiley, New York, 1989.
235   J.W. Dolan, L.R. Snyder, N.M. Djordjevic, D.W. Hill, D.L. Saunders, L. Van Heukelem and T.J. Waeghe, *J. Chromatogr. A*, 803 (1998) 1.
236   S.V. Galushko, A.A. Kamenchuk and G.L. Pit, *J. Chromatogr. A*, 660 (1994) 47.
237   R.M. Smith, *J. Chromatogr. A*, 1000 (2003) 3.

238   N.J.K. Simpson, *Solid-Phase Extraction Principles, Techniques and Applications*, Dekker, New York, 2000.

239   E.M. Thurman and M.S. Mills, *Solid Phase Extraction. Principles and Practice*, Wiley, New York, 1998.

240   J.S. Fritz, *Analytical Solid Phase Extraction*, Wiley, New York, 1999.

241   J. Pawlisyn, *Solid State Micro Extraction. Theory and Practice*, Wiley-VCH, New York, 1997.

242   J. Pawliszn, *Applications of Solid Phase Microextraction*, Royal Society of Chemistry, Cambridge, 1999.

243   B. Zygmunt, A. Jastrizebska and J. Namiesnik, *Crit. Rev. Anal. Chem.*, 31 (2001) 1.

244   S.A.S. Wercinski, *Solid Phase Microextraction. A Practical Guide*, Dekker, New York, 1999.

245   E. Baltussen, P. Sandra, F. David and C.J. Cramers, *J. Microcol. Sep.*, 11 (1999) 737.

246   W.R. Melander, B.-K. Chen and Cs. Horvath, *J. Chromatogr.*, 185 (1979) 99.

247   J.D. Thompson, J.S. Brown and P.W. Carr, *Anal. Chem.*, 73 (2001) 3340.

248   R.G. Wolcott, J.W. Dolan, L.R. Snyder, S.R. Bakalyar, M.A. Arnold and J.A. Nichols, *J. Chromatogr. A*, 869 (2000) 211.

249   J.W. Dolan, *J. Chromatogr. A*, 965 (2002) 195.

250   T. Greibrokk, *Anal. Chem.*, (2002) 374A.

251   R.M. Smith and R.J. Burgess, *J. Chromatogr. A*, 785 (1997) 49.

252   R.M. Smith, O. Chienthavorn, I.D. Wilson, B. Wright and S.D. Taylor, *Anal. Chem.*, 71 (1999) 4493.

253   B.A. Ingelse, H.G. Janssen and C.A. Cramers, *J. High Resolut. Chromatogr.*, 21 (1998) 613.

254   D.J. Miller and S.B. Hawthorne, *Anal. Chem.*, 69 (1997) 623.

255   E.W.J. Hooijschuur, C.E. Kientz and U.A.T. Brinkman, *J. High Resolut. Chromatogr.*, 23 (2000) 309.

256   T.M. Zimina, R.M. Smith, P. Myers and B.W. King, *Chromatographia*, 40 (1995) 662.

257   M.G. Cikalo, K.D. Bartle, M.M. Robson, P. Myers and M.R. Euerby, *Analyst*, 123 (1998) R87.

258   Z. Deyl and F. Svec, *Capillary Electrochromatography*, Elsevier, Amsterdam, 2001.

259   K.D. Bartle and P. Myers, *Capillary Electrochromatography*, Royal Society of Chemistry, Cambridge, 2001.

260   K. Hostettmann, A. Marston and M. Hostettmann, *Preparative Chromatography Techniques: Applications in Natural Product Isolation*, 2nd Edn. Springer, Berlin, 1998.

261   G. Guiochon, *J. Chromatogr. A*, 965 (2002) 129.

262   G. Subramanian, *Process Scale Liquid Chromatography*, VCH, Weinheim, 1995.

263   K. Jones, *Chromatographia*, 25 (1988) 437.

264   A.S. Rathore and A. Velayudhan, *Scale-up and Optimisation in Preparative Chromatography*, Dekker, New York, 2002.

265   G. Ganetsos and P.E. Barker, *Preparative and Production Scale Chromatography*, Dekker, New York, 1992.

266   K.K. Unger and R. Janzen, *J. Chromatogr.*, 373 (1987) 227.

267   Y. Qi and J.X. Huang, *Prog. Chem.*, 13 (2001) 294.

268   S.M. Lai and R.R. Loh, *J. Liq. Chromatogr.*, 25 (2002) 345.

269   G.M. Zhong and G. Guiochon, *Adv. Chromatogr.*, 39 (1998) 351.

270   S.M. Lai and R.R. Loh, *J. Liq. Chromatogr.*, 25 (2002) 345.

271   O. Di Giovanni, M. Mazzotti, M. Morbidelli, F. Denet, W. Hauck and R.M. Nicoud, *J. Chromatogr. A*, 919 (2001) 1.

272   M. Mazzotti, G. Storti and M. Morbidelli, *J. Chromatogr. A*, 786 (1997) 309.

Erich Heftmann (Editor)
Chromatography, 6th edition
Journal of Chromatography Library, Vol. 69A
© 2004 Elsevier B.V. All rights reserved

Chapter 3

Affinity chromatography

FRIEDRICH BIRGER ANSPACH

CONTENTS

3.1   Introduction 139
3.2   Chromatographic operations 141
3.3   Affinity interactions at solid interfaces 142
    3.3.1   Immobilization of affinity ligands 142
    3.3.2   Specific *vs.* non-specific interactions 143
    3.3.3   Mono-specific ligands 144
    3.3.4   Group-specific ligands 145
    3.3.5   Affinity tags 145
    3.3.6   Size of the affinity ligand 149
    3.3.7   Thermodynamics and kinetics 150
    3.3.8   Matrix effects 151
3.4   Affinity ligands 153
    3.4.1   Biochemical interactions 153
        3.4.1.1   Protein A 153
        3.4.1.2   Heparin 154
        3.4.1.3   Lectins 154
    3.4.2   Immuno-affinity ligands 155
    3.4.3   Synthetic ligands 155
        3.4.3.1   Dye ligands 155
        3.4.3.2   Phenylboronic acid 157
        3.4.3.3   Immobilized metal ions 157
    3.4.4   Combinatorial libraries 159
    3.4.5   Purification schemes 160
3.5   Summary 160
References 161

3.1 INTRODUCTION

In the majority of chromatographic separations, the overall retention and selectivity are primarily determined by a single type of interaction or property of the system, which is composed of the solute and the mobile, and stationary phases. Generally, these interactions are not very selective stereochemically or topologically. Since any given protein in a cell extract is not unique with respect to one of these chromatographic interactions, obtaining material of high purity requires that a number of different chromatographic steps be performed. However, each protein is unique with regard to the surface distribution of charges and hydrophobic and hydrophilic amino acid residues. Making use of the types of binding and recognition processes provided by nature, an adsorbent can be created which interacts in a complementary manner with these features and will selectively adsorb this protein from a cell extract. Such biological recognition processes include antigen/antibody recognition, hormone/receptor binding, enzyme/substrate interactions, and the various interactions involved in nucleic acid transcription. In nature, many mechanisms and structures have evolved which permit the discrimination of extraordinarily subtle topological differences between molecules. A minor change in chemical structure can have a profound effect on the binding constant. In these systems, selectivity relies on highly specific biological recognition, where multi-point interactions between the partners takes place and steric constraints will prevent other proteins with similarly oriented electrostatic and hydrophobic moieties from binding; this is often described as the "lock-and-key principle". Since each of the various interactions can be ionic, hydrogen-bonding, Van der Waals, or solvophobic in nature, the net effect is usually multimodal in terms of the nature of the binding mechanism.

The idea of exploiting nature's own selectivity to isolate a specific material from a complex biological mixture by "affinity chromatography", in contrast to generating columns with thousands of theoretical plates, as is done in HPLC, goes back to Cuatrecasas *et al.* [1]. They described a one-step isolation of hydrolases by means of adsorbents with immobilized enzyme inhibitors [2]. However, some publications on the successful application of specific adsorbents (mainly antibodies and antigens) for the one-step isolation of biologically active proteins appeared some years earlier [3,4]. Today, it is recognized that the development of affinity chromatography was a major advance in biochemical methodology. What was for a long time understood by the term "affinity chromatography" was retention based on this naturally occurring recognition process, being a shorthand for "bioselective affinity retention". This strong context was undermined in the last 15 to 20 years, as other modes of chromatography that involve the formation of complexes or covalent bonds between an immobilized ligand and a solute were often included under this term. Although there is much to be said about it, this chapter will concentrate on bioselective and group-specific affinity chromatography. Both are designated as a type of adsorption chromatography and cannot – practically – be operated under isocratic conditions. Since it is clearly impossible to cover all known affinity ligands and affinity techniques, only general issues are covered by means of some representative examples. Developments in the field of genetically engineered affinity tags as well as synthetic ligands from various libraries will also be considered. A collection of reviews on affinity chromatography can be found in Ref. 5.

3.2 CHROMATOGRAPHIC OPERATIONS

The immobilization on a solid phase of an active ligand that will recognize, selectively retain, and thereby ideally isolate the product of interest from all other components of the sample is an absolute necessity in affinity chromatography. Thus, retention is based on adsorption, and all unwanted components should freely pass through the column. Then, the adsorbed product should be released as a sharp peak – free of all contaminants – by a specific change of the mobile phase. Owing to the high association constants, affinity chromatography cannot be carried out by isocratic elution, except in the case of weak affinity ligands [6]. Instead, some type of step or gradient elution is commonly employed. The operational steps are as follows:

1. *Loading (adsorption) phase*: The sample, which contains one member of a complementary pair of interacting partners (*e.g.*, the antigen of an antibody or the receptor of a hormone, which are immobilized on the column support), is loaded onto the column in a buffer which promotes adsorption of the target molecule (the binding buffer). A solute that "recognizes" the immobilized affinity ligand is reversibly adsorbed on the stationary phase. Additional interactions of other solutes with the affinity ligand or the chromatographic support, often called "non-specific" interactions, may contribute to this binding or to "non-specific" adsorption of some of those solutes.

2. *Washing phase*: Non-binding or weakly bound species are washed off the column with either the binding buffer or another suitable, weak mobile phase, which promotes desorption of only the weakly bound species.

3. *Elution (desorption) phase*: Adsorbed species are eluted from the column by either completely or gradually replacing the binding buffer with the eluting buffer. To this end, a component can be added to the buffer which specifically competes for the ligand or product-binding sites. Alternatively, the ionic strength or the pH can be changed, or an organic solvent or chaotropic agent (urea, guanidine hydrochloride) can be added so as to decrease the strength of the association. The purpose of an elution buffer is to dissociate the various bonds that make up ligand/solute interactions and to return the target protein to the mobile phase in active form. For very high affinities ($K_A > 10^8\, M^{-1}$), the elution buffer may need to be very specific, unless a specific competitor is available. Prevention of denaturation is a priority [7], but this is not always possible [8] and a loss of activity of the released protein often occurs [9]. The purification of oligomeric proteins may present special problems. If non-specific elution conditions are used to disrupt the complex between a multi-unit protein and its affinity sorbent, the concomitant dissociation of its subunits may result in its inactivation. Such problems explain why affinity chromatography on immobilized antibodies is never used for the purification of oligomeric proteins.

4. *Regeneration phase*: Before the next sample is applied, the column must be re-equilibrated with the binding buffer. When a competing component that binds to the ligand was added during elution, several steps with various eluent strengths may be necessary, and re-equilibration may take considerable time.

3.3 AFFINITY INTERACTIONS AT SOLID INTERFACES

3.3.1 Immobilization of affinity ligands

An affinity sorbent is produced by chemically coupling molecules – the so-called affinity ligands – to an inert solid support. Proteins possessing complementary binding sites for the affinity ligand will then be adsorbed onto the affinity sorbent. A major requirement in this immobilization process is that there be no change in either the biological recognition or the interaction strength. After approaching each other in solution, target proteins and ligands may freely rotate to align their complementary binding sites and to establish the affinity complex. This "free" affinity system is usually characterized in *in vivo* or *in vitro* experiments by its equilibrium constant and stability in different environments. The affinity constant, K_A, is the reciprocal of K_D, the dissociation constant.

Immobilization requires covalent attachment *via* at least one functional group of the ligand. Ideally, this does not affect the properties of the ligand and causes its binding site to be unconditionally accessible to the target protein. In practice, various scenarios are possible during immobilization, which undermine this concept. Thinking of a plain, rigid surface, we may visualize:

(a) Spatial orientation of the immobilized ligand, which is statistical, with its binding site exposed in any direction, including the support surface.

(b) The functional group employed for covalent fixation is part of the binding site, and this causes partial or complete loss of recognition (Fig. 3.1).

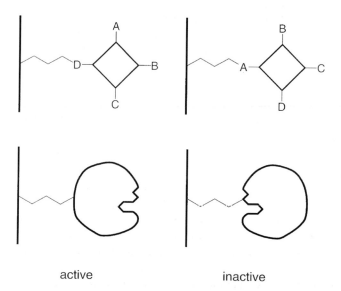

active   inactive

Fig. 3.1. Immobilization effects: Covalent fixation *via* an important functional group of a small ligand (A on top) or at the binding site of a large ligands (bottom) may lead to a non-functional ligand.

(c) Immobilization causes partial or complete loss of biological activity of the ligand due to denaturation.

(d) The density of immobilized ligands varies from batch to batch and may even vary in one batch at different sites of the support surface.

Any of these events may combine, so that the percentage of accessible binding sites may decrease with increasing ligand density, simply due to reduced accessibility at higher ligand densities of those binding sites, oriented more toward the support surface than toward the mobile phase; the immobilized ligand is no longer able to rotate. This is particularly noticeable with large affinity ligands. Small affinity ligands are prone to a loss of binding strength, if an important functional group is modified by the chemical reaction or selective binding is completely lost due to false matrix orientation (Fig. 3.1). In this case, a different immobilization strategy must be chosen, utilizing another functional group of the ligand [10]. However, the ligand density may still be subject to variations, having clustered ligands besides a more diffuse arrangement on the support surface. This may cause a heterogenous distribution of surface energies as an expression of a varying selectivity [11].

Proper orientation of a ligand may be achieved through a sandwich affinity approach, common in enzyme-linked immunosorbent (ELISA) or radio-immuno- (RIA) assay and also known in affinity chromatography [12,13]. If the condition of opposite binding sites for immobilization of the ligand and binding of the target molecule is fulfilled, ligand accessibility may be close to 100% [14]. However, it must also be kept in mind that immobilization is reversible. At low binding strength this may lead to ligand detachment during chromatography [15] and, consequently, to contamination of the product with the affinity ligand. Any of those scenarios will disabuse us of the view of a perfectly immobilized ligand and should remind us that ligand immobilization may require several attempts to be successful. It also shows quite plainly that, to keep changes in the properties of the affinity sorbent at a minimum, immobilization procedures should never be varied. One may also state that the immobilization SOP (standard operation procedure) should be minutely followed – if available.

3.3.2 Specific *vs.* non-specific interactions

Clearly, all forms of chromatography involve the differential interaction of a solute with the mobile and stationary phases. These interactions are described by a "chemical affinity" between the molecules. A special recognition mechanism is not a necessary requirement for an affinity interaction and, consequently, a differentiation between specific and non-specific binding, as often used to characterize a stationary phase, is questionable and has been an object of contention. Yet, in the context of affinity interactions, specific binding is understood to originate from the interactions observed with the dissolved ligand and solute (recognition in the "free affinity system"). Ideally, those interactions should be transferred to the immobilized affinity system without any change of recognition. Any unwanted binding, introduced by the immobilization process, is termed non-specific interaction. The latter may originate from the chromatographic matrix, the spacer, or the functional group employed for covalent binding of the ligand (Fig. 3.1). The binding strength of non-specific interactions, as described by an apparent equilibrium constant from thermodynamic investigations, can be higher or lower than a specific interaction. Thus, non-specific interactions may originate from the dominating selectivity in a chromatographic process.

However, non-specific binding may also originate from the ligand itself, for example, from the side-chains of amino acids and from charged or hydrophobic functional groups. These can be located anywhere on the ligand and may lead to binding of other molecules in the crude sample. Depending on the washing and elution conditions, these molecules may be eluted together with the target product. Another, mostly overlooked source of non-specific interactions is the target product. As it binds to the ligand, it gradually changes the characteristic of the stationary phase for newly arriving solutes. If the stationary phase has binding sites for other substances in the mixture, these interactions will lead to a more pronounced retention of these substances on the column; if the interactions are strong, they may even contaminate the target molecules [16,17]. This is particularly important when crude mixtures are used in a preparative purification, where the stationary-phase capacity is utilized as much as possible.

3.3.3 Mono-specific ligands

In principle, any of the molecules involved in biological recognition can be employed as an affinity ligand, and the corresponding affinity sorbent can be used to bind the complementary molecule. This is known from interactions between enzymes and their substrates and inhibitors, antibodies and antigens, and some special protein-binding families, such as the IGF/IGFBP (insulin-like growth factor/insulin-like growth-factor-binding protein) system [18]. Commonly, peptides or proteins are involved, and the corresponding interactions are strong and highly specific. Such protein/protein interactions are probably the most specific of all kinds of interactions that can be exploited for affinity chromatography [19–21] and are therefore very attractive, if one plans a one-step process for the purification of a target molecule from a complex mixture. Purification of proteins by this approach, especially on a larger scale, is limited to some well-characterized affinity systems, such as based on immunoglobulins or Protein A. The main problems stem from the fact that both the immobilized protein and the protein to be purified are subject to complicated unfolding and refolding processes, if harsh mobile-phase additives are required to dissociate the affinity complex. Due to the strong interactions, such elution conditions are not uncommon. Since unfolding is usually irreversible and changes the characteristic of an affinity sorbent, affinity systems in which the ligand is so affected are restricted to analytical applications; the sorbent may be used only once.

Competing with such chromatographic affinity systems are typical analytical application, such as ELISA or RIA techniques, where only detection of a specifically bound protein is accomplished and at least one antibody is involved. The technique of using microtiter well-plates with immobilized affinity ligands in ligand-binding assays is less known, but relies on a similar principle. Since only the specific binding is of interest, such an approach has advantages over chromatographic processes, where elution is required for detection. Such ligand-binding assays are increasingly employed in the field of proteomics, where arrays of immobilized ligands are screened to investigate the complex regulations inside a cell [22]. Because it is necessary to elute the bound solute from affinity columns, affinity systems employed in chromatography must be clearly differentiated from those in ELISA, RIA, or ligand-binding assays.

3.3.4 Group-specific ligands

Group-specific ligands, also termed general ligands, may bind a wide variety of proteins. For example, the co-factor NAD^+ binds NAD^+-dependent dehydrogenases [23]. Naturally occurring group-specific ligands, such as Protein A [24] or heparin [25] as well as synthetic compounds, such as phenylboronic acid [26], triazine dyes [27], or immobilized metal ions [28] have found widespread application in affinity purification. Especially the synthetic compounds are often not classified as affinity sorbents, as they do not display selectivity for only one protein. However, they do display site-selective binding, based upon specific interactions with defined pockets or functional groups on protein surfaces. On the other hand, the transition to other general purification techniques, such as ion-exchange or hydrophobic interactions is undefined; the latter was termed an affinity technique for a long time. As the ligand interaction becomes more generalized, the binding of a particular protein in a protein mixture becomes less selective and the product becomes more contaminated. Yet, by combining selective elution techniques and optimizing the purification procedure, excellent results can be achieved, as exemplified by a 12,600-fold purification of bovine lens aldose reductase by means of dye-ligand chromatography [29]. Group specificity has a major advantage over mono-specificity, because such affinity sorbents can be synthesized on a larger scale and become commercially available; meanwhile many of these affinity sorbents are certified according to GMP (good manufacturing practice) guidelines, like the standard chromatographic packing materials used in industrial operations.

3.3.5 Affinity tags

Lately, various techniques have been employed to genetically engineer ligands, affinity tails, and eluting agents to improve the efficiency of affinity chromatography and to allow a more general use of some affinity sorbents for a broad spectrum of recombinant proteins. These techniques include various extensions of the target protein, such that the properties of the added tag facilitates identification and provides an opportunity for a one-step purification process of the fusion protein [30]. A variety of expression vectors with different tag sequences has been designed for fusion to almost any target protein that can be cloned and expressed in a microbial host [31,32]. The size of these tags can range from small peptides to complete proteins, which in some cases consist of several subunits and can be attached to either the *N*- or *C*-terminus of the target protein or to both (Table 3.1).

Common tags include enzymes [33], peptides with protein-binding specificity [34–36], carbohydrate-binding proteins or domains [37,38], biotin-binding domains [39,40], antigenic epitopes [41], charged amino acids [42], poly(His) residues for binding to metal chelate supports [43,44] or a poly(acrylic acid) gel [45], and some short peptide chains composed of one amino acid, *e.g.*, poly(Phe) [46]. Still newer affinity tags keep appearing, such as a choline-binding tag [47], a tag based on *bis*-arsenical fluorescein [48], streptavidin-binding RNA ligands (aptamers) [49], and metal-chelate-binding tags with a

TABLE 3.1

COMMON AFFINITY TAGS

Type	Size	C- or N-terminal fusion	References
Proteins			
β-Galactosidase	116 kDa	N, C	[61]
Glutathione-S-transferase	26 kDa	N	[33,62]
Chloramphenicol acetyl transferase	24 kDa	N	[63]
TrpE	27 kDa	N	[64]
Calmodulin	17 kDa	N	[65,66]
Polypeptides			
Z- and ZZ-domain of Protein A	14, 31 kDa	N	[35,36,67,68]
Calmodulin-binding peptide	26 aa	N	[69,70]
Carbohydrate-binding domains			
Maltose-binding protein	40 kDa	N	[38,71]
Cellulose binding protein	111 aa	N	[37,72]
Biotin-binding domain	8 kDa	N	[39,40,73]
Antigenic epitopes			
recA	144 aa	C	[74]
FLAGJ	8 aa	N	[75]
Poly amino acids			
Poly(Arg)	5–15 aa	C	[42,76]
Poly(Glu)	6 aa	N	[77]
Poly(Asp)	5–16 aa	C	[78]
Poly(His)	1–9 aa	N, C	[79,80]
Poly(Phe)	11 aa	N	[46]
Poly(Cys)	4 aa	N	[80]

aa = amino acids.

selectivity similar to His-tags, *e.g.*, a natural poly(His) [50] and a peptide from *H. pylori* [51]. Affinity tags with carbohydrate specificity, such as a cellulose-binding domain or a chitin-binding domain, do not require special affinity sorbents, as they bind to cellulose or chitin chromatographic supports and utilize affordable purification methods [52]. Hydrophilic tags may improve the solubility of the target protein, as in the case of maltose-binding protein [53]. Usually, affinity tags are selected or optimized to allow mild elution conditions, such as salt elution or the addition of a competing molecule, such as imidazole or EDTA [54]. Also, specific competing peptides may be produced by a recombinant process to provide mild elution conditions [55].

A functional affinity tag requires correct folding, especially of a fused protein tag, and a location at an accessible site of the target protein. But these conditions are not always met, and some proteins may not be purified this way, even if a His-tag is employed, which allows purification in denatured condition [56,57]. Typically, the fusion partner is located at the 5′-position of the gene, where a better translational initiation is ensured than at the 3′-end. However, if an *N*-terminally fused tag is not accessible, due to crowding of peptide chains in its close neighborhood, positioning at the 3′-end should be envisaged, although in this configuration, expression yields may be more variable. High expression levels (up to 40% of the total protein content is routinely reported for *E. coli*) can be achieved, even with short affinity tags [58,59], which do not display protective properties, as is often the case with fused host-specific proteins. Just as with untagged proteins, high expression levels often cause the formation of inclusion bodies, which need to be solubilized by denaturing agents. Some affinity ligands, *e.g.*, IMAC (immobilized-metal affinity chromatography) sorbents, still work in the presence of 6 *M* urea, thus allowing on-column refolding of an adsorbed denatured His-tagged protein [57,60]. The His-tag has some other benefits, making it the most often used affinity tag today. Routinely, one-step affinity purification of target proteins is possible, achieving purities of >95% [81–83] – even >98% purity has been reported [84] – in a single chromatographic step or in series with other chromatographic modes, such as ion exchange [85]. Protein recoveries are also high with this affinity chromatographic method, normally exceeding 70% [86,87] and sometimes attaining 95% [88]. The high selectivity of IMAC in combination with His-tags has been exploited for direct affinity adsorption from a cell homogenate by using expanded-bed technology, again yielding high purities and recoveries in one step [89,90].

If it is necessary to remove the affinity tag, specific cleavage sites can be engineered between tag and target protein (Table 3.2). After cleavage of the fusion tag, the mixture of target protein and cleaved tag can again be passed through the affinity column. In this "negative" chromatographic mode, the cleaved target protein freely passes through the column, and the tag as well as non-cleaved target proteins are removed by affinity adsorption. Chemical cleavage methods are rare, as their specificity is lower than that of

TABLE 3.2

EXAMPLES OF CLEAVAGE STRATEGIES FOR REMOVING AFFINITY TAGS

Type of removal strategy	Cleavage specificity[*]	References
Enzymatic cleavage		
Carboxypeptidase B	R or K at *C*-terminus	[93]
Enterokinase	X-D-D-D-K-↓-X	[94]
Factor Xa	X-I-E-G-R-↓-X	[95]
Collagenase	P-X-↓-G-P-X-X	[96]
Thrombin	X-G-V-R-G-P-R-↓-X	[97]
Chemical cleavage		
Cyanogen bromide	X-M-↓-X	[98]
Hydroxylamine	X-N-↓-G-X	[67]

[*] One-letter code of amino acids, X = any amino acid, ↓ indicates cleavage site.

the common enzymatic methods. For small proteins, devoid of methionine, the cyanogen bromide method seems to be appropriate [91,92,98]. Current cleavage procedures based on enzymatic cleavage are expensive and may produce additional amino acids at either end of the proteins and contaminate the preparation with protease. In addition to specific enzymatic cleavage at the designated cleavage site (cleavage rates up to 95% are reported) some non-specific cleavage always occurs at other sites, creating fragments of the target protein [99]. The efficiency of proteases may also be reduced by some adventitious substances, such as detergents used to promote the solubility of membrane proteins [100]. Some enzymes are more specific than others and do not produce additional amino acids, *e.g.*, Factor Xa. Because this enzyme is expensive and available only from plasma, recombinant sources with engineered forms of this enzyme are being developed [101]. Post-cleavage removal of contaminating proteases can be achieved by introducing the affinity tag also into the proteolytic enzyme (without a cleavage site) [99], thus allowing simultaneous removal of the cleaved tag and the enzyme. Cleavage may also cause solubility problems in the case of proteins whose solubility is enhanced by the tag [53]. An interesting alternative to classical cleavage strategies is the incorporation of a modified intein, which is devoid of endonuclease activity [102], a so-called mini-intein [103–105], in the fusion system. For this, the target protein is localized at the *N*-terminus of the intein and the affinity tag at the *C*-terminus. This arrangement allows self-splicing of the target protein by the intein component of the fusion protein after induction by a thiol compound, *e.g.*, 1,4-dithiothreitol or β-mercaptoethanol. Then there is also pH- or temperature-induced splicing [106,107]. After affinity binding of fusion proteins to the sorbent, elution of target proteins can be accomplished under cleavage conditions [108]. These systems seem very attractive because they are easily implemented, but close scrutiny of cleavage selectivities is still missing.

Purification of therapeutic proteins requires additional procedures to remove fragments which may cause immunological reactions. Additional analytical procedures are also required to assure the absence of both the proteolytic enzyme employed for cleavage and any unwanted fragments having immunological potential. These are most properly the reasons why affinity tags are not widely used – if at all – industrially for the purification of therapeutics. An interesting strategy is to leave the tag on the therapeutic protein in order to improve its *in vivo* stability or to add a feature which improves the proteins delivery. This was shown in the case of a hydrophobic tag, which improved protein binding to an adjuvant [109], but such applications are rare in the accessible literature. In most cases, affinity tags are employed as analytical tools to follow expression levels of a recombinant protein, for which the tag may not be removed. The tag may also remain part of the target product for *in vitro* diagnostic tools or technical enzymes. In these cases, the affinity tag may interfere with protein functions [110,111]; interference with the structure of a protein must also be considered when crystallographic properties are investigated [112]. However, functional loss may occur at only one terminal site [113,114], so that the other site may be employed for tagging. IMAC (Sec. 3.4.3.3) also provides an opportunity for reversible immobilization of an enzyme as a diagnostic tool through binding of the tag to the affinity sorbent [115,116]. This immobilization process may also be employed to study affinity interactions *via* surface plasmon resonance [117]. A similar approach with histaminyl-purine residues has recently been used to purify DNA [118] by IMAC.

3.3.6 Size of the affinity ligand

Experience with immobilized small ligands suggests that their performance is more sensitive with regard to the immobilization procedure, matrix composition, and type of spacer than with regard to high molecular mass. This is also true for ligands from combinatorial libraries, which belong to the smaller ligands. As the ligand is immobilized on a solid support with a finite surface area, the maximum ligand density, as measured in $nmol/m^2$ for rigid supports or nmol/mL for soft gels, depends on the size of the ligand. Therefore, an affinity support with a macromolecular ligand may have a ligand density of only a few nanomoles per square meter. Only if the size of the solute is smaller or at most equal than the ligand can there be a 1:1 interaction between them without interference by neighboring ligands. With increasing solute size, a 1:1 saturation of ligands with solutes cannot occur, as partial or complete coverage of ligands occurs in close vicinity to a bound solute. Complete coverage of several ligands can be observed particularly with small affinity ligands, such as those used in dye-ligand or metal-chelate affinity chromatography (Fig. 3.2) (Sec. 3.4.3). As indicated above, binding of a solute is caused by electrostatic, hydrophobic, or some other, special interaction, and this interaction itself is not specific. Therefore, coverage of unused ligands, as shown in Fig. 3.2, induces non-specific interactions, their extent being related to the ligand. For example, dye ligands are composed of anionic and cationic functional groups (depending on the pH), giving rise to non-specific ionic interactions at the low ionic strengths commonly employed. A similar phenomenon was observed with IMAC sorbents, where a binding-site heterogeneity was linked to the adsorption thermodynamics of different proteins [11,119,120]. One could consider reducing the ligand density with small ligands. For example, it may be useful to limit the coverage to $50-100$ $nmol/m^2$, in order to avoid a large extent of non-specific binding. However, sorbents with low ligand densities tend to be less retentive, so that protein adsorption may no longer be possible. This phenomenon may be related to the number of available sites for binding during the very slow migration of the solute through the column while it is being loaded, provided that an equilibrium exists. The other extreme is that very strong elution conditions may be required at high ligand densities [121]. With dyes, it is generally accepted that non specific binding sites contribute largely to the retention, and thus it is meaningful to reduce the ligand density.

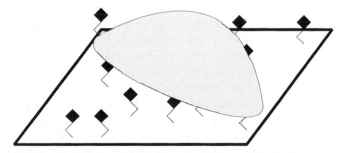

Fig. 3.2. Extent of coverage of a small ligand after binding a large protein through a single affinity bond.

If solutes with multiple binding sites exist, *e.g.*, multi-meric proteins, multi-valent binding can occur. This phenomenon is most prominent with small ligands, and it has been studied closely by several authors [122,123]. It generally gives rise to very strong interactions, as multiple binding of one solute raises the interaction strength [124]. However, in other cases, a limited ligand density may occur inadvertently. Especially, when low-molecular-weight ligands are used, the limitations on surface coverage are not only steric. For example, when a highly charged ligand is immobilized in a solution of low ionic strength, the ligand surface density may be controlled by Donnan exclusion. This is particularly known for dye-affinity sorbents, and therefore a high ionic strength is recommended to optimize ligand density [125]. One further difference exists between large and small ligands, which also depends on the ligand density. If the binding site of large affinity ligands is not optimally oriented, large ligand densities may cause this site to be blocked by the neighboring ligand, causing some decrease in relative ligand accessibility. This is often observed in immunoaffinity systems [126]. Clearly, no general advice for small or large affinity ligands can be given. The user must be prepared to tackle problems, such as low binding strength or low accessibility or the need for strong elution conditions.

3.3.7 Thermodynamics and kinetics

Thermodynamic investigations provide answers regarding specific and non-specific interactions and the heterogeneity of ligand densities. Although such an investigation is quite time-consuming and sometimes costly, it can be used to optimize ligand accessibility, the matrix, and immobilization. For adsorption of a protein on a packed-bed affinity column, the affinity constant, K_A, for the interaction must, for most practical purposes, exceed or equal 10^5 or $10^6\ M^{-1}$. If a very short packed bed or a single sheet of an affinity membrane is used, K_A should exceed $10^6\ M^{-1}$ [14]. If $K_A < 10^5\ M^{-1}$, there will also be some retardation, provided the ligand density is reasonable. With very weak affinity ligands ($K_A < 10^3\ M^{-1}$) isocratic elution chromatography can be useful [6]. It is sometimes impossible to use interactions at $K_A > 10^{10}\ M^{-1}$, as the conditions required to dissociate the affinity complex may unfold the protein.

By comparing experimentally derived equilibrium adsorption isotherms with the shape of isotherms calculated by different isotherm models, such as the monolayer or multilayer Langmuir model [127], protein/protein binding in multi-layers at high protein concentrations can be found with certain buffers. Also, high-affinity and low-affinity binding sites may be clearly distinguished, if the isotherm fits to a bi-Langmuir model, such as in the case of endotoxin-selective sorbents [128]. The data may be obtained from either batch experiments or from frontal analysis [129]. Both methods yield apparent equilibrium constants and a maximum binding capacity. These parameters may be used further to mathematically describe and model an affinity system [130,131]. Bergold *et al.* [132] have crudely modeled the effect of specific and non-specific interactions of the solute with the affinity stationary phase on the retention factor. They stated that even a small non-specific contribution to the binding free energy of *ca.* 10% can have a major effect on the capacity factor, although a test solute would appear to be completely unretained if non-specific interactions would be the sole interactions. Hence, the lack of retention of test solutes cannot be used as an indicator of the absence of non-specific interactions for a species that does not bind. In affinity chromatography, slow chemical equilibria in binding and desorption are often observed. In the case of non-specific elution conditions, such as a pH shift or the addition of a chaotropic agent, the kinetics of the chemical dissociation of a solute from a ligand tend to be so fast that band-broadening is dominated by mass-transport and flow-dispersion processes. However, when specific elution techniques, such as a competitive ligand or a competitive solute, are used, the rate-limiting step is often the chemical desorption. Slow chemical desorption leads to excessive tailing.

The complex adsorption and desorption process may be better understood in terms of the multi-dimensional nature of the affinity process. In order to transiently retain a solute at an immobilized ligand, only a favorable orientation will work. During the approach, orientation of the proper site must occur in order to achieve a fit. Since not all contacts of solutes with the surface will fulfill this prerequisite, several re-orientation steps may be necessary before they achieve an orientation favorable for adsorption. When ligand densities are high or non-specific binding with the chromatographic support is possible, other sites or chemical functionalities become favorably oriented, and a concerted multi-point adsorption results – perhaps also with neighboring adsorbed solutes – due to protein/protein interaction. During desorption, all of these interactions must be simultaneously disrupted in order for the solute to be released from the surface. Using soft chromatographic supports with large particle size, slower flow-rates and larger particle sizes cause mass transport to be the chief factor in broadening, and reaction rates are often immaterial. However, when HPLC monoliths, perfusion columns, or micro-filtration membrane are used as supports, these chemical reaction rates are often limiting [133–136]. That does not mean that all affinity separations are reaction-rate-limited. Dye systems often produce sharp elution profiles for proteins, as can be seen best in affinity membrane chromatography [137]. In cases of slow dissociation, it may be best to use a modified ligand that has a lower intrinsic adsorption constant and to adjust mobile-phase conditions so as to simultaneously reduce all types of interactions. The concanavalin A affinity ligand is well characterized with respect to the reaction kinetics [138–140]. Due to the slow dissociation kinetics, stopped-flow and pulsed-elution techniques have been proposed to enhance peak profiles [141]. This way, the mobile phase in the column is quickly replaced by the elution buffer, and the flow is stopped for several minutes. After several minutes of desorption, the flow is started again and the product can be eluted in a relatively small volume.

3.3.8 Matrix effects

As in other modes of chromatography, the quality of affinity chromatography is directly related to the quality of the packing media. It depends on their mechanical stability, flow properties, and a well-defined, uniform structure. Prior to ligand attachment, the matrix should be as inert as possible to minimize non-specific binding. Other desirable features include physical and chemical integrity and short diffusion paths. Because it is difficult to satisfy all of these requirements, priority is likely to be given to capacity, robustness, and throughput. Therefore, inexpensive and robust matrices, such as agarose, cellulose, and synthetic media may be considered [142]. Any affinity sorbent intended for process application must show not only stability of the matrix as well as the ligand and stable linkage under all process conditions, but also target-specificity and reproducibility [143]. In cases where reaction kinetics do not dominate the desorption process, the use of small chromatographic particles, common in HPLC, or of perfusive or monolithic supports will result in more efficient mass transport properties, as would be expected from general considerations of mass transport in porous supports. High-performance supports may also be used to quickly investigate the affinity properties of various ligands in the course of method development [144]. It must be kept in mind that the surface of these matrices may add interactions, such as the ion-exchange properties of the silanol groups in silica gel, which can be effective over a certain distance (even behind a small affinity ligand). Those non-specific interactions may need to be counterbalanced by the composition of the mobile phase, which, however, may affect the strength of the affinity interaction.

Some matrices lack a flat surface where the ligands can be anchored *via* a spacer, while in others the surface area available for binding is rather low. In order to increase the protein-binding capacity, polymers are immobilized in a first step, and then the ligands are immobilized in the polymer network (Fig. 3.3). With small ligands, a linear increase in the apparent association constant with ligand density is often observed [145], and the binding capacity may be also increased, as is expected [146]. However, large

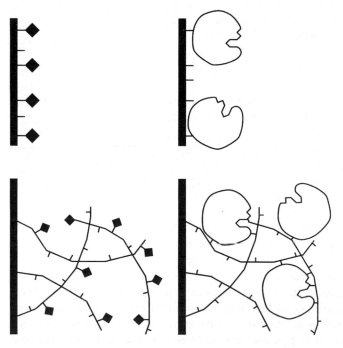

Fig. 3.3. Effect of a polymer network (coating): With a small ligand a higher binding capacity can often be achieved in a network (left), whereas a large ligand may not be accessible (right).

protein ligands, such as Protein A, may be interlocked through multiple covalent bonds in such a network, so that the binding of a solute may be strongly hindered [147].

3.4 AFFINITY LIGANDS

Countless affinity ligands are now being used in affinity chromatography, and there is no need to discuss all of them. However, some types of affinity ligands are more commonly employed and commercially available. Their development and, especially, their use in analytical and preparative applications will be reviewed here.

3.4.1 Biochemical interactions

3.4.1.1 Protein A

The most common laboratory and industrial application of affinity chromatography is the isolation of monoclonal antibodies. Classically, antibodies are isolated from ascites fluid, mammalian cells and, more recently, from animal and plant transgenic sources. The proteins are bound by immobilized Protein A, a protein obtained from *Staphylococcus aureus* which recognizes the Fc region of immunoglobulin G (IgG) in man and some other species. Protein A affinity chromatography can be employed on a large scale for the industrial purification of IgGs [148]. A recombinant Protein A has been engineered for greater binding specificity and is the most commonly used ligand. However, it cannot be used to recover all immunoglobulin sub-classes from blood plasma; *e.g.*, human IgG_1 binds to Protein A, but IgG_3 does not. Binding of murine IgG_1 is weak at neutral pH [149]; selective binding requires alkaline conditions and high ionic strength, *i.e.*, $0.1\,M$ glycine + 1 M NaCl (pH 8.9) [150].

A streptococcal Protein G was developed as affinity ligand to complement the selectivity for other IgG subclasses and additional species [151,152]. As the native protein also shows strong albumin binding, recombinant mono-functional forms are available, which show either albumin- or IgG-binding properties [144]. A Protein G with enhanced alkaline stability was introduced recently [153]. Some other bacterial cell-wall proteins may also find wider applications, *e.g.*, Protein L [154], which binds to the kappa light-chain region of IgGs, and Protein LG, a fusion protein containing binding domains from Proteins L and G [155].

As with other protein ligands, leakage from columns is considered a problem, as is the long-term stability under alternating pH (>pH 9 for binding, <pH 4 for elution) and cleaning-in-place conditions. Protein A sorbents can be very stable under normal operating conditions, though exceptions were also found [156]. They are stable in applications of chaotropic agents, such as 6 M guanidinium chloride, and show no loss of capacity [157]. Alkaline conditions (0.1 or 1 M NaOH) quickly destroy the affinity ligand [127], as would be expected from the action of hydroxide ions on peptide bonds; however, contrary results are also reported [158]. Therefore, when they are used on a production scale, a sensitive analytical technique should be applied to trace even minute amounts of a leaked ligand [159]. Protein G sorbents showed an increased affinity toward

IgG when alkaline conditions were employed as regeneration method, and this made elution of IgG difficult [160]. Use of a combination of urea and acetic acid was suggested to ensure efficient cleaning in the large-scale purification of a pharmaceutical monoclonal antibody.

Particularly in the case of the Protein A ligand, the user can choose among various sorbents, providing different performance and applicability to special applications [161], such as expanded-bed adsorption [162,163] or high-flow-rate applications with the use of superporous agarose with convective pores [164]. Protein A sorbents have also been used for quantifying polyclonal and monoclonal antibodies in 3-min analyses [165]; stability was increased when microbial growth was inhibited by overnight treatment with methanol. Protein A sorbents can also be employed for oriented immobilization of IgG, which, after further cross-linking of the affinity complex, can be used to isolate antigens from crude mixtures [166].

3.4.1.2 Heparin

Heparin is a negatively charged glycosaminoglycan of variable molecular mass (6000 to 20,000, depending on the preparation) containing carboxylic and sulfonic acid groups. It has long been used in affinity chromatography with varying success regarding the stability of affinity sorbents and the accessibility of the ligand. Despite a rigid chemical structure, some affinity sorbents are not stable under relatively mild cleaning-in-place conditions (0.1 M NaOH) [167]. Some stability problems may be ascribed to the heterogeneous composition of some heparin preparations, containing even protein constituents [168]. Also, the number of immobilization procedures employed proves the pitfalls of trial-and-error [169]. Yet, heparin preparations have been improved and so have heparin-based sorbents, as demonstrated with a HyperD-based matrix [170]. Aside from a countless number of applications to the purification of growth factors, such as fibroblast growth factor [170,171], chemokines [172], and coagulation proteins, such as thrombin [173], the use of heparin sorbents can be seen as an important development in the analytical and preparative techniques for lipases, phospholipases, and kinases [174]. It has also been employed for the purification of viruses, such as foot-and-mouth disease virus [175]. Due to the strongly negative charge of heparin, non-specific ionic interactions with positively charged by-products may be observed, if the target protein requires low ionic strength for binding. At intermediate ionic strength, interactions can be very selective, and effective purification methods have been developed [176]. However, when the selectivity is low, a cationic exchanger should be considered instead.

3.4.1.3 Lectins

Lectins are glycoproteins, used mainly for the structural determination of oligosac-charides [177–179] (Chap. 18). Recently, they have also found application in proteomics [180]. Several reviews have been published on the properties of lectins [181,182]. The most prominent lectin is concanavalin A, which is available commercially. It has been used, among many other applications, for the isolation of Protein C [183], lipophosphoglycans [184], and for the immobilization of sub-mitochondrial particles [185]. Compared to other affinity ligands, lectins show generally weaker interactions, which are less dependent on ionic charges. As a consequence, adsorption and desorption kinetics may be slow and may demand special chromatographic conditions, such as stopped-flow techniques. This effect is well known for concanavalin A, where it was closely examined [141].

3.4.2 Immuno-affinity ligands

Affinity ligands based on monoclonal antibodies are, in addition to polyclonal antibodies, indispensable in such analytical techniques as ELISA and RIA. The binding strength of antigen/antibody interactions is typically very high ($K_D < 10^{-11} M$), but antibodies with lower affinities are preferable for the purification of proteins, since strong binding interactions may require very stringent elution conditions, such as pH 2–3, urea, guanidine-HCl [8,186], or alkaline conditions [187], often accompanied by denaturation of the antigen [188]. An affinity of antibodies with a K_D of 10^{-7} to 10^{-8} M is sufficient for the isolation of antigens, even from crude extracts. At the same time, elution is considerably facilitated [189].

Immuno-affinity ligands have been used successfully in the purification of enzymes of the blood clotting cascade, such as Protein C [190,191], Factor IX, and Factor X [126,192]. Because the concentrations of these proteins in human blood plasma are low relative to other proteins, highly selective binding and high concentrating factors are required and can be obtained by this chromatographic method [192]. Immuno-affinity chromatography can also be used as an analytical method and compares favorably with ELISA and RIA with regard to speed and sensitivity [193,194]. Antibodies with very low affinity find application in the field of weak-affinity chromatography [6,195]. Binding strength in the millimolar range of K_D allows isocratic elution of low-molecular-weight compounds, such as various saccharides. In view of the weak retention, HPLC supports are recommended in order to reduce the peak-width and to improve the efficiency of this method.

3.4.3 Synthetic ligands

Synthetic ligands have significant advantages over ligands derived from natural products because of their relatively low cost and their robustness in industrial chromatographic process cycles, which include sterilization and sanitation steps. These ligands provide the additional advantage that they are not derived from intact animal, human or cell culture sources, which may be contaminated with an infectious particle or protein. In the wake of the concerns about transmissible diseases, traceability of such materials has become an industry requirement, enforced by the regulatory agencies.

3.4.3.1 Dye ligands

Dye-ligand affinity chromatography, also called biomimetic chromatography, was developed in the 1970s following the discovery that certain reactive textile dyes act like analogs of adenylate-containing co-factors and bind to the active sites of many proteins (Table 3.3). Screening of the many textile dyes available (*e.g.*, the Cibacron series from Ciba-Geigy, Procion series from ICI, and Remazol series from Hoechst) and some empirical ligand development [196] resulted in the introduction of many dyes into preparative fractionation procedures [197–199] and into the purification of HSA [200]. Spectroscopic techniques, including UV/visible, FTIR, NMR, ESR, and CD, have been utilized to scrutinize dye/protein interactions, the existence of competitive ligands (*e.g.*, substrates and coenzymes) and perturbing solutes (*e.g.*, salts and organic solvents) [201–204]. These studies revealed that interacting functional groups of dyes and enzymes must be at defined positions, and interactions are a mixture of electrostatic and hydrophobic forces at discrete sites of enzymes [205].

In the past, dye leakage and the often quite heterogeneous chemical composition of the dyes was a lasting problem and led to setbacks, particularly in preparative applications [206,207]. It seemed to be the cause of low apparent ligand accessibility of dye adsorbents [208]. Toxicity studies were carried out to account for the leaked dye [209]. However, these problems did not arise from the dye molecules but from weak covalent linkages at the chromatographic matrix and the heterogeneous composition of dye batches, as these dyes were intended for use in the textile industry rather than for the purification of pharmaceuticals. A large number of reactive dyes also bind to unrelated proteins (Table 3.3). Since most recorded examples of dye ligand chromatography required low ionic strength for adsorption [222,223], non-specific ionic or hydrophobic binding seemed common. These findings led to the designation of dye-ligand affinity chromatography as pseudo-affinity chromatography [224]. It is often considered to be at the fringe of what can be termed affinity chromatography [129]. Yet, it may be a problem of definition, as some pseudo-affinity sorbents show quite a high selectivity, as

TABLE 3.3

DYE-LIGAND AFFINITY SYSTEMS

Protein	Source	Immobilized ligand	References
Lactate dehydrogenase	Rabbit muscle	Procion Red	[27,210]
Malate dehydrogenase	Rat liver/heart	Cibacron Blue F3GA	[211]
	Rhodopseudomonas sp.	Procion Red H-3B	[212]
Glucose-6-P dehydrogenase	Yeast	Procion Red H8BN	[213]
Fructokinase	*Zymomonas mobilis*	Procion Yellow MX-GR	[214]
Aldehyde reductase	Rat liver	Procion Red HE-3B	[215]
Albumin	Human/bovine serum	Cibacron Blue F3GA	[216,217]
Polyclonal antibodies	Rabbit antiserum	Cibacron Blue F3GA	[218]
Enterotoxins	*Staphylococcus aureus*	Procion Red HE-3B	[219]
Plasminogen	Human serum	Procion Red HE-3B	[220]
Interferon	Human fibroblasts	Cibacron Blue F3GA	[221]

exemplified in the purification of penicillin acylase [225]. Anyhow, on the basis of the discrete interactions observed with textile dyes, new ligand variants were introduced [226]. These biomimetic dyes are specifically designed to mimic bioaffinity interactions, and they are claimed to show superior salt tolerance and alkaline stability [227–229]. They are applied mainly to enzymes, such as trypsin-like proteases [230] or glutathione-recognizing enzymes [231]. A major application may be seen in the purification of rHSA, as expressed in *Pichia pastoris* [232]. Because these ligands are compatible with CIP (cleaning-in-place) treatments, they have significant potential for capture. Reproducible use over many purification cycles is anticipated [233]. However, since the biomimetic ligand is specific for the target protein and must be redesigned for a different product, ligand development and validation requirements could restrict industrial use.

3.4.3.2 Phenylboronic acid

Cyclic complexes of *cis*-diols with boronic acid, displaying pK values of *ca.* 8.9, have been considered as affinity systems [234]. Since boronic acid cannot be incorporated into a matrix without losing its complexing abilities, meta-aminophenylboronic acid (PBA) is employed, which displays characteristics similar to boronic acid [235,236]. The phenylboronic acid/diol complex is relatively stable above pH 8, displaying retention factors between 1 and 20 for carbohydrate monomers [237]. Lower pK values of the complex are observed if pyrimidine or purine bases are integrated into the sugar, as in nucleosides and nucleotides, resulting in better stabilization of the complex and, consequently, higher retention factors. Proteins containing sugar side-chains with *cis*-diol groups can be purified like lectins [238]. If additional nitro groups are introduced into the phenyl ring of PBA, complexes are obtained which are stable at pH 6–7 and are especially practical for isolating highly pH-sensitive glycoprotein hormones [239,240].

Silica-immobilized PBA was employed for the quantitative determination of nucleosides in biological fluids [241]. A similar method was reported for the pre-fractionation of catecholamines from urine samples [242]. The isolation of RNA from DNA [243] as well as the separation of various oligoribonucleotides was described by several authors [244, 245]. Elution can be enforced by lowering the pH to 5 or 3. This dissociates the complex, freeing the bound molecule. However, in the isolation of glycoprotein hormones, care must be taken in applying such low pH conditions, since dissociation into subunits may occur. The major use of PBA sorbents is in diagnostic kits for the quantitation of glycosylated hemoglobins.

3.4.3.3 Immobilized metal ions

Immobilized metal-ion affinity chromatography (IMAC) – or metal chelate affinity (MCA), as it was named in the pioneering work by Porath *et al.* [246], or metal chelate interaction chromatography (MCIC), as named by El Rassi and Horváth [247] – was developed as a broad-spectrum adsorption method for proteins carrying accessible histidine residues at their surface [248]. IMAC is regarded as a special case of ligand-exchange chromatography (LEC), which was introduced by Helferich [249] for the separation of small molecules, such as amino acids [250] under isocratic conditions. Contrary to LEC, where the metal ion (usually Cu^{2+}) is bound weakly, incorporation of the metal ion into a metal chelator leads to very strong binding [251]. Especially transition metal ions, such as Cu^{2+} or Ni^{2+}, exhibit electron configurations creating very stable metal chelates. Traditionally iminodiacetic acid (IDA) is employed as immobilized-metal chelator in IMAC. Sorbents are commercially available from several manufacturers. The metal ion can be chosen as required; it is bound to the immobilized-metal chelator before the chromatographic process is started, in order to bring the metal-ion affinity sorbent into working conditions. The binding strength of proteins to such sorbents increases with the amount and spatial location of histidyl residues in the secondary structure, and it decreases in the order $Cu^{2+} > Ni^{2+} > Co^{2+} \approx Zn^{2+}$ [54,248,252,253]. The binding constant of an average protein with a single histidyl residue has been determined to be $K_A = 4.5 \times 10^3\,M^{-1}$, which is in agreement with data obtained for Cu^{2+}-IDA and histidine derivatives [254]. However, apparent binding constants are several orders of magnitudes higher and reflect the large number of potential binding sites at moderate to high ligand densities of this relatively small ligand [14,146].

For selective binding, buffers of pH 5–8 with added 0.5–1.0 M NaCl should be used to suppress ionic interactions, but, to avoid "bleeding" of metal ions, chelating buffers, based on Tris or imidazole, should not be used. Elution is accomplished at pH 3–4, leading to destabilization of the chelate. Competing substrates, such as 20 mM imidazole or up to 1 M ammonium chloride is recommended for elution at neutral pH. Proteins which cannot be eluted under these conditions, *e.g.*, those tagged with the poly(His) tail (Sec. 3.3.5), may be desorbed at very high imidazole concentrations or with 50 mM EDTA (a stronger chelator) + 0.5 M NaCl (to disrupt ionic binding) at pH 6–7. Non-linear elution methods are also described, with imidazole [255] and poly(His) displacers [256]. Divalent metal ions and chelating agents in the feedstock may affect adsorption. Metal leaching can be a problem, and removal of eluting agents may be required before further processing. Some proteins with high affinity binding sites may strip the metal ion off the chelate and will not be retained unless a second, "normal" binding site exists, a process called metal-ion transfer [54]. IMAC has become one of the most popular separation techniques, rivaling such techniques as ion-exchange and hydrophobic-interaction chromatography. Large-scale applications have been published, such as the application to serum albumin [257, 258], and it is a promising approach to the isolation of the anticoagulant Protein C [259]. IMAC is a very powerful method, because very high selectivity can be obtained using histidine tags [43,260]. It allows one-step purification on small columns or with affinity membranes on an analytical scale.

Incorporation of Fe^{3+} into immobilized IDA chelators allows the separation of phosphorylated from dephosphorylated proteins [261,262]. The adsorption mechanism is based on the interaction of phosphorus with Fe^{3+} ions. Proteins with binding sites for iron may also be purified with immobilized Fe^{3+} [263]. In addition to IDA, several other chelators have been introduced in the search for alternative binding mechanisms, *e.g.*, through "hard" metal ions (according to the Pearson classification). While the versatility of IMAC is its most appealing feature for developers, it means significant complexity for the ordinary user. At present, IDA, nitrilotriacetic acid (NTA), carboxymethylated aspartic acid (CM-Asp), and triscarboxymethyl ethylenediamine (TED) are available commercially; a more detailed description can be found in the literature [43,264,265]. A number of other chelating ligands [266–270] have also been used successfully for the separation of proteins. While these ligands introduce additional selectivity and could be better suited for immobilizing particular types of metal ions, such as Fe^{3+}, they are not commercially available.

3.4.4 Combinatorial libraries

Increased access to protein structural data, computer-aided molecular design, and novel combinatorial techniques and molecular modeling have assisted the design of new synthetic ligands. These ligands are often seen to open new avenues for the development of more efficient, less expensive and safer procedures for product purification at the industrial level; one separate, but very important aspect is the patent situation. According to their developers, synthetic affinity ligands are ideally suited for the purification of high-value therapeutic proteins. Unfortunately, little can be learned from a user about success and failure of such a ligand with respect to selectivity and efficiency for a particular product, after the sorbent has crossed the company border. Therefore, success is bound to the length of time a "new" ligand is referred to in the literature. Monoclonal antibodies are prime targets for these new ligands, as these proteins are becoming an important class of therapeutic agents; many of them are waiting for FDA approval. As these proteins do not act like hormones and are produced at relatively low titers by cell culture technologies, large masses must be purified from dilute, crude mixtures. There is still no solution to how this purification problem can be managed at low cost.

The tetrameric peptide TG19318 was identified after screening a multimeric library [271] and synthesized by solid-phase chemistry, for immunoglobulin affinity. The same adsorption and elution conditions can be employed for IgG as for Protein A sorbents. However, in contrast to Protein A sorbents, TG19318 affinity sorbents are not affected by the presence of denaturants, detergents, or other sanitizing reagents, and a reduction of 10–30% capacity after sterilization was ascribed to instability of the support rather than of the ligand [272]. This ligand purifies IgG from a wide variety of sources and also IgY from egg yolk [272], IgM [273], IgA [274], and IgE [275]. Results of a close examination demonstrated strong interactions with human IgM and rat IgG only, but weak interactions $(K_A > 10^5 \, M^{-1})$ with human IgG [276]. As TG19318 is strongly positively charged, one must be aware of non-specific electrostatic interactions. Another example is a Protein A analog peptide with 12 amino acids, which was isolated from two phage libraries, using a combination of phage display and bio-informatics. The peptides were selected from 20 peptide sequences, identified from two linear phage-display libraries, containing 3.4×10^9 initial sequences, by best affinity binding to a humanized antibody [277].

A mixture of defined and combinatorial chemical synthesis is another approach. The *de novo* design process is based on peptide templates, complementary to surface-exposed residues of target proteins and mimicking natural biological recognition. These ligands can even be tailored to the source to eliminate impurities specific to the source [278]. Based on a peptidal template of the dipeptide substrate of trypsin-like proteases, a ligand with high selectivity for kallikrein was designed [230]. A ligand designed to mimic a key motif of Protein A was designed and used to purify IgG from human plasma, ascites fluid, and fetal calf serum [279]. This ligand and the corresponding sorbent were able to withstand incubation in 1 M NaOH [280]. Complementary functionality to the target residues was employed to design a ligand for the insulin precursor MI3 [280]. Furthermore, a mixed apolar/metal-chelating ligand was designed, which adsorbed clotting Factor VII in a Ca^{2+}-dependent manner [281]. Another example of using a peptide library is exemplified by the large-scale fibrinogen purification [282]. The blood coagulation Factor VIII was purified by using special monoliths with peptides from a combinatorial library [283]. These examples prove that the future is apparently wide open for new affinity sorbents with ligands derived from combinatorial libraries.

3.4.5 Purification schemes

Use of recombinant microbial and animal cell-culture technologies for the production of highly effective pharmaceuticals requires a product which is free of bacterial DNA, endotoxins from Gram-negative bacteria, and other contaminants and impurities, such as proteolytic enzymes and viruses. Technologies that allow products to be purified in only a few steps will reduce their overall costs dramatically. The high purification factors frequently obtained by affinity chromatography recommend their use as one of the first steps in the purification sequence, possibly before separation of cell debris, as is demonstrated by the expanded-bed adsorption technique [284,285]. With this technique, expenses for centrifuges and filtration units can be largely eliminated. Furthermore, losses of enzymatic activity and overall recovery, which are frequently observed in multi-step processes due to proteolysis and non-specific adsorption, could be further minimized. Despite problems which can be ascribed to cell interactions, affinity ligands have been shown to serve with success, *e.g.*, Protein A in the direct affinity adsorption of IgG from cell culture supernatants [286,287], the dye ligand Procion Red HE3B in the isolation of recombinant formate dehydrogenase from *E. coli* [288], and a chelating ligand in the isolation of alcohol dehydrogenase from yeast [289]. However, there are also reasons for using chromatographic processes in purification schemes as late as possible. The risk of clogging of the column packing material by agglomerated or precipitated proteins from crude extracts is considerable. This is observed especially in large-scale processes and leads to reduced capacities and lifetimes of columns. Moreover, crude extracts may contain proteins that rapidly overload the column or complex with the proteins of interest to prevent their interactions with the sorbent. Generally, it is not necessarily the ligand with the highest purification factor that ought to be employed. Especially in large-scale purification schemes for pharmaceutical products, the stability of the affinity sorbent is a much more important factor, because appreciable amounts of the ligand or parts of it cannot be tolerated in the final product.

3.5 SUMMARY

Among the classical techniques for protein purification, affinity chromatography allows a several-fold purification from dilute preparations in a single step. At this time, the user can choose among a large variety of sorbents with different affinity ligands for almost any protein purification problem. However, we must clearly distinguish between applications in laboratory experiments and on an industrial scale. Whereas no restrictions apply in laboratory applications as to the type of ligand, its stability, or its origin, industrial applications require validated and certified affinity sorbents which are robust under process conditions, including cleaning-in-place. Future developments will concentrate on highly selective affinity ligands, tailored for a particular application, which are suitable for the elimination of by-products. For industrial applications, especially in the pharmaceutical sector, validation will be a problem, but one that can be solved during product development. Thus, the future seems bright for affinity sorbents based on synthetic ligands, whatever their chemical structure will be, as long as they are stable under industrial operating conditions.

REFERENCES

1   P. Cuatrecasas, M. Wilchek and C.B. Anfinsen, *Proc. Natl Acad. Sci. USA*, 61 (1968) 636.
2   P. Cuatrecasas, M. Wilchek and C.B. Anfinsen, *Biochemistry*, 61 (1968) 636.
3   G. Manecke and K.-E. Gillert, *Naturwissenschaften*, 42 (1955) 212.
4   I.M. Silman and E. Katchalsky, *Annu. Rev. Biochem.*, 35 (1966) 873.
5   E. Heftmann (Ed.), Affinity Chromatography, *J. Biochem. Biophys. Methods*, 49 (2001), Elsevier, Amsterdam.
6   S. Ohlson, M. Bergstrom, P. Pahlsson and A. Lundblad, *J. Chromatogr. A*, 758 (1997) 199.
7   S. Cochet, M. Hasnaoui, A. Debbia, Y. Kroviarski, P. Lambin, J. Carton and O. Bertrand, *J. Chromatogr. A*, 663 (1994) 175.
8   L.O. Narhi, D.J. Caughey, T. Horan, Y. Kita, D. Chang and T. Arakawa, *Anal. Biochem.*, 253 (1997) 236.
9   Y. Yarden, I. Harari and J. Schlessinger, *J. Biol. Chem.*, 260 (1985) 315.
10 P.-O. Larsson, M. Glad, L. Hansson, M.-O. Mansson and K. Mosbach, *Adv. Chromatogr.*, 21 (1982) 41.
11 R.D. Johnson and F.H. Arnold, *Biochim. Biophys. Acta*, 1247 (1995) 293.
12 Z. El Rassi, Y. Truei, Y.-F. Maa and C. Horváth, *Anal. Biochem.*, 169 (1988) 172.
13 J. Turková, L. Petkov, J. Sajdok, J. Kás and M.J. Benes, *J. Chromatogr.*, 500 (1990) 585.
14 T.C. Beeskow, W. Kusharyoto, F.B. Anspach, K.H. Kroner and W.D. Deckwer, *J. Chromatogr. A*, 715 (1995) 49.
15 F.B. Anspach and G. Altmann Haase, *Biotechnol. Appl. Biochem.*, 20 (1994) 313.
16 B. Ghebrehiwet, P.D. Lu, W. Zhang, S.A. Keilbaugh, L.E. Leigh, P. Eggleton, K.B. Reid and E.I. Peerschke, *J. Immunol.*, 159 (1997) 1429.
17 M. Sissler, C. Delorme, J. Bond, S.D. Ehrlich, P. Renault and C. Francklyn, *Proc. Natl Acad. Sci. USA*, 96 (1999) 8985.
18 D.R. Clemmons, *Cytokine Growth Factor Rev.*, 8 (1997) 45.
19 L.J. Ransone, *Methods Enzymol.*, 254 (1995) 491.
20 E.M. Phizicky and S. Fields, *Microbiol. Rev.*, 59 (1995) 94.
21 S. Beeckmans, *Methods*, 19 (1999) 278.
22 D. Stoll, M.F. Templin, M. Schrenk, P.C. Traub, C.F. Vohringer and T.O. Joos, *Front. Biosci.*, 7 (2002) c13–c32.

23   C.R. Lowe, M.J. Harvey, D.B. Craven, M.A. Kerfoot, M.E. Hollows and P.D.G. Dean, *Biochem. J.*, 133 (1973) 507.
24   R.A. Kagel, G.W. Kagel and V.K. Garg, *BioChromatography*, 4 (1989) 246.
25   B.A. Brennessel, D.P. Buhrer and A.A. Gottlieb, *Anal. Biochem.*, 87 (1978) 411.
26   X.C. Liu and W.H. Scouten, *J. Chromatogr.*, 687 (1994) 61.
27   D.A.P. Small, T. Atkinson and C.R. Lowe, *J. Chromatogr.*, 216 (1981) 175.
28   E. Sulkowski, *Trends Biotechnol.*, 3 (1985) 1.
29   K. Inagaki, I. Miwa and J. Okuda, *Arch. Biochem. Biophys.*, 216 (1982) 337.
30   R. Terpe, *Appl. Microbiol. Biotechnol.*, 60 (2003) 523.
31   P.A. Nygren, S. Stahl and M. Uhlen, *Trends Biotechnol.*, 12 (1994) 184.
32   J. Nilsson, S. Stahl, J. Lundeberg, M. Uhlén and P.A. Nygren, *Protein Expr. Purif.*, 11 (1997) 1.
33   K. Guan and J.E. Dixon, *Anal. Biochem.*, 192 (1991) 262.
34   M.C. Smith, T.C. Furman, T.D. Ingolia and C. Pidgeon, *J. Biol. Chem.*, 263 (1988) 7211.
35   B. Nilsson, L. Abrahmsen and M. Uhlén, *EMBO J.*, 4 (1985) 1075.
36   J. Nilsson, P. Jonasson, E. Samuelsson, S. Stahl and M. Uhlen, *J. Biotechnol.*, 48 (1996) 241.
37   J.M. Greenwood, E. Ong, N.R. Gilkes, R.A.J. Warren, R.C. Miller and D.G. Kilburn, *Protein Eng.*, 5 (1992) 361.
38   C. Guan, P. Li, P.D. Riggs and H. Inouye, *Gene*, 67 (1988) 21.
39   K.-L. Tsao, B. DeBarbieri, H. Michel and D.S. Waugh, *Gene*, 169 (1996) 59.
40   P.A. Smith, B.C. Tripp, E.A. DiBlasio-Smith, Z. Lu, E.R. LaVallie and J.M. McCoy, *Nucleic Acids Res.*, 26 (1998) 1414.
41   R. Hernan and K. Heuermann, *BioTechniques*, 28 (2000) 789.
42   S.J. Brewer and H.M. Sassenfeld, *Trends Biotechnol.*, 3 (1985) 119.
43   E. Hochuli, H. Döbeli and A. Schacher, *J. Chromatogr.*, 411 (1987) 177.
44   S.M. OBrien, R.P. Sloane, O.R.T. Thomas and P. Dunnill, *J. Biotechnol.*, 54 (1997) 53.
45   Q. Zeng, J. Xu, R. Fu and Q. Ye, *J. Chromatogr. A*, 921 (2001) 197.
46   M. Persson, M.G. Bergstrand, L. Bülow and K. Mosbach, *Anal. Biochem.*, 172 (1988) 330.
47   J. Caubin, H. Martin, A. Roa, I. Cosano, M. Pozuelo, J.M. de La Fuente, J.M. Sanchez-Puelles, M. Molina and C. Nombela, *Biotechnol. Bioeng.*, 74 (2001) 164.
48   K.S. Thorn, N. Naber, M. Matuska, R.D. Vale and R. Cooke, *Protein Sci.*, 9 (2000) 213.
49   C. Srisawat and D.R. Engelke, *RNA*, 7 (2001) 632.
50   G. Chaga, J. Hopp and P. Nelson, *Biotechnol. Appl. Biochem.*, 29 (1999) 19.
51   M.M. Enzelberger, S. Minning and R.D. Schmid, *J. Chromatogr. A*, 898 (2000) 83.
52   E. Shpigel, D. Elias, I.R. Cohen and O. Shoseyov, *Protein Expr. Purif.*, 14 (1998) 185.
53   J.D. Fox, K.M. Routzahn, M.H. Bucher and D.S. Waugh, *FEBS Lett.*, 537 (2003) 53.
54   E. Sulkowski, *BioEssays*, 10 (1989) 170.
55   J. Nilsson, P. Nilsson, Y. Williams, L. Pettersson, M. Uhlen and P.A. Nygren, *Eur. J. Biochem.*, 224 (1994) 103.
56   P. Braun, Y. Hu, B. Shen, A. Halleck, M. Koundinya, E. Harlow and J. LaBaer, *Proc. Natl Acad. Sci. USA*, 99 (2002) 2654.
57   G. Lemercier, N. Bakalara and X. Santarelli, *J. Chromatogr. B*, 786 (2003) 305.
58   V.M. Quintana Flores, R.C. Campos de Souza Fernandes, Z. Sousa de Macedo and E. Medina-Acosta, *Protein Expr. Purif.*, 25 (2002) 2516.
59   G.S. Choi, J.Y. Kim, J.H. Kim, Y.W. Ryu and G.J. Kim, *Protein Expr. Purif.*, 29 (2003) 85.
60   W. Tang, Z.Y. Sun, R. Pannell, V. Gurewich and J.N. Liu, *Protein Expr. Purif.*, 11 (1997) 279.
61   A. Ullmann, *Gene*, 29 (1984) 27.
62   D.B. Smith and K.S. Johnson, *Gene*, 67 (1988) 31.

63   J.A. Knott, C.A. Sullivan and A. Weston, *Eur. J. Biochem.*, 174 (1988) 405.

64   M. Hummel, H. Herbst and H. Stein, *Eur. J. Biochem.*, 180 (1989) 555.

65   U.A. Desai, G. Sur, S. Daunert, R. Babbitt and Q. Li, *Protein Expr. Purif.*, 25 (2002) 195.

66   S. Melkko and D. Neri, *Methods Mol. Biol.*, 205 (2003) 69.

67   T. Moks, L. Abrahmsen, E. Holmgren, M. Bilich, A. Olsson, M. Uhlén, G. Pohl, C. Sterky, H. Hultberg and S. Josephson, *Biochemistry*, 26 (1987) 5239.

68   J. Nilsson, P. Jonasson, E. Samuelsson, S. Stahl and M. Uhlen, *J. Biotechnol.*, 48 (1996) 241.

69   C.F. Zheng, T. Simcox, L. Xu and P. Vaillancourt, *Gene*, 186 (1987) 55.

70   W. Klein, *Methods Mol. Biol.*, 205 (2003) 79.

71   A. Blondel and H. Bedouelle, *Eur. J. Biochem.*, 193 (1990) 325.

72   E. Ong, J.M. Greenwood, N.R. Gilkes, D.G. Kilborn, R.C. Miller and R.A.J. Warren, *Trends Biotechnol.*, 7 (1989) 239.

73   J.E. Cronan, *J. Biol. Chem.*, 265 (1990) 10327.

74   G.G. Krivi, M.L. Bittner, E. Rowold, E.Y. Wong, K.C. Glenn, K.S. Rose and D.C. Tiemeier, *J. Biol. Chem.*, 260 (1985) 10263.

75   T.P. Hopp, B. Gallis and K.S. Prickett, *Bio/Technology*, 6 (1988) 1204.

76   H.M. Sassenfeld and S.J. Brewer, *Bio/Technology*, 2 (1984) 76.

77   S. LeBorgne, M. Graber and J.S. Condoret, *Bioseparations*, 5 (1995) 53.

78   J. Zhao, C.F. Ford, C.E. Glatz, M.A. Rougvie and S.M. Gendel, *J. Biotechnol.*, 14 (1990) 273.

79   M.C. Smith, T.C. Furman, T.D. Ingolia and C. Pidgeon, *J. Biol. Chem.*, 263 (1988) 7211.

80   E. Hochuli, W. Bannwarth, H. Döbeli, R. Gentz and D. Stüber, *Bio/Technology*, 6 (1988) 1321.

81   M. Masi, J.M. Pages and E. Pradel, *J. Chromatogr. B*, 786 (2003) 197.

82   W.T. Doerrler and C.R. Raetz, *J. Biol. Chem.*, 277 (2002) 36697.

83   G.K. Smith, D.G. Barrett, K. Blackburn, M. Cory, W.S. Dallas, R. Davis, D. Hassler, R. McConnell, M. Moyer and K. Weaver, *Arch. Biochem. Biophys.*, 399 (2002) 195.

84   Y. Hou, D. Vocadlo, S. Withers and D. Mahuran, *Biochemistry*, 39 (2000) 6219.

85   H. Jung, S. Tebbe, R. Schmid and K. Jung, *Biochemistry*, 37 (1998) 11083.

86   H. Wajant and F. Effenberger, *Eur. J. Biochem.*, 269 (2002) 680.

87   M. Pietzsch, A. Wiese, K. Ragnitz, B. Wilms, J. Altenbuchner, R. Mattes and C. Syldatk, *J. Chromatogr. B*, 737 (2000) 179.

88   C. Fiore, V. Trezeguet, P. Roux, A. Le Saux, F. Noel, C. Schwimmer, D. Arlot, A.C. Dianoux, G.J. Lauquin and G. Brandolin, *Protein Expr. Purif.*, 19 (2000) 57.

89   S. Gibert, N. Bakalara and X. Santarelli, *J. Chromatogr. B*, 737 (2000) 143.

90   W.S. Choe, R.H. Clemmitt, H.A. Chase and A.P. Middelberg, *Biotechnol. Bioeng.*, 81 (2003) 221.

91   C. Rais-Beghdadi, M.A. Roggero, N. Fasel and C.D. Reymond, *Appl. Biochem. Biotechnol.*, 74 (1998) 95.

92   S.G. Chang, D.Y. Kim, K.D. Choi, J.M. Shin and H.C. Shin, *Biochem. J.*, 329 (1998) 631.

93   J.C. Smith, R.B. Derbyshire, E. Cook, L. Dunthorne, J. Viney, J.J. Brewer, H.M. Sassenfeld and L.D. Bell, *Gene*, 32 (1984) 321.

94   K.S. Prickett, D.C. Amberg and T.P. Hopp, *BioTechniques*, 7 (1989) 580.

95   L. Morganti, M. Huyer, P.W. Gout and P. Bartolini, *Biotechnol. Appl. Biochem.*, 23 (1996) 67.

96   A.M. Shimabuku, T. Saeki, K. Ueda and T. Komano, *Agric. Biol. Chem.*, 55 (1991) 1075.

97   D.J. Hakes and J.E. Dixon, *Anal. Biochem.*, 202 (1992) 293.

98   D. Hüsken, T. Beckers and J.W. Engels, *Eur. J. Biochem.*, 193 (1990) 387.

99   S.I. Choi, H.W. Song, J.W. Moon and B.L. Seong, *Biotechnol. Bioeng.*, 76 (2001) 718.

100   A.K. Mohanty, C.R. Simmons and M.C. Wiener, *Protein Expr. Purif.*, 27 (2003) 109.

101   M.M. Guarna, H.C. Cote, E.M. Kwan, G.L. Rintoul, B. Meyhack, J. Heim, R.T. MacGillivray, R.A. Warren and D.G. Kilburn, *Protein Expr. Purif.*, 20 (2000) 133.

102   V. Derbyshire, D.W. Wood, W. Wu, J.T. Dansereau, J.Z. Dalgaard and M. Belfort, *Proc. Natl Acad. Sci. USA*, 94 (1997) 11466.

103   A. Zhang, S.M. Gonzalez, E.J. Cantor and S. Chong, *Gene*, 275 (2001) 241.

104   D.W. Wood, V. Derbyshire, W. Wu, M. Chartrain, M. Belfort and G. Belfort, *Biotechnol. Prog.*, 16 (2000) 1055.

105   J.P. Gangopadhyay, S.Q. Jiang, P. van Berkel and H. Paulus, *Biochim. Biophys. Acta*, 1619 (2003) 193.

106   D.W. Wood, W. Wu, G. Belfort, V. Derbyshire and M. Belfort, *Nat. Biotechnol.*, 17 (1999) 889.

107   M.W. Southworth, K. Amaya, T.C. Evans, M.Q. Xu and F.B. Perler, *Biotechniques*, 27 (1999) 118.

108   D.M. Myscofski, E.K. Dutton, E. Cantor, A. Zhang and D.E. Hruby, *Prep. Biochem. Biotechnol.*, 31 (2001) 275.

109   C. Andersson, L. Sandberg, H. Wernerus, M. Johansson, K. Lovgren-Bengtsson and S. Stahl, *J. Immunol. Methods*, 238 (2000) 181.

110   C.M. Kowolik and W. Hengstenberg, *Eur. J. Biochem.*, 257 (1998) 389.

111   K. Ragnitz, C. Syldatk and M. Pietzsch, *Enzyme Microb. Technol.*, 28 (2001) 713.

112   M.H. Bucher, A.G. Evdokimov and D.S. Waugh, *Acta Crystallogr. D Biol. Crystallogr.*, 58 (2002) 392.

113   A.E. Frankel, J. Ramage, M. Kiser, R. Alexander, G. Kucera and M.S. Miller, *Protein Eng.*, 13 (2000) 575.

114   A. Goel, D. Colcher, J.S. Koo, B.J. Booth, G. Pavlinkova and S.K. Batra, *Biochim. Biophys. Acta*, 1523 (2000) 13.

115   P. Loetscher, L. Mottlau and E. Hochuli, *J. Chromatogr.*, 595 (1992) 113.

116   S. Piesecki, W.Y. Teng and E. Hochuli, *Biotechnol. Bioeng.*, 42 (1993) 178.

117   L. Nieba, S.E. Nieba Axmann, A. Persson, M. Hamalainen, F. Edebratt, A. Hansson, J. Lidholm, K. Magnusson, A.F. Karlsson and A. Plückthun, *Anal. Biochem.*, 252 (1997) 217.

118   C.H. Min and G.L. Verdine, *Nucleic Acids Res.*, 24 (1996) 3806.

119   R.D. Johnson and F.H. Arnold, *Biotechnol. Bioeng.*, 48 (1995) 437.

120   H.J. Wirth, K.K. Unger and M.T.W. Hearn, *Anal. Biochem.*, 208 (1993) 16.

121   L.-Z. He, Y.-R. Gan and Y. Sun, *Bioprocess Eng.*, 17 (1997) 301.

122   D.J. Winzor and R.J. Yon, *Biochem. J.*, 217 (1984) 867.

123   S.J. Harris, C.M. Jackson and D.J. Winzor, *Arch. Biochem. Biophys.*, 316 (1995) 20.

124   J. Lyklema, in T.C.J. Gribnau, J. Visser and R.J.F. Nivard (Eds.), *Affinity Chromatography and Related Techniques*, Elsevier, Amsterdam, 1982, pp. 11–27.

125   D.A.P. Small, T. Atkinson and C.R. Lowe, *J. Chromatogr.*, 266 (1983) 151.

126   J.P. Tharakan, D.B. Clark and W.N. Drohan, *J. Chromatogr.*, 522 (1990) 153.

127   F.B. Anspach, D. Petsch and W.-D. Deckwer, *Bioseparation*, 6 (1996) 165.

128   D. Petsch, W.-D. Deckwer and F.B. Anspach, *Can. J. Chem. Eng.*, 77 (1999) 921.

129   D.J. Winzor and J. de Jersey, *J. Chromatogr.*, 492 (1989) 377.

130   A.I. Liapis, *J. Biotechnol.*, 11 (1989) 143.

131   M.A. McCoy and A.I. Liapis, *J. Chromatogr.*, 548 (1991) 25.

132   A.F. Bergold, A.J. Muller, D.A. Hanggi and P.W. Carr, in C. Horváth (Ed.), *High-performance Liquid Chromatography: Advances and Perspectives*, Academic Press, New York, 1988, pp. 96–209.

133   A.I. Liapis, *Sep. Purif. Methods*, 19 (1990) 133.

134   D. Josić and A. Buchacher, *J. Biochem. Biophys. Methods*, 49 (2001) 153.

135 Z. Yan and J. Huang, *J. Chromatogr. B*, 738 (2000) 149.

136 S.-Y. Suen, M. Caracotsios and M.R. Etzel, *Chem. Eng. Sci.*, 48 (1993) 1801.

137 B. Champluvier and M.R. Kula, *Biotechnol. Bioeng.*, 40 (1992) 33.

138 D.J. Anderson and R.R. Walters, *J. Chromatogr.*, 376 (1986) 69.

139 A.J. Muller and P.W. Carr, *J. Chromatogr.*, 357 (1986) 11.

140 P.D. Munro, D.J. Winzor and J.R. Cann, *J. Chromatogr. A*, 659 (1994) 267.

141 A.J. Muller and P.W. Carr, *J. Chromatogr.*, 294 (1984) 235.

142 S.R. Narayanan, *J. Chromatogr. A*, 658 (1944) 237.

143 K. Jones, *Chromatographia*, 32 (1991) 469.

144 T.V. Gupalova, O.V. Lojkina, V.G. Palagnuk, A.A. Totolian and T.B. Tennikov, *J. Chromatogr. A*, 949 (2002) 185.

145 P.M. Boyer and J.T. Hsu, *Chem. Eng. Sci.*, 47 (1992) 241.

146 T.C. Beeskow, K.H. Kroner and F.B. Anspach, *J. Colloid Interface Sci.*, 196 (1997) 278.

147 L.R. Castilho, W.-D. Deckwer and F.B. Anspach, *J. Membr. Sci.*, 172 (2000) 269.

148 A.C. Kenney and H.A. Chase, *J. Chem. Technol. Biotechnol.*, 39 (1987) 173.

149 M.R. Mackenzie, N.L. Warner and G.F. Mitchell, *J. Immunol.*, 120 (1978) 1493.

150 G. Schuler and M. Reinacher, *J. Chromatogr.*, 587 (1991) 61.

151 L. Björk and G. Kronvall, *J. Immunol.*, 133 (1984) 969.

152 P. Cassulis, M.V. Magasic and V.A. Debari, *Clin. Chem.*, 37 (1991) 882.

153 S. Gulich, M. Linhult, S. Stahl and S. Hober, *Protein Eng.*, 15 (2002) 835.

154 B.H. Nilson, L. Logdberg, W. Kastern, L. Bjorck and B. Akerstrom, *J. Immunol. Methods*, 164 (1993) 33.

155 R. Vola, A. Lombardi, L. Tarditi, L. Bjorck and M. Mariani, *J. Chromatogr. B*, 668 (1995) 209.

156 P. Füglistaller, *J. Immunol. Methods*, 124 (1989) 171.

157 R.M. Baker, A.-M. Brady, B.S. Combridge, L.J. Ejim, S.L. Kingsland, D.A. Lloyd and P.L. Roberts, in D.L. Pyle (Ed.), *Separations for Biotechnology*, Vol. 3, SCI/Royal Society of Chemistry, Cambridge, 1994, pp. 53–59.

158 A. Tejeda Mansir, R. Espinoza, R.M. Montesinos and R. Guzman, *Bioprocess Eng.*, 17 (1997) 39.

159 M.A.J. Godfrey, P. Kwasowski, R. Clift and V. Marks, *J. Immunol. Methods*, 149 (1992) 21.

160 E. Bill, U. Lutz, B.M. Karlsson, M. Sparrman and H. Allgaier, *J. Mol. Recognit.*, 8 (1995) 90.

161 R.L. Fahrner, D.H. Whitney, M. Vanderlaan and G.S. Blank, *Appl. Biochem.*, 30 (1999) 121.

162 N. Ameskamp, D. Lütkemeyer, H. Tebbe and J. Lehmann, *Bioscope*, 5 (1997) 14.

163 R.L. Fahrner, G.S. Blank and G.A. Zapata, *J. Biotechnol.*, 75 (1999) 273.

164 P.E. Gustavsson, K. Mosbach, K. Nilsson and P.O. Larsson, *J. Chromatogr. A*, 776 (1997) 197.

165 B.J. Compton, M.A. Lewis, F. Whigham, J. Shores Gerald and G.E. Countryman, *Anal. Chem.*, 61 (1989) 1314.

166 T.H. Sisson and C.W. Castor, *J. Immunol. Methods*, 127 (1990) 215.

167 F.B. Anspach, H. Spille and U. Rinas, *J. Chromatogr. A*, 711 (1995) 129.

168 B. Casu, *Ann. NY Acad. Sci.*, 556 (1989) 1.

169 H. Sasaki, A. Hayashi, H. Kitagaki-Ogawa, I. Matsumo-To and N. Seno, *J. Chromatogr.*, 400 (1987) 123.

170 G. Garke, W.-D. Deckwer and F.B. Anspach, *J. Chromatogr. B*, 737 (2000) 25.

171 M.A. Jacquot-Dourges, F.L. Zhou, D. Muller and J. Jozefonvicz, *J. Chromatogr.*, 539 (1991) 417.

172 G.S. Kuschert, A.J. Hoogewerf, A.E. Proudfoot, C.W. Chung, R.M. Cooke, R.E. Hubbard, T.N. Wells and P.N. Sanderson, *Biochemistry*, 37 (1998) 11193.

173   M. Bjorklund and M.T.W. Hearn, *J. Chromatogr. A*, 762 (1997) 113.

174   A.A. Farooqui, H.C. Yang and L.A. Horrocks, *J. Chromatogr. A*, 673 (1994) 149.

175   A.A.N. del Canizo, M. Mazza, R. Bellinzoni and O. Cascone, *Arch. Biochem. Biophys.*, 61 (1996) 399.

176   G. Garke, W.-D. Deckwer and F.B. Anspach, *Sep. Sci. Technol.*, 37 (2002) 1.

177   D.F. Smith and B.V. Torres, *Methods Enzymol.*, 179 (1990) 30.

178   A. Kobata and T. Endo, *J. Chromatogr.*, 597 (1992) 111.

179   M.F.A. Bierhuizen, M. Hansson, P. Odin, H. Debray, B. Öbrink and W. van Dijk, *Glycoconjugate J.*, 6 (1989) 195.

180   M. Geng, X. Zhang, M. Bina and F. Regnier, *J. Chromatogr. B*, 752 (2001) 293.

181   J.F. Kennedy, P.M.G. Palva, M.T.S. Corella, M.S.M. Cavalcanti and L.C.B.B. Coelho, *Carbohydr. Polym.*, 26 (1995) 219.

182   I. West and O. Goldring, in S. Doonan (Ed.), *Protein Purification Protocols*, Humana Press, Totowa, *Methods Mol. Biol.*, 59 (1996) 177.

183   E. PerezCampos, F. Cordoba, E. PerezOrtega, M. Martinez and E. Zenteno, *Prep. Biochem. Biotechnol.*, 26 (1996) 183.

184   P. Gorocica, A. Monroy, B. Rivas, S. EstradaParra, Y. Leroy, R. Chavez and E. Zenteno, *Prep. Biochem. Biotechnol.*, 27 (1997) 1.

185   M. Habibi Rezaei and M. Nemat Gorgani, *Appl. Biochem. Biotechnol.*, 67 (1997) 165.

186   M.L. Yarmush, K.P. Antonsen, S. Sundaram and D.M. Yarmush, *Biotechnol. Prog.*, 8 (1992) 168.

187   N. Ibarra, A. Caballero, E. Gonzalez and R. Valdes, *J. Chromatogr. B, Biomed. Sci. Appl.*, 735 (1999) 271.

188   A. Kummer and E.C. Li-Chan, *J. Immunol. Methods*, 211 (1998) 125.

189   P. Bailon and S.K. Roy, *ACS Symp. Ser.*, 427 (1990) 150.

190   K. Kang, D. Ryu, W.N. Drohan and C.L. Orthner, *Biotechnol. Bioeng.*, 39 (1992) 1086.

191   S.B. Yan, *J. Mol. Recognit.*, 9 (1996) 211.

192   S.S. Ahmad, R. Rawala-Sheikh, A.R. Thompson and P.N. Walsh, *Thromb. Res.*, 55 (1989) 121.

193   L.J. Janis and F.E. Regnier, *J. Chromatogr.*, 444 (1988) 1.

194   N.B. Afeyan, N.F. Gordon and F.E. Regnier, *Nature*, 358 (1992) 603.

195   L. Leickt, M. Bergstrom, D. Zopf and S. Ohlson, *Anal. Biochem.*, 253 (1997) 135.

196   N.E. Labrou and Y.D. Clonis, *Biochemistry*, 45 (1996) 185.

197   M.D. Scawen, J. Derbyshire, M.J. Harvey and A. Atkinson, *Biochem. J.*, 203 (1982) 699.

198   K.D. Kulbe and R. Schuer, *Anal. Biochem.*, 93 (1979) 46.

199   M.D. Scawen and T. Atkinson, in Y.D. Clonis, A. Atkinson, C. Bruton, and C.R. Lowe (Eds.), *Reactive Dyes in Protein and Enzyme Technology*, Macmillan, Basingstoke, 1987, pp. 51–85.

200   J. Travis, J. Bowen, D. Tewksbury, D. Johnson and R. Pennell, *Biochem. J.*, 157 (1976) 301.

201   L. Lascu, H. Porumb, T. Porumb, I. Abrudan, C. Tarmure, I. Petrescu, E. Presecan, I. Proinov and M. Telia, *J. Chromatogr.*, 283 (1984) 199.

202   T. Skotland, *Biochim. Biophys. Acta*, 659 (1981) 312.

203   S. Subramanian, *CRC Crit. Rev. Biochem.*, 16 (1984) 169.

204   K.I. Shimazaki, K. Nitta, T. Sato, T. Tomimura and M. Tomita, *Comp. Biochem. Physiol.*, 101B (1992) 541.

205   S.J. Burton, S.B. McLoughlin, C.V. Stead and C.R. Lowe, *J. Chromatogr.*, 435 (1988) 127.

206   D. Hanggi and P. Carr, *Anal. Biochem.*, 149 (1985) 91.

207   J.-F. Biellmann, J.-P. Samama, C.I. Bränden and H. Eklund, *Eur. J. Biochem.*, 102 (1979) 107.

208   Y.C. Liu, R. Ledger and E. Stellwagen, *J. Biol. Chem.*, 259 (1984) 3796.

209   O. Bertrand, E. Boschetti, S. Cochet, P. Girot, E. Hebert, M. Monsigny, A.-C. Roche, P. Santambien and N. Sdiqui, *Bioseparation*, 4 (1994) 299.

210   J. Kirchberger, F. Cadelis, G. Kopperschläger and M.A. Vijayalakshmi, *J. Chromatogr.*, 483 (1989) 289.

211   K.N. Kuan, G.L. Jones and C.S. Vestling, *Biochemistry*, 18 (1979) 4366.

212   K. Smith, T.K. Sundaram, M. Kernick and A.E. Wilkinson, *Biochim. Biophys. Acta*, 708 (1982) 17.

213   E.E. Farmer and J.S. Easterby, *Anal. Biochem.*, 141 (1984) 79.

214   R.K. Scopes, V. Testolin, A. Stoter, K. Griffiths-Smith and E.M. Algar, *Biochem. J.*, 228 (1985) 627.

215   A.J. Turner and J. Hryszko, *Biochim. Biophys. Acta*, 613 (1980) 256.

216   M.J. Harvey, in J.M. Curling (Ed.), *Methods of Plasma Protein Fractionation*, Academic Press, London, 1980, p. 189.

217   B.J. Horstmann, C.N. Kenney and H.A. Chase, *J. Chromatogr.*, 361 (1986) 179.

218   G. Giraudi and M. Petrarulo, *Ann. Chim. (Rome)*, 79 (1989) 231.

219   R.D. Brehm, H.S. Tranter, P. Hambleton and J. Melling, *Appl. Environ. Microbiol.*, 56 (1990) 1067.

220   N.D. Harris and P.G.H. Byfield, *FEBS Lett.*, 103 (1979) 162.

221   E. Knight and D. Fahey, *J. Biol. Chem.*, 256 (1981) 3609.

222   R.K. Scopes, *J. Chromatogr.*, 376 (1986) 131.

223   P.M. Boyer and J.T. Hsu, *Biotechnol. Tech.*, 4 (1990) 61.

224   Y. Clonis, *CRC Crit. Rev. Biotechnol.*, 7 (1988) 263.

225   X. Santarelli, V. Fitton, N. Verdoni and C. Cassagne, *J. Chromatogr. B*, 739 (2000) 63.

226   C.R. Lowe, S.J. Burton, N.P. Burton, W.K. Alderton, J.M. Pitts and J.A. Thomas, *Trends Biotechnol.*, 10 (1992) 442.

227   Y.D. Clonis, N.E. Labrou, V.P. Kotsira, C. Mazitsos, S. Melissis and G. Gogolas, *J. Chromatogr. A*, 89 (2000) 33.

228   J.A. Noriega, A. Tejeda, I. Magana, J. Ortega and R. Guzman, *Biotechnol. Prog.*, 13 (1997) 296.

229   K. Sproule, P. Morrill, J.C. Pearson, S.J. Burton, K.R. Hejnaes and H. Valore, *J. Chromatogr. B*, 740 (2000) 17.

230   N.P. Burton and C.R. Lowe, *J. Mol. Recognit.*, 5 (1992) 55.

231   S.C. Melissis, D.J. Rigden and Y.D. Clonis, *J. Chromatogr. A*, 917 (2001) 29.

232   H. Watanabe, K. Yamasaki, U. Kragh-Hansen, S. Tanase, K. Harada, A. Suenaga and M. Otagiri, *Pharm. Res.*, 18 (2001) 1775.

233   S.F. Teng, K. Sproule, A. Husain and C.R. Lowe, *J. Chromatogr. B*, 740 (2000) 1.

234   H.L. Weith, J.L. Wiebers and P.T. Gilham, *Biochemistry*, 9 (1970) 4396.

235   M. Glad, S. Ohlson, L. Hansson, M.O. Mansson and K. Mosbach, *J. Chromatogr.*, 200 (1980) 254.

236   L. Hansson, M. Glad and C. Hansson, *J. Chromatogr.*, 265 (1983) 37.

237   B. Anspach, K.K. Unger, P. Stanton and M.T.W. Hearn, *Anal. Biochem.*, 179 (1989) 171.

238   Y.C. Li, U. Pfüller, E.L. Larsson, H. Jungvid, I.Y. Galaev and B. Mattiasson, *J. Chromatogr. A*, 925 (2001) 115.

239   S. Soundararajan, M. Badawi, C. Montano Kohlrust and J.H. Hageman, *Anal. Biochem.*, 178 (1989) 125.

240   R.P. Singhal, B. Ramamurthy, N. Govindraj and Y. Sarwar, *J. Chromatogr.*, 543 (1991) 17.

241   C.W. Gehrke, K.C. Kuo, G.E. Davis, R.D. Suits, T.P. Waalkes and E. Borek, *J. Chromatogr.*, 150 (1978) 455.

242   K. Kemper, E. Hagemeier, D. Ahrens, K.S. Boos and E. Schlimme, *Chromatographia*, 19 (1984) 288.
243   S. Ackerman, B. Cool and J.J. Furth, *Anal. Biochem.*, 100 (1979) 174.
244   B. Pace and N.R. Pace, *Anal. Biochem.*, 107 (1980) 128.
245   T.F. McCutchan and P.T. Gilham, *Biochemistry*, 12 (1973) 4840.
246   J. Porath, J. Carlsson, I. Olsson and G. Belfrage, *Nature*, 258 (1975) 598.
247   Z. El Rassi and C. Horváth, *J. Chromatogr.*, 359 (1986) 241.
248   J. Porath, *Protein Expr. Purif.*, 3 (1992) 263.
249   F. Helfferich, *Nature*, 189 (1961) 1001.
250   P. Roumeliotis, K.K. Unger, A.A. Kurganov and V. Davankow, *J. Chromatogr.*, 255 (1983) 51.
251   L. Sundberg and J. Porath, *J. Chromatogr.*, 90 (1974) 87.
252   Y.J. Zhao, E. Sulkowski and J. Porath, *Eur. J. Biochem.*, 202 (1991) 1115.
253   G.S. Chaga, *J. Biochem. Biophys. Methods*, 49 (2001) 313.
254   F.H. Arnold, *Biotechnology*, 9 (1991) 151.
255   S. Vunnum and S. Cramer, *Biotechnol. Bioeng.*, 54 (1997) 373.
256   P. Arvidsson, A.E. Ivanov, I.Y. Galaev and B. Mattiasson, *J. Chromatogr. B*, 753 (2001) 279.
257   L. Andersson, E. Sulkowski and J. Porath, *J. Chromatogr.*, 421 (1987) 141.
258   L. Andersson, E. Sulkowski and J. Porath, *Bioseparation*, 2 (1991) 15.
259   H. Wu and D.F. Bruley, *Biotechnol. Prog.*, 15 (1999) 928.
260   T. Oswald, G. Hornbostel, U. Rinas and F.B. Anspach, *Biotechnol. Appl. Biochem.*, 25 (1997) 109.
261   G. Muszynska, G. Dobrowolska, A. Medin, P. Ekman and J.O. Porath, *J. Chromatogr.*, 604 (1992) 19.
262   L.D. Holmes and M.R. Schiller, *J. Liq. Chromatogr. Rel. Technol.*, 20 (1997) 123.
263   R.F. Boyer, S.M. Generous, T.J. Nieuwenhuis and R.A. Ettinger, *Biotechnol. Appl. Biochem.*, 12 (1990) 79.
264   T. Mantovaara, H. Pertoft and J. Porath, *Biotechnol. Appl. Biochem.*, 11 (1989) 564.
265   J. Porath, B. Olin and B. Granstrand, *Arch. Biochem. Biophys.*, 225 (1983) 543.
266   J.J. Winzerling, D.Q.D. Pham, S. Kunz, P. Samaraweera, J.H. Law and J. Porath, *Protein Expr. Purif.*, 7 (1996) 137.
267   M.C. Millot, F. Herve and B. Sebille, *J. Chromatogr. B*, 664 (1995) 55.
268   H. Chaouk and M.T.W. Hearn, *J. Biochem. Biophys. Methods*, 39 (1999) 161.
269   M. Zachariou, I. Traverso and M.T.W. Hearn, *J. Chromatogr.*, 646 (1993) 107.
270   F.B. Anspach, *J. Chromatogr. A*, 672 (1994) 35.
271   G. Fassina, A. Verdoliva, M.R. Odierna, M. Ruvo and G. Cassini, *J. Mol. Recognit.*, 9 (1996) 564.
272   A. Verdoliva, G. Basile and G. Fassina, *J. Chromatogr. B*, 749 (2000) 233.
273   G. Palombo, A. Verdoliva and G. Fassina, *J. Chromatogr. B*, 715 (1998) 137.
274   G. Palombo, S. De Falco, M. Tortora, G. Cassani and G. Fassina, *J. Mol. Recognit.*, 11 (1998) 243.
275   G. Palombo, M. Rosi, G. Cassani and G. Fassina, *J. Mol. Recognit.*, 11 (1998) 247.
276   L.R. Castilho, F.B. Anspach and W.-D. Deckwer, *J. Membr. Sci.*, 207 (2002) 253.
277   G.K. Ehrlich and P. Bailon, *J. Biochim. Biophys. Methods*, 49 (2001) 443.
278   C.R. Lowe, A.R. Lowe and G. Gupta, *J. Biochem. Biophys. Methods*, 49 (2001) 561.
279   R.X. Li, V. Dowd, D.J. Stewart, S.J. Burton and C.R. Lowe, *Nat. Biotechnol.*, 16 (1998) 190.
280   S.f. Teng, K. Sproule, A. Hussain and C.R. Lowe, *J. Chromatogr. B*, 740 (2000) 1.
281   P.R. Morrill, G. Gupta, K. Sproule, D. Winzor, J. Christensen, I. Mollerup and C.R. Lowe, *J. Chromatogr. B*, 774 (2002) 1.

282 D.B. Kaufman, M.E. Hentsch, G.A. Baumbach, J.A. Buettner, C.A. Dadd, P.Y. Huang, D.J. Hammond and R.G. Carbonell, *Biotechnol. Bioeng.*, 77 (2002) 278.

283 K. Amatschek, R. Necina, R. Hahn, E. Schallaun, H. Schwinn and D. Josic, *J. High Resolut. Chromatogr.*, 23 (2000) 47.

284 J. Thömmes, *Adv. Biochem. Eng.*, 58 (1997) 185.

285 F.B. Anspach, D. Curbelo, R. Hartmann, G. Garke and W.-D. Deckwer, *J. Chromatogr. A*, 865 (1999) 129.

286 R.L. Fahrner, G.S. Blank and G.A. Zapata, *J. Biotechnol.*, 75 (1999) 273.

287 J. Thömmes, A. Bader, M. Halfar, A. Karau and M.-R. Kula, *J. Chromatogr. A*, 752 (1996) 111.

288 U. Reichert, E. Knieps, H. Slusarczyk, M.-R. Kula and J. Thömmes, *J. Biochem. Biophys. Methods*, 49 (2001) 533.

289 N.A. Willoughby, T. Kirschner, M.P. Smith, R. Hjorth and N.J. Titchener-Hooker, *J. Chromatogr. A*, 840 (1999) 195.

Erich Heftmann (Editor)
Chromatography, 6th edition
Journal of Chromatography Library, Vol. 69A
© 2004 Elsevier B.V. All rights reserved

Chapter 4

Ion chromatography

CHARLES A. LUCY and PANOS HATSIS

CONTENTS

4.1   Introduction 172
4.2   Instrumentation 173
      4.2.1   General instrumentation 173
      4.2.2   Eluent 174
           4.2.2.1   Brief description of ion exchange 174
           4.2.2.2   Common eluents 174
           4.2.2.3   Eluent generation 175
      4.2.3   Suppressors 176
           4.2.3.1   Theory of suppression 176
           4.2.3.2   Packed-bed suppressors 177
           4.2.3.3   Membrane suppressors 178
           4.2.3.4   Electrochemical regeneration of suppressors 178
      4.2.4   Detection 180
           4.2.4.1   Conductivity detection 180
           4.2.4.2   Amperometric detection 183
           4.2.4.3   Ultraviolet-absorbance detection 184
           4.2.4.4   Post-column reaction detection 185
           4.2.4.5   Mass-spectrometric detection 186
4.3   Selectivity in ion chromatography 186
      4.3.1   Importance of selectivity 186
      4.3.2   Properties of analyte and eluent ions affecting selectivity 187
      4.3.3   Effect of mobile phase on selectivity 187
           4.3.3.1   Choice of eluent ion 188
           4.3.3.2   Effect of eluent on selectivity 188
           4.3.3.3   Use of organic additives 190
           4.3.3.4   Temperature 190
      4.3.4   Effect of stationary phase on selectivity 191
           4.3.4.1   Structure 191
           4.3.4.2   Role of column material 193

        4.3.4.3   Role of the type of ion-exchange site 194

        4.3.4.4   Role of the structure of the ion-exchange site 196

    4.3.5   Alternative modes of separation 198

        4.3.5.1   Cryptand columns 198

        4.3.5.2   Ion exclusion 198

        4.3.5.3   Electrostatic ion chromatography 201

4.4   Sample preparation 203

    4.4.1   Concentration of trace analytes 203

        4.4.1.1   High-volume direct injection 203

        4.4.1.2   Pre-concentrator column 204

    4.4.2   Matrix removal 205

        4.4.2.1   Solid-phase extraction 205

    4.4.3   On-line sample pre-treatment 206

4.5   Future directions 207

    4.5.1   High-speed analysis 207

    4.5.2   Process and field analysis 208

Acknowledgments 208

References 209

4.1 INTRODUCTION

Ion chromatography (IC) refers to the trace analysis of ions on low-capacity high-efficiency columns, possessing fixed ion-exchange sites. Most commonly, these columns are teamed with suppressed-conductivity detection to yield parts-per-billion limits of detection (LOD) of the seven common inorganic anions (F^-, Cl^-, NO_2^-, Br^-, NO_3^-, HPO_4^{2-}, and SO_4^{2-}) and, to a lesser extent, carboxylic acids, as well as the six common cations (Li^+, Na^+, NH_4^+, K^+, Mg^{2+}, and Ca^{2+}) and small amines. This chapter will focus on the fundamental and instrumental aspects of such analyses. Classical high-capacity ion-exchange resins (*e.g.*, Dowex) will not be discussed.

Ion chromatography had its beginnings in the seminal works of Hamish Small and co-workers in 1975 [1] and Gjerde *et al.*, in 1979 [2]. These papers demonstrated that conductivity could be used as a sensitive and universal on-line detector for ions. Low-capacity ion-exchange columns allowed dilute ionic solutions to be used as the eluent. These eluents possess a sufficiently low conductivity background that the conductivity of the analyte ions can be monitored directly [2]. This approach is generally referred to as non-suppressed IC. Alternately, the eluent can be dilute enough so that it can be converted to a neutral form after the separation. This approach is referred to as suppressed IC. Both suppressed and non-suppressed IC have demonstrated the ability to perform sensitive parts-per-million determinations of multiple ions in a few minutes. Since 1975, there have been steady developments and refinements in IC that have led to improved separation speed, detection limits, and ion selectivity. Currently, IC accounts for over $165 million of the $3 billion worldwide liquid-chromatography market, with over 2500 IC units sold in 2002. Environmental and contract laboratories account for over a third of the IC users and focus on the determination of anions in drinking water and wastewater, the chemical industry monitors inorganic anions, small amines, and carboxylates in process streams, the power utilities and the semiconductor industry determine trace ionic impurities in ultra-pure water, and the pharmaceutical companies determine small amines.

This article is intended to give an overview of the fundamental principles and characteristics of IC and to introduce recent developments. For a more detailed discussion, the reader is referred to the key monograph in the field, *Ion Chromatography: Principles and Applications* [3].

4.2 INSTRUMENTATION

4.2.1 General instrumentation

A list of current IC instrumentation is available in a recent product review in *Analytical Chemistry* [4] (Chap. 12). An IC system differs from a HPLC system in several respects: Firstly, the entire flow-path (pump, injector, column, tubing, and detector cell) must be metal-free. Even Standard 304 and 316 stainless steel can release significant levels of metal contamination, particularly when corrosive eluents are used. Metal ions can alter the column retention characteristics, can poison the cation-exchange sites on the suppressor and can interfere with detection. As a consequence, IC systems are preferably made of polymeric materials. While early polymeric systems had pressure limits that were significantly inferior to contemporary HPLC systems, since the mid-1980s, IC systems have been constructed of polyether ether ketone (PEEK). PEEK is chemically extremely inert, flexible, inexpensive, and capable of withstanding high pressures, so that current IC systems are rated to >4000 psi (28 MPa).

Dedicated or process IC units typically use isocratic pumps, while higher-end IC systems can perform gradient elution. Hydroxide gradients can also be performed, using an isocratic pump and on-line hydroxide generation (Sec. 4.2.2.3). The flow-range of IC pumps mirrors that of modern microbore (2-mm-ID) or conventional (4.6-mm-ID) columns, but no capillary IC systems are commercially available. Pump pulsation is an important consideration with non-suppressed-conductivity or suppressed-conductivity detectors and CO_3^{2-}/HCO_3^- eluents, where the background conductivity is high, but is less important with OH^- eluents, particularly those generated on-line. Degassing modules are available, but they are not as common or as necessary as in RP-LC, because pure aqueous eluents are typically used in IC. IC injectors are constructed of PEEK. The typical injection volume is 20 μL, with 1-mL injections and pre-concentrator columns used for trace (sub-ppb) analysis (Sec. 4.4.1). Temperature control is recommended for maintaining retention-time reproducibility; it can significantly improve the baseline, and this is particularly important for trace analysis.

IC stationary phases are low-capacity ion exchangers, typically 0.006–0.06 meq/mL. This is far below the 0.3–0.5 meq/mL of silica-based bonded-phase ion exchangers used for biochemical analysis or the 0.8–2.4 meq/mL of classical gel-type (polystyrene/divinylbenzene) ion exchangers. The low ion-exchange capacity of IC columns allows dilute electrolyte eluents to be used, which can subsequently be monitored directly or fully neutralized by the suppressor. Silica-based columns (*e.g.*, Vydac) can be used in non-suppressed methods where neutral eluents, such as phthalate are used. However, the strongly alkaline nature of suppressed IC eluents precludes the use of silica-based columns. Thus, most suppressed-IC packings are made from polymeric materials, such as ethylvinylbenzene (Dionex), methacrylate (Dionex, Metrohm, Alltech), polyvinyl alcohol (Metrohm), and polystyrene (Metrohm, Alltech), and are either nonporous or macroporous (200 nm), with a diameter of 5–15 μm. Efficiencies are typically 4000–7000 plates. Separation selectivity is governed to a greater extent by the column and to a lesser extent by the eluent than in RP-LC. The effects of column and eluent on ion selectivity are discussed in Secs. 4.3.3 and 4.3.4, respectively.

4.2.2 Eluent

4.2.2.1 Brief description of ion exchange

The primary equilibrium in ion chromatography is the ion-exchange displacement of an eluent ion, E, from the stationary or resin phase (denoted by the subscript r) by an analyte, A, initially in the mobile phase (denoted by the subscript m). For anion exchange of singly charged ions this equilibrium is

$$A_m^- + E_r^- \rightleftharpoons E_m^- + A_r^- \tag{4.1}$$

The charge of the eluent and analyte is balanced by cations (of the same charge as the exchange site on the ion exchanger) in the mobile phase. However, these cations play no role in the anion-exchange process, and so they are not generally shown.

4.2.2.2 Common eluents

As can be seen in Eqn. 4.1, the eluent ion is necessary for the process of ion exchange and affects retention through its nature and concentration. (More will be said about this in Sec. 4.3.3) Further, the eluent should be the salt of a weak acid ($pK_a > 7$) to benefit from suppression and should have adequate buffering capacity to ensure method robustness. The general order of eluent strength in anion exchange is

$$OH^- < BO_3^- < HCO_3^- < CO_3^{2-} \tag{4.2}$$

Hydroxide is the weakest eluent, although its relative eluent strength increases when the ion-exchange site on the stationary phase possesses alkanol functionalities (Sec. 4.3.4.4). Borate and HCO_3^- are also weak eluents, due to their monovalent charge. Carbonate is a powerful eluting ion. Indeed, CO_3^{2-} contamination of OH^- eluents originally made retention times with hydroxide eluents less reproducible. The advent of eluent generation (Sec. 4.2.2.3) has circumvented this problem. Commonly, HCO_3^- and CO_3^{2-} are used together, as this provides a well-buffered eluent over the pH range 8–11, having an eluent strength that is easily varied by altering the ratio of the two anions. Currently, OH^- and HCO_3^-/CO_3^{2-} are the most commonly used eluents in suppressed IC of anions, with an increasing trend towards OH^-, given its superior detection limits and linearity in suppressed-conductivity detection and its enhanced convenience and reliability in on-line eluent generation. Aromatic carboxylic acid buffers are the most commonly used eluents for nonsuppressed IC. They have low equivalent conductances (λ_{E-}, Eqn. 4.7), and so provide a low background conductivity. Further, the aromatic moiety is an excellent UV chromophore, and so aromatic carboxylates are ideal eluents for indirect photometric detection (Sec. 4.2.4.3). In practice, most anions are eluted effectively by an eluent with a -2 charge. Thus, *o*-phthalate ($pK_{a,1} = 2.95$, $pK_{a,2} = 5.41$) near its second pK_a is most commonly used. Eluents for nonsuppressed IC are prepared by dissolving the acid form of the eluent in water and adjusting the pH with LiOH. The lower conductance of Li^+ (λ_{E+}, Eqn. 4.7) relative to Na^+ or K^+ yields a reduced conductance background for nonsuppressed detection. The usual eluent ion for cation exchange is H^+. The acid used can vary considerably (*e.g.*, sulfuric acid, nitric acid, citric acid, tartaric acid), but the typical eluents for suppressed conductivity detection are sulfuric acid or methanesulfonic acid (MSA).

4.2.2.3 Eluent generation

There are many benefits in using OH^- as an eluent for suppressed IC. However, manual preparation of OH^- eluents for IC is not easy. They should be prepared daily from 50% NaOH solution, which has a reduced CO_3^{2-} solubility. Commercial 50% NaOH solution contains $0.06-0.17$ mol% CO_3^{2-} [6]. This CO_3^{2-} contributes to the background conductivity, causing substantial baseline shifts in gradient elution. Ionic impurities in commercial OH^- solutions further aggravate the challenges of using OH^- gradients. This is why, in suppressed IC, isocratic HCO_3^-/CO_3^{2-} was the preferred eluent for the first 20 years, despite the clear advantages of OH^- and gradient elution.

In the early 1990s Dasgupta and co-workers [6,7] pioneered the on-line generation of hydroxide eluents by electrodialysis. A commercial automated Eluent Generation Module (EG40) for on-line generation of KOH or MSA eluents was introduced by Dionex in 1997 [8]. A schematic diagram of the Eluent Generation Model for KOH production is shown in Fig. 4.1. It consists of a KOH generation chamber and a K^+ ion electrolyte reservoir. These are separated by a cation-exchange connector. To generate a KOH eluent, deionized water is pumped through the KOH generation chamber and a DC current is applied between the anode and cathode of the cartridge. Electrolysis of water occurs, generating H^+ and OH^-. The H^+ ions generated at the anode are displaced by K^+ within the cation-exchange connector in essentially a reverse-suppression process. The K^+ from the electrolyte reservoir and the OH^- generated at the cathode produce KOH. The eluent then passes through on-line degassing tubing to eliminate the electrolysis gases. The concentration of the resultant KOH eluent is directly related to the electrical current and inversely proportional to the eluent flow-rate. At 1.0 mL/min, the EG40 can generate up to 100 mM of KOH, and can operate at pressures up to 3000 psi (21 MPa). This eluent is carbonate-free and, thus, has a reduced background conductivity. Manually prepared NaOH exhibits a conductivity of $2-5$ μS/cm after suppression, and this results in significant baseline drift during gradient elution, as is evident in Fig. 4.5. With on-line KOH generation, background conductivity is about 0.3 μS/cm after suppression [7,8]. For comparison, pure water has a conductivity of 0.06 μS/cm. This lower background

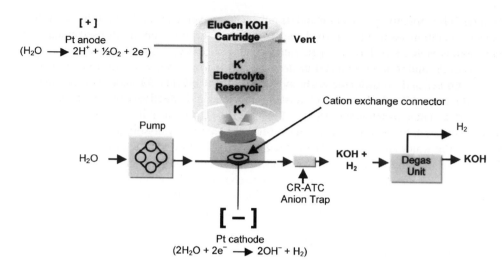

Fig. 4.1. Schematic diagram of electrodialytic on-line KOH eluent generator. Based on the Dionex Eluent Generation Model EG50. (Courtesy of Dionex Corporation.)

conductivity allows gradient elution to be performed with minimal baseline shift. The eluent generator for cation-exchange chromatography also operates in an electrodialytic fashion. Distilled water is pumped into the device and electrolyzed to generate OH^- at the cathode. This OH^- displaces MSA^- ions in the MSA^- electrolyte reservoir. The displaced ions migrate across the anion exchange into the MSA generation chamber to produce a MSA solution. The concentration of MSA is determined by the current applied to the MSA generator and the eluent flow-rate. Up to 100 meq/min can be generated, *i.e.*, 100 mM at 1.0 mL/min or 200 mM at 0.5 L/min.

4.2.3 Suppressors

4.2.3.1 Theory of suppression

Suppression is a post-column reaction primarily designed to enhance the sensitivity of conductivity detection by reducing the background conductivity through neutralization of the eluent ion. If, *e.g.*, a typical eluent anion, E^-, is the conjugate base of the weak acid HE, exchange of the counter-ion (*e.g.*, Na^+) of the eluent with H^+ will result in the formation of the weak acid HE.

$$(Suppressor)-H^+ + Na^+E^- \rightleftharpoons (Suppressor)-Na^+ + HE \tag{4.3}$$

HE will partially dissociate, generating a background conductivity.

$$HE \rightleftharpoons H^+ + E^- \tag{4.4}$$

The magnitude of this background conductivity correlates with the strength of the acid HE. For instance, typical background conductivities after suppression are 12–15 μS/cm for

HCO_3^-/CO_3^{2-} eluents ($pK_{a,1} = 6.35$) and, ideally, 0.06 μS/cm for OH^- ($pK_w = 14$). It is for this reason that eluents in IC should possess a pK_a greater than 7. The clear benefit of OH^- as an eluent for suppressed-conductivity detection has been the driving force in many of the developments in IC over the past decade. With less completely suppressed eluents (*e.g.*, carbonate) the residual conductivity after suppression can lead to nonlinearity of calibration, as will be discussed in Sec. 4.2.4.1. Another manifestation of the background conductivity is the appearance of a negative peak at the void volume. This peak is known as the *water dip*, and results from the absence of the eluent ion (E^-) from the sample solvent. The unretained solvent zone does not possess the residual conductivity resultant from Eqn. 4.4, and therefore has a lower conductivity. Thus, the magnitude of the water dip is directly proportional to the background conductivity of the eluent after suppression.

Simultaneous with the neutralization of the eluent, the counter-ion (Na^+) associated with the analyte anion (A^-) is also exchanged for H^+.

$$(Suppressor)-H^+ + Na^+A^- \rightleftharpoons (Suppressor)-Na^+ + HA \qquad (4.5)$$

If HA is a strong acid, as it is for the seven common anions (F^-, Cl^-, NO_2^-, Br^-, NO_3^-, HPO_4^{2-}, and SO_4^{2-}), it will fully dissociate, yielding A^- and H^+. Since the conductivity of H^+ (350 S cm² eq⁻¹) is very high, relative to Na^+ (50 S cm² eq⁻¹) and all other cations, there is an overall enhancement in the conductivity signal due to A^-. The significance of HA being a weak acid is discussed in Sec. 4.2.4.1. The use of suppression for cation-exchange chromatography is controversial, some people believing that suppression results in a significant gain in sensitivity, whereas others do not. The principles of the suppression of eluents in cation exchange are analogous to those in anion exchange. Briefly, in cation exchange the eluent ion is usually H^+ and the suppressor usually contains OH^-. Exchange of the counter ion of H^+ (*e.g.*, MSA^-) with OH^- results in the formation of water, and the eluent is effectively neutralized.

4.2.3.2 Packed-bed suppressors

Since the original patent on suppression technology expired in the mid-1990s, there has been a proliferation of different types of suppressors. Each manufacturer makes a different style of suppressor [5], which can be classified with regard to format (packed bed *vs.* membrane) and mode of regeneration (chemical *vs.* electrochemical). In the original work of Small *et al.* [1], the suppressor was a 9 mm × 250-mm column, packed with 200- to 400-mesh Dowex 50W-X8 in the H^+ form. This suppressor column revolutionized anion analysis by enabling for the first time rapid parts-per-million detection of anions in complex matrices. Alltech now markets a disposable 4.6 mm × 100-mm column suppressor (Model 335) with sufficient capacity for a 7-h operation. However, column suppression has a number of drawbacks. Firstly, the suppressor column has a limited capacity, and it must be regenerated or replaced periodically. Secondly, the large volume of the suppressor adds considerably to peak-broadening. Thirdly, retention times of weak acid analytes vary, due to variation in Donnan exclusion of these analytes, as the cation-exchange bed of the suppressor becomes exhausted. The drawbacks of column suppression are circumvented in the Metrohm Model 753 micro packed-bed suppressor.

In this device three small packed beds are housed in a three-position rotary valve. At any given time, one micro bed is suppressing the effluent, the second is being chemically regenerated with 100 mM H$_2$SO$_4$, and the third is being rinsed prior to returning to service. Each micro packed bed has sufficient capacity for 2 h of operation with a standard carbonate/bicarbonate eluent. In a typical operation, the valve is rotated prior to each chromatographic run. The micro-packed bed device is particularly robust, able to withstand pressures up to 2 MPa.

4.2.3.3 Membrane suppressors

Membrane suppressors provide continuous regeneration, low dead-volumes, and reproducible retention times for weak-acid analytes [9]. The heart of a Micromembrane Suppressor (MMS) for anion determination is a cation-exchange membrane, as shown in Fig. 4.2a [10]. Effluent (Na$^+$A$^-$, Na$^+$OH$^-$) flows on one side of the membrane, and sulfuric acid regenerant (H$^+$HSO$_4^-$) flows in the opposite direction on the other side. Cations move freely through the cation-exchange membrane. Thus, H$^+$ transfers from the regenerant into the eluent to generate H$^+$A$^-$ and H$_2$O, while the eluent cation (Na$^+$) passes simultaneously into the regenerant. The eluent is sandwiched between two cation-exchange membranes to maximize the cation flux and thereby the suppression capacity. Meanwhile, anions experience Donnan exclusion and do not pass through the membrane. The current version of the Micromembrane Suppressor (MMS III, Dionex) has a void volume of <50 or <15 μL in the 4- or 2-mm formats, respectively, and can suppress 150 or 37.5 μeq/min, respectively. In theory, membrane suppressors should achieve background conductivities of $12-15$ μS/cm for HCO$_3^-$/CO$_3^{2-}$ and 0.06 μS/cm for OH$^-$ eluents. However, in practice, with OH$^-$ eluents the background is limited by impurities in the eluent and by regenerant penetration through the membrane. The MMS III has a higher-capacity cation-exchange membrane than previous models, and not only improves the cation flux (thus increasing the amount of eluent that can be suppressed), but also lowers the background caused by leakage of regenerant ions into the eluent. When chemical suppression is used, the anion MMS III provides the lowest background levels (and thus the best detection limits) of the Dionex suppressors, and is compatible with eluents containing CO$_3^{2-}$ and OH$^-$ and eluents containing organic solvents. MMS III suppressors are also available for cation separations, with MSA- and sulfuric acid-containing eluents.

4.2.3.4 Electrochemical regeneration of suppressors

The primary drawback of membrane suppressors is that the regenerant flow must be three to ten times the eluent flow to ensure complete suppression and thus optimal sensitivity. For the anion MMS III (4-mm format) this translates to $5-10$ mL/min of 100 mM H$_2$SO$_4$. Thus, a second classification of suppressors is based on their mode of regeneration. All of the above suppressors are "chemically suppressed". That is, regeneration is achieved by using a solution of regenerant. An alternative is "electrolytic" or "self-regeneration" suppression. In these devices regeneration is accomplished by

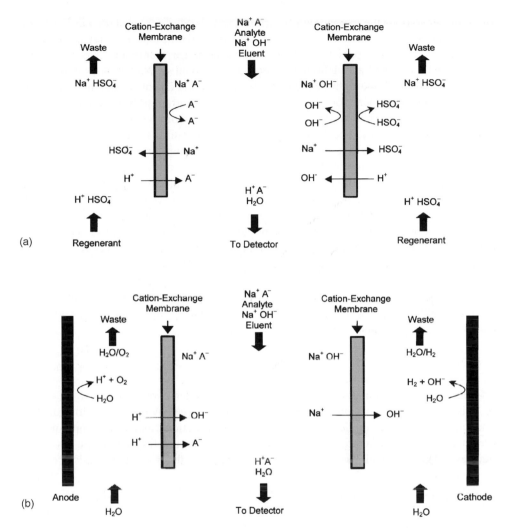

Fig. 4.2. Processes occurring within: (a) a membrane suppressor and (b) an electrolytic membrane suppressor. (Adapted from Ref. 10. Courtesy of Dionex Corporation.)

producing the regenerant ion *in situ* through electrolysis of water at electrodes within the suppressor, as shown in Fig. 4.2b [11]. Thus, the need for large volumes of regenerant and for a separate pumping system for the regenerant is eliminated.

When electrolytic regeneration is performed in a membrane suppressor [12], platinum electrodes are placed just outside the membrane. The effluent from the column flows in one direction on the inside of the ion-exchange membrane, while the outer stream, flowing in the opposite direction, is electrolyzed to produce H^+ at the anode and OH^- at the cathode:

Anode $H_2O \rightarrow 2H^+ + 1/2\,O_2 + 2e^-$ $\hspace{4cm}$ (4.6a)

Cathode $2H_2O + 2e^- \rightarrow 2OH^- + H_2$ $\hspace{4cm}$ (4.6b)

For anion separations the membrane is a strong-cation exchanger, as shown in Fig. 4.2b, and the neutralization reactions are similar to those shown in Fig. 4.2a. Hydrogen ions, generated at the anode, pass through the cation-exchange membrane into the eluent stream. Simultaneously, a stoichiometrically equal amount of Na^+ ions pass from the eluent stream through the right membrane into the cathode chamber. Again, as with the chemical membrane suppressor (Fig. 4.2a), Donnan equilibrium prevents co-anions from passing through the membrane. The capacity of the suppressor is controlled by the current applied to the electrodes. The Dionex Ultra self-regenerating suppressor (SRS) has a suppression capacity of 200 or 50 μeq/min in the 4- or 2-mm formats, respectively.

The choice of the source of the regenerant water is a compromise of sensitivity and convenience. Greatest sensitivity and tolerance to organic solvents are achieved by using fresh water as the regenerant, but it is more convenient to recycle the suppressed eluent through the outer chamber of the suppressor. A disadvantage of membrane-based suppressors is the fragile nature of the membranes. They are thin so as to maximize the suppression capacity, but this compromises their mechanical strength. Recent suppressor devices, such as the Atlas Electrolytic Suppressor (AES, Dionex), the DS-Plus (Alltech), and the Model 828 (Metrohm) combine the ruggedness of packed-bed suppressors with the ease of operation of electrolytic regeneration. The DS-Plus and Model 828 suppressors also incorporate on-line degassing to remove carbonic acid [13,14], and this results in a reduction of the background from 20 to $1-2$ μS [15].

4.2.4 Detection

Conductivity detection remains the most widely used detection scheme for IC, due to the universal response of conductivity detection to all ions. We will therefore discuss conductivity detection in detail. Other forms of detection, including amperometric detection, mass-spectrometric detection and post-column-reaction detection have also been used. These detection schemes will be discussed briefly at the end of this section.

4.2.4.1 Conductivity detection

Conductivity detection is performed with a detector cell, consisting of two electrodes to which an electric potential is applied [16]. The ions passing through the detector cell move in response to the electrical field, *i.e.*, anions move to the anode and cations to the cathode. The moving ions generate a current and it is this current that is the analytical signal. The generated current is dependent on the ionic conductance (λ) and concentration of the ions passing through the detector cell. To avoid polarization of the ions, the potential is an alternating current (AC) in the range of 50 to 10,000 Hz. The amplitude of this AC potential should be as high as possible to maximize the resultant current, but it must be low enough to avoid redox reactions at the electrode surfaces. In the absence of analyte ions, the conductivity signal is due entirely to the eluent ions. During elution, analyte ions displace eluent co-ions, so that when analyte ions pass through the conductivity cell, there is an increased concentration of analyte ions and a decreased concentration of eluent ions. The resultant detection signal is then dependent on the difference in ionic conductance between the eluent ions and the analyte ions. The change in conductivity, ΔG, during elution is given by Eqn. 4.7 [17]

$$\Delta G = \frac{(\lambda_{E^+} + \lambda_{A^-})\alpha_A - (\lambda_{E^+} + \lambda_{E^-})\alpha_E \alpha_A}{10^{-3}K} c_A \qquad (4.7)$$

where E^+ and E^- are the cation and anion of the eluent, λ is the ionic equivalent conductance of an ion, α_A and α_E are the degree of dissociation of the analyte and eluent ions, K is the cell constant of the detector and c_A is the analyte concentration passing through the detector. Therefore, Eqn. 4.7 predicts that an enhancement in signal can be obtained, if the conductivity of the eluent is reduced or essentially eliminated, and if the eluent is not fully dissociated. In the special case of complete dissociation of the analyte and eluent ions Eqn. 4.7 becomes

$$\Delta G = \frac{(\lambda_{A^-} - \lambda_{E^-})}{10^{-3}K} c_A \qquad (4.8)$$

There are a number of ways to reduce the background conductivity of the eluent. Firstly, and most obviously, a suppressor can be used to reduce the background conductivity and simultaneously enhance the conductivity of the analyte ion (Sec. 4.2.3). In this case, α_E in Eqn. 4.7 becomes 0 and the change in conductivity will arise from the analyte ion and an equivalent amount of H^+

$$\Delta G = \frac{(\lambda_{H^+} + \lambda_{A^-})\alpha_A}{10^{-3}K} c_A \qquad (4.9)$$

Alternatively, the eluent contribution to the background conductivity can be decreased by employing small concentrations of weakly conducting eluent ions [2]. Typically, suitable eluent ions for nonsuppressed anion exchange are organic aromatic acids, such as salicylate, phthalate, or benzoate, whereas, for cation exchange, organic bases, such as aniline, benzylamine, or methylpyridine, can be used. In this way, detection can be accomplished without a suppressor and with a substantial simplification of the IC instrument.

The linearity of calibration plots has been an issue with suppressed-conductivity detection when eluents are used that partially dissociate after suppression (*e.g.*, CO_3^{2-}/HCO_3^-). The nonlinearity is a result of the protonation of dissociated eluent ions by analyte ions after suppression, depicted in Fig. 4.3 [18]. Therefore, the background conductivity will be slightly lower in the presence of an analyte ion than in its absence, when the eluent alone is flowing through the detector cell (Fig. 4.3a). This decreases the slope of the calibration curve at the low-concentration end compared to the high-concentration end (Fig. 4.3b). This phenomenon is not seen when nonsuppressed conductivity detection is used.

The decision whether to suppress or not to suppress was a source of heated controversy in the early years of IC, when suppression was a patented technique. Currently, it is acknowledged that suppressed conductivity detection is preferred for anion exchange, since it results in a considerable decrease in the contribution of the eluent (Eqn. 4.7). Detection limits in anion exchange with suppressed-conductivity detection are at the part-per-billion (ppb) level, whereas with nonsuppressed detection, limits are about an order of

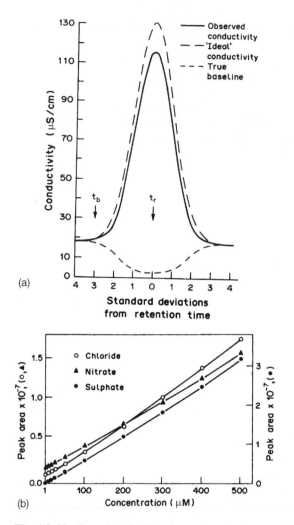

Fig. 4.3. Nonlinearity of calibration curves in suppressed IC. (a) Reduction in analyte signal due to protonation of dissociated eluent ions. (b) Effect on calibration curve. (Adapted from Ref. 8 with permission.)

magnitude higher. The use of suppressed conductivity detection for cation exchange is not as advantageous, since the terms $(\lambda_{A^+} - \lambda_{E^+})$ and $(\lambda_{OH^-} + \lambda_{A^+})$ (Eqns. 4.8 and 4.9) are similar in magnitude. It is best to view suppressed and nonsuppressed conductivity detection as complementary rather than competitive techniques. Suppression results in superior detection limits for analytes with a pK_a of less than 5 (*e.g.*, Cl^-, Br^-, NO_3^-, SO_4^{2-}). However, analytes with a pK_a greater than 5 (*e.g.*, CN^-) exhibit reduced detectability, and those with a pK_a greater than 7 are virtually undetectable. Clearly, in these cases, nonsuppressed-conductivity detection has a distinct advantage over suppression. Alternatively, suppressed and nonsuppressed conductivity detection can be combined [19]. In this technique, after suppressed conductivity detection is performed

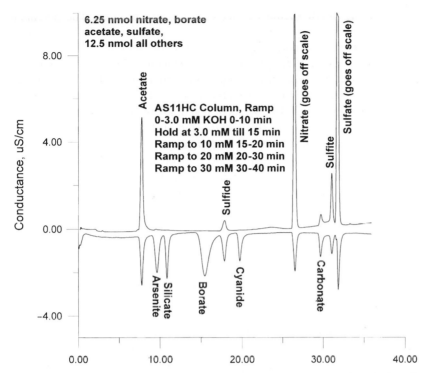

Fig. 4.4. Analytes determined with suppressed-conductivity detection (top) and nonsuppressed-conductivity detection (bottom). Experimental conditions: sample, 6.25 nmol nitrate, borate, acetate, and sulfate, all others 12.5 nmol; column, Dionex IonPac AS11-HC; injection volume, 25 μL; gradient program: from 0 to 3 mM KOH in 10 min, hold at 3.0 mM until 15 min, up to 10 mM 15–20 min, up to 20 mM 20–30 min, up to 30 mM 30–40 min. (Reproduced from Ref. 19 with permission.)

(using a hydroxide eluent), the eluent is passed into a membrane device, where potassium hydroxide is introduced into the eluent stream. In this way, the weak-acid analytes are converted to fully ionized potassium salts and are detected against a weak-hydroxide background as negative peaks. All ionic species are detected, regardless of pK$_a$ (Fig. 4.4). This detection scheme not only offers detection of strong- and weak-acid analytes, but it can also be used to detect unresolved peaks, estimate pK$_a$ values, calculate analyte equivalent conductance for peak identification, and even for universal calibration.

4.2.4.2 Amperometric detection

Amperometric detection is performed in a flow-cell that contains a working electrode, counter electrode, and reference electrode [16,20]. A potential is applied to the working electrode to oxidize or reduce analyte molecules at its surface. The resultant current is then measured and serves as the analytical signal. Usually, amperometric detection is carried out in the direct mode, which means that the analyte should be electrochemically active, whereas the mobile-phase co-ion should be electrochemically inactive. The potential applied to the working electrode may be constant (DC) or it may be pulsed. In pulsed methods, a repeating sequence of potentials is applied to the working electrode, and current is measured only at a specified time during the sequence. Pulsed potentials are advantageous when the products of the electrochemical reaction poison the surface of the working electrode. Typically, a measuring potential is applied, followed by a cleaning potential and a conditioning potential. This series of potentials serves to clean the surface of the working electrode electrochemically, making the surface reproducible for the detection of analytes. Amperometric detection is primarily used for the sensitive and selective detection of organic molecules, which are difficult to detect by other means. Examples include amines, amino acids, and carbohydrates [21]. Typically, these analytes are separated by anion exchange at high pH and detected under alkaline conditions. Detection limits are in the picomole to femtomole range.

4.2.4.3 Ultraviolet-absorbance detection

Direct monitoring of ultraviolet (UV) absorbance is the most straightforward mode of detection. Many analytes exhibit useful UV absorbance (Table 4.1). The sensitivity of UV detection can be enhanced by incorporation of a suppressor when the eluent is UV-absorbing in its deprotonated state. For example, OH^- has significant UV absorbance at 214 nm. If a suppressor is used before UV detection, then the OH^- ion is converted to water, and this dramatically reduces the background absorbance of the eluent. For ions without UV absorption, indirect UV detection can be used [22]. In this technique, the wavelength of detection is chosen such that the absorbance of analyte ions is zero, whereas the absorbance of the eluent ion is very high. Thus, when an analyte ion passes through the

TABLE 4.1

ULTRAVIOLET ABSORPTION OF ANIONS (TYPICAL OPERATING WAVELENGTHS)

Ion	nm
Br^-	215
BrO_3^-	210
$C_2O_4^{2-}$	205
CrO_4^{2-}	365
$HCOO^-$	190
IO_3^-	210
NO_2^-	210
NO_3^-	215
$S_2O_3^{2-}$	205
SCN^-	195

detector cell, it displaces a certain amount of eluent co-ions, the absorbance of the eluent decreases, and this results in negative peaks. Typical eluents for indirect UV detection are aromatic carboxylic acids (like those used for nonsuppressed conductivity) or sulfonic acids for anion separations, and copper, cobalt, or cerium salts or aromatic amines for cation separations.

4.2.4.4 Post-column reaction detection

Post-column reaction (PCR) detection involves monitoring of the product of an on-line reaction of an analyte, eluted from the separation column, with a reagent (or series of reagents) [23,24]. Advantages of PCR as a detection scheme include:
- (a) analytes that are not easily detected are converted into products that are more easily detected;
- (b) sensitivity and specificity of detection are improved; and
- (c) a wide array of reagents are available for PCR.

However, these advantages are balanced by an increased instrumental complexity and a broadening of the chromatographic peaks. Table 4.2 shows some of the most commonly used reagents for IC [25–27]. The post-column reagent is added after the separation column *via* a porous membrane, like those used for suppressors (Sec. 4.2.3), or a simple T-junction [28]. Complete mixing is essential for minimizing background noise. Pumping reagents under gas pressure produce much lower baseline "noise" than using either a syringe or a reciprocating pump. The reaction time is governed by the length and diameter of the reaction tube and the combined flow-rate of the eluent and reagent. The length of tubing is a compromise between the enhancement in sensitivity resulting from allowing the reaction to go further toward completion in a longer tube and the concomitant increased band-broadening. Reagents, such as PAR or Arsenazo III, react rapidly, so that the reaction is completed within the connecting tubing between the mixing device and the detector. Other reactions, such as that with BrO_3^-, are slower. In such cases, heating in conjunction with long, knitted, or potted tubing is used [23,27].

TABLE 4.2

COMMON POST-COLUMN REAGENTS IN ION CHROMATOGRAPHY

Reagent	Analytes	Detection (nm)	Ref.
2-(pyridylazo) resorcinol (PAR)	Transition metals and lanthanides (low µg/L)	520–530	
Arsenazo III	Lanthanides, U, Th, Y (100 µg/L)	658	[26]
1,5-diphenylcarbazide	CrO_4^{2-} (0.3 µg/L)	530	[25]
o-dianisidine dihydrochloride (ODA)	BrO_3^- (0.12 µg/L)	450	[27]

4.2.4.5 Mass-spectrometric detection

The coupling of IC with mass spectrometry (MS) meets the need for the determination of analytes other than the common anions/cations with increased sensitivity (sub-ppb level) and the need for speciation, *i.e.*, identification of the actual chemical form of an element. Coupling of IC to MS can be performed with an inductively coupled plasma (ICP) as an ionization source or with electrospray ionization. The former results in an element-specific detector, whereas the latter can be used to provide structural information on the analytes. The details of interfacing IC to MS can be found in a number of recent review articles [16,29,30] and Chap. 10. There has been a decline in the number of papers dealing with the coupling of IC with atomic emission spectroscopy (AES) as it is gradually being replaced by ICP/MS. However, a brief description of IC/ICP-AES is given in Ref. 30. IC/MS is now commonly used to determine such ions as perchlorate, bromate, haloacetates, selenium, arsenic, and chromium. Applications of IC/MS to the determination of inorganic ions can be found in Chap. 13 of this book and in review articles by Seubert [30], Sutton [29], and Buchberger [16].

4.3 SELECTIVITY IN ION CHROMATOGRAPHY

4.3.1 Importance of selectivity

Proper control of the selectivity of a separation is of utmost importance in method development. Common ways of altering selectivity are through changes in the mobile phase (*e.g.*, eluent type, eluent strength, pH, and additives) and in the stationary phase (*e.g.*, support and ion-exchange site). In IC, selectivity is most often altered through changes in the chemical nature of the stationary phase. The reason for this is primarily that IC separations performed with suppressed conductivity detection are restricted to eluents that can be suppressed (Secs. 4.2.2 and 4.2.3). Moreover, as will be shown in Sec. 4.3.3.2, the selectivity for ions of similar charge cannot be influenced through changes in the eluent concentration. Consequently, manufacturers of IC columns have invested a great deal of time and effort into producing columns with varying selectivity (Chap. 12). Selectivity in IC (or more specifically, ion exchange, which will be discussed exclusively until Sec. 4.3.5) can be qualitatively described as the likelihood of exchange occurring between two ions, one being the analyte of interest and the other being the eluent ion in the stationary phase [31]. Consider, *e.g.*, the exchange of ion A^{x-} with the eluent ion E^{y-}

$$yA_m^{x-} + xE_r^{y-} \rightleftharpoons yA_r^{x-} + xE_m^{y-} \tag{4.10}$$

where the subscript m denotes the mobile phase and r denotes the resin (stationary) phase. The selectivity coefficient, $K_{A,E}$, is the equilibrium constant for Eqn. 4.10 (assuming that activity coefficients are approximately equal to 1)

$$K_{A,E} = \frac{[A_r^{x-}]^y [E_m^{y-}]^x}{[A_m^{x-}]^y [E_r^{y-}]^x} \tag{4.11}$$

The selectivity coefficient, being a function of the analyte, the eluent ion, and the stationary phase, can be moderated in a number of ways.

4.3.2 Properties of analyte and eluent ions affecting selectivity

Perhaps the most intuitively obvious factor that affects retention of ions in ion exchange is the charge of the ion. As might be expected, doubly charged ions (*e.g.*, SO_4^{2-}) are much more strongly retained than singly charged ions (*e.g.*, Cl^-). Consequently, there is a gap in chromatograms between ions of differing charge, which can lead to long analysis times. One way to reduce the gap between ions of differing charge is to increase the eluent strength, as will be discussed in Sec. 4.3.3.2. Alternatively, if an ion is the conjugate base of a weak acid, the pH of the eluent may be used to control the degree of ionization of the ion and, therefore, its retention. However, in most cases this is only possible when nonsuppressed-detection modes are used. Suppression necessitates that the eluent be acidic for cation separations and basic for anion separations (Sec. 4.2.3.1). Apart from the charge of ions, solvation of ions also plays a significant role in retention. Since ion exchange is a process that occurs in aqueous solution, ions are surrounded by solvation spheres [32]. If an ion is to leave the solution phase (*i.e.*, the mobile phase) and enter the ion exchanger, it must rearrange and eventually partially shed its solvation sphere in order to come close to the ion-exchange site. The more an ion sheds its solvation sphere, the closer it can come to the ion-exchange site and the more strongly it will be bound to the site (*i.e.*, retained). The eluent ion undergoes similar processes. Generally speaking, the ion exchanger preferentially binds the ions with the smallest hydration sphere. It is for this reason that elution in anion exchange follows the order F^-, Cl^-, Br^- and in cation exchange it is Li^+, Na^+, K^+, Rb^+.

Polarizable ions display unique selectivity in ion exchange. They are very strongly retained on anion exchangers due to the secondary interactions they undergo. Such ions as I^-, SCN^-, and especially ClO_4^-, being relatively large and poorly hydrated, are unable to form a proper solvation sphere in the solution phase. They disturb the structure of water in the mobile phase and enter the stationary phase, where the structure of water is less ordered [33]. According to Diamond [34], polarizable ions may participate in water-structure-enforced ion pairing with the anion-exchange site, and this further increases their retention. This type of interaction is strong between two large, poorly hydrated ions (*e.g.*, tetraalkylammonium and ClO_4^-) and thus, polarizable ions are very strongly retained on most anion exchangers. Since the strong interaction of polarizable ions with anion exchangers stems from the disruption of the water structure in the mobile phase, a better term for them is *chaotropic*. The relative chaotropic nature of various ions is illustrated by the Hofmeister series (Eqn. 4.17a, Sec. 4.3.5.3).

4.3.3 Effect of mobile phase on selectivity

As will be shown in Sec. 4.3.4, the stationary phase has the strongest impact on selectivity in ion exchange. This does not mean that the mobile phase can be ignored in method development; "fine-tuning" of the selectivity can be accomplished through changes in the mobile phase. The most important and obvious way in which this is accomplished is through changes in the eluent type and concentration, although other variables, such as organic modifiers and temperature can also be used.

4.3.3.1 Choice of eluent ion

The choice of eluent is an important consideration, especially when separating multiply charged species. Weak eluents (*e.g.*, hydroxide or borate for anion separations and H^+ for cation separations) are effective for singly charged analytes, but retention times for multiply charged species are prohibitively long. Eluents commonly used for the separation of multiply charged species are listed in Table 4.3 [35,36]. The eluent strength is adjusted to fine-tune the resolution. Multivalent ions can also be eluted in a reasonable time by high concentrations of OH^-, as shown in Fig. 4.5. In cation exchange, alkaline-earth metals can be separated with a tartaric acid/dipicolinic acid eluent when using non-suppressed conductivity detection or with MSA when suppressed-conductivity detection is used. However, both separations are performed on carboxylated stationary phases for reasons given in Sec. 4.3.4.3.

4.3.3.2 Effect of eluent on selectivity

Based on Eqn. 4.10 and general chromatographic principles (Chap. 1), the linear-solvent-strength model [37–39] predicts that ion retention is governed by

$$\log k_A = \frac{1}{y}\log K_{A,E} + \frac{x}{y}\log\left(\frac{Q}{y}\right) + \log\left(\frac{w}{V_M}\right) - \frac{x}{y}\log[E_m^{y-}] \qquad (4.12)$$

TABLE 4.3

ELUENTS USED FOR CERTAIN DIVALENT ANALYTES AND THE CORRESPONDING COLUMNS [35,36]

Analytes	Eluent	Column (from Table 4.4)
Sulfate, phosphate	Hydroxide	Dionex IonPac AS10
	Bicarbonate/carbonate	Metrohm Metrosep A Supp 1, 4, and 5
Alkaline-earth metals	Tartaric acid–dipicolinic acid	Metrohm Metrosep A Supp 4
	Methanesulfonic acid or sulfuric acid	Dionex IonPac CS12A
Transition metals	Tartaric acid/citric acid or oxalic acid	Metrohm Metrosep C 2-250
	Oxalic acid or pyridine-2,6-dicarboxylic acid	Dionex IonPac CS5

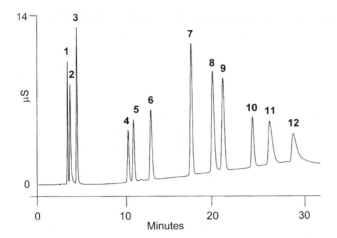

Fig. 4.5. Elution of multivalent anions with a OH⁻ eluent. Experimental conditions: column, Dionex IonPac AS10 (4 mm), 170 μeq capacity; eluent, 50–124 m*M* NaOH in 30 min; injection, 20 μL; detection, suppressed conductivity. Peaks (in mg/L): 1. fluoride (1.2); 2. acetate (30); 3. formate (5); 4. selenite (10); 5. chloride (1.8); 6. nitrite (9); 7. sulfate (10); 8. oxalate (10); 9. selenate (10); 10. phosphate (9); 11. bromide (10); 12. nitrate (6). (Courtesy of Dionex Corporation.)

where k_A is the retention factor of the analyte, $K_{A,E}$ is the ion-exchange selectivity constant for the analyte and the eluent competing ion (E^{y-}), Q is the effective ion-exchange capacity of the stationary phase, w is the mass of the stationary phase, and V_m is the volume of the mobile phase. The linear-solvent-strength model (Eqn. 4.12) can adequately describe retention behavior for eluents containing a single competing ion (*e.g.*, OH⁻ [38]), but is not effective for eluents containing multiple competing ions (*e.g.*, phthalate [37] or HCO₃⁻/CO₃²⁻ [39]). Further, secondary interactions, such as adsorption of the analyte on portions of the stationary phase that are not functionalized, or steric hindrance also influence retention. Nonetheless Eqn. 4.12 provides a useful framework for a discussion of the factors that influence selectivity in IC. For a given column and given eluent ion, Eqn. 4.12 is reduced to

$$\log k_A = \text{const} - \frac{x}{y} \log[E_m^{y-}] \tag{4.13}$$

Thus, increasing the eluent concentration results in a dramatic decrease in retention. For instance, doubling the concentration of NaOH will cut the retention time of Cl⁻ and NO₃⁻ in half and the retention time of SO₄²⁻ and C₂O₄²⁻ to a fourth. Thus, the use of high concentrations of OH⁻ compensates for its weak eluent strength. Modern suppressors are capable of neutralizing up to 200 meq/min OH⁻ (Sec. 4.2.3). While the eluent concentration strongly affects retention, it has little effect on the selectivity of ions of the same charge. However, the eluent concentration can have dramatic effects on selectivity among ions of different charge. If the eluent concentration is high enough, it is possible to elute multiply charged analytes before singly charged analytes (Fig. 4.5). Thus, manipulation of the eluent concentration is extremely useful for bridging the gap between analytes of different charge in chromatograms. However, this approach to selectivity modification only works with high-capacity columns. In the example shown in Fig. 4.5, the capacity of the Dionex IonPac AS10 column is 170 μeq. For low-capacity columns, high eluent strength produces inadequate separation of early-eluted ions, but with a higher capacity it is possible to use high eluent strengths to modulate selectivity without sacrificing the resolution of early-eluted anions.

4.3.3.3 Use of organic additives

The addition of organic solvents to the eluent (*e.g.*, methanol, acetonitrile) can have a marked effect on selectivity, peak shape, and efficiency in IC. However, care must be taken to ensure that the stationary phase is solvent-compatible. If the polymer making up the core of the agglomerated phases is lightly cross-linked (<5%), as it was in early IC phases, the beads will swell dramatically in the presence of even a small amount of solvent. This causes increased back-pressure and voids. Solvent-compatible columns are often more than 50% cross-linked. Organic modifiers affect selectivity through changes in solvation of the analyte and eluent. Examples of the use of organic solvents for selectivity control in anion separations can be found in Ref. 32. The use of an organic modifier is most often necessary in cation separations, *e.g.*, in the determination of amines [40]. In the case of the more hydrophobic amines, 100% aqueous eluents produce excessively long retention times (due to hydrophobic adsorption of the amine on the polymer backbone) and extreme peak-tailing. The organic additive attenuates hydrophobic interactions with the stationary phase and reduces retention time and peak-tailing. An added benefit of solvent-compatible packings is that column clean-up is more effective. Many compounds are strongly adsorbed on the stationary phase. The combined use of solvents and high ionic strength allows the removal of a large variety of contaminants, thus restoring column performance to nearly original levels.

4.3.3.4 Temperature

The effect of temperature has attracted much attention in RP-LC method development, but, until recently, the use of temperature to modify selectivity in IC has been relatively neglected [40–43]. In anion-exchange chromatography, the effect of temperature on selectivity can be quite complex [42]. Unfortunately, with the basic eluents typically used in anion exchange, the quaternary ammonium sites of the stationary phase (Sec. 4.3.4.4) are prone to nucleophilic substitution of OH^- on the benzylic carbon, and to a lesser extent the methyl carbon, resulting in decreased capacity and retention at high pH [44]. Conversely, cation exchangers are temperature-stable. Retention usually decreases with increasing temperature, but different classes of analytes, (*e.g.*, alkali metal, alkaline-earth metal, or amine) show decreases to varying extent [40,43]. Thus, selectivity is altered significantly through a change in temperature, especially in the separation of metals and amines. Temperature was found to have a marked effect on peak symmetry and efficiency in both anion and cation exchange. This is especially true for the separation of chaotropic anions in anion-exchange [42] and the separation of amines in cation-exchange chromatography [40,43].

4.3.4 Effect of stationary phase on selectivity

Undoubtedly, the stationary phase exerts the most pronounced effect on selectivity in ion exchange [31]. It is for this reason that a wide range of ion-exchange columns for IC has been manufactured (Chap. 12). Some of these IC columns are listed in Table 4.4 [35,36,45]. The selectivity of a separation can be altered by changing
   (a) the materials used to construct the stationary phase;
   (b) the degree of cross-linking of the resin;
   (c) the ion-exchange capacity of the resin;
   (d) the functional group on the ion exchanger.

4.3.4.1 Structure

Most stationary phases for IC are made from polymers (Sec. 4.2.1). The particles may be "agglomerated", *i.e.*, composed of an essentially nonporous solid inner core, on the surface of which a thin layer of stationary phase is deposited [36]. More specifically, a monolayer of charged latex particles (which determine the stationary-phase functionality/selectivity) is electrostatically attached to a surface-functionalized internal-core particle. Such stationary phases (manufactured by Dionex) produce higher efficiencies than completely porous ion exchangers, because exchange is faster and the pellicular layer is high permeable. Examples of such columns are the Dionex IonPac AS10 and the IonPac AS16 (Table 4.4). The capacity of agglomerated stationary phases is determined by the diameter of the latex particle. Latex particles of larger diameter have a larger capacity. However, there is a limit. When the latex particles become too large, chromatographic efficiency is sacrificed. Another way to construct polymeric particles is to chemically graft a functionality directly onto the particles. This approach is preferred when a stationary phase is to have a higher capacity. A grafted film is typically very thin (1–5 nm) and makes high-efficiency and high-capacity stationary phases. The Dionex IonPac AS14-A and IonPac CS12A (Table 4.4) employ this technology.

Typically, stationary phases are constructed from three different types of monomers:
   (a) a functional (or base) monomer for creating the ion-exchange site;
   (b) a cross-linking monomer to control the water content; and
   (c) a nonfunctional monomer can be used to control charge density and secondary selectivity interactions. Common base materials are ethylvinylbenzene (Dionex), methacrylate (Dionex, Metrohm, Alltech), polyvinyl alcohol (Metrohm), and polystyrene (Metrohm, Alltech) (Sec. 4.2.1). These base monomers need to be cross-linked to control their water content, which plays a role in stationary-phase selectivity. The choice of cross-linker depends on the base material used. Divinylbenzene is the most popular cross-linking material for aromatic base materials (*e.g.*, ethylvinylbenzene or polystyrene), whereas ethyleneglycol dimethacrylate (among others) can be used for methacrylate-based resins. Typically, methacrylate-based resins require a higher degree of cross-linking (8–40%) to achieve the same water content as aromatic resins (0.2–5%). Functionalization of the base monomer, which is usually a proprietary process, is accomplished in a number of ways.

TABLE 4.4

IC COLUMNS FROM DIONEX, METROHM, AND ALLTECH [35,36,45]

Column	Functional group	Stationary-phase material	Capacity (μeq/col.)	Particle size (μm)	Typical analytes
Dionex IonPac AS9-HC	Alkyl quaternary ammonium	EVB/DVB	170	9	Oxyhalides, common anions
Dionex IonPac AS14-A	Alkyl quaternary ammonium	EVB/DVB	120	5	7 common anions and low-MW acids
Dionex IonPac AS16	Alkanol quaternary ammonium	EVB/DVB	170	9	Chaotropic anions
Dionex IonPac CS5A	Sulfonate/quaternary ammonium	EVB/DVB	20/40	9	Transition metals and lanthanides
Dionex IonPac CS12A	Carboxylate/phosphonate	EVB/DVB	2800	5	Inorganic cations, alkylamines, alkanolamines
Metrohm Metrosep A Supp 1	Quaternary ammonium	PS/DVB	34	7	Common anions, oxyhalides
Metrohm Metrosep A Supp 4	Quaternary ammonium	PVA	46	9	7 common anions
Metrohm Metrosep A Supp 5	Quaternary ammonium	PVA	57	5	Common anions, oxyhalides, perchlorate
Metrohm Metrosep C 2-250	Carboxylate	Silica	194	7	Inorganic cations, alkylamines, alkanolamines
Metrohm Nucleosil 5SA	Sulfonate	Silica	186	5	Divalent cations, transition metals
Alltech Allsep Anion	Alkyl quaternary ammonium	Methacrylate	–	7	7 common anions
Alltech Wescan Anion/R	Trimethyl ammonium	PS/DVB	190	10	7 common anions, thiocyanate, thiosulfate
Alltech Universal Cation	Carboxylate	Silica	–	7	Inorganic cations

EVB/DVB = ethylvinylbenzene/divinylbenzene, PS/DVB = poly(styrene)/divinylbenzene, PVA = poly(vinylalcohol).

Typical examples include [31]:
  (a) for cation exchangers, reaction of polystyrene with sulfuric acid, chlorosulfonic acid, or sulfur trioxide to produce sulfonate groups on the base material;
  (b) for anion exchangers, reaction of the base material with a tertiary amine to produce a quaternary ammonium group.

4.3.4.2 Role of column material

The materials used in the synthesis of a stationary phase can have a pronounced effect on the selectivity of a separation. Fig. 4.6 shows the separation of seven common anions on each of three stationary phases. Stationary phases with high water content

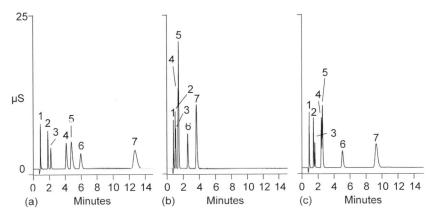

Fig. 4.6. Effects of stationary-phase composition on anion-exchange selectivity. (a) 5% cross-linked vinylbenzyl chloride (VBC) latex (75% water by mass), (b) 1% VBC (91% water by mass), (c) 5% cross-linked glycidylmethacrylate (GM) latex (87% water by mass). All stationary phases have a trimethylamine functional group. Conditions: eluent, 3 mM sodium carbonate; flow-rate, 2 mL/min; suppressed conductivity detection. Peaks: 1. fluoride, 2. chloride, 3. nitrite, 4. bromide, 5. nitrate, 6. phosphate, 7. sulfate. (Adapted from Ref. 31 with permission. Courtesy of Dionex.)

(*i.e.*, Latex B and Latex C) cannot completely separate Br^- from NO_3^-. Latex A (low water content) gives baseline resolution of Br^- from NO_3^-. However, this does not mean that resins with high water content always behave identically. This is especially true for the separation of chaotropic anions. Fig. 4.7 shows a separation of I^-, BF_4^-, and SCN^- on Latex B and Latex C. Latex C shows considerably more retention and resolution of these anions than Latex B. Stationary phases of high water content produce greater retention changes for chaotropic anions than stationary phases of lower water content.

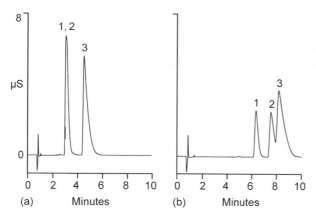

Fig. 4.7. Effect of stationary-phase composition on the separation selectivity of chaotropic anions. (a) 1% cross-linked vinylbenzyl chloride (VBC) latex (91% water by mass); (b) 5% cross-linked glycidylmethacrylate (GM) latex (87% water by mass). Conditions: eluent, 3 mM sodium carbonate; flow-rate, 2 mL/min; suppressed-conductivity detection. Peaks: 1. iodide, 2. tetrafluoroborate, 3. thiocyanate. (Adapted from Ref. 31 with permission. Courtesy of Dionex.)

4.3.4.3 Role of the type of ion-exchange site

Different types of ion-exchange sites can be produced through the functionalization of the base monomers. By convention, ion exchangers are classified according to their type of functionality (*i.e.*, ion-exchange site). Cation exchangers are broadly classified into strong acid (*e.g.*, sulfonate) and weak acid (*e.g.*, carboxylate), whereas anion exchangers can be classified into strong base (*e.g.*, quaternary ammonium) or weak base (*e.g.*, tertiary amine) types. Strong-acid/base ion exchangers retain their capacity over a wide pH range, since the ion-exchange site remains fully charged as the pH is varied. Conversely, weak-acid/base ion exchangers undergo changes in capacity with pH. Anion-exchange separations are usually performed on quaternary ammonium stationary phases. Selectivity variations in these materials are introduced through changes in the structure of the ion-exchange site, rather than through the type of ion-exchange site (Sec. 4.3.4.4). However, cation-exchange selectivity is usually altered through the use of different types of functional groups (*e.g.*, sulfonate, carboxylate, phosphonate).

The effect of the type of ion-exchange site on the selectivity in cation-exchange chromatography is illustrated by the retention data obtained on three cation exchangers, differing only in their functionality. Table 4.5 shows the retention observed on sulfonated, phosphonated, and carboxylated cation exchangers as a function of the eluent concentration. An eluent of 5 mM MSA can elute only the singly charged metals, regardless of the functionality of the cation exchanger. However, there are some subtle selectivity changes between the different functionalities. For example, NH_4^+ is eluted near Na^+ from the carboxylated resin, closer to K^+ from the sulfonated

TABLE 4.5

RETENTION DATA FOR CATIONS ON THREE GRAFTED CATION EXCHANGERS WITH METHANESULFONIC ACID (MSA) ELUENTS OF VARYING CONCENTRATION. N, NOT ELUTED. (ADAPTED FROM REF. 31.)

Eluent	Column	Retention time (min)					
		Li^+	Na^+	NH_4^+	K^+	Mg^{2+}	Ca^{2+}
5 mM MSA	Sulfonated	11.7	13.8	19.8	23.1	N	N
	Phosphonated	4.4	4.4	5.7	5.2	N	N
	Carboxylated	5.0	6.3	7.2	9.8	N	N
10 mM MSA	Sulfonated	5.5	6.4	8.2	9.3	N	N
	Phosphonated	3.0	3.0	3.6	3.4	N	N
	Carboxylated	2.8	3.3	3.5	4.2	6.5	7.6
25 mM MSA	Sulfonated	3.1	3.6	4.4	5.0	N	N
	Phosphonated	2.5	2.5	2.7	2.7	9.7	13.2
	Carboxylated	2.2	2.5	2.5	2.7	2.7	2.7
100 mM MSA	Sulfonated	2.2	2.2	2.4	2.4	13.2	22.8
	Phosphonated	2.2	2.2	2.2	2.2	2.7	2.9
	Carboxylated	2.0	2.0	2.0	2.0	2.0	2.4

resin, and after K^+ from the phosphonated cation exchanger. One of the primary uses of cation exchange is the separation of NH_4^+ from Na^+ and K^+ [46]. Elution of divalent species can be accelerated by increasing the eluent strength, but success of this approach is highly dependent on the functionality of the cation exchanger. For example, divalent cations are not eluted from the sulfonated cation exchanger until an eluent strength of 100 mM MSA is used. However, resolution of the singly charged ions is completely sacrificed in the process. Clearly, sulfonated cation exchangers are ill-suited for the simultaneous determination of singly and doubly charged cations. The divalent cations are eluted at a much lower eluent strength (25 mM MSA) from the phosphonated cation exchanger. However, as is the case with the sulfonated cation exchanger, resolution of the singly charged ions is unsatisfactory. The best overall separation is obtained with the carboxylated cation exchanger at an eluent strength of only 10 mM MSA. Thus, carboxylated cation exchangers are frequently used for the simultaneous determination of singly and doubly charged cations. This is seen for the Dionex IonPac CS12A and Metrohm Metrosep C 2-250 columns in Table 4.4. A separation of the six common cations on a mixed carboxylated/phosphonated stationary phase is shown in Fig. 4.8. Sulfonated cation exchangers are usually employed for the separation of transition metals and lanthanides with complexing eluents (Sec. 4.3.3.1). Examples of sulfonated cation exchangers for the separation of transition metals in Table 4.4 include the Metrohm Nucleosil 5SA and the Dionex IonPac CS5.

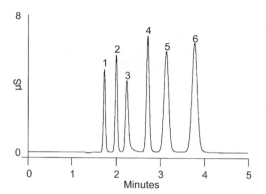

Fig. 4.8. Separation of univalent and divalent cations on a carboxylated stationary phase. Experimental conditions: column, Dionex IonPac CS12A (150 × 3 mm); eluent, 33 mM MSA; flow-rate, 0.5 mL/min; injection, 25 μL; suppressed-conductivity detection. Peaks (in mg/L): 1. Lithium (0.12), 2. sodium (0.5), 3. ammonium (0.62), 4. potassium (0.25), 5. magnesium (0.62), 6. calcium (1.25). (Courtesy of Dionex Corporation.)

4.3.4.4 Role of the structure of the ion-exchange site

The previous section showed that the functionality of an ion exchanger has a profound effect on the selectivity of a separation. However, the structure of an ion-exchange site can also have a significant effect on selectivity. This is especially true for anion exchangers, since there are a number of possibilities for varying the structure of the ion-exchange site. Changes in the structure of quaternary ammonium functional groups ($-NR_4^+$) are introduced through their alkyl chains. For example, a quaternary ammonium group can be made bulkier and more hydrophobic by lengthening some or all of the alkyl chains. Alternatively, an ion-exchange site can be made hydrophilic through the incorporation of an alcohol functionality (*e.g.*, an alkanolamine). The possible variations in the structure of an anion-exchange site may seem endless, but fortunately, a few generalizations can be made. Firstly, the retention of hydrophilic, polyvalent anions (*e.g.*, SO_4^{2-} and HPO_4^{2-}) decreases as the size of the functional group increases (*e.g.*, on going from a methyl to an ethyl substituent). This is because the charge density of the ion-exchange site decreases as it becomes larger. Thus, coulombic attraction between the solute and ion-exchange site is not as strong. Hydrophilic monovalent anions (*e.g.*, Cl^-) show a small increase in retention as the size of the ion-exchange site increases. For the separation of chaotropic anions (*e.g.*, I^-, SCN^-, ClO_4^-), the hydration of the ion-exchange site is of utmost importance. Usually, there is an increase in the retention of these ions as the ion-exchange site becomes more hydrophobic (less hydrated). However, the opposite is true if the eluting ion is more chaotropic than the solute (*e.g.*, *p*-cyanophenol).

Another variation of the structure of an ion-exchange site is seen in the hydroxide-selective stationary phases, such as the Dionex IonPac AS16 in Table 4.4. This variation is produced through the incorporation of alkanolamine functionalities in the ion-exchange site. Hydroxide is the preferred eluting ion for suppressed-conductivity detection, owing to its very low conductance after suppression (Sec. 4.2.3) and the ability to generate OH^- on-line (Sec. 4.2.2.3). However, OH^- is one of the weakest eluting ions for conventional [*e.g.*, $-N(CH_3)_3^+$] anion exchangers. The reason for this is that OH^- is a highly hydrated ion and, as such, prefers to stay in the mobile phase where water of hydration is relatively accessible [47]. The incorporation of alkanolamine functionalities in the ion-exchange site makes it more hydrophilic, and increased amounts of water will form hydrogen bonds with the site. Consequently, more water of hydration is now available in the stationary phase, and the hydration of hydroxide is not as extensively disturbed when it enters the stationary phase. This phenomenon makes hydroxide a much stronger eluting ion for these hydrophilic stationary phases. This is illustrated in Table 4.6 [47]. Retention factors for the

TABLE 4.6

RETENTION FACTORS WITH 100 mM SODIUM HYDROXIDE AS ELUENT. (ADAPTED FROM REF. 47.)

Column	Analyte						
	F^-	Cl^-	Br^-	NO_3^-	ClO_3^-	SO_4^{2-}	PO_4^{3-}
Methyldiethanolamine	0.06	0.24	0.92	1.1	1.0	0.2	0.31
Dimethylethanolamine	0.14	1.1	4.5	5.0	4.9	3.0	6.7
Trimethylamine	0.30	4.4	19.2	22.5	21.2	51.4	>100
Triethylamine	0.30	5.8	26.1	55.8	35.7	24.9	>100

conventional stationary phases (*i.e.*, trimethylamine and triethylamine functionality) with 100 mM OH^- as eluent are rather large. As the anion exchange-site becomes more hydrophilic, *i.e.*, in going from the triethylamine to methyldiethanolamine functionality, the retention factors for all anions show a significant decrease. This is due to the enhanced selectivity for OH^- shown by these stationary phases, thus effectively making hydroxide a stronger eluting ion.

Hydrophilic stationary phases (*e.g.*, methyldiethanolamine and dimethylethanolamine) are particularly well suited to the separation of chaotropic ions (*e.g.*, I^-, SCN^- and ClO_4^-), since they are hydrated to a greater extent than trialkylammonium stationary phases. This results in reduced retention of chaotropic anions for several reasons: Firstly, these stationary phases show a higher selectivity for hydroxide than for analyte ions. Secondly, the interaction of chaotropic anions with the anion-exchange site through water structure-enforced ion pairing (Sec. 4.3.2) is greatly diminished as the hydration of the stationary phase ion is increased [34].

4.3.5 Alternative modes of separation

4.3.5.1 Cryptand columns

Stationary phases made from macrocyclic ligands (cryptands) offer new possibilities for selectivity control in IC [48]. The unique property of these cryptand columns is their ability to vary the column capacity (and therefore the retention of ions) on the fly, by simply changing the mobile-phase cation from Na^+ to Li^+ to K^+. In this column the positively charged functional group is created by complexation of the metal with the cryptands on the stationary phase. Different cations are complexed by the cryptand to differing degrees ($Li^+ < Na^+ < K^+$). Switching from Na^+ to Li^+, for example, reduces the ion-exchange capacity of the stationary phase, since there are fewer complexed metal ions and therefore fewer anion-exchange sites. This approach is analogous to gradient elution chromatography, and is termed "capacity gradient". This column technology has recently been commercialized by Dionex Corporation [49]. The cryptand-based columns are expected to have advantages over conventional anion exchangers in several key areas, *e.g.*, the simultaneous separation of the common inorganic anions and chaotropic anions and the separation of common anions in concentrated acids [49].

4.3.5.2 Ion exclusion

Ion-exclusion chromatography is predominantly based on electrostatic repulsion, rather than electrostatic attraction, which is typically used in ion-exchange chromatography [50]. Thus, ion exclusion provides a significantly different selectivity, strong-acid anions (Cl^-, SO_4^{2-}, NO_3^-) being eluted in the void-volume of the column and weak-acid anions (*e.g.*, acetate, lactate) being retained. The degree of retention is determined by the analyte pK_a and size and by secondary adsorption on the stationary phase. Within an ion-exchange column there are three phasic regions: The first is the flowing eluent that passes through the interstitial channels between the particles. This is the mobile phase in ion exclusion. The second is the polymeric network of the resin material itself. For a cation exchanger, this network will possess a strong, fixed anion charge. The third phase is the liquid occluded within the pores of the resin. The occluded (stagnant) liquid (and to a secondary degree the polymer resin network) is the stationary phase in ion exclusion. The ion exchanger carries fixed, nonmoving charges (anionic for a cation exchanger). To maintain electroneutrality, these fixed charges are balanced by an equal number of freely diffusing counter-ions (cations in this case). Thus, inside an ion-exchange resin bead there is in essence an extremely concentrated electrolyte solution. Outside the bead, there is a dilute electrolyte solution. The concentration gradient generates a driving force for the counter-ions to diffuse out of the resin bead. However, the diffusion of counter-ions to the solvent outside the bead quickly sets up a strong (anionic) *Donnan potential* which draws the counter-ions (cations) back into the bead. A secondary effect of this Donnan potential, however, is to repel ions of the same charge as those fixed on the ion exchanger (anions).

Thus, if a cation-exchange resin is used for ion exclusion, strong-acid anions, such as Cl^-, SO_4^{2-}, and NO_3^-, are strongly repelled by the Donnan potential and cannot enter the stagnant liquid within the bead pores. As a consequence, these ions are eluted in the void volume, which corresponds to the interstitial volume of the column ($V_{interstitial}$). Alternatively, small neutral analytes experience no Donnan exclusion and therefore diffuse freely between the moving interstitial liquid and the stagnant occluded liquid. Thus, neutral analytes are eluted in the total permeation volume, which is equivalent to the dead-volume of a normal chromatographic column, ($V_{permeation}$), consisting of both the interstitial and the occluded liquid volumes. Alternatively, if the analyte is a weak acid ($pK_a = 2.5-6.5$) and the eluent pH is near the analyte pK_a, a portion of the analyte will be in the neutral, protonated form. This neutral acid (HA) is not repelled by the Donnan potential, and can therefore diffuse into the occluded liquid within the bead. Thus, migration of the weak acid is retarded relative to that of a strong-acid anion, which remains in the moving interstitial liquid. The ionized (A^-) and neutral (HA) forms of the acid are in constant equilibrium, so that the effective negative charge of the acid is a function of the acid dissociation constant of the analyte (pK_a) and the pH. On this basis, retention in ion exclusion is given by [51]

$$V_{eluted} = V_{interstitial} + K_d V_{occluded} \tag{4.14}$$

where K_d is the distribution coefficient, which ranges from 0 to 1. This is the same equation as the one that governs retention in size-exclusion chromatography (Chap. 5). Thus, as in size-exclusion chromatography, the columns used for ion-exclusion chromatography are larger than other IC columns (*e.g.*, Dionex IonPac ICE-AS6 in Fig. 4.9 is 9×250 mm). Assuming that elution behavior is based solely on ion exclusion, the distribution

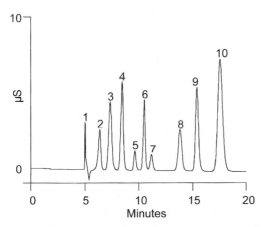

Fig. 4.9. Ion-exclusion separation of low-molecular-weight carboxylic acids. Experimental conditions: column, Dionex IonPac ICE-AS6; eluent, 0.4 m*M* heptafluorobutyric acid; flow-rate, 1.0 mL/min; injection volume, 50 µL; detection, suppressed conductivity. Peaks (pK_a): 1. oxalic, 5.0 mg/L (1.23, 4.19); 2. tartaric, 10 (2.98, 4.34); 3. citric, 15 (3.14, 4.77, 6.39); 4. malic, 20 (3.40, 5.11); 5. glycolic, 10 (3.83); 6. formic, 10 (3.75); 7. lactic, 10 (3.86); 8. hydroxyisobutyric (HIBA), 30 (4.71); 9. acetic, 25 (4.76); 10. succinic, 25 (4.16, 5.61). (Courtesy of Dionex Corporation.)

coefficient in Eqn. 4.14 is given by

$$K_d = \frac{C_{H^+}}{C_{H^+} + K_a}$$ (4.15)

where C_{H^+} is the hydronium concentration provided by the strong-acid eluent (*e.g.*, sulfuric acid, octanesulfonic acid, or heptafluorobutyric acid) and K_a is the first-dissociation constant of the analyte. To a first approximation, then, the elution order of an ion-exclusion column is in reverse order of analyte pK_a. This general trend can be seen in Fig. 4.9. This separation was performed at pH 3.4, dictated by the 0.4 m*M* heptabutyric acid eluent. Oxalic acid ($pK_a = 1.23$) is fully deprotonated and eluted in the dead-volume, along with any other strong-acid anions. The rest of the carboxylic acids are partially ionized and thus are somewhat retained by the column and generally follow the inverse dependence on pK_a, predicted by Eqn. 4.15. Acid strength can be used to alter the position where an acid is eluted within the elution window ($V_{interstitial}$ to $V_{permeation}$), but generally pH-based selectivity changes are only observed for multiprotic acids with second-dissociation constants close to their pK_{a1} (*e.g.*, fumaric $pK_{a1} = 3.03$, $pK_{a2} = 4.44$).

Secondary effects that alter retention in ion-exclusion chromatography are size-exclusion effects and adsorption on the stationary phase. Size exclusion (Chap. 5) restricts the access of larger analytes to the occluded liquid in the pores of the stationary phase. This results in lower retention of larger solutes than expected on the basis of pK_a alone. The lower the percent cross-linkage of the stationary phase, the more open the structure and the more permeable it is to higher-molecular-weight substances. For small-molecule separations, an 8% cross-linked polystyrene/divinylbenzene resin is generally used (*e.g.*, Dionex IonPac ICE-AS6 or Aminex HPX-87), whereas for oligosaccharides, size-exclusion effects can become dominant and therefore lower degrees of cross-linking (*e.g.* 4%) are more effective. Adsorption is generally significant only for the neutral form of the analyte, given the much greater surface area which it can access. The distribution coefficient is then modified by an additional partition coefficient, K_P

$$K_d' = K_d \cdot K_p$$ (4.16)

This adsorption results in an increase in retention of the analytes beyond that predicted on the basis of pK_a, potentially even beyond the total permeation volume. Typically, adsorption is based on hydrophobicity, so that longer-chain carboxylic acids show greater retention than shorter-chain acids of comparable pK_a [50]. Addition of organic modifiers, such as methanol or acetonitrile will decrease the retention of large hydrophobic acids, while leaving smaller acids unaffected. Alternatively, some manufacturers add a hydrophilic functionality to the ion-exclusion column to minimize this secondary retention mode and potentially induce hydrogen-bonding interactions with hydroxylated acids.

As illustrated in Fig. 4.9, the primary application of ion-exclusion chromatography is the determination of mixtures of carboxylic acids. In addition, ion exclusion can be used for separating other weak-acid analytes, such as sulfite ($pK_{a1} = 1.91$), phosphate

(2.15), fluoride (3.17), borate (9.02), and cyanide (9.21). Alternatively, ion-exclusion chromatography can be used as a pre-treatment step in the determination of contaminant anions in concentrated solutions of weak acids, such as H_3PO_4, HF, and glycolic acid [52]. Strong-acid anions in a concentrated weak-acid solution will be eluted at the excluded volume of an ion-exclusion separation (*i.e.*, with oxalic acid in Fig. 4.9). In contrast, the weak-acid matrix will enter the resin beads and be somewhat retained. Collection of the early-eluted ions and re-injection into an ion-exchange column allowed determination of 100- to 1000-μg/L levels of Cl^-, NO_3^-, and SO_4^{2-} in 85% phosphoric acid and 10–300 μg/L of Cl^-, NO_3^-, SO_4^{2-}, and HPO_4^{2-} in 24.5% HF [52], although subsequently problems with SO_4^{2-} contamination from the column and unstable HPO_4^{2-} recovery from HF solutions have been reported [53].

4.3.5.3 Electrostatic ion chromatography

Electrostatic ion chromatography (EIC) demonstrates a selectivity that is distinct from that of typical IC [54]. The heart of EIC is a bifunctional stationary phase where cationic and anionic functional groups are in close proximity on the stationary phase, as shown in Zwittergent 3–14. The stationary phase may either possess permanent and balanced positive and negative functionalities [55], or the zwitterionic character can be induced by dynamically coating a reversed-phase column with a zwitterionic surfactant, such as 3-(*N*,*N*-dimethylmyristylammonio)-propanesulfonate (Zwittergent 3–14) [54]. The elution order in EIC is determined by chaotropic, rather than electrostatic effects [56]. In essence, these ions interact with water less than water does with itself. The chaotropic nature of ions is empirically ranked in the Hofmeister series [57]. Strongly chaotropic ions are typically hydrophobic (*e.g.*, *p*-cyanophenol) or polarizable (*e.g.*, I^-), anions exhibiting much stronger chaotropic effects than cations. In EIC, the separation selectivity (Eqn. 4.17) follows the order of increasing chaotropic nature (Hofmeister series) [55,56,58].

$$SO_4^{2-} < F^- < Cl^- < NO_2^- < CNO^- < Br^- < NO_3^- \ll ClO_3^- \ll I^- \ll SCN^- \ll ClO_4^- \tag{4.17a}$$

$$Rb^+ \approx Cs^+ \approx K^+ \approx 0 < Na^+ < Li^+ < Mg^{2+} < Ce^{3+} \tag{4.17b}$$

The more chaotropic the anion, the greater is its retention. This is evident in Fig. 4.10 [59]. Cations are less chaotropic, so that the ion association constant for trivalent Ce^{3+} is about the same as that of monovalent NO_3^- [58].

The eluent also plays a unique role in EIC. Pure water can be used as an eluent [60]. With pure water, ions are eluted as "ion pairs", in every possible combination. For instance, injection of NaCl and KNO_3 would result in four peaks (Na^+Cl^-, K^+Cl^-, $Na^+NO_3^-$, $K^+NO_3^-$). Nonetheless, useful analytical results may be obtained with pure water as mobile phase, by converting all samples to a constant cation or by using a suitable electrolyte as eluent [60]. Addition of as little as 1 mM electrolyte (*e.g.*, Na_2SO_4, NaCl, or $NaClO_4$) is sufficient to change retention in EIC [56]. However, there is no further change in ion retention, as the electrolyte concentration in the eluent is further increased. This is

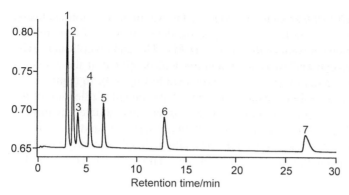

Fig. 4.10. Electrostatic ion chromatography of inorganic anions. Experimental conditions: column, 4.6×250 mm, 5-μm polymeric ODS material, pre-equilibrated with 30 mM Zwittergent 3–14 for 1 h; eluent, 10 mM NaHCO$_3$; flow-rate, 1.0 ml/min; injection, 100 μL of 0.1 mM of each anion in the Na$^+$ form; detection, suppressed conductivity. Peaks: 1, sulfate; 2, chloride; 3, nitrite; 4, bromide; 5, nitrate; 6, chlorate; and 7; iodide. The peak due to thiocyanate was eluted at a retention time of 378 min. (Reprinted from Ref. 59 with permission.)

contrary to typical ion-exchange behavior (Eqn. 4.12), but consistent with a mechanism whereby retention is governed by chaotropic behavior rather than electrostatics. Retention in EIC is determined by the nature of the eluent electrolyte rather than the eluent concentration. It has been proposed that the eluent electrolyte affects retention through ion-exclusion effects [56]. The sulfonate group on the outer part of the zwitterionic stationary phase (Fig. 4.11a) [56] contributes a negative charge that repels analyte anions by acting as a Donnan membrane. However, the magnitude of this negative charge (and thus the degree of repulsion created by the Donnan membrane) depends on the uptake of eluent ions by the zwitterionic surface. If the electrolyte contains a strongly chaotropic cation (divalent or trivalent), this will diminish the charge of the Donnan membrane, resulting in an increase in anion retention. Alternatively, if the electrolyte contains a strongly chaotropic anion (*e.g.*, ClO$_4^-$), the Donnan membrane potential will increase and anions will experience greater ion exclusion and thus less retention. The primary

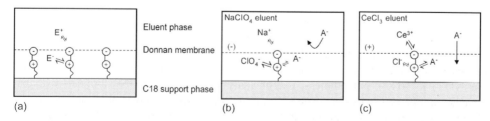

Fig. 4.11. Stationary phase and mechanism of retention in electrostatic ion chromatography. (Reproduced from Ref. 56 with permission.)

consequences of retention being governed by a chaotropic rather than an electrostatic mechanism are:

(a) a unique selectivity, where hydrophilic anions, such as Cl^- and SO_4^{2-}, are weakly retained, while ions high in the Hofmeister series (*e.g.*, I^-, SCN^-, ClO_4^-) are strongly retained; and

(b) retention which is insensitive to the ionic strength of the sample.

As an example, traces of Br^-, NO_3^-, and I^- in sea water were determined by using Zwittergent 3–14 as eluent [61].

4.4 SAMPLE PREPARATION

4.4.1 Concentration of trace analytes

Ions in the mid-μg/L to mg/L (ppb to ppm) range can be determined using sample injections of 10–50 μL. To achieve detection limits below this level, sample pre-concentration is required. This may be accomplished by direct injection of large volumes that are concentrated at the head of the column (Sec. 4.4.1.1) or by using a small pre-concentration column. However, regardless of the pre-concentration technique used, successful determination of trace analytes is critically dependent upon the precautions taken to minimize contamination. The de-ionized water used for preparing rinse solutions, eluents, and standards should be free of measurable levels of ionic impurities, organic material, microorganisms, and particulate matter larger than 0.2 μm. Containers and volumetric ware should be made of polyethylene or some other inert material. Even brief exposure to conventional laboratory glassware results in significant contamination, particularly by sodium and silicate. All containers should be soaked for at least 24 h in deionized water and rinsed several times prior to use. Disposable gloves, suitable for clean-room electronics, must be worn when handling apparatus that makes contact with eluent, standard, or samples. The total baseline shift between the beginning and the end of the gradient analysis should be no more than 1 μS. To ensure this, sodium hydroxide eluents should be prepared with minimal carbonate contamination or generated on-line (Sec. 4.2.2.3). In addition, a high-capacity anion-exchange column (1–3 meq) should be placed between the pump and the injector to trap anion contaminants. Finally, some systems require up to 5 h to achieve the baseline stability needed for the determination of trace analytes. Many users thus find it best to keep the system running continually.

4.4.1.1 High-volume direct injection

For ultra-clean samples, *e.g.*, from the power or semi-conductor industries, sensitivities at the low- to sub-μg/L levels can be achieved by high-volume, direct-injection techniques [62]. This approach is attractive, as the sensitivity enhancement is achieved without the need for a concentrator column or loading pump and valve. Water alone is a very weak eluent in IC. Thus, when ultra-pure water is injected into an ion-exchange column, the trace ions are strongly retained at the head of the separation column, a process known as *zone compression*. Only after all of the sample matrix (water) has passed through does the eluent finally reach the head of the column and starts eluting the ions. Direct injection of 900 μL into a 2-mm-ID anion-exchange column and employing an isocratic hydroxide eluent yields detection limits ranging from 0.04 μg/L for F$^-$ to 0.35 μg/L for oxalate [62]. However, when high-volume direct injection is used, the peak corresponding to the water dip becomes very large. This makes quantification of early-eluted peaks by suppressed-conductivity detection challenging. For ions absorbing UV, direct absorbance detection can circumvent this problem. Nitrite determination has been performed in this manner, with injection volumes as large as 10–50 mL. However, a more general solution is the use of an OH$^-$ gradient. In this manner, a 750-μL sample provides sub-μg/L detection limits with linear calibration curves (R^2 > 0.99) [62,63]. Addition of corrosion inhibitors, such as ethanolamine (8 mg/L), morpholine (8 mg/L), boric acid (8 mg/L), or ammonium hydroxide (0.3 mg/mL) had no significant effect on peak efficiency, retention time, or calibration [62,63]. Trace cations in high-purity water can also be determined with high-volume direct injection. Injection of 1000 μL into a 2-mm-ID Dionex IonPac CS12A column yielded detection limits ranging from 0.007 μg/L for Li$^+$ to 0.05 μg/L for Ca^{2+} [62].

4.4.1.2 Pre-concentrator column

The most popular means of performing pre-concentration in IC is to use a pre-column, designed to retain trace levels of analytes from a large volume of sample [64]. This method is simple, convenient, and easy to automate. Furthermore, it enables performing routine analysis for ions at μg/L (ppb) to ng/L (ppt) levels. However, it does require additional equipment (a concentrator column, sample pump, and an additional valve) not needed in direct injection. For pre-concentration, an accurately known volume of sample is passed through a small ion-exchange column (*e.g.*, 2 × 15 mm) with a syringe or a metered-sample pump. Analyte ions are trapped selectively on the concentrator column. A valve then switches the concentrator column to connect with the separation column. The capacity of the concentrator column is generally only a few percent of that on the separation column. Thus, the same eluent provides both rapid elution from the concentrator column and sufficient retention for separation on the analytical column.

In theory, the greater the volume of sample loaded onto the concentrator, the greater the sensitivity. However, if too large a volume is loaded onto the concentrator column, the sample is eluted from the concentrator by the sample matrix. The maximum sample volume that can be concentrated will thus be a function of the sample matrix. For instance, more than 45 mL of deionized water or 1 ppm ammonium in water can be reliably concentrated on an anion concentrator, but 20 mL of 1.2% boric acid is the maximum which can be concentrated [65]. Two procedures can be used to establish the maximum sample volume for a new application [65,66]. The first procedure involves performing a series of analyses with increasing volumes of injected sample. The peak height of the most weakly retained component is plotted against the volume concentrated. A negative deviation in this plot shows the volume at which the sample matrix starts to elute analytes from the column. The second method requires that the retention volume of the least-retained compound be calculated using a simulated sample matrix as the eluent.

This retention volume corresponds to the volume at which the matrix is eluting analytes from the column. Trace anions have also been determined in organic solvents such as 2-propanol, acetone, and *N*-methylpyrrolidone using a pre-concentration column [67]. Five mL of organic solvent was passed through the pre-concentration column, which was then rinsed with distilled water to remove the organic matrix prior to a standard IC separation. The detection limits for Cl^-, NO_3^-, SO_4^{2-}, and HPO_4^{2-} were in the sub-μg/L range.

4.4.2 Matrix removal

4.4.2.1 Solid-phase extraction

Small, disposable cartridges, containing 0.5–1.5 mL of highly selective, functionalized polystyrene resin, are a practical tool for matrix elimination. Samples are passed through these cartridges, using a disposable syringe, and the effluent is injected directly into the IC. Hydrophobic matrix components can interfere with IC analysis either by interfering with detection, or by altering retention, if they are strongly adsorbed on the column. Instead of ODS, polystyrene/divinyl benzene resins are used in solid-phase extraction (SPE) cartridges to remove hydrophobic species (*e.g.*, surfactants, organic acids, proteins) because of the extreme pH of many of the samples. Strongly acidic or basic samples can cause baseline disturbances and retention-time changes by protonating/deprotonating analytes and ion-exchange sites. A strong-anion-exchange column in the OH^- form can be used to reduce the pH of acidic samples prior to injection. Since OH^- is weakly retained by anion exchangers, it is readily displaced by anions within the sample. The liberated OH^- then neutralizes H^+ in the sample. However, the sample anions are retained on anion-exchange SPE cartridges. Thus, they can be used to remove or concentrate anions from the sample, to adjust the pH of samples for cation analysis, or to remove cations that form insoluble hydroxide salts. A strong-cation-exchange column in the H^+-form can be used to reduce the pH of alkaline samples prior to analysis. Cations in the sample displace weakly retained H^+ from the column. The H^+ is consumed in the neutralization reaction with OH^- from the sample. The net result is removal of hydroxide and an equivalent amount of cations from the sample. In effect, this is a column-based suppressor applied prior to injection. The use of membrane-based suppressors for neutralization is discussed in the next section. A cation exchanger in the H^+ form can be used in conjunction with a reversed-phase SPE cartridge to improve removal of organic acids. However, loss of weak-acid anions, such as nitrite (30–40%) or phosphate (13%), can occur upon passage through a H^+-SPE column [68]. A cation-exchange SPE cartridge in the Na^+ form can be used to remove cations from the sample without altering the pH. Alternatively, more selective trapping of transition-metal cations can be achieved by using an iminodiacetate resin cartridge. This type of resin chelates transition metals, such as Cd^{2+}, Mn^{2+}, Fe^{2+}, Fe^{3+}, Co^{2+}, Pb^{2+}, Ni^{2+}, Al^{3+}, Cu^{2+}, Zn^{2+}, and Ag^+, at pH > 4, even in the presence of high amounts of sodium [69].

Removal of chloride and sulfate from the sample can be achieved by precipitating these anions with counter-ions provided by a cation-exchange SPE cartridge. For instance, if a halide solution is passed through a cation exchange SPE cartridge in the Ag^+ form, the following equilibria occur

$$M_m^+ + Ag_r^+ \rightleftharpoons M_r^+ + Ag_m^+ \tag{4.18a}$$

$$Ag^+ + Cl^- \rightleftharpoons AgCl_{(s)} \qquad \log K_{SP} = 9.7 \tag{4.18b}$$

The metal (M_m) in the *mobile* phase (sample) displaces the weakly retained Ag^+ from the stationary (*Resin*) phase, which can combine with the anion to form a precipitate (SP). Those ions which form an insoluble salt (log $K_{SP} > 8$) with Ag^+ (*e.g.*, Cl^- and Br^-) are quantitatively removed, whereas ions such as F^-, NO_3^-, phosphate, and SO_4^{2-} are recovered in $>95\%$ yield [68]. Since silver hydroxide readily forms under basic conditions, the pH of the samples must be below 8 for effective removal of halides. As with the H^+ form of the SPE resin, some losses in NO_3^- are experienced. Silver-containing SPE cartridges (SPE-Ag^+) are not effective for removal of phosphate, although silver phosphate has a high K_{SP} (log $K_{SP} = 17.6$), since PO_4^{3-} is prevalent only at high pH. For the trace analysis of HPO_4^{2-}, recoveries of phosphate upon passage through a SPE-Ag^+ should be determined by standard addition. To avoid contamination of the separation column with Ag^+, the SPE-Ag^+ is usually followed by a cation-exchange column in the H^+ form. Some two-layer disposable cartridges are now available which contain Ag^+-form resin on top of H^+ resin for both, removal of halide and conversion of the sample cation.

Similarly, sulfate can be removed by using a cation-exchange SPE cartridge in the Ba^{2+} form

$$2M_m^+ + Ba_r^{2+} \rightleftharpoons 2M_r^+ + Ba_m^{2+} \tag{4.19a}$$

$$Ba^+ + SO_4^{2-} \rightleftharpoons BaSO_{4(s)} \qquad \log K_{SP} = 10.0 \tag{4.19b}$$

However, unlike Ag^+ in SPE-Ag^+, Ba^{2+} is relatively well retained by the cation-exchange resin. This means that there must be sufficient cation in the sample matrix to displace the barium ion, *i.e.*, *ca.* 200 mg/L Na^+ or 100 mg/L Ca^{2+}. If the cation content of the matrix is low, the sample can be spiked with calcium chloride and then successively passed through Ba^{2+}-form, Ag^+-form, and H^+-form resin. The SPE-Ba^{2+} removes any SO_4^{2-}, the SPE-Ag^+ removes Cl^-, and the SPE-H^+ traps any re-dissolved Ag^+ [70]. The $BaSO_4$ precipitate is soluble at low pH, and thus, SPE-Ba^{2+} cartridges are not very effective for removal of SO_4^{2-} from sulfuric acid. The recovery of Cl^-, NO_2^-, and Br^- was only 80–85%, when samples, containing 1000 mg/L SO_4^{2-}, were passed through SPE-Ba^{2+}, presumably due to occlusion in the $BaSO_4$ crystals [68]. Since these recoveries are reproducible, they can be corrected by preparing standards containing the same SO_4^{2-} concentration as the samples.

4.4.3 On-line sample pre-treatment

The SPE cartridges described above provide an effective means of dealing with a wide array of sample matrices. However SPE is performed manually off-line. Thus, it is both labor-intensive and unsuitable for automation. A number of other procedures have been developed for sample pre-treatment [71–73]. Acidic or alkaline samples can be neutralized by passage through an on-line neutralizer prior to direct injection or pre-concentration [71,72]. These neutralizers are based on the suppressor technologies described in Sec. 4.2.3. Using such neutralizers, single-digit $\mu g/L$ detection limits have been achieved for Cl^-, NO_2^-, Br^-, NO_3^-, SO_4^{2-}, and HPO_4^{2-} in 29% (w/w) ammonium hydroxide [53,74] and 25% trimethylammonium hydroxide (TMAOH) [74]. Weak-acid anions, such as F^- and NO_2^-, would not be expected to be amenable to this approach, as HF and HNO_2 would be lost *via* diffusion through the membrane. However, recoveries reported for NO_2^- have been near-quantitative. Leaching of Cl^-, and particularly SO_4^{2-}, can result in elevated blanks for these anions. However, cleaning of the neutralizer with $0.5\,M$ NaOH for several hours, followed by a 1-h rinse with distilled water minimizes this background. Similarly, cations at the low- to mid-$\mu g/L$ level can be determined in concentrated (up to 25%) HF, H_2SO_4, and H_3PO_4, using a self-regenerating (electrochemical) membrane suppressor [71]. HCl, $HClO_4$, and HNO_3 matrices damage the membrane either directly or indirectly (*via* electrolysis products). Weak-base analytes, such as ammonium and amines, yield low recoveries (50%), due to diffusion of the non-ionized weak base through the membrane. Near-quantitative recoveries have been achieved for NH_4^+ in HCl and H_2SO_4 using column-based suppressors [72–74].

Passive diffusion can be used to eliminate macromolecular matrix components. The sample, pumped into a flow-through dialysis cell, passes over a cellulose acetate membrane with 0.2-μm pores. Species below a certain mass diffuse across the membrane into a stationary acceptor solution. After about 5 min of dialysis time, the ion content of the acceptor matches that of the sample, but the matrix is much simplified. This acceptor solution is then introduced into the column. In this manner, standard anions and cations were readily determined in complex matrices, such as milk, untreated wastewater, engine coolant, and multi-vitamin tablets, with >87% recovery and no long-term decrease in column performance [73]. "Two-dimensional" or "heart-cut" chromatography can also be used to remove the sample matrix [75]. The sample is injected into a chromatographic column and, at a specific time, the effluent containing the component of interest is passed into a second column through a column-switching valve. This procedure works best when the sample matrix is of a consistent composition and does not permanently affect the column performance. Enhanced performance can be achieved by using two different separation modes in the two columns [52].

4.5 FUTURE DIRECTIONS

4.5.1 High-speed analysis

In recent years, high-speed separations have been of great interest in reversed-phase liquid chromatography, and another subject of intense development is the on-line monitoring of chemical processes (Sec. 4.5.2). Savings in analysis time have been achieved in HPLC mainly through shortening of the column with a simultaneous decrease in particle size (3 μm or less). However, IC columns are packed with rather large (>5-μm) particles.

This is the main reason why IC routinely cannot be performed in less than a minute – yet. Nevertheless, significant reductions in analysis time have already been achieved by Connolly and Paull [76–78]. They employed a 30 × 4.6-mm-ID reversed-phase column, packed with 3-μm particles for the ion-interaction separation of nitrite and nitrate in less than 45 s. Refinements in their technology resulted in a 2.5-min separation of 9 common anions. Furthermore, Hatsis and Lucy [79,80] have shown the applicability of monolithic columns in IC with separations of the common anions in less than 1 min.

4.5.2 Process and field analysis

On-line HPLC never achieved widespread acceptance in industry, partly due to the cost of purchase and disposal of solvents, and hazards of flammable solvents. In contrast, eluents in IC are nonflammable aqueous solutions, and the waste (after suppression) is essentially distilled water. Further simplification of operations can be achieved with on-line eluent generation (Sec. 4.2.2.3). Thus, it is not surprising that on-line IC has been much more readily accepted by industry than its RP-LC counterpart. The Dionex DX-800 and Metrohm-Peak Models 811 and 821 are IC systems specifically designed for production environments. These systems can monitor one or up to 10–21 different process streams and offer low-μg/L detection limits for anion and cation analyses with direct injection. Samples with lower concentrations can be monitored by using a pre-concentrator column (Sec. 4.4.1.2), whereas those with higher concentration are diluted on-line. On-line IC has been used for a variety of process applications [81], including monitoring: ultra-pure water in nuclear power plants for corrosive ions, such as chloride and sulfate [82]; various process and wastewater streams for anions and cations [83]; organic acids that are starting materials in the production of nylon [84]; and organic additives in plating baths used for manufacturing semi-conductor wafers [84].

A fully automated instrument for field measurements of acid gases and soluble anionic constituents of atmospheric particulate matter has recently been developed [85]. Acid gases include SO_2, HCl, HF, HONO, HNO_3, CH_3SO_3H, and various organic acids. The instrument incorporates two sample-collection manifolds (for soluble acid gases and particulates) and an IC system. Detection limits in the low- to sub-ng/m^3 range of concentrations of most gaseous and particulate constituents are readily attained, and the suitability of the instrument for field analysis has been extensively tested. A computer-controlled, field-portable, capillary IC system has also been developed [86]. The entire instrument (excluding laptop controller) fits into a standard briefcase and weighs only 10 kg. Detection limits are comparable to those of laboratory IC systems, while mass detection limits are more than 100 times lower than for standard systems. Pre-concentration can also be performed.

ACKNOWLEDGMENTS

The authors would like to thank Christopher Pohl and Jeff Rohrer of Dionex Corporation for their help in the preparation of this manuscript. IonPac, MMS, and SRS are registered trademarks of the Dionex Corporation and Allsep is a registered trademark of Alltech Associates Inc. This chapter was based on research supported by the Natural Sciences and Engineering Research Council of Canada (NSERC), the University of Alberta, and the Dow Chemical Company.

REFERENCES

1   H. Small, T.S. Stevens and W.C. Bauman, *Anal. Chem.*, 47 (1975) 1801.
2   D.T. Gjerde, J.S. Fritz and G. Schmuckler, *J. Chromatogr.*, 186 (1979) 509.
3   Dionex IonPac AS14-A Brochure, Dionex, Sunnyvale, 2001. http://www.dionex.com.
4   P.R. Haddad and P.E. Jackson, *Ion Chromatography: Principles and Applications*, Elsevier, Amsterdam, 1990.
5   B.E. Erickson, *Anal. Chem.*, 71 (1999) 465A.
6   D.L. Strong, C.U. Joung and P.K. Dasgupta, *J. Chromatogr.*, 546 (1991) 159.
7   D.L. Strong, P.K. Dasgupta, K. Friedman and J.R. Stillian, *Anal. Chem.*, 63 (1991) 480.
8   Y. Liu, N. Avdalovic, C. Pohl, R. Matt, H. Dhillon and R. Kiser, *Am. Lab.*, 30 (1998) 8.
9   T.S. Stevens, J.C. Davis and H. Small, *Anal. Chem.*, 53 (1981) 1488.
10 K.K. Haak, S. Carson and G. Lee, *Am. Lab.*, 18 (1986) 50.
11 D.L. Strong and P.K. Dasgupta, *Anal. Chem.*, 61 (1989) 939.
12 S. Rabin, J. Stillian, V. Barreto, K. Friedman and M. Toofan, *J. Chromatogr.*, 640 (1993) 97.
13 T. Sunden, A. Cedergren and D.D. Siemer, *Anal. Chem.*, 56 (1984) 1085.
14 D.D. Siemer and V.J. Johnson, *Anal. Chem.*, 56 (1984) 1033
15 R. Saari-Nordhaus and J.M. Anderson, Jr., *J. Chromatogr. A*, 956 (2002) 15.
16 W.W. Buchberger, *J. Chromatogr. A*, 884 (2000) 3.
17 P.R. Haddad, M.J. Shaw and G.W. Dicinoski, *J. Chromatogr. A*, 956 (2002) 59.
18 D. Midgley and R.L. Parker, *Talanta*, 36 (1989) 1277.
19 R. Al-Horr, P.K. Dasgupta and R.L. Adams, *Anal. Chem.*, 73 (2001) 4694.
20 R.D. Rocklin, *J. Chromatogr.*, 546 (1991) 175.
21 H. Yu, Y.S. Ding, S.-F. Mou, P. Jandik and J. Cheng, *J. Chromatogr. A*, 966 (2002) 89.
22 H. Small and T.E. Miller, Jr., *Anal. Chem.*, 54 (1982) 462.
23 P.K. Dasgupta, *J. Chromatogr. Sci.*, 29 (1989) 422.
24 R.M. Cassidy and B.D. Karcher, in I.S. Krull (Ed.), *Reaction Detection in Liquid Chromatography*, Dekker, New York, 1986.
25 Method 7199, United States Environmental Protection Agency, 1996, http://www.epa.gov/epaoswer/hazwaste/test/pdfs/7199.pdf.
26 D.J. Barkley, M. Blanchette, R.M. Cassidy and S. Elchuk, *Anal. Chem.*, 58 (1986) 2222.
27 H.P. Wagner, B.V. Pepich, D.P. Hautman and D.J. Munch, Method 317.0, United States Environmental Protection Agency, Cincinnati, OH, 2001, http://www.epa.gov/safewater/methods/met317rev2.pdf.
28 R.M. Cassidy, S. Elchuk and P.K. Dasgupta, *Anal. Chem.*, 59 (1987) 85.
29 K.L. Sutton and J.A. Caruso, *J. Chromatogr. A*, 856 (1999) 243.
30 A. Seubert, *Trends Anal. Chem.*, 20 (2001) 274.
31 C.A. Pohl, J.R. Stillian and P.E. Jackson, *J. Chromatogr. A*, 789 (1997) 29.
32 S. Rabin and J. Stillian, *J. Chromatogr. A*, 671 (1994) 63.
33 B. Chu, D.C. Whitney and R.M. Diamond, *J. Inorg. Nucl. Chem.*, 24 (1962) 1405.
34 R.M. Diamond, *J. Phys. Chem.*, 67 (1963) 2513.
35 Metrohm Column Selection Guide, Metrohm, Herisau, 2001, http://www.metrohm.ch.

36   1997–1998 Dionex Product Selection Guide, Dionex, Sunnyvale, 1998, http://www.dionex. com.

37   J.E. Madden and P.R. Haddad, *J. Chromatogr. A*, 829 (1998) 65.

38   J.E. Madden, N. Avdalovic, P.E. Jackson and P.R. Haddad, *J. Chromatogr. A*, 837 (1999) 65.

39   J.E. Madden and P.R. Haddad, *J. Chromatogr. A*, 850 (1999) 29.

40   M.A. Rey and C.A. Pohl, *J. Chromatogr. A*, 739 (1996) 87.

41   N.E. Fortier and J.S. Fritz, *Talanta*, 34 (1987) 415.

42   P. Hatsis and C.A. Lucy, *J. Chromatogr. A*, 920 (2001) 3.

43   P. Hatsis and C.A. Lucy, *Analyst*, 126 (2001) 2113.

44   M. Tomoi, K. Yamamguchi, R. Ando, Y. Kantake, Y. Aosaki and H. Kubota, *J. Appl. Polym. Sci.*, 64 (1997) 1161.

45   Alltech Chromatography Source Book, Alltech, Deerfield, 2001, http://www.alltechweb.com.

46   M.A. Rey, *J. Chromatogr. A*, 920 (2001) 61.

47   R.W. Slingsby and C.A. Pohl, *J. Chromatogr.*, 458 (1988) 241.

48   R.G. Smith, P.A. Drake and J.D. Lamb, *J. Chromatogr.*, 546 (1991) 139.

49   A. Woodruff, C.A. Pohl, A. Bordunov and N. Avdalovic, *J. Chromatogr. A*, 956 (2002) 35.

50   K.L. Ng, B.K. Glod, G.W. Dicinoski and P.R. Haddad, *J. Chromatogr. A*, 920 (2001) 41.

51   K. Tanaka, T. Ishizuka and H. Sunahara, *J. Chromatogr.*, 174 (1979) 153.

52   E. Kaiser, J.S. Rohrer and K. Watanabe, *J. Chromatogr. A*, 850 (1999) 167.

53   K.F. Wang, Y. Lei, M. Eitel and S. Tan, *J. Chromatogr. A*, 956 (2002) 109.

54   W.Z. Hu, T. Takeuchi and H. Haraguchi, *Anal. Chem.*, 65 (1993) 2204.

55   W. Jiang and K. Irgum, *Anal. Chem.*, 71 (1999) 333.

56   H.A. Cook, W.Z. Hu, J.S. Fritz and P.R. Haddad, *Anal. Chem.*, 73 (2001) 3022.

57   M.G. Cacace, E.M. Landau and J.J. Ramsden, *Q. Rev. Biophys.*, 30 (1997) 241.

58   T. Yokoyama, M. Macka and P.R. Haddad, *Anal. Chim. Acta*, 442 (2001) 221.

59   W.Z. Hu and P.R. Haddad, *Anal. Commun.*, 35 (1998) 317.

60   E. Twohill and B. Paull, *J. Chromatogr. A*, 973 (2002) 103.

61   W.Z. Hu, P.R. Haddad, K. Hasebe, K. Tanaka, P. Tong and C. Khoo, *Anal. Chem.*, 71 (1999) 1617.

62   E. Kaiser, J. Riviello, M. Rey, J. Statler and S. Heberling, *J. Chromatogr. A*, 739 (1996) 71.

63   Dionex Application Note 113, Determination of Trace Anions in High Purity Waters by High Volume/Direct Injection Ion Chromatography, Dionex, Sunnyvale, (1996), http://www.dionex. com/servletwl1/FileDownloader/slot114/8925/AN113.PDF.

64   P.R. Haddad, P. Doble and M. Macka, *J. Chromatogr. A*, 856 (1999) 145.

65   Dionex Application Note 56, Determination of Trace Anions and Key Organic Acids in High Purity, Ammoniated and Borated Water Found in Steam Cycle Power Plants, Dionex, Sunnyvale, 2001, http://www.dionex.com/servletwl1/FileDownloader/slot114/8925/AN56. PDF.

66   Method D5542-94, Vol. 11.01 Water (I): Water and Environmental Technology, American Society for Testing and Materials, Philadelphia, 1995.

67   E. Kaiser and J. Rohrer, *J. Chromatogr. A*, 858 (1999) 55.

68   I.K. Henderson, R. Saari-Nordhaus and J.M. Anderson, *J. Chromatogr.*, 546 (1991) 61.

69   Y. Liu and J.D. Ingle, Jr., *Anal. Chem.*, 61 (1989) 525.

70   R.W. Slingsby and C.A. Pohl, *J. Chromatogr. A*, 739 (1996) 49.

71   A. Siriraks and J. Stillian, *J. Chromatogr.*, 640 (1993) 151.

72   R.M. Montgomery, R. Saari-Nordhaus, L.M. Nair and J.M. Anderson, Jr., *J. Chromatogr. A*, 804 (1998) 55.

73   B.M. DeBorba, J.M. Brewer and J. Camarda, *J. Chromatogr. A*, 919 (2001) 59.

74   M.L. Wu and J.G. Chen, *Micro*, 15 (1997) 65.

75 S.R. Villaseñor, *Anal. Chem.*, 63 (1991) 1362.
76 D. Connolly and B. Paull, *J. Chromatogr. A*, 917 (2001) 353.
77 D. Connolly and B. Paull, *Anal. Chim. Acta*, 441 (2001) 53.
78 D. Connolly and B. Paull, *J. Chromatogr. A*, 953 (2002) 299.
79 P. Hatsis and C.A. Lucy, *Analyst*, 127 (2002) 451.
80 P. Hatsis and C.A. Lucy, *Anal. Chem.*, 75 (2003) 995.
81 M.J. Doyle and B.J. Newton, *CAST*, Jan/Feb (2002) 9.
82 T.O. Passell, *J. Chromatogr. A*, 671 (1994) 331.
83 E.M. Hodge, *J. Chromatogr. Sci.*, 38 (2000) 353.
84 J.C. Thompsen and R.H. Smith, *Process Contr. Qual.*, 2 (1992) 55.
85 C.B. Boring, R. Al-Horr, Z. Genfa, P.K. Dasgupta, M.W. Martin and W.F. Smith, *Anal. Chem.*, 74 (2002) 1256.
86 C.B. Boring, P.K. Dasgupta and A. Sjogren, *J. Chromatogr. A*, 804 (1998) 45.

Erich Heftmann (Editor)
Chromatography, 6th edition
Journal of Chromatography Library, Vol. 69A

Chapter 5

Size-exclusion chromatography

J. SILBERRING, M. KOWALCZUK, J. BERGQUIST, A. KRAJ, P. SUDER, T. DYLAG, M. SMOLUCH, J.-P. CHERVET and R. EKMAN

CONTENTS

5.1   Introduction 213
5.2   Theory 215
5.3   Columns 217
        5.3.1   Monolithic materials 218
        5.3.2   Capillary columns 219
        5.3.3   Microsystems 219
5.4   Mobile phases 219
5.5   Detectors 222
        5.5.1   Mass spectrometry 224
        5.5.2   Nuclear magnetic resonance and surface-enhanced Raman
                spectroscopy 227
5.6   Calibration 228
5.7   Applications 230
        5.7.1   Polymers 230
        5.7.2   Proteins and polypeptides 237
        5.7.3   Molecular biology 239
        5.7.4   Carbohydrates 239
        5.7.5   Other applications 240
Acknowledgments 241
References 241

5.1 INTRODUCTION

Size-exclusion chromatography (SEC), also called gel-permeation chromatography, molecular-sieve chromatography, or gel filtration, separates molecules according to their size. The term "gel permeation" is commonly used in polymer science for non-aqueous separations. Molecules of different size penetrate well-defined pores of column beads, where they are retained to various degrees. Smaller molecules are retarded on the column, while larger ones are eluted more rapidly. As the retention time can be directly correlated with the size of molecules, this method is particularly useful for the determination of molecular weight. Since the current IUPAC recommendation still accepts terms such as molecular weight, molar mass, and molecular mass, they will equally appear in this text.

The use of size-exclusion phenomena for the separation of natural polymers had been already discussed in the late 1940s when materials, such as zeolites, charcoal, and ion-exchange resins were developed as sorbents. However, it was not until the middle of the 1950s that size-exclusion chromatography appeared in the literature [1,2], with starch as the column material. This was followed by a period of intensive activity at Uppsala University [3–8], where in 1959 Porath and Flodin [9] produced cross-linked dextran, which was commercialized by Pharmacia as Sephadex, and in 1962 Hjertén and Mosbach [10,11] developed cross-linked polyacrylamide (Biogel P) and agarose (Sepharose). During the same period, Polson [12] reported that agar was also a suitable column material for separating natural polymers. Since then, a large number of separation media have been developed, and these will be described in greater detail in this chapter. General reviews on SEC, including theory, equipment, and instrumentation can be found in Refs. 13–17.

SEC is now generally applicable to the separation of molecules in the range between 0.5 and 1000 kDa, but larger proteins or other giant molecules can also be separated [18]. Moreover, SEC can also physically separate folded, from unfolded macromolecules, particularly in the case of slow equilibria [19]. The major applications of SEC include the determination of molecular weight and molecular-weight distribution (polydispersity) of natural and synthetic polymers. Separation strongly depends on many factors, among them column packing, column dimensions, flow-rate, sample volume, and mobile-phase composition. These parameters may substantially contribute to the quality of separation and possible errors, such as band-broadening and decreased resolution. This effect was observed by Busnel *et al.* [20] with standards having very narrow molar mass distribution. Mapping of band-broadening was observed for different column sets. These aspects were also studied in detail by Ricker and Sandoval [21], who presented practical guidelines for the development of reproducible SEC methods, based upon optimized sample volume, flow-rate, column length, and mobile-phase conditions that reduce nonideal SEC behavior – parameters often ignored in this type of separations. Adjustment of these factors frequently results in more accurate elution times, more precise determination of molar mass, sharper peaks for improved resolution and shorter run times for increased throughput. In general, sample volume and flow-rate should be kept to a minimum for optimal resolution. Increasing column length improves resolution and may be achieved by connecting columns in tandem. Adjustment of the mobile-phase conditions can significantly enhance resolution, but results are difficult to predict, because unique sample properties play a major role in this interaction, as does the column packing. Whenever possible, ionic strength and pH of the mobile phase should be adjusted until the peak(s) of interest are eluted at the expected time and with good peak shape. Finally, the use of smaller-diameter columns (*i.e.*, 4.6 mm rather than 9.4 mm ID) and packings of smaller particles (4–5 µm) may also be considered. These factors will be described below.

5.2 THEORY

SEC is defined as the differential elution of solutes, from a bed of a porous chromatographic medium, caused by different degrees of steric exclusion of sample molecules from the pore volume of the molecular size of solutes. Thus, in SEC, solutes are eluted strictly according to decreasing molecular size, and the maximum available volume for separation is equal to the total pore volume of the packing medium. The retention volume, V_R, is given by

$$V_R = K_D V_P + V_O \tag{5.1}$$

where K_D is the distribution coefficient, *i.e.*, the available pore fraction (ranging from 0 to 1), V_P is the pore volume of the packing medium, and V_O is the extra-particle void volume. The relative pore volumes of packing media vary from 50% for silica-based materials to over 95% for semi-rigid polymer-based media. Since the separation volume is limited in SEC, the calculation of peak capacity can be reduced to

$$nR_s = 1 + V_P/V_t(N/16)^{1/2}/R_s \tag{5.2}$$

where nR_s is the number of peaks separated with a resolution of R_S, V_t is the total liquid volume of the column (*i.e.*, $V_P + V_O$), and N is the maximum number of theoretical plates of the column. The peak capacity of SEC columns is often cited as

$$n = 1 + 0.2N^{1/2} \tag{5.3}$$

However, this equation is valid only if N has the same value for all solutes, which is not the case for macromolecules. That is why Giddings [22] suggested that an average plate count could be assumed in the calculation. From these equations and the statements above one can conclude that the peak capacity of a SEC column is much smaller than that for many other liquid chromatography (LC) columns. There is also little or no advantage, with respect to separation efficiency, in using a smaller-sized packing material in SEC, since the pore volume then decreases and the column length is often shorter. The only gain would thus be in separation speed. This could, of course, be very important in multiple separations, where processing speed needs to be high. Another point to consider is that zone-broadening in SEC is diffusion-controlled. That is why factors affecting the total diffusion time (*i.e.*, column length, the diffusivity of the solute, which is 10–100 times slower for macromolecules than for small inorganic molecules) and the diffusion distance (*i.e.*, pore and particle size of the packing media) are crucial. Other factors to be considered include the sample volume, which should typically be smaller than 0.2% of the total bed volume in order to avoid peak-broadening. There are also some simple "rules of thumb" regarding sample concentration in SEC, which generally should not exceed *ca.* 70 mg/mL for a globular protein, 5 mg/mL for a dextran with an average molar mass of 100 kDa, and 10 mg/mL for a dextran of 10 kDa.

Practical aspects of SEC, including instrumentation and troubleshooting, have been described in a recent review by Titterton [23]. Since no gradients are used in SEC, the chromatographic equipment can be rather simple. For optimal results, packing of the column must be carried out very carefully. The volume of the loaded sample should not exceed 5% of the column volume for preparative runs and 1% for analytical applications. For production-scale separations, column diameters up to 30 cm and gel-bed lengths up to 120 cm are recommended.

In protein separations, the resolution is not correlated with the total amount of protein loaded onto the column. Thus, highly concentrated protein samples will give the best separations. However, in the case of carbohydrates, resolution is likely to be affected by molecular interactions if the sample concentration is too high. It should be noted that SEC columns are not able to concentrate samples, in contrast to some other LC columns. On-line sample treatment for or by column liquid chromatography has been discussed by Brinkman [24].

A recent review by Winzor [25] summarized the theory of SEC and described the development of chromatographic techniques for the determination of reaction stoichio-metry and equilibrium constants for solute interactions of biological importance. Gel chromatography, affinity chromatography, and studies of interactions by biosensor technology were reviewed in detail. A general stochastic theory of SEC, which can account for size dependence on both pore ingress and egress processes, moving-zone dispersion, and pore-size distribution, has recently been developed by Dondi *et al.* [26]. A unified theory for gel electrophoresis and gel filtration was presented earlier by Rodbard and Chrambach [27], who later published on kinetics of hormone/receptor and antigen/antibody interactions. A quantitative theory for gel-exclusion chromatography also arose from the studies of Polson and Katz [28].

Calculations of the hydrodynamic permeability of gels and gel-filled microporous membranes have been published by Mika and Childs [29]. A model was developed by Rill *et al.* [30] to simulate SEC separations of globular proteins on templated gels. In this study, it was assumed that the partition coefficient for sieving of a protein is equal to the fraction of gel volume accessible to a sphere with a radius equal to the Stokes radius of the protein. An interpretation of virial coefficients, reflecting thermodynamic nonideality in incompressible solutions of a single macromolecular species for which there is no volume change on mixing, has been reported by Wills and co-workers [31]. The findings were discussed in relation to the results obtained by osmometry, isopiestic measurements, equilibrium dialysis, gel chromatography, and sedimentation equilibrium. This group has also reviewed the quantitative characterization of biospecific complex formation [32]. The merits of frontal gel chromatography, electrophoretic methods, and affinity chromato-graphy were discussed and theoretical and experimental studies have been made on the advancing elution profile in frontal gel chromatography [33]. A study of multiple polymerization equilibria by glass-bead exclusion chromatography with allowance for thermodynamic nonideality effects has also been reported [34]. Thermodynamic nonideality and the dependence of partition coefficients upon solute concentration were studied by Minton [35]. A gel-chromatographic procedure that corrects for Donnan effects in studies of ligand binding was discussed by Jordan *et al.* [36]. The origin and consequences of concentration dependence in gel chromatography were reported by Nichol and co-workers [37] and Winzor and Nichol [38], while Hibberd *et al.* [39] discussed an experimental and theoretical investigation of boundary spreading in gel chromatography. An interesting paper has recently been published by Brooks *et al.* [40], where an alternate picture proposes that the partition coefficient can be calculated from a thermodynamic model for the free energy of mixing of the solute with the gel phase. Size-dependent exclusion, caused by the unfavorable entropy of mixing, associated with the partition was predicted, the magnitude of the effect being modified by enthalpic interactions between the solute and the gel phase. This concept has been extended to describe the partition of macromolecules into a layer of terminally attached polymer chains, grafted onto a solid bead.

5.3 COLUMNS

Several excellent reviews summarize the properties and applications of various column packings, both inorganic-based and polymeric-based stationary phases [41,42]. Porous glass was one of the first new-generation rigid packings for SEC [43], as opposed to the soft gels. However, there were problems in its use, not the least of which was a tendency to irreversibly adsorb polar solutes, due to the presence of charged sites in the silicate matrix of the glass. Such packings were later superseded by silica-based materials, in which the charged sites had been deactivated by the use of an inactive bonded phase. Commercially available SEC particles are not homogenous in their size. Therefore, their separating capability is lower than expected. Ekman *et al.* [44] have described a simple and efficient method, called dried elutriation, to separate Sephadex beads of narrow size distribution.

Hjertén and Eriksson [45] studied the high-performance molecular-sieve chromatography of proteins on agarose columns, and investigated the relation between concentration and porosity of the gel. Curves showing the relationship between the logarithm of molecular weight and distribution coefficient were presented for proteins subjected to chromatography on cross-linked and non-cross-linked agarose gels of different concentrations. Plate numbers were determined for columns of 20% agarose at different flow-rates and bead sizes. Anspach *et al.* [46] subsequently performed comparative study of the application of the silica-based packings (Zorbax Bio Series GF 250 and GF 450 and TSK-Gel 3000 SW and SWXL columns) in the chromatography of proteins. It was found that reduction of the mean particle diameter of the silica-based packings in the SEC of proteins to about 5 μm generated the expected increase in column plate number over the traditional 10-μm SEC columns. Slightly lower column efficiency of the TSK-Gel 3000 SWXL compared with the GF 250 column was compensated by the fact that the phase ratio of the 3000 SWXL column is higher by a factor of two. Both columns showed nearly the same peak capacity of about 20–30 in this application.

Santarelli *et al.* [47] investigated dextran-coated silica packings for high-performance SEC of proteins. Porous silica beads have excellent mechanical properties, but non-specific interactions between silanols on the silica surface with the proteins require modification of these media before they can be used as stationary phases for SEC. Silica beads were coated with dextran, bearing a small number of positive charges in order to neutralize the negatively charged silanol groups. For this purpose, diethylaminoethyl-dextrans (DEAE-dextrans) with a relatively low percentage of dextran units, bearing DEAE functions, were layered on the silica beads. The effects of these packings on chromatographic performance were studied in order to determine the optimal conditions for the SEC of proteins. An investigation of the physical, chemical, and functional properties of one of the first semi-rigid polymer packings for SEC, Sephacryl HR, was described by Hagel *et al.* [48]. High-performance SEC of some standard proteins, peptides, and amino acids on another hydrophilic packing material, obtained by chemical transformation of a cross-linked polystyrene/divinylbenzene copolymer, was also studied [48]. The characteristics of columns filled with 4- and 7-µm particles were compared. The influence of acetonitrile, 2-propanol and trifluoroacetic acid at various concentrations in the mobile phase on the chromatographic performance of the columns was investigated. A linear calibration graph, covering the molecular-weight range from 200 to 700,000, was obtained under optimal conditions [49].

The mechanical stability of the chromatographic media marketed as Fractogel EMD BioSEC, permits higher flow-rates than softer gels. This feature appeared to be helpful during the regeneration and equilibration of columns, as well as in chromatographic runs. The pressure stability also facilitates packing of larger columns, which are sometimes necessary for refining pharmaceutical proteins. High stability toward alkali treatment enables production-scale separations on Fractogel EMD BioSEC. The separation media that have been commercialized under the names Superdex and Superose are based on dextran, cross-linked with agarose to various extent. The name Superose applies to packings of artificially cross-linked agarose, as opposed to the older agarose media, cross-linked only by natural hydrogen bonding. This type of matrix has a very good stability over a broad pH range. A Superdex Peptide column is a variant capable of separating small molecules (peptides and single amino acids) within the 100- to 7000 Da range. Based on the improved performance in the speed of chromatographic separation on Superdex-type materials, compared to conventional media, such as Sephadex and Bio Gel, a rapid SEC method was developed for the separation and analysis of carrageenan oligosaccharides [50]. Vilenchik and co-workers [51] have recently presented evidence that useful microporous materials can be obtained from protein crystals (CLPC). The CLPC materials can be made chemically and mechanically stable and are capable of separating molecules by size, chemical structure, and chirality. This allows estimation of the apparent pore-size and pore-size distribution in solid and soft hydrated porous sorbents by SEC.

5.3.1 Monolithic materials

Monolithic materials are continuous rods of solid, porous polymers rather than separate beads. Molecular-weight determination for polymers can be improved by using complementary techniques, such as chromatography on the monolithic columns, thus providing valuable information on the composition of co-polymers [52]. Preparation of molded, porous, polymer monoliths with controlled pore structure was described by Peters and co-workers [53], and the history of the development of macroporous columns (though mostly concerned with ion exchangers) was published by Abrams and Miller [54]. The use of monolithic high-performance SEC media for screening of polymers was reported recently [55]. The rapid development of this type of columns was intensified following publications by Hjertén's group [56,57]. Another methodology for the preparation of polymer gels with variable pore architecture was demonstrated by Antonietti *et al.* [58]. Electro-osmotically driven SEC of polystyrene standards with molecular weights of up to 10^6 Da has been described by Peters and co-workers [59]. Mayr *et al.* [60] studied the influence of polymerization conditions on the separation of proteins by inverse SEC. SEC was also applied to control the pore-size distribution of the monolithic poly-TRIM grafting substrate [61]. Separation of styrene oligomers by several modes with this technique has been reported [62]. It seems that further miniaturization of capillary columns (below 300 μm ID), filled with "traditional" gel-type stationary phases will eventually be replaced by the use of monolithic columns, as they are much easier and faster to prepare, providing suitable reproducibility and control over a broad mass range.

5.3.2 Capillary columns

Capillary SEC affords an alternative separation strategy for microscale purification [63–71]. Capillary SEC columns are best suited for direct coupling to electrospray-ionization mass spectrometry (ESI-MS), because they have comparable flow-rates that can be delivered to the ESI source without splitting. Because no split is necessary, the sensitivity of such separations is much higher, and less material can be applied on the column. For example, scaling down the diameter of a Superdex Peptide column from 10 to 0.3 mm causes a significant decrease in sample consumption from 5 μg to 50 ng. Various successful applications of SEC microcolumns have been described [72–74]. Micro-SEC separations can also be indirectly linked with matrix-assisted laser desorption ionization/time-of-flight mass spectrometry (MALDI-TOF-MS) by using robotic interfaces [75]. Other references to these topics are cited in Secs. 5.5 and 5.7.

5.3.3 Microsystems

There are still very few references dealing with microchip (micro-Total Analysis Systems, μTAS) technology, used together with packed channels, though there is a rapidly growing number of papers describing capillary electrophoresis on microchips in conjunction with MS and other detection techniques (Chap. 11). Rapid progress in the development of monolithic media can extend these applications [76], because preparation of channels filled with these stationary phases is much easier than filling them with "classical" column packings. Two papers giving a general and complete overview of the present status of μTAS were published recently [77,78]. A prototypic microchip has been constructed for the size separation of macromolecules and particles by hydrodynamic chromatography [79]. The device has been applied to size characterization of macromolecules.

5.4 MOBILE PHASES

SEC is a very mild separation method, because buffer systems are used as the mobile phase. Thus, optimal conditions with respect to the stability of solute molecules can be selected. This aspect was extensively studied by Garcia *et al.* [80]. Aqueous SEC was used to analyze the elution behavior of several standard ionic polymers as a function of the pH and ionic strength of the eluent. Two organic-based hydrophilic packings, Spherogel TSK PW4000 and Ultrahydrogel 250, were tested in order to select the optimal conditions for macromolecular separation. Deviations from ideal elution behavior have been attributed to ion-exclusion and hydrophobic effects, as a consequence of the repulsive or attractive interactions between the ionizable groups of the poly-electrolyte and the residual surface charge of the support. Quantitative evaluation of elution volumes of poly-electrolytes in salt-containing eluents was performed, taking into account electrical double-layer effects and the effective radius of poly-ions, and assuming that poly-electrolytes behave as rigid hydrodynamic spheres and that the geometry of gel pores is cylindrical [81]. Reported data on the elution of sodium polystyrene sulfonate and polyglutamic acid from both organic- and silica-based packings were used to test the accuracy of the predictions [82].

The effect of column and mobile-phase modifications on retention behavior in SEC of polycyclic aromatic hydrocarbons on poly(divinylbenzene) has also been described [83]. Molecular-mass distribution analysis of ethyl(hydroxyethyl)cellulose was studied by SEC with dual light-scattering and refractometric detection [84]. The polymer aggregates showed variations in behavior that were dependent on the flow-rate and ionic strength of the mobile phase. Molecular characterization of cellulose is technically difficult because of the limited number of appropriate non-degrading solvents. SEC of cellulose was described by Hasegawa *et al.* [85]. Schult *et al.* [86] applied SEC on macroporous, monodisperse poly(styrene/co-divinylbenzene) particles and LiCl in *N,N*-dimethylacetamide to dissolve cellulose. Phillips and Olesik [87] performed studies of LC with enhanced-fluidity liquids. This new technique has led to the continued development of LC at the critical condition (LC-CC) or liquid chromatography at the critical adsorption point (LC-CAP). LC-CC allows isolation of one area of the polymer matrix so that other areas of the polymer can be probed with size-exclusion or adsorptive chromatographic modes.

SEC analysis in aqueous systems has provided information on solubility and aggregation of xylans [88]. Xylan samples from different sources were investigated, using a multi-detector SEC system with two chromatographic column sets and mobile phases consisting of dimethylsulfoxide (DMSO)/water mixtures in various proportions. Molar-mass distribution could be best analyzed by using a mobile phase of DMSO/water (9:1) with the addition of 0.05 M LiBr, a system offering good solubilization of the polymers and an effective chromatographic separation. The use of tetrahydrofuran (THF) as mobile phase in the SEC characterization of a liquefaction extract and its hydrocracking products has been found to cause partial loss of sample and to give anomalous results [89]. However, the problem was solved by using 1-methyl-2-pyrrolidinone as the mobile phase. Li and co-workers [90] have described a method for the determination of iodide in seawater and urine by SEC with the iodine/starch complex. Iodide was converted to iodine, then sequestered with starch, and separated from the matrix, using a Shim-pack DIOL-150 size-exclusion column with methanol/0.01 M aq. phosphoric acid (1:9) as mobile phase at 1.2 mL/min. Batas and Chaudhuri [91] have described a mechanism for SEC-based protein-refolding, another novel effect in SEC. This model considers the steps of loading denatured proteins onto a column and its elution. The predictions were compared with results obtained by SEC of lysozyme on Superdex 75 HR with a refolding buffer. The main collapse in protein structure occurred immediately after loading, where the partition coefficient increased from 0.1 for unfolded lysozyme to 0.48 for the partially folded molecule. The use of a refolding buffer as the mobile phase resulted in complete refolding of lysozyme. The effect of mobile phase on the oligomerization state of α-helical coiled coil peptides during high-performance size-exclusion chromatography (HPSEC) was studied by Mant *et al.* [92]. HPSEC appeared to be useful for examining both the oligomerization state of coiled coils and the stability of such motifs, due to facile manipulation of the mobile phase and the lack of interaction of the peptide solutes with the stationary phase.

The influence of three experimental parameters – temperature, pH, and ionic strength of the eluent – on the retention of heparin samples and polysaccharides as calibration standards in SEC was investigated by Bergman *et al.* [93]. Silvestre *et al.* [94] separated protein hydrolysates on poly(2-hydroxyethylaspartamide)-silica columns with 0.05 *M* formic acid as eluent. Concentrated formic acid (70%) was applied for elution of Aβ fragments present in human brains, and a Superose-12 column was used without significant deterioration of the stationary phase due to the acidic eluent [95]. Nylander *et al.* [96] reported that the mobile-phase composition may influence adsorption and separation of neuropeptides on the Superdex-75 column. These studies were extended by Hedlund and co-workers [97]. In general, both acetic and formic acids, commonly used in chromatography and mass spectrometry, contain C=O groups in their structures. The double bond accounts for a high absorbance in the UV detectors up to 240 nm, a factor that should be carefully considered in the detection of, *e.g.*, a peptide bond (usually at 210 214 nm). Under overload conditions, the resolution in SEC of proteins can be compromised due to non-uniform flow, caused by the viscous-fingering flow instability. In the work of Fernandez *et al.* [98], the non-uniform flow under these conditions was analyzed by numerical simulation and magnetic resonance imaging, and a new column design was postulated. Another recent aspect of protein chemistry was investigated concerning speciation of zinc in complexation with proteins of low molecular mass that occurs in breast milk and infant-feeding formulas. SEC was used in conjunction with flame atomic-absorption spectroscopy [99]. After ultracentrifugation of the milk, the sample was injected into a TSK-Gel G2000 glass column and eluted with 0.2 *M* NH_4NO_3/NH_4OH (pH 6.7). This was followed by inductively-coupled plasma mass spectrometry (ICP-MS) analysis of the eluate.

An analytical HPLC method has been reported for the simultaneous measurement of low concentrations of dextran-methylprednisolone succinate and its degradation products, methylprednisolone hemisuccinate (MPS) and methylprednisolone (MP) [100]. The analytes were detected at 250 nm after resolution on a size-exclusion column with a mobile phase of 10 m*M* KH_2PO_4/McCN (3:1) at a flow-rate of 1 mL/min. The resolution of MP and MPS peaks was substantially affected by the pH of the mobile phase; the degradation products were not resolved at pH 3.4. Hyaluronic acid has been analyzed by HPSEC on a TSK-Gel 6000 PW column and eluted with 100 m*M* $NaNO_3$ [101]. A method for analytical and preparative SEC of large, water-insoluble, protected peptides in an organic solvent was developed by Karnoup and co-workers [102]. This method was applied to the analysis and separation of protected synthetic peptide tandem repeats and to control the peptide fragment coupling. Toyopearl HW-40, HW-50, HW-55, and HW-60 columns of fine grade were used, and the selectivity of each sorbent, as well as the chromatographic behavior of the peptides were examined. Fractionation ranges of these gels in N,N-dimethylformamide (DMF) were shown to extend over much smaller molecular masses (*ca.* 400–14,000 Da) than those of the same gels applied to separate proteins in aqueous buffers (100–1,000,000 Da).

Mobile-phase composition, pH, and flow-rate can significantly affect separation in SEC. Special care must be taken to optimize separations before sample application in order to minimize variations in retention times and possible sample loss due to, *e.g.*, non-specific adsorption on the gel. In certain cases, addition of NaCl is necessary to reduce non-specific interactions, but salts may influence further analysis by mass spectrometry. Moreover, phosphate buffers are not recommended, as they can polymerize in heated capillaries (ESI-MS), thus forming insoluble polyphosphates.

5.5 DETECTORS

Detection of separated components is a crucial step in the chromatographic procedure, as it provides information on the character of the molecules. The major features of an effective detector are: selectivity, specificity, sensitivity, stability, and linearity. It is also desirable that the detector will not destroy the sample, but this does not apply to some mass spectrometers, *e.g.*, those equipped with an ESI ion source. Samples from MALDI or Fast Atom Bombardment (FAB) targets can possibly be recovered for further analysis, such as radioimmunoassay (RIA), radio-receptor assay (RRA), enzyme-linked immunosorbent assay (ELISA), etc. Anyhow, modern mass spectrometers utilize only minute amounts of sample, and can give unambiguous results that outweigh these disadvantages. The characteristics of the ideal detector for liquid chromatography have been summarized by Lemiere [103], and several fundamental reviews on this topic have also been published in recent years [104–108]. The "ideal detector" should be able to identify the molecule of interest, quantify it (possibly in the presence of other components), and provide complete structural information for the unambiguous identification of the isolated compound. From this point of view, mass spectrometry, nuclear magnetic resonance (NMR), and Raman spectroscopy, including the flow-injection surface-enhancement Raman scattering (SERS) variant [109–111], are the methods of choice for either on-line or off-line detection. These are often used in conjunction with other techniques, such as RIA, ELISA or other, less specific detectors (UV, IR, electrochemical detectors). For example, Fourier-transform infrared (FTIR) spectrometry in the mid-infrared region is becoming more and more important in SEC. It is a powerful and potentially very widely applicable method for obtaining information on the chemical functional groups in each molecular-size fraction [112]. Quantitative evaluation of polymer composition across the SEC chromatogram can provide a more accurate characterization of heterogeneous polymer samples, which is necessary for problem solving and material specification. Detection limits, dilution factors, and technique compatibility in multi-dimensional separations have been discussed by Schure [113].

SEC with coupled multi-angle light-scattering- (MALS), and differential-refractometry detectors have been used to obtain molecular mass and radius of gyration distributions of polydisperse polymer samples [114]. From these data, the scaling relation between dimensions and absolute molecular mass was obtained with one sample of each polymer. The molecular mass (M_r) of the complexes of monoclonal anti-bovine serum albumin

(BSA) and monomer BSA were determined on-line [115] by means of SEC, coupled with a low-angle laser light-scattering (LALLS) detector and two concentration detectors, ultraviolet (UV) and refractive index (RI) (SEC/LALLS/UV/RI system). Also, the size and M_r of the complexes were evaluated by the SEC/LALLS/UV/viscometer system. The principle of the differential viscometer detector and its use in constructing a universal calibration were explained by Titterton [116]. The pioneers in the field of viscometry detection in SEC were Fishman and co-workers [117], who first reported this technique in 1989 in studies of pectins. The use of SEC with on-line light-scattering, absorbance, and RI detectors for studying proteins and their interactions, was the topic of a review by Wen *et al.* [118]. Aqueous SEC with on-line argon-ion MALLS photometry and differential viscometry detectors was reported by Muller and co-workers [119]. Starch was characterized by HPSEC with detection by both MALLS and RI [120].

HPSEC with UV absorbance and on-line dissolved-organic-carbon (DOC) detectors have also been adopted and optimized under various conditions [121]. For example, an enhanced HPSEC/UV system with a modified DOC detector provides an improved understanding of the qualitative and quantitative natural-organic-matter (NOM) properties in water samples by detecting aromatic and non-aromatic fractions of NOM as a function of molecular weight. The "retention analysis method", which is based on SEC in conjunction with an arsenic-specific detector (graphite furnace atomic-absorption spectrometer) was reported by Gailer and Lindner [122].

An interesting approach to on-line combination of SEC and GC has been designed, using LC/GC apparatus. This has been applied to determine organophosphorus pesticides in olive oil [123]. Per-*O*-sulfonated polysaccharides, including glycosaminoglycans and hyaluronan oligosaccharides have been analyzed by HPSEC with suppressed-conductivity detection [124]. The sensitivity of this method was compared to that of HPSEC with UV or fluorescence detection after reaction with 2-cyanoacetamide in strongly alkaline solution. The use of conductivity detection without derivatization and under isocratic conditions gave a limit of detection in the picogram range. Such a detection system is desirable for HPSEC of all polyelectrolytes.

HPSEC, combined with multispectral detection by a photodiode-array (PDA) UV detector was applied to the analysis of proteins and peptides in human cerebrospinal fluid (CSF) [125] and to the characterization of proteinergic profiles in the CSF of alcoholics [126]. Molecular components of the CSF were identified, and their purity was tested. The PDA detector, recently reviewed by several authors [127–133], provides significant savings of biological samples, as the entire spectrum can be registered simultaneously in a single run. Moreover, this technique eliminates possible errors caused by, *e.g.*, a shift in retention time, which may occur when manual injectors are used. Several papers were published on the identification of various compounds and on purity tests performed with this type of detectors [134–137]. The application of multi-wavelength detection in the study of unfolding equilibrium of growth hormones in urea by SEC has also been investigated [138]. On-line PDA instrumentation and comparison of spectral ratios figured in monitoring tertiary and quaternary structural changes associated with protein denaturation. Stationary-phase-induced effects on protein conformation were monitored by changes in the maximum-to-minimum ratio of the second-derivative spectrum. Commercial protein preparations were analyzed for purity on a size-exclusion column, coupled to a PDA detector [139]. Dupont [140] studied degradation of gelatin in paper upon aging by means of aqueous SEC and UV-PDA detection. Spectroscopic characterization by PDA detection of urinary and amniotic-fluid proteins, fractionated by anion-exchange and SEC/HPLC, was described by Calero *et al.* [141]. A similar technique was applied by de Vries [142] for the analysis of heparins. Moreover, a rapid quantitative determination of aromatic groups in lubricant oils by SEC combined with PDA detection was described by Varotsis and Pasadakis [143].

5.5.1 Mass spectrometry

As a growing number of applications and developments of multi-dimensional LC separations for proteomics are expected in the future, combinations of various modes of LC (including SEC) and MS are currently of widespread interest. Two-dimensional LC (2D-LC) is now being applied more and more in high-resolution separations. This approach has been used mainly in proteomics studies, where a large number of components must be separated, and as a technique complementary to 2D-electrophoresis. Due to technical difficulties, 2D-electrophoresis still cannot achieve the reproducibility required for unambiguous comparison of two independent separations. Many combinations of 2D-LC have been investigated and their applicability has been demonstrated. One of these approaches utilizes SEC, *e.g.*, in the separation of polyethylene glycols and surfactants [144]. Other combinations have been applied in the separation of peptides [145,146]. Interfacing LC techniques with MS is a rapidly growing strategy in analytical sciences (Chap. 10) [147]. However, some precautions must be taken before linking these two techniques in order to make them fully compatible with respect to the mobile-phase composition, flow-rates, size of molecules, scanning time, etc. Sandra and co-workers [148] and Lemiere [149] have provided some very useful practical tips. The instrumental set-up for nucleic acid analysis has been reviewed by Huber and Oberacher [150], for clinical and forensic toxicology by Marquet [151], for drugs of abuse by Moeller and Kraemer [152], for analysis and screening of combinatorial libraries by Shin and Van Breemen [153], for proteomics by Peng and Gygi [154], for metabolite identification in drug discovery by Clarke *et al.* [155], for general HPLC/MS by Erickson [156] and Niessen [157], for high-throughput quantitative analysis of biological material by Jemal [158], for pharmaceutical analysis by Ermer and Vogel [159], and in nanotechnologies by Guetens *et al.* [160]. Complex SEC assays for screening combinatorial libraries were reported by Schurdak *et al.* [161]. This list is a selection from only the recent literature, describing the application of various chromatographic techniques linked to mass spectrometers equipped with various ionization sources.

Capillary and nano-chromatography columns, such as capillaries for electrophoresis, produce separations within minutes. The major problem in fast separations stems from the relationship between peak-width and scan duration. If the scan speed is too slow, some components might be overlooked in the mass spectrum. Moreover, an MS/MS (or MS^n) experiment, even if performed automatically, requires several seconds for completion. A compromise between separation quality and the limitations of the mass spectrometer can be achieved by using a peak-parking method [162,163]. This technique substantially prolongs the signal in a mass spectrometer and also enhances its sensitivity without significantly affecting the resolution. As the peaks are retained longer on the column by decreasing the flow-rate of the mobile phase, the retention time cannot be considered as constant. Post-run data obtained with ESI-MS, coupled to the chromatographic system, require a deconvolution procedure to reveal the actual masses of the detected components. Often, the presence of complex mixtures of substances and/or the presence of contaminants may affect this algorithm, leading to false mass assignments [105]. Therefore, pre-separation of molecules according to their size may contribute to the unequivocal identification of all ions.

A method has been developed for on-line pseudo-cell SEC/MS (PsC/SEC/MS), providing rapid, real-time analyses of non-covalently bound protein complexes [164]. The methodology can be used to determine components of such complexes, as well as their exact stoichiometry. Furthermore, it enables the efficient determination of gross conformational changes upon complexation. The power of this new approach was demonstrated in the analysis of the global transition-state regulator AbrB and its complex with a target DNA sequence from the promoter sinIR. Non-covalent interactions were studied with the aid of SEC/coordinated ion spray (CIS)-MS [165]. Characterization of non-covalent complexes of antigen and recombinant human monoclonal antibody by cation exchange, SEC, and Biacore (detection and monitoring of the binding of biological molecules by surface-plasmon resonance technology) was reported by Santora *et al.* [166]. The effect of enzyme inhibitors on protein quaternary structure, determined by on-line SEC/micro-ESI/MS, has been described [167]. Blom *et al.* [168] have reported the determination of affinity-selected ligands and of binding affinities by on-line SEC/LC/MS. The use of SEC, linked to ESI-MS for the separation of neuropeptides and their fragments has been the subject of several papers [105,169–173]. A SMART™ System for micropurification was applied, using a Superdex Peptide column of 3.1 mm ID. Combination of such a system with MS required a volatile buffer, acceptable by the ESI source. In this case, a mobile phase consisting of 0.1% aq. TFA was found optimal. TFA generally suppresses the MS signal and decreases the sensitivity of measurements. Dilute formic acid gives much better results in this respect, but it was difficult to elute components from the column using this solvent. Capillary columns of Superdex Peptide have been applied for the identification of peptides and proteins by LC/MS [174]. Further details of this method have been described by Suder *et al.* [175]. An off-line identification of endogenous LVV-hemorphin-7 from CSF was performed on the Superdex Peptide column, followed by ESI-MS analysis of collected fractions [176].

Characterization of arsenic species in clams by multi-dimensional LC, linked to ICP-MS and ESI-TOF-MS/MS has been described by McSheehy *et al.* [177]. Such combinations of ICP-MS with LC have been receiving increasing attention in recent years. ICP-MS can also be applied to the detection of labile biological molecules, such as DNA fragments, where phosphorus can be monitored. Fractionation of phosphorus and trace elements in soybean flour and bean seeds by SEC, linked to ICP-MS was reported by Koplik *et al.* [178]. This group also discussed the application of various SEC columns to the separation of several other elements. Metal distribution patterns in cytosols from the mussel *Mytilus edulis* were demonstrated by Ferrarello *et al.* [179], who used SEC and double-focusing ICP-MS detection. Quantitative analysis of iron speciation in meat by using a combination of spectrophotometric methods and HPLC, coupled to sector

ICP-MS, has also been reported [180]. Wang and co-workers [181] have described the determination of trace elements in liver proteins by SEC/ICP-MS with a magnetic-sector mass spectrometer. A multi-dimensional approach, combining LC with parallel ICP-MS and ESI-MS/MS for the characterization of arsenic species in algae was reported by McSheehy *et al.* [182]. The identities of all the species were doubly checked, by matching the retention times of chromatographically pure species with standards and by ESI-MS/MS. The same group applied a similar approach in their investigation of arsenic speciation in oyster test reference material by multi-dimensional HPLC/ICP-MS and ESI-MS/MS [183].

Liu *et al.* [184] have recently reviewed the multi-dimensional separations of proteins and peptides. Micro-SEC on MicroSpin Sephadex G-25 columns was applied in combination with capillary LC/atmospheric-pressure chemical-ionization (APCI)-MS and used for the screening of potential drugs in recently discovered pharmaceutical compounds [185]. Harris *et al.* [186] have developed a size-exclusion-based system for the rapid isolation of plasmid DNA in a 96-well microplate format. A method for analyzing polysaccharide materials has been described, which employs SEC, followed by detection by on-line ESI-MS and off-line MALDI-TOF-MS [187]. The well-established method of two-dimensional gel electrophoresis is far too slow for screening in proteomics [188]. A new methodology has been developed, combining capillary electrophoresis/isoelectric focusing (IEF) with MS with sequential fragmentation (CIEF/MS^n), and preparative IEF followed by SEC, combined with MS. Isotope ratio mass spectrometry was used to detect very low alterations in ^{13}C abundance in analyte species that cannot be volatilized [189]. Examples were given of proteins, carbohydrates, and nucleotides, eluted from various types of HPLC columns.

Recent developments in MALDI-TOF techniques have enhanced the opportunities of linking HPLC with this ionization method. The MALDI source differs from the ESI source in that it operates in high vacuum. The high-vacuum source is capable of accommodating the LC mobile phase at a maximum flow-rate of $5-7$ μL/min. Introduction of a liquid at higher flow-rates would cause an immediate shutdown of the instrument. Two different approaches to the on-line linking of SEC to the MALDI-TOF-MS, continuous-flow and aerosol, were discussed by Fei and Murray [190]. Details on linking capillary SEC to MALDI-TOF-MS can be found, *e.g.*, in Ref. 193. In the aerosol method, the sample eluted from the column is mixed with matrix before nebulization. Typically, flow-rates are maintained between 0.5 and 1.0 mL/min. The problems with this type of interface have recently been solved by the use of a rotating-ball inlet [191], which prevents clogging of the vacuum interface by matrix crystals or frozen solvents. An alternative strategy utilizes a picoliter sampling onto the MALDI target plate with the help of a flow-through piezo-electric micro-dispenser [192], but larger robotic systems can be connected as well. Another promising methodology is based on the development of a MALDI source, operating at atmospheric pressure. Such a device can be mounted instead of an ESI sprayer and coupled to any of the known detectors (*e.g.* ion-trap, TOF, quadrupole). Promising developments in atmospheric-pressure MALDI (AP-MALDI) [193–196] and the application of an infrared laser at 3 μm may be a turning point for on-line LC/MALDI-MS^n applications. The additional advantage of the IR laser is that ions can be generated from water solutions. This makes the system compatible with the LC separations of biological molecules. On the other hand, Blais *et al.* [197] have discussed some limitations of the MALDI-TOF method in comparison with SEC for the characterization of phosphorus-containing dendrimers.

5.5.2 Nuclear magnetic resonance and surface-enhanced Raman spectroscopy

Another approach to unambiguous identification in LC is linking the chromatographic system with other structurally specific methods, such as NMR or surface-enhanced Raman scattering (SERS) [198–201]. The feasibility of interfacing flow-injection-based SERS methods with HPLC for the detection of individual components in a complex mixture was proposed in 1990 [202], but this approach has been studied more extensively in conjunction with capillary electrophoresis rather than with liquid chromatography. A microcoil NMR probe for coupling microscale HPLC with on-line NMR spectroscopy was described by Subramanian *et al.* [203] and adapted to the nanoliter scale by Behnke *et al.* [204]. These methods were developed for reversed-phase columns, but the same approach can also be applied to SEC. Application of LC, coupled on-line to ^{1}H-NMR, was reported for the investigation of flavonoids [205]. The eluent was a mixture of MeOH and D$_2$O. Separation and identification of terpenoids were achieved by coupling a commercial capillary HPLC system with a PDA detector and a custom-built NMR flow microprobe [206]. The eluent from a 3-μm-diameter C$_{18}$ HPLC column was linked to a 500-MHz ^{1}H-NMR microcoil probe with a volume of 1.1 μL. Direct on-line LC/NMR/MS/MS for the rapid screening of natural products was described by Sandvoss *et al.* [207]. An off-line MS/NMR procedure has been developed for rapid screening of small organic molecules and their ability to bind a target protein [208]. With this methodology it was also possible to obtain structure-related information as part of a structure-based drug discovery-and-design program. The methodology combines the inherent strengths of SEC, MS, and NMR to identify bound complexes in a relatively universal high-throughput screening approach. Another indirect technique was presented by Venter *et al.* [209] for the identification of membrane-transport proteins of *E. coli*. With a number of polymer additives as model compounds, practical problems encountered with multiple combinations were described for the coupling of HPLC with UV detection, on-line NMR spectroscopy, and MS, combined with a dedicated interface to collect chromatographic eluents for subsequent FT-IR [210]. SEC was performed with deuterated chloroform as eluent, the separation being monitored on-line by UV detection at 254 nm and on-flow ^{1}H-NMR and MS. Interfacing LC with NMR and other detection techniques has recently been reviewed by Wilson [211]. A separate chapter of his review was devoted to SEC/NMR/IR instrumentation. A fully integrated system, involving HPLC, with superheated D$_2$O as a mobile phase, and combined with on-line PDA-UV, ^{1}H NMR, FT-IR, and APCI-MS has been applied for the analysis of ecdysteroid-containing plant extracts [212]. The potential of such "multiple hyphenation" can yield a complete and unambiguous structural elucidation from a single experiment. The methods and theoretical basis for quantitative measurements in continuous-flow HPLC/NMR were presented by Godejohann *et al.* [213].

5.6 CALIBRATION

Deviations from the Benoit principle of universal calibration were observed by Belenkii *et al.* [214] when flexible-chain polymers were chromatographed on macro-porous swelling sorbents. These peculiarities were caused by different degrees of thermodynamic compatibility of the polymers with the sorbent matrix. More recently, a new SEC method for the estimation of the weight-to-number-average molecular-weight ratio M_w/M_n of polymers with a narrow molecular-weight distribution, approximated by a log-normal distribution, was proposed [215]. The method was applied to a series of polystyrene standards of narrow molecular-weight distribution. Guillaume *et al.* [216] have proposed a mathematical model for hydrodynamic and SEC of polymers on porous particles. The model described constitutes an attractive method for enhancing these two chromatographic techniques for separating biological or synthetic macromolecules. Different polynomial models of calibration curves have been evaluated and compared with respect to their predictive properties [217]. The best model across the effective fractionation range (linear range) was not always found to give a straight line. Polycyclic aromatic hydrocarbon standards were applied in studies on the polymerization of anthracene oil with $AlCl_3$ as the catalyst [218]. SEC separations have been carried out on a stationary phase of polystyrene/polydivinylbenzene with 1-methyl-2-pyrrolidinone at 80°C as eluent. Endogenous calibrants were used by Tsao *et al.* [219] for the quantitation of various molecular forms of chromogranin A in serum and urine. Oliva *et al.* [220] performed a comparative study of protein molecular weights by SEC and laser-light scattering. The results obtained by the two methods were compared, using samples of recombinant human growth hormone and β-lactoglobulin as test substances. The effect of peak-broadening and of the error in inter-detector volume on the calibration curve and experimental molecular-mass averages obtained by SEC were investigated by Netopilik [221]. The parameters affecting the fractionation performance in SEC of broad polymer samples were studied by Lou *et al.* [222]. Two different modes were considered, *i.e.*, using MALDI-MS to provide an absolute calibration curve for SEC, and using SEC as a sample preparation step for MALDI-MS measurements. The latter combination was demonstrated to be more reliable, because most problems inherent in SEC can be circumvented.

The pore dimensions, pore-size distributions, and phase ratios were determined [223] for a set of cation exchangers, using inverse SEC (ISEC). This technique [224] is alternatively called macromolecular porosimetry [225] and is used for the characterization of the porosity of materials [226]. The adsorbents examined represent a diverse set of materials from Pharmacia, TosoHaas, BioSepra, and EM Industries. The ISEC was carried out using dextran standards. This technique provided a comparative characterization of the accessible internal pore-surface area, as a function of solute size. ISEC was adopted to measure the permeability of microcapsules (hollow hydrogel spheres with diameter <1 mm), using dextran molecular-weight standards [227]. Alginate/poly(L-lysine)/ alginate microcapsules were chosen as a column substrate. Polysaccharides of known molecular weight were used as standards for HPSEC of humic substances [228]. Calibration curves were equivalent for both columns, whereas analytical parameters revealed that a TSK column was only slightly more efficient in separating polysaccharide standards. Variations between columns in the results obtained for the molar masses of humic substances were attributed to intrinsic properties of these substances, such as stability of conformational structures.

A universal calibrator has been applied to the determination of molecular masses of heparin samples and has been referred to as the Heparin Molecular Mass Calibrant [229]. This calibrator replaces the large number of calibrators (19 in the authors' previous work) that are required for molecular-mass analyses. Polysaccharide standards having average molar masses in the range of 180–100,000 Da were used in HPSEC, applied as a screening technique for the determination of average molar masses of polygalacturonic acid samples for use in pharmaceutical applications [230]. The method was employed in screening commercially available polygalacturonic acid raw materials with respect to both average molar mass and polydispersity. The analysis of poly(bisphenol A carbonate) by SEC/MALDI was reported by Puglisi *et al.* [231]. Their results show that MALDI spectra of the SEC fractions allow not only the detection of linear, and cyclic oligomers, but also the simultaneous determination of their average molar masses. Two slightly differing SEC calibration plots were obtained, due to the smaller hydrodynamic volume of the polycarbonate cyclic chains relative to the linear ones. The possibility of standardizing calibrants for SEC was investigated by Yomota and Okada [232]. Their review article (unfortunately in Japanese only) focuses on the water-soluble polysaccharides, such as dextran, hyaluronate, and chitosan. A HPSEC/RI method has been described for the quantification and molecular-weight determination of extractable water-soluble poly-amines in a novel, proprietary, polymeric pharmaceutical compound [233]. The extracted polyamines were synthetic impurities as well as potential degradation products of the polymer. Potzschke *et al.* [234] studied the molar masses and structure of hemoglobin hyperpolymers, commonly used in calibrations for SEC of these artificial oxygen carriers. The calibration curve was found to differ significantly from that given by native globular proteins due to a less compact structure of hemoglobin hyperpolymers. Therefore, the calibration of SEC with globular proteins for the determination of molar masses of hemoglobin polymers would be erroneous.

A method for the calibration of SEC columns, suggested by Harlan *et al.* [235], takes into account the nonlinear dependence of the Stokes radius, R_s, upon the partition coefficient, K_D. An application of this method, in which aggregation states of the membrane protein prostaglandin H2 synthase, solubilized in nonionic detergents, was reported by Duggleby [236]. He described a method and software with appropriate statistical operations to select the best straight or curved calibration line. The size of the unknown is then interpolated and an estimation of the error is calculated. Le Maire *et al.* [237] tested the hypothesis that porous media used in protein SEC are surface fractals. The data obtained in the calibration of "classical" gels (Sephacryl and Sepharose) and of HPLC phases (TSK SW and PW) with proteins having a wide range of molecular sizes were analyzed within the framework of this theory. While the model does not apply to "classical" gels, it seems that HPLC phases can be described as fractals over the range of protein sizes. The same group [238] also tested the hypothesis [239] that the elution position of macromolecules in gel chromatography is better correlated with the viscosity-based Stokes radius ($R\eta$), rather than with the Stokes radius (R_S) calculated from the frictional coefficient. By the use of different gel matrices (agarose, Sephadex, and TSK silica gel columns) it was found that the elution positions of dextran fractions and reduced proteins, denatured with guanidine hydrochloride, were in accordance with their $R\eta$. In this case, water-soluble, globular proteins were used for gel calibration. A simple routine for nonlinear least-square analysis was described [240] and applied to small-zone scanning data, where calibration of the gel column requires the use of fully characterized markers of known molecular size. Application of nonlinear least-squares analysis eliminates the difficulty encountered due to the scarcity of calibrating markers for gels whose porosities span a certain size range. A multilevel calibration has been discussed by Hinshaw [241]. By misapplying multilevel nonlinear calibration to compensate for gross systematic errors, one can mask serious malfunction of the analysis, leading to erroneous results. The author concludes that "it's never a good idea to assume that a chromatograph delivers equal responses to all components or at all levels". This is obviously true and should always be considered, not only with regard to chromatographic data, but also in MS and other techniques. An overview of the chemometric methods for calibration, optimization, and statistics can also be found in Ref. 242. Lazaro *et al.* [243] investigated the elution behavior of coal-derived materials on a polystyrene/divinylbenzene SEC column with 1-methyl-2-pyrrolidinone as the mobile phase and use of polystyrene calibration. This calibration was also correlated by mass spectrometry. The importance of calibration was emphasized in the measurement of dextran clearance in clinical studies [244].

5.7 APPLICATIONS

5.7.1 Polymers

Determination of molecular weight and molecular-weight distribution is crucial for understanding the relationship between the properties of polymeric materials and their molecular structures. The molecular-weight averages and distributions are given by the principal relationships defined in terms of number-average molecular weight (M_n), weight-average molecular weight (M_w) and z-average molecular weight (M_z). Further parameters to be considered are the ($z + 1$)-average molecular weight as well as the viscosity-average molecular weight (M_η). These averages are all connected with the population defined by the number of moles (n_i), or weight (w_i) of individual macromolecules possessing molecular weight M_i, contained in the polymer population. The distribution of polymer molecular weight is described by various mathematical relations, including Gaussian, Poisson, Flory-Schulz, and log-normal molecular weight distribution functions. The polydispersity of polymers, Q, is traditionally defined as the relation of their weight-average to number-average molecular weights

$$Q = M_w/M_n = U + 1 \qquad\qquad (5.4)$$

where U, the molecular inhomogeneity, has a numerical value of one less than Q. Polymers are usually distributed in more than one direction of inhomogeneity, which comprises, in addition to molecular-mass distribution (MWD), the distribution with respect to chemical composition (CCD), functionality (FTD) and topology (MAD). The chemical composition of polymers is related to the macromolecular chain structure, including the number of co-monomer units for co-polymers, their sequence of incorporation, configuration, and conformation. The topology of synthetic polymers concerns the polymer-chain architecture (linear, comb-like, branched, cyclic, etc.), and is connected with functionality, *i.e.*, with the number and localization of the functional groups.

Several approaches have been used for the separation of polymers according to both MWD and CCD. In contrast to homopolymers, which are monodisperse with respect to CCD for each retention volume, different combinations of molar masses, composition, and sequence length can be found in co-polymer fractions having the same hydrodynamic volume. In order to address this problem, several selective detection techniques have been combined with SEC. Hyphenated techniques in SEC have been reviewed recently [245]. Two-dimensional chromatography is another approach frequently used for the characterization of molar mass, composition, and functionality distributions of polymer macromolecules. The hyphenation of SEC in a two-dimensional "orthogonal" mode has been applied to the characterization of amphiphilic copolymers [246–248] Full co-polymer characterization can be achieved by measuring the bivariate distribution, *i.e.*, with respect to masses and composition. This measurement is particularly important for co-polymers obtained at high conversion of co-monomers, because, usually, at that stage of co-polymerization, one of the two monomers is preferentially incorporated into the co-polymer chain. Therefore, the spread in composition increases as conversion progresses, and the resulting co-polymer displays compositional heterogeneity. Several methods have been suggested to determine the bivariate distribution of masses and composition in co-polymers. For these purposes, the co-polymer is usually fractionated by SEC, and the collected fractions are subsequently analyzed by various spectrometric techniques [249]. For example, the fractions can be analyzed by NMR to determine the co-polymer abundance and composition and then by MALDI-MS to determine the molar mass of each fraction. Finally, NMR and MALDI-MS data are combined with bivariate distribution models to yield bivariate distribution maps [249].

The molecular size of a homo/co-polymer molecule in solution is a function of its chain length and chemical composition, solvent, and temperature. It is, therefore possible to selectively separate polymers with respect to hydrodynamic volume, chemical composition, and functionality by the use of different modes of LC. When the adsorption effects between the polymer molecule and the column packing particles are dominated by the enthalpic interactions, $K_D > 1$ in the SEC general retention equation (Eqn. 5.1), and chromatography takes place in the adsorption mode. It is generally assumed that adsorption chromatography is more suitable than SEC for the characterization of chemically heterogeneous polymers. This can be performed in either isocratic or gradient elution modes. Gradient polymer elution chromatography (GPEC) in both reversed-phase (RP) and normal-phase (NP) modes have been applied in microstructural characterization of aromatic co-polymers [250]. Several examples of GPEC and its combination with SEC for the determination of CCD of co-polymers as well as FTD of telechelic polymers, polymer blends, and resins have been reported [251]. Determination of polymer structure by SEC is, besides protein research, one of the largest field of applications. SEC is capable of providing both, the molecular mass distribution and the polydispersity index. On-line linking with ESI-MS or combination with MALDI-TOF-MS as specific detectors provides full structural information, including data on the nature of the end groups. Several reviews describe the application of SEC in the field of synthetic polymers [252].

A strategy for controlling the polymer topology through transition-metal catalysis has been described by Guan [253]. The molecular weight and intrinsic viscosity of the hyperbranched polymers were measured by MALDI-MS and SEC with triple detectors. Characterization of polyether and polyester polyurethane soft blocks was performed by MALDI-MS and compared with SEC/MALDI [254]. The SEC/MALDI results provide significantly larger values of M_w and polydispersity than MALDI alone. SEC is now considered to be a standard tool in the molecular characterization of commercial polymers. In this respect, two factors, *i.e.* repeatability and reproducibility, are important. In order to determine the variance in reproducibility, several interlaboratory studies (*i.e.*, round-robin experiments) have been undertaken, and the round-robin experiments conducted in Europe have done much to raise awareness of quality issues in SEC.

Recently, the results of an interlaboratory experiment in high-temperature SEC have been published [255]. Fifteen laboratories performed analyses of five polyethylene samples and two standards. It was found that the reproducibility, measured by the inter-laboratory standard deviation (sLAB) was greatly influenced by the width of the molecular-weight distribution (MWD) and branching. The sLAB values for M_w of linear polyethylenes of narrow and broad MWD were 4 and 14%, respectively. For branched polymers, SEC/viscometry methods were shown to measure significantly higher molecular weights than the non-hyphenated SEC method. For single-site polyethylene, only a couple of laboratories reported the MWD that closely matched the Flory distribution. It was concluded that many variations in instruments and analytical methods exist among laboratories, and that this technique must yet undergo many refinements before a truly standard method is widely accepted and implemented. Nevertheless, wide-ranging applications of SEC to the characterization of synthetic polymers, resins, coatings and paints were demonstrated, and some of the recent examples are given below. Calculations of the η/M relationship for ethylene/propylene co-polymers have been performed, and the results were used for a theoretical examination of the effect of co-polymer composition on the calibration of SEC columns [256]. The kinetics of peroxide-induced degradation of polypropylene has been investigated by comparing the molecular-weight distributions, as determined by SEC measurement, and the impact of imperfect mixing on the MWD shift has been examined [257].

The application of SEC to the quantitative measurement of adhesion between polypropylene blends and paints, self-nucleation behavior of the polyethylene block in polystyrene/β-polyethylene/β-polycaprolactone triblock co-polymers, characterization of the molecular structure of highly isotactic polypropylene, as well as the use of SEC for the determination of chain-scission distribution function for polypropylene degradation, constitute some further examples of the use of this technique in studies of polyolefins [258–261]. Polypropylene oligomers were isolated from the polymer matrix and have been characterized by a combination of SEC with FTIR spectrometry and HPLC with UV detection [262]. An analytical strategy recently developed [263] for the analysis of polyisobutylenes, partially functionalized with isothiocyanate groups, involved coupling of capillary SEC, SEC, and RP-LC. By comparing the results obtained with these techniques, a complete characterization of the polymer was achieved, and the degree of polymerization and relative quantity of the different series of macromolecular chains were estimated. SEC has been further utilized for the characterization of isobutylene co-polymers of various topology, such as poly(isobutylene/β-styrene) block co-polymers, triblock co-polymers with densely grafted styrenic end blocks (prepared from a poly-isobutylene macro-initiator), novel multiarm-star polyisobutylene/polystyrene thermo-plastic elastomers, linear and star-shaped block co-polymers of isobutylene, methacrylates (obtained by combination of living cationic and anionic polymerizations), and multiarm-star polyisobutylenes [264–268]. The application of SEC to PVC (polyvinyl chloride) characterization includes the determination of the MWD of commercial PVC, and was recently reported [269–271].

A comparison of the SEC performance with that of size-exclusion electrochromato-graphy (SEEC) for the mass distribution analysis of synthetic polymers, such as poly(methylmethacrylate) (PMMA), polycarbonate, polycaprolactam, poly(ethylene terephthalate), and polystyrene (PS) was recently published [272]. The repeatability of electro-osmotic flow control within-day, day-to-day and column-to-column was determined for SEEC with respect to retention and separation efficiency. It was shown that by using the retention ratio instead of the migration time, the precision of the mass distribution was sufficiently high, and that similar distributions were obtained for a sample analyzed by pressure-driven SEC and by SEEC. With the aid of direct on-line coupling of SEC with depolarization (D) multi-angle light scattering (SEC/DMALS), a method for studying optical anisotropy of polymers as a function of their molar mass has been developed [273]. The effects of tactic heavy-atom substitution on the main chain in the depolarization behavior of polymers were studied, using atactic and isotactic PMMA, atactic and brominated PS, and the semi-flexible polypeptide. An introduction to the theory of SEC/DMALS was also given. By combining gradient HPLC and SEC in a fully automated 2D chromatography setup, it was possible to simultaneously fingerprint the chemical composition and molar mass during investigations of the grafting of methyl-methacrylate onto ethene/propene/diene rubber [274]. Preparative SEC was used for fractionation of highly branched PMMA with an estimated degree of branching (3.7 branch-points per 100 monomer units). The fractions were characterized in solution by SEC/viscosity coupling and in the melt by visco-elastic spectroscopy [275].

Two experimental techniques, namely, SEC with fluorescence detection and SEC, coupled to forward recoil spectrometry (FRES), were used to monitor independently the extent of reaction between model end-functional polymers at a PS/PMMA interface [276]. This is of importance for investigations of polymer/polymer reaction kinetics and interfacial segregation during *in situ* reactive polymer compatibilization. Possibilities and limitations of photon correlation spectroscopy (PCS) in determining polymer molecular-weight distributions were studied in comparison with the results obtained by SEC for samples of PMMA having monomodal and bimodal distribution functions [277]. Evaluation of a single-capillary viscometer detector, coupled on-line to a SEC system, was performed, using various polymers that were soluble in organic (PS, PMMA, polyvinyl acetate (PVAc), PVC, polyalkylthiophene), and aqueous solvents (PEO, PEG, pullulan, and hyaluronan). Molar-mass distribution, intrinsic-viscosity distribution, and constants of the Mark-Houwink-Sakurada relationship were determined [278].

Using polymers of complex macromolecular architectures, Stogiou *et al.* [279] recently tested the validity of the universal calibration curve (UCC) in SEC. The polymers studied included the microarm stars, H-shaped, and pi-shaped, as well as a model linear tetrablock co-polymer of the PS/PI/PS/polyisoprene type. It was found that the universality of the relation of log $M\eta$ *vs.* peak elution volume is also valid for these complex molecules. However, the determination of the molecular weight of a polymer with the UCC was found to be very sensitive to the molecular and compositional homogeneity of the sample. A qualitative analysis of secondary mechanisms in SEC of polymers through the mean value of the viscosimetric exponent has been reported by Gomez *et al.* [280]. A computer simulation study for estimating the biases induced by branching under ideal fractionation and detection conditions was performed for the analysis of a styrene/butadiene graft co-polymer by SEC [281,282]. A novel polymerization model was developed for predicting the MWD, DBD (degree-of-branching distribution), and CCD for the total co-polymer and for each of its different branched topologies. To simulate the molecular-weight calibrations in SEC, the Zimm-Stockmayer equation was applied to each co-polymer topology. Negligible deviations due to branching were found in the MWD and the DBD with respect to the theoretical predictions. However, errors in the CCD were intolerably large and, consequently, it was concluded that the CCD cannot be estimated by SEC.

Park *et al.* [283] have recently investigated the reaction products of polystyryllithium with air by SEC, temperature-gradient interaction chromatography (TGIC) and MALDI-TOF-MS. It was confirmed that polystyryl ketone, polystyryl alcohol, and directly coupled polystyrene were the major products, in addition to the normally terminated polystyrene. Moreover, polystyrenes, end-capped with methoxy and carboxylic acid groups, as well as dipolystyryl ether, were also identified as minor products. Comparative studies on polymer characterization by TGIC and by SEC have been reported by Chang *et al.* [284]. TGIC is a form of HPLC in which the column temperature is changed in a programmed manner to control the retention of polymeric species during isocratic elution. The polymers were separated by TGIC in terms of their molecular weights. TGIC was considered to be superior to SEC with respect to resolution and sample loading capacity, and was found to possess higher sensitivity to molecular weight in the analysis of nonlinear polymers. The advantage of TGIC over solvent-gradient HPLC arises from the fact that the former permits the use of RI detection methods, such as differential refractometry and light scattering due to the use of isocratic elution. In addition, temperature variation provides finer and more reproducible values of retention volume than does variation in solvent composition. Such control is important when the MWD is determined by secondary calibration methods. TGIC was successfully applied to the characterization of star-shaped polystyrene, and kinetics of linking between living polystyrene anions and chlorosilane linking agent were investigated in detail. Thermodynamic principles and some applications of the TGIC analysis were reported, and it was found that the MWD of anionically polymerized species is much narrower than generally assumed from SEC analysis. The TGIC separation conditions for PS, PI, PMMA, PVC, and PVAc over a wide molecular-weight range were established. Studies of band-broadening occurring in SEC were performed, using very narrow PS standards, obtained and characterized by TGIC. Recently, TGIC was applied in the separation of stereo-regular polyethyl methacrylate (PEMA) according to the polymer tacticity. To isolate the tacticity effect from the molecular-weight effect on TGIC retention, the PEMA samples were fractionated by TGIC, and the accurate molecular weights of the fractions were determined by MALDI-TOF-MS. It was concluded that the retention in TGIC was affected by both the tacticity and the molecular weight [285,286].

A review on the recent progress in combining a full adsorption/desorption procedure with SEC (FAD/SEC) to the separation and characterization of co-polymers has been published by Nguyen and Berek [287]. FAD includes complete and selective adsorption of the polymer sample to be separated from a solvent promoting adsorption onto an appropriate adsorbent, which is packed into a designed LC-type microcolumn. In the following steps, macromolecules were displaced from the adsorbent by different eluents with increasing desorbing strength. The fractionation of polymers according to their MWD and CCD can be accomplished in the course of the FAD process [287]. Another review, covering the application of SEC to lipophils and biopolymers was also published by Berek [288]. A theoretical model and data analysis methods were recently proposed for SEC of step-growth polymers with cyclic species [289]. The measurement of MWD of many step-growth polymers is complicated due to the presence of cyclic oligomers, formed during polymerization. When SEC is used to determine the MWD, the cyclic oligomers are generally only partly separated from the linear polymer and, hence, distort the measured linear MWD. Moreover, the cyclic oligomers require a different calibration curve, in contrast to the linear species. Therefore, in general, their molecular weights are not accurately measured. The proposed model of the SEC separation of step-growth polymers with cyclic species was used for the characterization of these polymers by conventional and multi-detector SEC. The results were compared with experimental data for nylon 6, nylon 6.6, and polyethylene terephthalate.

The molar mass distribution of several functional oligo-amides has been studied by SEC [290]. The molar masses determined by SEC were compared with the values obtained by chemical titration and by NMR analysis. The kinetics of the reaction of tosyl isocyanate with polyhexamethylene/pentamethylene carbonate diol has also been evaluated by SEC [291]. Studies on polybisphenol A carbonate (PC), by SEC/MALDI have been conducted by Montaudo and co-workers [292,293]. The investigations included end-group and molar-mass determination, as well as examination of self-association, and the mechanism of thermal oxidation. The results showed that MALDI spectra of the SEC fractions allow not only the detection of linear and cyclic oligomers in the polymer, but also the simultaneous determination of their average molar masses. Two slightly differing SEC calibration plots were obtained, due to the smaller hydrodynamic volume of the polycarbonate cyclic chains, compared to the linear ones. In agreement with theory, the (M-cycle/M-linear)v(e) ratio at a fixed elution volume was found to be 1.22, independent of the molar mass values. Thermal oxidation products of PC, generated by heating at 300 and 350°C in air, were detected by SEC/MALDI. The SEC curves of the thermally oxidized samples showed extensive degradation as a function of heating time, up to the formation of oligomers having very low masses. Oxidized PC samples were subjected to SEC fractionation with collection of several fractions that were further analyzed off-line by MALDI/TOF. The mechanisms accounting for the formation of thermal oxidation products of PC involved several simultaneous reactions:

(a) hydrolysis of carbonate groups of PC to form free bisphenol A end groups,

(b) oxidation of the isopropenyl groups of PC, and

(c) oxidative coupling of phenolic end-groups to form biphenyl groups. The presence of biphenyl units among the thermal oxidation products confirmed the occurrence of cross-linking processes, which are responsible for the formation of the insoluble gel fraction.

The biological polyesters have been systematically investigated [294], and SEC and ESI-MS^n have been used to determine the sequence distribution and chemical structure of mass-selected macromolecules of poly-3-hydroxybutyrate-co-3-hydroxyvalerate (PHBV), a natural polyester macro-initiator, obtained by partial alkaline depolymerization of natural PHBV [295]. The microstructure of this bacterial co-polyester was assessed, starting from the dimer up to the oligomer containing 22 repeat units, and the results obtained were compared with those described previously involving other "soft" ionization MS techniques. The poly-3-hydroxybutyrate (HB)/co-ε-caprolactone (CL) co-polyesters, derived *via* acid-catalyzed trans-esterification of natural PHB have been characterized by SEC and MALDI-TOF [296] with regard to their molecular weights, molar compositions, and average block length of repeating units. The MALDI-TOF mass spectra of samples fractionated by SEC made it possible to ascertain that co-polymers rich in HB units had mostly hydroxyl and carboxyl end-groups, while copolymers rich in CL units had mostly tosyl-, and carboxyl end groups. The structure of biomimetic PHB and poly-2-methyl-3-hydroxyoctanoate was established by ESI/MS^n and SEC analyses. The addition/elimination mechanism of the polymerization of β-lactones containing α-hydrogen by alkoxide anion was demonstrated to be true for β-lactones having alkyl substituents in both α- and β-positions [297]. The comparison of SEC and ESI-MS results for the analysis of water-soluble racemic PHB oligomers (from dimer to dodecamer) was performed, based on ESI-selective ion-display patterns of individual macromolecular ions [298]. The structure of the water-soluble oligomers, composed of racemic PHB, covalently conjugated to L-alanine and Ala-Ala-Ala oligopeptide has also been assessed by ESI/MS^n, and the respective structural information on each has been complemented by SEC analyses [299]. SEC and MALDI-TOF-MS have recently been used to determine product molecular weight and end-group structure in mass-selective lipase-catalyzed poly(ε-caprolactone) transesterification reactions [300]. These studies have shown how enzymatic trans-esterification reactions can be further developed to provide oligomers with well-defined length and end-group structure.

Novel block co-polymers, based on 2-vinylpyridine (2VP) and ε-caprolactone (CL), which are expected to be useful pigment-dispersing agents for TiO$_2$ in, *e.g.*, polyester powder coatings, have been characterized by SEC [301]. It was found that part of the living 2VP chains was deactivated immediately after addition of CL, which yielded bimodal MWD. A comparison of SEC with ESI-MS and with MALDI-TOF-MS for the characterization of polyester resins has been reported. A series of 17 polyester paint resins has been analyzed in order to compare the structural and molecular-weight data derived from each technique [302]. The optimization of sample preparation and laser power were found to be important factors in obtaining constant MWD by MALDI. However, the use of SEC for the fractionation of polyester samples prior to ESI or MALDI analysis showed that both techniques significantly underestimated the average molecular weights of the polydisperse polyesters. Differences in the relative abundances of branched and cyclic species in ESI *vs.* MALDI mass spectra were also noted. It was concluded that it would be useful to employ both techniques to ensure complete characterization of complex polymer samples by SEC. A novel analytical method, called two-dimensional correlation gel permeation chromatography, has recently been introduced in a study of the octyltriethoxysilane sol/gel polymerization process [303]. This technique, based on the combination of 2D-correlation analysis and time-resolved SEC, is a useful method for studying polymerization processes. A combination of LC with ESI orthogonal acceleration time-of-flight (oaTOF) MS has been applied in on-line polymer analysis [304]. In one experimental setup, three different LC modes were interfaced with MS: SEC/MS, GPEC/MS, and LC-CAP/MS.

5.7.2 Proteins and polypeptides

One of the challenges of the last decade has been the identification of critical cellular markers of neurodegenerative diseases. Such markers include amyloid beta (Aβ) protein and its fragments, prion proteins, etc. Cells and tissues contain thousands of proteins (an average of *ca.* 10,000 proteins per cell). A complete analysis of the entire protein content, as well as their structures and interactions with other molecules, is an objective of an increasing importance, bearing in mind the long-term goal of developing novel drugs. Handling of such complex mixtures has been the subject of many reviews, and despite several methodological breakthroughs [3,305–309], a consensus on the most suitable and reproducible strategy has not been achieved. A polymerized form of recombinant mouse prion protein (mPrP) domain 23–231 [mPrP-(23–231)], designated mPrP-z, was recently generated [310] and isolated by RP-HPLC or size-exclusion HPLC. Transgenic mouse Aβ peptides were purified by sequential SEC and RP-LC, and subjected to amino acid sequencing and MS [311]. Analysis of the *in vivo*-derived Aβ polypeptides by on-line 2D-chromatography/MS was performed to detect Aβ 1–40 and Aβ 1–42 directly in cell lysates [312]. The method consisted of on-line SEC to provide initial separation of analytes from the sample (based on their molecular mass), coupled with sample concentration prior to analysis by microbore LC/MS.

The chromatographic system Äkta-Purifier 10, scaled up for preparative HPLC, was used for the purification of substance P (SP) endopeptidase activity in the ventral tegmental area (VTA) of the rat brain [313]. By use of this strategy, it was possible to achieve a purification factor of almost 7500, based on the specific activity. Righetti and Verzola [19] discussed the usefulness of present methodologies in the study of folding/unfolding/refolding of proteins in comparison with the capabilities of capillary zone electrophoresis. The effect of enzyme inhibitors on protein quaternary structure has been determined by on-line SEC/MS [314]. SEC and electrophoretic separation techniques have been used by Underberg *et al.* [315] to investigate the physical stability of peptides and proteins. These authors also presented an overview of various separation and detection techniques for peptides and proteins that have been used in stability research and biochemical analysis. Assembling of γ- with α-globin chains to form human fetal hemoglobin *in vitro* and *in vivo* was verified by SEC [316]. Analytical-scale SEC was applied to the large-scale production of recombinant hepatitis B surface antigen from *Pichia pastoris* [317].

Direct interaction of a high-affinity complex between the bacterial outer membrane protein, FhuA, and the phage T5 protein, pb5, has been demonstrated by isolating a pb5/FhuA complex by SEC [318]. Kim and Park [319] used this method to test the aggregation stability of encapsulated recombinant human growth hormone. The behavior of gelatin in dilute aqueous solution has been studied with the objective of designing a nanoparticulate formulation by using SEC under various conditions of time, temperature, pH, and ethanol concentration [320]. Kinetics and thermodynamics of dimer formation and dissociation for a recombinant humanized monoclonal antibody against vascular endothelial growth factor were investigated as a function of pH, temperature, and ionic strength by SEC, using the concentration jump method [321]. The relevance of techniques, such as analytical ultracentrifugation, SEC, and MS in the structural investigation of detergent-solubilized membrane proteins was discussed by LeMaire *et al.* [322]. A simple chromatographic assay for Rab geranylgeranyl transferase has been developed [323]. The method involves separation of the reaction mixture on a Sephadex G-25 Superfine minicolumn.

The proteome determines the cellular phenotype, and the regulation of the entire inter- and intracellular network requires simultaneous monitoring [324–326]. Data from 2D-separation techniques, such as 2D-gas chromatography (GC × GC), liquid chromatography/liquid chromatography (LC × LC), and liquid chromatography/capillary electrophoresis (LC × CE) can be readily analyzed by various chemometric methods to increase the information content of chemical analysis [327]. Proteolysis of whole-cell extracts with application of immobilized-enzyme columns as a part of multi-dimensional chromatography, was investigated by Wang and Regnier [328]. The effectiveness of proteolysis was evaluated with extracts of *E. coli*, the extent of degradation being monitored by SEC. Hille *et al.* [329] proposed a combined technique, involving capillary isoelectric focusing coupled to mass spectrometry (CIEF/MS^n) and preparative IEF, followed by SEC linked to MS (PIEF/SEC/MS) to improve automation, speed, and precision of proteome analysis. Peptide mapping with combinations of SEC, RP-LC, and CE was reported by Stromqvist [330]. Recombinant extracellular superoxide dismutase was proteolytically degraded by trypsin, and the digest was fractionated by three different separation techniques, among them SEC. Zhang *et al.* [331] showed how several preparative steps were essential for obtaining information about modified human lens β-crystallins. The preparative techniques prior to MS included SEC, RP-LC, 2D-PAGE, *in situ* digestion of the proteins, and peptide trapping before the final LC/MS analysis. To understand the structural properties of buffalo growth hormone, an equilibrium denaturation with guanidine chloride was carried out and was monitored by UV spectroscopy, intrinsic fluorescence spectroscopy, far-UV circular dichroism, and SEC [332]. Native, unfolded/refolded frutalin and a distinct molecular form denoted "misfolded", were separated on Superdex 75 [333]. On-line coupling of SEC with imaged capillary isoelectric focusing and a membrane interface for proteins separation have been described by Tragas and Pawliszyn [334]. The system is equivalent to 2D-PAGE, transferring the principle of 2D-separation to the capillary format.

5.7.3 Molecular biology

SEC and denaturing gel electrophoresis of a recombinant enzyme was applied [335] to characterize NADH-dependent methylenetetrahydrofolate reductase from higher plants and to prove that it exists as a dimer of *ca.* 66-kDa subunits. SEC analysis of adenovirus particles has also been reported [336]. DNA binding properties of basic helix-loop-helix fusion proteins of Tal and E47 have been investigated [337]. It has been concluded from the SEC studies that all mutant and fusion proteins are dimeric. Pacek *et al.* [338] studied DNA box sequences as the site for helicase delivery during plasmid RK2 replication initiation in *E. coli*, and Shirakawa *et al.* [339] investigated the targeting of high-mobility proteins in chromatin. Meyer and co-workers [340] observed a difference with respect to dimer formation between native PrP(C) and recombinant PrP prion proteins. Application of various strategies, including SEC, to the purification of tumor-specific immuno-therapeutics, obtained by recombinant DNA technology, was reviewed by Matthey *et al.* [341]. SEC was also applied in an automated one-step DNA sequencing technique [342], based on the nanoliter reaction volumes and capillary electrophoresis. The reaction products were purified by SEC, followed by an on-line injection of the DNA fragments into a capillary. Over 450 bases of DNA could be separated and identified by this technique. A review including descriptions of various separation techniques used in nucleic acid research was published by Takenaka and Kondo [343]. For large natural polymers, such as double-stranded DNA molecules or synthetic polymers, slalom chromatography (SC) or hydrodynamic chromatography (HDC) can be applied. The elution order in HDC is the same as in SEC. On the other hand, the observed elution order of double-stranded DNA molecules is the opposite of that expected for HDC or SEC [344]. The theory and validation for these methods have been described [345–347]. All three modes can be treated as complementary and linked in a global separation mechanism, utilizing a nonequilibrium chromatographic principle.

5.7.4 Carbohydrates

Methods for the preparation and characterization of hyaluronan oligosaccharides of defined length have been recently reported [348]. The preparations obtained by SEC were characterized by a combination of ESI-MS, MALDI-TOF MS, and fluorophore-assisted carbohydrate electrophoresis. Another approach to the analysis of polysaccharide materials employs SEC [349], followed by detection by on-line ESI-MS and off-line MALDI-TOF-MS. It was demonstrated that formation of the multiply charged oligomers that bind up to five sodium cations permits the rapid analysis of polysaccharides with molecular masses in excess of 9 kDa. Isolation and purification of proteoglycans by SEC was the subject of a review by Fedarko [350]. Characterization of the molar masses of hemicelluloses from wood and pulps by SEC and MALDI-TOF-MS has been reported by Jacobs *et al.* [351]. Negative-ion (NI)-ESI-MS and SEC have been used to reveal structural heterogeneity in κ-carrageenan oligosaccharides [352]. HP-SEC, combined with isotope-ratio MS, was applied by Abramson *et al.* [353] to measure low levels of underivatized materials, such as proteins, carbohydrates, and nucleotides. This combination produces a device capable of measuring very low alterations in

[13]C abundance from analyte species that cannot be volatilized. The occurrence of internally $(1 \rightarrow 5)$-linked arabinofuranose and arabinopyranose residues in arabinogalactan side-chains from soybean pectic substances was investigated by Huisman *et al.* [354], and the compositional analysis of glycosaminoglycans by SEC/MS has recently been described by Zaia and Costello [355]. The distribution of 4-*O*-methylglucuronic acid residues along the polysaccharide chains of xylans, isolated from various trees, has been studied by analysing the oligosaccharide mixtures obtained by partial acid hydrolysis [356]. The hydrolysates thus obtained were analyzed by MALDI-MS or by capillary electrophoresis as well as by SEC in combination with MALDI-MS. Fructans (fructo-oligosaccharides and inulin) originally extracted from chicory roots have been separated by continuous annular and fixed-bed conventional gel chromatography [357]. Both columns were packed with Toyopearl HW 40 (S) and eluted with de-ionized water. The productivity of the annular system was found to be 25 times higher than the conventional system. More detailed information on SEC analysis of oligosaccharides and polysaccharides can be found in Chap. 18 of this book.

5.7.5 Other applications

SEC is a very useful tool for investigating the self-association of many substances, such as surfactants, chlorpromazine hydrochloride, Methylene Blue, and a sulfobetaine derivative (CHAPS) of cholic acid [358]. The SEC of lipids has been described in several reviews [359,360]. More recently, its use in the detection of diacylglycerols and other compounds to verify the quality and authenticity of olive oil has been reported by Dauwe *et al.* [361]. MS of myelin proteolipids, pre-separated by means of SEC with an organic solvent as the mobile phase, has been described [362]. Humic substances have been analyzed by APCI and ESI-MS in positive and negative modes, in combination with SEC [363]. The effects of ozone, chlorine, hydrogen peroxide, and permanganate on aquatic humic matter with different molecular-size fractions, and the formation of organic acids in drinking water have been studied by SEC [364]. Aquatic humus in lake water, artificially recharged groundwater, and purified, artificially recharged groundwater were fractionated by HP-SEC/UV before and after oxidation. Fractionation of natural organic matter in drinking water by SEC and characterization by ^{13}C cross-polarization magic-angle spinning NMR spectroscopy and SEC have been achieved by Wong *et al.* [365]. Studies of specific interactions of organic substances, such as alcohols, mono- and dicarboxylic acids, aromatic acids, and amino acids were conducted by Specht and Frimmel [366]. Adsorption of several different organic polyelectrolytes from aqueous solution by activated carbon was investigated by Kilduff *et al.* [367]. The polyelectrolytes studied included humic acids, extracted from peat and soil, polymaleic acid, a synthetic polymer identified as a fulvic acid surrogate, and natural organic matter in river water.

A method has been developed [368] for analyzing pesticides in dust. For non-acidic pesticides, the extract, after centrifugation and filtration, was purified by SEC and then analyzed by GC/MS. Coupling of SEC to LC/MS for the determination of trace levels of thifensulfuron-methyl and tribenuron-methyl in cottonseed and cotton gin trash has been described [369]. An on-line SEC/GC method for the detection and quantification of organophosphorous pesticides in crude edible oils was described by Jongenotter and Janssen [370]. Other compounds, such as sterols and wax esters were also identified by an off-line technique. SEC on a Bio-Beads SX-3 column, followed by a dual GC determination has been developed by Jover and Bayona [371] for a multi-class pesticide determination in lanolin. The effluent from the analytical column (50% diphenylmethyl- or 14% cyanopropylphenylpolysiloxane) was split into an electron capture detection (ECD) and a nitrogen phosphorous detection (NPD) system. That system was optimized for 28 pesticides commonly used to control sheep pests. The molecular-mass distributions of organic poly-electrolytes remaining in solution after equilibration with various amounts of activated carbon were determined by HP-SEC. The shape separations of suspended gold nanoparticles were investigated by Wei *et al.*, using SEC [372]. A sample of Athabasca bitumen was fractionated by preparative SEC, and the MWD of five fractions and of the original sample were determined by SEC, using a calibration based on polystyrene standards [373]. SEC on Sephacryl S-1000 has recently been used by Loa *et al.* to purify turkey coronavirus (TCoV) from infected turkey embryos [374]

ACKNOWLEDGMENTS

This work was supported by the Eivind and Elsa K:son Foundation, by the Swedish Research Council (Grant 13123), and by a grant from the Polish Committee for Scientific Research (KBN 3 P04 B02024).

REFERENCES

1   B. Lindqvist and T. Storgårds, *Nature*, 175 (1955) 511.
2   G.H. Lathe and C.R.J. Ruthven, *Biochem. J.*, 62 (1956) 665.
3   J. Porath, *Biochem. Soc. Trans.*, 7 (1979) 1197.
4   J. Porath, *Curr. Content*, 19 (1981) 21.
5   J. Porath, *J. Chromatogr.*, 218 (1981) 241.
6   J.-C. Janson, *Chromatographia*, 23 (1987) 361.
7   T.C. Laurent, *J. Chromatogr.*, 633 (1993) 1.
8   J. Porath, *J. Protein Chem.*, 16(5) (1997) 463.
9   J. Porath and P. Flodin, *Nature*, 183 (1959) 1657.
10 S. Hjertén and R. Mosbach, *Anal. Biochem.*, 3 (1962) 109.
11 S. Hjertén, *Arch. Biochem. Biophys.*, 99 (1962) 466.
12 A. Polson, *Biochim. Biophys. Acta*, 50 (1961) 565.
13 W.R. LaCourse, *Anal. Chem.*, 72 (2000) 37R.
14 J.G. Dorsey, W.T. Cooper, B.A. Siles, J.P. Foley and H.G. Barth, *Anal. Chem.*, 70 (1998) 591R.
15 J. Cazes (Ed.), *Encyclopedia of Chromatography*, Dekker, New York, 2001.
16 W.W. Yau, J.J. Kirkland and D.D. Bly (Eds.), *Modern Size-exclusion Liquid Chromatography: Practice of Gel Permeation and Gel Filtration Chromatography*, Wiley, New York, 1979.
17 M. Potschka and P.L. Dubin (Eds.), *Strategies in Size-exclusion Chromatography*, ACS, Washington DC, 1996.

18   H. Parvez, Y. Kato and S. Parvez (Eds.), *Chromatography of Proteins and Peptides*, VSP, Zeist, 1985.
19   P.G. Righetti and B. Verzola, *Electrophoresis*, 22 (2001) 2359.
20   J.P. Busnel, F. Foucault, L. Denis, W. Lee and T. Chang, *J. Chromatogr. A*, 930 (2001) 61.
21   R.D. Ricker and L.A. Sandoval, *J. Chromatogr. A*, 743 (1996) 43.
22   J.C. Giddings, *Anal. Chem.*, 39 (1967) 1072.
23   A. Titterton, *LC-GC*, January 1 (2002).
24   U.A.T. Brinkman, *J. Chromatogr. A*, 665(2) (1994) 217.
25   D.J. Winzor, *J. Mol. Recognit.*, 13(5) (2000) 279.
26   F. Dondi, A. Cavazzini, M. Remelli, A. Felinger and M. Martin, *J. Chromatogr. A*, 943 (2002) 185.
27   D. Rodbard and A. Chrambach, *Proc. Natl Acad. Sci. USA*, 65 (1970) 970.
28   A. Polson and W. Katz, *Biochem. J.*, 12 (1969) 387.
29   A.M. Mika and R.F. Childs, *Ind. Eng. Chem. Res.*, 40 (2001) 1694.
30   R.L. Rill, D.H. Van Winkle and B.R. Locke, *Anal. Chem.*, 70 (1998) 2433.
31   P.R. Wills, W.D. Comper and D.J. Winzor, *Arch. Biochem. Biophys.*, 300 (1993) 206.
32   D.J. Winzor and J. De Jersey, *J. Chromatogr.*, 492 (1989) 377.
33   J.R. Cann and D.J. Winzor, *Arch. Biochem. Biophys.*, 256 (1987) 78.
34   L.W. Nichol, R.J. Siezen and D.J. Winzor, *Biophys. Chem.*, 9 (1978) 47.
35   A.P. Minton, *Biophys. Chem.*, 18(2) (1983) 139.
36   D.O. Jordan, S.J. Lovell, D.R. Phillips and D.J. Winzor, *Biochemistry*, 13 (1974) 1832.
37   L.W. Nichol, M. Janado and D.J. Winzor, *Biochem. J.*, 133 (1973) 15.
38   D.J. Winzor and L.W. Nichol, *Biochim. Biophys. Acta*, 104 (1965) 1.
39   G.E. Hibberd, A.G. Ogston and D.J. Winzor, *J. Chromatogr.*, 48 (1970) 393.
40   D.E. Brooks, C.A. Haynes, D. Hritcu, B.M. Steels and W. Muller, *Proc. Natl Acad. Sci. USA*, 97 (2000) 7064.
41   H. Barth, B. Boyes and C. Jackson, *Anal. Chem.*, 70 (1998) 251R.
42   J.G. Dorsey, W.T. Cooper, B.A. Siles, J.P. Foley and H.G. Barth, *Anal. Chem.*, 70 (1998) 591.
43   A.M. Basedow and K.H. Ebert, *Infusionsther Klin Ernahr.*, 2 (1975) 261.
44   R. Ekman, B.G. Johansson and U. Ravnskov, *Anal. Biochem.*, 70 (1976) 628.
45   S. Hjertén and K.O. Eriksson, *Anal. Biochem.*, 137 (1984) 313.
46   A. Anspach, H.U. Gierlich and K.K. Unger, *J. Chromatogr.*, 443 (1988) 45.
47   A. Santarelli, D. Muller and J. Jozefonvicz, *J. Chromatogr.*, 443 (1988) 55.
48   L. Hagel, H. Lundstrom, T. Andersson and H. Lindblom, *J. Chromatogr.*, 476 (1989) 329.
49   Y.B. Yang and M. Verzele, *J. Chromatogr.*, 391 (1987) 383.
50   S.H. Knutsen, M. Sletmoen, T. Kristensen, T. Barbeyron, B. Kloareg and P. Potin, *Carbohydr. Res.*, 331 (2001) 101.
51   L.Z. Vilenchik, J.P. Griffith, N.St. Clair, M.A. Navia and A.L. Margolin, *J. Am. Chem. Soc.*, 120 (1998) 4290.
52   M. Petro, F. Svec, I. Gitsov and J. Frechet, *Anal. Chem.*, 68 (1996) 315.
53   E.C. Peters, F. Svec and J.M.J. Frechet, *Chem. Mater.*, 9 (1997) 1898.
54   I.M. Abrams and J.R. Millar, *React. Funct. Polym.*, 35 (1997) 7.
55   S. Lubbad and M.R. Buchmeiser, *Macromol. Rapid Commun.*, 23 (2002) 617.
56   S. Hjertén, Y.-M. Li, J.L. Liao, J. Mohammad, K. Nakazato and G. Pettersson, *Nature*, 356 (1992) 810.
57   Y.M. Li, P. Brostedt, S. Hjertén, F. Nyberg and J. Silberring, *J. Chromatogr. B*, 664 (1995) 426.
58   M. Antonietti, R.A. Caruso, Ch.G. Göltner and M.C. Weissenberger, *Macromolecules*, 32 (1999) 1383.

59 E.C. Peters, M. Petro, F. Svec and J. Frechet, *Anal. Chem.*, 70 (1998) 2296.

60 A. Mayr, R. Tessadri, E. Post and M. Buchmeiser, *Anal. Chem.*, 73 (2001) 4071.

61 B. Viklund and K. Irgum, *Anal. Chem.*, 33 (2000) 2539.

62 F. Svec and J. Frechet, *Ind. Eng. Chem. Res.*, 38 (1999) 34.

63 A. Hirose and D. Ishii, *J. Chromatogr.*, 411 (1987) 221.

64 C.L. Flurer, C. Borra, F. Andreolini and M. Novotny, *J. Chromatogr.*, 448 (1988) 73.

65 H.J. Cortes, B.M. Bell, C.D. Pfeiffer and D. Graham, *J. Microcol. Sep.*, 1 (1989) 278.

66 R.T. Kennedy and J.W. Jorgensson, *J. Microcol. Sep.*, 2 (1990) 120.

67 K. Jinno and M. Nishibara, *Anal. Lett.*, 13 (1980) 673.

68 A. Ishii and T. Takeuchi, *J. Chromatogr.*, 255 (1983) 349.

69 T. Takeuchi and D. Ishii, *J. Chromatogr.*, 257 (1983) 327.

70 H. Yun, S.V. Olesik and E.H. Marti, *Anal. Chem.*, 70 (1998) 3298.

71 M.T. Smoluch, P. Mak, J.-P. Chervet, G. Hohne and J. Silberring, *J. Chromatogr. B*, 726 (1999) 37.

72 J.J. Keve, B.G. Belenkii, E.S. Gankina, L.Z. Vilenchik, O.I. Kurenbin and T.P. Zhmakina, *J. High Res. Chromatogr. Commun.*, 4 (1981) 425.

73 J.J. Keve, B.G. Belenkii, E.S. Gankina, L.Z. Vilenchik, O.I. Kurenbin and T.P. Zhmakina, *J. Chromatogr.*, 207 (1981) 145.

74 T. Schenk, H. Irth, G. Marko-Varga, L.E. Edholm, U.R. Tjaden and J. van der Greef, *J. Pharm. Biomed. Anal.*, 26 (2001) 975.

75 M.W. Nielen, *Anal. Chem.*, 70 (1998) 1563.

76 A. Ericson, J. Holm, T. Ericson and S. Hjertén, *Anal. Chem.*, 72 (2000) 81.

77 D.R. Reyes, D. Iossifidis, P.-A. Auroux and A. Manz, *Anal. Chem.*, 74 (2002) 2623.

78 P.-A. Auroux, D. Iossifidis, D.R. Reyes and A. Manz, *Anal. Chem.*, 74 (2002) 2637.

79 A. Chmela, R. Tijssen, M. Blom, J.G. Gardeniers and A. van den Berg, *Anal. Chem.*, 74 (2002) 3470.

80 R. Garcia, I. Porcar, A. Campos, V. Soria and J.E. Figueruelo, *J. Chromatogr. A*, 655 (1993) 191.

81 R. Garcia, I. Porcar, A. Campos, V. Soria and J.E. Figueruelo, *J. Chromatogr. A*, 655 (1993) 3.

82 R. Garcia, I. Porcar, A. Campos, V. Soria and J.E. Figueruelo, *J. Chromatogr. A*, 662 (1994) 61.

83 A.L. Lafleur and E.F. Plummer, *J. Chromatogr. Sci.*, 29 (1991) 532.

84 A. Porsch, M. Andersson, B. Wittgren and K.G. Wahlund, *J. Chromatogr. A*, 946 (2002) 69.

85 M. Hasegawa, A. Isogai and F. Onabe, *J. Chromatogr.*, 635 (1993) 334.

86 T. Schult, S.T. Moe, T. Hjerde and B.E. Christensen, *J. Liquid Chromatogr.*, 23 (2000) 2277.

87 S. Phillips and S.V. Olesik, *Anal. Chem.*, 74 (2002) 799.

88 A. Saake, T. Kruse and J. Puls, *Bioresour. Technol.*, 80 (2001) 195.

89 A.A. Herod, S.-F. Zhang, B.R. Johnson, K.D. Bartle and R. Kandiyoti, *Energy Fuels*, 10 (1996) 743.

90 H.B. Li, F. Chen and X.R. Xu, *J. Chromatogr. A*, 918 (2001) 335.

91 B. Batas and J.B. Chaudhuri, *J. Chromatogr. A*, 864 (1999) 229.

92 C.T. Mant, H. Chao and R.S. Hodges, *J. Chromatogr. A*, 791 (1997) 85.

93 M. Bergman, R. Dohmen, H.A. Claessens and C.A. Cramers, *J. Chromatogr. A*, 657 (1993) 33.

94 M.P.C. Silvestre, M. Hamon and M. Yvon, *J. Agric. Food Chem.*, 42 (1994) 2778.

95 J. Näslund, A. Schierhorn, U. Hellman, L. Lannfelt, A.D. Roses, L.O. Tjärnberg, J. Silberring, S.E. Gandy, B. Winblad, P. Greengard *et al. Proc. Natl Acad. Sci. USA*, 91 (1994) 8378.

96 I. Nylander, K. Tan-No, A. Winter and J. Silberring, *Life Sci.*, 57 (1995) 123.

97 H. Hedlund, L. Kärf, T. Nyhammar and A. Winter, http://www.amershambiosciences.com/product/publication/brochure/ST31p12_13.pdf.

 98   E.J. Fernandez, T.T. Norton, W.C. Jung and J.G. Tsavalas, *Biotechnol. Prog.*, 12 (1996) 480.
 99   P. Bermejo, E.M. Pena, D. Fompedrina, R. Dominguez, A. Bermejo, J.A. Cocho, J.R. Fernandez and J.M. Fraga, *J. AOAC Int.*, 84 (2001) 847.
100   R. Mehvar, *J. Pharm. Biomed. Anal.*, 19 (1999) 785.
101   Wyatt Technology Corp., Application note. *LC-GC*, 2 (2002) 33.
102   A.S. Karnoup, V.M. Shiryaev, V.N. Medvedkin and Y.V. Mitin, *J. Pept. Res.*, 49 (1997) 232.
103   F. Lemiere, *LC-GC Europe*, December (2001) 2.
104   W.R. LaCourse and C.O. Dasenbrock, *Anal. Chem.*, 70 (1998) 37R.
105   L.D. Rothman, *Anal. Chem.*, 68 (1996) 587.
106   A.J. Anderson, *Anal. Chem.*, 71 (1999) 314.
107   L.G. Hargis, J.A. Howell and R.E. Sutton, *Anal. Chem.*, 68 (1996) 169.
108   O.Y. Al-Dirbashi and K. Nakashima, *Biomed. Chromatogr.*, 14 (2000) 406.
109   A. Ni, R.S. Sheng and T.M. Cotton, *Anal. Chem.*, 62 (1990) 1958.
110   R.M. Seifar, M.A.F. Altelaar, R.J. Dijkstra, F. Ariese, U.A.T. Brinkman and C. Gooijer, *Anal. Chem.*, 72 (2000) 5718.
111   S.A. Korhammer and A. Benreuther, *Fresenius' J. Anal. Chem.*, 354/2 (1996) 131.
112   K. Torabi, A. Karami, S.T. Balke and T.C. Schunk, *J. Chromatogr. A*, 910 (2001) 19.
113   M. Schure, *Anal. Chem.*, 71 (1999) 1645.
114   M. Teresa, R. Laguna, R. Medrano, M.P. Plana and M.P. Tarazona, *J. Chromatogr. A*, 919 (2001) 13.
115   R.L. Qian, R. Mhatre and I.S. Krull, *J. Chromatogr. A*, 787 (1997) 101.
116   A. Titterton, *Laboratory News (Europe)*, 11–12, November 1989.
117   M.L. Fishman, D.T. Gillespie, S.M. Sondey and R.A. Barford, *J. Agric. Food Chem.*, 37 (1989) 584.
118   J. Wen, T. Arakawa and J.S. Philo, *Anal. Biochem.*, 240 (1996) 155.
119   A. Muller, H.A. Pretus, R.B. McNamee, E.L. Jones, I.W. Browder and D.L. Williams, *J. Chromatogr. B*, 666 (1995) 283.
120   M.L. Fishman, L. Rodriguez and H.K. Chau, *J. Agric. Food Chem.*, 44 (1996) 3182.
121   N. Her, G. Amy, D. Foss, J. Cho, Y. Yoon and P. Kosenka, *Environ. Sci. Technol.*, 36 (2002) 1069.
122   J. Gailer and W. Lindner, *J. Chromatogr. B*, 716 (1998) 83.
123   J.J. Vreuls, R.J. Swen, V.P. Goudriaan, M.A. Kerkhoff, G.A. Jongenotter and U.A. Brinkman, *J. Chromatogr. A*, 750 (1996) 275.
124   A. Chaidedgumjorn, A. Suzuki, H. Toyoda, T. Toida, T. Imanari and R.J. Linhardt, *J. Chromatogr. A*, 959 (2002) 95.
125   J. Silberring, S. Lyrenas and F. Nyberg, *Biomed. Chromatogr.*, 3 (1989) 203.
126   J. Silberring, P. Brostedt, J. Neiman, U. Hellman, S. Liljequist and L. Terenius, *Biomed. Chromatogr.*, 8 (1994) 137.
127   H.H. Maurer, *Comb. Chem. High Throughput Screen*, 3 (2000) 467.
128   M. Rizzo, *J. Chromatogr. B*, 747 (2000) 203.
129   X.G. He, *J. Chromatogr. A*, 880 (2000) 203.
130   M.J. Bogusz, *J. Chromatogr. B*, 733 (1999) 65.
131   A. Tracqui, P. Kintz and P. Mangin, *J. Forensic Sci.*, 40 (1995) 254.
132   E.H. Jansen, L.A. van Ginkel, R.H. van den Berg and R.W. Stephany, *J. Chromatogr.*, 580 (1992) 111.
133   W.E. Lambert, J.F. Van Bocxlaer and A.P. De Leenheer, *J. Chromatogr. B*, 689(1) (1997) 45.
134   A.F. Fell, B.J. Clark and H.P. Scott, *J. Chromatogr.*, 316 (1984) 423.
135   J.B. Castledine, A.F. Fell, R. Modin and B. Sellberg, *J. Pharm. Biomed. Anal.*, 9 (1991) 619.

136 D. Lincoln, A.F. Fell, N.H. Anderson and D. England, *J. Pharm. Biomed. Anal.*, 10 (1992) 837.

137 J.B. Castledine and A.F. Fell, *J. Pharm. Biomed. Anal.*, 11 (1993) 1.

138 M.T. Hearn, M.I. Aguilar, T. Nguyen and M. Fridman, *J. Chromatogr.*, 435 (1988) 271.

139 J. Frank, Jr., A. Braat and J.A. Duine, *Anal. Biochem.*, 162 (1987) 65.

140 A.L. Dupont, *J. Chromatogr. A*, 950 (2002) 113.

141 M. Calero, J. Escribano, F. Soriano, A. Grubb, K. Brew and E. Mendez, *J. Chromatogr. A*, 719 (1996) 149.

142 J.X. de Vries, *J. Chromatogr.*, 465 (1989) 297.

143 N. Varotsis and N. Pasadakis, *Ind. Eng. Chem. Res.*, 36 (1997) 5516.

144 R.E. Murphy, M.R. Schure and J.P. Foley, *Anal. Chem.*, 70 (1998) 1585.

145 G.J. Opiteck and J.W. Jorgenson, *Anal. Chem.*, 69 (1997) 2283.

146 T. Stroink, G. Wiese, H. Lingeman, A. Bult and W. Underberg, *Anal. Chim. Acta*, 444 (2001) 193.

147 O.Y. Al-Dirbashi and K. Nakashima, *Biomed. Chromatogr.*, 14 (2000) 406.

148 P. Sandra, G. Vanhoenacker, F. Lynen, L. Li and M. Schelfaut, *LC-GC Europe*, December (2001) 2.

149 F. Lemiere, *LC-GC Europe*, December (2001) 2.

150 C.G. Huber and H. Oberacher, *Mass Spectrom. Rev.*, 20 (2001) 310.

151 P. Marquet, *Ther. Drug Monit.*, 24 (2002) 255.

152 M.R. Moeller and T. Kraemer, *Ther. Drug Monit.*, 24 (2002) 210.

153 Y.G. Shin and R.B. van Breemen, *Biopharm. Drug Dispos.*, 22 (2001) 353.

154 J. Peng and S.P. Gygi, *J. Mass Spectrom.*, 36 (2001) 1083.

155 N.J. Clarke, D. Rindgen, W.A. Korfmacher and K.A. Cox, *Anal. Chem.*, 73 (2001) 430A.

156 B.E. Erickson, *Anal. Chem.*, 72 (2000) 711A.

157 W.M. Niessen, *J. Chromatogr. A*, 856 (1999) 179.

158 M. Jemal, *Biomed. Chromatogr.*, 14 (2000) 422.

159 J. Ermer and M. Vogel, *Biomed. Chromatogr.*, 14 (2000) 373.

160 D. Guetens, K. Van Cauwenberghe, G. De Boeck, R. Maes, U.R. Tjaden, J. van der Greef, M. Highley, A.T. van Oosterom and E.A. de Bruijn, *J. Chromatogr. B*, 739 (2000) 139.

161 M.E. Schurdak, M.J. Voorbach, L. Gao, X. Cheng, K.M. Comess, S.M. Rottinghaus, U. Warrior, H.N. Truong, D.J. Burns and B.A. Beutel, *J. Biomol. Screen*, 6 (2001) 313.

162 M.T. Davis, D.C. Stahl, S.A. Hefta and T.D. Lee, *Anal. Chem.*, 67 (1995) 4549.

163 J.P. Vissers, R.K. Blackburn and M.A. Moseley, *J. Am. Soc. Mass Spectrom.*, 13 (2002) 760.

164 J. Cavanagh, R. Thompson, B. Bobay, L.M. Benson and S. Naylor, *Biochemistry*, 41 (2002) 7859.

165 B. Ganem, Y.T. Li and J.D. Henion, *J. Am. Chem. Soc.*, 113 (1991) 6294.

166 L.C. Santora, Z. Kaymakcalan, P. Sakorafas, I.S. Krull and K. Grant, *Anal. Biochem.*, 299 (2001) 119.

167 M.L. Shen, L.M. Benson, K.L. Johnson, J.J. Lipsky and S. Naylor, *J. Am. Soc. Mass Spectrom.*, 12 (2001) 97.

168 K.F. Blom, B.S. Larsen and C.N. McEwen, *J. Comb. Chem.*, 1 (1999) 82.

169 J. Silberring, in C. Williams and B. Irvine (Eds.), *Methods in Molecular Biology*, Humana Press, New Jersey, 1997, pp. 129–140.

170 J. Sandin, J. Georgieva, J. Silberring and L. Terenius, *Neuroreport*, 10 (1999) 71.

171 M. Vlaskovska, L. Kasakov, P. Suder, J. Silberring and L. Terenius, *Brain Res.*, 818 (1999) 212.

172 J. Sandin, I. Nylander and J. Silberring, *Regul. Pept.*, 73 (1998) 67.

173   J. Sandin, K. Tan-No, L. Kasakov, I. Nylander, A. Winter, J. Silberring and L. Terenius, *Peptides*, 18 (1997) 949.

174   M.T. Smoluch, P. Mak, J.P. Chervet, G. Hohne and J. Silberring, *J. Chromatogr. B*, 726 (1999) 37.

175   P. Suder, J. Kotlinska, A. Legowska, M. Smoluch, G. Hohne, J.-P. Chervet, K. Rolka and J. Silberring, *Brain Res. Brain Res. Protoc.*, 6 (2000) 40.

176   J. Silberring and F. Nyberg, *J. Chromatogr. A*, 777 (1997) 41.

177   S. McSheehy, J. Szpunar, R. Lobinski, V. Haldys, J. Tortajada and J.S. Edmonds, *Anal. Chem.*, 74 (2002) 2370.

178   R. Koplik, H. Pavelkova, J. Cincibuchova, O. Mestek, F. Kvasnicka and M. Suchanek, *J. Chromatogr. B*, 770 (2002) 261.

179   C.N. Ferrarello, M.R. Fernandez de la Campa, C. Sariego Muniz and A. Sanz-Medel, *Analyst*, 125 (2000) 2223.

180   C.F. Harrington, S. Elahi, S.A. Merson and P. Ponnampalavanar, *Anal. Chem.*, 73 (2001) 4422.

181   J. Wang, D. Dreessen, D.R. Wiederin and R.S. Houk, *Anal. Biochem.*, 288 (2001) 89.

182   S. McSheehy, P. Pohl, D. Velez and J. Szpunar, *Anal. Bioanal. Chem.*, 372 (2002) 457.

183   S. McSheehy, P. Pohl, R. Lobinski and J. Szpunar, *Analyst*, 126(7) (2001) 1055.

184   A. Liu, D. Lin and J.R. Yates, 3rd, *Biotechniques*, 32 (2002) 898, 900, 902 passim.

185   P.A. Wabnitz and J.A. Loo, *Rapid Commun. Mass Spectrom.*, 16 (2002) 85.

186   A. Harris, M. Engelstein, R. Parry, J. Smith, M. Mabuchi and J.L. Millipore, *Biotechniques*, 32 (2002) 626.

187   M.J. Deery, E. Stimson and C.G. Chappell, *Rapid Commun. Mass Spectrom.*, 15 (2001) 2273.

188   J.M. Hille, A.L. Freed and H. Watzig, *Electrophoresis*, 22 (2001) 4035.

189   F.P. Abramson, G.E. Black and P. Lecchi, *J. Chromatogr. A*, 913 (2001) 269.

190   X. Fei and K.K. Murray, *Anal. Chem.*, 68 (1996) 3555.

191   M.W. Nielen, *Anal. Chem.*, 70 (1998) 1563.

192   A. Orsnes, T. Graf, H. Degn and K.K. Murray, *Anal. Chem.*, 72 (2000) 251.

193   P. Onnerfjord, S. Ekstrom, J. Bergquist, J. Nilsson, T. Laurell and G. Marko-Varga, *Rapid Commun. Mass Spectrom.*, 13 (1999) 315.

194   V.V. Laiko and A.L. Burlingame, in Abstracts of the 4th Int. Symp. Mass Spectrom. Health Life Sci., August 25–29, 1998, San Francisco, CA, 1998, p. 72.

195   V.V. Laiko, M.A. Boldwin and A.L. Burlingame, *Anal. Chem.*, 72 (2000) 652.

196   V.V. Laiko, N.I. Taranenko, V.D. Berkout, M.A. Yakshin, C.R. Prasad, H.S. Lee and V.M. Doroshenko, *J. Am. Soc. Mass Spectrom.*, 13 (2000) 354.

197   J.-C. Blais, C.-O. Turrin, A.-M. Caminade and J.-P. Majoral, *Anal. Chem.*, 72 (2000) 5097.

198   S.P. Mulvaney and C.D. Keating, *Anal. Chem.*, 72 (2000) 145.

199   W.E. Smith, *Methods Enzymol.*, 226 (1993) 482.

200   K. Kneipp, H. Kneipp, I. Itzkan, R.R. Dasari and M.S. Feld, *Chem. Rev.*, 99 (1999) 2957.

201   B.J. Marquardt, P.G. Vahey, R.E. Synovec and L.W. Burgess, *Anal. Chem.*, 71 (1999) 4808.

202   A. Ni, R.S. Sheng and T.M. Cotton, *Anal. Chem.*, 62 (1990) 1958.

203   R. Subramanian, W.P. Kelley, P.D. Floyd, Z.J. Tan, A.G. Webb and J.V. Sweedler, *Anal. Chem.*, 71 (1999) 5335.

204   A. Behnke, G. Schlotterbeck, U. Tallarek, S. Strohschein, L.-H. Tseng, T. Keller, K. Albert and E. Bayer, *Anal. Chem.*, 68 (1996) 1110.

205   F.D. Andrade, L.C. Santos, M. Datchler, K. Albert and W. Vilegas, *J. Chromatogr. A*, 953 (2002) 287.

206   M.E. Lacey, Z.J. Tan, A.G. Webb and J.V. Sweedle, *J. Chromatogr. A*, 922 (2001) 139.

207 M. Sandvoss, A. Weltring, A. Preiss, K. Levsen and G. Wuensch, *J. Chromatogr. A*, 917 (2001) 75.

208 F.J. Moy, K. Haraki, D. Mobilio, G. Walker, R. Powers, K. Tabei, H. Tong and M.M. Siegel, *Anal. Chem.*, 73 (2001) 571.

209 A. Venter, A.E. Ashcroft, J.N. Keen, P.J. Henderson and R.B. Herbert, *Biochem. J.*, 363 (2002) 243.

210 M. Ludlow, D. Louden, A. Handley, S. Taylor, B. Wright and I.D. Wilson, *J. Chromatogr. A*, 857 (1999) 89.

211 I.D. Wilson, *J. Chromatogr. A*, 892 (2000) 315.

212 A. Louden, A. Handley, R. Lafont, S. Taylor, I. Sinclair, E. Lenz, T. Orton and I.D. Wilson, *Anal. Chem.*, 74 (2002) 288.

213 M. Godejohann, A. Preiss and C. Mugge, *Anal. Chem.*, 70 (1998) 590.

214 B.G. Belenkii, L.Z. Vilenchik, V.V. Nesterov, V.J. Kolegov and S.Y. Frenkel, *J. Chromatogr.*, 109 (1975) 233.

215 M. Netopilik, S. Podzimek and P. Kratochvil, *J. Chromatogr. A*, 922 (2001) 25.

216 Y.C. Guillaume, J.F. Robert and C. Guinchard, *Anal. Chem.*, 73 (2001) 3059.

217 Y.V. Heyden, S.T. Popovici and P.J. Schoenmaker, *J. Chromatogr. A*, 957 (2002) 127.

218 J. Bermejo, A.L. Fernandez, M. Granda, I. Suelves, A.A. Herod, R. Kandiyoti and R. Menendez, *J. Chromatogr. A*, 919 (2001) 255.

219 K.C. Tsao, G.H. Liu, P.Y. Chang, C.N. Lin, T.L. Wu, C.F. Sun and J.T. Wu, *J. Clin. Lab. Anal.*, 15 (2001) 193.

220 A. Oliva, M. Llabres and J.B. Farina, *J. Pharm. Biomed. Anal.*, 25 (2001) 833.

221 M. Netopilik, *J. Chromatogr. A*, 915 (2001) 15.

222 X. Lou, J.L. van Dongen and E.W. Meijer, *J. Chromatogr. A*, 896 (2000) 19.

223 P. DePhillips and A.M. Lenhoff, *J. Chromatogr. A*, 883 (2000) 39.

224 H.G. Barth, B.E. Boyes and C. Jacson, *Anal. Chem.*, 66 (1994) 595R.

225 L.Z. Vilenchik, J. Asrar, R.C. Ayotte, L. Ternorutsky and C.J. Hardimann, *J. Chromatogr. A*, 648 (1993) 9.

226 A. Kurganov, K. Unger and T. Issaeva, *J. Chromatogr. A*, 753 (1996) 177.

227 M. Brissova, M. Petro, I. Lacik, A.C. Powers and T. Wang, *Anal. Biochem.*, 242 (1996) 104.

228 P. Conte and A. Piccolo, *Chemosphere*, 38 (1999) 517.

229 A. Ahsan, W. Jeske, H. Wolf and J. Fareed, *Clin. Appl. Thromb. Hemost.*, 6 (2000) 169.

230 G.W. White, T. Katona and J.P. Zodda, *J. Pharm. Biomed. Anal.*, 20 (1999) 905.

231 A. Puglisi, F. Samperi, S. Carroccio and G. Montaudo, *Rapid Commun. Mass Spectrom.*, 13 (1999) 2260.

232 A. Yomota and S. Okada, *Kokuritsu Iyakuhin Shokuhin Eisei Kenkyusho Hokoku*, 115 (1997) 213.

233 K. Juliano, W.L. Champion, Jr., M.A. Schreiber and J.A. Blackwell, *J. Pharm. Biomed. Anal.*, 16 (1997) 499.

234 H. Potzschke, W.K. Barnikol, U. Domack, S. Dinkelmann and S. Guth, *Artif. Cells Blood Substit. Immobil. Biotechnol.*, 25 (1997) 527.

235 J.E. Harlan, D. Picot, P.J. Loll and R.M. Garavito, *Anal. Biochem.*, 224 (1995) 557.

236 R.G. Duggleby, *Comput. Appl. Biosci.*, 10 (1994) 133.

237 M. le Maire, A. Ghazi, M. Martin and F. Brochard, *J. Biochem. (Tokyo)*, 106 (1989) 814.

238 M. le Maire, A. Viel and J.V. Moller, *Anal. Biochem.*, 177 (1989) 50.

239 M. Potschka, *Anal. Biochem.*, 162 (1987) 47.

240 E.E. Brumbaugh, M.L. Hilt, R.B. Shireman, A.J. Espinosa and P.W. Chun, *Anal. Biochem.*, 154 (1986) 287.

241 J.V. Hinshaw, *LC-GC*, 20 (2002) 350.

242   S.D. Brown, S.T. Sum, F. Despagne and B.K. Lavine, *Anal. Chem.*, 68 (1996) 21.
243   M.J. Lazaro, C.A. Islas, A.A. Herod and R. Kandiyoti, *Energy Fuels*, 13 (1999) 1212.
244   M.H. Hemmelder, P.E. de Jong and D. de Zeeuw, *J. Lab. Clin. Med.*, 132 (1998) 390.
245   H. Pasch, in M. Schmidt (Ed.), *Advances in Polymer Sciences*, Vol. 150, Springer, Berlin, 2000, pp. 5–66.
246   D. Berek, S.H. Nguyen and G. Hild, *Eur. Polym. J.*, 36 (2000) 1101.
247   J. Falkenhagen, H. Much, W. Stauf and A.H.E. Muller, *Macromolecules*, 33 (2000) 3687.
248   D. Lee, I. Teraoka, T. Fujiwara and Y. Kimura, *J. Chromatogr. A*, 966 (2002) 41.
249   M.S. Montaudo, *Polymer*, 43 (2002) 1587.
250   H.J.A. Philipsen, F.P.C. Wubbe, B. Klumperman and A.L. German, *J. Appl. Polym. Sci.*, 72 (1999) 183.
251   H.J.A. Philipsen, *Mechanisms of Gradient Polymer Elution Chromatography and its Application to (Co)polyesters*, Technical University Eindhoven, 1998.
252   K.J. Clevett, *Bioprocess. Technol.*, 6 (1990) 47.
253   Z. Guan, *J. Am. Chem. Soc.*, 124 (2002) 5616.
254   J.T. Mehl, R. Murgasova, X. Dong, D.M. Hercules and H. Nefzger, *Anal. Chem.*, 72 (2000) 2490.
255   L. D'Agnillo, J.B.P. Soares and A. Penlidis, *J. Polym. Sci. Pol. Phys.*, 40 (2002) 905.
256   L. Simek, J. Dostal and M. Bohdanecky, *Polimery*, 46 (2001) 817.
257   P.D. Iedema, C. Willems, G. van Vliet, W. Bunge, S.M.P. Mutsers and H.C.J. Hoefsloot, *Chem. Eng. Sci.*, 56 (2001) 3659.
258   H.X. Tang, B. Foran and D.C. Martin, *Polym. Eng. Sci.*, 41 (2001) 440.
259   V. Balsamo, Y. Paolini, G. Ronca and A.J. Muller, *Macromol. Chem. Phys.*, 201 (2000) 2711.
260   P. Viville, D. Daoust, A.M. Jonas, B. Nysten, R. Legras, M. Dupire, J. Michel and G. Debras, *Polymer*, 42 (2001) 1953.
261   S.V. Canevarolo, *Polym. Degrad. Stabil.*, 70 (2000) 71.
262   H. El Mansouri, N. Yagoubi, D. Scholler, A. Feigenbaum and D. Ferrier, *J. Appl. Polym. Sci.*, 71 (1999) 371.
263   W. Buchmann, H.A. Nguyen, H. Cheradame, J.-P. Morizur and B. Desmazieres, *Chromatographia*, 55 (2002) 483.
264   D. Cunliffe, J.E. Lockley, J.R. Ebdon, S. Rimmer and B.J. Tabner, *Macromolecules*, 34 (2001) 3882.
265   J.H. Truelsen, J. Kops and W. Batsberg, *Macromol. Rapid Commun.*, 21 (2002) 98.
266   J.E. Puskas, W. Pattern, P.M. Wetmore and V. Krukonis, *Rubber Chem. Technol.*, 72 (1999) 559.
267   J. Feldthusen, B. Ivan and A.H.E. Muller, *Macromolecules*, 31 (1998) 578.
268   J.E. Puskas and C.J. Wilds, *J. Polym. Sci. Polym. Chem.*, 36 (1998) 85.
269   G. Pepperl, *J. Vinyl Addit. Technol.*, 6 (2000) 88.
270   G.D. Longeway and D.E. Witenhafer, *J. Vinyl Addit. Technol.*, 6 (2000) 100.
271   N. Manabe, K. Kawamura, T. Toyoda, H. Minami, M. Ishikawa and S. Mori, *J. Appl. Polym. Sci.*, 68 (1998) 1801.
272   F. Ding, R. Stol, W.T. Kok and H. Poppe, *J. Chromatogr. A*, 924 (2001) 239.
273   A.M. Striegel, *Anal. Chem.*, 74 (2002) 3013.
274   A. Siewing, J. Schierholz, D. Braun, G. Hellmann and H. Pasch, *Macromol. Chem. Phys.*, 202 (2001) 2890.
275   P.F.W. Simon, A.H.E. Muller and T. Pakula, *Macromolecules*, 34 (2001) 1677.
276   J.S. Schulze, B. Moon, T.P. Lodge and C.W. Macosko, *Macromolecules*, 34 (2001) 200.
277   A. Faraone, S. Magazu, G. Maisano, V. Villari and G. Maschio, *Macromol. Chem. Phys.*, 200 (1999) 1134.

278 R. Mendichi and A.G. Schieroni, *J. Appl. Polym. Sci.*, 68 (1998) 1651.
279 M. Stogiou, C. Kapetanaki and H. Iatrou, *Int. J. Polym. Anal. Chem.*, 7 (2002) 273.
280 C.M. Gomez, R. Garcia, I. Recalde and A. Codoner, *Int. J. Polym. Anal. Chem.*, 6 (2001) 365.
281 D.A. Estenoz, J.R. Vega, H.M. Oliva and G.R. Meira, *Int. J. Polym. Anal. Chem.*, 6 (2001) 315.
282 J.R. Vega, D.A. Estenoz, H.M. Oliva and G.R. Meira, *Int. J. Polym. Anal. Chem.*, 6 (2001) 339.
283 S. Park, D. Cho, J. Ryu, K. Kwon, T. Chang and J. Park, *J. Chromatogr. A*, 958 (2002) 183.
284 T.Y. Chang, H.C. Lee, W. Lee, S. Park and C.H. Ko, *Macromol. Chem. Phys.*, 200 (1999) 2188.
285 J.P. Busnel, F. Foucault, L. Denis, W. Lee and T. Chang, *J. Chromatogr. A*, 930 (2001) 61.
286 D. Cho, S. Park, T. Chang, K. Ute, I. Fukuda and T. Kitayama, *Anal. Chem.*, 74 (2002) 1928.
287 S.H. Nguyen and D. Berek, *Int. J. Polym. Anal. Chem.*, 7 (2002) 52.
288 D. Berek, *Polimery*, 46 (2001) 777.
289 D.E. Niehaus and C. Jackson, *Polymer*, 41 (2000) 259.
290 A. Eceiza, J. Zabala, J.L. Egiburu, M.A. Corcuera, I. Mondragon and J.P. Pascault, *Eur. Polym. J.*, 35 (1999) 1949.
291 V. Girardon, M. Tessier and E. Marechal, *Eur. Polym. J.*, 34 (1998) 1325.
292 C. Puglisi, F. Samperi, S. Carroccio and G. Montaudo, *Rapid Commun. Mass Spectrom.*, 13 (1999) 2260.
293 G. Montaudo, S. Carroccio and C. Puglisi, *Polym. Degrad. Stabil.*, 77 (2002) 137.
294 G. Montaudo and R.P. Lattimer, *Mass Spectrometry of Polymers*, CRC Press, Boca Raton, 2002.
295 G. Adamus, W. Sikorska, M. Kowalczuk, M. Montaudo and M. Scandola, *Macromolecules*, 33 (2000) 5797.
296 G. Impallomeni, M. Giuffrida, T. Barbuzzi, G. Musumarra and A. Ballistreri, *Biomacromolecules*, 3 (2002) 835.
297 A.H. Arkin, B. Hazer, G. Adamus, M. Kowalczuk, Z. Jedlinski and R.W. Lenz, *Biomacromolecules*, 2 (2001) 623.
298 M.L. Focarete, A. Scandola, D. Jendrossek, G. Adamus, W. Sikorska and M. Kowalczuk, *Macromolecules*, 32 (1999) 4814.
299 Z.G. Arkin, J. Rydz, G. Adamus and M. Kowalczuk, *J. Biomater. Sci. Polym. Ed.*, 12 (2001) 297.
300 M. Bankova, A. Kumar, G. Impallomeni, A. Ballistreri and R.A. Gross, *Macromolecules*, 35 (2002) 6858.
301 F.L. Duivenvoorde, J.J.G.S. van Es, C.F. van Nostrum and R. van der Linde, *Macromol. Chem. Phys.*, 201 (2000) 656.
302 S.M. Hunt and M.M. Sheil, *Eur. Mass Spectrom.*, 4 (1998) 475.
303 K. Izawa, T. Ogasawara, H. Masuda, H. Okabayashi and I. Noda, *Macromolecules*, 35 (2002) 92.
304 M.W.F. Nielen and F.A. Buijtenhuijs, *Anal. Chem.*, 71 (1999) 1809.
305 N. Funasaki, *Adv. Colloid Interface Sci.*, 43 (1993) 87.
306 P. Andrews, *Methods Biochem. Anal.*, 18 (1970) 1.
307 F.E. Regnier, *Science*, 222 (1983) 245.
308 F.E. Regnier, *J. Chromatogr.*, 418 (1987) 115.
309 A.C. Gavin, M. Bosche, R. Krause, P. Grandi *et al. Nature*, 415 (2002) 141.
310 B.Y. Lu and J.Y. Chang, *Biochem. J.*, 364 (2002) 81.
311 W. Kalback, M.D. Watson, T.A. Kokjohn, Y.M. Kuo, N. Weiss, D.C. Luehrs, J. Lopez, D. Brune, S.S. Sisodia, M. Staufenbiel, M. Emmerling and A.E. Roher, *Biochemistry*, 41 (2002) 922.

312   N.J. Clarke, F.W. Crow, S. Younkin and S. Naylor, *Anal. Biochem.*, 298 (2001) 32.

313   A. Karlsson and F. Nyberg, *J. Chromatogr. A*, 893 (2000) 107.

314   M.L. Shen, L.M. Benson, K.L. Johnson, J.J. Lipsky and S. Naylor, *J. Am. Soc. Mass Spectrom.*, 12 (2001) 97.

315   W.J. Underberg, M.A. Hoitink, J.L. Reubsaet and J.C. Waterval, *J. Chromatogr. B*, 742 (2000) 401.

316   A. Adachi, Y. Zhao, T. Yamaguchi and S. Surrey, *J. Biol. Chem.*, 275 (2000) 12424.

317   A. Hardy, E. Martinez, D. Diago, R. Diaz, D. Gonzalez and L. Herrera, *J. Biotechnol.*, 77 (2000) 157.

318   B. Plancon, C. Janmot, M. le Maire, M. Desmadril, M. Bonhivers, L. Letellier and P. Boulanger, *J. Mol. Biol.*, 318 (2002) 557.

319   H.K. Kim and T.G. Park, *Biotechnol. Bioeng.*, 65 (1999) 659.

320   C.A. Farrugia and M.J. Groves, *J. Pharm. Pharmacol.*, 51 (1999) 643.

321   J.M. Moore, T.W. Patapoff and M.E. Cromwell, *Biochemistry*, 38 (1999) 13960.

322   M. Le Maire, P. Champeil and J.V. Moller, *Biochim. Biophys. Acta*, 1508 (2000) 86.

323   V.S. Hung, P. Low and E. Swiezewska, *Anal. Biochem.*, 289 (2001) 36.

324   B. Pradet-Balade, F. Boulme, H. Beug, E.W. Mullner and J.A. Garcia-Sanz, *Trends Biochem. Sci.*, 26 (2001) 225.

325   A. Kumar and M. Snyder, *Nature*, 415 (2002) 123.

326   E.A. Golemis, K.D. Tew and D. Dadke, *Biotechniques*, 32 (2002) 636–638, 640, 642 passim.

327   C.G. Fraga, B.J. Prazen and R.E. Synovec, *Anal. Chem.*, 73 (2001) 5833.

328   S. Wang and F.E. Regnier, *J. Chromatogr. A*, 913 (2001) 429.

329   J.M. Hille, A.L. Freed and H. Watzig, *Electrophoresis*, 22 (2001) 4035.

330   M. Stromqvist, *J. Chromatogr. A*, 667 (1994) 304.

331   Z. Zhang, D.L. Smith and J.B. Smith, *Proteomics*, 1 (2001) 1001.

332   K. Maithal, H.G. Krishnamurty and K. Muralidhar, *Indian J. Biochem. Biophys.*, 38 (2001) 368.

333   P.T. Campana, D.I. Moraes, A.C. Monteiro-Moreira and L.M. Beltramini, *Eur. J. Biochem.*, 268 (2001) 5647.

334   C. Tragas and J. Pawliszyn, *Electrophoresis*, 21 (2000) 227.

335   S. Roje, H. Wang, S.D. McNeil, R.K. Raymond, D.R. Appling, Y. Shachar-Hill, H.J. Bohnert and A.D. Hanson, *J. Biol. Chem.*, 274 (1999) 36089.

336   V. Klyushnichenko, A. Bernier, A. Kamen and E. Harmsen, *J. Chromatogr. B*, 755 (2001) 27.

337   I. Ghosh, P. Bishop and J. Chmielewski, *J. Pept. Res.*, 57 (2001) 354.

338   M. Pacek, G. Konopa and I. Konieczny, *J. Biol. Chem.*, 276 (2001) 23639.

339   H. Shirakawa, J.E. Herrera, M. Bustin and Y. Postnikov, *J. Biol. Chem.*, 275 (2000) 37937.

340   R.K. Meyer, A. Lustig, B. Oesch, R. Fatzer, A. Zurbriggen and M. Vandevelde, *J. Biol. Chem.*, 275 (2000) 38081.

341   B. Matthey, A. Engert and S. Barth, *Int. J. Mol. Med.*, 6 (2000) 509.

342   H.M. Pang and E.S. Yeung, *Nucleic Acids Res.*, 28 (2000) E73.

343   S. Takenaka and H. Kondo, in A.M. Griffin and H.G. Griffin (Eds.), *Molecular Biology: Current Innovations and Future Trends*, Horizon Scientific, Norfolk, 1995.

344   Y.C. Guillaume, F.X. Perrin, Ch. Guinchard, L. Nicod, T.T. Truong, A. Xicluna, J. Millet and M. Thomassin, *Anal. Chem.*, 74 (2002) 1217.

345   Y.C. Guillaume, J.F. Robert and C. Guinchard, *Anal. Chem.*, 73 (2001) 3059.

346   Y.C. Guillaume, M. Thomessin and C. Guinchard, *J. Chromatogr. Sci.*, 39 (2001) 361.

347   K. Kasai, *J. Chromatogr.*, 618 (1993) 203.

348   D.J. Mahoney, R.T. Aplin, A. Calabro, V.C. Hascall and A.J. Day, *Glycobiology*, 11 (2001) 1025.

349 M.J. Deery, E. Stimson and C.G. Chappell, *Rapid Commun. Mass Spectrom.*, 15 (2001) 2273.

350 N.S. Fedarko, *EXS*, 70 (1994) 9.

351 A. Jacobs and O. Dahlman, *Biomacromolecules*, 2 (2001) 894.

352 D. Ekeberg, S.H. Knutsen and M. Sletmoen, *Carbohydr. Res.*, 334 (2001) 49.

353 F.P. Abramson, G.E. Black and P. Lecchi, *J. Chromatogr. A*, 913 (2001) 269.

354 M.M. Huisman, L.P. Brul, J.E. Thomas-Oates, J. Haverkamp, H.A. Schols and A.G. Voragen, *Carbohydr. Res.*, 330 (2001) 103.

355 J. Zaia and C.E. Costello, *Anal. Chem.*, 73 (2001) 233.

356 A. Jacobs, P.T. Larsson and O. Dahlman, *Biomacromolecules*, 2 (2001) 979.

357 B. Finke, B. Stahl, M. Pritschet, D. Facius, J. Wolfgang and G. Boehm, *J. Agric. Food Chem.*, 50 (2002) 4743.

358 R. Walkenhorst, *LC-GC Europe*, November 1 (2001) 2.

359 W.W. Christie, *Lipid Technol.*, 7 (1995) 17.

360 C.M. Dobarganes and G. Marquez-Ruiz, in W.W. Christie (Ed.), *Advances in Lipid Methodology-Two*, Oily Press, Dundee, 1993, pp. 113–137.

361 C. Dauwe, G. Reinhold and O. Okogeri, *LC-GC Europe*, December (2001) 22.

362 O.A. Bizzozero, S.P. Malkoski, C. Mobarak, H.A. Bixler and J.E. Evans, *J. Neurochem.*, 81 (2002) 636.

363 T. Pfeifer, U. Klaus, R. Hoffmann and M. Spiteller, *J. Chromatogr. A*, 926 (2001) 151.

364 T. Myllykangas, T.K. Nissinen, P. Rantakokko, P.J. Martikainen and T. Vartiainen, *Water Res.*, 36 (2002) 3045.

365 S. Wong, J.V. Hanna, S. King, T.J. Carroll, R.J. Eldridge, D.R. Dixon, B.A Bolto, S. Hesse, G. Abbt-Braun and F.H. Frimmel, *Environ. Sci. Technol.*, 36 (2002) 3497.

366 Ch.H. Specht and F.H. Frimmel, *Environ. Sci. Technol.*, 34 (2000) 2361.

367 J.E. Kilduff, T. Karanfil, Y.-P. Chin and W.J. Weber, *J. Environ. Sci. Technol.*, 30 (1996) 1336.

368 S. Wong, J. Kim, A.T. Lemley, S.K. Obendorf and A. Hedge, *J. Chromatogr. Sci.*, 39 (2001) 101.

369 J.J. Stry, J.S. Amoo, S.W. George, T. Hamilton-Johnson and E. Stetser, *J. AOAC Int.*, 83 (2000) 651.

370 B. Jongenotter and H.-G. Janssen, *LC-GC Europe*, 1 June (2002).

371 E. Jover and J.M. Bayona, *J. Chromatogr. A*, 950 (2002) 213.

372 G.-T. Wei, F.-K. Liu and C.R.C. Wang, *Anal. Chem.*, 71 (1999) 2085.

373 M. Domin, A. Herod, R. Kandiyoti, J.W. Larsen, M-J. Lazaro, S. Li and P. Rahimi, *Energy Fuels*, 13 (1999) 552.

374 C.C. Loa, T.L. Lin, C.C. Wu, T.A. Bryan, H.L. Thacker, T. Hooper and D. Schrader, *J. Virol. Methods*, 104 (2002) 187.

Erich Heftmann (Editor)
Chromatography, 6th edition
Journal of Chromatography Library, Vol. 69A
253

Chapter 6

Planar chromatography

SZABOLCS NYIREDY

CONTENTS

6.1 Introduction 254
6.2 Classification of planar chromatographic techniques 255
6.3 Principles of planar chromatographic methods 257
    6.3.1 Capillary flow-controlled planar chromatography 257
    6.3.2 Forced-flow planar chromatography 257
        6.3.2.1 Overpressured-layer chromatography 257
        6.3.2.2 Rotation planar chromatography 257
        6.3.2.3 Electro planar chromatography 258
6.4 Principal factors in planar chromatography 258
    6.4.1 Stationary phase 258
    6.4.2 Sample amount and application 260
    6.4.3 Solvent system and its optimization 261
    6.4.4 Chamber type and vapor phase 262
        6.4.4.1 N-chambers 264
        6.4.4.2 S-chambers 264
        6.4.4.3 Chambers for forced flow 266
    6.4.5 Flow velocity 266
    6.4.6 Development mode 267
        6.4.6.1 Linear development mode 267
        6.4.6.2 Circular development mode 268
        6.4.6.3 Anticircular development mode 269
        6.4.6.4 Multiple development modes 269
        6.4.6.5 Multidimensional development mode 271
    6.4.7 Separation distance and time 272
    6.4.8 Temperature 273
6.5 Instrumentation 273
    6.5.1 Overpressured-layer chromatography 273
    6.5.2 Rotation planar chromatography 274
    6.5.3 Electro planar chromatography 276
    6.5.4 Automated multiple development 276

6.6   Qualitative and Quantitative Analysis 277
      6.6.1   Identification of separated compounds 277
              6.6.1.1   Chromatographic data 277
              6.6.1.2   Pre- and post-chromatographic derivatization 279
              6.6.1.3   Optical spectroscopy 279
              6.6.1.4   Infrared and Raman spectroscopy 280
              6.6.1.5   Mass spectrometry 281
              6.6.1.6   Nuclear magnetic resonance spectroscopy 281
              6.6.1.7   Digital autoradiography 281
      6.6.2   Quantitative determination 282
              6.6.2.1   Measurements after elution 282
              6.6.2.2   *In situ* measurements 282
6.7   Preparative planar chromatography 282
      6.7.1   Off-line separation 283
      6.7.2   On-line separation 284
      6.7.3   Selection of the appropriate method 284
6.8   Special planar chromatographic techniques 285
      6.8.1   Combination with flame-ionization detection 285
      6.8.2   Sequential techniques 286
      6.8.3   Mobile-phase gradient 286
      6.8.4   Layer-thickness gradient 286
6.9   Comparison of various planar chromatographic techniques 287
6.10 Trends in planar chromatography 287
      6.10.1 Development of instrumentation 287
      6.10.2 Development of forced-flow planar chromatographic methods . . 289
      6.10.3 Multidimensional planar chromatography 290
      6.10.4 Multimodal separations by planar chromatography 290
References 291

6.1 INTRODUCTION

Planar chromatography (PC) is a collective term including all analytical, micropreparative, and preparative separation methods where the mobile phase moves through the stationary phase (porous sorbent) in a planar arrangement. The movement of compounds to be separated by PC is the result of two opposing forces, the driving force of the solvent system and the retarding action of the stationary phase [1]. The planar geometry (flat bed) has several advantages, such as simplicity, flexibility, parallel analysis of a large number of samples, various development modes, as well as the applicability of selective and specific chemical and biological detection methods. A disadvantage of PC is that a certain amount of skill and experience is required to derive full benefit from its possibilities, due to the relatively large number of parameters influencing the result [2].

In the present chapter on PC only a short mention is made of paper chromatography, since this technique has been eclipsed by the faster, more efficient, and more sensitive method of thin-layer chromatography (TLC), and because both TLC and paper chromatography are very closely related in their technical aspects [3]. Because it is difficult to find a sharp borderline between classical TLC and high-performance TLC (HPTLC), both are discussed side by side in this chapter. The term mobile phase is more appropriate in forced-flow planar chromatography (FFPC), where it migrates in a closed system; hence the term solvent system [2] should be used for classical TLC/HPTLC. Three topics are discussed in greater detail in this chapter: The various chamber types and development modes, which are only practicable in PC, and the newer FFPC methods. The topic of PC has been surveyed in detail for paper chromatography [4,5], classical TLC [6–12], and HPTLC [13–16]. The principles and possibilities of the FFPC techniques and the state of the art of modern planar chromatography have also been reviewed recently [18–22]. The fundamentals of modern PC are discussed in detail in the textbook by Geiss [1]. The state of art of planar chromatography – a retrospective view for the third millennium was recently summarized in 29 chapters [23].

6.2 CLASSIFICATION OF PLANAR CHROMATOGRAPHIC TECHNIQUES

Planar chromatography can be classified from different points of view, *e.g.*, (a) the flow of the solvent system/mobile phase, (b) the nature of the stationary phase, (c) the mechanism of the separation, (d) the polarity relation between the mobile and stationary phase, (e) the aim of the separation, and (f) whether the principal steps are performed as separate operations or not.

(a) The solvent can migrate through the stationary phase by capillary action or under the influence of forced flow. Forced flow can be achieved either by application of external pressure [overpressured-layer chromatography (OPLC)] [24–28], an electric field [electro planar chromatography (EPC)] [17,29–31], or by centrifugal force [rotation planar chromatography (RPC)] [32–36]. The superior efficiency of FFPC techniques, comparing their analytical properties with those of classical TLC and HPTLC [25], are schematically demonstrated in Fig. 6.1. FFPC techniques permit the advantage of optimal mobile-phase velocity to be exploited over practically the whole separation distance without loss of resolution. This effect is independent of the type of forced flow and the layer thickness used.

(b) PC can also be classified as TLC and HPTLC, based on the average particle size. Pre-coated TLC uses an average particle size of *ca.* 11 μm, ranging from 3 to 18 μm. The average particle size of HPTLC plates is now 5–6 μm with a very small range of particle sizes. In both categories, the solvent migrates through a layer of porous stationary phase, which is generally bound to an inert support in a planar configuration.

(c) Depending on its nature, the stationary phase promotes the separation of compounds by adsorption, partition, ion-exchange, or size-exclusion separation processes. Some types of stationary phases cannot be classified into these four basic classes, and separations actually involve a combination of these four basic mechanisms.

(d) In normal-phase (NP) PC the sorbent is more polar than the solvent system, whereas in reversed-phase (RP) PC the stationary phase is less polar than the mobile phase.

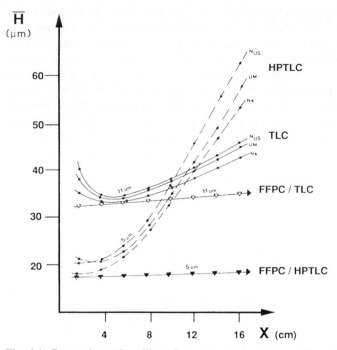

Fig. 6.1. Comparison of capillary flow-controlled and forced-flow planar chromatographic methods for different plates and chamber types. The developing distance from the start to the solvent front (X) is plotted against plate height (H). N_{US} = normal, unsaturated chamber, UM = ultramicrochamber, N_S = normal, saturated chamber.

    (e) An aim of a separation may be analysis for the qualitative assay or quantitative determination of the separated compounds. Another goal of PC may be a preparative separation for the isolation or purification of substances.

    (f) PC may be also classified as an off-line or an on-line separation technique. Classical TLC for analytical purposes is a typical fully off-line process, where the principal steps of sample application, development, evaporation of the solvent system, and densitometric evaluation are performed as separate operation steps. In preparative applications the separated substances are not subjected to *in situ* quantification, but instead, the zones are scratched off the support and the separated compounds are eluted from the sorbent, using a solvent of high solvent strength.

In the fully on-line mode the principal steps are not performed as separate operations; rather, the separated compounds are drained from the layer [37,38]. If an FFPC system is equipped with an injector and a detector, the various off-line and on-line operations can also be combined. Possible combinations are off-line sample application and on-line separation/detection, or on-line sample application/separation and off-line detection. OPLC can be used for fully on-line analytical separations, while for preparative separations not only OPLC but also RPC may be used [14] by connection to a flow-through detector, recorder and/or by collection of isolated compounds with a fraction collector.

6.3 PRINCIPLES OF PLANAR CHROMATOGRAPHIC METHODS

6.3.1 Capillary flow-controlled planar chromatography

The driving force for solvent migration in classical TLC is the decrease in free energy of the liquid as it enters the porous structure of the layer; the transport mechanism is the result of the capillary effects [39]. Under these conditions the velocity at which the solvent moves is a function of the distance of the solvent front from the solvent entry position: the velocity declines at this distance increases. The consequences of this effect are the following:

- the solvent system velocity varies as a function of time and migration distance, therefore, capillary forces are inadequate for achieving the desired optimum velocity;
- the solvent system velocity is established by the system variables and is otherwise beyond experimental control. The solvent system velocity can only be reproduced by careful attention to experimental details.

Classical TLC or HPTLC is a fully off-line process, because the principal steps are performed as separate operations; quantitative determination is usually performed *in situ* on the plate.

6.3.2 Forced-flow planar chromatography

6.3.2.1 Overpressured-layer chromatography

Apart from capillary action, the driving force for solvent migration in OPLC is external pressure [24]. Depending on the desired mobile-phase velocity, low (2–5 bar), medium (10–30 bar), and high (50–100 bar) operating pressures can be used [31]. In OPLC the vapor phase is completely eliminated, the chromatoplate being covered with an elastic membrane under external pressure; thus, the separation can be carried out under controllable conditions. The principle of the method is shown in the schematic drawing of the instrument in Sec. 6.5.1. The separation may be started with a dry layer, as in classical TLC, but the closed system also permits performance of a fully on-line separation, where the separation can be started with a mobile-phase-equilibrated stationary-phase system, as in high-performance liquid chromatography [40].

6.3.2.2 Rotation planar chromatography

In RPC the driving force for solvent migration is the centrifugal force, in addition to capillary action. The samples are applied to the rotating stationary phase near the center. The centrifugal force drives the mobile phase through the sorbent from the center to the periphery of the plate. For analytical purposes (Fig. 6.1a), up to 72 sample can be applied, and quantification can be carried out *in situ* on the plate [34–36]. For micropreparative and

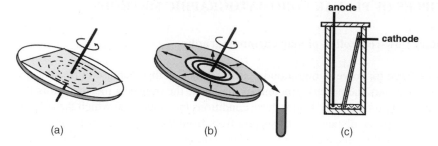

Fig. 6.2. Principle of (a) analytical RPC, (b) preparative RPC, and (c) analytical EPC.

preparative purposes only one sample is applied as a circle; the separations can be carried out either in the off-line or on-line mode. In the latter, the separated compounds are eluted from the stationary phase by the centrifugal force and collected in a fraction collector, as shown in Fig. 6.2b [41].

6.3.2.3 Electro planar chromatography

In EPC – earlier name: high-speed TLC [29] – in addition to capillary action, the driving force for solvent migration is the electric field, with a voltage gradient greater than 1 kV/cm [42]. During EPC, the components of the sample are separated simultaneously by two processes, electrophoresis, and adsorption. The use of an electric field results in the reduction of analysis time and a higher theoretical-plate number; therefore better separation can be achieved [30]. The principle of EPC is shown in Fig. 6.2c.

6.4 PRINCIPAL FACTORS IN PLANAR CHROMATOGRAPHY

The chromatographic processes in analytical or preparative PC are basically the same, regardless of whether the driving force is capillary action alone or augmented by additional forces. Among the factors which may influence PC separations, the chamber types and development modes deserve special treatment.

6.4.1 Stationary phase

Most users prefer the commercially available precoated layers to preparing their own plates. Besides saving time, precoated layers have the advantage of much higher reproducibility than can be obtained with self-prepared plates. Generally, two types of pre-coated plates are available, *viz.* unmodified sorbents and modified silicas in layer thicknesses of 0.1–2 mm. For analytical separations the layer thickness should be between 0.1–0.25 mm, whereas for preparative purposes 0.5-mm, 1-mm, and 2-mm layer thicknesses are available.

The unmodified stationary phases include silicas, aluminas, Kieselguhr, silicates, controlled-porosity glass, cellulose, starch, gypsum, polyamides, and chitin [43]. For TLC and HPTLC the most frequently used stationary phase is silica. It is prepared by spontaneous polymerization and dehydration of aqueous silicic acid, which is generated by adding acid to a solution of sodium silicate. The product of this process is an amorphous, porous solid, the specific surface area of which can vary over a wide range (200 to more than 1000 m²/g), as can the average pore diameter (10–1500 Å) [2,44]. Modified silicas may be nonpolar or polar sorbents. The former class includes silicas bearing alkane chains or phenyl groups, while the polar modified silicas contain cyano, diol, amino, or thiol groups or substance-specific complexing ligands [45,46]. The structures of some chemically modified silicas are shown in Fig. 6.3.

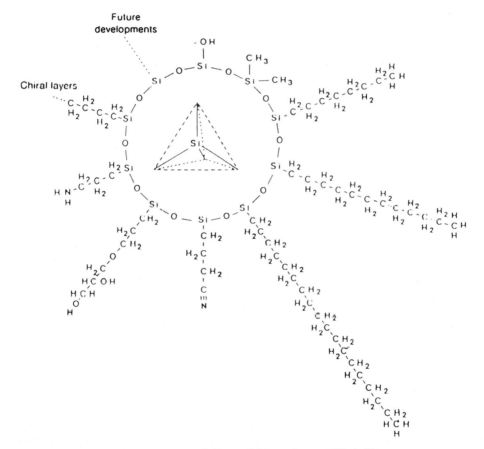

Fig. 6.3. Structures of some commercially available surface-modified silicas.

It is generally accepted that resolution is higher on a thinner layer (0.1 mm), but this effect is also dependent on the detection mode [47]. Commercially available analytical thin-layer plates have dimensions of 10 × 10, 10 × 20, or 20 × 20 cm. The silica materials commonly used for pre-coated plates have an average particle size of *ca.* 11 μm, ranging from 3 to 18 μm; for self-prepared analytical layers the average particle size is 15 μm, and the range of particle sizes is much greater. The average particle size of pre-coated HPTLC plates is now 5–6 μm, with a very small range of particle sizes. Various pre-coated analytical layers with a pre-adsorbent zone are also commercially available for linear development. This zone serves to hold the sample until development begins. Compounds soluble in the solvent system pass through the pre-adsorbent zone and are concentrated in a narrow band before entering the chromatographic layer proper, and this improves their resolution. Just about all the stationary phases used in NP- and RP-HPLC are now also available for PC [1,48].

6.4.2 Sample amount and application

The process of sample application is one of the most important steps for a successful PC separation [49]. The aim of sample application is quantitative transfer of the sample to the layer in such a way that the sample penetrates the layer to form a compact zone of minimum size without causing damage to the layer. The sample should be dissolved in a nonpolar, volatile solvent in such a concentration that the components of the sample are adsorbed throughout the whole thickness of the plate and not only on the surface of the layer. Local overloading may distort the applied bands, since the rate of dissolution of the components in the solvent system will then become a limiting factor [11].

Samples can be applied by contact transfer as spots or by spraying as bands or rectangles. Narrow bands as starting zones ensure the highest resolution, even when a large number of substances with minimal differences between migration characteristics must be distinguished. For spot-wise application, a peak capacity value of 9.5 is calculated for a track length of 50 mm, compared with a peak capacity value of 14.4 for band-wise application. Because mass distribution is uniform over the full length of the bands, densitometric evaluation can be performed by aliquot scanning; this ensures maximum quantitative accuracy. Spot broadening in the direction of development is smaller for bands than for spots. Overloading of the stationary phase, which can lead to spot tailing, can thus be avoided by use of this technique [50]. Spraying in the form of rectangles enables the application of large volumes [51]. Before developing the plate, the rectangles are focused into narrow bands by means of a solvent of high solvent strength. Spraying with this technique enables the application of larger volumes.

Among the various applicators designed for the deposition of small sample volumes on the chromatoplates, the microcapillary (especially Pt-Ir capillary) is one of the simplest and most useful. An alternative to these capillaries is the microsyringe, which offers greater flexibility in the choice of sample volume. Use of automatic applicators enables rapid and reproducible sample application. In automated instruments for sample streaking the plate moves beneath a syringe, containing the sample solution. An atomizer, operating with a controlled stream of nitrogen, sprays the sample from the syringe, forming narrow spots (with a stationary plate) or homogeneous bands (with a moving plate) on the stationary phase [50].

6.4.3 Solvent system and its optimization

Solvent system optimization can be performed by trial and error, solvent selection being based both on the analyst's experience and intuition and on modifications of published data. However, as the sample composition becomes more complex, systematic solvent optimization becomes more important. The methods used for optimizing isocratic mobile phases in HPLC (Chap. 1) are generally also applicable, with some modifications, to PC. Window diagrams have been successfully applied to the optimization of PC solvent systems [52–54]. Similarly, overlapping-resolution maps were used as criterion by Issaq *et al.* [55] and by Nurok *et al.* [56], who also extended this approach to continuous development. The fruitful application of the sequential simplex method was also reported [57,58].

Geiss [2] has suggested a structural approach, which assumes that selectivity and solvent strength are independent variables. For this optimization process he used the Vario KS chamber (Sec. 6.4.4.2), and three strong solvents (methyl *tert*-butyl ether, acetonitrile, and methanol). All three solvents were diluted with a suitable amount of a weaker fourth solvent (F-113 or 1,2-dichloroethane) to obtain a series of solutions spanning the solvent strength (e^0) range from 0.0 to 0.70 in increments of 0.05 e^0. In the next step, the appropriate solvent strength must be determined. Once this has been identified, fine-tuning is accomplished by blending solvent mixtures of this strength but of different selectivity. Many elegant separations have been achieved in this manner, but this method reduces the number of solvents available for optimization.

On the basis of Snyder's solvent characterization [59], the solvent optimization method, called the "PRISMA" system [60–65] has been developed. It consists of three parts: In the first part, the basic parameters, such as the stationary phase and the individual solvents are selected by TLC. In the second part, the optimal combination of these solvents is selected by means of the "PRISMA" model. The third part of the system includes selection of the appropriate FFPC technique (OPLC or RPC) and HPTLC plates, selection of the development mode, and finally, application of the optimized mobile phase in the various analytical and preparative chromatographic techniques. The tripartite "PRISMA" system is shown in Fig. 6.4.

After selection of the stationary phase, the recommended initial optimization step in the first part of the "PRISMA" system is TLC with each of ten neat solvents – characterized by their solvent strength (s_i) and selectivity value (s_v) [109] – selected on the basis of their miscibility with hexane and representing at least one member of each of Snyder's eight selectivity groups (Table 6.1). For the separation of nonpolar compounds, the solvent strength can be decreased with hexane; for the separation of polar compounds the solvent strength can be increased by adding water or another polar solvent in low concentration, to bring the R_F values of the compounds to be separated into a range of 0.2–0.8. Generally, the solvents giving the best resolutions are then selected for further optimization. After selection of the solvents, the second part of the system, the construction of the "PRISMA" model, is begun. Generally, between two and five solvents may be selected; modifiers may also be added. The tripartite "PRISMA" model is a three-dimensional geometric construction, which correlates the solvent strength with the selectivity of the solvent system. As a rule, in normal-phase PC, the

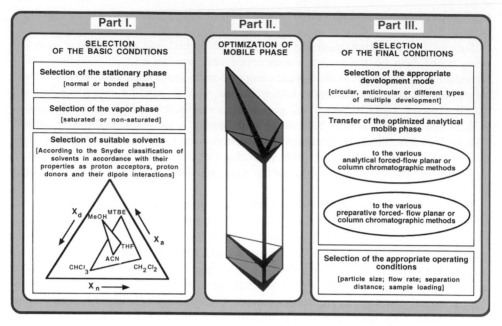

Fig. 6.4. The "PRISMA" system in planar chromatography. (Reprinted from Ref. 70 with permission.)

upper frustrum is used for solvent optimization for polar compounds, the center portion of the prism is used for solvent optimization for nonpolar compounds, and the lower part symbolizes the modifiers. The third part of the procedure includes selection of the development mode (circular, linear, or anticircular), the appropriate forced-flow chromatographic technique (OPLC or RPC) with HPTLC plates, transfer of the optimized mobile phase to the various analytical and/or planar or column preparative liquid-chromatographic techniques, and selection of the operating conditions. The "PRISMA" system offers special possibilities for solving difficult PC separation problems. For details the reader is referred to Ref. 55, where the whole optimization process is presented with the help of flow charts. Strategies for optimizing the solvent systems for PC, including the two-dimensional TLC separations, have been summarized by Geiss [2], Nurok [66,67], as well as by Siouffi and Abbou [68].

6.4.4 Chamber type and vapor phase

Selection of the chamber type and vapor space is a variable offered only by PC, because the separation process occurs in a three-phase system of stationary, mobile, and vapor phases, all of which interact with one another until equilibrium is reached [69]. Basically, one can distinguish between the normal (N) chamber and the sandwich (S) chamber. In the common N-chamber there is a distance of more than 3 mm between the layer and the wall (or between the layer and the lid of the chamber in horizontal development) of the

TABLE 6.1

SOLVENTS PROPOSED FOR MOBILE-PHASE OPTIMIZATION IN PC

Group	Solvent	Solvent strength (S_i)	Selectivity value (S_v)
–	n-Hexane	*ca.* 0	*ca.* 0.01
I	n-Butyl ether	2.1	2.44
	Diisopropyl ether	2.4	3.43
	Methyl butyl ether	2.7	3.50
	Diethyl ether*	2.8	4.08
II	n-Butanol	3.9	3.11
	2-Propanol*	3.9	2.89
	1-Propanol	4.0	2.84
	Ethanol*	4.3	2.74
	Methanol	5.1	2.18
III	Tetrahydrofuran*	4.0	1.90
	Pyridine	5.3	1.85
	Methoxyethanol	5.5	1.59
	Dimethylformamide	6.4	1.86
IV	Acetic acid*	6.0	1.26
	Formamide	9.6	1.57
V	Dichloromethane*	3.1	1.61
	1,1-Dichloroethane	3.5	1.43
VI	Ethyl acetate*	4.4	1.48
	Methyl ethyl ketone	4.7	1.59
	Dioxane*	4.8	1.50
	Acetone	5.1	1.52
	Acetonitrile	5.8	1.15
VII	Toluene*	2.4	0.89
	Benzene	2.7	0.72
VIII	Chloroform*	4.1	0.61
	Water	10.2	1.00

*Proposed solvents for the first experiments for mobile-phase optimization wher using the "PRISMA" optimization system.

chromatographic tank. If this distance is smaller, the chamber is said to have the S-configuration [70]. Both types of chromatographic chamber can be used for unsaturated or saturated systems. Although the chambers used for FFPC separation can also be assigned to the above two categories, their special features warrant a separate discussion of FFPC-chambers.

6.4.4.1 N-chambers

Although a wide variety of chromatographic chambers are available, rectangular glass N-chambers with internal dimensions of $23 \times 23 \times 8$ cm or $13 \times 13 \times 5$ cm are most frequently used in PC for development of two 20×20-cm or 10×10-cm plates, respectively. Starting the separation with unsaturated tanks generally gives higher R_F values for NP systems because of the evaporation of the solvents from the surface of the layer. The disadvantages of using unsaturated tanks is that the reproducibility of the R_F values may be poor and that a concave solvent front can occur, leading to higher R_F values for solutes near the edges [70]. If the layer is placed in the chamber immediately after introducing the solvent systems into the chromatographic tank, separation starts in an unsaturated system (Fig. 6.5a) which will become progressively more saturated in the course of the separation. A chamber is saturated when all components of the solvent are in equilibrium with the entire vapor (gas) space before and during the separation [2]. Therefore, the N-chamber is lined on all four sides with filter paper, thoroughly soaked with the solvent systems (Fig. 6.5b). The prepared tank should stand for 60–90 min to allow the internal atmosphere to become saturated with the vapor of the solvent system. Each plate must lean against a side wall, such a way that plates do not touch each other. The saturated tanks have the advantage, that a front (the first front of a multicomponent mobile phase) is much more regular, and also the separation efficiency is higher when a development distance of 18 cm is used.

A versatile version of the N-tank is the twin-trough chamber [71], which has a raised glass ridge along the center that effectively separates the chromatographic tank into two separate compartments (Fig. 6.5c). To obtain a saturated vapor space, the solvent system is placed in one of the compartments, while the plate to be developed is put in the other one. When the layer has become equilibrated with the vapor phase, the tank is carefully tipped to transfer the solvent system from one compartment to the other and to start the separation [2]. The short-bed continuous development (SB-CD) chamber is a flat N-chamber [72,73], designed for short separation distances and high solvent velocities. Low-strength mobile phases will then resolve slow-moving solutes in a reasonable time. The bottom of the SB-CD chamber has five ridges, serving as stop positions for the plate which leans against the rim, protruding from the chamber (Fig. 6.5d). The solvents migrating up the chromatoplate evaporate from the protruding portion. Because the migration distance, which depends on the angle of the plate, is short and migration is slow, diffusion is decreased and extremely high resolution can be obtained.

6.4.4.2 S-chambers

S-chambers are very narrow, unsaturated tanks; the plate with the layer is usually sandwiched with a glass cover plate (Fig. 6.5e). Saturation can be established with a facing chromatographic plate that has been soaked with the mobile phase (Fig. 6.5f). Part of the stationary phase of the plate to be developed is removed by scraping, so that, initially, the mobile phase can only reach the level of the facing plate. After sorptive and capillary saturation of this plate, the depth of the mobile phase is

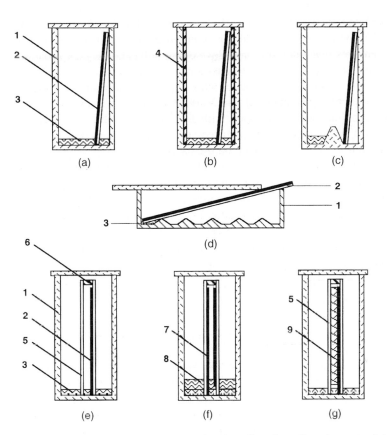

Fig. 6.5. Chromatographic chambers. 1 = Chamber, 2 = chromatoplate, 3 = mobile phase, 4 = soaked filter paper, 5 = glass cover plate, 6 = spacer, 7 = counter chromatoplate, 8 = additional mobile phase, 9 = elastic inert material.

increased to start the separation. The Brenner and Niederwieser (BN) chamber [74,75] is a special case of the S-chamber for linear, continuous development, where the solvent system and the compounds advance on the plate practically continuously at a constant velocity. The top part of the chromatoplate does not face another plate but rests on a heated metal block. Once the α-front has reached the heated block, the solvent system starts to evaporate. The BN-chamber is suitable for the separation of compounds which otherwise would have very low R_F values. The ultra-micro (UM) chamber is a special variety of S-chamber, in which the vapor space is reduced [76]. Between the chromatographic plate and the sandwiched glass cover plate an elastic inert material (e.g., Teflon) is placed, and the plates are pressed together (Fig. 6.5g).

The most versatile S-chamber, which can also be used in the N-chamber configuration, is the Vario-KS chamber, devised by Geiss and Schlitt [2,77]. It is suitable for evaluation of the effects of different solvents, solvent vapors, and relative humidities. The chromatoplate is placed face down over a glass conditioning tray, containing 5, 10, or 25 compartments to hold the required conditioning solvents. The design of the device ensures that saturating and developing solvents are kept separate. A major advantage of this chamber is that up to ten activity and/or saturating conditions can be compared on the same chromatoplate with the same solvent for solvent optimization purposes. Another advantage of the Vario-KS chamber is that the same chromatoplate can be simultaneously developed with five different solvents. Additionally, the chamber can also be used with a heating accessory for continuous development. The vapor-phase saturation can be characterized on the basis of the saturation grade (S_G) [78]. The S_G value of a given chromatographic chamber can be calculated by dividing the sum of the hR_F values of the three furthest-migrating substances (preferably test dye mixtures) by the sum of the hR_F values of all the components, subtracting the result from 1, and multiplying the answer by 100. The saturation grade can be used as a measure of the reproducibility of separations with given stationary and mobile phases and at different temperatures and humidity; this enables transfer of the mobile-phase data to other vapor-phase conditions.

6.4.4.3 Chambers for forced flow

The chambers used for OPLC separations are unsaturated S-chambers, theoretically and practically devoid of any vapor space. This must be considered in the optimization of the solvent system, especially in connection with the disturbing zone and multifront effect, which are specific features of the absence of a vapor phase [79,80]. The main difference between the chamber types used in RPC lies in the size of the vapor space, which is an essential criterion in rotation planar chromatography [35]. Therefore, we use an additional symbol to indicate the vapor space [normal-chamber, micro-chamber, ultramicro-chamber, and column RPC (N-RPC, M-RPC, U-RPC, and C-RPC, respectively)].

In N-chamber RPC the layer rotates in a stationary N-chamber, where the vapor space is extremely large. Due to extensive evaporation, this chamber is practically unsaturated. The M- and U-chambers in RPC belong to the S-chamber type, the difference between these two chambers being that the former is saturated, while the latter is unsaturated. Since in M-chamber RPC the chromatoplate rotates together with the small chromatographic chamber, where the distance between the layer and the lid of the chamber is smaller than 2 mm, the vapor space is rapidly saturated. In the case of the ultramicrochamber the lid of the rotating chamber is placed directly on the chromatoplate [41] so that practically no vapor space exists.

6.4.5 Flow velocity

In capillary-flow-controlled PC the solvent velocity is the parameter which, in principle, cannot be influenced by the chromatographer. In TLC this depends entirely on the characteristics of the stationary phase and the solvent chosen. The solvent velocity constant can be determined from the time required for the solvent to travel between fixed points, marked on the TLC plate, as the slope of a plot of Z_f^2 against time, *i.e.* based on the linear equation [2,39]

$$Z_f^2 = k \times t \tag{6.1}$$

where: Z_f is the distance migrated by the solvent front from the solvent entry (cm), k is the solvent system velocity constant (cm^2/sec), and t is the time (sec). When FFPC is used in the linear development mode, with a constant-flow pump, a linear relationship exists between the migration distance, the mobile phase velocity, and the time required [40].

$$Z_f = u \times t \tag{6.2}$$

where: Z_f is the distance migrated by the solvent front from the solvent entry (cm), u is the mobile-phase velocity constant (cm/sec), and t is the time (sec).

One possibility of exerting an influence on the flow velocity is to avoid solvents of higher viscosity in solvent optimization. Also, saturated chromatographic systems have the advantage that development is much faster than in unsaturated chambers. This means that the solvent velocity increases in going from the unsaturated N-chamber *via* the unsaturated S(U)-chamber to the saturated S-chamber. The local mobile-phase velocity can be influenced by the selection of the development mode (Sec. 6.4.6). In continuous development (SB-CD chambers), migration of the solvent system starts off relatively fast, but always falls off along the separation distance. Once the α front can be evaporated, the migration rate of the solvent system becomes constant. The highest mobile-phase velocity can be achieved by using FFPC techniques. In OPLC, the upper limit of velocity depends on the applied external pressure, besides the viscosity [81]. In RPC, the higher the rotational speed, the faster is the migration of the mobile phase. The flow-rate is limited by the amount of solvent that may be kept in the layer without skimming over the surface. The amount of migrating solvent can also be increased by scratching a round hole into the center of the layer, varying its perimeter according to the optimal mobile-phase velocity [41,82].

6.4.6 Development mode

PC differs from CLC not only in having a vapor space, but also in permitting a selection of the optimal development mode.

6.4.6.1 Linear development mode

The ascending mode is most frequently used for capillary-flow-controlled PC. Because ascending development has no theoretical advantage over horizontal development, the latter, being more adaptable, has become increasingly common in the recent years. Continuous development can be achieved by allowing the end of the plate to remain uncovered during development (BN-chamber) or to protrude from a slot in the cover of the chamber in the ascending mode (SB-CD-chamber). In either case, the solvent system flows continuously and evaporates from the uncovered area. In OPLC, the most frequent development modes are the linear one- and two-directional

(Fig. 6.6a and b) modes [1]. However, the linear-type OPLC technique requires a special chromatoplate, sealed along the edge by impregnation to prevent the solvent from flowing off the layer. In RPC with a M- or an U-chamber the mobile-phase movement can be also linearized (linear development mode) by scraping the layer to form lanes (Fig. 6.6c) [30].

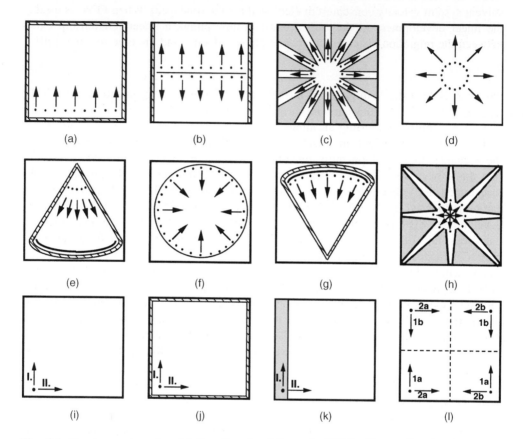

Fig. 6.6. Development modes. (a) One-directional linear in OPLC, (b) two-directional linear in OPLC, (c) linear in RPC, (d) circular in TLC, OPLC, and RPC (e) circular in OPLC, (f) anticircular in TLC, (g) anticircular in OPLC, (h) anticircular in RPC, (i) two-dimensional TLC, (j) two-dimensional OPLC, (k) two-dimensional TLC on a bilayer plate, (l) two-dimensional TLC of 4 samples.

6.4.6.2 Circular development mode

The advantage of circular development (Fig. 6.6d), where the solvent system migrates radially from the center of the plate to the periphery, is well known for the separation of compounds in the lower R_F range [2,70]. Using the same solvent system, the resolution, particularly in the lower R_F range, is about 4−5 times greater in circular than in linear development mode. The separating power of the circular development mode can be better exploited, if the samples are spotted near the center [35]. As the distance between the mobile-phase inlet and sample application increases, the resolution begins to approach that of linear development. For off-line circular OPLC and RPC, no preparation of the plate is necessary (Fig. 6.6d); for on-line circular OPLC, a sector must be isolated by scratching off the surrounding sorbent and impregnating the sector (Fig. 6.6e).

6.4.6.3 Anticircular development mode

Anticircular development can be used in analytical TLC, if the resolution must be increased in the higher R_F range [2,70]. In this development mode, the solvent system enters the layer at a circular line and flows towards the center (Fig. 6.6f). Since the solvent flow velocity decreases with the square of the distance, but the area wetted also decreases with the square of the distance traveled, the speed of solvent system migration is practically constant [83]. Therefore this developing mode is the fastest with respect to separation distance. However, the flow-rate cannot be controlled. Anticircular on-line OPLC development is carried out similar to the circular mode, except the mobile phase enters the plate from the opposite side (Fig. 6.6g). For anticircular ultramicro-chamber RPC separations the plate must be prepared as shown in Fig. 6.6h.

6.4.6.4 Multiple development modes

Multiple development (^nD) includes all development procedures, where development is repeated after a development is completed and the mobile phase has been carefully evaporated [1,84]. From the point of view of development distance and solvent system composition the ^nD techniques can be classified into five basic categories [86].
1) Unidimensional MD (U^nD) means the repeated development of the chromatoplate on the same development distance (D) in the same (S_{T1}, S_{V1}) solvent system (Fig. 6.7). The removal of the solvent system between development steps is performed by careful drying of the chromatoplate. The dried layer is returned to the development chamber for repeated development under the same chromatographic conditions as for earlier development steps. U^nD can be used to increase the resolution. The location of the compounds to be separated, and hence the R_F value, is influenced by the number of developments. After the first development (^1R_F) the R_F values of a multiple developed solute can be predicted by the following equation [72]:

$$^n(R_F) = 1 - (1 - {}^1R_F)^n \qquad (6.3)$$

where n is the number of developments. In this way, the R_F values and thus also the DR_F values can be calculated for all compounds of interest. Needless to say, unidimensional ^nD can also be carried out with FFPC techniques. In this case, the spot capacity increases linearly as a function of the square root of the development distance, instead of going through a maximum, as it does for capillary-controlled flow conditions.

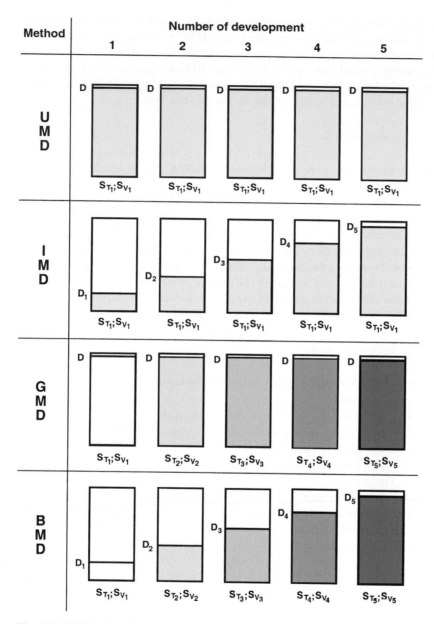

Fig. 6.7. Multiple development techniques.

2) Incremental MD (I^nD) means that the re-chromatography is performed in increasing development distances ($D_1 \rightarrow D_n$) with the same (S_{T1}, S_{V1}) mobile phase (Fig. 6.7). Using the ^nD technique, the first development length is the shortest, the subsequent development steps are accomplished in a (usually by an equal distance) longer

development distance, and the last front migration distance is the longest, corresponding to the useful development length of the chromatoplate and to the feature of the solvent system [85,86].

3) Gradient MD (G^nD), called the multiple development technique [85,86], where the successive chromatographic development steps are performed with a change in solvent strength and selectivity ($S_{T1}, S_{V1}; \rightarrow S_{Tn}, S_{Vn}$) on the same chromatographic length (D = const.) (Fig. 6.7).

4) Bivariate MD (B^nD) is the most complex multiple development technique [85,86], where the development distance and the solvent system composition are varied simultaneously ($D_1, S_{T1}, S_{V1}; \rightarrow D_n, S_{Tn}, S_{Vn}$) during the successive chromatographic runs (Fig. 6.7).

Based on the idea of Perry *et al.* [87–88], Burger [89–91] derived the automated multiple development (A^nD or AMD) mode, which is a special case of B^nD. In AMD the chromatoplate is developed repeatedly in the same direction over an increasing migration distance ($D_1 \rightarrow D_n$). The developing solvent system for each successive run differs from the one used before, so that a stepwise negative gradient can be obtained. In contrast to column LC, in PC the gradient used starts with the most polar solvent or solvent system, for which the shortest developing distance is employed, and it is changed towards decreasing polarity. The longest migration distance is used with the most nonpolar solvent or solvent system. Another unique feature of the multiple development technique, which is also used in AMD, is the zone re-concentration mechanism [92,93], which ensures that if a sufficiently large number of developments are used, the zone will be compressed to a thin band. The five basic variants can also be varied. The most powerful ^nD technique is the use of AMD in two directions [96].

6.4.6.5 Multidimensional development mode

The two-directional ^nD (2^nD) technique is used for the separation of complex mixtures [94,95]. The sample is spotted at the corner of the layer and developed linearly (Fig. 6.6i). After evaporation of the first solvent system, the second development is carried out at right angle to the first one. For this, different types of solvent systems are used in the two directions. If the same solvent system is used for both directions, the sample components will be distributed along a line diagonally from the origin to the farthest corner. In this case, the resolution will only be increased by a factor of $\sqrt{2}$, corresponding to the increased migration distance for the sample, but this method is useful for finding out whether the analyte has undergone some chemical change. Compounds that are unchanged by the separation process will lie on the diagonal of the plate.

According to Giddings [97], the correct definition of multidimensional (MD) chromatography includes two conditions. "First, it is one in which the components of a mixture are subjected to two or more separation steps in which their displacements depend on different factors. The second criterion is that when two components are substantially separated in any single step, they always remain separated until the completion of the separative operation." This latter condition, therefore, precludes simple tandem arrangements in which compounds separated in the first separation system can re-merge in the second [98]. The following modes have most frequently been used for MD separations involving PC [96]:

(a) 2-D development on the same monolayer stationary phase with solvent systems characterized by different total solvent strength (S_T) and selectivity values (S_V);

(b) 2-D development on the same bilayer stationary phase either with the same solvent system or with solvent systems of different composition;

(c) multiple development (^nD) in one or two directions on the same monolayer stationary phase with solvent systems characterized by different solvent strength and selectivity values;

(d) coupled layers with stationary phases of decreasing polarity developed with solvent systems of constant composition;

(e) a combination of at least two of the above-mentioned modes;

(f) automated coupling of two chromatographic techniques, in which PC is used as the second dimension and another separation method, *e.g.* gas chromatography (GC), high-performance liquid chromatography (HPLC), etc., as the first.

MD-PC is discussed in detail in Ref. 99, and attention is drawn to the different possibilities, as well as advantages and limitations of the various modes of the technique.

6.4.7 Separation distance and time

In PC, the separation distance improves with the square root of the migration distance. However, the optimum depends on the quality of the plate, the vapor space, the development mode, and the properties of the compounds to be separated [2]. In capillary flow-controlled PC, the maximal length of commercially available precoated plates is 20 cm. Commercially available U-chambers can accommodate 10×10-cm chromatoplates for analyses by the circular and anticircular development modes in capillary flow-controlled PC. Therefore, the maximal separation distance is 5 cm and 4 cm, respectively. In linear OPLC, the maximum separation distance is 18 cm for the 20×20-cm chromatoplates. Working with off-line circular OPLC, the maximal separation distance is 10 cm; in this case, only one sample can be analyzed. If the distance between the mobile phase inlet and the sample application is 2 cm, then a separation distance of 8 cm can be achieved, which allows more samples to be applied. To carry out conventional anticircular separation is difficult, due to the large perimeter (*ca.* 60 cm for a 20×20-cm plate) of the mobile-phase inlet. After a suitable preparation of the plate (Fig. 6.6g) by scraping a segment out of the layer and sealing the segment with a polymer suspension, circular as well as anticircular fully off-line and on-line separations can be carried out over a separation distance of 18 cm.

In analytical RPC, the separation distance and number of samples depend on the development mode. In the radial mode, which is the most commonly used, up to 36 samples may be applied as points on a circle of 2-cm radius; in this case, the separation distance is 8 cm. In the separation of enantiomers, baseline separation of 72 samples on a single chromatoplate with a 5-cm separation distance was reported [100]. Since special preparation of the plate is necessary (by scribing lines into the layer), for linear and anticircular RPC, the sample number is correspondingly reduced [29]. Using capillary controlled PC, the development time depends on the solvent system, particle size of the stationary phase, and development modes. Generally, it can be stated that for the separation of nonpolar compounds, by FFPC techniques and with silica as the stationary phase, an extremely short separation time (1–2.5 min, over a separation distance of 18 cm) can be used without great loss in resolution. In contrast, longer separation times are extremely important when silica is used for the separation of polar compounds [101].

6.4.8 Temperature

Temperature is not an effective parameter for modifying selectivity and maximizing resolution under normal circumstances [2]. Generally, it can be stated that in saturated chromatographic chambers, which are the most commonly used, the temperature does not exert a great influence on separations. A change of $\pm 5\,°C$ results in a change of the ΔR_F of less than 0.03 [2]. Nevertheless, in the interest of reproducibility in repetitive separations it is important to note the working temperature. Normally, if two compounds are unresolved at a given temperature, they will remain unseparated at other temperatures, regardless of whether N- or S-chambers are used for the separations. Remarkably, temperature turns out to play an important role in the selectivity and efficiency of OPLC separations [102]. Experiments show that development by temperature-programmed OPLC with a suitable control unit can effect further improvement in PC separations.

6.5 INSTRUMENTATION

6.5.1 Overpressured-layer chromatography

OPLC separations can be performed on a commercially available instrument, the P-OPLC 50 (OPLC-NIT). The equipment offers 18-cm separation distances in linear, circular, and anticircular on-line mode and, in addition, off-line circular development, on a maximal 10-cm development distance. The P-OPLC 50 accommodates a high cushion pressure (50 bar), which allows the use of more viscous mobile phases and/or higher mobile-phase velocities. The OPLC chamber (Fig. 6.8a) consists of four main units [40]: holding unit, hydraulic unit, tray-like layer casette, and attached drain valve. The prepared layer (Fig. 6.8b) is placed in the tray-like layer casette, in which the chromatoplate is covered with a Teflon sheet. To ensure that mobile-phase migration forms a linear front, a channel is scratched into the Teflon sheet. A second channel of the Teflon sheet cut at a distance of 18 cm (for a 20 × 20-cm plate) from the inlet channel permits on-line separations. The separation chamber with two mobile-phase connections is built to withstand the 5 MPa maximum external pressure. The microprocessor-controlled liquid-delivery system includes a two-in-one hydraulic pump and a mobile-phase delivery pump. All conditions for single or repeated development can be programmed and stored in the delivery system software [40]. External pressure (max 5 MPa), mobile-phase volume, and mobile-phase flow rate can be selected, and development time is calculated automatically [103]. After one separation is finished, in the tray-like layer casette the next chromatoplate

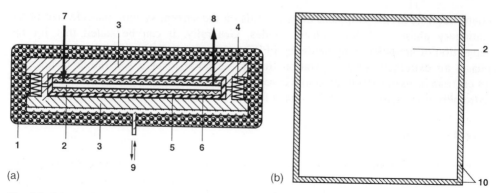

Fig. 6.8. Schematic drawing of casette-type OPLC instrument (A) and the prepared chromatoplate (B) 1 = support of the instrument, 2 = chromatoplate, 3 = inner part of the instrument, 4 = spring, 5 = casette system for the chromatoplate between two Teflon sheets, 6 = teflon sheet, 7 = mobile phase inlet, 8 = mobile phase outlet, 9 = hydraulic system, 10 = polymer suspension for the protection of the chromatoplate.

is put in place, and the equipment is ready for the next separation. Linear separations require specially prepared plates, with edges that are chamfered off and impregnated with a suitable polymer suspension, in order to prevent solvent leakage at overpressure.

6.5.2 Rotation planar chromatography

Presently two instruments are commercially available for RPC; the Chromatotron Model 7924 (Harrison) and the Extrachrom rotation planar separator (RIMP). The Chromatotron consists of an annular N-chamber, inclined at an angle and fixed on a pedestal [104,105]. A flat glass rotor, covered with stationary phase, is mounted on an axle with the aid of a retaining screw. The motor-driven glass disk of 24-cm diameter rotates at a constant speed of 750 rpm. The mobile phase inlet eccentrically pierces a quartz lid, which covers the chromatographic chamber. The chamber is provided with a circular channel for the collection of eluate. The solvent outlet is placed at the lowermost point of the collection channel. An inlet tube is mounted on the side of the chamber for flushing with nitrogen or other inert gases. The preparation of the preparative layers is detailed in Ref. 104.

A schematic drawing of the Extrachrom in the preparative normal-chamber RPC mode is shown in Fig. 6.9a. The upper part of the instrument is a standing chromatographic chamber, which is covered with an inspection window. The two solvent delivery devices are horizontally and vertically adjustable; vertical adjustment is necessary because the layers have different heights, the horizontal adjustment is especially necessary for sequential RPC. The delivery needles lead from the solvent delivery systems to the center of the chromatographic chamber. For preparative RPC separations the prepared stationary phase on the rotor is fixed at the center of the collector. M-RPC is performed in a special chamber where the vapor space is reduced and defined [1]; therefore, M-RPC resembles

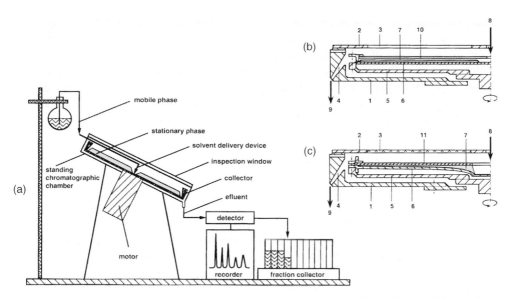

Fig. 6.9. Schematic diagram of on-line preparative RPC separations. Flow scheme of the method: (a) schematic diagrams of the chamber for M-RPC (b) and C-RPC (c). 1 — Lower part of the stationary chamber, 2 = upper part of the stationary chamber, 3 = inspection window, 4 = collector, 5 = motor shaft with the rotating disk, 6 = rotor, 7 = stationary phase, 8 = mobile phase inlet, 9 = eluent outlet, 10 = quartz glass cover plate, 11 = stainless-steel cover plate.

a saturated system (Fig. 6.9b). Since the vapor space is reduced, this method may be used for off-line analytical, off-line micropreparative, and on-line preparative purposes. In normal-chamber RPC and sequential RPC the layer rotates together with the collector in the instrument; no quartz glass cover plate is used [35]. For preparative separations the rotor with a 4 mm preparative layer is placed in the collector, thus eliminating the vapor space. This chamber can be adapted for analytical and preparative micro- and ultramicro RPC separations [34]. The collector is used for collection of the effluents in the various preparative modes.

Column rotation planar chromatography (C-RPC) is an on-line preparative technique, in which the stationary phase is placed in a closed circular chamber (Fig. 6.9c). The volume of stationary phase stays constant along the separation distance, hence the name "column" RPC [13]. Since it is a closed system, there is no vapor space, and any stationary phase may be used in fine particle size (3–5 mm) with or without binder. The rotating planar column has a special geometric design, which may be described by the following function:

$$h = K/(a + br + cr^2) \tag{6.4}$$

where h is the actual height of the planar column at radius r, r is the radius of the planar column, and a, b, c, and K are constants. The surface of the rotor has the special geometrical design described above and forms the lower part of the planar column.

Thus, the column consists of the rotor with the special design and stainless-steel cover plate. To enable the filling of the column and for solvent delivery this cover plate has a central hole of large diameter. Due to the column properties, the design creates a linear mobile-phase velocity and eliminates the extreme band-broadening, combining the advantages of both planar and column chromatography. The front panel of the instrument casing carries the adjusting and controlling units. The microprocessor-controlled rotation speed may be varied between 20 and 2000 rpm, in steps of 10 and 100 rpm.

6.5.3 Electro planar chromatography

Electro planar chromatography (EPC) can be performed on wetted or nonwetted layers in vertical or horizontal chambers. Starting with a wetted layer, the flow-rate is different at the beginning of the development of the chromatogram. Until it reaches the wetted part of the chromatoplate, the solvent system migrates under the influence of capillary forces and the electro-osmotic flow generated by the electric field; subsequent migration is a result of electro-osmotic flow only [42]. Starting the separation with a nonwetted layer, the chromatograms can be developed from both sides (the cathode and anode sides) simultaneously. The flow-rate is the sum of migration resulting from capillary activity and that induced by the electric field [42]. The flow-rate depends on the applied voltage, the type of stationary phase, the solvent system, and the degree of saturation of the chamber. A schematic diagram of EPC is shown in Fig. 6.10. Details concerning the principle and applicability of EPC are given in Ref. 42.

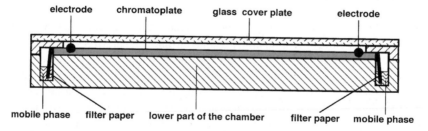

Fig. 6.10. Schematic drawing of horizontal chamber for PEC.

6.5.4 Automated multiple development

A schematic flow diagram of the AMD system (Camag) is shown in Fig. 6.11. Its central component is a closed N-chamber with connections for feeding and withdrawing developing solvents, and for pumping gas in and out. The solvent system is made up from up to six reservoir bottles, containing the neat solvents, which are passed *via* a motorized valve to the gradient mixer. The gas phase is made up externally by passing nitrogen through the wash bottle into the reservoir, from where it is pumped into the chromatographic chamber at the appropriate time. After the plate has been placed in the chromatographic chamber and the solvent system is fed into the gradient mixer, the

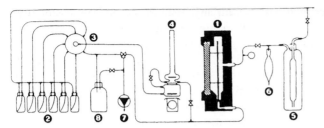

Fig. 6.11. Schematic drawing of AMD instrument. 1 = Chromatographic chamber, 2 = reservoir bottles, 3 = motor valve, 4 = gradient mixer, 5 = wash bottle, 6 = mobile phase reservoir, 7 = vacuum pump, 8 = waste collecting bottle.

separation is started. When the pre-programmed time determining the running distance has elapsed, the solvent system is withdrawn from the chamber, first to the waste-collection bottle, and then, after the solvent system has been completely removed, vacuum is applied by a pump, to dry the chromatoplates [91]. The duration of the drying cycle can be programmed. Before the next developing cycle is started, the chromatoplate is reconditioned by feeding a gas from the reservoir into the chamber. During the drying phase, the solvent system is prepared for the next chromatographic step. More details about the highly developed microprocessor-controlled AMD system are given in Refs. 22, 90, and 106.

6.6 QUALITATIVE AND QUANTITATIVE ANALYSIS

6.6.1 Identification of separated compounds

Apart from chromatographic mobility data, the list of analytical methods that can be brought to bear upon the analysis of a chromatoplate includes visual inspection, UV/visible spectrophotometry, fluorescence spectrophotometry, optical and electron microscopy techniques, Auger, reflectance IR spectroscopy, radio-imaging methods, near-IR analyses, and mass spectrometry in various forms, including secondary-ion mass spectrometry, fast-atom bombardment, and laser-desorption ionization [1,107]. It should be remembered that the chromatographic retention data are not enough for correct identification; at least one spectroscopic method is necessary for a definitive statement.

6.6.1.1 Chromatographic data

In PC, compounds are identified primarily on the basis of their mobility in a suitable solvent system. The mobility of separated zones is described by the R_F value of each compound, where

$$R_F = \frac{\text{distance of spot migration, } Z_x}{\text{distance start to front, } Z_F - Z_0} \qquad (6.5)$$

The R_F value bears no linear relationship with any basic PC parameter or structural element of the analyte. Such a linearity can be achieved with the R_M value, which is a logarithmic function of the R_F value:

$$R_M = \log \frac{1}{R_F} - 1 \tag{6.6}$$

For continuous and multiple development, where the solvent front is not measurable, the R_x value can be used, which is defined by Eqn 6.7 [12]

$$R_{x_a} = \frac{\text{distance traveled by Solute } a}{\text{distance traveled by Standard } x} \tag{6.7}$$

Instead of R_F values, it is more convenient to use the hR_F ($= 100 \times R_F$) values. The problems surrounding the determination of correct R_F values for linear, radial, and antiradial development, and calculation of other retention data are discussed in detail in Ref. 2. Identification by comparison with R_F, hR_F, R_M, or R_x values in the literature is fraught with uncertainty, because they depend on many factors, such as quality of stationary phase, saturation grade of the vapor phase, humidity, layer thickness, development distance, and temperature [108]. Comparison of the retention data with those of reference substances gives better results, but still does not provide unequivocal proof of identity. If the retention data of the compound to be identified are identical with those of the reference substance in three different solvent systems but the same stationary phase or with the same solvent system but three different types of stationary phases, the two compounds can be regarded as identical.

A new identification method said to be applicable to the identification of known compounds by TLC has been published [109]. The method is based on the retention data measured. If the hR_F values of the compound to be identified have practically the same value (± 3) as the reference substance in three mobile phases with different total solvent strength and total selectivity value, these values can be depicted as a triangle in a coordinate system. The area of the triangle, the $I_{P(Chr)}$ value, characterizes the goodness of the chromatographic identification. This dimensionless value can be calculated according to the rule of coordinate geometry. For any chromatographic identification, sample and the reference substances must always be applied on the same chromatoplate. The hR_F values must be calculated from the densitograms. Two compounds can be considered identical, if the variance of the DhR_F values of the compound to be identified and the standard substance is less than 3. The higher the value of $I_{P(Chr)}$, the greater is the probability that two compounds are identical. If the $I_{P(Chr)}$ value is less than 0.1, the identification is inadequate. For routine laboratory work, the $I_{P(Chr)}$ value should be between 0.1 and 0.5. If the value of $I_{P(Chr)}$ is higher than 0.5, the substances are chromatographically identical with a high degree of probability. In the identification of new natural compounds, the value of $I_{P(Chr)}$ must be higher than 0.5 [109]. The probability can be increased by *in situ* UV and/or VIS spectra, by a color reaction, or with off-line VIS spectra.

6.6.1.2 Pre- and post-chromatographic derivatization

For the visualization of compounds, one can use physical, chemical, or biological detection methods. Physical detection methods are based on substance-specific properties. The most commonly employed methods are the absorption or emission of electromagnetic radiation, which are measured by detectors (Sec. 6.6.1.3). The β-radiation of radioactively labeled compounds can also be detected directly on the plate. These nondestructive detection methods allow subsequent isolation and can also be followed by microchemical and/or biological detection methods [110]. Since physical detection methods are frequently not sufficient to establish identity, they must be complemented by specific chemical reactions (derivatization). These reactions may be carried out either before or after chromatography. Pre-chromatographic derivatization can be performed either during sample preparation or on the chromatoplate at the origin. It is generally used to introduce a chromophore, leading to the formation of strongly absorbing or fluorescent derivatives, to increase the selectivity of the separation, enhance the sensitivity of detection, and improve the linearity. It includes oxidation, reduction, hydrolysis, halogenation, nitration, diazotization, esterification, etherification, hydrazone formation, and dansylation [110].

The primary aim of post-chromatographic derivatization is the detection of the chromatographically separated compounds for better visual evaluation of the chromatogram. This step generally also improves the selectivity and the detection sensitivity. Post-chromatographic reactions can be carried out by spraying reagents onto the chromatoplate, by dipping the layer in reagent solutions, or exposing the plate to vapors. The reagent can also be in the solvent system or in the adsorbent. In most instances, subsequent heating is necessary. Post-chromatographic derivatizations have been extensively reviewed [111, 112]. For biological/physiological detection, the separated compounds can be transferred to the biological system. Alternatively, bio-autographic analysis, reprint methods, and enzymatic test may also be applied. Pre- and post-chromatographic derivatization methods have been summarized recently by Cimpan [113].

6.6.1.3 Optical spectroscopy

Regardless of whether they are colored or colorless compounds, absorbing in the UV range, they can be detected by direct scanning at the wavelength of maximum absorption [1]. Some types of compound are naturally fluorescent, while others can be converted to fluorescent derivatives through pre- and/or post-chromatographic reactions; this is a highly specific and sensitive method. Fluorescence quenching is limited to compounds that readily adsorb in the wavelength range of maximum excitation of a phosphor incorporated in the stationary phase [9]. Instruments for scanning densitometry after the above treatments can be operated in the reflectance, transmission, or combined (reflectance/transmission) mode. Usually, the sample beam is fixed and the plate is scanned, mounted on a movable stage, which is controlled by a stepper motor. The geometry of the light beam of scanners can have the form of a slit or a spot. The most common method is slit scanning, in which the sample beam illuminates a rectangular area on the chromatoplate surface, while the plate is transported in the direction of development.

With the other type, a small light spot moves in two dimensions. In this case, three different kinds of movement are possible: meandering, zig-zag, and flying-spot scanning, the last-mentioned with a sinusoidal type of movement.

Different principal optical geometries are used in scanning densitometry: the single-beam methods, which can be used in the reflectance, transmittance, or simultaneous mode, and the double-beam methods [1]. In the latter case the two beams can be either separated in time at the same point on the chromatoplate or separated in space and recorded simultaneously by two detectors. Absorption spectra are rarely sufficient for identification, except when directly compared with the spectrum of a standard substance, measured on the same chromatoplate. If possible, preference should be given to the fluorescence mode, which has a much higher spectroscopic selectivity because two different wavelengths, an excitation and an emission wavelength, are used for each measurement. Recently, diode-array scanners [114,115] as well as image analysis [116–119] have been introduced for evaluating planar chromatograms. The uses of densitometry in PC have been summarized by Dammertz and Reich [120].

6.6.1.4 Infrared and Raman spectroscopy

Depending on the nature of the compound, the IR absorption of a TLC spot can be measured either after transfer to an IR-transparent substrate or *in situ*. The off-line method involves elution of *ca.* 5 µg substance, evaporation of the solvent, and pressing into a micropellet suitable for recording spectra. IR spectra can also be recorded *in situ* by DRIFT-IR spectroscopy. To obtain suitable IR spectra, all solvents must be carefully removed prior to measurement, and a background correction must be applied for absorption by the chromatoplate. Kovar *et al.* [121,122] reported a new method for on-line coupling of TLC and FTIR spectroscopy; a commercially available special HPTLC-DRIFT unit was constructed for this purpose. Achievements in this field have been summarized by Rager and Kovar [123].

Analysis of TLC spots by photoacoustic spectrometry (PAS) has been preferred over DRIFT analysis of strongly IR-absorbing samples [124]. The spot must be physically removed from the chromatoplate and, after some preparation, it is placed in the photoacoustic cell for measurement. Recently, investigation of the in-depth distribution of the analytes in the sorbent layer was studied by means of PAS [125]. Vovk and Močnik wrote a comprehensive review of photoacoustic and photothermal methods used in PC [125]. Reasonable spectra can also be obtained by surface-enhanced Raman spectroscopy (SERS) [126,127]. Recently, a new method was reported for preparing SERS-active surfaces, in which colloidal silver spheres are deposited on HPTLC plates. The sensitivity of theses activated HPTLC plates is so high that *in situ* vibrational investigations of spots are possible at the picogram level [128]. Comparison of the relative intensities of HPTLC-SERS spectra and normal Raman spectra was reported by Koglin [129]. The results demonstrate that variations in electric charge density and hydrophilic character of the silver sol-activated nano-TLC plate may strongly influence the HPTLC-SERS detection limit.

6.6.1.5 Mass spectrometry

Different methods [130–139] have been described for obtaining mass spectra of TLC spots. The zone of the compound to be identified can be scraped from the layer, and after elution and solvent evaporation, the residue is inserted into the mass spectrometer. Alternatively, the sample, together with the stationary phase, can be inserted directly. In fast atom bombardment (FAB) and secondary-ion mass spectrometry (SIMS), a high-energy ion or atom beam is used to sputter molecules from the condensed phase into the gas phase for mass spectrometric analysis. However, most mass spectrometric measurements are destructive in nature. FAB-MS and SIMS are surface-sensitive methods in which the material actually consumed in the analysis is sputtered only from the top layers of the sample spot. By using a special ion-source housing and an appropriate direct TLC-MS probe [131], one-dimensional mass spectra and mass chromatograms can be recorded from one track of the developed chromatographic plate.

Busch *et al.* [132,133] have described an interface for the combination of PC and MS. Using this device, 2-dimensional mass-imaging was carried out. The FAB/SIMS liquid matrices that are usually used limit the time for imaging, because of spot diffusion. To overcome this limitation, a solid matrix was used [134], thus preserving the spatial resolution of the chromatogram for a longer time and enabling the time-consuming 2D-MS mapping. Also, a charge-coupled device for optical detection of sample bands was applied [134]. A matrix-assisted laser-desorption time-of-flight mass-spectrometric (MALDI-TOF-MS) system was also described [135]. Wilson *et al.* [136–139] described the application of tandem MS (TLC/MS/MS). They demonstrated that FAB-MS alone gave spectra which were dominated by ions from the matrix. When MS/MS was applied to the same samples, the resulting spectra were devoid of matrix interferences and contained only ions from the compounds of interest. Recently, Busch [140] outlined the state of the art of TLC/MS coupling.

6.6.1.6 Nuclear magnetic resonance spectroscopy

Wilson *et al.* [141] reported the solid-state NMR spectroscopy (high-resolution magic-angle spinning) for identification in PC. The substances were separated on a C_{18} stationary phase, and the spots were located by UV illumination. Usable spectra could be obtained when the appropriate zone was scraped off the TLC plate, slurried with D_2O, and placed in the NMR tube without eluting the substance from the stationary phase.

6.6.1.7 Digital autoradiography

Digital autoradiography (DAR) offers higher sensitivity than contact autoradiography, saves time in the detection of radiolabeled compounds, and/or reduces the quantity of the analyte. Implementation of OPLC separation has significantly improved the performance of the method. Klebovich [142] has recently reviewed the applications of DAR in planar PC.

6.6.2 Quantitative determination

Methods for the quantitative evaluation of TLC/HPTLC may be divided into two categories. In the first, solutes are eluted from the stationary phase before being examined further; in the second, solutes are assayed directly on the layer.

6.6.2.1 Measurements after elution

Quantitation may be performed after scraping off the separated zones and recovery of the substances by elution from the stationary phase. Thereafter, the methods used for quantitative analysis are essentially the current methods of microanalysis (e.g., GC, HPLC, liquid scintillation counting). A special accomplishment is on-line OPLC, where only one sample can be analyzed at a time, but all compounds are eluted from the chromatoplate and can be measured with the help of a flow-through detector, as in HPLC [143].

6.6.2.2 In situ *measurements*

Modern optical densitometric scanners (Sec. 6.6.1.3) are linked to computers and are equipped for automated peak location, multiple-wavelength scanning, and spectral comparison of fractions, and are capable of measurement in any operating mode (reflectance, absorption, transmission, fluorescence) [120]. In electronic scanning densitometry, the stationary chromatoplate is scanned electronically. This type of densitometer for image analysis requires a computer with video digitizer, light source and appropriate optics, such as lenses, filters, and monochromators, and a vidicon tube or charge-coupled video camera. Electronic scanning always operates in the point-scanning mode, as opposed to devices with electro-mechanical scanning, which mostly use the slit-scanning mode. For 2D separations, modern pattern recognition techniques can be employed. A powerful microcomputer of the IBM-AT class is usually employed. For the quantification of separated radioactive substances, autoradiography, liquid scintillation counting, and direct scanning with radiation detectors can be used. Photothermal deflection densitometers measure the refractive index gradient formed in the gas phase over a solid sample, heated by a laser [144]. In the absorption mode, the detection limit can be similar to that of optical scanning densitometers. The applications of flame-ionization detectors are summarized in Sec. 6.8.1. Many books, book chapters and review articles [11,12,16,145,146] on quantitative PC are available.

6.7 PREPARATIVE PLANAR CHROMATOGRAPHY

The aim of preparative PC is the isolation of compounds in amounts of 10–1000 mg for structure elucidation (MS, ^{1}H-NMR, ^{13}C-NMR, IR, UV, etc.), for various analytical purposes (further chromatography), or for the determination of biological activity. The preparative chromatographic processes operative in capillary flow-controlled PC, OPLC, and RPC basically resemble those in their analytical counterparts [104]. However, some special characteristics need to be considered, such as the average particle size and thickness of the stationary phase, application of a large amount of sample, detection of the separated compounds, as well as their removal. In preparative capillary-flow PC and in micropreparative RPC, only off-line methods can be used. This means that once the analytical or preparative plate has been developed and the solvent system is evaporated, the separated bands must be located and the desired compounds removed from the plate. For OPLC and RPC also on-line elution may be used, where all compounds migrate over the whole separation distance. This results in a better separation in the lower R_F range, and additionally allows the separated compounds to be directly eluted from the chromatoplate [105].

For self-made preparative plates the stationary phases most often applied are silica, alumina, cellulose, and gypsum. To produce layers with a thickness between 0.5 and 2 mm, so-called P-type sorbents may be used, which do not contain $CaSO_4$. Sorbents designated "$P + CaSO_4$" are suitable for preparing layers up to 10 mm thick. The advantage of preparing one's own plates is that any desired thickness or composition of plates becomes feasible. Besides saving time, precoated layers have the advantage of much better reproducibility than self-made plates. Generally, four types of pre-coated preparative plates are commercially available: silica, alumina, RP-2, and RP-18, in layer thicknesses of 0.5–2 mm. It is generally accepted that high resolution requires relatively thin layers (0.5–1 mm). On a high-capacity (1.5- to 2-mm) layer the resolution is much more limited. The load capacity of a preparative layer without loss of separating power increases with the square root of the thickness. The load capacity of a 0.5-mm layer is approximately half that of a plate with a layer thickness of 2 mm [105]. In addition to the 20 × 20-cm or 20 × 40-cm plates, 20 × 100-cm plates are also available for the separation of larger amounts of sample. The silica materials commonly used for preparative separations have excessively coarse particles (average of *ca.* 25 μm), and their distribution is also excessive (5–40 μm). Various precoated preparative layers with a pre-adsorbent zone are also commercially available.

The process of sample application is an important step in preparative PC. The preferred method of placing a sample on a preparative layer is to apply it as a narrow streak across the plate. It is convenient to use pre-coated preparative layers with a concentrating zone [105]. A solid-phase sample application method by Botz *et al.* [147] permits uniform sample application over the whole cross-section of the preparative layer with the advantage of *in situ* sample concentration and clean-up and an extremely sharp edge. This device can be applied in both capillary-action and forced-flow PC.

6.7.1 Off-line separation

For off-line separations of 5- to 15-mg samples, analytical TLC and/or HPTLC plates can be used, but where the total amount of substances to be separated lies between 50 and 1000 mg preparative plates must be used [148]. Many methods are available for the location/detection of the separated components. Pre-coated plates containing indicators fluorescing at 254 nm or 365 nm provide a general nondestructive mode of detection. If the compounds themselves are not visible or fluorescent, they can be detected by applying specific reagents. After development, a vertical channel is scraped in the layer, *ca.* 0.5 cm from the beginning of the applied streak. After covering the rest of the layer with a glass plate, the uncovered part of the layer, sprayed with a suitable reagent, serves as a guide for locating the zones on the remainder. The mechanical removal of the zones is followed by elution of the compounds from the stationary phase with a suitable solvent, separation from the residual adsorbent, and concentration of the solvent. Several commercially available devices and individually developed methods exist for eluting the compounds from the stationary phase [148].

6.7.2 On-line separation

Mincsovics *et al.* summarized [149] all possible combinations of off- and on-line OPLC separation techniques, which are also valid for RPC. The fastest FFPC separation can be obtained in the fully on-line mode. This operating mode is also the simplest and most economic preparative FFPC method, because, after cleaning and re-equilibration, the same plate may be used several times without loss of resolution. Since OPLC and RPC may be used not only for on-line preparative separations, but also for analytical and micropreparative purposes, both analytical methods allow a direct scale-up to preparative OPLC and RPC, respectively [148].

From the TLC separation in unsaturated or saturated chromatographic tanks, the solvent system can be transferred *via* analytical OPLC, microchamber and ultra-microchamber RPC to preparative OPLC, microchamber and ultra-microchamber RPC, respectively [35]. For scale-up, the sample may be applied to an analytical TLC plate and the amount of sample increased stepwise in subsequent separations. The resulting plates are scanned (off-line) to see whether the resolution is satisfactory. Thus, the maximum amount of sample for the on-line preparative separation is determined, considering the particle size and the volume of the stationary phase. The flow-rate of the mobile phase must be adapted to preparative separation, so that the migration of the α-front is as fast as in the analytical separation. Mixed sorbents can be used for on-line separation in the column RPC [79]. The stationary phase occupies a closed radial chamber (column), which has a special geometric design, hence the name "column" RPC. The volume of stationary phase stays constant along the separation distance; the flow is accelerated linearly by centrifugal force [147]. The simplest method is to fill the major part of the planar column with silica and then for the last 1 cm separation distance with kieselguhr, as a preadsorbent zone. It is also possible to fill the same planar column with more than two successive stationary phases in order of increasing or decreasing polarity.

6.7.3 Selection of the appropriate method

Whether the use of forced-flow techniques is necessary or not depends on the kind of sample to be separated. Instrumental methods will increase preparation time and cost but also significantly improve efficiency. As a rule of thumb, if the sample contains more than 5 substances, up to 10 mg of sample can be separated by a micropreparative method and up to 500 mg by a preparative method. If the sample contains fewer than 5 substances, the amounts may be increased up to 50 mg and 1000 mg, respectively. If no more than

6 compounds are to be separated, distributed over the whole R_F range, present in more or less the same amounts, and if the total amount of sample exceeds 150 mg, preparative capillary flow-controlled PC can be used successfully. This is the simplest and therefore the most widely used method [150].

The potential of on-line linear OPLC on 20 × 20-cm plates with a separation distance of 18 cm as a preparative method is considerable. Because the particle size of the precoated plates is so large, not all advantages of this method can be realized yet. Generally, on-line OPLC can be used for the separation of 6 to 8 compounds in amounts up to 300 mg [38]. The oldest forced-flow planar chromatographic method uses centrifugal force for on-line purification and isolation [32,33]. Generally, 15-μm particle size is used in the stationary phase with all the advantages of free selection of the size of the vapor space and development mode. Up to 10 compounds in amounts up to 500 mg can be isolated by the appropriate RPC method [148].

6.8 SPECIAL PLANAR CHROMATOGRAPHIC TECHNIQUES

6.8.1 Combination with flame-ionization detection

TLC can also be carried out on permanent, rod-shaped layers, having mechanical and chemical properties that permit detection of the separated compounds with a flame-ionization detector (FID) [151]. It should be noted that this is a planar but not a "flat-bed" technique, *i.e.*, the term flat-bed chromatography does not include all PC methods. A TLC/FID system commercially available is the Iatroscan TH-10 (Iatron Laboratories). The layer, composed of a suitable sintered mixture of glass powder and adsorbent, is coated on a quartz rod with a diameter of 0.9 mm and a length of 15 cm. The glass powder (1–10 μm) is mixed with the stationary phase (5–10 μm) in a ratio between 2 and 10 to 1. At present, three types of rods are commercially available; Chromarod S (10-μm silica; layer thickness, 100 μm), Chromarod S II (5-μm silica; layer thickness, 50 μm), and Chromarod A (10-μm aluminium oxide; layer thickness, 35 μm).

Before the rods are used, they are cleaned and activated in the flame of the detector. After they have been placed in a holder, the samples are applied (1- to 50-μg samples in 0.1–1 μl), and the chromatogram is developed in a saturated N-chamber. After development, the solvent system is evaporated and the rods are placed in a sliding frame of the instrument, which passes through the FID at constant speed. The individual zones are ionized in a hydrogen flame, and the ionization current produced is amplified and fed into the integrator and recorder. Some recent developments, involving the use of novel detectors, such as the flame-thermionic ionization detector (FTID), which responds to substances containing nitrogen and halogen atoms, and the flame-emission photometric detector, which detects compounds containing sulfur and/or phosphorous as well as the chemiluminescent nitrogen detector, coupled on-line with FID, should be able to widen the range of possible applications of the coated-rod TLC-FID systems. The method, instruments, and applications have been surveyed by Mukherjee [152].

6.8.2 Sequential techniques

The sequential development of analytical and preparative plates has the advantage that the solvent system supply is fully variable in time and location. Thus, the resolution can be improved and the separation time reduced. The principle of the sequential technique is based on the fact that the solvent system velocity is much higher at the beginning of the separation than later on. After a first separation, the layer is dried, and either the same or a different suitable solvent system is applied. The supply of solvent system may be stopped at any time in order to transfer it directly to the area of the compound zones to be separated. Therefore, the high initial velocity of the solvent system is always used, and this substantially shortens the analysis time and increases the resolution. Sequential TLC can be carried out with the S-chamber-type Mobil-R_F chamber, developed by Buncak [153,154].

Combination of the circular and anticircular development modes is possible by sequential development in RPC. In this technique, the mobile phase can be introduced at any desired place on the plate and at any time. In sequential RPC, the solvent application system – a sequential solvent delivery device – operates by centrifugal force and with the aid of capillary action against a reduced centrifugal force (antiradial mode). Generally, the circular mode is used for the separation of zones and the anticircular mode for pushing zones back towards the center with a stronger solvent (*e.g.*, ethanol). After the plate has been dried at a high rotational speed, the next development with another mobile phase may be started. This combination of two operating modes makes the separation pathways in sequential RPC theoretically unlimited [41,79].

6.8.3 Mobile-phase gradient

In PC, a true mobile-phase gradient can only be used with the FFPC techniques. Whereas all forms of gradient are possible with these methods, so far only the application of step gradients has been reported. The positive effects of step gradients were demonstrated not only for analytical [155] and preparative OPLC separations [156], but also for analytical and preparative RPC separations of various plant extracts [82].

6.8.4 Layer-thickness gradient

Use of preparative taper plates [157] greatly reduces spot elongation and overlapping, due to the gradient effect of layer thickness. The improved performance of the taper plate is similar to the improved resolution in the lower R_F range observed with radial TLC. In the taper plate, the cross sectional area traversed by the solvent increases as development progress. Therefore, the cross sectional flow per unit stationary phase area is always highest at the beginning of the layer, decreasing toward the mobile-phase front. As a result, the tail end of a zone moves faster than the front end, thus keeping each component focused in a narrow band. Band-broadening is significantly reduced, especially for compounds with higher R_F values. Compounds with lower R_F values are subject to greater mobile-phase velocity, because of the increase in the amount of solvent at the front with migration distance. Because of this, the distance between bands at lower R_F values is increased, resulting in better separations.

6.9 COMPARISON OF VARIOUS PLANAR CHROMATOGRAPHIC TECHNIQUES

PC methods may be compared with respect to simplicity, reproducibility, efficiency, rapidity, and speed TLC and HPTLC plates and techniques are compared in Refs. 2 and 20. In this context, it should be noted that at smaller average particle size of the stationary phase and, therefore, at shorter separation distance, the detection limit is as much as 5–10 times lower in the absorption and in the fluorescence mode. Chamber saturation is one of the most important factors in achieving reproducibility in capillary flow-controlled PC. Therefore, the type of chromatographic chamber and the degree of saturation of the vapor phase should be stated when reporting results [2,108]. The possibilities of transferring optimized solvent systems between various PC methods are discussed in Ref. 158.

Capillary flow-controlled PC is much simpler to use than the forced-flow techniques. As a rule, FFPC techniques are of value only with the small-particle-size HPTLC plates. The linear development mode (over a 18-cm separation distance) can be recommended when pairs of peaks show incipient separation on a TLC plate with capillary-controlled flow, but not enough resolution for quantitative evaluation. If the separation problem is in the upper R_F range, anticircular development should be tried; if it is in the lower R_F region, circular development is preferred [65]. In both cases, the resolution can be increased further by FFPC techniques. If many compounds are to be separated (up to 72), ultra-microchamber RPC or OPLC in the circular development mode is preferred. A comparison of the analytical PC methods is given in Table 6.2. In capillary flow-controlled PC under optimized operating conditions, the spot capacity lies between 15 and 25. For $2^{n}D$ development this value can be increased to 400, but it is almost impossible to exceed 500, except under very favorable conditions [20]. Using FFPC, the spot capacity can be between 60 and 100. If OPLC development is used in the first direction and the elution of the compounds in the second direction, a spot capacity of a few thousand could be achieved theoretically [159]. However, this has not yet been accomplished in practice, due to technical difficulties. At the moment, the fastest separations can be achieved with OPLC. A separation time of less than 5 min for 72 samples of enantiomers has been reported [36]. A comparison of the different RPC methods has been published [35]. Classical preparative-layer chromatography was also compared with the various preparative FFPC techniques [105].

6.10 TRENDS IN PLANAR CHROMATOGRAPHY

6.10.1 Development of instrumentation

There is a strong trend towards instrumentation of the individual steps in PC. A significant improvement is the Camag sample applicator, which enables automatic sample

TABLE 6.2

COMPARISON OF ANALYTICAL PC METHODS

Basis of comparison	Methods				
	TLC	HPTLC	OPLC	U-RPC	M-RPC
Flow-rate	Capillary action		Depending on overpressure	Depending on centrifugal force	
Vapor space	Variable		Absent	Practically absent	Saturated
Development mode	Linear, circular, anticircular two-directional, multiple		Linear, circular	Circular, multiple (linear, anticircular)	
Separation distance	Max. 18 cm	Max. 7 cm	18 cm (54 cm)		8 cm (10 cm)
Number of samples	Up to 36		Up to 72		Up to 72
Temperature	Controllable		Programmable		Controllable
Disadvantages	Low separation power	Short migration distance	Disturbing zone Multifront effect		Overflow effect
Special possibilities	Over running		On-line detection	Observation during separation	Sequential technique

application in linear, circular, and anticircular PC by selection of the x, y, and z coordinates. Solvent front detection is now available with a double-beam optical sensor [160] in automated developing chromatographic tanks [161] with a multiple plate holder. Based on the computer "desk jet" printer, a new instrument for pre- and post-chromatographic derivatization was also developed [162]. A further improvement of multidimensional detection systems can be expected soon in the field of identification by spectroscopic methods, based on results achieved so far. The most recently developed generation of diode-array densitometers enables greater accuracy and more rapid quantitative determinations [114,115]. This new type of densitometer allows the on-line detection on TLC/HPTLC chromatoplates [163].

6.10.2 Development of forced-flow planar chromatographic methods

It is expected that future research will concentrate on the positive effects of forced flow, *e.g.*, the applied pressure in OPLC and the centrifugal force in RPC. As a consequence, smaller particle size and a narrower distribution range will be needed for the stationary phases in order to achieve maximum resolution. The advantage of combining on-line and off-line separations [149] as well as two-dimensional development can also be exploited in OPLC. This development mode for planar columns was first described by Guiochon *et al.* [159,164]. After development with the first mobile phase, zones are eluted in perpendicular direction with a second mobile phase into a diode-array detector. Under optimized conditions, a separation number (SN) > 1000 could be achieved [1,2]. This would open up new vistas in the separating power of PC, but the experimental difficulties in implementing such a separation process are excessive. A very realistic possibility of increasing the efficiency and rapidity of the PC separation of complex samples is the use of multilayer OPLC. In the proposed version of Tyihák *et al.* [27] the same or different types of stationary phases can be used for the simultaneous development of several chromatoplates (Fig. 6.12a,b,c) [158]. This version is not only excellent for rapid off-line analytical OPLC, but also suitable for RPC. The efficiency of the method was recently demonstrated by Botz *et al.* [165], who developed five HPTLC plates simultaneously. By circular OPLC, 360 samples of plant extracts could be separated in 150 sec.

A novel category of multilayer OPLC is the long-distance OPLC, where the efficiency of the separation is increased significantly [166,167]. The end of the first chromatoplate

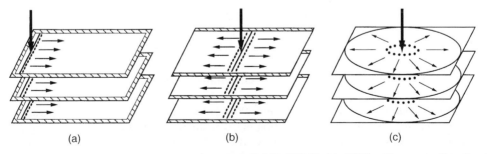

(a)   (b)   (c)

Fig. 6.12. Schematic drawing of multilayer OPLC (ML-OPLC). (a) Off-line linear one-directional development, (b) off-line linear two-directional development, (c) off-line circular development.

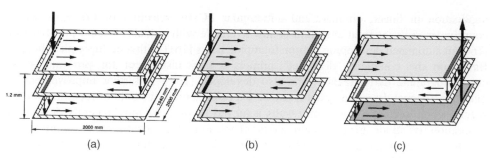

(a)   (b)   (c)

Fig. 6.13. Schematic drawing of long-distance OPLC (LD-OPLC). (a) Principle of the method, (b) fully off-line LD-OPLC using homolayers, (c) fully on-line LD-OPLC on heterolayers.

has a slit-like perforation to permit the mobile phase to migrate to a second layer. On this basis, a very long separation distance can be achieved by adding one chromatoplate to another, as it is shown in Fig. 6.13a. Also, different stationary phases can be used so that each part of a complex mixture can reach a suitable stationary phase. The method is applicable for off-line (Fig. 6.13b) and on-line determination (Fig. 6.13c) of analytical samples and perhaps a combination of off-line and on-line analysis of complex samples.

6.10.3 Multidimensional planar chromatography

Multidimensional planar chromatography (MD-PC) requires not only a multiplicity of separation stages, but also that the integrity of separation achieved in one stage be transferred to the others. The process of separation on a two-dimensional plane is the clearest example of multidimensional separations [96]. The greatest strength of MD-PC – when properly applied – is that compounds are distributed widely over a two-dimensional space of high peak (zone) capacity. The best possibility to achieve optimal separation is the parallel combination of stationary phases and solvent systems/mobile phases. To take advantage of the double effect of MD-PC it is recommended that fine-particle-size bilayer chromatoplates be combined with multiple-development techniques, in which total solvent strength and mobile-phase selectivity are changed simultaneously. On the basis of theory and experimental observations, it can be predicted that a zone capacity of *ca.* 1500 could be achieved by MD development. Because the same result can be achieved by application of two-dimensional forced-flow development on HPTLC plates, it can be stated that with the combination of finer (3-μm)-particle-size stationary phases, FFPC and ^{n}D promises a bright future for modern PC [96]. For quantitation in MD-PC we are close to the time when slit-scanning densitometry will be replaced by quantitative image analyzers.

6.10.4 Multimodal separations by planar chromatography

In multimodal separations [22], where PC is one of the modes, two separation techniques having complementary retention mechanism are used. It is generally accepted that in multimodal separations PC is the second step. The first separation technique can be a GC or a CLC separation. Although, GC/PC was used successfully earlier, it is no longer popular. Boshoff *et al.* [168] were the first to describe an interface for depositing the effluent from conventional CLC on a TLC plate. Hofstraat *et al.* [169,170] reported the coupling of narrow-bore HPLC and PC using HPTLC plates. Due to the low mobile-phase velocity (10–100 µl/min), the complete column effluent can be deposited on the chromatoplate, using a modified spray-jet band applicator. Using the advantage of the coupled RP microbore HPLC and normal-phase AMD techniques, Jänchen [171] reported the separation of more than 50 compounds on a single chromatoplate. Though only a few papers have been published so far concerning this type of multimodal separations, gradual application of this technique can be expected in the near future. Another possibility that can be envisioned is that the effluent from HPLC may be coupled with various types of FFPC, like OPLC and RPC. Furthermore, the combination of on-line OPLC and AMD techniques is expected to result in extremely high resolutions, while on-line multi-channel OPLC, combined with diode-array detection, will open up new vistas in PC. For the analysis of complex samples on-line multi-channel detection can be combined with the diode-array scanner [114,115] as well as image analysis [116–119] for compounds remaining on the stationary phase. These possibilities are also summarized in Ref. 172.

REFERENCES

1 Sz. Nyiredy, in E. Heftmann (Ed.), *Chromatography, Fundamentals and Applications of Chromatographic and Electrophoretic Methods, part A*, Elsevier, Amsterdam, 1992, pp. 109–150.

2 F. Geiss, *Fundamentals of Thin Layer Chromatography (Planar Chromatography)*, Hüthig, Heidelberg, 1987.

3 K. Macek, in E. Heftmann (Ed.), *Chromatography, Fundamentals and Applications of Chromatographic and Electrophoretic Methods*, Elsevier, Amsterdam, 1983, pp. 162–194.

4 I.M. Hais and K. Macek, *Handbuch der Papierchromatographie, Band I, II, and III*, VEB Gustav Fischer Verlag, Jena, 1958, 1960, 1963.

5 J. Gasparic and J. Churacek, *Laboratory Handbook of Paper and Thin-Layer Chromatography*, Wiley, New York, 1978.

6 E. Stahl (Ed.), *Dünnschicht-Chromatographie, Ein Laboratoriumshandbuch*, Springer, Berlin, 1967.

7 G. Zweig and J. Sherma (Editors-in-Chief), *Handbook of Chromatography*, CRC Press, Boca Raton, FL, 1972.

8 J. Kirchner and E.S. Perry, *Thin-Layer Chromatography*, Wiley, New York, 2nd Edn., 1978.

9 J.C. Touchstone and M.F. Dobbins, *Practice of Thin Layer Chromatography*, Wiley, New York, 1983.

10 C.F. Poole and S.A. Schuette, *Contemporary Practice of Chromatography*, Elsevier, Amsterdam, 1989.

11 J.C. Touchstone and J. Sherma (Eds.), *Techniques and Applications of Thin Layer Chromatography*, Wiley, New York, 1985.

12 B. Fried and J. Sherma, *Thin-Layer Chromatography, Techniques and Applications,* 2nd Edn., Revised and Expanded, Dekker, New York, 1986.

13 R.E. Kaiser (Ed.), *Einführung in die Hochleistung-Dünnschicht-Chromatographie*, Institut für Chromatographie, Bad Dürkheim, 1976.

14 R.E. Kaiser and A. Zlatkis, *High Performance Thin Layer Chromatography*, Elsevier, Amsterdam, 1977.

15   W. Bertsch, S. Hara, R.E. Kaiser and A. Zlatkis (Eds.), *Instrumental HPTLC*, Hüthig, Heidelberg, 1980.
16   R.E. Kaiser (Ed.), *Planar Chromatography*, Vol. 1, Hüthig, Heidelberg, 1986.
17   E. Tyihák and E. Mincsovics, *J. Planar Chromatogr.*, 1 (1988) 6.
18   Z. Witkiewicz and J. Bladek, *J. Chromatogr.*, 373 (1986) 111.
19   D.E. Jänchen and H.J. Issaq, *J. Liquid Chromatogr.*, 11 (1988) 1941.
20   C.F. Poole and S.K. Poole, *Anal. Chem.*, 61 (1989) 1257A.
21   A.M. Siouffi, E. Mincsovics and E. Tyihák, *J. Chromatogr.*, 492 (1989) 471.
22   C.F. Poole, S.K. Poole, W.P.N. Fernando, T.A. Dean, H.D. Ahmed and J.A. Berndt, *J. Planar Chromatogr.*, 2 (1989) 336.
23   Sz. Nyiredy (Ed.), *Planar Chromatography – A Retrospective View for the Third Millennium*, Springer, Berlin, 2001.
24   E. Tyihák, E. Mincsovics and H. Kalász, *J. Chromatogr.*, 174 (1979) 75.
25   E. Mincsovics, E. Tyihák and H. Kalász, *J. Chromatogr.*, 191 (1980) 293.
26   E. Tyihák, E. Mincsovics, H. Kalász and J. Nagy, *J. Chromatogr.*, 211 (1981) 45.
27   E. Tyihák, E. Mincsovics and T.J. Székely, *J. Chromatogr.*, 471 (1989) 250.
28   Sz. Nyiredy, *Trends Anal. Chem.*, 20 (2001) 91.
29   V. Pretorius, B.J. Hopkins and J.D. Schicke, *J. Chromatogr.*, 99 (1974) 23.
30   D. Nurok, M.C. Frost, C.L. Pritchard and D.M. Chenoweth, *J. Planar Chromatogr.*, 11 (1998) 244.
31   D. Nurok, M.C. Frost and D.M. Chenoweth, *J. Chromatogr. A*, 903 (2000) 211.
32   Z. Deyl, J. Rosmus and M. Pavlicek, *Chromatogr. Rev.*, 6 (1964) 19.
33   E. Heftmann, J.M. Krochta, D.F. Farkas and S. Schwimmer, *J. Chromatogr.*, 66 (1972) 365.
34   Sz. Nyiredy, S.Y. Mészáros, K. Dallenbach-Tölke, K. Nyiredy-Mikita and O. Sticher, *J. Planar Chromatogr.*, 1 (1988) 54.
35   Sz. Nyiredy, L. Botz and O. Sticher, *J. Planar Chromatogr.*, 2 (1989) 53.
36   Sz. Nyiredy, L. Botz and O. Sticher, *Am. Biotechnol. Lab.*, 8 (1990) 9.
37   E. Mincsovics and E. Tyihák, *J. Planar Chromatogr.*, 1 (1988) 309.
38   Sz. Nyiredy, C.A.J. Erdelmeier, K. Dallenbach-Toelke, K. Nyiredy-Mikita and O. Sticher, *J. Nat. Prod.*, 49 (1986) 885.
39   C.F. Poole, in Sz. Nyiredy (Ed.), *Planar Chromatography – A Retrospective View for the Third Millennium*, Springer, Berlin, 2001, pp. 13–32.
40   E. Tyihák and E. Mincsovics, in Sz. Nyiredy (Ed.), *Planar Chromatography – A Retrospective View for the Third Millennium*, Springer, Berlin, 2001, pp. 137–176.
41   Sz. Nyiredy, in Sz. Nyiredy (Ed.), *Planar Chromatography – A Retrospective View for the Third Millennium*, Springer, Berlin, 2001, pp. 177–199.
42   J.K. Rozylo and I. Malinowska, in Sz. Nyiredy (Ed.), *Planar Chromatography – A Retrospective View for the Third Millennium*, Springer, Berlin, 2001, pp. 200–219.
43   J.K. Rozylo, I. Malinowska and A.V. Musheghyan, *J. Planar Chromatogr.*, 2 (1989) 374.
44   K.K. Unger, *Porous Silica*, Elsevier, Amsterdam, 1979.
45   K. Günther, *J. Chromatogr.*, 450 (1988) 11.
46   M. Mack and H.-E. Hauck, *J. Planar Chromatogr.*, 2 (1989) 190.
47   S.H. Han and D.W. Armstrong, in N. Gringerg (Ed.), *Modern Thin-Layer Chromatography*, Dekker, New York, 1990, pp. 398–427.
48   T. Kowalska, in Sz. Nyiredy (Ed.), *Planar Chromatography – A Retrospective View for the Third Millennium*, Springer, Berlin, 2001, pp. 33–46.
49   R.E. Kaiser, *J. Planar Chromatogr.*, 1 (1988) 182.

50   T. Omori, in Sz. Nyiredy (Ed.), *Planar Chromatography – A Retrospective View for the Third Millennium*, Springer, Berlin, 2001, pp. 120–136.

51   E. Hahn-Deinstrop, *Dünnschichtchromatographie*, Wiley-VCH, Weinheim, 1998, pp. 48–67.

52   D. Nurok and M.J. Richard, *Anal. Chem.*, 53 (1981) 563.

53   D. Nurok, R.M. Becker, M.J. Richard, P.D. Cunningham, W.B. Gorman and C.L. Bush, *J. High. Resolut. Chromatogr. Chromatogr. Commun.*, 5 (1982) 373.

54   Q.S. Wang, Z.P. Zhan and C.X. Wang, *J. Planar Chromatogr.*, 4 (1991) 442.

55   H.J. Issaq, J.R. Klose, K.L. McNitt, J.E. Haky and G.M. Muschik, *J. Liquid Chromatogr.*, 4 (1981) 2091.

56   D. Nurok, R.M. Becker and K.A. Sassic, *Anal. Chem.*, 54 (1982) 1955.

57   S. Turina, in R.E. Kaiser (Ed.), *Planar Chromatography*, Vol. 1, Hüthig, Heidelberg, 1986, p. 15.

58   B.M.J. De Spiegeleer, P.H.M. De Moerloose and G.A.S. Slegers, *Anal. Chem.*, 59 (1987) 59.

59   L.R. Snyder, *J. Chromatogr. Sci.*, 16 (1978) 223.

60   Sz. Nyiredy, C.A.J. Erdelmeier, B. Meier and O. Sticher, *Planta Med.*, (1985) 241.

61   K. Dallenbach-Toelke, Sz. Nyiredy, B. Meier and O. Sticher, *J. Chromatogr.*, 365 (1986) 63.

62   Sz. Nyiredy, *Application of the "PRISMA" model for the selection of eluent-systems in Overpressure Layer Chromatography (OPLC)*, Labor MIM, Hungary, 1987.

63   Sz. Nyiredy, K. Dallenbach-Tölke and O. Sticher, *J. Planar Chromatogr.*, 1 (1988) 336.

64   Sz. Nyiredy, K. Dallenbach-Toelke and O. Sticher, *J. Liquid Chromatogr.*, 12 (1989) 95.

65   Sz. Nyiredy, *J. Chromatogr. Sci.*, 40 (2002) 553.

66   D. Nurok, *LC–GC*, 6 (1988) 310.

67   D. Nurok, *Chem. Rev.*, 89 (1989) 363.

68   A.-M. Siouffi and M. Abbou, in Sz. Nyiredy (Ed.), *Planar Chromatography – A Retrospective View for the Third Millennium*, Springer, Berlin, 2001, pp. 47–67.

69   F. Geiss, *J. Planar Chromatogr.*, 1 (1988) 102.

70   Sz. Nyiredy, in I.D. Wilson, E.R. Adlard, M. Cooke and C.F. Poole (Eds.), *Encyclopedia of Separation Science*, Vol. 10, Academic Press, London, 2000, pp. 4652–4666.

71   P. Petrin, *J. Chromatogr.*, 123 (1972) 65.

72   J.A. Perry, *J. Chromatogr.*, 165 (1979) 117.

73   R.E. Tecklenburg, R.M. Becker, E.K. Johnson and D. Nurok, *Anal. Chem.*, 55 (1983) 2196.

74   A. Niederwieser, *Chromatographia*, 2 (1969) 519.

75   A. Niederwieser and C.C. Honegger, in J.C. Giddings and R.A. Keller (Eds.), *Advances in Chromatography*, Vol. 2, Dekker, New York, 1966, pp. 123–131.

76   D.E. Jänchen, in W. Bertsch, S. Hara, R.E. Kaiser and A. Zlatkis (Eds.), *Instrumental HPTLC*, Hüthig, Heidelberg, 1980, pp. 133–164.

77   F. Geiss, H. Schlitt and A. Klose, *Z. Anal. Chem.*, 213 (1965) 331.

78   Sz. Nyiredy, Zs. Fatér, L. Botz and O. Sticher, *J. Planar Chromatogr.*, 5 (1992) 308.

79   Sz. Nyiredy, S.Y. Mészáros, K. Nyiredy-Mikita, K. Dallenbach-Toelke and O. Sticher, *J. High Resolut. Chromatogr. Chromatogr. Commun.*, 9 (1986) 605.

80   Sz. Nyiredy and Zs. Fatér, *J. Planar Chromatogr.*, 7 (1994) 329.

81   E. Tyihák and E. Mincsovics, *Hung. Sci. Instr.*, 57 (1984) 1.

82   Sz. Nyiredy, K. Dallenbach-Toelke and O. Sticher, in F.A.A. Dallas, H. Read, R.J. Ruane and I. Wilson (Eds.), *Recent Advances in Thin Layer Chromatography*, Plenum Press, London, 1988, pp. 45–54.

83 R.E. Kaiser, *J. Planar Chromatogr.*, 1 (1988) 265.

84 J.A. Perry, T.H. Jupille and L.H. Glunz, *Anal. Chem.*, 47 (1975) 65A.

85 B. Szabady and Sz. Nyiredy, in R.E. Kaiser, W. Günther, H. Gunz and G. Wulff (Eds.), *Dünnschicht-Chromatographie in memoriam Prof. Dr. Hellmut Jork*, InCom Sonderband, Düsseldorf, 1996, pp. 212–224.

86 B. Szabady, in Sz. Nyiredy (Ed.), *Planar Chromatography – A Retrospective View for the Third Millennium*, Springer, Berlin, 2001, pp. 88–102.

87 J.A. Perry, K.W. Haag and L.H. Glunz, *J. Chromatogr. Sci.*, 11 (1973) 447.

88 J.A. Perry, *J. Chromatogr.*, 113 (1975) 267.

89 K. Burger, *Z. Anal. Chem.*, 318 (1984) 228.

90 K. Burger, *GIT Suppl. Chromatogr.*, 4 (1984) 29.

91 K.D. Burger and H. Tengler, in R.E. Kaiser (Ed.), *Planar Chromatography*, Vol. 1, Hüthig, Heidelberg, 1986, pp. 193–205.

92 T.H. Jupille and J.A. Perry, *J. Chromatogr. Sci.*, 13 (1975) 163.

93 D.C. Fenimore and C.M. Davis, *Anal. Chem.*, 53 (1981) 252A.

94 G. Guiochon, M.F. Gonnord, A. Siouffi and M. Zakaria, *J. Chromatogr.*, 250 (1982) 1.

95 M. Zakaria, M.F. Gonnord and G. Guiochon, *J. Chromatogr.*, 271 (1983) 127.

96 Sz. Nyiredy, in L. Mondello, A.C. Lewis and K.D. Bartle (Eds.), *Multidimensional Chromatography*, Wiley, Chichester, 2001, pp. 171–196.

97 J.C. Giddings, in H.J. Cortes (Ed.), *Multidimensional Chromatography*, Dekker, New York, 1990, pp. 1–27.

98 J.C. Giddings, *J. High Resolut. Chromatogr. Chromatogr. Commun.*, 10 (1987) 319.

99 Sz. Nyiredy, in Sz. Nyiredy (Ed.), *Planar Chromatography – A Retrospective View for the Third Millennium*, Springer, Berlin, 2001, pp. 103–119.

100 Sz. Nyiredy, K. Dallenbach-Tölke and Sz. Nyiredy, *J. Chromatogr.*, 450 (1988) 241.

101 K. Dallenbach-Tölke, Sz. Nyiredy, S.Y. Mészáros and O. Sticher, *J. High Resolut. Chromatogr. Chromatogr. Commun.*, 10 (1987) 362.

102 E. Tyihák, Sz. Nyiredy, G. Verzár-Petri, S.Y. Mészáros, I. Farkas-Tompa, A. Nagy, L. Szepesy, L. Vida, E. Mincsovics, G. Kemény and Z. Baranyi, *Hung. Pat. No. 189.737; German Pat. No. 3.512.547* (1986)

103 E. Mincsovics, M. Garami, L. Kecskés, B. Tapa, Z. Végh, Gy. Kátay and E. Tyihák, *J. AOAC Int.*, 83 (1999) 587.

104 Sz. Nyiredy, C.A.J. Erdelmeier and O. Sticher, in R.E. Kaiser (Ed.), *Planar Chromatography*, Vol. 1, Hüthig, Heidelberg, 1986, pp. 119–164.

105 Sz. Nyiredy, in J. Sherma and B. Fried (Eds.), *Handbook of Thin Layer Chromatography*, Dekker, New York, 1996, pp. 307–340.

106 D.E. Jänchen, *Int. Lab.*, March (1987) 66.

107 C.F. Poole and S.K. Poole, *J. Chromatogr.*, 492 (1989) 539.

108 Sz. Nyiredy, *J. Planar Chromatogr.*, 9 (1996) 403.

109 Sz. Nyiredy, Zs. Fatér and B. Szabady, *J. Planar Chromatogr.*, 7 (1994) 406.

110 H. Jork, W. Funk, W. Fischer, H. Wimmer, *Thin-Layer Chromatography, Reagents and Detection Methods, Physical and Chemical Detection Methods, Reagents I*, Vol. 1a, VCH, Weinheim, 1990.

111 W. Funk, *Z. Anal. Chem.*, 318 (1984) 206.

112 L.R. Treiber (Ed.), *Quantitative Thin Layer Chromatography and Its Industrial Applications*, Marcel Dekker, New York, 1987.

113 G. Cimpan, in Sz. Nyiredy (Ed.), *Planar Chromatography – A Retrospective View for the Third Millennium*, Springer, Berlin, 2001, pp. 410–445.

114 B. Spangenberg and K.-F. Klein, *J. Chromatogr. A*, 898 (2000) 265.

115 B. Spangenberg and K.-F. Klein, *J. Planar Chromatogr.*, 14 (2001) 260.

116 I. Vovk, M. Prosek and R.E. Kaiser, in Sz. Nyiredy (Ed.), *Planar Chromatography – A Retrospective View for the Third Millennium*, Springer, Berlin, 2001, pp. 464–488.

117 J. Stroka, T. Peschel, G. Tittelbach, G. Weidner, R. van Otterkijk and E. Anklam, *J. Planar Chromatogr.*, 14 (2001) 109.

118 S. Mustoe and S. McCrossen, *J. Planar Chromatogr.*, 14 (2001) 252.

119 S. Essig and K.-A. Kovar, *Chromatographia*, 53 (2001) 321.

120 W. Dammertz and E. Reich, in Sz. Nyiredy (Ed.), *Planar Chromatography – A Retrospective View for the Third Millennium*, Springer, Berlin, 2001, pp. 235–246.

121 G. Glauninger, K.-A. Kovar and V. Hoffmann, *Z. Anal. Chem.*, 338 (1990) 710.

122 K.-A. Kovar, H.K. Ensslin, O.R. Frey, S. Rienas and S.C. Wolff, *J. Planar Chromatogr.*, 4 (1991) 246.

123 I.O.C. Rager and K.-A. Kovar, in Sz. Nyiredy (Ed.), *Planar Chromatography – A Retrospective View for the Third Millennium*, Springer, Berlin, 2001, pp. 247–260.

124 R.L. White, *Anal. Chem.*, 57 (1985) 1819.

125 I. Vovk and G. Mocnik, in Sz. Nyiredy (Ed.), *Planar Chromatography – A Retrospective View for the Third Millennium*, Springer, Berlin, 2001, pp. 312–335.

126 E. Koglin, *J. Mol. Struct.*, 173 (1988) 369.

127 E. Koglin, *J. Planar Chromatogr.*, 2 (1989) 194.

128 E. Koglin, *J. Planar Chromatogr.*, 3 (1990) 117.

129 E. Koglin, *J. Planar Chromatogr.*, 6 (1993) 88.

130 J.W. Fiola, G.C. Didonato and K.L. Busch, *Rev. Sci. Instrum.*, 57 (1986) 2294.

131 Y. Nakagawa and K. Iwatani, *J. Chromatogr.*, 562 (1991) 99.

132 S.M. Brown, H. Schurz and K.L. Busch, *J. Planar Chromatogr.*, 3 (1990) 222.

133 K.L. Busch, *J. Planar Chromatogr.*, 5 (1992) 72.

134 S.M. Brown and K.L. Busch, *J. Planar Chromatogr.*, 5 (1990) 338.

135 A. Crecelius, M.R. Clench, D.S. Richards, J. Mather and V. Parr, *J. Planar Chromatogr.*, 13 (2000) 76.

136 I.D. Wilson and W. Morden, *J. Planar Chromatogr.*, 4 (1991) 226.

137 P. Martin, W. Morden, P. Wall and I. Wilson, *J. Planar Chromatogr.*, 5 (1992) 255.

138 W. Morden and I.D. Wilson, *J. Planar Chromatogr.*, 8 (1995) 98.

139 I.D. Wilson and W. Morden, *LC–GC*, 12 (1999) 72.

140 K.L. Busch, in Sz. Nyiredy (Ed.), *Planar Chromatography – A Retrospective View for the Third Millennium*, Springer, Berlin, 2001, pp. 261–277.

141 I.D. Wilson, M. Spraul and E. Humpfer, *J. Planar Chromatogr.*, 10 (1997) 217.

142 I. Klebovich, in Sz. Nyiredy (Ed.), *Planar Chromatography – A Retrospective View for the Third Millennium*, Springer, Berlin, 2001, pp. 293–311.

143 G.C. Zogg, Sz. Nyiredy and O. Sticher, *J. Planar Chromatogr.*, 1 (1988) 351.

144 C.F. Poole and S.K. Poole, *J. Chromatogr.*, 492 (1989) 539.

145 S. Ebel, in F. Geiss (Ed.), *Fundamentals of Thin Layer Chromatography (Planar Chromatography)*, Hüthig, Heidelberg, 1987, pp. 420–436.

146 S. Ebel, in Sz. Nyiredy (Ed.), *Planar Chromatography – A Retrospective View for the Third Millennium*, Springer, Berlin, 2001, pp. 353–385.

147 L. Botz, Sz. Nyiredy and O. Sticher, *J. Planar Chromatogr.*, 3 (1990) 10.

148 Sz. Nyiredy, in I.D. Wilson, E.R. Adlard, M. Cooke and C.F. Poole (Eds.), *Encyclopedia of Separation Science*, Vol. 2, Academic Press, London, 2000, pp. 888–899.

149 E. Mincsovics, E. Tyihák and A.M. Siouffi, *J. Planar Chromatogr.*, 1 (1988) 141.

150   Sz. Nyiredy, *Anal. Chim. Acta*, 236 (1990) 83.

151   M. Ranny, *Thin-Layer Chromatography with Flame Ionization Detection*, D. Riedel, Dordrecht, 1987.

152   K.D. Mukherjee, in J. Sherma and B. Fried (Eds.), *Handbook of Thin-Layer Chromatography, Second Edition, Revised and Expanded*, Dekker, New York, 1996, pp. 361–372.

153   P. Buncak, *GIT Suppl. Chromatogr.*, 3 (1982) 1.

154   P. Buncak, *Z. Anal. Chem.*, 318 (1984) 291.

155   J. Vajda, L. Leisztner, J. Pick and N. Anh-Tuan, *Chromatographia*, 21 (1986) 152.

156   Sz. Nyiredy, C.A.J. Erdelmeier, K. Dallenbach-Tölke, K. Nyiredy-Mikita and O. Sticher, *J. Nat. Prod.*, 49 (1986) 885.

157   UNIPLATE™ Taper Plate, *Analtech Technical Report No. 8202*, Newark, DE, 1985.

158   Sz. Nyiredy, in I.D. Wilson, E.R. Adlard, M. Cooke and C.F. Poole (Eds.), *Encyclopedia of Separation Science*, Vol. 10, Academic Press, London, 2000, pp. 4652–4666.

159   G. Guiochon, M.F. Gonnord, M. Zakaria, L.A. Beaver and A.M. Siouffi, *Chromatographia*, 17 (1983) 121.

160   T. Omori, *J. Planar Chromatogr.*, 1 (1988) 66.

161   B. Szabady, Zs. Fatér and Sz. Nyiredy, *J. Planar Chromatogr.*, 12 (1999) 82.

162   Sz. Nyiredy, *Swiss Patent No. 674314*.

163   Sz. Nyiredy, *Hungarian Patent Application No. P0202968*.

164   M.F. Gonnord and G. Guiochon, in E. Tyihák (Ed.), *Proceedings of the International Symposium on TLC with Special Emphasis on OPLC*, Labor MIM, Budapest, 1986, pp. 241–250.

165   L. Botz, Sz. Nyiredy and O. Sticher, *37th Annual Congress on Medicinal Plant Research*, Braunschweig, FRG, 5–9 September 1989, Abstr. No. 1–20.

166   L. Botz, Sz. Nyiredy and O. Sticher, *J. Planar Chromatogr.*, 3 (1990) 352.

167   L. Botz, Sz. Nyiredy and O. Sticher, *J. Planar Chromatogr.*, 4 (1991) 115.

168   P.R. Boshoff, B.J. Hopkins and V. Pretorius, *J. Chromatogr.*, 126 (1976) 35.

169   J.W. Hofstraat, M. Engelsma, R.J. van de Nesse, C. Gooijer, N.H. Velthorst and U.A.Th. Brinkman, *Anal. Chim. Acta*, 186 (1986) 247.

170   J.W. Hofstraat, S. Griffioen, R.J. van de Nesse, U.A.Th. Brinkman, C. Gooijer and N.H. Velthorst, *J. Planar Chromatogr.*, 186 (1986) 247.

171   D.E. Jänchen and H.J. Issaq, *J. Liquid Chromatogr.*, 11 (1988) 1941.

172   Sz. Nyiredy and E. Tyihák (Eds.), *Forced-Flow Planar Chromatography*, Elsevier, Amsterdam, in press.

Erich Heftmann (Editor)
Chromatography, 6th edition
Journal of Chromatography Library, Vol. 69A
© 2004 Elsevier B.V. All rights reserved

Chapter 7

Electrokinetic chromatography

ERNST KENNDLER and ANDREAS RIZZI

CONTENTS

7.1 Introduction 298
   7.1.1 Modes of electro(kinetic) chromatography 298
7.2 Electro-osmotic flow in open and packed capillaries 301
   7.2.1 Electro-osmotic flow in open tubes 302
   7.2.2 Electro-osmotic flow in porous beds 303
7.3 Electrochromatography with stationary phases 304
   7.3.1 Migration 304
   7.3.2 Peak dispersion 305
   7.3.3 Open tubes 306
   7.3.4 Packed beds and monoliths 306
      7.3.4.1 Selectivity 307
      7.3.4.2 Peak dispersion 307
7.4 Electrokinetic chromatography with pseudo-stationary phases 310
   7.4.1 Micelles and micro-emulsions 310
      7.4.1.1 Migration 311
      7.4.1.2 Peak dispersion 312
   7.4.2 Soluble charged polymers and other additives 313
   7.4.3 Charged, small-molecular-mass additives 314
7.5 Electrically driven *vs.* pressure-driven chromatography 315
   7.5.1 Validation 315
   7.5.2 Gradient elution 315
   7.5.3 Overall sensitivity 315
   7.5.4 Liquid/liquid partitioning 316
7.6 Conclusions 316
References 317

7.1 INTRODUCTION

Electrokinetic chromatography (EC) comprises separation systems in which (a) distribution (partitioning or adsorption) of analytes takes place between two "phases" and (b) an electroosmotic flow (EOF) serves for the transport of a liquid phase through the separation system. This EOF is established after the application of an electric field on the separation capillary. Mobile-phase movement is thus based on electro-osmosis instead of on a pressure gradient, as in classical chromatography. The dissolved analytes migrate through the separation system in the mobile phase, and the different migration velocities of the analytes are implemented by utilizing interactions with a second "phase", which is a real phase in the simplest case. However, there is no difference as a matter of principle in the distribution equilibrium (which is, by the way, established solely at the interface between the two phases) whether a real phase is considered or a species, which is simply dissolved in the liquid as a soluble additive. Indeed, in electrokinetic chromatography we can span the range from real stationary solid or liquid phases to small molecules that form an adduct with the analytes. Common to all electrokinetic chromatography methods is a different velocity of the two "phases" in relation to each other, and the operation of separation selectivity due to a different extent of interaction of the analytes with these two "phases".

Conceptually, one of the main advantages of EC is the significant reduction of plate height due to the plug-type profile of the electro-osmotically driven bulk flow. The lower plate height and the fact that column length and particle size are not restricted by a maximum applicable pressure drop across the column – as it is in pressure-driven liquid chromatography (pd LC) – results in EC separation systems with very high plate numbers. It is evident that high plate numbers significantly improve the resolution at a given selectivity in general; they are of particular benefit when dealing with complex mixtures of many components, as they enable attainment of a high peak capacity, *i.e.*, an increase in the maximum number of distinguishable peaks.

Electrokinetically driven (ekd) chromatographic systems do not require pumps. This is an important aspect when working with miniaturized chromatographic beds, *e.g.*, in edged chips, for which a pump system might not be appropriate. Further miniaturization of separation systems will utilize the EOF even more.

7.1.1 Modes of electro(kinetic) chromatography

The general scheme which depicts the situation in all EC systems is shown in Fig. 7.1. The two "phases" between which the analyte is distributed are indicated by (1) and (2), respectively; they migrate with the different velocities $v^{(1)}$ – phase (1) – and $v^{(2)}$ – phase (2). Phase (1) is always driven by the EOF, and consequently $v^{(1)} = v^{EOF}$.

From Fig. 7.1, the migration velocity of analyte, i, can be derived as follows. The total velocity, v_i, is composed of the velocities of the analyte in the particular "phases", each weighted by the probability of being present there. This probability is equivalent to the fraction, x_i, of the mole number, n_i, of i in the respective phase related to the total mole

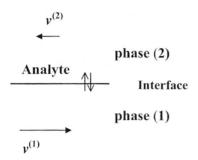

Fig. 7.1. General scheme of electrokinetic chromatography with partitioning of the analyte between two "phases", (1) and (2). $v^{(1)} = v^{EOF}$ and $v^{(2)}$ are the velocities of the two "phases".

number, of i present in both phases. In Phase (1) this fraction, $x_i^{(1)}$, is given by

$$x_i^{(1)} = \frac{n_i^{(1)}}{n_i^{(1)} + n_i^{(2)}} \qquad (7.1)$$

Eqn. 7.1 can be rewritten as

$$x_i^{(1)} = \frac{1}{1 + \dfrac{n_i^{(2)}}{n_i^{(1)}}} = \frac{1}{1 + k_i} \qquad (7.2)$$

where k_i is the distribution coefficient of analyte i between phases (2) and (1) in terms of mole numbers. In chromatography, k_i is usually referred to as retention factor (in the older literature also as capacity factor). The retention factor is related to the Nernst distribution coefficient, K_i, defined as the ratio of the molar concentrations, c_i, by the phase ratio, $q = V^{(2)}/V^{(1)}$, V being the phase volume

$$k_i = \frac{n_i^{(2)}}{n_i^{(1)}} = \frac{c_i^{(2)} V^{(2)}}{c_i^{(1)} V^{(1)}} - K_i q \qquad (7.3)$$

In case both phases are bulk phases, K_i is dimensionless. In systems in which one phase is a surface, q is defined as $A^{(2)}/V^{(1)}$, A being the surface area, and K_i then has the dimension of an inverse length. The probability of finding a molecule during a certain time period in Phase (2) is equivalent to $x_i^{(2)}$, being the difference between unity and $1/(1 + k_i)$ (see Eqn. 7.1)

$$x_i^{(2)} = \frac{k_i}{1 + k_i} \qquad (7.4)$$

as the molecules are either in (1) or in (2). The total velocity of the analyte is thus the sum of its weighted velocities in the two phases

$$v_i = \frac{1}{1 + k_i} v_i^{(1)} + \frac{k_i}{1 + k_i} v_i^{(2)} \qquad (7.5)$$

Assuming the absence of any pd hydrodynamic flow, the velocity can now be derived for the individual electrokinetic method simply by taking into account the velocity the analyte has in the two phases. Dividing by the electric field strength, $E = U/L$ (L is the capillary length, U the applied voltage) yields the relation of the respective mobility, μ. The mobility is the velocity in an electric field with unit strength, $\mu = v/E$.

$$\mu_i = \frac{1}{1 + k_i} \mu_i^{(1)} + \frac{k_i}{1 + k_i} \mu_i^{(2)} \tag{7.6}$$

where μ_i is the total mobility, which can be derived directly from the measured velocity, v_i.

It should be mentioned that all mobilities in Eqn. 7.6 are those determined from the velocity relative to the capillary wall as reference. This reference is obvious when dealing with phases that are actually stationary, but is not the only possible reference when dealing with moving pseudophases. Eqn. 7.6 can be transformed into Eqn. 7.7, which enables the retention factor values to be calculated from the experimentally determined mobilities

$$k_i = \frac{\mu_i^{(1)} - \mu_i}{\mu_i - \mu_i^{(2)}} \tag{7.7}$$

and which is valid for all EC methods. Note that the velocities are vectors, with the convention that their sign is positive if directed to the cathode, and negative when directed to the anode. The same holds for the mobilities, which are also signed quantities, but this is differently treated in the theory of conductance.

Electrically neutral solutes migrate in the liquid phase with the velocity v^{EOF} of the electro-osmotic flow. Ionic analytes, which have their own charge, exhibit an additional velocity vector, v_i^{eph}, under the influence of the applied electric field in this phase, which is proportional to their electrophoretic mobility, μ_i^{eph}. The total migration velocity of analyte i in Phase (1), $v_i^{(1)}$, thus reads

$$v_i^{(1)} = v_i^{eph} + v^{EOF} \tag{7.8}$$

The particular form of the "phases" (1) and (2) leads to the different modes of electrokinetic chromatography:

(a) *Electrochromatography with solid phases.* The stationary phase, (2), is the surface of either a porous material in the form of a bed of packed particles (packed-bed EC), or of a continuous, homogeneous, porous, so-called "monolithic" bed (monolithic-bed EC), or it is the wall of the chromatographic capillary (open-tube EC). This stationary phase is normally fixed in the system, but it could move as well (for instance, as charged colloidal particles), as it is only the relative difference in migration of the two phases which is the prerequisite for analyte separation. When the stationary phase is fixed, it is obvious that its migration velocity, $v^{(2)}$, as well as $v_i^{(2)}$ is zero.

(b) *Micellar electrokinetic chromatography (MEKC).* Phase (2) is a pseudo-stationary phase, consisting of (charged) micelles, suspended in the liquid phase – Phase (1). Due to their own charge, the micelles migrate in the electric field with a total velocity that differs from that of the EOF. Pseudophases may also consist of non-charged micelles in those cases where the analytes are charged; such systems are considered as rather belonging to capillary zone electrophoresis (CZE).

(c) *Micro-emulsion electrokinetic chromatography (MEEKC).* The pseudo-stationary phase consists of charged micro-droplets, dispersed in the liquid phase. As in MEKC, the micro-droplets have their own migration velocity.

(d) *Electrokinetic chromatography with soluble polymers.* Charged soluble polymers (linear or branched polymers, dendrimers, calixarenes) serve as Phase (2). The pseudo-stationary phase does not have to consist of discrete micelles or micro-droplets.

(e) *Electrokinetic chromatography based on (hydrophobic) interaction or complex formation with oligomeric or monomeric additives.* Charged low-molecular-weight additives, which undergo a kind of complexation with the analyte serve as Phase (2). As analytes can interact also with these molecularly dispersed partners, there is no reason to limit EC to large particles. Indeed, low-molecular-weight additives represent a borderline case of the EC systems. The additive can be neutral, if the analyte is charged; this is, however, usually considered as a variant of CZE.

7.2 ELECTRO-OSMOTIC FLOW IN OPEN AND PACKED CAPILLARIES

The two electrokinetic phenomena, electro-osmosis and electrophoresis, share the common feature that a relative movement of the adjacent liquid occurs tangentially to a charged surface after the application of an electric field [1,2]. Electro-osmosis and electrophoresis are mirror images of the same phenomenon. In electro-osmosis, it is the liquid that is transported through a capillary, an immobilized set of particles, a membrane, or a porous plug. Electro-osmosis is the result of an electric double layer, formed between the liquid and the charged surface, which are in contact with each other. The electric potential falls off perpendicular to the surface, with the characteristic ζ potential at the slip plane, the plane of shear [1,2]. The electro-osmotic velocity of the liquid is proportional to the field strength and depends on the ζ potential, the relative permittivity (dielectric constant) of the solvent, ε_r, and its dynamic viscosity, η.

For a spherical, non-conducting particle with radius r, the electrokinetic mobility (*i.e.*, μ^{EOF} or μ^{eph}, the electrophoretic mobility of a particle with a certain ζ potential) is expressed in a general manner by the Henry equation

$$\mu = \frac{2}{3} \frac{\varepsilon_r \varepsilon_0 \zeta}{\eta} f(\kappa r) \tag{7.9}$$

where ε_0 is the permittivity of vacuum and $f(\kappa r)$ is Henry's correction factor. $f(\kappa r)$ depends on the relation between the thickness (Debye length) of the double layer, κ^{-1}, to the particle radius, r. The Debye length, κ^{-1}, is inversely proportional to the ionic strength, $I = 0.5 \sum z_i^2 c_i$, of the solution: $\kappa^{-1} = \sqrt{(\varepsilon_r \varepsilon_0 RT)/F^2 \sum z_i^2 c_i}$; where R is the gas constant, F the Faraday constant, T the absolute temperature, z_i the charge number of the ion, i, in solution, and c_i its molar concentration.

Depending on the size ratio of the Debye length to the radius, we can differentiate two extreme cases of the Henry formula. The one, the Hückel–Onsager case (relevant for the electrophoretic migration of small molecules), is given when the particle is considered as point charge, that is when $r \ll \kappa^{-1}$, and consequently $\kappa r \ll 1$; the Henry coefficient

$f(\kappa r) = 1$. The Hückel–Onsager case ignores the deformation of the applied field around the particle. More relevant for the evaluation of the electro-osmotic flow mobility, μ^{EOF} and thus more interesting for CEC is the Helmholtz–Smoluchowsky case, where the surface is considered a flat plane and the field is assumed to be uniform and parallel to the surface. Here, the particle (or capillary) radius is large compared to the Debye length: $\kappa r \gg 1$. For the limiting Helmholtz–Smoluchowsky case, the Henry correction factor $f(\kappa r) = 3/2$. The electro-osmotic mobility is then given by

$$\mu = \mu^{EOF} = \frac{\varepsilon_r \varepsilon_0 \zeta}{\eta} \tag{7.10}$$

The zeta potential is related to the charge density of the surface, σ, as

$$\zeta = \frac{\sigma}{\varepsilon_r \varepsilon_0 \kappa} \tag{7.11}$$

Combination of Eqn. 7.10 with that for the Debye length gives

$$\mu = A \frac{\sigma}{\eta} \sqrt{\frac{\varepsilon_r T}{I}} \tag{7.12}$$

where A is a proportionality constant.

Organic solvents, either pure or as constituents of the background electrolyte, affect the electro-osmotic flow severely [3,4]. According to Eqn. 7.10, the electro-osmotic mobility should depend on the solvent-specific parameters, ε_r and η, actually on ε_r/η, given that the ζ potential does not change upon variation of the solvent. However, such changes can hardly be excluded, and changes of the surface charge groups, of pK_a values of the buffer constituents, etc., have to be considered when varying the solvent composition. In addition, changes in mobilities of the buffer constituents, and occurrence of ion association should be accounted for. For these reasons, the actually observed influence of the solvent will in most cases deviate from that predicted by theory.

7.2.1 Electro-osmotic flow in open tubes

Under the conditions of $\kappa r \gg 1$ (when $\kappa^{-1} \ll r$) the EOF-mobility is independent of the particle or capillary radius (Eqn. 7.10). As a consequence, outside the double layer distance, the electro-osmotic velocity of the liquid is constant over the cross-section of an open capillary. It is clear that within the double layer the mobility changes with the radial ordinate. Normally, in EC the Debye thickness is small compared to the radius of the capillaries, as can be seen upon insertion of the applicable constants: $\kappa^{-1} = 0.304/\sqrt{I}$ at 25°C for aqueous solutions (with κ^{-1} being in nm, when the ionic strength is in M). For a 10 mM solution, the thickness is about 3 nm, which is orders of magnitude smaller than the usual capillary radii. It further decreases with increasing ionic strength. When the capillary diameter does not exceed the double layer thickness significantly (say, at least 50-fold) the EOF velocity is smaller than that calculated from Eqn. 7.10, and the flow is no longer plug-like; it resembles a parabolic profile, as described by Rice and Whitehead [5]. However, down to a capillary or channel diameter of 20 times the Debye length the effect will not be dramatic.

7.2.2 Electro-osmotic flow in porous beds

Plugs formed from solid particles have irregular shapes. Therefore, the equations for impulse transport cannot be solved without making some geometric models. Two models were introduced for the theoretical treatment of the EOF in porous systems: the cell model, and the equivalent-cylinder model [1]. The latter, which is more common, considers the porous plug as a bundle of parallel cylinders of a certain mean diameter. If the radius is much larger than the double-layer thickness ($\kappa r \gg 1$), the Smoluchowsky equation can be applied, in which the electro-osmotic mobility is independent of the radius. For the packed columns used in EC we consider the channels between the particles to be capillaries with an average diameter of *ca.* 1/4 of the particle diameter [6,7]. For HPLC material with particle diameters in the range of $1-5$ μm, the average channel diameter is then between 0.25 and 1.25 μm. With monolithic columns, the diameters of the macro-porous channels can vary considerably, depending on the porosity of the material; they can be assumed to range between 0.15 and 5 μm. In both cases, r is significantly larger than κ^{-1}.

There are a number of restrictions that limit the applicability of the theory in practice. One is given by the fact that the pore diameter *inside* the porous particles can be comparable to the Debye length, even when the entire particle diameter is large. It was shown above that the Debye length is in the range of a few nm in mobile phases with several 10 to 100 mmol/L ionic strengths. The pore diameters of such particles are, however, in the 8- to 12-nm range, which is only 2 to 3 times larger than the double-layer thickness. Within the pores, double-layer overlap occurs, which reduces the EOF mobility [8,9], given that the pores are accessible to flow. In that case the EOF profile is no longer plug-like.

In tubes filled with particles typically used in CEC, the column-to-particle diameter ratio, the so-called aspect ratio, is relatively small (it is the order of 2 to 20). In such systems, the local total porosity of the packed bed is not constant, but varies with radial position. It is unity in the immediate vicinity of the wall, and reaches a constant value, typically 0.6 to 0.7, far from the wall, displaying a damped oscillation over a distance of several particle diameters [10–13]. As the EOF in the packed capillary has a radial velocity distribution which depends on the aspect ratio and on the ratio of the ζ potentials of the wall to that of the particle surface, we find a radial EOF mobility distribution that is not consistently described by the Smoluchowsky theory.

Another refinement of the theoretical treatment becomes necessary, if particles exhibit a certain surface conductance, *i.e.* an excess conduction tangential to the charged surface. As the mobile phases or background electrolytes (BGE) usually applied in CEC exhibit a relatively high electric conductance, surface conductance most often does not play a relevant role. However, it might become significant in systems with low bulk conductance and large κr (>25). Then the zeta potential, as related to the electro-osmotic mobility, is underestimated. Under these conditions, the Dukhin theory [14–16] offers a more adequate relation between mobility and ζ-potential.

In CEC practice another point must be taken into account, if the column consists of a packed and an open segment. Even given that the ζ potential at the capillary wall is identical along these segments (and assuming the theoretical case that the particles do not contribute to the EOF), the electric conductances and, therefore, the field strengths in both segments will not be equal. The ratio of the conductance of a packed capillary, κ_p, with interstitial porosity, ε_i, to the conductance of an open capillary, κ_o, of the same dimension, and filled with the same electrolyte solution, is [17]

$$\kappa_p / \kappa_o = (\varepsilon_i)^m \tag{7.13}$$

Typical values for CEC columns with spherical packings are: ε_i between 0.3 and 0.4 and the exponent m 1.5 [18,19]. This leads to a *ca.* 4 times larger conductance in the open segment, thus to a different field strength and, consequently, to a smaller EOF. As the liquid is non-compressible, the volume flow-rate must be the same in both segments. Therefore, a hydrodynamic pressure occurs at the border between the open and the packed segments, and a laminar flow is generated. As this additional flow will occur most probably mainly through the open segment, due to its higher permeability, it often does not influence significantly the analyte migration in packed columns [20]. Considering charged particle surfaces as a source of EOF, the EOF in the packed segment will be even larger due to the higher charge density there.

7.3 ELECTROCHROMATOGRAPHY WITH STATIONARY PHASES

7.3.1 Migration

Commonly, when we speak of CEC, we assume the use of a stationary phase, fixed in the system. This means that velocity $v^{(2)}$ is zero, and Eqn. 7.5 for the analyte velocity reduces to

$$v_i = \frac{1}{1 + k_i} v_i^{(1)} \tag{7.14}$$

The migration velocity of analytes in the mobile phase, $v_i^{(1)}$, equals v^{EOF} in case the analytes are electrically neutral. If they possess an electric charge, their velocity in the mobile phase will be given by Eqn. 7.8. Thus, the total migration velocity, v_i, (which is often called apparent velocity in the context of electrophoresis) depends on the analyte velocity in the mobile phase and on the extent of partitioning between the two phases. Neutral analytes migrate with the EOF velocity when k_i is zero, but charged analytes can move even faster than the EOF, even though they have partition coefficients different from zero.

Insertion of Eqn. 7.8 for the analyte velocity into Eqn. 7.14, and division by the field strength gives the analyte mobility

$$\mu_i = \frac{1}{1 + k_i} (\mu_i^{eph} + \mu^{EOF}) \tag{7.15}$$

and the equation by which the retention factor can be evaluated from the measured mobilities as

$$k_i = \frac{\mu_i^{eph} + \mu^{EOF} - \mu_i}{\mu_i} \tag{7.16}$$

For a neutral analyte ($\mu_i^{eph} = 0$) Eqn. 7.16 yields the classical relation between retention factor and migration time – mobility and migration time are inversely proportional; $\mu_i = v_i/E = l/(t_iE)$, where l is the migration distance

$$k_i = \frac{t_i - t_{EOF}}{t_{EOF}} \qquad (7.17)$$

This allows simple assessment of the retention factor from the residence time of the analyte, t_i, and the residence time, t_{EOF}, of the EOF marker (a compound which is unretained). For charged solutes it is necessary to determine the electrophoretic mobility of the analyte under conditions where it is not retained (*e.g.*, in free solution of the same pH, ionic strength, and solvent composition).

7.3.2 Peak dispersion

The theoretical treatment of peak dispersion in CEC follows the plate-height concept used in pd HPLC. Commonly, five processes are considered as being predominantly responsible for peak-broadening in HPLC with packed beds: longitudinal diffusion, convective mixing, eddy dispersion, broadening due to the kinetics of mass exchange in the mobile and the stationary phase (including kinetic effects from the adsorption/desorption process), respectively [21,22] (Chap. 1). Given that the system is linear, the total plate-height is simply the sum of the plate-heights of the different contributions.

Electrokinetic chromatography differs from the corresponding pd system significantly and particularly with respect to the following contributions:

(a) *Convective mixing.* In contrast to pd HPLC, where the mobile-phase flow profile is parabolic, the EOF, transporting the mobile phase in CEC, does not form a radial velocity gradient (at least for channel diameters which are significantly larger than the thickness of the electric double layer). In porous chromatographic beds the plug-like profile of the EOF largely reduces a possible radial flow-velocity distribution, and most importantly, makes the flow-velocity independent of the channel diameter. It is particularly this contribution which is usually significant in pd LC and is essentially reduced in CEC due to the characteristic features of EOF flow profiles.

(b) *Joule heating.* This adventitious contribution common to all electrokinetic methods is due to the heat generated in the ionic solution upon passage of an electric current. There is no equivalent in pd HPLC. The increased temperature of the solution leads to a temperature gradient between the center of the channel and the wall. The additional increment to dispersion based on this effect has been discussed by Knox *et al.* [7,23–25] and Horváth *et al.* [26–31]. For CZE, it has been demonstrated that it is often an overestimated effect on peak broadening, although the temperature increase inside the column could be drastic [32].

As in classical column chromatography, two main techniques can be differentiated, depending on how the stationary phase is brought into the separation system. In one method the stationary phase is fixed on the inner wall of the capillary (open-tube chromatography), and in the second technique the stationary phase is at the surface of a porous material, of either particles or a monolithic rod.

7.3.3 Open tubes

Open-tubular capillary electrochromatography is performed with the same separation capillary as pd open-tube LC. As such systems have a well-defined geometry, which allows a clear theoretical description of the dispersion processes, the processes which contribute to peak-broadening here can be expressed analytically. The k_i dependence of the plate-height increment of the combined term, originating from convective mixing and mobile-phase mass exchange with EOF transport in the open cylindrical capillary reads [33]

$$H_{c,m} = \frac{k_i^2 r^2}{4D_i^{(1)}(1 + k_i)^2} v^{(1)} \tag{7.18}$$

r being the capillary inner radius, and $D_i^{(1)}$ the diffusion coefficient of i in the mobile phase; $v^{(1)} = v^{EOF}$ is the mobile-phase velocity.

$H_{c,m}$ in CEC is lower than in the case of open-tubular chromatography with laminar flow (Taylor dispersion). The corresponding equation for the latter case reads

$$H_{c,m} = \frac{(1 + 6k_i + 11k_i^2)r^2}{24D_i^{(1)}(1 + k_i)^2} v^{(1)} \tag{7.19}$$

In this equation, $v^{(1)}$ is the pressure-caused mean flow-velocity of the mobile phase. Note that the term for EOF transport reaches a limiting value of $(1/4)[r^2 v^{(1)}/D_i^{(1)}]$ for analytes with very large k_i values, whereas that for the pd flow has a limiting value of $(11/24)[r^2 v^{(1)}/D_i^{(1)}]$, which is nearly twice as large [34].

Open-tube CEC faces the analogous problems as open-tube HPLC: due to the small diffusion coefficients of the analytes in the liquid phase, very low mobile-phase velocities are needed to approach the Van Deemter minimum, and this results in very long analysis times. Another way to overcome this limitation is to shorten the diffusional path length by reducing the inner diameter of the capillary. This strategy, however, leads to low detection sensitivity and a high demand on miniaturization in order to suppress extra-column effects.

7.3.4 Packed beds and monoliths

In CEC with solid beds, the driving force for the EOF is associated to the ζ potential *via* the charged surfaces of the particles and (to a minor extent) of the capillary wall. With silica-based alkyl-modified support material, the surface charge results either from dissociated residual silanol groups still present after derivatization or, in addition, from immobilized charged groups, fixed on the support material *via* a short spacer. When strongly or weakly acidic groups, like sulfonic acids or carboxylic acids, are immobilized, a cathodic EOF will be generated, while binding basic groups to the surface yields an anodic EOF (given that the anodic flow resulting from the positively charged surface groups prevails over the cathodic flow of the silanolates). When polymeric support materials, like polystyrene/divinylbenzene (PS/DVB) co-polymers or poly(methacrylate) (PMA)-based support particles, are used, the incorporation of ion-exchange groups is the method of choice for obtaining a charged surface. In this case, the ligand density alone determines the surface charge density and thus the strength of EOF. The immobilization of strongly acidic ligands which produce an EOF even at very low pH values, offers an opportunity to perform CEC separations under highly acidic conditions, at which the dissociation of the silanol groups is insufficient. It is noteworthy that even surfaces of relatively inert organic polymeric particles, like polyethylene, exhibit a manifest ζ potential [4], often due to the adsorption of ionic species from the solution.

7.3.4.1 Selectivity

Most applications of CEC are based on a reversed-phase (RP) mode of operation with predominantly hydrophobic surfaces and aqueous mobile phases containing a certain amount of organic modifier. The addition of organic solvent components to the mobile phase significantly influences the EOF. Charge density, viscosity, and relative permittivity (Eqns. 7.9–7.12) of mixed organic/aqueous solvents are considerably different from the corresponding values in water. Acetonitrile is most frequently employed as the organic solvent constituent, as it does not affect the EOF so pronouncedly like methanol and ethanol; the ε_r/η ratio of the mobile phase changes much less over the composition range in case of water/acetonitrile mixtures [3]. Usually, the EOF in mixed organic/aqueous solvents is lower than in pure aqueous solution.

Considering the retention mechanism with regard to non-charged analytes, EC, performed in this RP mode, proved to be entirely comparable, in most aspects, to the RP-LC mechanism observed in pd LC systems. Particularly, the dependence of the (chromatographic) retention factor, k_i, on the volume fraction (% v/v) of the organic component in the mixed mobile phase (usually acetonitrile) shows – within a certain range of solvent composition – in many instances a fairly linear decrease of the logarithm of k_i with increasing modifier content, similar to classical RP-HPLC [34]. A similar concordance is also observed for the temperature-dependence of k_i, where the Van't Hoff plots exhibit fairly linear dependencies of $\ln k_i$ on $1/T$. However, in CEC one must always be aware that – in dependence of the applied voltage – the actual temperature in the chromatographic bed may drastically exceed the temperature of the thermostated surrounding. For charged analytes, clearly, the selectivity is governed, according to Eqn. 7.16, not only by the partition equilibrium but also by the differences in the mobilities of analytes.

A substantial number of materials have been developed in the form of beads and monolithic beds. Beyond the RP-type stationary phases, materials were made up containing various surface-immobilized ligands that can act as selectors to provide specific selectivity [35,36]. Especially in the context of chiral separations, non-aqueous mobile phases (*e.g.*, mixtures of acetonitrile with methanol, containing mM concentrations of acetic acid or triethylamine) have been employed for this [37,38].

7.3.4.2 Peak dispersion

Considering aspects of efficiency in packed-bed CEC, it is the substantially reduced contribution of convective mixing in particular which is responsible for the very high efficiency (Fig. 7.2). With monolithic beds – briefly discussed below – an additional and

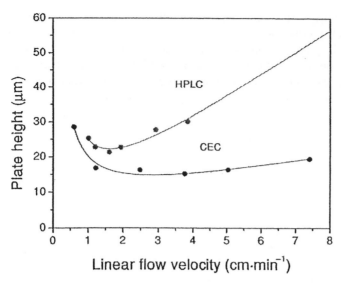

Fig. 7.2. Comparison of the *H vs. v* curves for pressure- and electro-driven LC. Column: Monolith, produced by co-polymerization of butyl methacrylate, ethyleneglycol dimethacrylate, and 2-acrylamido-2methyl-1-propane sulfonic acid. Total length, 41.5 cm; effective length, 33 cm; 100 μm ID. Mobile phase, 20% (v/v) aq. sodium phosphate (5 mM, pH 7)/80% acetonitrile. Temperature, 21°C. (Reproduced from Ref. 64 with permission.)

significant improvement in the mass-exchange term is achieved. Finally, in CEC the column length is not as much limited as in pd HPLC, since the packing itself is the source of the flow and does not cause a pressure drop. It is obvious that with long capillaries higher plate numbers can be obtained.

Optimized packing techniques yield plate-height values of nearly half of those obtained in classical pd HPLC systems. This means that the reduced plate height, h, ($h = H/d_p$, where d_p is the particle diameter) for very well packed beds could be established at a value of *ca.* 1 for CEC, compared to *ca.* 2 for pd systems [34]. Plate numbers as large as 200,000 are thus attained with capillaries of 50-cm length, packed with 3-μm (monodisperse) particles (corresponding to *ca.* 400,000 plates/meter). For comparison, a typical capillary of 25-cm length, packed with 3-μm particles yields only about 25,000 plates when operated in the reversed-phase pd LC mode (this is *ca.* 100,000 plates/meter). Due to the large pressure drop, the column length cannot be extended much over 25 cm in practice. With well-packed and, especially, with monolithic EC capillaries, there is a potential for achieving or even exceeding the plate numbers usually reserved for capillary gel electrophoresis or CZE (*i.e.*, several hundred thousands up to a million plates, depending on the charge number of the analytes).

Silica and, particularly, synthetic organic polymer materials, based on PS/DVB and various methacrylates are increasingly employed as monolithic phases, which are synthesized directly in the tube (capillary) and form uniform, continuous chromatographic beds. Monolithic capillaries provide a further reduction of the plate-height due to a reduced mass-exchange term. This is based on the fact that in the totally porous bed an in-pore flow is established, which results in the mass transfer no longer being controlled by analyte diffusion. Monolithic capillaries are particularly well suited to CEC [30,39]. Whereas polymeric monoliths can offer up to 150,000 plates/meter in pd HPLC, they yield up to 400,000 plates/meter (or 80–150,000 plates/column) in CEC [39,40]. With silica-based monoliths, values up to 220,000 plates/meter have been reported for CEC [39]. This means that the efficiency of polymeric monoliths in CEC is higher than that of HPLC by a factor of 2–3. Monolithic capillaries have a great potential in CEC of biological polymers [30,41] (Fig. 7.3), which exhibit low diffusion coefficients and for which the mass-exchange term in packed columns or capillaries is significant.

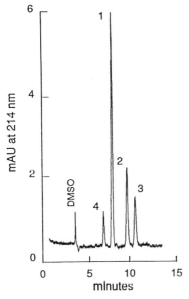

Fig. 7.3. Electrochromatogram of acidic and basic polypeptides: 1. angiotensin II; 2. angiotensin I; 3. [Sar[1], Ala[8]]-angiotensin II; 4. insulin. Column, porous styrene monolith with quaternary ammonium and octyl groups. Total length, 31 cm; effective length, 21 cm; 75 μm ID. Mobile phase, 5 mM phosphate/50 mM NaCl, 25% acetonitrile (pH 3). Voltage, 5 kV. (Reproduced from Ref. 31 with permission.)

A discussion of the flow properties in monolithic beds, instructions for preparing various monolithic materials for CEC, as well as a history of this technique can be found in the comprehensive reviews by Svec *et al.* [39,42,43]. A discussion of silica-based monoliths is presented in Ref. 44. Methods for controlling the porous properties of polymeric monoliths during polymerization by balancing the temperature, the concentration of the cross-linking monomer and the solvent components acting as micro- and macro-porogens are given in Refs. 31,42, and 44. In principle, a solvent mixture is used for dissolving the monomer solution, one solvent (A) being a good solvent for both, the monomer and the polymeric reaction product. Solvent B is a much poorer solvent for the polymer formed, and causes the precipitation of the polymer when it has attained a certain chain length. Solvent B is called porogen, because the chain-length of the polymer formed, and consequently the porosity of the polymeric material depends on its concentration. Higher concentrations of the porogen yield materials with larger pores (and higher permeability) and *vice versa*. The selection of the type and concentration of the porogen is critical for controlling monolithic column porosity. Through-pores in monolithic materials usually have diameters between 1 and 3 μm and narrow mesopores between 10 and 25 nm [39]. At these dimensions, double-layer overlap cannot be expected to occur in the through-pores. The finding that in CEC with monoliths the EOF increases linearly with increasing pore size is most probably not related to double-layer overlap [42]. It seems to indicate that pore structure and flow resistance in the chromatographic bed also play a role in CEC – which in some way contradicts expectations from theory. Monolithic beds are chemically bound to the inner wall of the capillary. As they do not need frits at the outlet, some problems frequently associated with frits, *e.g.*, bubble formation, are avoided. This is an advantage, because the preparation of frits by sintering packing materials – as it is often done with packed-particle beds – needs some practice. However, the reproducible production of monolithic-bed capillaries also requires experience, and there is a demand for commercially available materials.

7.4 ELECTROKINETIC CHROMATOGRAPHY WITH PSEUDO-STATIONARY PHASES

CEC with pseudo-phases can be regarded as being situated in-between chromatography and capillary electrophoresis. Whereas in MEKC and MEEKC the chromatographic origins can easily be recognized, CEC with low-molecular-weight additives is usually considered as a variant of classical CZE. Thus, many of the definitions and concepts commonly used in CZE are basic to the discussion of migration and efficiency in CEC with pseudo-phases.

7.4.1 Micelles and micro-emulsions

Instead of anchoring the stationary phase in the system, substances can be added to the BGE to form a stable suspension of micelles or micro-droplets. Sodium dodecylsulfate (SDS) was introduced as such a pseudo-stationary phase by Terabe [45], who invented the very popular technique of MEKC. Other micelle-forming compounds, either anionic or cationic, have been used as well. These pseudo-stationary phases consist of a lipophilic chain and an ionic end-group. The chain can be, *e.g.*, an alkyl group (octyl, decyl, and dodecyl are the most common), combined with, *e.g.*, the anionic sulfate, carboxylate, or cholate or the cationic tetraalkylammonium group. Low-molecular-weight surfactants have a critical micellar concentration above which micelles are formed. Others, like polymeric surfactants (so-called micelle polymers) always form a kind of micelle, even in mixed aqueous/organic solvents with a high content of organic solvent, due to their covalently stabilized structure.

Watarai [46] used stable oil-in-water emulsions as electrokinetic separation systems [47]. Typically, a micro-emulsion consists of (a) a lipophilic organic solvent; (b) the additive that imparts a charge to the micro-droplets; (c) a so-called co-surfactant (a more hydrophilic organic solvent), added in order to decrease the surface tension between the micro-droplets and the bulk aqueous phase; and (d) the aqueous phase, which is a buffer having a certain pH, ionic strength, or salinity. Various anionic or cationic detergents are applied as surfactants; the droplet-forming organic solvents are alkanes, ethers, ketones, esters, or long-chain alcohols; short-chain alcohols are used as co-surfactants.

Pseudo-stationary phases have several practical advantages: they are replaceable and can be removed and renewed after each run, thus producing a new column for each analysis. The capillary can be rinsed after each run with aggressive chemicals in order to remove impurities and to re-establish EOF reproducibly without damaging the chromatographic column. The system also obviates the problems associated with frits, used to retain particulate stationary-phase materials in the capillary.

7.4.1.1 Migration

The mobility of the analytes is in accordance with Eqn. 7.6, where $\mu_i^{(1)}$ is the EOF mobility in case of a neutral analyte, and it consists of both, μ_i^{eph} plus μ^{EOF}, in case of electrically charged analytes (see Eqn. 7.8). Due to their charge, the particles of the pseudo-stationary phase have their own mobility, μ_{PS}^{eph}. Under the assumption that the electrophoretic mobility, μ_{PS}^{eph}, of a relatively large micelle or droplet is the same, whether the analyte is dissolved in it or not, $\mu_i^{(2)}$ will be identical with the apparent mobility of the pseudophase, $\mu_{PS}^{app} = \mu_{PS}^{eph} + \mu^{EOF}$. Eqn. 7.6 will then read

$$\mu_i^{app} = \frac{1}{1 + k_i}\left(\mu_i^{eph} + \mu^{EOF}\right) + \frac{k_i}{1 + k_i}\left(\mu_{PS}^{eph} + \mu^{EOF}\right) \tag{7.20}$$

The retention factor can then be expressed by

$$k_i = \frac{\mu_i^{eph} + \mu^{EOF} - \mu_i^{app}}{\mu_i^{app} - \mu_{PS}^{eph} - \mu^{EOF}} \tag{7.21}$$

A neutral analyte ($\mu_i^{eph} = 0$) with $k_i = 0$ has the mobility of the EOF marker; a compound with infinitely large k_i is eluted with the mobility of the pseudo-stationary phase. The corresponding migration times confine the operational window in the MEK chromatogram. For charged analytes the situation is more versatile, as implied by Eqn. 7.21. Decisive for the migration order are the sign of the charge and the magnitude of the electrophoretic mobility of the analyte, the pseudo-stationary phase, and the EOF. Analytes can, in principle, migrate ahead of the EOF marker, even when they are partitioning into the micelle. This could happen, *e.g.*, in a system with an EOF directed towards the cathode (positive μ^{EOF}), negatively charged pseudo-phases, and weakly retained cationic analytes having a high electrophoretic mobility of their own. One could expect also peaks with retention times longer than that of the micelle marker, *e.g.*, in the system just mentioned, with anionic analytes that have a high electrophoretic mobility and are weakly partitioning into the pseudo-stationary phase.

7.4.1.2 Peak dispersion

MEKC has proved to be a very efficient electrophoretic method. Plate numbers up to several hundred thousand have been achieved, and this is in many instances better than (or at least equivalent to) classical CZE. Separation efficiency in systems with pseudo-stationary phases is usually described by the plate-height model, as applied to chromatography and CZE. Longitudinal diffusion is always present, as in elution chromatography and in CZE. It is relevant also in the presence of pseudo-stationary phases. The contributions in CZE due to Joule heating and wall adsorption can be significant in EC systems with pseudo-stationary phases as well. The kinetics of mass transfer between the two phases must also be considered as in chromatographic systems. Some contributions are specific for the EC systems, *e.g.*, those stemming from the polydispersivity of the micelles or from micellar overload. The particular effects are, briefly:

(a) *Joule heating-related dispersion.* In principle, the contribution of Joule heating is generally encountered in electroseparation techniques (in CZE as in CEC with stationary or pseudo-stationary phases) in the effect of temperature on mobility, solvent viscosity, and secondary equilibria, *e.g.*, in the extent of partitioning into the pseudo-phase. The radial temperature gradient causes a radial velocity gradient in the capillary and a radial dependence of the retention factor, k_i. It is still not clear whether or not thermal peak broadening plays a significant role in the context of MEKC [48]. In MEKC there are still a number of specific processes that influence peak-broadening, in addition to those described earlier, and some contributions need to be treated differently compared to classical CZE.

(b) *Longitudinal diffusion.* For MEKC, the classical chromatographic and CZE term $H_{diff} = 2D_i^{(1)}/v^{(1)}$, should be modified. It is obvious that in the mobile phase (the electrolyte solution) the zone diffuses proportional to the free-solution diffusion coefficient, $D_i^{(1)}$, weighted by the analyte fraction, $x_i^{(1)} = 1/(1 + k_i)$, present in this phase. Accordingly, micelle diffusion is governed by the micelle diffusion coefficient, $D_{(PS)}$, weighted by the fraction in the pseudo-stationary phase (in classical chromatography diffusion in the stationary phase is neglected). The total diffusion coefficient, D_i, is then expressed by a relation analogous to that given for the mobility (Eqn. 7.6) [49,50]

$$D_i = x_i^{(1)} D_i^{(1)} + x_i^{(2)} D_{PS}^{(1)} \tag{7.22}$$

Note that both, mobility and diffusion coefficient, are parameters describing the extent of mass transport across gradients – that of the electric and the chemical potential, respectively. For polymeric pseudophases the dependence of the overall diffusion coefficient on the extent of partitioning has, in fact, been demonstrated [51].

(c) *Micellar overload.* A contribution specific for MEKC is due to micellar overload, which could occur when the analyte concentration is too high relative to the micelle concentration [52]. Such an effect can be assumed from the observation that higher plate numbers are often obtained upon reduction of the analyte concentration. Highly skewed peaks of neutral analytes can result from non-linear chromatographic partitioning [52] and not from electromigrative dispersion in the mobile phase, as observed for ionic analytes.

However, both effects result from the fact that the underlying parameters (partition coefficient or mobility and diffusion coefficient, respectively) are not constant and that the respective differential equations to be solved for the mass balance equations are not linear.

(d) *Kinetics of mass transfer.* The resistance to diffusion across the micelle or micro-emulsion/electrolyte interface as well as the mass transport between the micelles can be relevant. As the analyte must diffuse only over the short distances between the micelles, the minor significance of peak broadening caused by this effect has been demonstrated [49,54]. Considering the kinetics of adsorption/desorption processes, contributions to the plate-height are found in certain cases when solid stationary phases are involved and, particularly, when steric hindrance results in high free energies of the transition state. Such analogous contributions are expected to be less pronounced in liquid partition systems, as there are less steric constraints.

(e) *Polydispersity of the micelles* [53].

(f) *Wall-adsorption-related dispersion.* Peak dispersion based on wall adsorption is frequently underestimated. An increasing loss of plate numbers after repetitive injections could be an indication of adsorption. However, initial efficiency can often be re-established upon rinsing the capillary under conditions where the adsorbed contaminants are removed. Interestingly, the contribution of wall adsorption can be unexpectedly great, compared to CZE, despite the presence of a detergent.

Although many of these effects are understood, there is still no unequivocal quantitative theory of the separation efficiency in electrokinetic methods with micelles. Somewhat contradictory experimental results concerning the influence of the particular processes on band dispersion have been reported by different authors – some of these results support theory, some contradict it.

7.4.2 Soluble charged polymers and other additives

In principle, it is not necessary for Phase (2) to form distinct entities like micelles or droplets. Thus, high-molecular-weight additives, like polymers or dendrimers, can also be used, as long as they form a stable suspension or solution in the BGE. Applications of such additives have recently been reviewed [55]. What is needed for separation – in addition to interaction with the analytes – is that at least one, either the analyte or the additive, possess a charge. The retention of an analyte can be described in the same manner as for micelles, when it can be assumed that the adduct formed between the analyte and the pseudo-stationary phase has the same mobility as the pseudo-phase alone. A deviation from this assumption, *e.g.*, when small molecules are used as additive, will be discussed below.

Polymeric solutions often have a higher macroscopic viscosity than micellar solutions or micro-emulsions. This can be a disadvantage in the handling of the BGE, *e.g.*, in filling or rinsing the capillary. Interestingly, high macroscopic viscosity does not automatically cause low analyte mobility [56], because the macroscopic viscosity is a bulk-phase property, whereas the movement of a small analyte through the solution is affected on the molecular scale. Thus, it is the microscopic viscosity that determines the electrophoretic migration as well as the diffusion. As a consequence, Walden's rule, which states that the product of mobility and (macroscopic) viscosity is constant, is not obeyed in polymer solutions. In some highly viscous solutions, analytes can even exhibit a mobility which is about the same as in pure aqueous solutions with a viscosity one order of magnitude smaller [56].

7.4.3 Charged, small-molecular-mass additives

In principle, the chain-length of a large polymer can be reduced to units with smaller and smaller number of repeats. Then the question arises from what size on can an additive be considered as a pseudo-stationary phase. It is a fact that even monomeric units of the polymer can implement retention [57]. Therefore, we can, in principle, treat the system consisting of a BGE with dissolved oligo- or monomers, at least formally, like other electrokinetic chromatography systems. The difference lies in the fact that the mobility, μ_{cplx}^{app}, of the analyte/additive adduct (which we will call a "complex" for simplicity) will not be the same as that of the monomeric additive. This is because analyte and additive can be of similar size and molecular mass. Thus, we have to modify Eqn. 7.21 to

$$k_i = \frac{\mu_i^{eph} + \mu^{EOF} - \mu_i^{app}}{\mu_i^{app} - \mu_{cplx}^{eph} - \mu^{EOF}} \tag{7.23}$$

The term "complex" indicates that, in the case of low-molecular-size additives, the partition between the two "phases" is better described by a stoichiometric view of the interaction between the analyte, A, and the additive, C. This interaction (complexation) is then described by the (complexation) equilibrium

$$aA + cC = A_a C_c \tag{7.24}$$

a and c being the stoichiometric numbers. The (complexation) equilibrium constant is

$$K^{cplx} = \frac{[A_a C_c]}{[A]^a [C]^c} \tag{7.25}$$

For the simple case of a uni-univalent reaction (1:1 type complexation stoichiometry) the mobility of the analyte, expressed as a function of the additive concentration, reads

$$\mu_i^{app} = \frac{1}{1 + K^{cplx}[C]} \mu_i^{eph} + \frac{K^{cplx}[C]}{1 + K^{cplx}[C]} \mu_{cplx}^{eph} + \mu^{EOF} \tag{7.26}$$

Note the similarity with Eqn. 7.6. Eqn. 7.26 is the well-known relation applied in CZE separations (Chap. 9), used in particular when dealing with the separation of enantiomers, where C is an enantiomerically pure chiral selector. Using charged selectors (which are frequently also employed in the separation of charged analytes), we face a true counter-current system, and the selectivity can be markedly increased *via* migration and/or EOF counterbalancing. The selectivity is influenced by the strength of interaction, expressed by the equilibrium constant of complexation, K^{cplx}, as well as by the concentration of the additive, and depends on the electrophoretic mobilities of the free and the complexed analytes as well as the EOF. The selectivity coefficient for two analytes (the ratio of their apparent mobilities), plotted *vs.* the additive concentration, $[C]$, exhibits – in dependence on the numerical values of the individual mobilities and K^{cplx} – maximum values, may pass through zero or may exhibit discontinuities when approaching infinity [58,59]. Such systems are powerful tools in achieving high separation selectivity. When modeling such dependencies, one should consider that with increasing concentrations of charged additives the ionic strength increases too, and K^{cplx} is not longer constant [58,59].

7.5 ELECTRICALLY DRIVEN *VS.* PRESSURE-DRIVEN CHROMATOGRAPHY

In comparing the over-all potential of EC with that of pd HPLC in capillaries, it is necessary to address some further parameters of practical relevance, as for instance aspects of method validation, implementing solvent gradients, and over-all sensitivity.

7.5.1 Validation

With respect to method validation one must consider, among others, the reproducibility of elution time and resolution, the reproducibility of sample application, and the range of a fairly linear detector response. Since EOF is a surface phenomenon, it is obvious that its magnitude strongly depends on the actual condition of the electrified surface. It is well known that surfaces are subject to undesirable modifications, *e.g.*, due to the adsorption of sample or buffer constituents or contaminants. Certainly, in ekd systems these modifications have a negative influence on the reproducibility of elution time and resolution. With electrokinetic sample application, the reproducibility of the applied sample volume – which is very high in pd LC systems – may be significantly impaired and a systematic discrimination of charged and therefore electrophoretically migrating analytes may occur – as in CZE. Due to these factors, a considerably enhanced calibration effort is usually required in CEC.

7.5.2 Gradient elution

Gradients in mobile-phase composition can be produced in CEC by specialized technical equipment [60–62]. This is, however, not as straightforward and simple as in pd LC [63]. Upon changing the solvent composition in the column, the ζ potential, relative permittivity, and viscosity will gradually change, yielding progressively different electro-osmotic flow velocities. The application of a step gradient – effected, *e.g.*, by changing the "inlet" vial – will *gradually* change the EOF velocity in the column; to implement continuous gradients – as they are applied in pd LC – is quite difficult.

7.5.3 Overall sensitivity

In the absence of mobile-phase gradients, the maximum injection volume, V_{inj}^{max}, just small enough to avoid a decrease in resolution is limited to values of approximately twice the volume standard deviation, σ_v, of the analyte peaks produced by the column. Assuming a reduced plate-height value, h, of 2 (the limit for pd systems), the V_{inj}^{max} values are near 260 nL for a 300-μm-ID capillary, 25 cm in length, (assuming a retention factor of 3), and *ca.* 30 nL and *ca.* 7 nL for 100-μm- and 50-μm-ID capillaries, respectively.

Taking *h* values of as low as 1 for ekd systems, the corresponding V_{inj}^{max} values for comparable column dimensions are calculated as *ca.* 180, *ca.* 20, and *ca.* 5 nL. The values for the 50-μm capillary are comparable to the volumes usually injected into CZE systems without peak stacking. In these instances, EC faces analogous problems as CZE regarding the overall sensitivity of the method with respect to sample concentration and dynamic range of detection.

7.5.4 Liquid/liquid partitioning

Chromatography based on liquid/liquid partitioning equilibria (*e.g.*, liquid/liquid chromatography, LLC) is beset by certain problems in the pd mode. LLC with liquid stationary phases coated onto a solid support suffers from problems of instability and lack of reproducibility. LLC without a solid support, performed as counter-current chromatography, is a technique requiring a substantial instrumental investment. In contrast, the corresponding CEC techniques, with the use of charged micelles (MEKC) or micro-emulsions (MEEKC), are very reproducible, easily performed, and widely applicable.

7.6 CONCLUSIONS

Electrically driven chromatographic methods have been introduced as an useful alternative to pressure-driven methods, because they offer a number of advantages. In the CEC mode with stationary phases they are not restricted by the back-pressure of the column, and beds with much smaller particles or enhanced length can be utilized. This, together with the favorable geometric properties of the EOF, produces lower plate-heights and affords significantly larger plate-numbers and thus better separation efficiencies than pd methods. One must keep in mind that the EOF is a surface phenomenon, suffering some restrictions in practice. Unpredictable changes in surface properties can occur, *e.g.*, due to adsorption of sample or buffer constituents. This results in lower reproducibility of the flow than in pd chromatography. Moreover, due to the limited charge density of the surface, EOF velocities remain quite low and increase the analysis time. EC with pseudo-stationary phases have the great advantage that the column can be exhaustively cleaned after each run, and, as the phases are replaceable, the entire column can be renewed for each analysis.

Although the high plate numbers were doubtless the main motivation for the implementation of CEC, the achieved selectivity deserves some discussion. When dealing with charged analytes, two types of selective mechanisms are operative in CEC. One source of selectivity is the difference in electrophoretic mobilities, the second one is the chromatographic selectivity resulting from differences in partition coefficients. These two sources of selectivity can be synergistic or antagonistic, depending on the analyte structure, mobile-phase composition, and type of stationary or pseudo-stationary phase. In such systems the non-selective movement of the EOF, when migrating opposite to the analytes permits – at least in principle – very high selectivity values to be obtained by migration balancing. CEC with solid stationary phases and pseudo-phases combine features of chromatography and electrophoresis. It allows one – depending on the particular problem – to benefit, in part, from the great potential and advantages of both methods.

REFERENCES

1   J. Lyklema, *Fundamentals of Interface and Colloid Science*, Academic Press, San Diego, 1995.
2   R.J. Hunter, *Zeta Potential in Colloid Science*, Academic Press, London, 1981.
3   C. Schwer and E. Kenndler, *Anal. Chem.*, 63 (1991) 1801.
4   W. Schützner and E. Kenndler, *Anal. Chem.*, 64 (1992) 1991.
5   C.L. Rice and R. Whitehead, *J. Phys. Chem.*, 69 (1965) 4017.
6   T.S. Stevens and H.J. Cortes, *Anal. Chem.*, 55 (1983) 1365.
7   J.H. Knox and I.H. Grant, *Chromatographia*, 24 (1987) 135.
8   J.T.G. Overbeek, in H.R. Kruyt (Ed.), *Colloid Science*, Elsevier, New York, 1952, p. 194.
9   J.T.G. Overbeek and P.O.W. Wijga, *Rec. Trav. Chim.*, 65 (1946) 556.
10 A.S. Rathore and C. Horvath, *J. Chromatogr. A*, 781 (1997) 185.
11 U. Tallarek, T.W.J. Scheenen and H. Van As, *J. Phys. Chem. B*, 105 (2001) 8591.
12 R.F. Benenati and C.B. Brosilow, *AIChE J.*, 8 (1962) 359.
13 A.I. Liapis and B.A. Grimes, *J. Chromatogr. A*, 877 (2000) 181.
14 S.S. Dukhin and B.V. Derjaguin, *Surface and Colloid Science*, Wiley-Interscience, New York, 1974.
15 N.M. Semenikhin and S.S. Dukhin, *Kolloidn. Zh.*, 37 (1975) 1123.
16 S.S. Dukhin, *Adv. Colloid Interf. Sci.*, 61 (1995) 17.
17 G.E. Archie, *Trans. A.I.M.E.*, 146 (1941) 54.
18 R.E. de la Rue and C.W. Tobias, *J. Electrochem. Soc.*, 106 (1959) 827.
19 G. Choudhary and C. Horvàth, *J. Chromatogr. A*, 781 (1997) 161.
20 O.L. Sánchez Muñoz, E. Pérez Hernández, M. Lämmerhofer, E. Tobler, W. Lindner and E. Kenndler, *Electrophoresis*, 24 (2003) 390.
21 R. Tijssen, in E. Katz, R. Eksteen, P. Schoenmakers and N. Miller (Eds.), *Handbook of HPLC*, Dekker, New York, 1998, p. 55.
22 J.C. Giddings, *Unified Separation Science*, Wiley, New York, 1991.
23 J.H. Knox, *Chromatographia*, 26 (1988) 244.
24 J.H. Knox and I.H. Grant, *Chromatographia*, 32 (1991) 317.
25 J.H. Knox, *Chrom. Int.*, 1 (1996) 39.
26 A.S. Rathore and C. Horvath, *J. Chromatogr. A*, 743 (1996) 231.
27 A.S. Rathore and C. Horvath, *Anal. Chem.*, 70 (1998) 3271.
28 A.S. Rathore and C. Horvath, *Anal. Chem.*, 70 (1998) 3069.
29 A.S. Rathore and C. Horvath, *Electrophoresis*, 23 (2002) 1211.
30 S. Zhang, X. Huang, J. Zhang and C. Horvath, *J. Chromatogr. A*, 887 (2000) 465.
31 I. Gusev, X. Huang and C. Horvath, *J. Chromatogr. A*, 855 (1999) 273.
32 S.P. Porras, E. Marziali, B. Gas and E. Kenndler, *Electrophoresis*, 24 (2003) 1553.
33 G.J.M. Bruin, P.P.H. Tock, J.H. Kraak and H. Poppe, *J. Chromatogr.*, 517 (1990) 557.
34 K.D. Bartle and P. Myers, *J. Chromatogr. A*, 916 (2001) 3.
35 T. Koide and K. Ueno, *J. Chromatogr. A*, 909 (2001) 305.
36 M. Lämmerhofer, F. Svec, J.M.J. Frechet and W. Lindner, *Trends Anal. Chem.*, 19 (2000) 676.
37 M. Lämmerhofer, F. Svec, J.M.J. Frechet and W. Lindner, *J. Microcolumn Sep.*, 12 (2000) 597.
38 M. Lämmerhofer, F. Svec, J.M.J. Frechet and W. Lindner, *Anal. Chem.*, 72 (2000) 4623.

39   N. Tanaka and H. Kobayashi, *Anal. Chem.*, 73 (2001) 420A.
40   A. Palm and M.V. Novotny, *Anal. Chem.*, 69 (1997) 4499.
41   I.S. Krull, A. Sebag and R. Stevenson, *J. Chromatogr. A*, 887 (2000) 137.
42   F. Svec, E.C. Peters, D. Sykora and J.M.J. Frechet, *J. Chromatogr. A*, 887 (2000) 3.
43   F. Svec, E.C. Peters, D. Sykora, C. Yu and J.M.J. Frechet, *J. High Resolut. Chromatogr.*, 23 (2000) 3.
44   C. Viklund, F. Svec, J.M.J. Frechet and K. Irgum, *Chem. Mater.*, 8 (1996) 744.
45   S. Terabe, K. Otsuka, K. Ichikawa, A. Tsuchiya and T. Ando, *Anal. Chem.*, 56 (1984) 111.
46   J.H. Schulman, U. Stoeckenius and L. Prince, *J. Phys. Chem.*, 63 (1959) 1677.
47   H. Watarai, *Chem. Lett.*, 3 (1991) 391.
48   J.M. Davis, in M.G. Khaledi (Ed.), *High-Performance Capillary Electrophoresis. Theory, Techniques and Applications*, Wiley, New York, 1998, p. 141.
49   S. Terabe, K. Otsuka and T. Ando, *Anal. Chem.*, 61 (1989) 251.
50   J.M. Davis, *Anal. Chem.*, 61 (1989) 2455.
51   B. Maichel, B. Gas and E. Kenndler, *Electrophoresis*, 21 (2000) 1505.
52   K.W. Smith and J.M. Davis, *Anal. Chem.*, 74 (2002) 5969.
53   S. Terabe, N. Matsubara, Y. Ishinama and Y. Okado, *J. Chromatogr. A*, 608 (1992) 23.
54   L. Yu and J.M. Davis, *Electrophoresis*, 16 (1995) 2104.
55   I. Peric and E. Kenndler, *Electrophoresis*, 24 (2003) 2924.
56   T. Shimizu and E. Kenndler, *Electrophoresis*, 20 (1999) 3364.
57   B. Maichel, K. Gogova, B. Gas and E. Kenndler, *J. Chromatogr. A*, 894 (2000) 25.
58   B.A. Williams and G. Vigh, *J. Chromatogr. A*, 777 (1997) 295.
59   A. Rizzi, *Electrophoresis*, 22 (2001) 3079.
60   G.P. Rozing, A. Dermaux and P. Sandra, *Capillary Electrochromatography*, Elsevier, Amsterdam, 2001, p. 39.
61   Z. El Rassi, *Carbohydrate Analysis by Modern Chromatography and Electrophoresis*, Elsevier, Amsterdam, 2002, p. 597.
62   K. Mistry, I. Krull and N. Grinberg, *J. Sep. Sci.*, 25 (2002) 935.
63   J.L. Liao, C.M. Zeng, A. Palm and S. Hjerten, *Anal. Biochem.*, 241 (1996) 195.
64   T. Jiang, J. Jiskra, H.A. Claessens and C.A. Cramers, *J. Chromatogr. A*, 923 (2001) 215.

Erich Heftmann (Editor)
Chromatography, 6th edition
Journal of Chromatography Library, Vol. 69A
© 2004 Elsevier B.V. All rights reserved

Chapter 8

Gas chromatography

PHILIP J. MARRIOTT

CONTENTS

8.1 Introduction 320
8.2 Basic operating variables 321
    8.2.1 Column 321
    8.2.2 Carrier gas 323
    8.2.3 Stationary phase 325
    8.2.4 Phase types 327
8.3 Enhanced and fast separations 332
    8.3.1 Coupled columns 332
    8.3.2 Pressure tuning 332
    8.3.3 Multi-dimensional gas chromatography 333
    8.3.4 Fast gas chromatography 336
    8.3.5 Comprehensive two-dimensional gas chromatography 337
8.4 Sample introduction 340
    8.4.1 Split/splitless injection 341
    8.4.2 Cool on-column injection 343
    8.4.3 Programmed-temperature vaporizer and large-volume injection . . 345
    8.4.4 Headspace and solid-phase micro-extraction 346
    8.4.5 Coupling with liquid chromatography 347
8.5 Detection 348
    8.5.1 General principles 348
    8.5.2 General classifications of detectors 351
    8.5.3 Selected detectors 352
        8.5.3.1 Ionization detectors 352
        8.5.3.2 Bulk physical property detectors 358
        8.5.3.3 Photometric detectors 358
        8.5.3.4 Other detectors 362
    8.5.4 Dual detection 364
References 364

8.1 INTRODUCTION

Gas chromatography (GC), which is now 50 years old [1], provides separation and quantitative analysis for volatile, thermally stable compounds in a broad variety of mixtures, from the simplest (*e.g.*, purity tests of individual compounds) to the most complex (*e.g.*, petrochemical assays of samples comprising hundreds of individual components). For the analysis of complex samples, no other physical method offers both the broad analysis of total sample and the specific information on individual components of the sample as is produced by GC. This is a direct consequence of the resolving power of long, narrow-bore (capillary) columns, coated with a thin film of stationary phase, which maximize the ability to separate closely related chemical components. There has been little advance in extending the upper limit of GC; 350–425°C is normally accepted as the upper boundary (determined by stationary-phase stability). The lower temperature limit is often not of too much concern, but sub-ambient operation allows the most volatile analytes to be chromatographed. The mass range of GC may therefore be defined as being from 2 to *ca.* 1200–1500 Da of molar mass.

The retention of a compound depends on both, its boiling point and the specific interactions with the stationary phase. The carrier gas, for the most part, merely serves to move the compound in the gas phase from the injector (inlet) to the detector (outlet). The primary property of the gas which affects the chromatographic result is its flow-rate and viscosity (which affects the magnitude of diffusion coefficients of solutes in the gas). The stationary phase is the primary determinant in retention and separation. It may be an uncoated porous material, as in gas/solid chromatography (GSC) or a polymer-coated support material, as in gas/liquid chromatography (GLC). The GLC mode is by far the most commonly used method. The stationary phase may be a solid-phase column packing, as in packed-column chromatography, or it may be a liquid, coating the inner wall of the column, as in the open-tubular (OT) or capillary (C) GC methods. The liquid phase must be heat-stable and nonvolatile; hence, cross-linked polymeric materials, chemically bonded to the capillary column wall, are favored. Phase selectivity is a measure of how well the phase differentiates between different compounds and, hence, how well the compounds will be separated. The column performance measures its efficiency (*i.e.*, the narrowness of peaks).

In the low-efficiency, packed columns, phase selectivity is important in providing separation. It is common to classify phases according to polarity. This is a derived measure, based on how the phase alters the relative peak positions (retention). To a first approximation, a greater degree of retention leads to the phase being classified as more polar. The shift towards capillary GC, with its increased efficiency (narrow peaks), has made phase selectivity a less important parameter. That is fortunate, because the limitation of achieving good, thermally stable coatings on glass (fused silica) walls means that there are fewer phase choices for open-tubular (OT) than for packed columns. Specialty phases, such as those composed of enantioselective additives, or liquid crystal phases, provide specific interaction mechanisms for specific separation goals.

Sample introduction procedures have undergone a renaissance over the past few years, although it may remain the Achilles heal of some methods. Detection methods have not changed greatly in the past 15 years, although re-engineering of some detectors to allow faster detection signal response is now gaining pace, and this will support work in faster GC separation. Given the importance of GC/MS (Chap. 10) to routine analytical laboratories, the dominant position of, especially, quadrupole MS detection technology must be acknowledged. The sophistication of on-line library-searching identification, with quantification capabilities has given considerable power to the analysis of volatile samples. The technology of time-of-flight mass spectrometry (TOF-MS) has gone from being the first mass spectrometry technique, to oblivion, and is now back in vogue. The GC technique today is mature, reproducible and reliable – primarily due to active control enabled by microprocessors and sophisticated flow control – and provides the analyst with confidence in the identification and quantification of compounds in mixtures. The automated gas chromatograph delivers much greater retention reproducibility than manual control. Fully automated operation and reporting, with an array of supporting sample-handling options, and additional method variations that offer specific enhanced analysis capabilities are available. However, within this zone of sophisticated analytical performance, there are still some areas where new developments provide improved operation and performance.

8.2 BASIC OPERATING VARIABLES

8.2.1 Column

Packed columns range in size from capillary dimensions to preparative columns, having inner dimensions of centimeter proportions. The conventional analytical packed column is usually of dimensions 1/8 or 1/16 in. (1–3 mm) ID, packed with support particles of 60–230 mesh [0.1 to 0.4 mm particle diameter (dp)]; the phase loading is given in %mass/mass (*e.g.*, 5% mass); the film thickness is not usually quoted, since the specific area is not commonly reported. Packed columns are usually less than 5 m long. OTGC can only be performed in microcolumn (capillary) dimensions, otherwise diffusion to the wall will be too slow. The C term of the van Deemter equation (Chap. 1) indicates that insufficient gas-to-liquid mass transfer will take place, and this will significantly reduce column efficiency. The upper limit for the ID of capillary columns is *ca.* 1 mm. In practice, mega-bore columns of 0.53 mm are the widest offered by manufacturers, 0.1–0.33 mm ID being the most common. Capillary columns may be packed also, but these are not commonly used. Open-tubular capillaries are available in two formats – wall-coated (WCOT) or porous layer (PLOT). The capillary-array (multi-capillary) column is comprised of an array of hundreds (919) of 40-mm-ID capillaries, with a total length of only 1–2 m. It is intended to permit fast analysis on very-narrow-bore columns with reasonably high sample capacity [2]. This approach has not been widely adopted by chemists.

The above details are summarized in Table 8.1 [3]. The phase ratio (β) is the volume of mobile to stationary phases in the column, which is also the cross-sectional area ratio of

TABLE 8.1

PROPERTIES OF VARIOUS GC COLUMN TYPES

Column type	Typical dimensions		Phase ratio	H_{min}	$\hat{u}_{opt}$	Permeability
	Length (m)	ID (mm)				
Packed	1–5	1–3	4–200	0.5–2	5–15	1–50
Micro-packed	1–5	<1	50–200	0.02–1	5–10	1–100
Packed capillary	1–5	<0.5	10–300	0.05–2	5–25	5–50
SCOT	10–50	0.2–0.5	20–300	0.5–1	10–100	200–1000
WCOT	10–50	0.2–0.5	15–500	0.03–0.8	10–100	300–20000
Megabore	10–30	= 0.54	20–300	0.1–2	10–100	200–1000
Microbore	5–20	0.05–0.15	50–700	0.01–0.4	10–100	200–1000
Capillary array	1–2	0.04	50	2.0	10–100	200–1000

Reprinted, in part, from Ref. 3 with permission.

the phases. This is given for thin-film-phase OT columns by the simplified equation

$$\beta = r/2d_f \qquad (8.1)$$

where r = radius; d_f = stationary-phase film-thickness.

There are a number of aspects differentiating packed and capillary GC columns. Whereas the back-pressure in packed columns prohibits lengths in excess of 5–10 m, the ease of carrier-flow (permeability) in capillary columns allows much longer columns to be used. Their similar plate heights (H_{min}; Table 8.1) confers significantly greater efficiency to the capillary column. Thus, components with a similar separation factor (α) will be better resolved in the capillary column, while packed columns must rely on phase selectivity to provide resolution in some instances. Smaller stationary-phase (thin-film) loading in capillary columns will reduce the sample capacity (loadability) so that, to prevent overloading, lower concentrations or smaller volumes of samples must be injected to maintain linear chromatographic conditions. Packed columns can, in many cases, accommodate high-concentration solvent injections. Thus, a 10%-w/w load of stationary phase in a packed column of a total 4-g mass means 0.4 g of stationary phase, whereas a 10-m capillary column having 0.25 mm ID and a film thickness of 0.5 μm will have a total mass of *ca.* 0.01 g phase. While this is distributed over the total length of column, it is the local phase volume or amount that is important in deciding whether overloading will arise. The capillary column will have an even lower proportion of phase in a given separation element of the column. To maintain K (distribution constant = C_s/C_m; where C_s and C_m are solute concentrations in the stationary and mobile phases, respectively) in the linear region, the ratio of solute concentrations in each phase must be in equilibrium and equal to K, as the amount of total solute injected is altered. This means that the retention time (t_R) of the solute will also be constant, since K $\propto$ k and constant k means constant t_R. If there is less stationary phase, the maximum amount of sample injected must be reduced. Since the gas phase has a limited ability to accommodate increased amounts of solute, excess amounts will cause the concentration residing in the stationary phase to be increased such that K is no longer constant; K increases, k increases, and the zone maximum increases, along with the peak becoming increasingly asymmetric. This is commonly seen in capillary column chromatography, especially where a sample has a (few) major component(s), as triangular-shaped peaks with a broad leading edge and sharp return to baseline.

8.2.2 Carrier gas

The carrier gas was earlier stated to be a passive participant in the separation process. This is certainly true with respect to the mechanism of interaction between the solute and the stationary phase – the gas is merely a medium in which the solute vapor resides. However, when we refer to separation in respect of resolution of components, and hence the band-width of a solute (Chap. 1), then the efficiency of the column must be an active part of the resolution consideration. Once the column geometry (dimensions, film thickness) has been decided – and the column placed in the chromatograph – the efficiency is very much affected by carrier gas properties. While this largely arises from its flow velocity through the column, it is a result of the diffusion coefficient of solute in the carrier gas, and it has two distinct and seemingly opposite effects when one considers the microscopic forces that act on a solute molecule in the column. These may also be interpreted through the terms of the van Deemter equation:

(a) *Longitudinal diffusion*. Molecules in a gas will have a tendency to spread apart in that phase. This follows the Einstein–Stokes diffusion expression. Thus, a higher diffusion coefficient in the gas phase (D_M, promoted by either lower solute molar mass or its higher volatility, and a lower gas viscosity) will cause more spreading and so broader peaks. A zero carrier gas flow will still cause this longitudinal broadening – spreading along the column. A higher carrier flow-rate will cause the solute to reach the end of the column faster, so that there will be less broadening from this effect. Thus, high flow minimizes the relative contribution of longitudinal broadening. This gives rise to the van Deemter B term.

(b) *Mass transfer*. In order to achieve separation and retention, a solute must interact with the stationary phase. The solute in the gas phase therefore must reach the stationary phase. We wish to have rapid, and many, equilibration events with the stationary phase. This is enhanced by having a high diffusion coefficient. In this case a high D_M will give a better, more efficient result. We refer to this effect as a resistance to mass transfer. It is hoped that rapid equilibration occurs at the gas/liquid phase interface and that the solute stationary-phase profile along the column reflects (*i.e.*, is proportional to) that of the solute in the gas phase at all points. If the solute lags in the stationary phase, then this will delay the solute re-emerging from the stationary phase, and this will exacerbate the temporal difference between the gas- and stationary-phase solute distribution. Mass-transfer effects reside in the van Deemter C term.

(c) *Overall effect*. Since both of the above effects are apparently opposite, a high diffusion coefficient will favor mass transfer, but disadvantage longitudinal diffusion. This is an apparent dichotomy, and results in the familiar van Deemter curve, where the

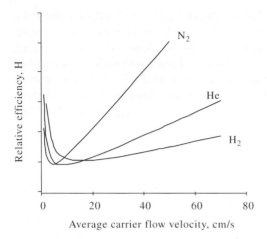

Fig. 8.1. Van Deemter curves for three different carrier gases: N_2, He, and H_2.

minimum represents the optimum flow rate, $\hat{u}_{opt}$. Comparative curves for the three gases are sketched in Fig. 8.1. It is clear that this minimum occurs at lower flow-rates for the gas which has a lower diffusion coefficient (nitrogen), but at the same time, a solute with lower diffusivity should have a lower optimum flow-rate. Another consequence of this is that the lighter gas – H_2, having a flatter curve, should be a better choice at high flow-rates, since there is less loss of efficiency; its lower C term is responsible for this. At high-flow operation, which may be useful for routine analysis, hydrogen will produce more plates per unit time of analysis. In the van Deemter equation, B causes longitudinal spreading, while C causes migration to and across the phase boundary. The solute must travel to the phase if separation is to occur, and this happens more effectively if narrow-bore columns are used; hence they provide greater efficiency.

The above discussion on gas viscosity needs to be considered along with the effect of pressure on viscosity. The more viscous gas will produce a higher inlet pressure for a given flow-rate. Again, H_2 will have the highest flow-rate for a given temperature, T, and pressure, P. This can be seen from the general expression for viscosity, η, where $\eta = aT + b$. For H_2, N_2, and He, the terms are: a = 0.183, b = 83.99; a = 0.384, b = 167.4; a = 0.399, b = 186.6, respectively. The practical effect of this is also reflected in the change in carrier flow during temperature-programmed operation. The carrier gas becomes *more viscous* as temperature increases, so an unretained solute will have a longer retention time at higher oven temperature with the same inlet pressure. Instruments often offer a correction for this effect by allowing constant-flow operation by suitable pressure adjustment during a temperature program. This is considerably aided by electronic pressure control. Also, since best performance is achieved with hydrogen (especially less loss of efficiency at higher flow-rates), H_2 is much more widely used in capillary GC than in packed-column GC. Carrier-gas leak detectors and low-cost H_2 generators have added to its acceptance as a carrier gas in capillary GC.

In the above discussion, the focus is on the conventional gases used for GC. It is important to acknowledge the work in the area of solvating GC (SGC), where gases capable of adding a solvating mechanism to the GC separation are used. Very little routine work is conducted by SGC. Wu *et al.* [4] contrasted fast GC by packed-column SGC with ordinary OTC. CO_2 was used as carrier gas for SGC experiment on short (10- to 50-cm) capillaries of 0.25 mm ID, packed with 5-μm particles. It was concluded that SGC shows advantages for fast separations of simple mixtures of low-retained compounds where retention factor and selectivity are important considerations. Otherwise, OTC is superior for achieving fast analysis.

8.2.3 Stationary phase

With a non-participating mobile phase, it is the stationary phase which is responsible for molecular separation. The solutes themselves bring molecular characteristics of boiling point and structural features to the consideration of retention. Separation therefore arises due to differences in solute vapor pressure and specific interactions between solute molecules and the stationary phase. The specific interactions may be classified broadly into dispersion and those interactions that depend upon polarity considerations, which may be considered as donor/acceptor, orientation, and induction. All molecules possess *dispersion force* interactions, and are molecular-size-dependent (and independent of temperature). Interactions arise from instantaneous atomic dipoles formed by electron motion, which interact with molecules in the stationary phase.

(a) *Donor/acceptor forces* are considered to be electron-based (partial) transfer of electron density from one molecule to another (*e.g.*, a donor orbital of one molecule to a vacant orbital of another). The hydrogen bond is a specific example, as would be the interaction between a ligand-type molecule and a metal acceptor.

(b) *Orientation forces* arise from the alignment of permanent dipoles in one molecule and another. If permanent dipoles are so aligned, they will contribute to a molecular interaction (attraction) that increases the association energy between the molecule and the phase, and increases retention. As the temperature increases, this energy will become less strong as molecular motion becomes more chaotic or randomized.

(c) *Induction forces* behave similarly with temperature, and arise from a similar effect. However, it is the ability of a permanent dipole on one molecule to induce a dipole on another that leads to an intermolecular attractive force. The other molecule must therefore be polarizable to allow an induced dipole to arise.

The dispersion force is the most important primary force, since it is common to all molecules. It explains, *e.g.*, why, as members of a class of compounds become larger, they are retained to a greater extent. This force must be the one explaining hydrocarbon retention behavior, and so we observe that the retention is correlated well with the boiling point of these molecules. Superimposed on the dispersion force-related retention effect are the effects of the other forces. If two molecules of different chemical class (polarity) have a similar boiling point, then their separation depends primarily on (the magnitude of) the specific interactions between the molecules and the stationary phase. Dipole moments, polarizabilities, induction effects, etc., come into play. We normally do not differentiate between such interactions and broadly classify them as "polarity", expecting that a more polar molecule will have stronger interactive forces with a stationary phase. This aids a general interpretation of relative retention. However, we cannot state that a more polar molecule has greater retention than a less polar molecule, because their relative dispersive forces must still be considered. Hence, a nonpolar molecule of higher molecular mass will most likely be eluted later than a polar molecule of lower molecular mass (*e.g.*, methanol is eluted before decane on both polar and nonpolar phases). If more polar phases are used, the polar solutes will progressively be retained more strongly than nonpolar solutes.

The above interactions are by no means the only ones that provide altered selectivity for various molecular separations. Here, selectivity refers to the ability of the phase to provide a mechanism for separating two particular solutes. Thus, if two molecules prove difficult to resolve with phases of varying polarity, just based on the above interactions, it will be necessary to try a different mechanism, based on different additives or molecular geometries in the stationary phase. Perhaps the most important of these classes is that provided by host/guest interactions within the stationary phase, as in cyclodextrin (CD) derivatives in the stationary phase. There may still be dipole/dipole associations, for instance, but the unique selectivity provided by the additives is important. We can refer to these as general shape–selectivity interactions, and hence, the phase can be used for molecular-geometry selectivity. One example of this is enantiomeric isomer separation. This has proven to be a very successful application of GC to, *e.g.*, essential oil components. The selective CD phase provides sufficiently different interaction energies to provide some degree of separation between the two enantiomers. Liquid-crystal phases, similarly, provide an alternative mechanism, where the shape of a molecule determines the extent of interaction, such as whether the molecule is planar or nonaxial.

Temperature limits range from the lowest temperature at which the stationary phase remains fluid, to the upper limit where thermal degradation of the phase starts. The upper limit is usually determined by the phenomenon of "bleed", where polymer phase components are released from the phase, often leading to an increase in the detector baseline towards the end of a temperature-programmed analysis. Some degree of bleed is acceptable (or unavoidable), but excessive bleed reduces column lifetime and stability. Cross-linked chemically bonded phases on capillary columns are favored, because of improved thermal stability. However, "thermal cracking" of the ends of polymer chains in the phase will still occur at some (high) temperature. Common bleed components are small cyclic siloxane molecules, which arise from the cyclization of polymer chain moieties, *e.g.*, those referred to as the D3, D4, or D5 components, consisting of 3, 4, or 5 siloxane units. This can be reduced, or the upper temperature limit increased, by reducing the polymer backbone flexibility, *e.g.*, by insertion of a phenyl group within the polymer backbone [5]. The phases originally developed by SGE and given the designation BPX are based on a polysilphenylene polymer, and incorporate such a stabilizing unit. Phases of these types are popular, being more stable with less background bleed, for use, especially, in mass spectrometry, *e.g.*, DB-5ms (J&W), and Rtx-5ms (Restek).

The stationary phase must coat the support, *e.g.*, the capillary wall, in a uniform, thin film. The process of capillary column manufacture can be simply summarized as follows: Column surface pre-treatment (*e.g.*, acid washing); column surface preparation (*e.g.*, treatment with an activating material); static coating of the surface with the stationary phase, which may contain additives to promote cross-linking, such as peroxides; thermal treatment to achieve reaction and bonding; rinsing; and finally, testing. The column is acceptable only when it passes a testing procedure, which usually covers aspects of phase activity (by chromatographing polar compounds to test tailing), resolution of test compounds, efficiency (a measure of phase uniformity), and peak asymmetry. Grob *et al.* [6] proposed a test that provided some measure of column phase quality, and most manufacturers base their testing on this. The net result should be a nonextractable (hence solvent-rinseable), thermally stable, uniformly coated phase.

8.2.4 Phase types

It is necessary to differentiate stationary phases for packed columns from those used in capillary columns. Packed columns have a wide range of polarities and, provided it can be dispersed on the support, any type of phase may be used, within the bounds of the temperature range. Only a few phase types are suitable for capillary columns, since it is necessary to use a polymer which can wet the surface and give a uniform coating. Early attempts at coating with polar phases were aided by roughening and modifying the glass surface with a mechanically stable support (*e.g.*, carbon whiskers, salt particles). For special separation mechanisms afforded by adsorbents, the surface may be prepared with an appropriate adsorbent. Adsorbent-coated capillaries are still available, and de Zeeuw and Luong [5] have described coatings of molecular sieves, alumina, porous polymers, deposited carbon, silica, and multi-layer phases. Sternberg *et al.* [7] have used capillary and micropacked columns for space exploration. These included a micro-packed carbon molecular sieve (2 m × 0.75 mm ID), a glassy carbon PLOT column (14 m × 0.18 mm ID) and a conventional cyanopropyl WCOT column (10 m × 0.18 mm ID). Ji *et al.* have reviewed the preparation and application of PLOT columns [8]. The most recent commercial phases are based on sol/gel processes [9].

In short, the primary characteristic of a phase is its polarity, or solvent strength. The solvent strength may be considered to be a measure of the stationary phase to interact with the solute, based on the previous interactions. Thus, it is the sum total of those interactions, each not readily measurable, that determines the polarity. The selectivity is the relative magnitude of specific interactions towards given solutes. These interactions are not readily quantifiable, but most practical decisions are made based on the basis of experience and rarely on considerations of primary interaction strengths. Since this implies a phenomenological approach to understanding phase properties, the approach of McReynolds constant, *i.e.*, the use of a series of molecular probes to gauge the strength of the relative total interaction of a molecule and the phase, appears to provide at least a scale of relative phase interaction. While that method is not difficult to implement, it is only a rule-of-thumb of phase properties. It is very unlikely that for routine analysis and optimization a phase will be selected on the basis of basic measurements of specific interaction energies. It is, rather, on the basis of knowing that a series of phases have different overall bulk properties (such as shown by the McReynolds approach) and testing these by direct experimentation. Phase polarity is compared with polarity of a squalane phase, which is estimated by retention index calculations of a set of five molecular probes – benzene, butanol, 2-pentanone, nitropropane, and pyridine. The summation of the difference in retention indices of these probes between the nonpolar squalene and the phase being tested is taken as a measure of phase polarity. The lowest-polarity polysiloxane phase is usually taken to be the 100% dimethylpolysiloxane. However, this phase has a McReynolds constant of about 170. A polar phase, such as polyethylene glycol has a McReynolds constant of about 2280. We call these ranges low and high polarity, respectively. Note that the Chrompack CPSil-2 CB phase has an additional, high-molar-mass hydrocarbon in its structure, which leads to a lower polarity that the 100% dimethylpolysiloxane phase.

Table 8.2 is a representative compilation of phase types available from selected manufacturers. This table is not exhaustive, and recent company literature should be consulted for a more complete product range. In some instances, the phases are close to those indicated, or are claimed by the manufacturers to be equivalent to other phases in the indicated category. The chemical nature of the phases is not known in all cases. Some manufacturers have novel or modified phase types, such as Restek Rt-TCEP, including 1,2,3-tris(2-cyanoethoxy)propane, Supelco PAG – poly(alkylene glycol), and SGE HT5/HT8 carborane phases. Some phases are proprietary, with no information given, such as Supelco MDN-12. Some phases claim low bleed, suited for mass spectrometry, such as the Agilent DB-XLB (eXtra Low Bleed), and the Chrompack "Factor Four" VF series, for which a bleed is claimed that is only 1/4 of that of other comparable phases. Chiral columns contain various derivatives, and the chiral selector is mixed (usually at 30% *w/w* load) with a phenyl or cyanopropyl phenyl dimethylpolysiloxane phase prior to column coating. There are many additional column manufacturers, such as those specializing in chiral columns (*e.g.*, Advanced Separation Technologies). The important observation here is that there are not many variations on the phase types that are used. Thus, to extend the polarity of the low-polarity 100% dimethylpolysiloxane, various levels of phenyl, cyanopropyl(phenyl), and trifluoropropyl groups are added to the base polymer, or the popular polar polyethyleneglycol phase may be used. The requirement is that the new phase type must still be suitably bonded/cross-linked to produce a stable phase coating. Increased polarity often corresponds to decreased phase or coating stability, and thus, the upper operating temperature will be reduced. The chiral phases serve the specific purpose of providing separation of enantiomers. Some compounds can be chromatographed on columns with low temperature limits (*e.g.*, for volatile essential oil components), but others, which require much higher temperature, are less studied, because chiral phases tend to be poorly stable at high temperature. They often contain a chiral selector (*e.g.*, a derivatized cyclodextrin) in a medium-polarity phase, such as 35% phenyl dimethyl polysiloxane or 14% cyanopropylphenyldimethylpolysiloxane. Being physically dispersed, and not chemically bonded, these additives lead to lower temperature stability (upper limits of 240–260°C). Other specific applications require other special columns, such as the BPX70 (70% cyanopropylphenyl) column for fatty acid methyl ester analysis, the terephthalic acid-treated polyethyleneglycol column for the analysis of volatile free acids, or the DB-1EVDA for drugs of abuse. Columns may also be promoted for use with specified EPA methods, in which case the column dimensions are more likely to be also specified as appropriate for that analysis.

TABLE 8.2

TYPICAL CAPILLARY GC PHASE TYPES

Phase chemical type	Manufacturer								
	Agilent[a] (J&W; HP)	Alltech	Chrompack	Phenomenex	Quadrex	Restek	SGE	Supelco
100% Dimethyl polysiloxane	DB-1 HP-1	AT-1	CP-Sil-5 CB	Zebron ZB-1	007-1	Rtx-1	BP1;sol Gel-1ms	SPB-1
5% Phenyl 95% dimethyl polysiloxane	DB-5 HP-5	AT-5	CP-Sil-8 CB	ZB-5	007-5	Rtx-5	BP5	PTE-5 PTE-5QTM
20% Phenyl 80% dimethyl polysiloxane			CP-Sil-13 CB (14% phenyl)			Rtx-20		SPB-20
35% Phenyl 65% dimethyl polysiloxane	DB-35	AT-35		ZB-35	007-35	Rtx-35		SPB-35 MDN-35
50% Phenyl 50% dimethyl polysiloxane	DB-17 HP-50 +		CP-Sil-24 CB	ZB-50	007-17	Rtx-50		SPB-2250 SPB-50
65% Phenyl 35% dimethyl polysiloxane					007-65HT	Rtx-65		
6% Cyanopropyl phenyl 94% dimethyl polysiloxane	DB-624 DB-1301	AT-624	CP-Sil-1301 CB	ZB-624	007-624	Rtx-1301	BP624	OVI-G43
14% Cyanopropyl phenyl 86% dimethyl polysiloxane	DB-1701	AT-1701	CP-Sil-19 CB	ZB-1701	007-1701	Rtx-1701	BP10	
50% Cyanopropyl phenyl 50% dimethyl polysiloxane	DB-225 HP-225	AT-225	CP-Sil-43 CB (25% cyano 25% phenyl)		007-225	Rtx-225	BP225	

(*Continued on next page*)

TABLE 8.2 (continued)

Phase chemical type	Manufacturer							
	Agilent* (J&W; HP)	Alltech	Chrompack	Phenomenex	Quadrex	Restek	SGE	Supelco
Polyethylene glycol	DB-WAX HP-20M INNOWax	AT-WAX	CP-Sil-88 CP-WAX 52 CB	ZB-Wax	007-CW BTR-CW	Rt-CW20M	BP20; SolGel-WAX	Nukol Omegawax
Polyethylene glycol, terephthalic acid treated	DB-FFAP HP-FFAP	AT-1000	CP-WAX 58(FFAP) CB	ZB-FFAP	007-FFAP		BP21	
Cyclodextrin (of various derivatizations) usually coated in either a phenyl dimethyl polysiloxane or a 14% cyanopropyl dimethyl polysiloxane	Cyclodex-B CycolSil-B HP-Chiralβ		CP-Chirasil-Dex – CB CP-Cyclodextrin β-2,3,6-M-19			Rt-BDEXm Rt-BDEXcst Rt-BDEXsa Rt-BDEXsm Rt-BDEXse Rt-BDEXsp	Cydex-B	α-Dex 120 β-Dex110 β-Dex 120 γ-Dex 120 Dex-225 Dex 325
5% (or 8%) Phenyl (equivalent) polysiloxane-carborane							HT5 or HT8	
5% Phenyl (equivalent) polysilphenylene-siloxane	DB-ms HP5ms					XTI-5	BPX5	
35% Phenyl (equivalent) polysilphenylene-siloxane					007-11; 007-608	Rtx-35	BPX35; BPX50 (50%)	

70% Cyanopropyl (equivalent) polysilphenylene-siloxane		AT-SILAR	CP SIL 88		Rtx-2330	BPX70	SPB-2330 80%/20% SPB-2380 90%/10%
Biscyanopropyl-cyanopropylphenyl polysiloxane	DB-23 (50% cyano)			007-23 (78% cyano)			
50% Trifluoropropyl 50% dimethyl polysiloxane	DB-200 (35%); DB-210			007-210	Rtx-200 (%tfp NA)		

* Both J&W and HP phases are now sold under the Agilent Technologies brand name.

8.3 ENHANCED AND FAST SEPARATIONS

8.3.1 Coupled columns

Stationary-phase selectivity (column selectivity) may be varied by use of a mixture of different stationary phases to provide an intermediate separation performance. This is readily achieved in packed-column chromatography, where the phases are simply dissolved in a suitable solvent, which is then mixed with the solid support. For capillary columns, preparing a mixed-phase coating is not so straightforward. Most manufacturers use a single polymer material to prepare the desired phase. Since there is a limited choice of capillary GC phases, the use of coupled capillary columns is a convenient way to achieve a mixed-phase selectivity. This certainly extends the range of apparent stationary phase polarity, but it is probably not a widely used strategy in routine laboratories. Jennings [10] relates that it was this type of study which led to the development of a new phase for specific applications in environmental analysis. The DB-1 phase and DB-1701 phases did not quite give the desired result, and a window-diagram approach indicated that an intermediate phase would be better; that is how the DB-1301 phase was born, which is now quite popular. The window diagram is simply a plot of separation factors against volume percent of Phase B in a binary phase (AB), or against the length of Column B in a coupled-column (AB) system. The simple technique of coupling two columns has also been called multi-chromatography. Hinshaw [11] has discussed various aspects of multi-chromatography. It is not surprising perhaps, that the order of coupling the columns gives a slightly different performance. The choice of lengths is also part of method development, as this will result in different net contributions of the individual phase selectivities to the overall separation. One application that immediately comes to mind is the coupling of two different CD columns for a particular analysis, where one of the chiral selectors is suited to the separation of some enantiomeric compounds, and the other is better for other enantiomeric compounds.

8.3.2 Pressure tuning

The coupled-column technique may be elegantly modified to provide a flexible approach to continuously variable solute selectivity, within given bounds, again intermediate between that of the two joined columns. This is done by providing a variable pressure at the confluence of the columns. The variation in pressure will affect the flow through each of the columns, and this modified the contribution of each column to the total coupled-column separation. Indeed, for two closely eluted solutes exhibiting quite different retention factors on each column, it is possible to demonstrate a situation where the two solutes fully swap positions, just by varying the mid point (or tuning) pressure. The effect of an increase in midpoint pressure is to reduce the pressure drop across Column 1; its flow-rate decreases and therefore the retention of a solute on this column increases. Conversely the pressure drop across Column 2 increases; the flow-rate increases and, therefore, a solute will travel across Column 2 much faster. This will lead to a greater contribution of Column 1 to the overall separation. This procedure was proposed

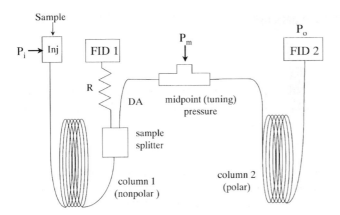

Fig. 8.2. Diagram of a pressure-tuning experiment. Columns of different polarity are used with a midpoint gas supply. R, flow restrictor; DA, deactivated tubing; P_i, injector pressure; P_m, midpoint (tuning) pressure; P_o, outlet pressure (1 atm); FID, monitor detector 1; FID, analytical detector 2. (Reproduced from Ref. 19 with permission.)

by Deans [12], and studied at some length by Sandra [13] and Hinshaw [11,14], and has been more recently re-introduced by Sacks' group [15,16]. They used short coupled columns, electronic pressure control for accurate control of carrier flows, and new methods for interpreting and planning the pressure-tuned experiment. In more recent cases, stop-flow and modulated operation of the carrier gas during the analysis was employed [17,18]. Fig. 8.2 shows a schematic diagram of the pressure-tuning experiment. The mid-point pressure gauge allows the pressure at this point to be gauged, when no supplement of extra carrier is provided. The FID 1 permits the separation performance at the end of Column 1 to be compared with that achieved at the end of Column 2, allowing the effect of Column 2 to be quantified.

8.3.3 Multi-dimensional gas chromatography

With an experimental set-up as in Fig. 8.2, an alternative mode of analysis, multi-dimensional gas chromatography (MDGC), can be devised, where again two columns are employed, but the object is now to allow only designated fractions to pass through the second column. Fig. 8.3 shows that an interface or valve permits solute to pass either from Column 1 to Column 2 or to Detector 1. The two columns may be housed in different ovens. This GC technique, also known (crudely) as "heart-cutting", has been acknow-ledged as giving the highest possible resolution [20,21]. A portion of the first chromatogram that is too complex and inadequately resolved is selectively passed into a second column, containing a different phase, where that fraction is more fully analyzed.

Giddings [22] and Schomburg [23] have contributed greatly to advances in this technique. Coupling of a packed column to a capillary column allows increased sample

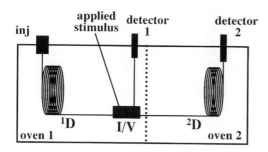

Fig. 8.3. Schematic setup for multi-dimensional gas chromatography. A second oven is optional, but may assist in system optimization. The stimulus is applied to the interface or valve (*e.g.*, a Deans' switch) to permit fractions of the effluent from Column 1 to pass into Column 2, where each fraction undergoes further fractionation.

introduction without overloading the capillary column. Back-flushing allows introduction of a sample with a wide boiling-point range, when the solute of interest is a more volatile component. Once that component has passed the midpoint valve, the midpoint flow is altered so that a reverse flow is introduced into the pre-column to flush out the low-volatility components. Since the components may not have migrated very far into the column, it is relatively easy to flush them out of the column and out the injection vent. Thus, low-volatility components need not be chromatographed through the total column system. The back-flush/venting procedure allows lower temperature operation, saves time, and avoids elution at higher temperature, thus increasing column life.

The primary fields of MDGC application are petroleum and essential oil analysis. For instance, for petroleum, the aromatic/aliphatic/cyclic aliphatic ratio may be a desired chemical characterization parameter of oil, and if these compound types can be measured as total classes, it simplifies the measurement and classification process. MS detection is widely employed, and so enhanced separation and identification are simultaneously achieved. Essential oil quality may be established by study of enantiomer ratios of various chiral compounds. If the enantiomeric excess (ee) of certain compounds in natural oil extracts is well known, the presence of adulterants (synthetic oils or cheaper oil substitutes, usually containing compounds of different ee) may be detected by MDGC with a chiral second column. Mondello *et al.* [24,25] used MDGC to estimate enantiomeric distributions in mandarin and bergamot oils, which enabled correlation of enantiomeric ratios with the technology of oil isolation, or the assessment of extraneous or contaminating oils in the products. The logical approach is to use an achiral pre-column and then to heart-cut the unresolved enantiomers of interest to the second, chiral column, where they are resolved. Fig. 8.4 represents a typical study of chiral compound analysis by using enantio-MDGC [26]. In this application the back-flush method was also used to clear the column of residual, higher-boiling material, once the target solute had been measured.

Many reviews and books have been devoted to MDGC [27–29]. Some examples of the use of MDGC for high-resolution analysis include the study of tobacco components [30]

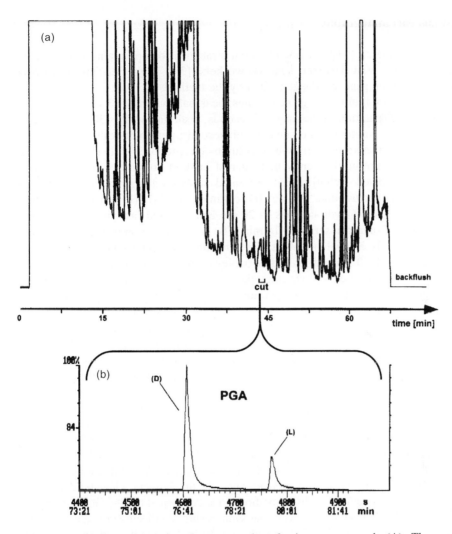

Fig. 8.4. Multi-dimensional gas chromatography of urinary compounds (A). The enantiomeric pyroglutamic acid is resolved on a chiral column (B) with MS detection. (Reproduced from Ref. 26 with permission.)

with 23 heart-cuts, each of *ca.* 1- to 3-min duration, separately eluted over 48 h, which revealed literally thousands of compounds, 308 of them identified and 80 found in tobacco for the first time. Wilkins' group [31] employed multiple parallel cryo-traps to store selected heart-cuts for a pre-column petroleum GC analysis. Each cryo-trap was individually eluted and analyzed on a second column with both FTIR and MS detectors. The key to MDGC is the switching system, coupling the two columns. The pressure-balanced Deans' switch [32] has been widely used, but other valves and on/off flow switches are also available.

8.3.4 Fast gas chromatography

Cramers and Leclercq [33] described general strategies for optimizing the speed of separations in GC. Van Deursen and co-workers [34] suggested the following classification, based on peak-widths generated by the method:

Fast GC = separation in the range of minutes; peak half-widths 1–3 sec

Very fast GC = separation in the range of seconds; peak half-widths 30–200 msec

Ultra-fast GC = separation in the range of sub-seconds; peak half-widths 5–30 msec.

Although the use of short and narrow-bore columns was proposed early in the history of capillary GC, it was not until recently that interest in short columns – and hence fast analysis – was rekindled. Schutjes [35] presented studies on high-speed profiling of complex mixtures in narrow-bore (50-μm ID) columns in 1981. He separated 24 alkyl aromatics in 6 min and 13 hydrocarbons in 20 sec and reported fast profiling of TMS-derivatized acids and sugars in serum in 14 min. The key observation was that separation performance effectively equivalent to that of a conventional capillary GC column of the day (25 m, 0.32 mm ID, 0.5 μm df) could be achieved in a fraction of the time.

Hinshaw [36] demonstrated various strategies for a faster GC analysis of a number of test mixtures: elevated isothermal temperature, increased temperature-program rates, shorter columns, and faster linear carrier-flow velocity. Analysis of a petroleum sample on a shorter, narrow-bore column showed that resolution performance equivalent to that of a longer column of conventional dimensions could be achieved in a shorter time. The process of using short, narrow-bore columns to speed up analysis has been adopted in an automated procedure called "method translation" or "retention-time locking", and developed by Agilent Technologies [37]. (The program is available as free ware on the Agilent Technologies web home page.) It allows input of various parameters of an existing method (column dimensions, carrier-flow, and temperature) to generate a new set of conditions on a "fast" column, which should, to the first approximation, give a separation of equal quality. A calculation of the speed gain is also provided. Generally, chromatography on the narrow-bore column will require a faster temperature program rate to produce the same separation performance as the longer, wider-bore column. Additionally, a major advantage is that a method, once developed on one GC system, can be "translated" to another, with effectively the same result. Van Deursen and co-workers [38] presented approaches to fast GC, based on narrow-bore multi-capillary columns, operating under vacuum conditions with vacuum outlet [39] and fast column heating [40]. Bicchi *et al.* [41] also used shorter capillary columns with conventional inner diameter for the analysis of samples of medium complexity. Understandably, separating power will be sacrificed, but this may be a useful strategy for some analyses. Blumberg [42–45] has published a series of papers on fast GC, in which he considered the effects of the major column operating parameters (column efficiency, speed of analysis, gas flow-rate, and liquid-film thickness) on a theoretical basis. When there is a high pressure-drop across a capillary column (*e.g.*, when a narrow-ID column is used), an extra term must be introduced into the van Deemter equation:

$$H = \frac{B}{\bar{u}^2} + C_1 \bar{u}^2 + C_2 \bar{u} \tag{8.2}$$

This shows that the effect of flow rate ($\hat{u}$) on column efficiency (H) is much greater than anticipated. The H vs. $\hat{u}$ plot is also much steeper for a longer capillary column, and this means that increasing the flow-rate to speed up an analysis will result in a greater loss of efficiency in a long column. This again suggests H_2 as a carrier-gas of choice.

An important factor in the acceptance of fast GC by routine laboratories is the availability of a range of columns with the required geometry. It is unlikely that a laboratory will introduce just one fast GC analysis to its range of analyses, and lack of chemical phases suitable for other applications may contribute to maintaining the *status quo*. However, there is every reason to believe that the trend to faster GC analysis will accelerate in the years ahead. Not surprisingly, fast GC demands both injection and detection techniques in keeping with rapid sample delivery and analysis of fast peaks. Less sample must be delivered to the column when narrow-bore columns are used (column loadability is reduced), hence, higher split ratios or more dilute solutions should be used. Peaks will be faster (narrower), and to maintain an acceptable number of data points per peak, the detection process may need to acquire data at a faster rate. Korytar *et al.* [46] have reviewed a range of practical routes towards fast GC analysis and presented a series of simulated chromatograms that were generated with various strategies.

8.3.5 Comprehensive two-dimensional gas chromatography

Comprehensive 2D GC (GC × GC) is a recently established mode that can be considered a (very) special case of MDGC. It requires a coupled-column system, with a column of conventional length in the first dimension (^{1}D), and a short, fast-elution column in the second dimension (^{2}D). (In conventional MDGC two standard-length capillary columns are used.) Thus, in GC × GC we use concepts from both, Sec. 8.3.3 and Sec. 8.3.4. The key to GC × GC is the use of a *modulation mechanism* that causes peaks eluted from Column 1 to be sliced into fractions, and each of these to be separately and sequentially injected into Column 2 (see below). Based on the two coupled columns, it achieves a two-column separation performance for *all compounds* in a chromatogram, as opposed to MDGC with heart-cutting, which only achieves a small increase in the overall separation process because it encompasses only a limited number of heart-cut fractions. It has been estimated statistically that the number of components which can be isolated in a GC separation is far less than the peak capacity (maximum number of resolvable components) of the column; a single column can achieve adequate separation when the number of peaks is no more than only 18% of the maximum capacity [47]. Giddings [48] recognized that a comprehensive GC experiment could have significantly greater separation power. In fact, it essentially multiplies the peak capacity of the first column by that of the second column, so that it can theoretically resolve thousands of separate compounds. For instance, if Column 1 has a peak capacity of 400 peaks and Column 2 has 20, the capacity is 8000 peaks! Column 2 is deliberately stated here as having a capacity of 20 peaks, since in GC × GC Column 2 is a short, fast-separation/elution column (*e.g.*, with a retention time of *ca.* 3–6 sec) and therefore has limited separating power.

Phillips and Beens [49] have reviewed the position of GC × GC up to 1999. At the time of writing this chapter, more than 110 papers have appeared on various aspects of GC × GC, including theoretical treatments, technical details, such as modulator design, and applications that demonstrate its use for complex sample analysis. Pursch *et al.* [50] reviewed the technical features of GC × GC, while Marriott and Shellie [51] reviewed application areas to which GC × GC has been applied. Fig. 8.5A is a generic schematic diagram of the GC × GC instrument, with the most important feature – the modulator – shown located between the two columns. A variety of devices are shown which are able to modulate the chromatographic peak as it emerges from Column 1, and pulse it in discrete packets to Column 2. The moving cryo-trap modulator developed by Marriott's group [52] is shown in Fig. 8.5B. The modulators deliver the performance that allows the unique two-dimensional separation of a sample mixture to be generated. Lee *et al.* [53] reviewed a range of modulator designs in 2001, but since then a number of other modulators have appeared. The modulators are essentially based on a rotating heater slot, the longitudinally modulated cryogenic trapping, which leads to cryogenic jets, and a number of valve [54] and diaphragm [55] systems. These all have the effect of "decoupling" the separation on ^{1}D from the elution on ^{2}D. Thus, whatever separation performance has been achieved on the first column is retained, and the second column gives a completely independent chromatographic analysis. To achieve this, the sample introduction into ^{2}D must be achieved in almost the same fashion as a normal injection, *i.e.*, it must be fast, and have no inherent bias against any of the components in that injection zone. If the modulator "collects" sample for a given (short) time from the end of ^{1}D, all the various compounds collected must be delivered to ^{2}D, both instantaneously and as one homogeneous pulse.

It is easy to see how valves can sample quickly and efficiently very small amounts of effluent from ^{1}D into ^{2}D, but they do not allow all sample to be delivered from ^{1}D to ^{2}D (this can only be done if the valve has a storage facility, such as a sample loop, and a back-flush operation pushes the collected effluent to ^{2}D). If sample is lost to waste during switching, not all sample is delivered to ^{2}D. The other modulators produce a "zone-compression" effect, and so concentrate the effluent into a discrete, sharp, narrow band

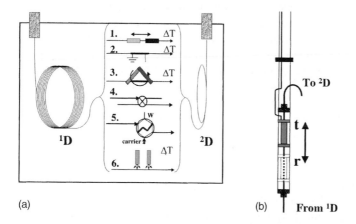

(a)   (b)

Fig. 8.5. (a). Schematic diagram of the comprehensive two-dimensional gas chromatography method. ^{2}D is a short (usually 1-m or shorter) column that gives very fast elution. (b). The longitudinally modulated cryogenic method, which incorporates a movable cryo-trap to trap then release solute traveling through a capillary column.

before rapid introduction (sometimes called re-injection) into ^{2}D. All this occurs in the capillary column, and it can therefore only be achieved by heating the solute in order to collect it into a leading zone, or by cooling it in order to effect trapping accumulation. The first of the cryogenic systems employed a movable trap that slides back and forth over less than a 3.5-cm length of capillary column. The solute in the effluent from ^{1}D is trapped into a narrow (a few mm) zone, and when the trap moves away (upstream), this band rapidly pulses into ^{2}D.

More recently, a cryogenic jet system was shown to give a similar performance [56,57]. The rotating slotted heater, which also achieved the band compression and modulation required in GC × GC [58], has effectively been superseded by the other modulation methods. Since each ^{1}D peak is pulsed into the second, fast-separation column, containing a different stationary phase, the aim is to "tune" or adjust experimental parameters such that unresolved solutes from the end of ^{1}D are resolved on ^{2}D. The requirements of Giddings that comprehensive 2D chromatography should maintain the first-column separation means that modulation be fast and that the ^{2}D chromatogram should be completed approximately within the time frame of the modulation time. Typical modulation times are between 2 and 10 sec. The cryo-trap procedure has been reported to offer the distinct mode of complete peak trapping with fast expulsion to the second

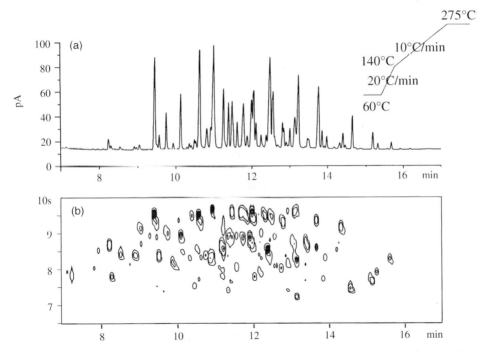

Fig. 8.6. Comprehensive two-dimensional gas chromatography of a polychlorinated biphenyl sample. Note that most PCB congeners are represented as individual, well-resolved contour spots. Modulation period, 3 sec; contour levels, 13, 20, 80, 120, 180, 220, 280 pA; base response, 10 pA. (Reproduced from P. Haglund, M. Harju, R. Ong and P. Marriott, *J. Microcol. Sep.* 13 (2001) 306 with permission.)

column. Since it is possible to target specific analytes in this mode, it was called *targeted multidimensional GC*. Features of a typical GC × GC analysis include enhanced peak capacity, and hence increased resolution, signal improvement over a conventional GC analysis performed under the same conditions, and "structured retention", which permits class separation of components of the sample [59]. These features clearly set GC × GC apart from conventional analysis. It is important that the two columns chosen for the GC × GC experiment allow "orthogonal separation" of the sample constituents. For the most part, a non-polar first column acts to provide discrimination of sample components on the basis of boiling points, and then a more polar, second column allows specific polarity differences of the unresolved components from ^{1}D to be exploited for their separation. The column phase chosen for ^{2}D will depend upon the nature of the compounds in the sample. For instance, a phase with a high phenyl content is useful for separating compounds on the basis of their aromaticity, elution being generally in the order: alkanes, cyclic alkanes, olefines, monoaromatics, and diaromatics. Thus, a "structural map" of compounds arises in the 2D plot. Applications in forensics [60], pesticides analysis [61], and essential oil analysis [62] attest to the unique utility of the 2D plot for the identification of solutes, based on dual-column retention. The application to polychlorinated biphenyl compounds in Fig. 8.6 demonstrates the power of the 2D GC × GC separation for such an environmental analysis. Given the present activity in GC × GC, it is anticipated that this will continue to be a fruitful area for gas chromatographic research, and will be accepted for and applied to a wide applications base.

8.4 SAMPLE INTRODUCTION

An earlier review on sample injection by Sandra [63] has been superseded by the comprehensive review of Grob [64], which can be regarded as the definitive reference source on injection. The introduction of a sample into a GC column probably calls for more attention of the analyst than any other part of the technique. Sample may be introduced into the injector in gas, liquid, or solid form, each involving specific operational aspects. Direct injection into a packed column is rather straightforward, since the high-flow rate into the column and its large sample capacity make special maneuvers unnecessary. In contrast, capillary column injection may involve critical decisions by the analyst. The solid-sample introduction procedure is perhaps the least used. In one technique, a syringe is used with a small depression in the injection needle to accommodate a finite mass of solid material, which can be injected through the septum into the heated injection zone. In another procedure, a small glass vessel, containing a solid sample, is inserted into the injector and then smashed to allow its contents to escape into the carrier gas. The solid, *e.g.*, a fiber filter on which solid particles have been collected, may also be inserted into a cool injector with programmable temperature operation to release volatiles from the sample (an arrangement available from the ATAS GL Company). These are merely three variants of solid-sample introduction, which imply that quantitative transfer of sample into the injector is achieved.

8.4.1 Split/splitless injection

The pioneers of capillary GC (Golay and Desty) recognized very early that rapid introduction of sample into a capillary column was essential for preserving the high-performance capability (efficiency) of the capillary column. Ettre [65], recounting the early history of capillary GC injection, credits Desty with the development of the dynamic split procedure. Grob [64] gave an excellent account of the rationale of the use and of the basic principles of the split/splitless injection method, and discussed the strategies for improving recoveries. Since the total peak variance is a summation of contributions by the injector, column, and detector, any deficiencies arising from the injector become increasingly significant as column efficiency improves. Therefore, the ideal injector will minimize band-spreading or injection zone-spreading during sample introduction. Since the flow-rate in a capillary column is low (at most, a few mL/min) and the injection of μL-volumes of solution leads to many mL of vapor, it is not possible to have this much vapor enter the column fast enough to give a narrow solute band without some intervention. The first practical way to inject a sample in solution was the split injection method, in which a large split-vent is used to purge most of the injected solvent and its contents. In this way, vaporization occurs instantaneously and (a small amount of) sample enters the column. Practical split ratios (ratio of vent-flow to column-flow) of from 10:1 to 200:1 are used, *i.e.*, only *ca.* 10 to 0.5% of the sample enters the column. It is often assumed that the split ratio and/or the solvent fractional ratio between vent and column is equal to the solute split ratio. Fig. 8.7 is a schematic diagram of the typical split/splitless injector. Common features of the injector include the glass liner, pre-heated carrier-gas inlet, vent-flow line, and septum purge. While the split injection method has served the user well, and is still widely used, it suffers from the very fact that most of the sample never reaches the column, and sample loss is undesirable. Also, the very real problem of discrimination, where certain – usually higher-boiling – components suffer greater reduction in response than would be expected simply on the basis of the split ratio, has been acknowledged for some time [66]. Discrimination effects inherent in various methods of injection increase in the order: on-column < fast hot needle < slow hot needle < solvent-flushing fast cold needle < slow cold needle.

The type of injector liner is believed to be critical. Some of the proposed modifications to injector liners include a glass-wool-packed liner, inverted-cup liner, cyclo-liner, and a liner with baffles or beads. Many of the strategies behind the use of different liners are related to the idea that a tortuous flow path and intimate mixing of sample and carrier gas will result in the most consistent and reproducible splitting of sample. Thus, inverted-cup arrangements and liners packed with glass-wool and similar material seem logical. Grob and Biedermann [67,68] have used a video of the injection process, with a glass injector construction, UV light, and a fluorescent dye (perylene) in the injected solution to permit tracing of the liquid flow during injection. It was seen that, contrary to expectations, solution could channel directly through glass wool, without the expected mixing and uniform vaporization. The conclusion was that an open injection liner is as effective as any of the above modifications and, in the case of sensitive samples, the reduction in active surfaces in the unpacked liner may be a real advantage. However, the glass-wool packing does keep some contaminants from reaching the column, and for this purpose the wool is

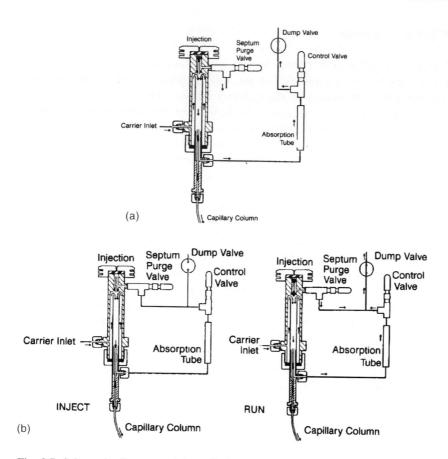

Fig. 8.7. Schematic diagrams of the split (a) and splitless (b) injection modes. (Reproduced with permission from SGE International.)

best positioned at the tip of the needle. The injected volume should be sufficient to give a volume of vapor consistent with the inner volume of the liner. An excess of injected solution may back-flush against the carrier-gas supply, or be lost through the septum purge. If the needle is too short, sample may evaporate too high up in the liner, and again be lost through the upper part of the liner.

The splitless injector method was introduced in order to overcome the problem of losing most of the sample out the vent. Fig. 8.7 illustrates the typical two-step injection process used in the splitless mode. If all (or most) of the sample can be injected, then the peak response will be maximized. This is important for routine trace analysis. The most critical consideration is that the analyte band must be introduced into the column in such a way that it does not spread or smear down the column in an excessively long band, and so certain steps must be taken to provide some degree of band focusing or trapping. Again, Grob studied this extensively, and provided the analyst with useful advice on implementing the splitless procedure. It may be summarized as follows: inject the sample into a hot injector liner to allow sample evaporation; keep the injector vent closed during injection transfer of sample to the column; use an oven temperature at least 30°C below the boiling point of the sample solvent; open the vent at a specified time after sample injection, *e.g.*, 45 sec to 1.5 min; use pressure pulsing to ensure that vapor does not back-flush up the inlet line. This may be explained as follows: Closing the vent retains the sample in the injector liner for the splitless period, and so having a higher concentration in the liner for a longer period ensures that more sample enters the column. After a given period, there is no extra benefit (*i.e.*, no further increase in response) of analyte transfer to the column. The low oven temperature permits the solvent effect to be realized. Solvent may be considered to form a condensed phase along the inner walls of the capillary. This effectively increases the instantaneous film thickness (reduces phase ratio), and this causes analyte to be trapped or move slowly in this region (a small phase ratio increases retention or reduces band mobility). This effect is in addition to the fact that the less volatile analyte itself will not be amenable to elution at the low oven temperature. The opening of the vent effectively flushes away mainly residual solvent and a small amount of sample that is still in the injector after a minute or more. The long time taken by sample transfer in splitless injection contributes to band-broadening of the more volatile components of a sample, if they are not effectively trapped. Thus, early-eluted peaks may be broader than they should be, and later peaks will tend to be more acceptable. An undergraduate laboratory experiment has been described that demonstrates the comparison of split and splitless injection, and emphasizes the problems that may arise if the splitless mode is not performed correctly [69], such as not opening the split vent or not cooling the oven during splitless injection. This has been included in a popular text on analytical chemistry by Harris [70].

8.4.2 Cool on-column injection

To address the concerns of discrimination and possible decomposition of thermally labile solutes into a high temperature split-splitless flash vaporization injector, the on-column capillary injector was developed as an alternative introduction method. The two primary features of on-column injection are the use of a fine needle and a method for sealing the injector about the needle as it is inserted into the column. The narrow needle is not robust enough for piercing the septum, used in conventional syringe injection. Some on-column injectors have a two-stage closure where the needle is pushed through a narrow channel which provides a seal about the needle, and then a second closure valve is opened to allow the needle to be introduced into the carrier-gas flow region. Other designs use a sealing procedure that permits the needle to gently push aside a ball-seal that normally prevents escape of carrier gas. The needle may be made of either fused silica or fine stainless steel. Once the needle is inside the column, the plunger is slowly depressed to permit a balance of vaporization of the solvent and its re-condensation inside the capillary to again achieve the solvent effect, to prevent the solvent and sample from spreading too far down the column during the injection step. This ensures narrow band formation. If flooding of solvent occurs, droplets of solvent may carry the analyte some distance down the capillary. Thus, the oven is held at a

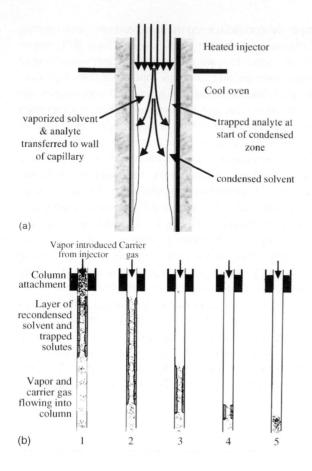

Heated injector

Cool oven

vaporized solvent
& analyte
transferred to wall
of capillary

trapped analyte at
start of condensed
zone

condensed solvent

(a)

Vapor introduced Carrier
from injector   gas

Column
attachment

Layer of
recondensed
solvent and
trapped
solutes

Vapor and
carrier gas
flowing into
column

(b)   1   2   3   4   5

Fig. 8.8. The solvent effect used in splitless and cool on-column injection to provide trapping and therefore zone focusing of analytes in a condensed solvent zone. (a). Initial process at time of injection. (b). Subsequent steps, illustrating the progressive evaporation of solvent and focusing of analyte. (Reproduced from Ref. 64 with permission.)

temperature just below the boiling point of the solvent. Fig. 8.8 shows the proposed effect of solvent condensation, which may occur in both splitless and on-column injection. Once injection is complete, the solvent vaporizes out of the condensation region to leave the solute in the stationary phase at the start of the column, ready to commence the chromatographic process. On-column injection methods were shown to overcome the bias against higher-boiling analytes. This method was used for porphyrins, which had retention indices up to 5500 [71,72] and for such other high-temperature applications as triglyceride gas chromatography [73], where the low-volatility components require column temperatures up to 400°C. On-column injection into a wider-bore capillary column is feasible, with downstream coupling of the wide-bore column to a narrow-bore analytical column.

8.4.3 Programmed-temperature vaporizer and large-volume injection

The programmed-temperature vaporizer (PTV) is an introduction procedure, invented 20 years ago, which is intended to decouple the process of solvent introduction into the injection region and the subsequent solvent evaporation and sample vaporization steps [74]. The problems of discrimination inside a heated injector and other uncertainties involved in injection are potentially tractable by using PTV injection. The injector temperature can be programmed so that the various steps may be individually optimized or controlled. Provided the required injection volume can be fully dispensed into a cool injector in which all the sample is retained, the excess solvent may be expelled from the injector (e.g., through the split vent), as the temperature is increased. The vent is then closed, and the increased temperature of the PTV will cause the analyte to be vaporized into the column. Some residual solvent may also be introduced in order to focus the sample. The description of a PTV method will include the PTV injector-temperature profile, the oven-temperature profile, and the valve-switching steps that permit either venting of volatiles or transfer of the evaporated volatiles to the column. Fig. 8.9 illustrates the PTV process.

This injection mode is very versatile, because it can also be used for the injection of large volumes of solution [75]. Volumes in excess of 100 μL may be injected, with a corresponding decrease in concentration sensitivity. If a normal splitless injection of 1 μL has a detection limit of 0.1 mg/L, a 100-μL PTV injection of the same analyte would have a detection limit of 1 μg/L. The problem of filling the PTV injector with over 100 μL of solution prior to evaporation is solved by concurrent evaporation of solvent while the

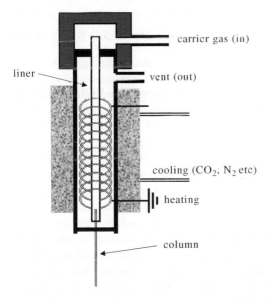

Fig. 8.9. Programmed temperature vaporizer injection unit, with supplementary heating and rapid cooling.

sample solution is slowly introduced. This normally requires a carefully measured introduction rate at the prevailing temperature to ensure that solvent elimination matches the rate of solution introduction. Once the total volume has been introduced, the vent is shut and the temperature is increased to permit analyte vaporization into the column.

Another large-volume injection procedure has been described, in which injection of the solution onto an uncoated pre-column permits some degree of solvent effect, but requires a short segment of retaining column to follow the pre-column. As sample solution travels into the retaining column, it is trapped by the stationary phase. Just after the retaining column and at its junction with the analytical column, there is a solvent vapor exit, which allows venting of the excess solvent before it reaches the analytical column. When almost all the solvent is vented and before any analyte is lost through the solvent vapor exit, the valve is closed to channel the analyte into the analytical column. The net effect is an increase in analyte response, which allows trace analysis.

8.4.4 Headspace and solid-phase micro-extraction

When the volatile headspace above a solid or liquid is sampled, it is expected that an unbiased sample of the headspace is collected and analyzed, *i.e.*, no sample components are lost in the transfer lines, and sampling does not disturb the equilibrium between headspace and sample. The sample is transferred directly to the GC injector with a gas-tight syringe. Headspace sampling is primarily suited to volatile analytes, which may either arise from volatile molecules released from a solid or liquid matrix or they may be part of the sample, such as in the analysis of atmospheric organic species. For the latter, sample collection is often performed in the field by use of canisters, which are a convenient vehicle for transporting the sample to the laboratory for analysis. There, the canister is placed in the instrumental sampling system to permit a given volume to be introduced into the gas chromatograph. On-line sample processing may involve the removal of ozone and water vapor from the gas by scrubbing the sample prior to cryo-trapping. Headspace introduction is best conducted with a cryo-focusing step at the head of the column to create a narrow sample band. Often, for studies of very volatile components, the chromatograph will be operated at sub-ambient (*e.g.*, $-50°C$) temperature to achieve adequate retention and resolution of light hydrocarbons or permanent gases.

The solid-phase micro-extraction (SPME) introduction method, developed by Pawliszyn [76], is a popular, convenient procedure for analyzing organic volatile and semi-volatile compounds, with direct transfer of the extracted solutes to a GC injector. It requires a small amount of sorptive phase, not unlike a stationary phase for GC, coated onto a fused-silica fiber, which is attached to a fine wire. The wire and fiber are retractable into a supporting needle so that they are protected during the insertion of the needle through a septum. There are now literally hundreds of applications [77] and theoretical studies based on the use of SPME for sampling. The process may involve the following steps: Clean and condition the fiber at high temperature in a flow of gas (*e.g.*, by insertion in a GC injector port). Retract the fiber and put the needle into the sample, which most often is in a sealed vessel to permit headspace equilibration. Insert the needle through a sampling septum and expose the fiber. Allow the analyte to be sorbed from the headspace for a given time. Retract the fiber into the needle. Transfer the needle to the GC injector and insert it through the GC septum. Push down the plunger to expose the fiber and desorb the analytes from the needle into the carrier gas. The fiber may be left in the injector until it is clean. Desorption should occur rapidly, and the injector may be used in split or splitless mode, depending on the amounts of analyte sorbed.

While SPME may be used for both qualitative and quantitative analysis, the latter requires very careful handling and control, with appropriate calibration. Baltussen *et al.* [78] have shown that a magnetic stir-bar, coated with a polydimethylsiloxane phase, can perform similar sorptive extractions of analytes from headspace and from solution. Stir-bar sorptive extraction (SBSE), having a larger volume of polymer phase than a fiber, apparently recovers much more analyte in the sample extraction step. Both SPME and SBSE may be employed directly in aqueous solution for an efficient concentration of analytes onto the polymer. The stir-bar is more difficult to introduce into the GC injector than the SPME fiber, and it is more suited to placement into a PTV-type injector for thermal desorption of analytes.

8.4.5 Coupling with liquid chromatography

The objective of liquid chromatography prior to gas chromatography is to simplify sample mixtures. Coupled LC/GC can be automated by on-line introduction of the eluate from LC into the GC injector. Mol *et al.* [79] have reviewed the technical aspects of on-line sample enrichment and large-volume injection for the analysis of traces of pollutants in aqueous samples. The most effective procedure involves normal-phase LC, with eluents rich in organic solvents, suitable for GC injection. Normal-phase LC is also most useful for class separation of complex mixtures into fractions of different polarity. For example, a mineral oil may be fractionated into aromatics, cyclic alkanes, unsaturated alkanes, and saturated alkanes. Each of these can then be separately introduced into the gas chromatograph to permit specific classes of compounds to be individually analyzed. Davies and coworkers [80] reported the automated chemical-class characterization of kerosene and diesel oil by on-line coupled microbore HPLC/capillary GC. The whole aromatic fraction was back-flushed as a single peak from the LC column, after being separated from aliphatics. Beens and Tijssen [81] demonstrated the HPLC fractionation of a middle-range distillate for quantitative characterization of oil fractions in an on-line HPLC/HRGC system. A total of 7 different chemical-class fractions (saturates, mono-aromatics, naphthalenes, biphenyls + benzothiophenes, fluorenes, dibenzothiophenes, and tri-aromatics) were isolated and quantified by summing the total FID response area and using an average response factor for each class. Mondello *et al.* [82] analyzed bergamot oil on a silica column with a step-gradient elution program of pentane and diethyl ether to obtain four fractions (F1–F4), which were separately analyzed by GC to reveal the following compositions: F1 = mono- and sesquiterpenes; F2 = aliphatic aldehydes and esters; F3 = monoterpene aldehydes, sesquiterpene alcohols and some monoterpene alcohols; F4 = the remaining monoterpene alcohols.

8.5 DETECTION

Several reviews [83–85] have covered the general principles of chromatographic detectors, but did not include more recent studies or newer detection methods.

8.5.1 General principles

The chromatographic detector provides means of "visualization" or recording of the chromatographic separation. Unlike thin-layer methods, where direct visual confirmation of a separation is possible, from the position and color of solute spots located on the TLC plate, it is not practical in GC methods to use direct visual monitoring of compounds emerging from a column. One exception would be to (carefully!) observe the flame color, produced by a flame-photometric detector, in order to confirm the presence of sulfur or phosphorus compounds entering the flame. The detector should respond quickly (on the time scale of compounds entering the detector), if the recorded signal is to faithfully represent the chromatographic peak distribution as it is eluted from the column. This is a matter of great importance, because we use the chromatographic peak profile to derive much information relating to the chromatographic process and performance of the chromatographic system, so that we rely completely on the integrity of the recorded peak. The variety of detectors available means that the analyst must consider the choice of appropriate detector for a particular application and take advantage of the extra information available from different detectors. Amirav [86] recently compared the role and application of GC detectors in a climate where mass spectrometry is becoming a more universally available and competitive detection tool. The detector is a chemical transducer that converts a given property of a compound into a measurable electrical signal. The chemical property may be its specific elemental composition (*e.g.*, the presence of a halogen, as sensed by the electron-capture detector), or it may be related to the molecular composition and structure of a compound (*e.g.*, producing a fingerprint mass spectrum in a mass spectrometer). Selected general operational principles of relevance to chromatographic detectors are listed as follows:

Sensitivity – The magnitude of signal response, *Sig*, for a given change in sample mass. This is essentially the slope of the calibration plot of mass (m) against response (R); $R = Sig\ m$.

Selectivity – The response of a detector towards a given compound, in comparison to its response to "carbon". This is essentially a measure of the response of a selective detector towards the (hetero)atom for which the detector exhibits a selective response. It is defined in this manner precisely because the response to "carbon" is a general response mechanism for the most widely used detector – the flame-ionization detector (FID) – and the response to a different element provides a different, and hence selective, response mechanism. Selectivity is therefore a response ratio, and may be of the order of 10^3 to 10^5.

Linearity – The classical detector responds linearly to changes in the amount of sample entering the detector. Linear range varies with detector type, and for quantitative analysis it is important to operate the detector within its linear range. The linear range is reported in terms of number of orders of magnitude over which the relationship $R \propto m$ holds. Not all detectors, inherently, give a linear response. Some suffer concentration effects that lead to non-linearity at higher amounts of solute. The one detector that stands apart for having a non-linear response is the flame-photometric detector (FPD) (S-mode).

Detection limit – This is defined as the mass of sample injected that leads to a signal equal to $3 \times$ (or $2 \times$, depending on the definition) the noise response level. This is an important analytical parameter, since it determines the solute mass which must be introduced into the column in order to obtain a detectable signal. This is then related to the analyte concentration and the volume that must be injected. In respect of the total analytical method, detection limit considerations may also dictate such factors as sample mass, final extract volume, and choice of injection method. The limit of quantification is often defined as that mass of sample which leads to a signal equal to 10 times the noise level.

Response time – The detector response time is the speed with which the detector responds to a change in mass of sample entering the detector. In order for the detector to have negligible effect on the peak profile (shape) and so preserve the performance of the column, the detector should be fast-responding – we might say that the detector should give an instantaneous response. For very narrow peaks, the detector time-constant may indeed not be fast enough, and lead to an apparent broadening of the recorded chromatographic band. For a detector, such as the electron-capture detector (ECD), which has an "active" internal volume, use of a low-flow-rate detector make-up gas will cause apparent peak-broadening for fast GC peaks.

Detection frequency – The frequency of sampling of the detector response is an electronic signal-processing parameter. The detector will normally produce an analog signal which is discretely sampled as a digital response for subsequent computer processing. As GC peaks have become narrower through application of faster GC methods, the detector response must be sampled faster to provide sufficient data points across a peak in order to adequately "describe" the peak. It is accepted that a minimum of about 10 data points per peak half-width is required (*i.e.*, 4 data points per peak standard deviation, or 17 points per peak base-width) for an accurate description of a peak, or more, if a shoulder is present [87]. Dyson [88] discussed the effect of data sampling frequency and pointed out that more errors will arise from asymmetric peaks than from symmetric peaks and that for the former a higher frequency should be used. In a further treatment, Dyson reported the number of data samples required for defined levels of measurement uncertainty, taking into account various levels of peak asymmetry. For 1.0 and 0.1% error, 27 and 85 data samples, respectively, per peak-width (defined here as $6 \times$ standard deviation) are required for a Gaussian peak (based on the trapezoidal rule). This then will permit accurate peak area and peak maximum response (*i.e.*, retention time) to be derived [89]. Most conventional capillary GC detectors will permit 10 Hz data sampling, but this is insufficient for very fast GC peaks which may have half-widths of 100 msec, for which, as the above discussion indicates, operation at 50–100 Hz is required. Recently, FID and ECD operation at these, and higher, frequencies has become available.

Response factors – The chromatographic detector response factor acknowledges the fact that the detector response is not an absolute relationship between response and amount for different compounds. This is equivalent to stating that the sensitivity value, S, varies from one compound to another. While for a given compound class, the response factor may be similar, for different compound classes, the response factor may vary considerably. This is directly related to the response mechanism of the detector, *i.e.*, the process by which the transducer produces an electrical signal arising from the compound. If the mechanism and magnitude of response can be determined precisely from the chemical structure of a compound, then the response factor for the compound may be precisely derived. More often than not, the response factor will be an experimentally determined value, calculated by using standard mixtures of known concentrations and the resulting responses of the compounds when the mixture is chromatographed. Since the response of a given compound is proportional to its injected mass (see above), its response factor under given detector conditions is constant. However, if the detector conditions vary, the response factor of a compound may vary. This reflects the need to calibrate the detector response frequently enough to ensure that there is no change to the response factor, and this is often the reason why calibration data should not be applied to the analysis of samples on subsequent days without re-calibrating the response. If the response mechanism of a detector is precisely known, its response may also be exactly predictable. Due to the uncertainty of the response mechanism, unfortunately this is often not the case. Continuing studies in this area include the use of artificial neural networks for predicting the response of various detectors [90]. It appears that reasonable prediction was possible for the FID and two photo-ionization detector (PID) conditions studied, although in some instances the predicted and experimental response factors differed sufficiently to suggest that precise analysis will still require direct experimental determination of response factors.

Response factor data are usually tabulated in the form of normalized responses, against the response of a compound that has a well-defined, dependable response. A typical procedure will be to prepare a standard mixture of known volumes or masses of compounds to be calibrated, and to record their relative response areas. Response factors may then be calculated, based on relative masses or moles of compound (to give mass response factors, or mole response factors, respectively), by calculating the ratio (relative response)/(relative mass). This value can then be normalized to the reference compound. Table 8.3 illustrates such a calculation. Here, toluene gives the best response. The response factor of ethyl acetate is about 1/3 that of toluene. Butanol has a response about 2/3 that of toluene. These recently calculated response factors are quite consistent with older compilations of response factors, such as the relative sensitivity data reported by Dietz [91], also listed in Table 8.3. The response is an indicator of the mechanism of the FID. By calculating various parameters for the mixture – volume %, mass %, mol % and "relative mass of carbon" in each compound – it can be seen that the FID response correlates best with the amount of carbon in each compound. Thus, ethyl acetate has the lowest response, and has proportionally less carbon than the other compounds. This is because each molecule of ethyl acetate has a considerable amount of oxygen. This reduces the relative amount of carbon, and does not, in itself, contribute to the response of the FID in generating the measured flame ionization. A similar comparison for a mixture of alcohols (Table 8.4) shows that, again, the FID response follows the relative proportion of carbon in each compound, but the depressive effect of the heteroatom in ethanol makes its response somewhat lower than its carbon mass %. In contrast, the trend in thermal-conductivity

TABLE 8.3

RESPONSE FACTOR CALCULATION FOR A SIMPLE STANDARD SOLUTION

Compound	Volume %	Mass %	Mole %	Mass % carbon	FID response area %	Area %/ mass %	Response factor[*]	Literature value[**]
Ethyl acetate	25	26.1	26.4	18.8	12.5	0.479	0.360	0.38
Toluene	25	25.0	24.2	30.3	33.3	1.332	1.000	1.07
Butanol	25	23.4	28.1	20.2	19.6	0.838	0.629	0.66
o-Xylene	25	25.5	21.3	30.7	30.7	1.204	0.904	1.02

[*] Response factors relative to toluene = 1.000.
[**] Literature values [91] relative to heptane = 1.00.

TABLE 8.4

RESPONSE COMPARISON OF FID AND TCD FOR AN ALCOHOL MIXTURE

Compound	Volume %	Mass %	Mol %	Carbon mass %	FID area %	TCD area %
Ethanol	25	24.5	33.8	20.8	17.1	26.9
Propanol	25	25.0	26.4	24.4	25.2	26.1
Butanol	25	25.2	21.6	26.6	27.5	24.2
Pentanol	25	25.3	18.2	28.1	30.3	22.8

detector (TCD) response is opposite that of the FID, and one could suspect that it may reflect the relative mol %.

8.5.2 General classifications of detectors

Almost any physical method that can be used as a transducer in the gas phase and has real-time (on-line) measurement capabilities may be used as a GC detector. This means that there are many possible detectors, some of which may be used routinely for particular applications. It is possible here to mention only those that are more readily available and/or used in analytical laboratories. One may also expect that for a detector to be widely accepted, it should be reliable, should have a reproducible run-to-run, and day-to-day response, should not require too much maintenance, and should exhibit little drift and/or noise. Dyson [88] discussed these and other aspects of measurement of chromatographic peak parameters in depth. Two broad classifications for GC detectors place them in either mass- or flow (concentration)-sensitive categories. For a mass-sensitive detector, an increase in carrier flow will increase the mass flux into the detector, and thereby increase the response proportionally. The flow (concentration)-sensitive detector will show no change in response, since the concentration of sample in the carrier will not change. A further classification of detectors, and one that is perhaps more useful, is to denote a detector as universal or selective (specific). This is directly related to the response mechanism of the detector, and thus determines whether the detector responds to a broad range of compound types (universal), or is strictly suited to only certain classes of compounds (selective). Both, the TCD and the Fourier transform infrared (FT-IR) detectors are universal, since almost any compound that can be chromatographed will possess the property of thermal conductivity and will have molecular structural features that lead to infrared activity. Likewise, the mass spectrometer will produce a mass spectrum for almost any compound that can be eluted from a GC column (provided the molecule is suitably ionized). However, the FT-IR and MS detectors will also be highly selective, since the spectral discrimination of the detector permits selection of a feature which will be indicative of certain molecular functional groups. We may refer to these detectors as having multi-channel detection capabilities, where the detector records a discrete response intensity at given molecular vibrational frequencies (FT-IR) or mass-to-charge ratios (MS); the selection of one or more desired channels then constitutes isolation of molecular-specific detection performance. Finally, the FID is considered a carbon counter, and while this may be thought of as giving a specific response mode, when we appreciate that almost all compounds that can be subjected to GC will likely contain carbon (or a C–H bond), the FID should respond to most compounds. The most important exceptions are the permanent gases and related molecules. Table 8.5 lists various classifications of GC detectors, and selected comments on their response characteristics. Further quantitative properties of some detectors with respect to sensitivity and selectivity are summarized in Table 8.6.

8.5.3 Selected detectors

8.5.3.1 Ionization detectors

The success of ionization detectors is related to the relatively low background response of the carrier-gas stream. The presence of an ionizable molecule generates an easily measurable signal, its magnitude depending on the amount of solute present. A variety of processes can lead to either ionization of a molecule, or produce ions from a molecule by reaction in the detector, and hence, a variety of ionization detectors are used in GC, as listed in Table 8.5. Clearly, if they all operated according to the same mechanism, there would be little reason for this variety; the detector names give a hint to their different mechanisms.

The *flame ionization detector*, now the most widely used detector, burns carbon-containing compounds in an air-rich hydrogen/air flame. The hydrogen is normally pre-mixed with the carrier gas, and hence enters at the base of the flame jet. The standing current may be as little as 10 pA, but it is believed that few molecules which enter the

TABLE 8.5

CLASSIFICATIONS AND EXAMPLES OF GAS CHROMATOGRAPHY DETECTORS

Detector type	Mass/Flow	Comments
Ionization		
Flame-ionization FID	Mass	Universal detector; responds to carbon; variation in response to different chemical forms of carbon; good sensitivity
Thermionic ionization TID or NPD	Mass	Selective; responds to N, P, S, B, X, Sb, As, Sn, Pb; very sensitive
Electron-capture ECD	Flow	Selective; chiefly for halogens; very sensitive
Photo-ionization PID	Mass	Selective; response arises from photon-induced ionization
Mass spectrometer MS; ICPMS	Mass	Universal; requires (molecular) ionization to permit mass measurement; fragmentation patterns permit molecular structural assignment. ICP produces primarily atomic ions for mass measurement
Pulse discharge PDD	Flow	Variety of modes – electron capture and photo-ionization – so may have specific or universal response; may also stimulate atomic emission
Bulk physical property		
Thermal conductivity TCD	Flow	Universal; responds to change in thermal conductivity of the carrier gas as it passes over a heated filament; not sensitive
Optical		
Flame photometric FPD	Mass	Selective; molecular emission produced in a cool hydrogen flame; good sensitivity; for P-,S-, halogen, and N-containing compounds
Atomic emission AED; ICPOES	Mass	Selective; atomic emission excited in various discharge types
Chemiluminescence	Mass	Selective; may be very sensitive; very useful for sulfur and nitrogen compounds
Infrared FT-IR	Flow	Universal; records a compound's infrared spectrum as is passes through a flow cell
Electrochemical		
Hall electrolytic conductivity HECD	Flow	Selective; sensitive for halogens; an operationally more complex detector but may be specified in some methods for pollutant analysis
Other		
Olfactory GC/O	Mass	Selective; odor of compounds eluted from a GC column and detected by human nose
Electro-antennography (EAG)	Mass	Insect antenna response to chemical stimulus

TABLE 8.6

TYPICAL FIGURES OF MERIT OF SELECTED DETECTORS

Detector	Sensitivity (g/sec)	Selectivity (g solute/g carbon)
FID	2×10^{-12}	Depends on compound's response factor
FPD	2×10^{-12} (P)	$6 \times 10^{4} - 1 \times 10^{5}$ (S/C)
	5×10^{-11} (S)	
NPD	1×10^{-13} (N)	5×10^{4} (N/C)
	5×10^{-14} (P)	2×10^{4} (P/C)
PID	2×10^{-12} (benzene)	Variable – compound-dependent
ECD	$\sim 10^{-13}$ (variable)	Variable but may be high
HECD	$2 \times 10^{-12} - 10^{-11}$(S)	10^{5} (S/C)
	$2 - 4 \times 10^{-12}$ (N)	$> 10^{6}$ (N/C)
	5×10^{-13} (Cl)	$> 10^{6}$ (Cl/C)

flame actually produce a measurable ion – perhaps one ion per million molecules. The ionization response may be optimized by altering the relative flows of air and hydrogen, while injecting a given standard solution. Fig. 8.10 is a schematic diagram of a typical contour plot of flame-composition optimization. It shows that, as hydrogen is varied, the response goes through a maximum, whereas as air is increased, the response reaches a plateau and thereafter does not diminish greatly. Organic compounds produce ions according to the representative exothermic reaction [92]:

$$\text{C-containing compound} + \text{flame} \rightarrow \text{CH·} + \text{O}^{*} \rightarrow \text{CHO}^{+} + \text{e}^{-}$$

This reaction explains why carbon is a key elemental requirement for producing the ionization response, why in particular the C–H bond is important, and why the response is

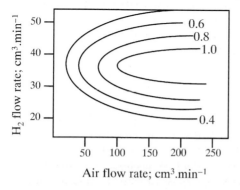

Fig. 8.10. Response contour plot, indicating typical effects of hydrogen and air flows on the response of the flame-ionization detector, approximately normalized to a maximum response of 1.0.

proportional to the amount of injected solute. The main charge-carrying species is believed to be H_3O^+ (and clusters of hydrated ions) [93]. However, there is still uncertainty as to the exact mechanism. The FID still attracts attention from those offering new understanding of its response. Holm [94] concluded that the initial generation of CH_4 from hydrocarbon compounds in cooler parts of the flame is critical to its response.

Due to the differences in the chemical bond of carbon in organic compounds, the relationship between the number of carbons and the resulting response is not the same for different compounds. This explains the different response factors observed for the FID (Sec. 8.5.1). For saturated organic compounds, we note a similar response per carbon, and so *x* mols of butane will give a response very similar to *x*/2 mols of octane. However, smaller responses for a given mass of compound are found for compounds that contain heteroatoms. CCl_4, CO, COS, CO_2, CS_2 will show very poor sensitivity in the FID, as will the permanent and inorganic gases.

The *thermionic ionization detector* (TID), also known as the nitrogen/phosphorus detector (NPD), was developed by Kolb [95] from studies on the alkali-FID, and may be considered to be a variation of the FID. However the unique conversion of the "universal" FID response into a sensitive and selective mode is put to significant advantage in the analysis of a wide range of nitrogen- and phosphorus-containing compounds, such as drugs and organophosphorus pesticides. The TID still employs a mixture of air and H_2 gases, which may be thought of as similar to the FID flame, but the gas composition will not by itself support a flame. The presence of an alkali-metal salt bead or ceramic material (initially based on rubidium, but cesium may also be used), located just above the detector jet tip and before the collector electrode, suggests that the gases form a plasma in the detector. The metal salt is formed at a ceramic bead on an electrical wire, the voltage of which can be separately controlled to maximize the signal and reduce the background response. The detector may also be considered as having a FID-like response, but this is suppressed by using a negative voltage on the bead to increase the detector specificity for the target elements. Whether the reaction between the alkali metal and the compound occurs in the plasma as a gas-phase process [96] or at the bead surface [97], is still open to interpretation. However, the degradation products of nitrogen and phosphorus compounds must become negatively charged by electron-extraction processes, and thus provide the selective response. The effective result is a response that is linear with the mass of solute, is sensitive to trace amounts of solute, and gives a specific response to the elements N and P. Enhanced selectivity of response to nitrogen compounds, in an inert gas, without H_2 or air, has been described recently [98]. It was found that compounds with acidic protons lead to strong structure-related responses, those with more acidic protons give a greater response, and gases with low thermal conductivity induce stronger responses.

The *electron-capture detector* (ECD) uses a radioactive source (typically [63]Ni, but other sources can be used) that produces beta-particles within the ionization chamber, incorporating two electrodes, across which ionization may be measured. The detector normally uses a make-up gas of methane-doped nitrogen, or just nitrogen. Within the detector, the beta-particles interact with the carrier and make-up gases to form a plasma, containing, among other species, thermal electrons. This sets up a standing current between the electrodes. When an electron-capturing species enters the chamber, it captures some of the thermal electrons, reducing the standing current, and this is measured as the detector response. The source is prone to contamination, and this makes the detector somewhat noisy or subject to drift. Also, strongly electron-capturing solvents, such as dichloromethane, and derivatization reagents that leave significant traces of residual halogenated reagent in reaction mixtures, should be avoided. This detector is widely used in trace analysis of chlorinated pollutants (Chap. 21), and the recent interest in high-resolution analysis makes miniaturization of the detector's internal volume important. The use of a counter-flow of make-up gas over the anode isolates it from the halogenated solutes and has been proposed to enhance detector stability. Greater make-up flow also reduces solute residence time in the ionization chamber and minimizes peak-width, however, at the expense of sensitivity.

The signal output of the constant-current variable-frequency ECD is in frequency units (Hz). The ECD is reportedly linear over up to 7 orders of magnitude (note that this may be greater than the concentration range that maintains linear-chromatography conditions). For a highly sensitive molecule, such as lindane, a micro-ECD is reported to have a detection limit of 0.01 pg/sec. The micro-ECD is capable of faster data acquisition, and is therefore compatible with faster (narrower) GC peaks. Classes of compounds that yield an ECD response include alcohols, ketones, amines, and anhydrides, especially in conjugated form, which exhibit electron-capturing groups. Selectivity is shown most clearly towards halogens, for which sensitivity is in the order of I > Br > Cl > F. The extent of response to halogenated compounds increases dramatically with the degree of halogenation. The ECD response to multiply halogenated compound mixtures, such as polychlorinated biphenyls (PCB), means that the response factors of each compound should be known in order to accurately quantify complex sample mixtures. Cochran and Frame [99] compiled diagrammatically the response factors of all 209 PCB, and listed those for the pentachlorobiphenyls.

Photoionization detection (PID) employs an energetic light source, which provides a supply of photons capable of ionizing chemical compounds. A typical design uses a UV light source, such as a hydrogen lamp, providing 10.2 eV energy photons. The cell also contains a polarizing and collecting electrode. The mechanism can be represented as $R + UV \rightarrow R^+ + e^-$, where R is the sample compound. Since the photon flux determines the extent of ionization, maximizing photon intensity should give the greatest sensitivity. Low-pressure operation aids this, but it is inconvenient. Isolation of the source and ionization chambers permits ion production and ion measurement to be individually optimized [100]. Improvements in design to make the PID more suited for use with capillary columns by reduction in chamber volume have been reported. Clearly, compounds with lower ionization potentials will be more favorably ionized, and this accounts for the enhanced sensitivity towards olefins and aromatics, with their C–C π-electrons. Thus, with the H_2 lamp, permanent gases and small organic molecules will be poorly ionized. The ratio of PID to FID response can be used to indicate the class of compounds detected, a higher PID/FID ratio being indicative of aromatics, but this is not an accurate or reliable rule. The non-destructive nature of the PID means that it can be located ahead of a FID, to provide dual detection without the loss of sensitivity that may arise from stream splitting. The responses with the FID and PID have been contrasted for a range of atmospheric hydrocarbons [101]. The PID/FID ratios were on the order of <1 for *n*-alkanes, *ca.* 5 for 1-alkenes, and 5–12 for mono-aromatics. Considerable variation within the classes was observed. Dissolved and sedimentary sulfur was measured, following a range of H_2S generation methods from various materials. With a cryogenic pre-concentration step, a detection limit of 0.13 nM was found [102]. Through hydride generation, As, Se, Sn, and Sb species in natural waters were sensitively and selectively analyzed by GC/PID [103,104]. Chlorobenzenes in air, blood, and urine were extracted and analyzed by GC/PID. It was observed that the detector response to any given molar concentration of any chlorobenzene is identical, and the response is linearly related to molar concentration over 5 orders of magnitude [105]. Detection limits were in the low-ppb range. Thus, the PID has been applied to a variety of analyses, in areas where either the FID offers a competitive performance, or where the FID cannot be applied.

The *pulse-discharge detector* (PDD) has been developed during the 1990s. There are two general modes by which pulse discharge may be used in a GC detector. One is the measurement of ionization produced by the discharge; the other is the generation of emission, which is then monitored spectrophotometrically. A pulsed direct-current (DC) discharge in helium is used as the ionization source. In an ionization mode, the PDD may be operated as an electron-capture device, or as a helium-photoionization device. The pulsed discharge generates high-energy (13.5- to 17.5-eV) photons, which photoionize a dopant gas (*e.g.*, methane). The energy is sufficient to photoionize all elements, except neon. The dopant acts as a source of photoelectrons, and in the PDECD (pulsed-discharge ECD) [106] mode these electrons are captured by an electronegative element. Thus, this bears some resemblance to the more familiar radioactive ECD. In the absence of the dopant, the photon emission from He is energetic enough to photoionize the compounds in the GC effluent directly, and hence leads to the pulsed-discharge photoionization detector PDPID. The noble gas used for photon generation is defined as the prefix (He-, Ar-, or Kr-PDPID). Compared with the radioactive ECD, the PDECD is cleaner, since there is no radioactive source that may become contaminated. Also, the chamber can be made smaller so that is it suited to lower make-up flows and potentially better maintains the performance of narrow GC peaks. Wentworth *et al.* [107] discussed the rationale for the choice of dopant – inert to electron-capture, low ionization potential, and possessing vibrational and rotational modes – and listed a range of dopants that may be used. For a variety of reasons, one of which is that it produces results that parallel those of the radioactive ECD, it appears that CH_4 is the preferable dopant. In general, sensitivity of the PDECD and radioactive ECD are similar.

Dojahn *et al.* [108] compared the He-PDPID in its photoionization mode with the FID in the analysis of a hydrocarbon mixture. While the FID is normally considered to be a carbon-counter detector, they concluded the He-PDPID to be more accurate for determining the percent composition of the mixture. Moreover, the latter detects even compounds that the FID cannot detect, and so it is a more universal detector. A pulse period of 220 μsec was used, and a pulse width of 28 μsec, during which time a DC voltage of up to 20 V was applied (adjusted for stability of the discharge). A collector electrode measured the photoinduced current that was produced. Judging from the chromatograms presented, the He-PDPID gave a more consistent area across the test mixture components, and a peak for CS_2 was noted for the PID mode. The He-PDPID was found to be of advantage in the simultaneous determination of dissolved gases (H_2, N_2, O_2, CO, CO_2, CH_4, C_2H_6, C_2H_2, C_3H_8) and moisture in mineral oils, following static headspace sampling and GC [109]. A dual parallel column/detection system was used (with GSC and GLC columns and a switching valve in the GSC separation line). In its emission mode, the PDD has been shown to be an element-specific detector, with wavelength selection by a monochromator [110]. For the halogens studied, specifically chlorine, vacuum UV wavelengths were used [111]. Comparison of the Cl-PDECD and He-PDPID chromatograms showed that the former exhibits no sensitivity towards hydrocarbons, and therefore gives a much simpler chromatogram [112].

The mass spectrometer occupies an important enough area to warrant a separate chapter (Chap. 10). Being arguably the most important GC detector today, and having specific relevant aspects related to its use in analytical GC/MS experiments (*e.g.*, selected-ion monitoring, selected- or multiple-reaction monitoring, extracted-ion chromatograms, collisional activation, correlation with retention indices for enhanced identification, etc.), there is considerable technical and interpretive information associated with MS detection in chromatography.

8.5.3.2 Bulk physical property detectors

The *thermal-conductivity detector* (TCD) is an excellent, robust, low-maintenance detector, ideally suited to remote as well as on-site analysis. It does not require additional detector gases (there is no flame) and relies purely on electrical means to heat a filament in the carrier-gas stream. While it suffers analytically from a poorer sensitivity than other detectors, its universal nature means that it can be used for mixtures that might not be suited to other detectors. The original design compared filaments in two arms of a Wheatstone bridge. When a compound enters the measurement arm, the thermal conductivity of the compound alters the resistivity of the filament, and the out-of-balance current of the bridge corresponds to the amount of compound present. The design was later modified to permit use of a single filament, with flow modulation switching the column effluent every 100 msec between the filament cell and a bypass channel. Hence, a reference reading is also taken every 100 msec. The response of the TCD was shown to follow the kinetic theory of gases, for a number of small molecules [113], and the prediction of response factors was studied [114]. Since TCD was essentially the first detector used for GC, it is not surprising that soon after the introduction of capillary GC, there was a need for modifying the detector by reducing the volume of the flow-cell [115]. Since most uses of the TCD are for permanent gases, its application range is limited and application-specific.

8.5.3.3 Photometric detectors

Photometric or optical detectors in GC employ either molecular or atomic emission phenomena as the sensing mechanism, with various spectrophotometric devices to record the emission intensity. Note that, in contrast to HPLC, absorption detectors are rarely used in GC, except for the infrared detector. Some isolated reports of GC/UV have appeared in the literature in the past. Lagesson *et al.* [116] presented a compelling study on high-resolution GC/UV with spectra taken over the range 168–330 nm. Three-dimensional and

2D contour plots of chromatograms of a flavor sample and a petroleum product illustrate the degree of sophistication that can be offered by GC/UV analysis. One of the earliest photometric detectors is the *flame-photometric detector* (FPD). Grant [117] suggested flame emissivity as a detection mode in 1958. The FPD is essentially reserved for molecular emissions, stimulated in a cool diffusion flame, as opposed to atomic emission. In 1966 Brody and Chaney [118] proposed the use of the emission from diatomic sulfur for the detection of sulfur compounds. The role of the FPD for sulfur analysis was reviewed by Farwell and Barinaga [119]. The combustion of a sulfur compound yields a range of sulfur-containing products (S, SO, SO_2, H_2S, HS, and S_2). The blue emission used analytically is from the S_2 species, and is due to a recombination reaction that produces the electronically excited diatomic molecule, S_2^*. Relaxation to the ground state is accompanied by a molecular band emission with a variety of maxima, the most intense of which is at 383 nm. The following reaction occurs:

$$\text{S-compound} + \text{flame} \rightarrow \text{S} + \text{S} \rightarrow [S_2]^*$$

and therefore

$$h\nu \text{ intensity} = k[S_2]^*$$

where $[S_2]^* \propto [S][S] \propto [S]^2$. Since the concentration of S in the flame, [S], is proportional to the amount of S-compound, but the amount of S_2 generated is proportional to the square of the amount, S, the response, R, is proportional to the square of the amount of sample.

$$R = k(\text{mass of S})^2$$

The theoretical value of 2 for a square-law response is often not observed, and the exponent may vary considerably, from 1.5 to 2.2. It also follows that the calibration graph will be nonlinear, and a plot of $\log(R)$ *vs.* $\log(S)$ will show a theoretical slope of 2 (or from 1.5 to 2.2, as the case may be). Also, just as the S-compound mass flux in the detector changes according to the Gaussian peak shape expected of the chromatographic peak, the peak shape produced by the FPD in the S_2 mode will not be Gaussian but rather "Gaussian^2". This affects the interpretation of chromatographic data, such as resolution and efficiency, when the FPD-S_2 mode is used [120]. Since organic contaminants interfere to some extent with production of the chemiluminescent species and cause deviations from the theoretical value, a dual-flame arrangement was developed. It provides a pre-combustion chamber, followed by an upper analytical flame, the emission of which is monitored.

The other major mode of the FPD is the phosphorus mode. It finds use in pesticide analysis, often in conjunction with a TID. The TID responds to both N- and phosphorus-containing compounds, while the FPD-P mode will indicate which of these residues contains just phosphorus. This mode is more sensitive than the S_2 mode, and also arises from a recombination reaction in the flame to produce an excited-state molecule, HPO^*. In this case, the response is linear with amount of phosphorus compound:

$$\text{P-compound} + \text{flame} \rightarrow \text{combustion products, P, PO, } PO_2 \rightarrow HPO^*$$

$$h\nu \text{ intensity} = k[HPO^*]$$

where [HPO*] $\propto$ P-compound. In this case, an emission is produced, centered on 525 nm. The emissions of HPO* and S$_2^*$ overlap to some degree, and so there is some overlap – a small emission intensity arising from HPO* may be recorded at the detection wavelength used for the S$_2$ mode, and from S$_2^*$ when the FPD-P mode is used.

More recently, the *pulsed-flame photometric detector* (PFPD) was introduced, where the gas flow-rates cannot sustain a continuous flame operation [121,122]. When ignited, the flame front propagates back to the combustible gas mixture and will self-terminate. Emission from an excitable species will therefore be pulsed. This isolates the combustion- and emission-generating processes in space and time. Fig. 8.11 is a schematic cross-section of the PFPD. Note that the igniter is in a position remote from the flame. The optical sensing elements are located just above the quartz combustor. This diagram may be considered representative of many different sorts of PFPD, with simple modifications to suit different transducer mechanisms. It has been reported that improved detection limits for S and P, higher selectivity against hydrocarbons, and reduced emission quenching were achieved with the PFPD. Additionally, selective detection of N, and simultaneous detection of S and C were reported. Detection limits of the order of 2×10^{-13} g S/sec, 1×10^{-14} g P/sec, 5×10^{-12} g N/sec and 6×10^{-11} g C/sec were quoted. The PFPD arrangement reduced the compound-dependency of the sulfur response. A typical application is represented by the selective determination of sulfur compounds in beer with the use of headspace SPME [123].

The *atomic emission detector* (AED) offers a novel opportunity to establish the elemental composition of a compound that is separated and eluted from a GC column. Van Stee and Brinkman [124] have reviewed the performance and applications of GC/AED.

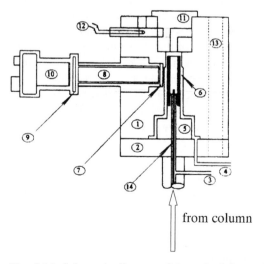

from column

Fig. 8.11. Schematic diagram of the pulsed flame-photometric detector. Important features are: 3, hydrogen/air mixture inlet to combustor; 4, outer bypass hydrogen/air mixture inlet; 5, combustion cell holder; 6, quartz combustion tube; 7, sapphire optical window; 8, quartz rod light guide; 9, glass optical filter; 10, photomultiplier; 12, igniter; 14, column. (Reproduced from Ref. 122 with permission.)

The first report describing the use of emission spectrometry (atomic lines and molecular bands) as a GC detector for organic compounds was that of McCormack *et al.* [125], who used a microwave-induced plasma (MIP). Both, excellent sensitivity and selectivity towards certain elements, were demonstrated, and fast response and wide dynamic range were also cited as advantages. Detection limits in pg/sec were observed, but selectivity against carbon was not good. Uden [126] summarized the developments in this area up to the early 1990s. The Beenakker TM_{010} cavity [127] enabled an atmospheric helium-supported plasma to be reliably used, with higher energy, that was more applicable to GC detection and became the basis for a commercial GC/AED device. The compound enters a high-temperature plasma source, which decomposes/atomizes it, and from which atomic emission is stimulated. Classical spectral methods are used to isolate atomic emission wavelengths and to report their intensities. Van Stee and Brinkman [124] compiled analytical figures of merit of various elements, which are reported in Table 8.7; a similar tabulation was provided by Uden [128]. Stan and Linkerhagner [129] have tabulated detection limits for 385 pesticides, using the diagnostic lines for the elements in each pesticide, as well as retention times under standard conditions. Large-volume injection ensured adequate sensitivity for the residues in food analysis. To screen for pesticides, two analyses were conducted, S, P, N, and C lines being selected in the first run, and Cl and Br in the second. The advantage of this selective detection mode was that matrix interferences were largely confined to the C, H, and O emission lines.

Chemiluminescent detectors occupy niche application areas for nitrogen and sulfur compounds, based on the reaction between them and ozone. Yan [130] has reviewed the

TABLE 8.7

TYPICAL SENSITIVITIES AND SELECTIVITIES REPORTED FOR THE ATOMIC EMISSION DETECTOR

Element	Wavelength (nm)	LOD (pg/sec)	Selectivity over carbon ($\times 10^{-3}$)
N	174.2	15–50	2–5
S	180.7	1–2	5–20
C	193.1	0.2–1	–
P	178.1	1–3	5–8
C	495.8	15	–
H	486.1	1–4	–
Cl	479.5	25–40	3–10
Br	478.6	30–60	2–6
F	685.6	60–80	20–50
O	777.2	50–120	10–30
Si	251.6	1–7	30
Hg	253.7	0.1–0.5	250
Pb	261	0.2–1	300
Sn	271	1	300

present state of these detectors. The respective reactions are:

$$NO + O_3 \rightarrow NO_2^* + O_2$$

$$NO_2^* \rightarrow NO_2 + h\nu \ (\lambda_{max} = 1200 \ nm)$$

and

$$SO + O_3 \rightarrow SO_2^* + O_2$$

$$SO_2^* \rightarrow SO_2 + h\nu \ (\lambda_{max} = 360 \ nm)$$

The reactive N and S species must be generated in the detector, often by a pyrolysis process, and then the products are transported to the reaction chamber, where ozone is supplied and the emission is monitored. These mechanisms are the bases for the *thermal energy analyzer* (TEA, for nitrogen) and the *sulfur chemiluminescence detector* (SCD). Sensitivities are of the order of 10^{-12} g (S or N)/sec, selectivity against C is *ca.* 10^7, and the linear range is 4 to 5 orders of magnitude.

   The only other detector of note under this category that will be discussed here is the *infrared detector*. Rather than being an emission-based detector of the above types, it operates by spectrophotometric absorption of radiation. The heyday of FT-IR detection was in the 1980s and early 1990s, but it has had considerable competition from MS detection more recently. The role of the FT-IR detector for various chromatographic methods was discussed by White [131]. It is almost the definitive universal detector, since almost every molecular species possesses IR activity, and it also has some degree of specificity, arising from the spectral fingerprint associated with modes of molecular vibrations [132]. Along with the mass spectrometer, IR will give molecular structural information, and these two spectroscopic detectors have been compared by Ragunathan *et al.* [133]. The only practical IR method to be used with chromatography is the Fourier transform mode, with rapid acquisition of the interferogram on the chromatography time-scale. The compound must pass through the light-pipe, in which its IR spectrum is measured in real time. Cryogenic trapping of eluate on a moving substrate with subsequent IR measurement, is another operational method [134]. Library searching and reconstructed chromatograms, based on selected frequencies of vibration, offer some unique advantages. The first is identification of geometric isomers, such as the *o-*, *m-* and *p*-isomers of mono-aromatic compounds (*e.g.*, dimethylbenzenes), which effectively have the same mass spectra, and two of which are eluted very close together (the *o-* and *p*-isomers). The second is the ability to reconstruct chromatograms of certain chemical classes from a mixture and present simplified chromatograms that make interpretation easier. The FT-IR detector is a powerful, but often overlooked detection mode, and this is probably due to the overwhelming reliance on the MS detector for situations where identification is required, while accepting the small number of cases where MS cannot yield a definitive structural identification.

8.5.3.4 Other detectors

   Two specialty GC detectors will be mentioned in this section, both of which are used in the areas of essential oils, flavor and fragrances, and semiochemicals. These two detectors are based on biochemical receptor responses. The first is the *olfactory detector*, and olfactometry has given rise to the term GC/O. Here, the human nose is the sensing element. The GC effluent is mixed with humid air, and the effluent is "continuously" monitored for the presence of compounds that elicit an odor response. This detector is best operated in parallel with a mass spectrometer *via* an effluent splitter, to provide identifications of the odorous compounds. The experimenter not only registers the time he senses certain odors, but also the type of odor (fruity, coconut, spicy, floral, malty, ethereal, sulfurous, cut grass). The intensity of the odor can also be indicated by using a signaling device, and the resulting chromatogram is a plot of retention time *vs.* the sensory response, which is proportional to the perceived intensity of the stimulant. The olfactogram can be compared directly with the GC/FID profile to correlate compound retentions and estimate concentrations, or with GC/MS to provide identification. Sometimes, a peak occurs on the olfactogram but not on the GC/FID tracing, for compounds that have a very low odor threshold but are below the FID detection limit.

Quantitation of an individual compound in GC/O is achieved by either the aroma extract-dilution analysis (AEDA) [135,136] or combined hedonic and response measurements (CHARM) [137] methods. The former presents trained sniffers with successive dilutions of the analyzed sample until no odor is perceived. This gives a semi-quantitative measure of the individual compounds. In CHARM, the sniffer records the concentration when it exceeds the odor threshold, and again when it drops below the threshold. A major peak on the CHARM diagram (chromatogram) may not necessarily be the biggest peak in the sample. The odor profile of a sample may then be presented on a wagon-wheel diagram, where each spoke is a specific odor descriptor. The magnitude of each odor in the given sample can then be plotted on the diagram for facile comparison. The sniffer port of the GC/O device now incorporates a supply of moisture to ensure that the sniffer's nasal cavity does not become dry and irritated [138]. The analysis is also often undertaken in a "sensory deprivation" environment, to ensure objective measurement of the effluent. Steinhart *et al.* [139] and Van Asten [140] recently reviewed advances in flavor research, with the emphasis on GC methods. The analysis of essential oils by GC, while often a challenging and complex task, is now a well-developed procedure that depends on correlation of linear temperature-programmed retention indices and mass spectrometry [141], for which there are literature compilations available [142], although there is still a need for improved resolution.

The second dedicated detector in this area is the electro-antennograph (EAG). It is a highly specific detector, which incorporates an organ of an insect – usually the antenna – that contains receptors to target compounds, usually allelochemicals or pheromones [143]. A preliminary separation by GC is almost indispensable, and the biological response then provides greatly increased identification power for studies of active compounds [144]. Since the target molecules elicit a biological response, which the transducer converts to an electrical signal as the potential across the antennae, the activity of the antennae towards an eluted chemical will be recorded as a peak. This permits identification of the chemicals that are responsible for a particular insect behavior. Studies were conducted with sample collection in open-tubular polysiloxane phase-coated traps, followed by thermal-desorption GC/EAG of dung beetle attractant [145]. A sensitivity of 15 pg was reported for the antennal response in a GC/EAD (electro-antennographic detection) system to the major component of the Egyptian armyworm sex pheromone, (*Z,E*)-9,11-tetradecadienyl acetate [144]. As the search for more highly targeted agricultural management chemicals continues, this area of analysis will likely play an increasingly important role in natural control measures for insects and other parasites. Current trends towards miniaturization (Chap. 11) have stimulated efforts directed towards incorporation of detector technologies on chip devices for GC. The obvious method will be miniaturized TCD, but DC plasma optical emission has also been reported [146].

8.5.4 Dual detection

While two detectors provide "double" the information, and potentially more than that of a single detector, rarely are three detectors used simultaneously (Chap. 12). The purpose of the dual detector is to provide improved quantification and/or identification from the GC experiment. This effectively confirms that retention time, or the peak position in a chromatogram, is insufficient in itself to provide confirmation of a compound's identity. Hence either chromatography on multiple columns (MDGC, GC × GC), or multiple detection, may be considered. For some applications, MS detection in full-scan mode is required to confirm peak identity. However, MS may not provide adequate quantitation, or it may be too complicated to set up for the quantification of a multitude of different components in a sample. In such cases, a parallel GC/FID analysis will be relied upon for quantitation, based on the FID response. GC/FTIR may also provide the required identification, but may not be as convenient for quantification; the FID channel can then be used for this.

For complex samples, it is often of advantage to obtain a simplified chromatogram, if the selective detector can offer this. Thus simultaneous PFPD and MS detection of pesticides has been described [147], in which the simplicity of dimethoate and benzothiazole detection in the PFPD channel is apparent, whereas in the MS (total-ion current) tracing they are almost unrecognizable. For malathion in rosemary essential oil, the PFPD P- and S-modes were contrasted with the MS-TIC tracing. For a multi-residue pesticide extract from a lanolin sample, dual detection with ECD and NPD provided an interesting application, where in one region of the chromatogram the NPD gave a simpler result, and in another, the ECD gave a simpler result [148]. However, the simplicity provided by the second detector is not always achieved. For a urine sample, the volatile fraction gave very complex chromatograms with both detectors, the FID and TID [149]. Complex sample fractionation by multi-dimensional GC, combined with parallel MS and FT-IR detection, has been used for petrochemical analysis, where the FT-IR is used for isomer characterization. The two detectors may be considered complementary in this respect [150]. Most of the sample effluent is split to the less sensitive FT-IR in order to obtain adequate sensitivity.

REFERENCES

1   A.T. James and A.J.P. Martin, *Biochem. J.* 50 (1952) 679.
2   M. van Lieshout, M. van Deursen, R. Derks, J.G.M. Janssen and C.A. Cramers, *J. Microcol. Sep.*, 11 (1999) 155.

3   C.F. Poole and S.K. Poole, *Chromatography Today*, Elsevier, Amsterdam, 1991.

4   N.-J. Wu, J.C. Medina and M.L. Lee, *J. Chromatogr. A*, 892 (2000) 3.

5   J. de Zeeuw and J. Luong, *Trends Anal. Chem.*, 21 (2002) 594.

6   K. Grob, Jr., G. Grob and K. Grob, *J. Chromatogr.*, 156 (1978) 1.

7   R. Sternberg, C. Szopa, D. Coacia, S. Zubrycki, F. Raulin, C. Vidal-Madjar, H. Niemann and G. Israel, *J. Chromatogr. A*, 846 (1999) 307.

8   Z.-H. Ji, R.E. Majors and E.J. Guthrie, *J. Chromatogr. A*, 842 (1999) 115.

9   D. Wang, S.-L. Chong and A. Malik, *Anal. Chem.*, 69 (1997) 4566.

10   W. Jennings, *Analytical Gas Chromatography*, Academic Press, Orlando, 1987, p. 173.

11   J.V. Hinshaw and L.S. Ettre, *Chromatographia*, 21 (1986) 561.

12   D.R. Deans and I. Scott, *Anal. Chem.*, 45 (1973) 1137.

13   P. Sandra, F. David, M. Proot, G. Diricks, M. Verstappe and M. Verzele, *J. High Resolut. Chromatogr. Chromatogr. Commun.*, 8 (1985) 782.

14   J.V. Hinshaw and L.S. Ettre, *Chromatographia*, 21 (1986) 669.

15   R. Sacks and M. Akard, *Environ. Sci. Technol.*, 28 (1994) 428A.

16   H. Smith and R. Sacks, *Anal. Chem.*, 69 (1997) 5159.

17   T. Veriotti, M. McGuigan and R. Sacks, *Anal. Chem.*, 73 (2001) 279.

18   T. Veriotti and R. Sacks, *Anal. Chem.*, 73 (2001) 3045.

19   B. Lorentzeas, in *Chemometric Studies on Optimization of Solute Separation in Pressure Tuned Serially Coupled Capillary Gas Chromatography Columns*, Honours Thesis, Applied Chemistry Department, RMIT University, Melbourne, 1998.

20   W. Bertsch, *J. High Resolut. Chromatogr.*, 22 (1999) 647.

21   W. Bertsch, *J. High Resolut. Chromatogr.*, 23 (2000) 167.

22   J.C. Giddings, *J. Chromatogr. A*, 703 (1995) 3.

23   G. Schomberg, in P. Sandra (Ed.), *Sample Introduction in Capillary Gas Chromatography*, Hüthig, Heidelberg, 1985, p. 235.

24   L. Mondello, M. Catalfamo, A.R. Proteggente, I. Bonaccorsi and G. Dugo, *J. Agric. Food Chem.*, 46 (1998) 54.

25   L. Mondello, A. Verzera, P. Previti, F. Crispo and G. Dugo, *J. Agric. Food Chem.*, 46 (1998) 4275.

26   M. Heil, F. Podebrad, T. Beck, A. Mosandl, A.C. Sewell and H. Bohles, *J. Chromatogr. A*, 714 (1998) 119.

27   L. Mondello, A.C. Lewis and K.D. Bartle, *Multidimensional Chromatography*, Wiley, Chichester, 2001.

28   H. Cortes, *Multidimensional Chromatography: Techniques and Applications*, Dekker, New York, 1990.

29   P.J. Marriott and R.M. Kinghorn, in A.J. Handley and E.R. Adlard (Eds.), *Gas Chromatographic Techniques and Applications*, Sheffield Academic Press, Sheffield, 2001.

30   B.M. Gordon, M.S. Uhrig, M.F. Borgerding, H.L. Chung, W.M. Coleman, III, J.F. Elder, Jr., J.A. Giles, D.S. Moore, C.E. Rix and E.L. White, *J. Chromatogr. Sci.*, 26 (1988) 174.

31   K.A. Krock, N. Ragunathan and C.L. Wilkins, *J. Chromatogr.*, 645 (1993) 153.

32   D.R. Deans, *Chromatographia*, 1 (1968) 18.

33   C.A. Cramers and P.A. Leclercq, *J. Chromatogr. A*, 842 (1999) 3.

34   M.M. van Deursen, J. Beens, H.-G. Janssen, P.A. Leclercq and C.A. Cramers, *J. Chromatogr. A*, 878 (2000) 205.

35   C.P.M. Schutjes, E.A. Vermeer, J.A. Rijks, C.A. Cramers, in R.E. Kaiser (Ed.), *Proceedings of the Fourth International Symposium on Capillary Chromatography*, Hindelang, Institute of Chromatography, Bad Durkheim, 1981, p. 687.

36   J.V. Hinshaw, *LC-GC*, 13 (1995) 944.

37   L.M. Blumberg and M.S. Klee, *Anal. Chem.*, 70 (1998) 3828.
38   M. van Deursen, M. van Leishout, R. Derks, H.-G. Janssen and C. Cramers, *J. High Resolut. Chromatogr.*, 22 (1999) 119.
39   M. van Deursen, H.-G. Janssen, J. Beens, P. Lipman, R. Reinierkens, G. Rutten and C. Cramers, *J. Microcol. Sep.*, 12 (2000) 613.
40   M. van Deursen, H.-G. Janssen, J. Beens, G. Rutten and C. Cramers, *J. Microcol. Sep.*, 13 (2001) 337.
41   C. Bicchi, C. Brunelli, M. Galli and A. Sironi, *J. Chromatogr. A*, 931 (2001) 129.
42   L.M. Blumberg, *J. High Resolut. Chromatogr.*, 20 (1997) 597.
43   L.M. Blumberg, *J. High Resolut. Chromatogr.*, 20 (1997) 679.
44   L.M. Blumberg, *J. High Resolut. Chromatogr.*, 22 (1999) 403.
45   L.M. Blumberg, *J. High Resolut. Chromatogr.*, 22 (1999) 501.
46   P. Korytár, H.-G. Janssen, E. Matisova and U.A.Th. Brinkman, *Trends Anal. Chem.*, 21 (2002) 558.
47   J.M. Davis and J.C. Giddings, *Anal. Chem.*, 55 (1983) 418.
48   J.C. Giddings, *Anal. Chem.*, 56 (1984) 1258A.
49   J.B. Phillips and J. Beens, *J. Chromatogr. A*, 856 (1999) 331.
50   M. Pursch, K. Sun, B. Winniford, H. Cortes, A. Weber, T. McCabe and J. Luong, *Anal. Bioanal. Chem.*, 373 (2002) 356.
51   P. Marriott and R. Shellie, *Trends Anal. Chem.*, 21 (2002) 573.
52   P.J. Marriott and R.M. Kinghorn, *Anal. Chem.*, 69 (1997) 2582.
53   A.L. Lee, A.C. Lewis, K.D. Bartle, J.B. McQuaid and P.J. Marriott, *J. Microcol. Sep.*, 12 (2000) 187.
54   J.V. Seeley, F. Kramp and C.J. Hicks, *Anal. Chem.*, 72 (2000) 4346.
55   C.G. Fraga, B.J. Prazen and R.E. Synovec, *J. High Resolut. Chromatogr.*, 23 (2000) 215.
56   E.B. Ledford, Jr. and C. Billesbach, *J. High Resolut. Chromatogr.*, 23 (2000) 202.
57   J. Beens, M. Adahchour, R.J.J. Vreuls, K. van Altena and U.A.Th. Brinkman, *J. Chromatogr. A*, 919 (2001) 127.
58   J.B. Phillips, R.B. Gaines, J. Blomberg, F.W.M. van der Wielen, J.-M. Dimandja, V. Green, J. Grainger, D. Patterson, L. Racovalis, H.-J. deGeus, J. deBoer, P. Haglund, J. Lipsky, V. Sinha and E.B. Ledford, Jr., *J. High Resolut. Chromatogr.*, 22 (1999) 3.
59   J. Blomberg, P.J. Schoenmakers, J. Beens and R. Tijssen, *J. High Resolut. Chromatogr.*, 20 (1997) 539.
60   G.S. Frysinger and R.B. Gaines, *J. Forensic Sci.*, 47 (2002) 471.
61   P. Korytár, P.E.G. Leonards, J. de Boer and U.A.Th. Brinkman, *J. Chromatogr. A*, 958 (2002) 203.
62   R. Shellie, P. Marriott and C. Cornwell, *J. High Resolut. Chromatogr.*, 23 (2000) 554.
63   P. Sandra, *Sample Introduction in Capillary Gas Chromatography*, Vol. 1, Hüthig, Heidelberg, 1985.
64   K. Grob, *Split and Splitless Injection for Quantitative Gas Chromatography. Concepts, Processes, Practical Guidelines, Sources of Error*, Wiley-VCH, Weinheim, 2001.
65   L.S. Ettre, in P. Sandra (Ed.), *Sample Introduction in Capillary Gas Chromatography*, Vol. 1, Hüthig, Heidelberg, 1985.
66   K. Grob and H.P. Neukom, *J. High Resolut. Chromatogr. Chromatogr. Commun.*, 2 (1979) 15.
67   K. Grob and M. Biedermann, *J. Chromatogr. A*, 897 (2000) 237.
68   K. Grob and M. Biedermann, *J. Chromatogr. A*, 897 (2000) 247.
69   P.J. Marriott and P.D. Carpenter, *J. Chem. Educ.*, 73 (1996) 96.
70   D.C. Harris, *Quantitative Chemical Analysis*, 6th Edn. Freeman, New York, 2002.

71   P.J. Marriott and G. Eglinton, *J. Chromatogr.*, 249 (1982) 311.

72   P.J. Marriott, J.P. Gill, R.P. Evershed, G. Eglinton and J.R. Maxwell, *Chromatographia*, 16 (1982) 304–308.

73   E. Geeraert, in P. Sandra (Ed.), *Sample Introduction in Capillary Gas Chromatography*, Vol. 1, Hüthig, Heidelberg, 1985, p. 133.

74   F. Poy and L. Cobelli, in P. Sandra (Ed.), *Sample Introduction in Capillary Gas Chromatography*, Vol. 1, Hüthig, Heidelberg, 1985, p. 77.

75   W. Engewald, J. Teske and J. Efer, *J. Chromatogr. A*, 842 (1999) 143.

76   H. Lord and J. Pawliszyn, *J. Chromatogr. A*, 885 (2000) 153.

77   J. Pawliszyn, *Applications of Solid Phase Microextraction*, The Royal Society of Chemistry, Cambridge, 1999.

78   E. Baltussen, P. Sandra, F. David, H.-G. Janssen and C. Cramers, *Anal. Chem.*, 71 (1999) 5213.

79   H.G.J. Mol, H.-G. Janssen, C.A. Cramers, J.J. Vreuls and U.A.Th. Brinkman, *J. Chromatogr. A*, 703 (1995) 277.

80   I.L. Davies, K.D. Bartle, G.E. Andrews and P.T. Williams, *J. Chromatogr. Sci.*, 26 (1988) 125.

81   J. Beens and R. Tijssen, *J. Microcol. Sep.*, 7 (1995) 345.

82   L. Mondello, K.D. Bartle, P. Dugo, P. Gans and G. Dugo, *J. Microcol. Sep.*, 6 (1994) 237.

83   D.J. David, *Gas Chromatographic Detectors*, Wiley, New York, 1974.

84   J. Sevčík, *Detectors in Gas Chromatography*, Elsevier, Amsterdam, 1976.

85   M. Dressler, *Selective Gas Chromatographic Detectors*, Elsevier, Amsterdam, 1986.

86   A. Amirav, *Am. Lab.*, 33 October, (2001) 28.

87   A. Braithwaite and F.J. Smith, *Chromatographic Methods*, Chapman & Hall, London, 1985.

88   N. Dyson, *Chromatographic Integration Methods*, Royal Society of Chemistry, Cambridge, 1990.

89   N. Dyson, *J. Chromatogr. A*, 842 (1999) 321.

90   M. Jalali-Heravi and Z. Garkani-Nejad, *J. Chromatogr. A*, 950 (2002) 183.

91   W.A. Dietz, *J. Gas Chromatogr.*, February, (1967) 68.

92   A.J.C. Nicholson, in A.J.C. Nicholson (Ed.), *Detectors and Chromatography*, Australian Scientific Industry Association, Melbourne, 1983, p. 21.

93   I.G. McWilliam, in A.J.C. Nicholson (Ed.), *Detectors and Chromatography*, Australian Scientific Industry Association, Melbourne, 1983, p. 5.

94   T. Holm and J.O. Madsen, *Anal. Chem.*, 68 (1996) 3607.

95   B. Kolb and J. Bischoff, *J. Chromatogr. Sci.*, 12 (1974) 625.

96   B. Kolb, M. Auer and P. Pospisil, *J. Chromatogr. Sci.*, 15 (1977) 53.

97   P.L. Patterson, *J. Chromatogr.*, 167 (1978) 381.

98   H. Carlsson, G. Robertsson and A. Colmsjo, *Anal. Chem.*, 73 (2001) 5698.

99   J.W. Cochran and G.M. Frame, *J. Chromatogr. A*, 843 (1999) 323.

100   J.N. Davenport and E.R. Adlard, *J. Chromatogr.*, 290 (1984) 13.

101   W. Nutmagul, D.R. Cronn and H.H.J. Hill, *Anal. Chem.*, 55 (1983) 2160.

102   G.A. Cutter and T.J. Oatts, *Anal. Chem.*, 59 (1987) 717.

103   L.S. Cutter, G.A. Cutter and M.L.C. San Diego-McGlone, *Anal. Chem.*, 63 (1991) 1138.

104   S.H. Vien and R.C. Fry, *Anal. Chem.*, 60 (1988) 465.

105   M.L. Langhorst and T.J. Nestrick, *Anal. Chem.*, 51 (1979) 2018.

106   W.E. Wentworth, J. Huang, K.-F. Sun, Y. Zhang, L. Rao, H.-M. Cai and S.D. Stearns, *J. Chromatogr. A*, 842 (1999) 229.

107   H.-M. Cai, W.E. Wentworth and S.D. Stearns, *Anal. Chem.*, 68 (1996) 1233.

108   J.G. Dojahn, W.E. Wentworth, S.N. Deming and S. Stearns, *J. Chromatogr. A*, 917 (2001) 187.

109   J. Jalbert, R. Gilbert and P. Tetreault, *Anal. Chem.*, 73 (2001) 3382.

110   W.E. Wentworth, K.-F. Sun, D. Zhang, J. Madabushi and S.D. Stearns, *J. Chromatogr. A*, 872 (2000) 119.

111   K.-F. Sun, W.E. Wentworth and S.D. Stearns, *J. Chromatogr. A*, 872 (2000) 141.

112   K.-F. Sun, W.E. Wentworth and S.D. Stearns, *J. Chromatogr. A*, 872 (2000) 179.

113   D.W. McMorris, *Anal. Chem.*, 46 (1974) 42.

114   B.D. Smith and W.W. Bowden, *Anal. Chem.*, 36 (1964) 82.

115   J.A. Petrocelli, *Anal. Chem.*, 35 (1963) 2220.

116   V. Lagesson, L. Lagesson-Andrasko, J. Andrasko and F. Baco, *J. Chromatogr. A*, 867 (2000) 187.

117   D.W. Grant, in D.H. Desty (Ed.), *Gas Chromatography 1958*, Butterworths, London, 1958.

118   S.S. Brody and J.E. Chaney, *J. Gas Chromatogr.*, 4 (1966) 42.

119   S.O. Farwell and C.J. Barinaga, *J. Chromatogr. Sci.*, 24 (1986) 483.

120   P.J. Marriott and T.J. Cardwell, *Chromatographia*, 14 (1981) 279.

121   S. Cheskis, E. Atar and A. Amirav, *Anal. Chem.*, 65 (1993) 539.

122   A. Amirav and H.-W. Jing, *Anal. Chem.*, 67 (1995) 3305.

123   P.G. Hill and R.M. Smith, *J. Chromatogr. A*, 872 (2000) 203.

124   L.L.P. van Stee, U.A.Th. Brinkman and H. Bagheri, *Trends Anal. Chem.*, 21 (2002) 618.

125   A.J. McCormack, S.C. Tong and W.D. Cooke, *Anal. Chem.*, 37 (1965) 1470.

126   P.C. Uden, *Element-specific Chromatographic Detection by Atomic Emission Spectroscopy*, American Chemical Society, Washington, DC, 1992.

127   C.I.M. Beenakker, *Spectrochim. Acta B*, 32 (1977) 173.

128   P.C. Uden, *J. Chromatogr. A*, 703 (1995) 393.

129   H.-J. Stan and M. Linkerhagner, *J. Chromatogr.*, 750 (1996) 369.

130   X.W. Yan, *J. Chromatogr. A*, 842 (1999) 267.

131   R. White, *Chromatography/Fourier Transform Infrared Spectroscopy and Its Applications*, Dekker, New York, 1990.

132   T. Visser, *Trends Anal. Chem.*, 21 (2002) 627.

133   N. Ragunathan, K.A. Krock, C. Klawun, T.A. Sasaki and C.L. Wilkins, *J. Chromatogr. A*, 856 (1999) 349.

134   A.M. Haefner, K.L. Norton and P.R. Griffiths, *Anal. Chem.*, 60 (1988) 2441.

135   F. Ullrich and W. Grosch, *Z. Lebensm. Unters. Forsch.*, 184 (1987) 277.

136   W. Grosch, *Flavour Fragr. J.*, 9 (1994) 147.

137   T.E. Acree, J. Barnard and D.G. Cunningham, *Food Chem.*, 14 (1984) 273.

138   T.E. Acree, R.M. Butts, R.R. Nelson and C.Y. Lee, *Anal. Chem.*, 48 (1976) 1821.

139   H. Steinhart, A. Stephan and M. Bucking, *J. High Resolut. Chromatogr.*, 23 (2000) 489.

140   A. van Asten, *Trends Anal. Chem.*, 21 (2002) 698.

141   R. Shellie, L. Mondello, P. Marriott and G. Dugo, *J. Chromatogr. A*, 970 (2002) 225.

142   R.P. Adams, *Identification of Essential Oil Components by Gas Chromatography/Mass Spectrometry*, Allured Publishing Corp., Carol Stream, IL, 1995.

143   D.L. Struble and H. Arn, in H.E. Hummel and T.A. Miller (Eds.), *Techniques in Pheromone Research*, Springer, New York, 1984, p. 161.

144   E.A. Malo, M. Renou and A. Guerrero, *Talanta*, 52 (2000) 525.

145   B.V. Burger, A.E. Nell and W.G.B. Petersen, *J. High Resolut. Chromatogr.*, 14 (1991) 718.

146   J.C.T. Eijkel, H. Stoeri and A. Manz, *Anal. Chem.*, 72 (2000) 2547.

147   A. Amirav and H.-W. Jing, *J. Chromatogr. A*, 814 (1998) 133.

148   E. Jover and J.M. Bayona, *J. Chromatogr. A*, 950 (2002) 213.

149   M.J. Hartigan, J.E. Purcell, M. Novotny, M.L. McConnell and M.L. Lee, *J. Chromatogr.*, 99 (1974) 339.

150   C.L. Wilkins, *Anal. Chem.*, 66 (1994) 295A.

Erich Heftmann (Editor)
Chromatography, 6th edition
Journal of Chromatography Library, Vol. 69A
© 2004 Elsevier B.V. All rights reserved

Chapter 9

Capillary zone electrophoresis

PIER GIORGIO RIGHETTI, ALESSANDRA BOSSI, LAURA CASTELLETTI
and BARBARA VERZOLA

CONTENTS

9.1   Introduction   369
9.2   The instrument   370
9.3   The capillary   374
9.4   How to modulate the EOF   377
      9.4.1   Buffer changes and additives   377
      9.4.2   Adsorbed polymers   379
      9.4.3   Adsorbed surfactants   380
      9.4.4   Covalently-bonded phases   381
9.5   The buffers   382
9.6   Modes of operation   384
      9.6.1   Open-tube capillary zone electrophoresis   384
      9.6.2   Capillary isoelectric focusing   385
      9.6.3   Capillary zone electrophoresis in sieving liquid polymers   387
      9.6.4   Capillary isotachophoresis   388
      9.6.5   Two-dimensional separations   389
9.7   Micellar electrokinetic chromatography   390
9.8   Biosensors   394
9.9   Conclusions   394
Acknowledgments   395
References   395

9.1 INTRODUCTION

In 1967, Hjertén [1] described a novel apparatus for free-zone electrophoresis. It consisted of three units: 1) a high-voltage power supply; 2) a detector, and 3) a unit holding a 1- to 3-mm-ID quartz tube, which was immersed in a cooling bath. In order to minimize solute adsorption and electro-osmotic flow (EOF), the internal surface of the separation tube was coated with methylcellulose. The instrument was based on a "Copernican principle": in order to prevent electrodecantion of separated macromolecular zones, the tube was continuously rotated about its axis. The rotating quartz cell and electrode reservoirs were mounted on a movable carriage; detection of separated components could thus be achieved by driving the capillary past a fixed UV monitor. In spite of the fact that Hjertén's instrument proved many of the concepts of modern capillary zone electrophoresis (CZE), it did not mark the birth date of present-day CZE. First of all, the apparatus was mammoth in size and cumbersome to operate, since it still lacked the automation of modern CZE instrumentation; secondly, no capillaries were available in those days, impeding the operations at high voltages that led to high-performance separations. Years later, Virtanen [2] demonstrated that the use of tubes approaching capillary dimensions (0.2 mm ID) obviated the need for capillary rotation to counteract sample decantation, as predicted by Hjertén in 1967. Yet, even with Virtanen's report, CZE lay dormant for most of the decade, although the Seventies witnessed the birth of isotachophoresis (ITP), with proper commercial instrumentation for its routine application [3]. However, ITP did not gain popular acceptance as an analytical technique, in part because the unusual data presentation was foreign to practitioners and in part because of the introduction of a plethora of discontinuous buffers by Jovin [4,5], throwing the users into a state of perennial confusion. Nevertheless, ITP is enjoying a renaissance as a sample pre-fractionation and concentration step prior to CZE separations [6–8].

The modern era of CZE is considered by many to have commenced with publication of a series of papers by Jorgenson and Lukacs [9–11]. These reports described a simple research instrument consisting of a fused-silica capillary, electrode reservoirs, a high-voltage power supply and, usually, a modified HPLC optical detector. Samples were easily introduced by dipping the capillary inlet into the analyte solution and applying voltage or raising the level of the sample vial. Polyimide-clad fused-silica capillaries, as adopted in gas chromatography [12], were employed with inside diameters of 25 to 100 µm, and a section of the polyimide was burned away to provide a detection "window".

9.2 THE INSTRUMENT

Countless publications offer the scheme of the standard CZE instrument, much as described in Refs. 9–11. To this classical drawing, one has to add, today, a thermostatic unit, present essentially in any available instrument and, of course, a computer for data acquisition, analysis, and apparatus control (see, *e.g.*, Fig. 3 of Oda and Landers [13]). The focus today has considerably changed from single-capillary units to multichannel or capillary array machines, and to this equipment we will turn our attention. As early as 1992 Mathies and Huang [14] developed the first capillary array electrophoresis (CAE) system, using a confocal fluorescence scanner. Its application to DNA sequencing was demonstrated by running 25 capillaries simultaneously, thus increasing the production rate of sequencing data 25-fold [15,16]. The search for a comprehensive CAE instrument was soon attempted by several other groups seeking equipment capable of fast, automated, sensitive, and rugged operation [17–26]. Today, there are three commercial versions of CAE instruments, while a few groups are still trying to offer new alternatives in system design. All CAE DNA sequencers feature 96 capillaries: MegaBACE 1000 from Molecular Dynamics (Sunnyvale, CA, USA), ABI 3700 from PE Biosystems

(Foster City, CA, USA), and SCE9600 from SpectrumMedix (State College, PA, USA). Their actual design originated from prototypes of Huang *et al.* [15,16], Kambara and Takahashi [17], and Ueno and Yeung [21], respectively. The instrument from Molecular Dynamics is based on confocal detection, consisting in a microscope objective for focusing the laser light inside the capillaries and, at the same time, for collecting the emitted light from the center of the column. In order to collect the data from all the capillaries, a scanning system is used. The confocal system utilizes a set of four photomultiplier tubes with proper filters and dichroic beam splitters. Columns coated with linear polyacrylamide (LPA) and filled with low-viscosity LPA solutions are used [27]. The equipment from Perkin-Elmer employs post-column detection with liquid sheath flow [18,20]. The capillary bundle is aligned inside a quartz cuvette. Along the dead-space between the capillaries and the walls of the cuvette a buffer solution is pumped through the cell. The liquid sheath, flowing on the outside, drags down the DNA zones being eluted from the columns and tapers them to a small diameter. A laser beam crosses all flow streams and excites the fluorescent molecules. Light is collected at 90° from the laser plane, thus reducing the light scatter. The fluorescent light is dispersed through a diffraction grating and imaged onto a cooled charge-coupled device camera (CCD camera). The SpectrumMedix instrument exploits on-column detection [21,28]. The laser beam crosses all 96 capillaries that are immersed in a refractive-index-matching liquid between glass plates; the latter also confine the laser light, minimizing its losses. The fluorescent light is collected at right angle from the laser axis by a lens and detected by a CCD camera able to perform multi-color detection. It is expected that this system can be expanded to 384 capillaries.

In addition to the commercial instrumentation just described, research on CZE instrumentation is continuing in different laboratories. Thus, Kambara's group [29] has recently improved its detection system by intercalating the capillaries with glass rods of the same external diameter as the capillaries. With this design, the light transmission with side illumination was ameliorated, but at the expense of the total number of capillaries, reduced from 96 to 48. The approach of Quesada and Zhang [23], who use optical fibers for illumination and light collection, has also been improved by adopting a single beam for illuminating all capillaries [30]. The capillary walls are used as lenses, an approach similar to that of Anazawa *et al.* [19]. In another approach, Mathies' group [31,32] has also continued CAE development. The latest version of their equipment, employs a confocal system with the capillaries set in a circular array. The microscope objective spins inside a drum, interrogating each one of the columns in turn. Currently, their wheel accommodates 96 channels, but a larger number of capillaries could easily be lodged in this geometry (up to 384 channels). Fig. 9.1 shows the CAE wheel devised by Medintz *et al.* [31] and Paegel *et al.* [32] (*cf.* Fig. 11.9). This design consists of 96 separation channels, arrayed in a radial format around a central common anode reservoir in a 10-cm- or 15-cm-diameter wafer. Each pair (or doublet) of microchannels shares a common waste and cathodic reservoir (Fig. 9.1b). The peculiarity of this design is that only a single doublet needs to be designed, because all doublets are rotationally equivalent. Channels are grouped into 4 quadrants of 12 pairs each. The mask design of a 96-channel chip with an extended separation geometry (16 cm in length) is presented in Fig. 9.1c, along with a close-up of the actual taper geometry (Fig. 9.1d). For image acquisition, a rotary scanning confocal fluorescence

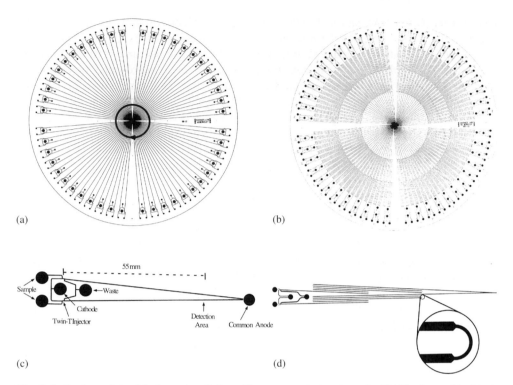

Fig. 9.1. Design of the 96-channel radial capillary array electrophoresis (CAE) plates. (a): Mask pattern used to form the 96-channel radial CAE plate on a 15-cm diameter disk. The black circle with an arrow in the center of the plate indicates where the laser beam from the rotary scanner interrogates the separations; (b): mask pattern used to form a 96-channel radial CAE plate with folded channels for increasing the separation length to 16 cm; (c): enlarged view of a doublet of individual straight channels showing the common cathode, waste, and anode reservoirs; (d): enlarged view of the folded-channel layout and the novel hyper-turn geometry. (Reproduced from Ref. 31, with permission.)

detector was developed, as illustrated in Fig. 9.2. The detection system consists of a rotating objective head (Fig. 9.2a), coupled to the four-color confocal detection unit (Fig. 9.2b). The 488-nm beam from an Ar laser is directed through a series of mirrors and dichroic beam splitters up to the rotating central hollow shaft of a stepper motor. The beam is focused onto the disk by a rotating microscope objective. Fluorescence from the sample zone is collected by this objective and passed back along the reverse optical path through the dichroic beam splitter to the four-color confocal fluorescence detector (see Fig. 9.2b). The latter separates light into four distinct channels: red (>600 nm), black (560–590 nm), green (530–560 nm) and blue (505–530 nm).

Another variant of CAE instruments comes from Heller's group [33]. These authors use a laser line generator for producing a uniform intensity profile over the array of 96 channels, yet they use a low-power laser (*ca.* 30 mW). In the detection set-up, they have adopted a holographic transmission grating, with the image of the array formed onto a back-illuminated CCD chip. A general problem with most of the CZE instruments is the

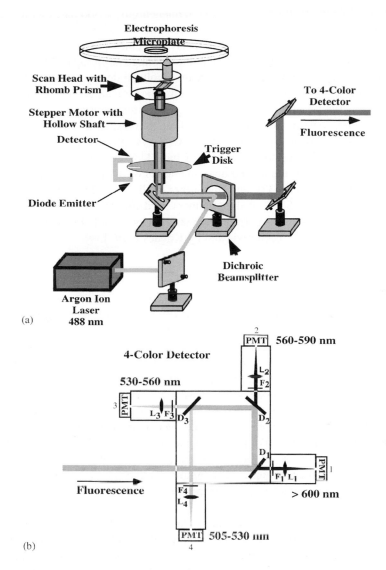

Fig. 9.2. (a): Schematic drawing of the Berkeley radial confocal fluorescence plate scanner; (b): four-color detector. Pinholes and apertures are indicated by blank white spaces between compartments. D (1 to 3) dichroic beam splitters; F (1 to 4) filters; L (1 to 4) lenses; PMT (1 to 4) photomultiplier tubes. Fluorescence from the scanner is directed sequentially by reflectance from D_1 (light <595 nm is reflected) to D_2 (reflects light <565 nm), and then to D_3 (reflects light <537 nm). Fluorescence light is filtered by a 600-nm long-pass filter in Channel 1, by a 580-nm band-pass filter in Channel 2, by a 550-nm band-pass filter in Channel 3 and by a 520-nm band-pass filter in Channel 4. The filter band-widths for each channel, as defined by the dichroics and filters, correspond to the 4 common DNA sequencing dye emission windows. (Reproduced from Ref. 31, with permission.)

reading length: it is in general difficult to achieve sequencing of >500 bases with $>99\%$ accuracy, although it would be highly desirable to extend this value to at least 800 bases. In order to address this problem, Endo *et al.* [34] have recently proposed electric-field-strength gradients with a typical duty cycle of an initial voltage ramp (up to 220 V/cm) for accelerating short fragments, followed by a plateau, a voltage decrement, and, finally, a constant, lower voltage of 90–130 V/cm for the separation of longer DNA fragments. These electric field strength gradients, coupled to column temperatures of 60°C, allow extension of the reading length, with almost linear transit times, up to 800 bases.

Quite a different platform for performing CAE is provided by microchips, micromachined and microfabricated devices (see Chap. 11). Two major procedures for preparing microchips should be mentioned here: one uses conventional photolithography on glass or silicon substrates [35] and the other utilizes molded plastic over a micromachined template [36]. Some of the polymers used for microchip fabrication are poly(dimethylsiloxane) [36,37] and poly(methylmethacrylate) [38]. Procedures for the fabrication of such devices have been reviewed by Kovacs *et al.* [39].

9.3 THE CAPILLARY

One of the main inventions, which is at the heart of present CZE technology, was the development of the fused-silica column at the Hewlett-Packard Company in the late 1970s [12]. This invention changed the field of gas chromatography, where until 1978 glass capillaries were used, which were fragile and difficult to manufacture as separation columns. The discovery of fused-silica capillaries eliminated the intermediate step of using flexible quartz, which was developed in those days for fiber-optic production. The final decision of adopting silica columns came form the need for using the purest possible material: quartz still contained 60 ppm metal oxides, whereas their levels in fused silica was <1 ppm. The second major breakthrough was the development of an exterior coating process, allowing full flexibility of the silica material and column coiling without breakage. For this purpose, the previous rubber column coating was abandoned in favor of a thin coating of polyimide, characterized by its inertness, high strength, and good thermal capacity.

Untreated fused silica consists of a number of different acidic surface silanols (see Fig. 9.3), which impart to its surface, when it is bathed under an electric field in any buffer above pH 2, some unique properties that dramatically affect separation efficiency in CZE. As shown in Fig. 9.3, there are at least three types of ionizing groups: isolated, vicinal, and geminal silanols. Interspersed among these ionogenic groups, one can envision inert siloxane bridges and highly acidic hydrogen-bonding sites [40,41]. The total density of ionogenic silanols on this surface is given as 8.31×10^{-6} M/m^2, corresponding to about 5 silanols per nm^2. There has been a serious debate on the assessment of the pK value of such silanols. While Schwer and Kenndler [42] reported a pK of 5.3, data by Huang *et al.* [43] and Bello *et al.* [44] suggest that the pK value is in fact 6.3, *i.e.* one pH unit higher. This latter value is consistent with the fact that the point of zero electro-osmotic flow (EOF) is found at precisely pH 2.3 and with the fact that the EOF does not plateau before reaching a pH of *ca.* 10 in the electrolyte solution bathing the wall. The pK value of 6.3, however,

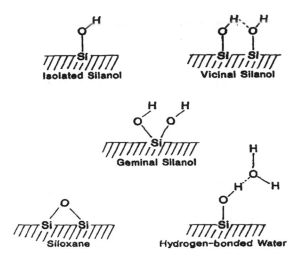

Fig. 9.3. Pictorial representation of surface silanols on a fused-silica capillary. At least three types of silanols can be envisioned: isolated, geminal, and vicinal, suggesting a microheterogeneity of pK values. (Reproduced from Ref. 12, with permission.)

should be taken to represent a mean pK value and does not in itself exclude the existence of a spectrum of micro pK values centered on 6.3, as one would expect from the heterogeneity of silanol groups, depicted in Fig. 9.3.

As a result of the progressive ionization of silanols at progressively higher pH values, an electro-osmotic flux is produced, increasing in magnitude from acidic to alkaline pH values until it reaches a plateau at a pH around 10. The electro-osmotic force in a capillary is produced by the electric field and transmitted by the drag of ions acting in a thin sheath of charged fluid adjacent to the silica wall (Chap. 7). The origin of the charge in this sheath is an imbalance between positive and negative ions in the diffuse double layer, extending for a few nm thickness away from the wall into the bulk solution. This excess of cations in the diffuse double layer is continuously carried away by the voltage gradient towards the cathode, where it migrates with its shell of hydration water: the macroscopic phenomenon observed is a net transport of liquid towards the cathode. As long as the electric field is applied, the EOF cannot cease, since the zeta potential (ζ) existing at the surface of shear (*i.e.* at the outer edge of the single stratum of counter-ions tightly clinging to the fixed negative charges on the silica) keeps regenerating the depleted excess of cations, which in turn are immediately harvested towards the cathode, in a endless and instantaneous depletion/replenishment cycle. The net liquid flux associated with the EOF can be quite strong. McCormick [45] has measured this flow-rate at different pH values and found that at pH 6 (when silanols are only about half ionized) it reaches 180 nL in a 50-µm-ID capillary. This means that, in a number of applications with buffers at pH > 7, the entire liquid content of a capillary could be replenished in a couple of minutes of operation, depending on the voltage applied and on the actual pH value. A typical EOF profile of untreated silica wall is shown in Fig. 9.4; it should be noted that the increment from the residual minimum values in strongly acidic medium to the plateau value in alkaline milieus could be as high as one order of magnitude. The velocity of EOF is

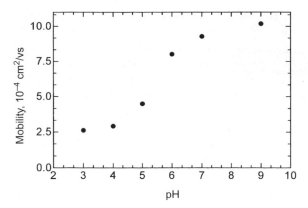

Fig. 9.4. Mobility of the electro-osmotic flow as a function of pH of the buffer, bathing the silica wall.

dependent on pH, ionic strength, buffer composition, and the chemical nature of the capillary wall. The reproducibility of the EOF is poor, particularly in the sigmoidal transition. This is due to the slow equilibration of the silica surface when cycling from alkaline to acidic conditions. This slow equilibration causes flux variations, which, in turn, affect run-to-run and day-to-day reproducibility of the transit times.

One might wonder if it is possible to find different types of surfaces which do not contain ionogenic groups. This problem has been addressed by Wanders *et al.* [46] and by Van de Goor *et al.* [47], who proposed a method for the on-line determination of EOF by using an analytical microbalance capable of monitoring minute transports of liquid. By coupling these measurements to a precise assessment of streaming potentials, they were able to construct a curve relating variations of the zeta potential with the pH of the buffer bathing the capillary wall. These experiments were unique in that they were performed on a plastic material, consisting of polyfluorocarbon (a neutral material, by definition). Yet, even this surface exhibited a strong ζ value, reaching a plateau of -45 mV at pH *ca.* 10 and reversing its sign to assume a value of 10 mV at a pH of *ca.* 3. Zeroing of the zeta potential was obtained at pH 3.5, which can be assumed to be a sort of "isoelectric point" of the wall. Later on, these data were confirmed by Schützner and Kenndler [48], who demonstrated that essentially all plastic materials (polyfluorocarbon, polyethylene, polyvinylchloride, even polystyrene and polyurethane) would exhibit a ζ value (reaching a plateau at -60 mV and pH > 7), possibly due to either intrinsic, dissociating carboxyl groups (incorporated during polymerization) or to ions adsorbed on the wall (or perhaps to both mechanisms).

Is EOF good or bad for a CZE separation or simply an "ugly" phenomenon one has to cope with? It depends. EOF is advantageous for the separation, in a single run, of basic, neutral, and acidic solutes. The separation of molecules based on differences in electrophoretic mobilities takes advantage of EOF, because this force, which influences the migration velocity and the direction of all analytes to the same extent, independent of their charge, modulates their transit times, a phenomenon which can be exploited to improve the separation. A case in point is micellar electrokinetic chromatography (MEC) (Chap. 7), which exploits a fine balance of opposite movements, those of the EOF towards the cathode and those of the anionic micelles retromigrating towards the anode: this offers an ever-expanding separation time window for sorting out uncharged compounds [49]. Thus, EOF is generally well accepted and most often exploited for the analysis of small-molecular-mass migrants. However, when dealing with macromolecules, especially DNA and proteins, it is a most unwanted phenomenon, since it impairs their separation. This is particularly evident with peptides and proteins which, due to their amphoteric character, can often be irreversibly adsorbed on the silica surface, even when present in solution above their pI value, which ought to ensure a net negative surface. Yet, even in this last case, regions of a polypeptide chain rich in positive charges can still stick to the negative charges on the silica wall. (It should be recalled that, to ensure a negative charge on a protein surface, one should generally work at quite high pH values, which, in turn, will automatically generate a maximum ionization of silanols, allowing co-operative binding with any positively charged stretch on a proteinaceous surface.)

9.4 HOW TO MODULATE THE EOF

For all the reasons given above, it is often necessary to modulate or even fully suppress, the EOF. There are several ways to control partially or fully the EOF and an unwanted adsorption of analytes: buffer changes and additives, use of organic solvents, adsorption of neutral and/or charged polymers (including surfactants) on the wall, and use of chemically bonded phases. Each technique has some merits and limitation, which will be evaluated below.

9.4.1 Buffer changes and additives

Quite a few research groups [50–52] have observed that the mobility of EOF (μ_{eo}) is inversely proportional to the buffer concentration and that a plot of log concentration versus μ_{eo} is linear. Conflicting reports on the effect of the buffer composition on EOF were resolved by VanOrman *et al.* [53] who noted that all buffers, independent of their chemical nature, would have the same effect on the EOF, provided that data were expressed not in terms of buffer molarity, but in terms of prevailing ionic strength at a given pH. Isoionic strength plots demonstrated that all buffers would indeed quench EOF in the same way. The same authors confirmed that viscosity increments would quench EOF in the same fashion, independent of the type of viscous agent added [54].

Also, the buffer cations seem to have an important role in modulating μ_{eo}. According to Atamna *et al.* [55] and Issaq *et al.* [56], in the series of monovalent cations, Li, Na, K, Rb, and Cs, the quenching of μ_{eo} seems to be directly correlated to the atomic radius: the larger it is, the higher is the effect of the cation in reducing μ_{eo}, Cs being the most, and Li the least, effective. It is hypothesized that the larger cations are more strongly adsorbed on the silica wall, thereby altering its charge and effectively reducing the EOF. Similar results were obtained by Green and Jorgenson [57], who used alkali cations for minimizing protein adsorption on the bare silica. Interestingly, the same inorganic, monovalent cations (Li, Na, K, Rb, Cs), when present in the Debye–Hückel layer of DNA, were found to have an opposite effect on DNA migration [58]. Its free mobility was seen to increase from Li- to Cs-equilibrated DNAs, since the last cation, having a weaker surface-charge distribution and a greater physical size (in the nonhydrated state), is more loosely bound to the DNA helix, thus providing less screening of its negative charges.

Another way of manipulating EOF and reducing analyte interaction with the silica wall is to add amines to the background electrolyte. Nahum and Horváth [59] first recommended them as additives to the mobile phases in reversed-phase chromatography of ionogenic substances, in order to suppress the untoward effect of the residual unmasked silanolic groups in the stationary phase. Amines make up a vast number of such compounds, *e.g.*, mono-amines, such as triethylamine and propylamine [60,61], morpholine [62], glucosamine and galactosamine [63], *N,N*-diethylethanolamine, *N*-ethyldiethanolamine, triethanolamine [64], ethanolamine [65], hydroxylamine, ethylamine [66], as well as the quaternary tetramethylammonium chloride [67]. Among the di-amines, there are 1,3-diaminopropane [68], 1,4-diaminobutane (putrescine) [69], 1,5-diaminopentane (cadaverine) [70], ethylenediamine [71], *N,N,N′,N′*-tetramethyl-1,3-butanediamine [72], and the α,ω-*bis*-quaternary ammonium alkanes, such as hexamethonium [73] and decamethonium bromides [74]. In the family of oligo-amines, we may mention diethylenetriamine and triethylenetetramine [75], *N,N′*-*bis*(3-aminopropyl)1,4-butanediamine (spermine) [66], and 1,4,7,10-tetraazocyclodecane (cyclen) [76]. Finally, among the polyamines, we may list chitosan [77], polyethylenimine [78,79], and polydimethylallyl ammonium chloride [80]. These dynamic wall modifiers have recently been re-evaluated by Verzola *et al.* [81], who could derive a scale of relative potencies. They specified two values for molarities: a value at 50% (a kind of a dissociation constant) and a value at 90% inhibition of binding of macromolecules to the silica surface. According to these figures of merit, mono- and diamines are rather poor quenchers of interaction with the wall, since the 50% values are of the order of 50–100 mM and the 90% values go as high as 560 mM. On the contrary, oligo-amines, especially spermine and tetraethylenepentamine (TEPA) are most effective, since the 50% molarities are in the sub-millimolar range and the 90% values are of the order of *ca.* 1 mM.

Another way of controlling EOF (without altering the buffer conductivity) is to add zwitter-ions to the running buffer. Ideally, such zwitter-ions should have a large ΔpK, so as to be isoelectric over the pH 3–10 range. Petersen and Merion [82] have proposed a number of such compounds: trimethylammoniumpropyl sulfonate, triethylammoniumpropyl sulfonate, and tripropylammoniumpropyl sulfonate. A similar approach was adopted by Zhu *et al.* [83], who suggested a number of zwitter-ions (called Z4A, Z6C, and Z6O, and described as "experimental multi-component mixtures of amphoteric species") for quenching protein interaction with the silica surface. Zwitter-ionic buffers have also been proposed for protein analysis by Bushey and Jorgenson [84] and by Chen *et al.* [85].

The solvent can also have an important role in modulating μ_{eo}. According to Camilleri and Okafo [86], even replacement of water by deuterium oxide can lower the EOF, since the latter compound has a higher viscosity (1.23 times greater) and an ionization constant one order of magnitude lower than water. Binary mixtures of water and a protic (methanol, ethanol, 2-propanol) or aprotic dipolar solvent (acetonitrile, acetone, dimethyl sulfoxide), all in a 1:1 (v/v) ratio, substantially lower the EOF and the ζ value by shifting the pK of silanols to higher pH ranges [42]. In addition, organic solvents have a unique effect on

MEC, in that they can be used to manipulate selectivity by decreasing the polarity of the mobile phase and thus altering the partition coefficient of analytes in the micellar phase [87].

9.4.2 Adsorbed polymers

A number of substance can be adsorbed on the silica wall through electrostatic interactions and/or hydrogen bonding. Moreover, when acidic protons of silanols are replaced by an appropriate Si-aryl group, the surface becomes hydrophobic and can thus adsorb substances through hydrophobic interactions. If the wall adsorbs a neutral polymer, EOF can be substantially reduced and the solutes may be sterically prevented from coming into contact with the wall and from being adsorbed; when cationic surfactants are added to the separation buffer in a hydrophobically derivatized capillary, the charge on the surface will be changed from negative to positive and the EOF direction reversed. We will here briefly review the behavior of two major classes of adsorbed compounds: neutral polymers and surfactants.

Although anionic, cationic and zwitter-ionic polymers have also been described as potential additives for inhibiting protein interaction with the silica wall, we will limit this digression to neutral polymers, which can be divided into two classes:

(a) *Neutral, hydrophilic polymers.* They are typically, but not exclusively, bonded to the silica wall. Among them are: polyacrylamide [88], poly(acryloylaminoethoxy ethanol) [89], poly(acryloylamino propanol) [90], celluloses and dextran [91–94], poly(vinyl alcohol) [95,96]; epoxy polymer [97–99], poly(ethylene-propylene glycol) [100,101], poly(ethylene oxide) [102], poly-*N*-(acryloylaminoethoxy)ethyl β-D-glycopyranoside [103], poly(glycidylmethacrylate-co-*N*-vinylpyrrolidone [104], cross-linked poly(vinyl alcohol) [105], and epoxy-poly(dimethylacrylamide) [106].

(b) *Neutral, hydrophobic polymers*: cellulose acetate [107], polytetrafluoroethylene (PTFE) [108], polydimethylsiloxane [109], and highly cross-linked poly(styrene-divinylbenzene) [110]. The latter forms a "tube-in-the-tube" structure, but it is so hydrophobic that it must be rendered hydrophilic with a second layer of, *e.g.*, polyoxyethylene oligomer.

A broad review covering all aspects of polymer wall coating has recently been published by Horváth and Dolnik [111]. A number of neutral polymers, used as potential quenchers of protein binding to the silica wall, were evaluated by Verzola *et al.* [112]. The data for the polymers studied show a very similar behavior for all, independent of the length and hydrophilicity/hydrophobicity of the different molecules. Contrary to data obtained in the case of amines, where an exponential decay is observed at increasing amine concentrations, here only *ca.* 50–60% inhibition of binding is obtained at the lower levels of polymers used. Nothing is gained when augmenting their amounts in solution, and a plateau level is quickly reached. The only polymer that shows higher inhibition (*ca.* 85%) is poly(dimethylacrylamide) (DMA), which was the only hydrophobic polymer of the series tested [112].

9.4.3 Adsorbed surfactants

Although a number of publications have dealt with adsorption of both, anionic [*e.g.*, sodium deoxycholate and sodium dodecylsulfate (SDS)] [113,114] and cationic [115,116] surfactants, these compounds are generally not optimal, due to their strong denaturing activity on proteins. Neutral and/or zwitter-ionic surfactants are much milder to proteins, and only these will be considered here. In the case of neutral detergents, and of their potential interactions with the silica wall, the strategy adopted by, *e.g.*, Town and Regnier [117] and Ng *et al.* [75,118], has been a coating mediated by hydrophobic interaction. The idea was to sequester covalently many of the surface silanols with octadecyl- or dimethylsilane and then to coat dynamically the hydrophobic surface of the column *via* nonionic surfactants. It had, in fact, previously been shown in RP-HPLC that such species are stereospecifically adsorbed on hydrophobic surfaces and create a hydrophilic surface layer [119]. Alkyl groups of the surfactant are directed into the alkylsilane layer, while the hydrophobic portion of the surfactant projects into the aqueous phase, thus shielding the surface from the approach of proteins. Adopting the strategy of hydrophobically mediated binding of surfactant to the treated silica wall, mostly the Tween (20, 40, 80) and Brij (35, 78) series were studied, although other reports have dealt with pluronic surfactants [118, 120]. Only in a few cases direct adsorption of neutral compounds (notably Triton X-100) on uncoated capillaries has been studied, but more in terms of their effect on the electro-endo-osmotic flow rather than of their potential capability of impeding analyte adsorption.

Recently, Castelletti *et al.* [121] have evaluated the efficacy of adsorbed neutral and zwitter-ionic surfactants in modulating EOF, by the procedure described by Verzola *et al.* [81,112]. It was shown that neutral surfactants are rather poor quenchers of protein binding to the wall and that 90% adsorption inhibition can be obtained at only very high levels (10%) of these species in solution. Such levels are unacceptable, since, among other problems, the solution viscosity becomes very high and handling of solutions becomes problematic. Among all the neutral surfactants, only Tween 20 appears to exhibit a reasonable inhibitory power, since it is already quite effective at a 3% concentration. On the contrary, zwitter-ionic surfactants exhibited much higher efficacy (except for SB-10, which resembles Tween 20) than that of the neutral surfactants, the best compound appearing to be the most hydrophobic one, SB-16.

As a general conclusion about all the data on dynamic coatings, it would appear that the efficacy of all such additives is closely related to their hydrophobicity [122]. This is quite clear in the case of amines, where the CH_2/N ratio seems to be the dominant factor [81] and it is apparent in the case of neutral polymers, where the most effective one was found to be poly(N,N-dimethylacrylamide) (quite a hydrophobic polymer) [112]. It is also true for the detergents, where it was clearly demonstrated that the shielding efficacy of surfactants closely follows a hydrophobicity scale [121]. But why should hydrophobic interaction occur between such a hydrophilic surface as silica and hydrophobic compounds approaching it? The mechanism of adsorption, at least in the case of polymers and surfactants, could be driven primarily by residual Van der Waals interactions. Although these interactions are extremely weak (they decay exponentially at the rate of $1/r^6$) [123], the situation could be more complex than that. This decay refers to interactions between two point-like objects. When dealing with two surfaces (silica on one side, polymers on the other) the rate becomes $1/r^2$, thus indicating a much stronger interaction. Therefore, it is quite possible that the mechanism of adsorption of polymers and surfactants on silica is driven by such residual Van der Waals forces and is then amplified by additional interaction mechanisms, once such compounds are deposited on the silica surface. Such interactions could be hydrophobically driven, but in the sense that they are vectorially oriented parallel to the silica surface, *i.e.* they occur among the various polymer molecules deposited on the surface, interlocking them *via* precisely hydrophobic bonding. Thus, layers of polymers and surfactants could build up on the silica surface, preventing or minimizing adsorption of proteins and peptides.

9.4.4 Covalently-bonded phases

Extensive reviews on this (as well as on most types of coatings) can be found in the papers of Chiari *et al.* [124] and of Horváth and Dolnik [111]. These phases could be simple organic molecules or polymeric material. In the first category, Swedberg [125] reported a coating composed of arylpentafluoro groups, Bruin *et al.* [126] maltose-bonded phases, Dougherty *et al.* [127] C_8 and C_{18} functionalities, and Capelli *et al.* [128] the Immobiline chemicals (acrylamido weak acids and bases). Of greater interest are polymeric coatings, since they should afford better coverage of the silica surface and thus better shielding of peptide/protein analytes. As early as 1985, Hjertén [88] proposed coating the inner surface of the capillary with covalently bonded polyacrylamide strings. He used a bifunctional chemical, Bind-Silane [(γ-methacryloxypropyl)trimethoxysilane] as a bridging agent [129]. Subsequently, Ganzler *et al.* [130] modified this coating by polymerizing onto it an additional hydrophilic layer, composed of polysaccharides (*e.g.*, dextran). A further improvement came from Cobb *et al.* [131], who substituted the siloxane bridge (labile even under mild alkaline conditions) with a stable, direct Si–C linkage, obtained *via* a Grignard reaction. The hydrophilicity of this coating has been markedly improved by substituting acrylamide with *N*-acryloylaminoethoxy ethanol [89] or with *N*-acryloyl aminopropanol [90], the last two monomers exhibiting also an extraordinary resistance to hydrolytic attack on the amido bond. The last-mentioned coatings guaranteed a reduction of the EOF by at least two orders of magnitude and were stable for >300 runs [90].

Although the main classes of coatings described so far are of two types, static or dynamic, there seems to be a third category emerging: dynamic coatings which become permanent. Recent reports have described this novel class of coatings, which exhibit the peculiar habit of first binding dynamically to the wall, *via* transient hydrogen and ionic bonds, to produce a covalent bond. This peculiar class of molecules belongs to a family of mono-quaternarized piperazines, the first one synthesized being an *N*(methyl-*N*-ω-iodobutyl),*N'*-methylpiperazine. It is hypothesized that at pH values above 7, three different types of bonds with the silica wall occur: (a) an ionic interaction with the quaternary amine; (b) a hydrogen bond interaction *via* the tertiary amine (deprotonated), and (c) a covalent link *via* reaction between the silanol on the wall and the terminal iodine molecule on the alkyl chain [132,133]. Unique separations of proteins/peptides and other organic molecules have been reported with this [134,135]. This class of compounds is dealt with in more detail in Chap. 15.

9.5 THE BUFFERS

Although we have described buffer additives in previous sections (*e.g.*, Sec. 9.4.1), here we must mention the various types of buffers that can be adopted in CZE and their basic properties. Buffers provide a conductive medium through which charged analyte molecules can migrate freely in an electric field; in addition, they provide selectivity by allowing manipulation of analyte mobility. The quality of a separation is dependent on the running current and the resulting Joule heating inside the capillary, whereas the reproducibility of a series of analyses is dependent on the buffering power of the chosen buffer at a given pH. Many different factors must be considered in choosing a good buffer, the most important being the ionic strength (conductivity) and the UV absorbance at low wavelengths (190–220 nm). An extensive chapter on buffer properties can be found in Ref. 54. The theory of CZE and how the mobility of analytes can be influenced by charge and by variation of pK values has been presented by Kenndler [136]. The type of solvent can also considerably influence migration and selectivity in CZE; a whole new field of CZE analyses is based on the use of nonaqueous solvents [137]. An entire issue of *Electrophoresis* has recently been devoted to these aspects [138].

Buffers used in CZE should provide good pH stability so that the pH does not increase or decrease due to ion depletion during a series of analyses, *e.g.*, up to 5 h of continuous operation. This implies that they have a good buffering capacity and are used within their buffering range. The acceptable buffering range can be determined as follows. If a weak protolyte is titrated within ± 0.572 pH units on either side of its pK value, its buffering power should still be 2/3 of the maximum (centered at pH $=$ pK). If one sets a reasonable lower limit for a useful buffering power at 1/3, this can be found at ± 0.996 pH units on either side of its pK value [139]. With this definition of an acceptable buffering power, the buffering range of a monoprotic weak protolyte becomes 2 pH units, although, whenever possible, one should limit this to only 1 pH unit across the pK value, for a more robust buffering power. The ionic strength and molar concentration of any buffer should also be sufficiently high to prevent pH drift due to the accumulation of opposing ions at the anode and cathode.

Table 9.1 lists the buffers most widely used in CZE, together with some of their important properties. It is worth recalling here some general properties of buffering/titrating ions that can greatly affect the separation quality and selectivity:

(a) Oligoprotic buffers: an anion like citrate, a tricarboxylic acid and strong chelator, can strongly ion pair with some analytes (*e.g.*, Lys/Arg-rich peptides) [140].

(b) Borate buffers interact strongly with molecules which contain polyhydroxyl groups, such as glycols, phenols, and saccharides [141,142]. This borate complexation imparts a charge on otherwise neutral carbohydrates, so that they will migrate in an electric field.

(c) An increase in the size of the counter-ion (non-buffering ion), in the case of monovalent cations, results in an increase in buffer conductivity and a decrease in electrophoretic mobility and EOF (μ_{eo}: $Li^+ > Na^+ > K^+ > Rb^+ > Cs^+$) [143]. In addition, the type of cation can strongly reduce the adsorption of analytes

TABLE 9.1

BUFFERS MOST FREQUENTLY USED IN CZE

Buffer		pK$_a$ (20°C)	Range	Abs* (mAU) (200/220 nm)
Phosphate	a	2.12	1.6–3.2	100/50
	b	7.21	5.9–7.8	1200/50
	c	12.32	10.8–13.0	
Citrate	a	3.06	2.1–6.5	>3000/2600
	b	4.74	2.1–6.5	
	c	5.40	2.1–6.5	
Formate		3.75	2.6–4.8	>3000/2600
Succinate	a	4.19	3.2–6.6	>3000/>3000
	b	5.57	3.2–6.6	
Acetate		4.75	3.4–5.8	2500/900
Acetate/triethylamine		4.75	3.4–5.8	2200/750
		10.72	10.0–11.5	
MES[1]		6.15	4.9–6.9	2100/190
BES[2]		7.15	5.9–7.9	2800/1000
Carbonate	a	6.35	6.2–10.8	1250/40
	b	10.33	6.2–10.8	
MOPS[3]		7.20	5.9–7.9	–
Tris		8.3	6.6–8.8	1750/10
Bicine		8.35	7.4–9.2	>3000/>3000
Borate		9.24	8.0–10.5	300/50
CHES[4]		9.50	8.6–10.0	2600/700
CAPS[5]		10.4	9.1–11.1	2700/740
Triethylamine		10.72	10.0–11.5	–

1 = 2-(*N*-morpholino)ethanesulfonic acid; 2 = *N,N-bis*(2-hydroxyethyl)-2-aminoethanesulfonic acid; 3 = 3-(*N*-morpholino)propanesulfonic acid; 4 = 2-(*N*-cyclohexylamino)ethanesulfonic acid; 5 = cyclohexylaminopropanesulfonic acid.
* Absorbance values at buffer concentrations of 40 m*M*, 1-cm path length.

(proteins) on the wall [57]; Cu^{2+} and Zn^{2+} can enhance peptide resolution *via* chelation [144].

(d) An increased buffer molarity is, in general, beneficial, since it will act to enhance competitive ion pairing of buffer cations with the sample molecules at the capillary wall, thus diminishing sample adsorption [85]. However, these benefits may be outweighed, if the ionic strength exceeds a safety value, by the increased sample dispersion that occurs due to excessive Joule heating [56]. When an Ohm's law plot (current *vs.* voltage) starts deviating from linearity, the voltage drop over the capillary should not be increased any further. In addition, at equal molarities, oligoprotic buffers will give substantially higher currents than monoprotic species.

(e) Phosphate buffers seem to have a unique effect on the capillary wall. It has been reported [45,145] that the phosphate ion reacts chemically at low pH with silanol groups to form a phosphate ester (Si–O–HPO$_4$). The phosphate group is bound to the capillary by flushing with a 1% solution of phosphoric acid for 15 min and can be removed by rinsing the capillary with 1 N KOH. This altered surface has a smaller ionic binding affinity for polycationic analytes (*e.g.*, peptides).

(f) The use of amine buffer salts, such as *tris*-(hydroxymethyl)aminomethane (Tris), 2-(*N*-morpholino)ethanesulfonic acid (MES), *N*-2-hydroxyethylpiperazine-*N'*-2-ethanesulfonic acid (HEPES) and oligo- or polyamines can induce a strong interaction with the wall, in some cases dynamically coating it. Chiesa and Horváth [146] have demonstrated the use of triethylammonium phosphate in the pH 2.5–3 range, for the separation of polysaccharides; in this buffer, the EOF could be controlled and even reversed. Good's buffers (amphoteric), alone or in a mixture, titrated around the pK value of the amino group with a weak counter-ion, often provide excellent separations due to their low conductivity, allowing delivery of high voltages [147].

(g) Amphoteric buffers, used as the sole ionic species at their isoelectric points, can have a unique effect on the separation of peptides and oligonucleotides, due to their very low conductivity and high buffering capacity [148–155]. They should not be confused with Good's buffers, which are very bad if used as isoelectric species (*i.e.* in the absence of a titrating ion). They are dealt with *in extenso* in Chap. 15.

9.6 MODES OF OPERATION

It is generally agreed that there are four main modes of operating in capillary electrophoresis: open-tube (free solution) CZE; isoelectric focusing (IEF); CZE in sieving liquid polymers, and isotachophoresis (ITP). We will briefly review these aspects below.

9.6.1 Open-tube capillary zone electrophoresis

In this mode, the capillary is filled with just plain buffers (with suitable additives, when necessary, provided that they are not polymeric or micellar phases) and the separation is conducted by exploiting the differential mobility of the analytes, as dictated, as a first approximation, by their charge-to-radius ratio (for small analytes). It is of interest to note that the liquid is maintained in the capillary tube by virtue of the anticonvective nature of the wall. On slabs or flat plates, the liquid would flow off; hence, a gel is necessary for maintaining the liquid on the plate, even if the porous network is not needed, *per se*, in the separation process. It should also be recalled here that the liquid becomes more difficult to maintain in the electrophoretic space as the surface-area-to-volume ratio decreases, *i.e.* in wider tubes. Thus, even small differences in the height of the two ends of a 200-μm-ID capillary could cause bulk liquid flow due to the small pressure drop. In free-solution CZE, as long as there is a zeta potential at the walls of the capillary and the resistance to flow is not significant, when an electric field is applied, bulk flow will occur, which is called electro-osmotic flow (*cf.* Secs. 9.3 and 9.4). The velocity, *v*, of this flow can be expressed as:

$$v = (\epsilon \zeta E)/\eta$$

Where ϵ and η are the permittivity and viscosity of the solution, respectively, ζ is the zeta potential at the wall, and *E* is the electric field strength (in V/cm). Thus, the bulk flow increases in direct proportion to the applied field. In addition, since η decreases by *ca.* 2% and the apparent electrophoretic mobility increases by *ca.* 2% for each degree centigrade, one needs a good column temperature control for achieving reproducible runs. An important characteristic of EOF, though, is that it is plug-like rather than parabolic, as is common for pressure-driven flow. In the case of "bare" fused-silica capillaries, moreover, bulk flow will occur toward the cathode, since the wall surface will be negatively charged. Thus, in principle, not only positively, but also negatively charged species will move in the same direction past the detector, provided that the mobility of the negatively charged species does not exceed the EOF.

There are innumerable examples of open-tube CZE in the literature. One application that is gaining interest is in clinical analysis [156–159], *e.g.*, the detection of normal and abnormal compounds in serum, urine, and cerebrospinal fluid. Clinical applications are published for the analysis of serum proteins and abnormal metabolites [160]. Immunoassay detection has also been shown to operate successfully in CZE [161], and the future of this approach appears bright. Since antibodies and competitors can be fluorescently labeled and detected by laser-induced fluorescence (LIF), exquisitely low detection levels are possible, at the level of attomoles to zeptomoles. Indeed, single-molecule detection by LIF has been claimed [162].

9.6.2 Capillary isoelectric focusing

Capillary isoelectric focusing (cIEF) stems from the analogous techniques, and know-how has been developing since the early seventies for both the gel tube and slab formats [163], by exploiting soluble carrier ampholytes (CAs). Only CA-IEF has been developed so far, the technique of IEF in immobilized pH gradients (IPGs) [164] being not so readily amenable to miniaturization in a capillary format. In its usual gel strip configuration, IEF (especially with IPGs) is widely adopted as the first dimension for two-dimensional (2-D) slab gel electrophoresis, the second dimension being a molecular mass separation by means of SDS/polyacrylamide gel electrophoresis (SDS-PAGE). The method is very important in cell component analysis, protein expression, and proteomics [165]. IEF has been extensively reviewed [166–178]. There are different ways of performing cIEF: in covalently coated capillaries and in dynamically coated tubes. In the former system, due to suppression of EOF, the focused stack of carrier ampholytes and proteins is arrested; thus, ways must be found to mobilize the stack past the detector. A number of procedures can be adopted:

    (a) applying a mechanical pump to the capillary and generating a hydrodynamic flow (at the end of the IEF process) (pressure as well as vacuum elution);

(b) replacing the base at the cathode with acid or the acid at the anode with base;

(c) salt mobilization.

This latter technique has attained wide popularity. When a salt (*e.g.*, NaCl) is added to the anolyte, mobilization will be towards the anode; conversely, if added to the catholyte, the train of bands will be eluted at the cathode. In general, the time required for mobilization is *ca.* 15 min at 360 V/cm. During mobilization, the current, which has reached a minimum at the end of the focusing stage (typically 1 μA), rises again to as high as 50 μA. It is during mobilization that the train of zones, titrated away from the pI by the cations or anions (other than protons or hydroxyl ions), enters the tube from one of the electrode reservoirs, transits in front of the detector, and is registered as a spectrum of bands. Alternatively, in the absence of elution, the whole train of focused protein bands can be visualized *in situ*, in straight, optically transparent capillaries, via optical absorption (with a linear photodiode array or a CCD camera) refractive-index gradient (schlieren shadowgraphy), or fluorescent imaging (LIF detector) [179]. In the dynamic coating approach, rather than eliminating EOF completely, one may try to reduce it to such an extent as to allow attainment of steady-state conditions; from there on, the bulk flow would keep mobilizing the "arrested" stack past the detection window. This approach would then obviate the need for performing salt, vacuum, or hydrodynamic mobilization, focusing and elution being accomplished in one step. A simple way to modulate EOF is to add viscous polymer solutions, such as methyl cellulose, hydroxypropylmethyl cellulose, polyvinyl alcohol, polyvinyl pyrrolidone, poly(dimethylacrylamide) (*cf.* Sec. 9.4.2) or even surfactants (*cf.* also Sec. 9.4.3) [180].

A vast body of applications of cIEF exists; here, we will show just some selected examples, where the resolving power is maximized by resorting to narrow, non-linear pH gradients. With umbilical cord blood, where only 3 major hemoglobin (Hb) components are present (Hb F, Hb A, and the acetylated Hb F, F_{ac}), one can screen for thalassemia, provided a good separation is obtained between Hb A and Hb F_{ac}. This is difficult because differences in pI values are minute. In order to improve the separation, the pH 6–8 Ampholine was mixed with an equimolar mixture of "separators", namely 0.2 *M* β-alanine and 0.2 *M* 6-aminocaproic acid, which would flatten the pH gradient in the focusing region of Hb A, Hb F, and F_{ac} [181] (Fig. 9.8). Using the same principle, Conti *et al.* [182] have also attempted an IEF separation of HbA from Hb A_{1c} (the glycated form of Hb A), the latter component being of diagnostic value for the long-term control of diabetic patients (Fig. 9.9). This was a challenging separation, since the pI difference between these two species is barely 0.01 pH unit. Of course, flat pH gradients will produce high resolution, but at the expense of the number of peaks that can be resolved. At the opposite extreme, for complex sample mixtures, one may wish to combine an acceptable resolution with a high peak capacity. This seems to have been achieved in proteome analysis by Smith *et al.* [183], who have reported the use of a wide pH (3–10) gradient and 50-cm-long capillaries for the resolution of *ca.* 210 peaks from the lysate of *Deinococcus radiodurans*. When the sample is not so terribly complex, the eluate from cIEF can be injected directly into a mass spectrometer (in the present case, a Fourier-transform ion-cyclotron-resonance machine). Smith's group [184] developed automated data analysis software which allows the visualization of the results in a 2-D format. The 2-D displays are produced by plotting molecular mass *vs.* scan number, which is correlated to protein pI.

9.6.3 Capillary zone electrophoresis in sieving liquid polymers

As practiced in the slab-gel format, electrophoresis is predominantly a size-based separation method. This can be seen in the two major applications of the technique: (a) DNA separation and analysis and (b) SDS-PAGE sorting-out of proteins. In the case of DNA, the electrophoretic mobility, above the critical size of 400 bp in length [185], has been shown to be independent of base number for both single- and double-stranded DNA. This is a general consequence of the compensation of additional charge from the phosphate group by the mass of each nucleotide added. In the case of proteins, it has been shown that fully unfolded proteins will add a fixed amount of SDS per unit mass of protein (in general 1.4 mg SDS per mg protein) [186]. As a consequence, the charge-to-mass ratio is (nearly) constant and the electrophoretic mobility will be constant in free solution. Initially, it seemed logical to use cross-linked, gel-filled matrices in CZE [187]. However, it was soon found that cross-linked gels were too rigid and could not respond properly to stresses caused by temperature, ionic strength, or pressure, when poured into a capillary. The next logical step was therefore to adopt polymers in a non-cross-linked format [188], particularly since earlier work on slab gels had shown that sieving was possible with entangled polymers [189]. For DNA analysis, in general, the best polymers affording the highest resolution, up to 500 bases, in sequencing belong to the family of polyacrylamides. Since DNA separations by CZE are dealt with in Chap. 19, we will not mention them here. Suffice it to say that extensive reviews exist (see, *e.g.*, [190]) and that the principles of size-based separations for DNAs in polymer solutions have been amply covered [191–196]. The unique separating capability of entangled polymers is exemplified by Fig. 9.5, which

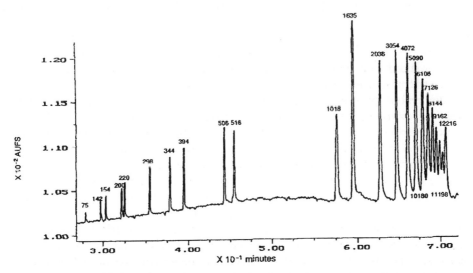

Fig. 9.5. Capillary zone electrophoresis of the 1 kb DNA ladder. Conditions, 39 cm-long-capillary of 100 μm ID; 100 m*M* Tris-borate buffer (pH 8.2) and 2 m*M* EDTA, containing 10%T poly(AAEE) (at 0%C); sample injection, 3 sec. at 4000 V; run, 4000 V at 8.8 μA; detection at 254 nm. Water's Quanta capillary unit with forced-air cooling. The numbers on each peak represent the fragment length (Reproduced from Ref. 197 with permission).

shows a representative run of DNA fragments in a novel poly(*N*-acryloylaminoethoxy-ethanol) (AAEE) polymer, offering excellent resolution up to at least 12,000 bp [197] (Fig. 9.5). Although for DNA analysis most polymers work quite well, for SDS/protein complexes, which are preferably monitored at 210 nm (a region of higher sensitivity for protein/peptide detection), polyacrylamides are not the polymers of choice, due to their rather strong absorbance at this wavelength. Preferably, UV-transparent polymers should be utilized [130,198]. As in the case of DNA analysis, a coating on the fused-silica capillary seems to offer better reproducibility, particularly in the screening of complex samples. The molecular mass of the proteins can be estimated within 10%, using this approach [199]; thus MS is clearly the method of choice when one needs correct molecular mass (*Mr*) values [200]. If one needs higher detection sensitivities than afforded by 210 nm monitoring, one can resort to labeling with fluorescent dyes. Sweedler *et al.* [201] have visualized trace levels of peptides *via* LIF detection after labeling them with Texas Red [202,203].

9.6.4 Capillary isotachophoresis

In isotachophoresis (ITP), a sample is separated in a discontinuous electrolyte system, formed by leading and terminating electrolytes [204]. In simple terms, the leading electrolyte contains the ion with the highest mobility, whereas the terminating electrolyte contains the species with the lowest mobility. When a sample is injected in between the leading and terminating electrolytes, the ions with intermediate mobilities will migrate isotachophoretically and create typical stacked isotachophoretic zones with sharp boundaries. The first real exploitation of this principle, in biochemical analysis, came from the work of Ornstein [205] and Davis [206], who developed disc electrophoresis. In that technique, the first part of the electrophoretic path, that occurring in the sample and stacking gels, consisted of an isotachophoretic race track. This had two extraordinary results: it provided an incredibly effective concentration of the macromolecular analyte (which had to adjust its molarity to that of the leading ion), of the order of 3 to 4 orders of magnitude, and it prevented boundary decay (a self-sharpening effect efficiently counteracting diffusion phenomena and peak-spreading). Too good to be true! In fact, there was a price to pay: as long as the ITP train was kept as such, detection of individual peaks, as the train zoomed past the UV detector, was very nearly impossible: there was no solution of continuity in between the separated zones, since no vacuum of ions could possibly be permitted, so that the position of peaks and their total number could not be estimated with certainty. Ornstein and Davis were aware of these shortcomings, and therefore, on purpose, they ran the train off the tracks, *i.e.* they ran it into a sieving gel in which each macromolecular species would now move according to both size and charge; the ITP run was thus abolished and the movement continued in the zonal mode, with individual analyte peaks separated by a zone of pure buffer components. Modern CZE analysis followed the same fate: Today, almost nobody utilizes ITP as a separation mechanism *per se*, *i.e.* by maintaining the entire run under the ITP principle. ITP is used only for handling relatively large sample volumes and for the separation and concentration adjustment of individual components, thus providing an ideal sampling method for CZE separations in the form of short zones with sharp boundaries. During the CZE step, the individual components will have to leave the stack and migrate independently with different velocities in the background electrolyte. There are many examples of the use of ITP as a transient step for concentrating a trace analyte above the detection limit. For instance, Tomlinson *et al.* [207–209] developed a unique method for loading, cleaning up, and separating peptides by using moving-boundary transient ITP conditions for membrane pre-concentration in small-diameter capillaries. This strategy worked very well for interfacing the CZE eluate with tandem MS for peptide sequencing. Plenty of other data on cITP can be found in a recent special issue of *Electrophoresis* [210].

9.6.5 Two-dimensional separations

Although this is not, strictly speaking, a *modus operandi* for CZE, we will deal with it here, since it bears some resemblance to the topics outlined above. In Sec. 9.6.2, we quoted Smith's work [183], displaying in a 2-D format the results of a cIEF separation, coupled to MS detection. This is, in fact, a typical example of coupling two procedures [211]. Yet, a true 2-D map, resembling the classical ones generated in proteome analysis [165], can also be generated in the CZE format, by coupling on-line liquid chromatography (LC) with CZE. The first comprehensive coupling of LC with CZE was achieved as early as 1990 by Bushey and Jorgenson [212]. For the first separation they utilized RP-LC, the LC column being 1-mm-ID and 25 cm long, operated under gradient conditions. The key part of the coupling was an electrically actuated six-port valve, fitted with a 10-µL loop. To be sure, this was not an easy job: peak elution volumes from the LC column were 20–40 µL, whereas the injection volumes for CZE capillary were of the order of nanoliters. Thus, only representative portions of the eluate could be transferred to the CZE capillary. The 2-D data were displayed as both a surface plot and a contour map. The surface plots allow for the intensities of the peaks to be seen, but not all peaks in the 2-D separation can be seen simultaneously. The contour plot allows the entire data set to be seen and is useful for protein fingerprinting or peptide mapping [213]. As there was a tremendous mismatch in volumes between these two dimensions, Lemmo and Jorgenson [214] resorted to microcolumns (*e.g.*, in the size-exclusion chromatography mode), in which the peak elution volumes are of the orders of a few hundred microliters, *i.e.* considerably closer to the order of volumes typically handled in CZE capillaries. Coupling of µRP-LC (micro-RP-LC) with CZE for the analysis of peptides was also demonstrated [215]. For high-sensitivity detection, the peptides were derivatized with tetramethylrodhamine isothiocyanate, which, having an absorption maximum at 548 nm, was an excellent match for LIF detection with the green helium/neon laser (543.5 nm). For those brave enough to venture in a multi-dimensional space, Jorgenson's group has also devised a three-dimensional separation scheme [216], in which SEC is coupled to RP-LC, which in turn is coupled to high-speed CZE. Their conclusion "the increased peak capacity of this system may not be worth the extra effort and added complexity that is entailed" [217] is seconded. It is hoped that this will slow down efforts by other groups to invent tetra-, penta- and higher orders of dimension in Separation Science!

9.7 MICELLAR ELECTROKINETIC CHROMATOGRAPHY

Micellar electrokinetic chromatography (variously abbreviated as MEC, MEKC, EKC), first reported by Terabe *et al.* [218], is now used for the separation of various compounds, because ionic and nonionic species can be separated at the same time, with high theoretical plate numbers. MEC is usually included in the field of electrophoresis, although the separation principle is more similar to RP-LC. In fact, whereas, in case of CZE, separation is based on differences in electrophoretic mobilities of analytes, in MEC it is mainly based on differences in distribution constants of analytes in the micellar pseudo-phase (Chap. 7). Since the partitioning into the micelles is determined by the strength of hydrophobic interactions between analytes and micelles, the migration order is similar to that of RPLC. Fig. 9.6 gives an example of the basic principle of MEC. The inner walls of the tube are

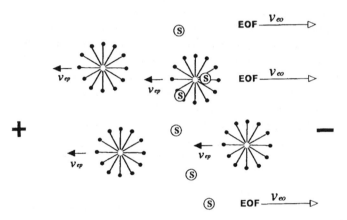

Fig. 9.6. Schematic drawing illustrating the separation principle of micellar electrokinetic chromatography with anionic surfactants (represented by a wheel-like motif). EOF = electroosmotic flow; S = analyte molecule (Reproduced from Ref. 224, with permission).

negatively charged, from weakly acidic to basic pH. Under these conditions, the EOF is in the direction of the negative electrode at the velocity v_{eo}. SDS, added to the solution above the critical micellar concentration, will migrate toward the positive electrode at the velocity v_{ep}. However, at pH above 5, SDS micelles will migrate toward the cathode at the net velocity $v_{mc} = (v_{eo} - v_{ep})$, since the absolute value of v_{eo} is larger than that of v_{ep} [219]. The migration time of an uncharged analyte will then depend on its distribution constant in the micellar phase. If it barely enters the micelle, it will migrate at v_{eo} with a migration time of t_0. If it is totally incorporated into the micelles, it migrates at their velocity (v_{mc}) with a migration time of t_{mc}. Examples of the former are methanol and acetonitrile, and of the latter are strongly hydrophobic azo dyes, such as Sudan III and Sudan IV. The difference ($t_{mc} - t_0$) is the time window available for the resolution of peaks in the electropherogram. There is a vast literature in the field of MEC, enough to fill up a few books on the topic. Here, we can only summarize some important references, such as reviews by Terabe's group [220–224] and Khaledi [49,225], and an entire book on the subject [226]. In addition, it should be appreciated that, in the last few years, not just pseudo-stationary phases, as in MEC, but true stationary phases, have been introduced in capillaries. These phases are in fact chromatographic beads (*e.g.*, reversed phases), used to pack the inner space of the capillary and then sealed inside with frits. Thus, a true chromatographic process takes place in the capillary, and EOF is typically used for eluting the bound substances [227]. Over the years, *Electrophoresis* has published a number of special issues (called paper symposia) under the general heading "Capillary Electrochromatography" and "Electrokinetic Capillary Chromatography", which give a broad view of the state of the art in the field [228–236], under the editorship of Z. El Rassi. This deluge does not seem to have subsided a bit. Thus, there is a strong suspicion that this prolific "guest editor" will strike again and again!

It would be impossible here to enter a deep review of the field. Suffice to say that MEC is based on pseudo-stationary phases that can be categorized into two general groups: the first and most widely used are charged micelles (*i.e.*, dynamic aggregates of charged surfactants); the other consists of covalently bonded or polymerized, charged, and organized assemblies. Among the first, are anionic (sodium dodecyl- or tetradecylsulfates; sodium dodecylsulfonate; lithium perfluorooctanesulfonate) and cationic (cetyltrimethylammonium bromide, cetyltrimethylammonium chloride, dodecyltrimethylammonium bromide) surfactants, as well as bile salts, such as cholate, deoxycholate, taurocholate, glycodcoxycholate, and taurodeoxycholate. The basic differences among these surfactants are in their micellar size, due to the number of monomers forming the micelle (called aggregation number). Cationic and anionic (as well as neutral and zwitter-ionic) surfactants form, in general, rather large micelles (quasi-spherical, with a diameter between 3 and 6 nm and aggregation numbers between 30 and 100). Bile salts, on the contrary, have a much smaller aggregation number (typically from 2 to 10), and form micelles presumed to have a helical structure. The second group (polymeric phases) is formed by covalently bonded, organized assemblies. Examples of this group are polymerized micelles (in which the monomer surfactants are covalently bonded together through a polymerization process), cascade macromolecules (dendrimers) and ionic polymers. They can be used in mixtures rich in organic solvents, and their primary application has been to the separation of highly hydrophobic compounds, requiring the addition of organic co-solvents.

We will offer here just an example of the tremendous separating power of MEC. Under standard MEC conditions, both the micelle and the surrounding aqueous phase migrate in the capillary. It had been hinted by Terabe that, if electro-osmosis were to be suppressed (in a well-coated capillary), the aqueous phase would remain in the capillary and the micelle would migrate through it. In other words, the only carrier of the analyte will be the micelle and not the concomitant EOF. Under these conditions, only analytes that effectively interact with the micelle will be carried past the detector (placed at the anodic end when using SDS). Although this separation mechanism was not implemented by Terabe, it was put into practice by Chiari *et al.* [237] in the separation of charged and neutral isotopes. These authors reported excellent separations of, *e.g.*, positively charged analytes, such as pyridine-h_5 and -d_5 or aniline-h_5 and -d_5, as well as of negatively charged compounds, such as benzoic-h_5 and -d_5 acids, either alone or in the presence of neutral (2% Nonidet-P40) or charged (50 mM SDS) surfactant micelles. In a way, such separations, although outstanding, could have been predicted, since oxygen isotopic

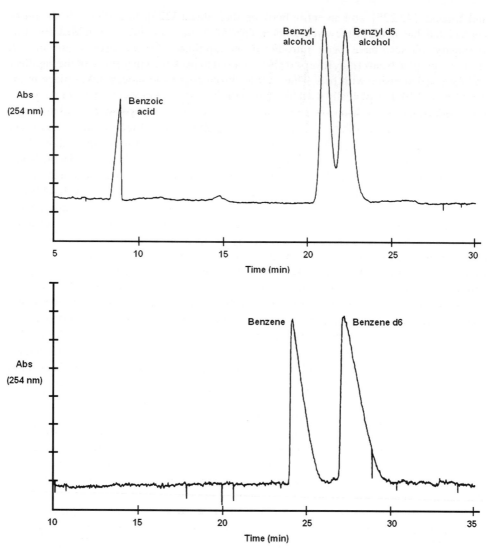

Fig. 9.7. Separation of (a) benzyl-h_5 and $-d_5$ alcohols and (b) of benzene-h_6 and $-d_6$ by Micellar Electrokinetic Chromatography. Conditions: coated capillary; background electrolyte, 50 mM MOPS (pH 7.2), containing 50 mM SDS; run, 13 kV at 48 μA towards the anode; sample injection, 10 sec., by hydrostatic pressure (Reproduced from Ref. 237, with permission).

benzoic acids had been already separated by Terabe *et al.* [238]. However, separations of neutral isotopic compounds had never been reported before. Figs. 9.7a and b give an example of such exquisite (and quite extraordinary) separations. Whereas, in the case of oxygen isotopic benzoic acids, the separation seems to be driven by the slightly different pK values of the weakly acidic carboxyl groups, due to oxygen replacement, the mechanism is not quite so obvious in the case of neutral compounds. Since such

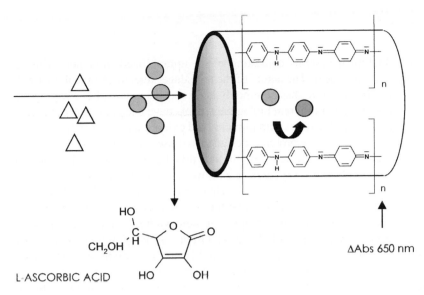

L-ASCORBIC ACID

ΔAbs 650 nm

Fig. 9.8. Example of a chemical biosensor relying upon a redox reaction over a polyaniline sensitive film. L-ascorbic acid molecules in an analyte mixture are entering the detection window of the capillary, where a layer of redox-sensitive PANI film is deposited (Reproduced from Ref. 249, with permission).

separations, as shown in Fig. 9.7, occur only in the presence of SDS micelles, it is reasonable to hypothesize that the deuterium substitution changes the affinity of such species for the SDS moiety by weakening it, since the first-eluted compound is, in all cases, the non-deuterated form.

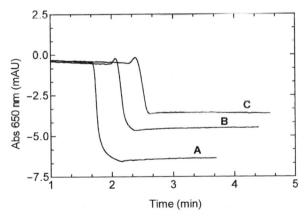

Time (min)

Fig. 9.9. Electropherograms of three different samples of ascorbic acid, obtained with the polyaniline/sensor: aniline/ammonium persulfate = 12/1. Riboflavin was used as internal standard (not shown) to correct for variations in time of migration and absorbance. Conditions: 50 mM aspartic acid buffer (pH 2.7), 20 kV, T = 25°C, λ = 650 nm. The concentrations of ascorbic acid injected were: (a) 75 mg/L; (b) 40 mg/L; (c) 15 mg/L (Reproduced from Ref. 249, with permission).

9.8 BIOSENSORS

Biosensors are analytical systems consisting of an immobilized biological sensing element and a physical transducer. The nature of the transducer (electrochemical, thermal, piezo-electric, optical, or magnetic) has commonly been used as a way of classifying the various devices. A wide range of different biological sensors have been developed, ranging from classical amperometric electrodes with immobilized oxidoreductases [239] to optical sensors, based on fluorescent-labeled proteins, designed to sense and report data concerning the dynamic distribution of specific reactants or reaction kinetics [240]. The features required by sensors are: high selectivity, high sensitivity, short analysis times, flexibility, reusability, and, in some cases, remote control. There are two major types of sensors: chemical, as reviewed by Janata *et al.* [241], and biological sensors, in which biological components, such as an enzyme, antibody, receptor, or their synthetic analogs, are devoted to the selective identification of analytes [242–244]. Often, a separation tool and a sensor element are brought together, as exemplified by liquid chromatography equipped with electrochemical detection [245], and, sometimes, additionally coupled with microdialysis sampling [246]. Even more complex assemblies, including the combination of liquid chromatography with a post-column enzymatic reactor, reflect the current trend in designing analytical instrumentation intended for the routine use in neuroscience and pharmacokinetics [247].

Although there appear to be quite a number of applications of biosensors coupled to column chromatography, their use in CZE has only recently begun [248]. Here, we will give an example of the application of a chemical biosensor, coupled to CZE [249], for the analysis of ascorbic acid. The capillary column was partially modified with a thin film of polyaniline (PANI) redox-sensitive material, deposited at the detection window. Ascorbic acid was detected by monitoring the changes in optical absorbance, occurring on the polyaniline film, following the reduction reaction. The sensor response (change in optical absorbance at 650 nm) was proportional to the concentration of ascorbic acid over a range of 2.5–250 mg/L, and the response range exhibited a clear dependence on the characteristics of the polymerized film. This type of sensor could find application not only in the food industry, but possibly also in the study of the relation between the content of L-ascorbic acid in body fluids and clinical parameters, such as cell ageing. Fig. 9.8 gives an example of the assembly of the chemical biosensor and of the reaction involved in ascorbic acid detection. Fig. 9.9 offers an example of the type of response obtained upon passage of the reducing agent over the PANI film.

9.9 CONCLUSIONS

This chapter contains a selection of old and new references that summarize the development of CZE, from early home-made instruments to the most sophisticated commercial equipment in use today, exploiting a battery of channels for parallel, massive operations, such as DNA sequencing. The accent is mostly on the fundamentals of the technique. Thus, it deals *in extenso* with basic concepts, such as silanol ionization and corresponding EOF, ways and means to control and/or suppress this flow, and buffer recipes and cocktails now in vogue for proper CZE analysis. The four basic modes of performing CZE are summarized as well, encompassing free CZE, CZE in sieving liquid polymers, CZE in the isoelectric focusing mode, and CZE under isotachophoretic conditions. The only applications dealt with in the present chapter are micellar electrokinetic chromatography and the use of biosensors in CZE. Topics, such as DNA analysis and sequencing, protein and peptide analysis are not covered since they are dealt with in Part B. It is nevertheless hoped that this chapter will convey to the readers the feeling of the unique *force de frappe* of the present CZE technique and of the tremendous impact it has had upon modern separation science technology in the last decade. In fact, whereas chromatographic techniques had already plateaued by the mid-1990s CZE was the only bright star rising over the horizon. It is felt that the CZE methodology is also plateauing now. So, perhaps we will have to look for a novel technique to accompany us into the third millennium.

ACKNOWLEDGMENTS

A.B. is supported by MURST 2001 (prot. 2001033797_003) and by MURST 2003 "Ingegneria di Sistemi di Separazione ad Elevate Prestazioni, Sensori ed Arrays".

REFERENCES

1   S. Hjertén, *Chromatogr. Rev.*, 9 (1967) 122.
2   R. Virtanen, *Acta Polytech. Scand.*, 123 (1974) 1.
3   F.M. Everaerts, J.L. Beckers and Th. P.E.M. Verheggen, *Isotachophoresis*, Elsevier, Amsterdam, 1976.
4   T. Jovin, *Biochemistry*, 12 (1972) 871; 879; 890
5   T. Jovin, *Ann. NY Acad. Sci.*, 209 (1973) 477.
6   V. Dolnick, K. Cobb and M. Novotny, *J. Microcol. Sep.*, 2 (1990) 127.
7   F. Foret, V. Sustacek and P. Bocek, *J. Microcol. Sep.*, 2 (1990) 229.
8   S. Hjertén, K. Elenbring, F. Kilar and J.L. Liao, *J. Chromatogr.*, 403 (1987) 47.
9   J.W. Jorgenson and K.D. Lukacs, *Anal. Chem.*, 53 (1981) 1298.
10 J.W. Jorgenson and K.D. Lukacs, *J. Chromatogr.*, 218 (1981) 209.
11 J.W. Jorgenson and K.D. Lukacs, *Science*, 222 (1983) 266.
12 R.D. Dandenau and E.H. Zerenner, *LC-GC*, 4 (1990) 10.
13 R.P. Oda and J.P. Landers, in J.P. Landers (Ed.), *Handbook of Capillary Electrophoresis*, CRC Press, Boca Raton, 1994, p. 9.
14 R.A. Mathies and X.C. Huang, *Nature*, 359 (1992) 167.
15 X.C. Huang, M.A. Quesada and R.A. Mathies, *Anal. Chem.*, 64 (1992) 967.
16 X.C. Huang, M.A. Quesada and R.A. Mathies, *Anal. Chem.*, 64 (1992) 2149.
17 H. Kambara and S. Takahashi, *Nature*, 361 (1993) 565.
18 S. Takahashi, K. Murakami, T. Anazawa and H. Kambara, *Anal. Chem.*, 66 (1994) 1021.
19 T. Anazawa, S. Takahashi and H. Kambara, *Anal. Chem.*, 68 (1996) 2699.
20 S. Bay, H. Strake, J.-Z. Zhang, L.D. Elliot and N.J. Dovichi, *J. Capillary Electrophor.*, 1 (1994) 121.
21 K. Ueno and E.S. Yeung, *Anal. Chem.*, 66 (1994) 1424.

22 E.S. Yeung and Q. Li, in M.G. Khaledi (Ed.), *High Performance Capillary Electrophoresis*, Wiley, New York, 1998, pp. 767.

23 M.A. Quesada and S.P. Zhang, *Electrophoresis*, 17 (1996) 1841.

24 I. Kheterpal, J.R. Scherer, S.M. Clark, A. Radhakrishnan, J. Ju, C.L. Ginther, G.F. Sensabaugh and R.A. Mathies, *Electrophoresis*, 17 (1996) 1852.

25 R.S. Madabushi, M. Vainer, V. Dolnik, S. Enad, D.L. Barker, D.W. Harris and E.S. Mansfield, *Electrophoresis*, 18 (1997) 104.

26 H.M. Pang, V. Pavski and E.S. Yeung, *J. Biochem. Biophys. Methods*, 41 (1999) 121.

27 V. Dolnik, D. Xu, A. Yadav, J. Bashkin, M. Marsh, O. Tu, E. Mansfield, M. Veiner, R. Madabushi, D. Barker and D. Harris, *J. Microcol. Sep.*, 10 (1998) 175.

28 Q. Li and E.S. Yeung, *Appl. Spectrosc.*, 49 (1995) 825.

29 T. Anazawa, S. Takahashi and H. Kambara, *Electrophoresis*, 20 (1999) 539.

30 M.A. Quesada, H.S. Dhadwal, D. Fisk and F.W. Studier, *Electrophoresis*, 19 (1998) 1415.

31 I.L. Medintz, B.M. Paegel, R.G. Blazej, C.A. Emrich, L. Berti, J.R. Scherer and R.A. Mathies, *Electrophoresis*, 22 (2001) 3845.

32 B.M. Paegel, C.A. Emrich, G.J. Wedemayer, J.R. Scherer and R.A. Mathies, *Proc. Natl Acad. Sci. USA*, 99 (2002) 574.

33 S. Behr, M. Matzig, A. Levin, H. Eickhoff and C. Heller, *Electrophoresis*, 20 (1999) 1492.

34 Y. Endo, C. Yoshida and Y. Baba, *J. Biochem. Biophys. Methods*, 41 (1999) 133.

35 S.C. Jacobson, R. Hergenroder, L.B. Koutny, R.J. Warmack and J.M. Ramsey, *Anal. Chem.*, 66 (1994) 1107.

36 D.C. Duffy, J.C. McDonald, O.J.A. Schueller and G.M. Whitesides, *Anal. Chem.*, 70 (1998) 4974.

37 C.S. Effenhauser, G.J.M. Bruin, A. Paulus and M. Ehrat, *Anal. Chem.*, 69 (1997) 3451.

38 S.M. Ford, B. Kar, S. McWhorter, J. Davies, S.A. Soper, M. Klopf, G. Calderon and V. Saile, *J. Microcol. Sep.*, 10 (1998) 413.

39 G.T.A. Kovacs, K. Petersen and M. Albin, *Anal. Chem.*, 68 (1996) 407A.

40 H.J. Ritchie, P. Ross and D.R. Woodward, *Int. Lab.*, 5 (1991) 54.

41 W.H. Wilson, H.M. McNair and K.J. Hyver, *J. Chromatogr.*, 540 (1991) 77.

42 C. Schwer and E. Kenndler, *Anal. Chem.*, 63 (1991) 1801.

43 T.L. Huang, P. Tsai, C.T. Wu and C.S. Lee, *Anal. Chem.*, 65 (1993) 1993.

44 M.S. Bello, L. Capelli and P.G. Righetti, *J. Chromatogr. A*, 684 (1994) 311.

45 R.M. McCormick, *Anal. Chem.*, 60 (1988) 2322.

46 B.J. Wanders, A.A.M. Van de Goor and F.M. Everaerts, *J. Chromatogr.*, 470 (1989) 89.

47 A.A.M. Van de Goor, B.J. Wanders and F.M. Everaerts, *J. Chromatogr.*, 470 (1989) 95.

48 W. Schützner and E. Kenndler, *Anal. Chem.*, 64 (1991) 1992.

49 M.G. Khaledi, in M.G. Khaledi (Ed.), *High Performance Capillary Electrophoresis*, Wiley, New York, 1998, pp. 77.

50 K.D. Altria and C.F. Simpson, *Chromatographia*, 24 (1987) 527.

51 G.J.M. Bruin, J.P. Chang, R.H. Kuhlman, K. Zegers, J.C. Kraak and H. Poppe, *J. Chromatogr.*, 471 (1989) 429.

52 W. Nashabeh and Z. El Rassi, *J. Chromatogr.*, 514 (1990) 57.

53 B.B. VanOrman, G.G. Liversidge, G.L. McIntire, T.M. Olefirowicz and A.G. Ewin, *J. Microcol. Sep.*, 2 (1990) 176.

54 E.S. Moring, in P.G. Righetti (Ed.), *Capillary Electrophoresis in Analytical Biotechnology*, CRC Press, Boca Raton, 1996, pp. 37.

55 I.Z. Atamna, H.J. Issaq, M. Muschik and G.M. Janini, *J. Chromatogr.*, 588 (1991) 315.

56 H.J. Issaq, I.Z. Atamna, M. Muschik and G.M. Janini, *Chromatographia*, 32 (1991) 155.

57 J.S. Green and J.W. Jorgenson, *J. Chromatogr.*, 478 (1989) 63.

58   P.G. Righetti, S. Magnusdottir, C. Gelfi and M. Perduca, *J. Chromatogr. A*, 920 (2001) 309.
59   A. Nahum and C. Horvath, *J. Chromatogr.*, 203 (1981) 53.
60   J.A. Bullock and L.C. Yuan, *J. Microcol. Sep.*, 3 (1991) 241.
61   A. Cifuentes, M.A. Rodriguez and F.J. Garcia-Montelongo, *J. Chromatogr. A*, 742 (1996) 257.
62   A. Cifuentes, M. de Frutos, J.M. Santos and J.C. Diez-Masa, *J. Chromatogr.*, 555 (1993) 63.
63   D. Corradini, A. Rhomberg and C. Corradini, *J. Chromatogr. A*, 661 (1994) 305.
64   D. Corradini, C. Cannarsa, E. Fabbri and C. Corradini, *J. Chromatogr. A*, 709 (1995) 127.
65   G.R. Paterson, J.P. Hill and D.E. Otter, *J. Chromatogr. A*, 700 (1995) 105.
66   M.E. Legaz and M.M. Pedrosa, *J. Chromatogr. A*, 719 (1996) 159.
67   N.A. Guzman, J. Moschera, K. Iqbal and W. Malick, *J. Chromatogr.*, 608 (1992) 197.
68   J. Bullock, *J. Chromatogr.*, 633 (1993) 235.
69   J.P. Landers, R.P. Oda, B.J. Madden and T.C. Spelsberg, *Anal. Biochem.*, 205 (1992) 115.
70   V. Rohlicek and Z. Deyl, *J. Chromatogr.*, 494 (1989) 87.
71   L. Song, Q. Ou and W. Wu, *J. Liq. Chromatogr.*, 17 (1994) 1953.
72   D. Corradini and G. Cannarsa, *Electrophoresis*, 16 (1995) 630.
73   R.P. Oda, B.J. Madden, T.C. Spelsberg and J.P. Landers, *J. Chromatogr. A*, 680 (1994) 82.
74   R.P. Oda and J.P. Landers, *Electrophoresis*, 17 (1996) 431.
75   F. Kalman, S. Ma, R.O. Fox and C. Horvath, *J. Chromatogr. A*, 705 (1994) 135.
76   A. Cifuentes, J.M. Santos, M. de Frutos and J.C. Diez-Masa, *J. Chromatogr.*, 652 (1993) 161.
77   Y.J. Yao and S.F.Y. Li, *J. Chromatogr. A*, 663 (1994) 97.
78   A. Cifuentes, H. Poppe and J.C. Kraak, *J. Chromatogr. B*, 681 (1996) 21.
79   J.K. Towns and F.E. Regnier, *J. Chromatogr.*, 516 (1990) 69.
80   N. Cohen and E. Grushka, *J. Capillary Electrophor.*, 1 (1994) 112.
81   B. Verzola, C. Gelfi and P.G. Righetti, *J. Chromatogr. A*, 868 (2000) 85.
82   J. Petersen and M. Merion, Eur. Patent No. 0494686A1, 1992
83   M. Zhu, R. Rodriguez, D. Hansen and T. Wher, *J. Chromatogr.*, 516 (1990) 123.
84   M.M. Bushey and J.W. Jorgenson, *J. Chromatogr.*, 480 (1989) 301.
85   F.A. Chen, L. Kelly, R. Palmieri, R. Biechler and H. Schwartz, *J. Liq. Chromatogr.*, 15 (1992) 1143.
86   P. Camilleri and G. Okafo, *J. Chromatogr.*, 541 (1991) 489.
87   S. Terabe, in N.A. Guzman (Ed.), *Capillary Electrophoresis Technology*, Dekker, New York, 1993, pp. 65.
88   S. Hjertén, *J. Chromatogr.*, 347 (1985) 191.
89   M. Chiari, C. Micheletti, M. Nesi, M. Fazio and P.G. Righetti, *Electrophoresis*, 15 (1994) 177.
90   C. Gelfi, M. Curcio, P.G. Righetti, A.R. Sebastiano, A. Citterio, H. Ahmadzadeh and N. Dovichi, *Electrophoresis*, 19 (1998) 1677.
91   S. Hjertén and K. Kubo, *Electrophoresis*, 14 (1993) 390.
92   M. Huang, J. Plocek and M.V. Novotny, *Electrophoresis*, 16 (1995) 396.
93   J.L. Liao, J. Abramson and S. Hjertén, *J. Capillary Electrophor.*, 2 (1995) 191.
94   J.T. Smith and Z. El Rassi, *Electrophoresis*, 14 (1993) 396.
95   M. Gilges, M.H. Kleemiss and G. Schomburg, *Anal. Chem.*, 66 (1994) 2038.
96   E. Simo-Alfonso, M. Conti, C. Gelfi and P.G. Righetti, *J. Chromatogr. A*, 689 (1995) 85.
97   J.K. Towns, J. Bao and F.E. Regnier, *J. Chromatogr.*, 599 (1992) 227.
98   Y. Liu, R. Fu and J. Gu, *J. Chromatogr. A*, 723 (1996) 157.
99   Y. Liu, R. Fu and J. Gu, *J. Chromatogr. A*, 694 (1995) 498.
100 X. Ren, Y. Shen and M.L. Lee, *J. Chromatogr. A*, 741 (1996) 115.
101 C.L. Ng, H.K. Lee and S.F.Y. Li, *J. Chromatogr. A*, 659 (1994) 427.
102 M. Iki and E.S. Yeung, *J. Chromatogr. A*, 731 (1996) 273.

103   M. Chiari, N. Dell'Orto and A. Gelain, *Anal. Chem.*, 668 (1996) 2731.

104   P. Porras, S.K. Wiedmer, S. Strandman, H. Tenhu and M.L. Riekkola, *Electrophoresis*, 22 (2001) 3805.

105   D. Belder, A. Deege, H. Husmann, F. Kohler and M. Ludwig, *Electrophoresis*, 22 (2001) 3813.

106   M. Chiari, M. Cretich, M. Stastna, S.P. Radko and A. Chrambach, *Electrophoresis*, 22 (2001) 656.

107   M.H.A. Busch, J.C. Kraak and H. Poppe, *J. Chromatogr. A*, 695 (1995) 287.

108   E. Drange and E. Lundanes, *J. Chromatogr. A*, 771 (1997) 301.

109   Y. Esaka, K. Yoshimura, M. Goto and K. Kano, *J. Chromatogr. A*, 822 (1998) 107.

110   X. Huang and C. Horvath, *J. Chromatogr. A*, 788 (1997) 155.

111   J. Horvath and V. Dolnik, *Electrophoresis*, 22 (2001) 644.

112   B. Verzola, C. Gelfi and P.G. Righetti, *J. Chromatogr. A*, 874 (2000) 293.

113   T. Tadey and W.C. Purdy, *J. Chromatogr.*, 652 (1993) 131.

114   M.R. Karim, S. Shinagawa and T. Takagi, *Electrophoresis*, 15 (1994) 1141.

115   A. Emmer, M. Jansson and J. Roeraade, *J. Chromatogr.*, 547 (1991) 544.

116   A. Emmer and J. Roeraade, *Electrophoresis*, 22 (2001) 660.

117   J.K. Town and F.E. Regnier, *Anal. Chem.*, 63 (1991) 1126.

118   C.L. Ng, H.K. Lee and S.F.Y. Li, *J. Chromatogr. A*, 659 (1994) 427.

119   C.P. Desilets, M.A. Rounds and F.E. Regnier, *J. Chromatogr.*, 544 (1991) 25.

120   X.W. Yao, D. Wu and F.E. Regnier, *J. Chromatogr.*, 636 (1993) 21.

121   L. Castelletti, B. Verzola, C. Gelfi, A. Stoyanov and P.G. Righetti, *J. Chromatogr. A*, 894 (2000) 281.

122   P.G. Righetti, C. Gelfi, B. Verzola and L. Castelletti, *Electrophoresis*, 22 (2001) 603.

123   K.E. Van Holde, W.C. Johnson and P.S. Ho, *Principles of Physical Biochemistry*, Prentice Hall, Upper Saddle River, 1998, pp. 9.

124   M. Chiari, M. Nesi and P.G. Righetti, in P.G. Righetti (Ed.), *Capillary Electrophoresis in Analytical Biotechnology*, CRC Press, Boca Raton, 1996, pp. 1.

125   S.A. Swedberg, *Anal. Biochem.*, 185 (1990) 51.

126   G.J.M. Bruin, R. Huischen, J.C. Kraak and H. Poppe, *J. Chromatogr.*, 480 (1989) 339.

127   A.M. Dougherty, C.I. Wooleey, D.L. Williams, D.F. Swaile, R.O. Cole and M.J. Sepaniak, *J. Liq. Chromatogr.*, 14 (1991) 907.

128   L. Capelli, S.V. Ermakov and P.G. Righetti, *J. Biochem. Biophys. Methods*, 32 (1996) 109.

129   S. Hjertén and M. Kiessling-Johansson, *J. Chromatogr.*, 550 (1991) 811.

130   K. Ganzler, K.S. Greve, A.S. Cohen and B.L. Karger, *Anal. Chem.*, 64 (1992) 2665.

131   K.A. Cobb, V. Dolnik and M. Novotny, *Anal. Chem.*, 62 (1990) 2478.

132   R. Sebastiano, C. Gelfi, P.G. Righetti and A. Citterio, *J. Chromatogr. A*, 894 (2000) 53.

133   R. Sebastiano, C. Gelfi, P.G. Righetti and A. Citterio, *J. Chromatogr. A*, 924 (2001) 71.

134   E. Olivieri, R. Sebastiano, A. Citterio, C. Gelfi and P.G. Righetti, *J. Chromatogr. A*, 894 (2000) 273.

135   C. Gelfi, A. Viganò, M. Ripamonti, P.G. Righetti, R. Sebastiano and A. Citterio, *Anal. Chem.*, 73 (2001) 3862.

136   E. Kenndler, in M.G. Khaledi (Ed.), *High Performance Capillary Electrophoresis*, Wiley, New York, 1998, pp. 25.

137   J.L. Miller and M.G. Khaledi, in M.G. Khaledi (Ed.), *High Performance Capillary Electrophoresis*, Wiley, New York, 1998, pp. 525.

138   M.L. Riekkola (Guest Ed.), Non-aqueous Capillary Electrophoresis, *Electrophoresis*, 23 (2002) 367

139   H. Rilbe, *pH and Buffer Theory*, Wiley, Chichester, 1996, p. 11.

140 A.D. Tran, S. Park, P.J. Lisi, O.T. Huynh, R.R. Ryall and P.A. Lane, *J. Chromatogr.*, 542 (1991) 459.

141 M. Van Druin, A. Peters, A.P.G. Kieboom and H. Van Bekkum, *Tetrahedron*, 41 (1985) 3411.

142 M. Stefanson and M. Novotny, *J. Am. Chem. Soc.*, 115 (1993) 11573.

143 I.Z. Atamna, C.J. Metral, M. Muschik and H.J. Issaq, *J. Liq. Chromatogr.*, 13 (1990) 2517.

144 R.A. Mosher, *Electrophoresis*, 11 (1990) 765.

145 B.M. Mitsyuk, *Russ. J. Inorg. Chem.*, 17 (1972) 471.

146 C. Chiesa and C. Horvath, *J. Chromatogr.*, 645 (1993) 337.

147 C. Gelfi, A. Viganò, M. Curcio, P.G. Righetti, S.C. Righetti, E. Corna and F. Zunino, *Electrophoresis*, 21 (2000) 785.

148 P.G. Righetti and F. Nembri, *J. Chromatogr. A*, 772 (1997) 203.

149 A. Bossi and P.G. Righetti, *Electrophoresis*, 18 (1997) 2012.

150 A. Bossi and P.G. Righetti, *J. Chromatogr. A*, 840 (1999) 117.

151 A. Bossi, E. Olivieri, L. Castelletti, C. Gelfi, M. Hamdan and P.G. Righetti, *J. Chromatogr. A*, 853 (1999) 71.

152 A.V. Stoyanov and P.G. Righetti, *J. Chromatogr. A*, 790 (1997) 169.

153 A.V. Stoyanov and P.G. Righetti, *Electrophoresis*, 19 (1998) 1674.

154 P.G. Righetti, C. Gelfi, M. Perego, A.V. Stoyanov and A. Bossi, *Electrophoresis*, 18 (1997) 2145.

155 P.G. Righetti, C. Gelfi, A. Bossi, E. Olivieri, L. Castelletti, B. Verzola and A. Stoyanov, *Electrophoresis*, 21 (2000) 4046.

156 S.M. Palfrey (Ed.), *Clinical Applications of Capillary Electrophoresis*, Humana Press, Totowa, 1999, pp. 1–250.

157 J.R. Petersen and A.M. Mohammad (Eds.), *Clinical and Forensic Applications of Capillary Electrophoresis*, Humana Press, Totowa, 2001, pp. 1–453.

158 J.P. Landers (Guest Ed.), Capillary and Gel Electrophoresis in Biomedicine, *Electrophoresis*, 21 (2000) 689–822

159 E. Verpoorte, J. Caslavska and W. Thormann (Guest Eds.), Chips in Clinical and Forensic Analysis, *Electrophoresis*, 23 (2002) 675–814

160 E. Jellum, H. Dollekamp and C. Blessum, *J. Chromatogr. B*, 683 (1996) 55.

161 L. Tao and R.T. Kennedy, *Anal. Chem.*, 68 (1996) 3899.

162 D.Y. Chen and N.J. Dovichi, *Anal. Chem.*, 68 (1996) 690.

163 P.G. Righetti, *Isoelectric Focusing: Theory, Methodology and Applications*, Elsevier, Amsterdam, 1983, pp. 1–386.

164 P.G. Righetti, *Immobilized pH Gradients: Theory and Methodology*, Elsevier, Amsterdam, 1990, pp. 1–397.

165 P.G. Righetti, A. Stoyanov and M. Zhukov, *The Proteome Revisited: Theory and Practice of the Relevant Electrophoretic Steps*, Elsevier, Amsterdam, 2001, pp. 1–400.

166 X. Liu, Z. Sosic and I.S. Krull, *J. Chromatogr. A*, 735 (1996) 165.

167 T. Pritchett, *Electrophoresis*, 17 (1996) 1195.

168 P.G. Righetti, C. Gelfi and M. Conti, *J. Chromatogr. B*, 699 (1997) 91.

169 P.G. Righetti, A. Bossi and C. Gelfi, *J. Capillary Electrophor.*, 4 (1997) 47.

170 P.G. Righetti and A. Bossi, *Anal. Chim. Acta*, 372 (1998) 1.

171 R. Rodriguez-Diaz, T. Wher and M. Zhu, *Electrophoresis*, 18 (1997) 2134.

172 M.A. Strege and L. Lagu, *Electrophoresis*, 18 (1997) 2343.

173 M. Taverna, N.T. Tran, T. Merry, E. Horvath and D. Ferrier, *Electrophoresis*, 19 (1998) 2572.

174 S. Hjertén, in P.D. Grosman and J.C. Colburn (Eds.), *Capillary Electrophoresis: Theory and Practice*, Academic Press, San Diego, 1992, pp. 191.

175 K. Shimura, *Electrophoresis*, 23 (2002) 3847.

176   F. Kilar, in J.P. Landers (Ed.), *Handbook of Capillary Electrophoresis*, CRC Press, Boca Raton, 1994, pp. 95–110.
177   T. Wher, M. Zhu and R. Rodriguez-Diaz, *Methods Enzymol.*, 270 (1996) 358.
178   J.R. Mazzeo and I.S. Krull, in N.A. Guzman (Ed.), *Capillary Electrophoresis Technology*, Dekker, New York, 1993, pp. 795–818.
179   X. Fang, C. Tragas, J. Wu, Q. Mao and J. Pawliszyn, *Electrophoresis*, 19 (1998) 2290.
180   L. Steinman, R. Mosher and W. Thormann, *J. Chromatogr. A*, 756 (1996) 219.
181   M. Conti, C. Gelfi and P.G. Righetti, *Electrophoresis*, 16 (1995) 1485.
182   M. Conti, C. Gelfi, A. Bianchi-Bosisio and P.G. Righetti, *Electrophoresis*, 17 (1996) 1590.
183   R.D. Smith, L. Pasa-Tolic, M.S. Lipton, P.K. Jensen, G.A. Anderson, Y. Shen, T.P. Conrads, H.R. Udseth, R. Harkewitcz, M.E. Belov, C. Masselon and T.D. Veenstra, *Electrophoresis*, 22 (2001) 1652.
184   P.K. Jensen, L. Pasa-Tolic, G.A. Anderson, J.A. Horner, M.S. Lipton, J.E. Bruce and R.D. Smith, *Anal. Chem.*, 71 (1999) 2076.
185   N. Stellwagen, C. Gelfi and P.G. Righetti, *Biopolymers*, 42 (1997) 687.
186   R. Pitt-Rivers and F.S.A. Impiombato, *Biochem. J.*, 109 (1968) 825.
187   A.S. Cohen, D.R. Najarian, A. Paulus, A. Guttman, J.A. Smith and B.L. Karger, *Proc. Natl Acad. Sci.*, 84 (1988) 9660.
188   D.N. Heiger, A.S. Cohen and B.L. Karger, *J. Chromatogr.*, 516 (1990) 33.
189   H.J. Bode, *Anal. Biochem.*, 83 (1977) 204.
190   P.G. Righetti and C. Gelfi, in P.G. Righetti (Ed.), *Capillary Electrophoresis in Analytical Biotechnology*, CRC Press, Boca Raton, 1996, pp. 431–476.
191   J.L. Viovy and C. Heller, in P.G. Righetti (Ed.), *Capillary Electrophoresis in Analytical Biotechnology*, CRC Press, Boca Raton, 1996, pp. 477–508.
192   C. Heller, *Electrophoresis*, 19 (1998) 3114.
193   C. Heller, *Electrophoresis*, 20 (1999) 1962.
194   C. Heller, *Electrophoresis*, 20 (1999) 1978.
195   C. Heller, *Electrophoresis*, 21 (2000) 593.
196   V. Dolnik, W.A. Gurske and A. Padua, *Electrophoresis*, 21 (2001) 692.
197   M. Chiari, M. Nesi and P.G. Righetti, *Electrophoresis*, 15 (1994) 616.
198   P. Shieh, N. Cooke and A. Guttman, in M.G. Khaledi (Ed.), *High Performance Capillary Electrophoresis*, Wiley, New York, 1998, pp. 185–221.
199   A. Guttman, J. Horvath and N. Cooke, *Anal. Chem.*, 65 (1993) 199.
200   K.B. Tomer, L.J. Deterding and C.E. Parker, in M.G. Khaledi (Ed.), *High Performance Capillary Electrophoresis*, Wiley, New York, 1998, pp. 405–448.
201   J.V. Sweedler, R. Fuller, S. Tracht, A.T. Timperman, V. Tma and K. Khatib, *J. Microcol. Sep.*, 5 (1993) 403.
202   G.M. McLaughlin and K.W. Anderson, in M.G. Khaledi (Ed.), *High Performance Capillary Electrophoresis*, Wiley, New York, 1998, pp. 637–681.
203   F.E. Regnier and S. Lin, in M.G. Khaledi (Ed.), *High Performance Capillary Electrophoresis*, Wiley, New York, 1998, pp. 683–728.
204   L. Krivankovà and P. Bocek, in M.G. Khaledi (Ed.), *High Performance Capillary Electrophoresis*, Wiley, New York, 1998, pp. 251–275.
205   L. Ornstein, *Ann. NY Acad. Sci.*, 121 (1964) 321.
206   B.J. Davis, *Ann. NY Acad. Sci.*, 121 (1964) 404.
207   A.J. Tomlinson, L.M. Benson, S. Jameson and S. Naylor, *Electrophoresis*, 17 (1996) 1801.
208   A.J. Tomlinson, S. Jameson and S. Naylor, *J. Chromatogr. A*, 744 (1996) 273.
209   A.J. Tomlinson, S. Jameson and S. Naylor, *Biomed. Chromatogr.*, 10 (1996) 325.

210 L. Krivankova and P. Bocek (Guest Eds.), Isotachophoresis and Stacking Phenomena, *Electrophoresis*, 21 (2000) 2745–3070

211 U.A.Th. Brinkman (Ed.), Hyphenation: Hype and Fascination, *J. Chromatogr. A*, 856 (1999) 1–532

212 M.M. Bushey and J.W. Jorgenson, *Anal. Chem.*, 62 (1990) 978.

213 M.M. Bushey and J.W. Jorgenson, *J. Microcol. Sep.*, 2 (1990) 293.

214 A.V. Lemmo and J.W. Jorgenson, *J. Chromatogr.*, 633 (1993) 213.

215 J.P. Larmamm, Jr., A.V. Lemmo, A.W. Moore Jr. and J.W. Jorgenson, *Electrophoresis*, 14 (1993) 439.

216 A.W. Moore, Jr. and J.W. Jorgenson, *Anal. Chem.*, 67 (1995) 3456.

217 T.S. Hooker, D.J. Jeffery and J.W. Jorgenson, in M.G. Khaledi (Ed.), *High Performance Capillary Electrophoresis*, Wiley, New York, 1998, pp. 581–612.

218 S. Terabe, K. Otsuka, K. Ichikawa, A. Tsuchiya and T. Ando, *Anal. Chem.*, 56 (1984) 111.

219 K. Otsuka and S. Terabe, *J. Microcol. Sep.*, 1 (1989) 150.

220 H. Nishi and S. Terabe, *Electrophoresis*, 11 (1990) 691.

221 H. Nishi and S. Terabe, *J. Pharm. Biomed. Anal.*, 11 (1993) 1277.

222 S. Terabe, in N.A. Guzman (Ed.), *Capillary Electrophoresis Technology*, Dekker, New York, 1993, pp. 65–87.

223 S. Terabe, N. Chen and K. Otsuka, in A. Chrambach, M.J. Dunn and B.J. Radola (Eds.), *Advances in Electrophoresis*, VCH, Weinheim, 1994, pp. 87–107.

224 N. Matsubara and S. Terabe, in P.G. Righetti (Ed.), *Capillary Electrophoresis in Analytical Biotechnology*, CRC Press, Boca Raton, 1996, pp. 155–182.

225 M.G. Khaledi, in J.P. Landers (Ed.), *Handbook of Capillary Electrophoresis*, CRC Press, Boca Raton, 1993, pp. 43–93.

226 J. Vindevogel and P. Sandra, *Introduction to Micellar Eelctrokinetic Chromatography*, Hüthig, Heidelberg, 1992.

227 Z. Deyl and F. Svec (Eds.), *Capillary Electrochromatography*, Elsevier, Amsterdam, 2001, pp. 1–400.

228 Z. El Rassi (Guest Ed.), Capillary Electrochromatography and Electrokinetic Capillary Chromatography, *Electrophoresis*, 20 (1999) 1–220

229 Z. El Rassi (Guest Ed.), Capillary Electrophoresis and Electrochromatography Innovations, *Electrophoresis*, 20 (1999) 2311–2567.

230 Z. El Rassi (Guest Ed.), Capillary Electrophoresis and Electrochromatography Reviews, *Electrophoresis*, 20 (1999) 2987–3330.

231 M.J. Sepaniak (Guest Ed.), Detection in Capillary Electrophoresis and Electrochromatography, *Electrophoresis*, 21 (2000) 1237–1434

232 Z. El Rassi (Guest Ed.), Capillary Electrophoresis and Electrochromatography Innovations, *Electrophoresis*, 21 (2000) 3071–3328.

233 Z. El Rassi (Guest Ed.), Capillary Electrophoresis and Electrochromatography Reviews: Theory and Methodology, *Electrophoresis*, 21 (2000) 3871–4192.

234 Z. El Rassi (Guest Ed.), Capillary Electrochromatography and Electrokinetic Capillary Chromatography, *Electrophoresis*, 22 (2001) 1249–1444.

235 Z. El Rassi (Guest Ed.), Capillary Electrophoresis and Electrochromatography Innovations, *Electrophoresis*, 22 (2001) 2357–2630.

236 Z. El Rassi (Guest Ed.), Capillary Electrophoresis and Electrochromatography Reviews: Applications, *Electrophoresis*, 22 (2001) 4033–4294.

237 M. Chiari, M. Nesi, G. Ottolina and P.G. Righetti, *J. Chromatogr. A*, 680 (1994) 571.

238 S. Terabe, T. Yashima, N. Tanaka and M. Araki, *Anal. Chem.*, 60 (1988) 1673.

239 J.H.T. Luong, A. Mulchandani and G.G. Guilbault, *Trends Biotechnol.*, 12 (1988) 310.

240   K.A. Giuliano and D.L. Taylor, *Trends Biotechnol.*, 16 (1998) 135.

241   J. Janata, M. Josowicz and D.M. Devaney, *Anal. Chem.*, 66 (1994) 207R.

242   A.P.F. Turner, I. Karube and G.S. Wilson (Eds.), *Biosensors: Fundamentals and Applications*, Oxford University Press, Oxford, 1987.

243   A.E.G. Cass, *Biosensors: A Practical Approach*, IRL Press, Oxford, 1990.

244   K. Mosbach and O. Ramström, *Biotechnology*, 14 (1996) 163.

245   P.T. Kissinger, *Electroanalysis*, 4 (1992) 359.

246   M.C. Linhares and P.T. Kissinger, *Pharm. Res.*, 10 (1993) 598.

247   T. Huang, L. Yang, J. Gitzen, P.T. Kissinger, M. Vreeke and A. Heller, *J. Chromatogr. B*, 670 (1995) 323.

248   A. Bossi, S.A. Piletsky, P.G. Righetti and A.P.F. Turner, *J. Chromatogr. A*, 892 (2000) 143.

249   L. Castelletti, S.A. Piletsky, A.P.F. Turner, P.G. Righetti and A. Bossi, *Electrophoresis*, 23 (2002) 209.

Erich Heftmann (Editor)
Chromatography, 6th edition
Journal of Chromatography Library, Vol. 69A
403

Chapter 10

Combined techniques

W.M.A. NIESSEN

CONTENTS

10.1 Introduction 404
10.2 Coupled columns 405
    10.2.1 Two-dimensional gas chromatography 405
        10.2.1.1 Principles and instrumentation 405
        10.2.1.2 Selected applications 407
    10.2.2 Liquid chromatography–gas chromatography 407
        10.2.2.1 Principle and instrumentation 408
        10.2.2.2 Selected applications 409
        10.2.2.3 On-line solid-phase extraction–gas chromatography . . 409
    10.2.3 Coupled-column liquid chromatography 410
        10.2.3.1 Principles and instrumentation 410
        10.2.3.2 Selected applications 411
        10.2.3.3 On-line solid-phase extraction–liquid
                chromatography 411
10.3 Chromatography–spectrometry 413
    10.3.1 Gas chromatography–mass spectrometry 413
    10.3.2 Gas chromatography–atomic emission detection 414
    10.3.3 Gas chromatography–Fourier-transform infrared spectroscopy . . 415
    10.3.4 Liquid chromatography–photodiode-array spectrometry 416
    10.3.5 Liquid chromatography–inductively coupled plasma
        mass spectrometry 416
    10.3.6 Liquid chromatography–Fourier-transform infrared
        spectroscopy 417
    10.3.7 Liquid chromatography–nuclear magnetic resonance
        spectroscopy 417
10.4 Liquid chromatography–mass spectrometry 418
    10.4.1 Principles and instrumentation 418
        10.4.1.1 Atmospheric-pressure ionization 418
        10.4.1.2 Developments in mass spectrometers 419

10.4.2   Applications 420
        10.4.2.1   Pharmaceutical applications   421
        10.4.2.2   Quantitative biochemical analysis   422
        10.4.2.3   Environmental applications 423
        10.4.2.4   Biochemical applications 423
10.4.3   Perspective   424
References 425

10.1 INTRODUCTION

Coupled, combined, or hyphenated chromatographic techniques are becoming increasingly important in analytical chemistry. These terms refer to techniques where chromatography is coupled either to a (scanning) spectrometric technique [*e.g.*, gas chromatography (GC)–mass spectrometry (MS) (GC–MS)], or to another chromatographic technique [*e.g.*, liquid chromatography (LC)–GC]. In the latter case, the term multi-dimensional chromatography is sometimes applied [1]. Although the term "hyphenated" is stylistically dubious, it has been widely used in this context. The scopes of coupled-column techniques on one hand and hyphenated chromatography– spectrometry are partly overlapping, they are also partly quite distinct. Coupled-column techniques take advantage of the probability that the use of different chromatographic modes improves selectivity and separation. In hyphenated chromatography–spectrometry, the rationale is to improve the information content with the detection step. The selectivity inherent in an improved information content supplements that of the separation technique.

The emergence of each new hyphenated technique raises the question whether the two techniques should be used on-line or combined off-line [2]. For instance, the initial difficulties in achieving efficient and robust LC–MS coupling prompted the use of fraction collection and subsequent mass analysis for many years. The possibilities of automation, of the reduction of artifacts and sample losses, and of improved efficiency in most cases turned opinions in favor of the on-line approaches. Returning to LC–MS as an example, at the present time the most efficient way of collecting fractions for manual MS would be LC–MS-directed fraction collection. Often, the on-line coupling of two chromatographic techniques or of a chromatographic and a spectrometric technique requires some kind of interface. Some general properties for such interfaces have been established [3]. Ideally, an interface should not in any way restrict or impair the operational parameters of the two techniques applied, *e.g.*, in terms of mobile-phase composition, flow-rate, selection of measurement conditions, detection limits and/or quantification capabilities. In addition, the interface should have a very high analyte transfer efficiency and should not introduce, or at least minimize, additional peak-broadening or losses in chromatographic resolution, and should not lead to uncontrolled chemical modifications of the analytes. Finally, the interface should be reasonably priced and easy to operate. This chapter provides a brief overview of the various coupled-column and chromatography–spectrometry techniques available, highlighting some of the areas where these technologies are applied. Subsequently, on-line LC–MS is discussed in somewhat more detail. LC–MS has been in the focus of attention for many years, and may be considered to be the success story of hyphenated techniques.

10.2 COUPLED COLUMNS

Two major objectives of coupled-column chromatography (multi-dimensional chromatography) are the improvement of selectivity in separation, *e.g.*, the creation of very high peak capacities in order to resolve a large number of components, and the analysis of one particular (group of) compound(s) in a complex matrix. The separation of all or selected groups of sample components is repeated in two or more columns of preferentially different properties (*e.g.*, polarity), which are coupled in series. The theory of multi-dimensional separation techniques aimed at increasing peak capacity has been treated by Giddings *et al.* [1,4,5]. They demonstrated that the practical effectiveness of high peak capacities is limited by a disordered distribution of component peaks. The general objectives of multi-dimensional separations are:

  (a) to increase the peak capacity in the analysis of complex samples,
  (b) to achieve high-resolution separation of isomers and enantiomers by using chiral selectivity in the second column,
  (c) to shorten analysis time and/or improve selectivity by analysis of selected cuts from a (fast) pre-separation of complex mixtures, and
  (d) to improve the determination of trace components eluted close to major components by heart-cutting.

10.2.1 Two-dimensional gas chromatography

Two-dimensional GC was pioneered by Deans [6], who first demonstrated the heart-cut method for resolving closely eluted components present in extremely disproportionate concentrations, and by Schomburg [7–9], who demonstrated an enhanced separation of complex mixtures *via* selected cuts. In comprehensive two-dimensional GC, pioneered by Phillips [10,11], a complex sample is analyzed completely in two dimensions of GC, using a combination of conventional GC and high-speed GC [12].

10.2.1.1 Principles and instrumentation

In conventional two-dimensional GC (GC–GC), two columns with different stationary phases are coupled in series. In some applications, especially in trace analysis, a packed column is preferred in the first separation step, because of its higher sample capacity. Preferably, each column is operated in a separate column oven. An intermediate detector is required for method development and optimization. An intermediate injector can be useful in some applications as well, *e.g.*, with highly diluted samples. In that case, the analytes are strongly retarded at the head of the first column, while the more volatile solvent is eluted. The analyte zone is subsequently transferred to the second column by back-flushing the first column. No valves should be used for switching the carrier gas to transfer cuts from one column to another, because sample components could come in contact with movable parts having potentially reactive surfaces at elevated temperatures. Miniaturized dead-volume coupling devices are required, especially for coupling narrow-bore open-tubular columns. Devices can be installed between the two columns for trapping and/or cryofocusing the components in the eluate fraction to be transferred. Without intermediate trapping, the components transferred would enter the second column with different starting times, determined by the selectivity of the first column, and zones would be broadened due to their transport through the first column.

In comprehensive two-dimensional GC (GC × GC), the effluent of the first column is separated into a large number of small fractions, and each of these is subsequently separated on the second column. The first column contains a film of nonpolar stationary phase in order to produce relatively broad peaks, while the second column of typically 100-μm ID contains a more polar stationary phase, *e.g.*, OV-1701 (14% cyanopropyl-phenyl in dimethylpolysiloxane). A modulator interface between the columns is required for the method to become two-dimensional. The modulator may be either thermal or cryogenic. Directly or indirectly heated capillaries have been used as thermal modulators [10,13,14]. The moving-heater modulator is a device for moving a separate, high-capacity heating element over the modulator capillary to heat locally a section of the modulator capillary and to pulse solute to the second column at a pulse width as short at 60 msec. Because a temperature difference of 100°C is required between retention in the modulator and desorption for re-injection, the maximum allowable oven temperature for the first column is lowered by 100°C when this device is used. In the longitudinal modulating cryogenic system or moving cryogenic trap [15] the modulator is shifted every few seconds from trap to release position (Fig. 10.1). In the release position, trapped fractions will be desorbed and separation on the second column can start immediately, while at the some time, the effluent from the second column is trapped. The jet-type cryogenic modulator [16] consists of two cold and two hot jets, mounted orthogonally at such a distance that a 6-sec modulation time is achieved. For comprehensive GC × GC, a fast-responding detection system is required (sampling rates in excess of 100 Hz). Flame-ionization detectors and time-of-flight MS have been used in most applications.

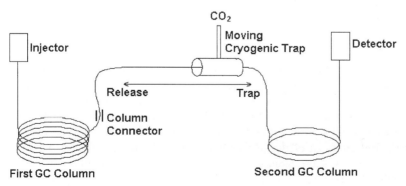

Fig. 10.1. Schematic diagram of a moving cryogenic trap for GC × GC.

10.2.1.2 Selected applications

Conventional GC–GC and comprehensive GC × GC occupy overlapping application areas, GC–GC being focused on specific areas of the chromatogram and GC × GC providing a more complete picture. Typical applications are in petroleum research, *e.g.*, the separation of coal-derived gasoline fractions [8,9], and the separation of benzene, toluene, ethylbenzene, and xylenes (BTEX) and total aromatics in gasoline [17], and in environmental analysis, *e.g.*, the separation of polychlorobiphenyls (PCB), polychloro-dibenzodioxins (PCDD), and/or polychlorodibenzofuran (PCDF) isomers [18], and the identification of pesticides in food extract [19]. As an example, a detail of the GC × GC chromatogram of a pesticide-spiked celeriac extract, acquired by full-scan TOF-MS, is shown in Fig. 10.2 [19]. Applications in the field of flavor and essential oil analysis have

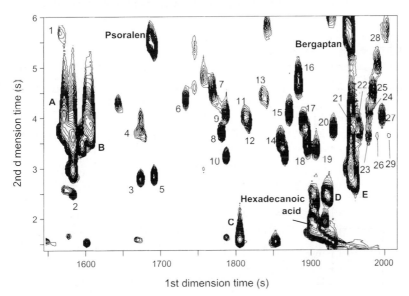

Fig. 10.2. Detail of the comprehensive GC × GC chromatogram, acquired with a time-of-flight mass spectrometer, of a plant extract, spiked with pesticides at a level of 0.32 mg/kg. (Reprinted from Ref. 19 with permission.)

been published as well, *e.g.*, the identification of methyl jasmonate stereoisomers in jasmine extracts [20]. More recently, the detailed analysis of cigarette smoke was described [16].

10.2.2 Liquid chromatography–gas chromatography

On-line liquid chromatography–gas chromatography (LC–GC), based on transfer of LC fractions *via* an autosampler, was first described by Majors in 1980 [21]. LC as a separation and sample pre-treatment technique is considered superior to other sample preparation techniques, *e.g.*, those involving SPE cartridges. Automated, on-line coupling of LC to GC eliminates manual sample preparation and, thus, the disadvantages of off-line operations, *i.e.*, sample loss, contamination, and/or artifact formation.

10.2.2.1 Principle and instrumentation

LC–GC entails the transfer of a LC fraction, comprising a whole LC peak or even a range of peaks, to a gas chromatograph. Since the volume of such fractions is typically in the range of 100 to 1000 µL, the development of large-volume introduction techniques for GC is the key to LC–GC. Since 1980, a variety of approaches to LC–GC have been proposed [22–24]. Two of them are still widely used: concurrent eluent evaporation with the loop-type interface and the retention-gap technique with the on-column interface (Fig. 10.3). Alternatively, a programmed-temperature vaporizer injector (PTV) can be applied.

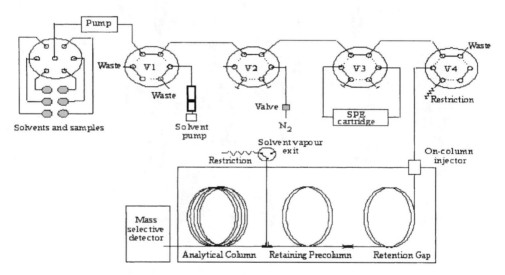

Fig. 10.3. Schematic diagram of a system for SPE/GC/MS. (Reproduced from Ref. 62 with permission.)

A retention gap [25] consists of an uncoated, deactivated, fused-silica capillary (15 m × 0.32 mm ID), installed ahead of the coated capillary column. The sample from the LC is injected into the retention gap at a temperature below the boiling point of the solvent. If the injected liquid can form a solvent film on the inner wall of the capillary, a flooded zone is formed, where analytes spread out. This precludes the use of most aqueous mobile phases. The solvent film evaporates from the rear to the front, and volatile analytes are re-concentrated by solvent-trapping effects. In addition, phase soaking at the top of the column affects re-concentration of analytes, because their retention is increased by the swelling of the stationary-phase layer. Less volatile components remain spread out over the retention gap, but they are re-concentrated upon entering the coated column. The retention gap allows volumes up to 100 μL to be introduced and is therefore compatible with microbore LC columns (0.25–1 mm ID). To protect the GC detector and to accelerate solvent evaporation, a solvent vapor exit is installed between the retention gap and the GC capillary column.

In concurrent eluent evaporation [26], the eluent is completely volatilized upon introduction into the GC pre-column at a temperature higher than the boiling point of the solvent. Volatile components are lost with the solvent by evaporation. This technique enables the introduction of relatively large volumes (≥1 mL), but it is only applicable to analytes with boiling points 80–100°C above the introduction temperature. Modifications of this approach, *i.e.*, partial concurrent solvent evaporation [27] or co-solvent trapping techniques [28], can be applied to avoid the loss of volatile sample components. A programmed-temperature vaporizer injector (PTV) is another coupling method applicable to LC–GC. Basically, a PTV is a split–splitless injector with a packed liner, which can be rapidly heated or cooled [29]. The sample is injected in the liner with an open split exit at an injector temperature below the boiling point of the solvent. Again, volatile components are lost with the solvent by evaporation. Subsequently, the analytes retained in the liner are transferred to the GC column. Alternatively, injection is performed with a closed split exit, and the solvent vapor is either vented *via* the GC column, where volatile sample components are trapped in the swollen stationary phase, or vented *via* an open septum purge.

10.2.2.2 Selected applications

LC–GC applications have usually involved normal-phase LC on silica, with an eluent of pentane or hexane plus a few percent of a more polar solvent. Applications have been reported for the petrochemical and food industry. A typical food-related example is the determination of sterols and squalene in edible oils in order to detect whether oils or fats have been refined or subjected to other thermal stress [30]. Petrochemical applications include the characterization of coal liquids and fuels, and the determination of PAH in various petroleum fractions and fuels. The complete quantitative characterization of oil fractions in the middle distillate range may serve as an example [31]. The on-line coupling of reversed-phase LC to GC *via* a retention gap or (partial) concurrent solvent evaporation is difficult. A PTV-type system consisting of a packed liner, in combination with a retaining pre-column and a solvent–vapor exit, was described and successfully applied to the determination of phthalates in water and the determination of pesticides in red wines [32,33].

10.2.2.3 On-line solid-phase extraction–gas chromatography

Various approaches to the on-line sample pre-treatment in combination with GC are technologically related to LC–GC coupling. Loop-type injections, in combination with concurrent solvent evaporation, have been used for on-line liquid–liquid extraction–GC in the analysis of hydrocarbons [34] and hexachlorocyclohexanes [35] in water. In a system for on-line solid-phase extraction (SPE), coupled to GC, the (large-volume) sample is injected into a short cartridge column (2–10 mm long and 1–4.6 mm ID). After clean-up, the pre-column is eluted with 50–100 μL of an organic solvent. The latter is introduced into the gas chromatograph *via* a large-volume introduction technique, previously described for LC–GC [35–37]. A schematic diagram of a SPE–GC–MS system is shown in Fig. 10.3. The liquid effluent of the SPE cartridge is injected into a retention gap, which is coupled to the analytical column *via* a retaining pre-column. SPE–GC has been optimized over the years. It is frequently used in environmental applications, *e.g.*, the determination of organic micropollutants in surface waters [38,39].

10.2.3 Coupled-column liquid chromatography

Coupled-column liquid chromatography (CCLC) is essentially based on the transfer of a heart-cut from a first LC column on to a second LC column [40]. Multi-dimensional chromatography with different phase systems, offering increased selectivity, seems attractive, but it is often complicated by mobile-phase compatibility problems. The application of CCLC is to some extent determined by the separating power of the first column. A poor resolution in the first step will result in the transfer of a multi-component fraction in the heart-cut, thus requiring complicated separation in the second step, whereas a high resolution in the first column will result in the transfer of a heart-cut containing just one or only a few closely related analytes [41,42]. Because separation times in LC are relatively long compared to GC, comprehensive two-dimensional LC (LC × LC) has not been used frequently. This calls for either a very low flow-rate in the first column or a stop-flow technique [43].

10.2.3.1 Principles and instrumentation

In CCLC, two LC columns are coupled *via* a valve-switching system. Part of the effluent from the first column is transferred to the second column. When only a small volume of column effluent from the first column must be transferred, direct transfer is permissible. Often, either an intermediate sample collection loop or a short trapping column is applied. A diagram of a simple CCLC system is shown in Fig. 10.4. The trapping column in most cases is eluted to the second column in back-flush mode.

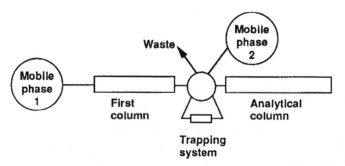

Fig. 10.4. Schematic diagram of a typical setup for coupled-column liquid chromatography.

10.2.3.2 Selected applications

A wide variety of applications of coupled-column chromatography have been described. A number of applications of reversed-phase LC coupled to reversed-phase LC have been reviewed [42]. The analysis of the very polar ethylenethiourea in ground water [44] may serve as an example. Immunoaffinity chromatography has also been exploited as the first separation or sample clean-up step. After initial reports, *e.g.*, on the determination of estrogens in plasma and urine and of aflatoxins in milk [45,46], this technique was applied to the selective determination of trace levels of pesticides, (*e.g.*, of carbofuran [47] and carbendazim [48]) in environmental samples, of steroidal estrogens in wastewaters [49], and of etheno-2′-deoxycytidine DNA adducts [50]. These applications are frequently supplemented with MS.

Chiral separations are another field of application for CCLC, in which a chiral and an achiral column are combined. If the enantiomers are well resolved on the chiral column, it is used as the first column. In the second step, analyte compression resulting from the better column efficiency of the achiral column compared to the chiral column is exploited, and separation from some interfering material can be achieved. An example of this technique is the determination of enantiomers of metoprolol, terbutaline, bupivacaine, and oxazepam in plasma [51]. If a near-baseline separation is achieved on the chiral column, the preliminary separation of the enantiomers from interfering compounds is performed on an achiral first column. The fraction containing the enantiomers is then transferred to the chiral column for final separation. An example of the second method is the quantitative analysis of enantiomers of chlorthalidone, oxazepam, and terbutaline from plasma, using β-cyclodextrin as a chiral selector in the mobile phase of the second column [52].

10.2.3.3 On-line solid-phase extraction–liquid chromatography

Instead of a full-size column, a short column can be used for solid-phase extraction (SPE) ahead of and on-line with the LC column. Initially, SPE–LC was developed for the trace enrichment of phthalates from water [53]. Subsequently, the same equipment was adopted in many application areas with a wide variety of sorbents. SPE and on-line SPE–LC have recently been reviewed [54]. Typically, a sample is injected or pumped through a short solid-phase column and the analytes are selectively adsorbed on the pre-column, while highly polar and macromolecular sample constituents are passed through unretained. After an optional washing step, the SPE column in eluted, either in forward or in back-flush mode, on to an analytical column, where the enriched analytes are separated prior to detection. Numerous examples are available in the literature [54]. The on-line SPE–LC approach has become very popular in environmental analysis, whereas in pharmaceutical laboratories the initial enthusiasm has given way to off-line SPE for quantitative analysis in most cases.

Typical sorbents used for SPE are conventional chemically bonded reversed-phase silicas, poly(styrene–divinylbenzene) co-polymers, carbon-based sorbents, and ion-pair and ion-exchange type sorbents [54]. Restricted-access packing materials (RAM) [55] combine size exclusion of proteins and other high-mass matrix components with the simultaneous enrichment of low-mass analytes at the hydrophobic inner pore surface. RAM columns are especially useful in the analysis of small molecules in samples containing proteins, *e.g.*, in the analysis of drugs in plasma or urine. A recent example of the use of RAM columns is the analysis of neuropeptide Y and its metabolites with a RAM column having sulfonic acid cation-exchange properties [56]. The immunoaffinity CCLC described above can be considered to be on-line SPE–LC as well, although it employs longer columns than SPE. The high selectivity of immunoaffinity separations has led to the development of molecularly imprinted polymers, containing specific recognition sites for certain molecules [57]. A recent example is the highly selective on-line clean-up of triazines from river water samples prior to LC–MS [58]. The use of a SPE pre-column, packed with somewhat smaller particles than in most conventional SPE applications, for both concentrating the sample and separating a limited number of pesticides has also been proposed. This single-short-column procedure is especially useful in combination with MS detection [59].

Another recent modification of SPE–LC, especially for use in LC–MS, is turbulent-flow chromatography [60]. A schematic diagram of the setup, which, in fact, is very similar to the set-up for conventional SPE–LC, is shown in Fig. 10.5. The 200-μL loop is filled with a solvent composition strong enough to elute the clean-up column. In the first step, an aqueous sample, *e.g.*, plasma after centrifugation, is injected into a special turbulent-flow clean-up column (50 × 1 mm, 50-μm-ID particles) at a flow-rate of 4 mL/min for 30 sec. Highly hydrophilic interfering material and proteins are efficiently removed under these conditions. In the second step, the 200-μL loop is switched in-line to transfer the analyte to the analytical column (0.3 mL/min for 90 sec). The effluent from the clean-up column is mixed with 1.2 mL/min aqueous mobile phase ahead of the analytical column. In the final step, a 2-min ballistic gradient, *e.g.*, 5–95% organic additive, is applied at 1 mL/min, and the column effluent is analyzed by LC–MS. Simultaneously, the clean-up column is regenerated and the 200-μL loop is refilled. The total analysis time per sample is about 5 min. In addition, systems are commercially available which enable staggered parallel operation, resulting in a four-fold increase in sample throughput.

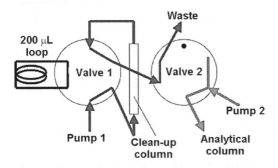

Fig. 10.5. Schematic diagram of turbulent-flow chromatography. Pump 1 delivers either an aqueous mobile phase for sample injection, or a hydro-organic mobile phase for regeneration of the cleanup column. Pump 2 delivers a fast ballistic gradient. (Reproduced from Ref. 60 with permission.)

10.3 CHROMATOGRAPHY–SPECTROMETRY

Both gas and liquid chromatography have been coupled to a variety of spectroscopic and spectrometric techniques. The prime objective of on-line spectroscopy is improvement of the information content concerning the compounds separated. For example, GC–MS has become indispensable for the identification of components, *e.g.*, contaminants in environmental or food samples, but it also is a powerful tool in the selective and sensitive determination of analytes in complex matrices. With respect to the latter, the routine analysis of tetrachlorodibenzodioxins (TCDD) and related compounds (toxaphenes, dibenzofurans, biphenyls, *etc.*) at ppt-levels in biological matrices (*e.g.*, meat and milk) may serve as an example. Obviously, each combination of GC or LC with one of the spectrometric or spectroscopic techniques will have distinct application areas and may also come with its own problems. A number of successful examples are briefly reviewed below. One of these combinations, *i.e.*, LC–MS is subsequently discussed in more detail in Sec. 10.4.

10.3.1 Gas chromatography–mass spectrometry

The on-line combination of GC and MS (GC–MS) is perhaps the best example for demonstrating the power of a combination of a highly efficient separation method and a spectrometric detection method that is both selective and sensitive [61,62]. Initially, the coupling of packed GC columns to the mass spectrometer resulted in some problems with respect to the gas flow from the chromatograph, which could not be handled completely by the vacuum system of the spectrometer. Gas-phase analyte-enrichment interfaces, such as the Watson–Biemann effusion interface, the membrane interface, and especially the jet-separator interface, have helped to overcome these problems. With the introduction of the open-capillary GC columns in the mid-1970s and the fused-silica columns with 0.25- to 0.32-mm ID in 1979, the vacuum problems were solved, because the typical gas flow from a capillary column (1 mL/min of He) fits the pumping capacity of the vacuum systems perfectly. In principle, the column outlet can be placed directly into the ion source of the spectrometer, to effect immediate ionization and minimize excessive band-broadening. Introduction of the capillary column simultaneously also stimulated the development of benchtop quadrupole mass spectrometers for dedicated GC–MS application.

The analytical power of GC–MS can be attributed to various features of the coupled system. GC is not only a highly efficient separation technique for volatile compounds, providing excellent resolution and high plate numbers, but it also presents the separated analytes in the vapor state, which is ideal for electron ionization (EI), the most widely used MS ionization technique. The features of EI for analyte detection after GC are similar to those of a flame-ionization detector (FID): there are no large differences in the ionization efficiency, and the detector response is similar for a wide variety of compounds. In addition, EI provides structure-informative fragmentation for a wide variety of substances that can be analyzed by GC. Due to the good reproducibility of EI mass spectra among instruments from various manufacturers, collecting mass spectra and building mass spectral libraries is feasible. Rapid computer searches with advanced algorithms of these commercially available libraries helps in the identification of unknown components of complex samples. Furthermore, the fragmentation of many compound classes under EI is well described and well understood [63,64], enabling also the identification of unknown compounds, not represented in mass-spectral libraries. EI–MS requires only minute amounts of analyte (less 1 ng) to give a useful mass spectrum. Whether identification of unknown compounds at this level is feasible depends mainly on the presence of interfering materials, which may obscure important spectral features needed for the interpretation of the spectrum. Moreover, selected-ion monitoring is available for monitoring selectively the intensity of just one or a limited number of ions from a particular analyte or set of analytes, and thereby significantly improves the lower limit of quantification in comparison to full-scan operation. Finally, a variety of more advanced MS techniques may help in solving analytical problems, *e.g.*, various mass analyzers, alternative analyte ionization techniques, and tandem MS [65]. Time-of-flight mass analyzers are becoming more prevalent in GC–MS, because they can offer either enhanced mass-spectrometric resolution, enabling accurate mass determination for the estimation of elemental composition, or very high spectrum acquisition rates (up to 500 spectra per second), which are widely used in fast-GC and GC × GC. GC–MS has found abundant applications in the petroleum industry, in environmental analysis of chlorinated organic compounds, pesticides, polycyclic aromatic hydrocarbons, in pharmaceutical and clinical applications, in food analysis, in forensic and toxicological studies, as well as in many other areas [62].

10.3.2 Gas chromatography–atomic emission detection

Microwave-induced plasmas (MIP) have been proposed for use in combination with GC since the mid 1960s [66,67] but never found broad application. An atomic-emission detector (AED), based on a microwave-induced plasma, was commercially introduced in 1989 [68]. The column effluent is fed into a MIP cavity, where the analytes are destroyed and atomized. The atoms are excited by the energy of the plasma. The emitted light is dispersed by an optical monochromator or polychromator system and measured *via* a photodiode array. The photodiode array can be positioned along the wavelength axis to enable the detection of various elements. A schematic diagram of the system is shown in Fig. 10.6. The AED enables the simultaneous determination of various elements. The success of GC–AED can be attributed to three features of the AED:
  (a) The AED shows excellent selectivity in the detection of heteroatoms.
  (b) Calibration of the AED is virtually compound-independent, since the elemental response is independent of the molecular structure. This also opens the way to estimate (partial) elemental composition from the data obtained.
  (c) The AED provides a response for any element that can be excited by a He plasma. The GC–AED system has been widely applied, *e.g.*, in the petrochemical industry, especially in the analysis of compounds containing sulfur [69], lead, vanadium [70], mercury, and palladium, in environmental analysis of pesticides [71], as well as of organometallic compounds containing tin [72], mercury, lead, as well as in the food and pharmaceutical industries.

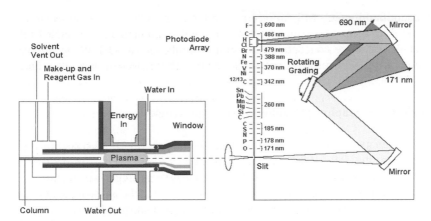

Fig. 10.6. Schematic diagram of an atomic-emission detector for GC. (Adapted from JAS Brochure on AED).

10.3.3 Gas chromatography–Fourier-transform infrared spectroscopy

Initially, there was little interest in an on-line combination of GC with Fourier-transform infrared spectroscopy (FT-IR). This can be attributed to the success of GC–MS and the comparatively poorer sensitivity of GC–FT-IR. The introduction of mercury cadmium telluride (MCT) photodetectors and gold-coated borosilicate glass lightpipes led to a substantial improvement in the performance of GC–FT-IR systems [65]. Next to the lightpipe, two other adjuncts to GC–FT-IR have been found useful, *i.e.*, matrix isolation and sub-ambient trapping. A lightpipe [73] is a 100- to 200-μm-ID borosilicate capillary, internally coated with a thin layer of gold. By reflecting the IR beam through the lightpipe, the path length of the cell is increased significantly, and this leads to a greatly increased sensitivity. The matrix isolation method [74] is based on freezing the column effluent, after mixing it with an inert gas, and depositing it on a rotating gold disk. The complete chromatogram is put on the disk, and later on, reflection–absorption spectra are obtained from the deposited material on the disk. The matrix isolation and cryogenic trapping of the analyte reduce the rotational line broadening in the IR spectra and thereby improve the resolution of the spectra. In subambient trapping [75], the column effluent is directly frozen on a moving IR-transparent zinc selenide window. The spectra are collected shortly after deposition in real time. The latter two methods show similar detection limits in the 10-pg range, which is about 100-fold better than those of the lightpipe instruments [65,76].

An important field of application of GC–FT-IR is the differentiation between positional isomers, something which is frequently difficult to obtain from the EI mass spectrum. Some examples are: the differentiation of PCB [77] and BTEX [78] in environmental analysis and of amphetamine and methamphetamine in forensic applications [79]. In addition to the ability to discriminate between structural isomers, the FT-IR detector has the virtues of being nondestructive and providing direct functional-group information. Given the importance of information from FT-IR in the structure elucidation of compounds unidentified by GC–MS, an on-line combination of GC–FT-IR–MS would be extremely useful for some applications. Such a system is commercially available [80]. Improvement of the effective concentration detection limits for GC–FT-IR by the on-line combination with SPE pre-concentrating sample pre-treatment has also been reported [81].

10.3.4 Liquid chromatography–photodiode-array spectrometry

The on-line combination of LC with a UV–VIS photodiode-array (PDA) spectrometer became available in the late 1980s [82] and is now widely used. Technological improvements have yielded PDA detectors that are as sensitive as single-wavelength UV detectors. The most important application areas of LC–PDA are in the assessment of peak purity, identification of unknown compounds and/or confirmation of their identity. Significant progress has been made in the assessment of peak purity through the development of intelligent algorithms [83–86] and their application to actual problems, *e.g.*, in drug impurity profiling [87,88].

Identification of unknown compounds and confirmation of their identity is important in forensic analysis and toxicological screening [89–93]. Examples of such studies are the rapid screening for 100 basic drugs and metabolites in urine by cation-exchange SPE and LC–PDA [90], and emergency toxicology testing to provide both detection, confirmation, and quantification of basic drugs in serum [91]. Computer searching of spectral libraries has become an important and powerful tool in work of this kind [92, 93]. Similar procedures have been adopted for multi-residue analysis in environmental monitoring programs related to quality control of surface water. A typical procedure employs an integrated system, called SAMOS [94,95], which enables the automated, unattended analysis of filtered 100-mL samples of surface water by means of on-line SPE on a short cartridge column, and subsequent gradient HPLC with UV-PDA detection. Compounds with sufficient UV activity can be quantified. Moreover, a UV spectral library allows provisional identification of compounds or compound classes, at least when the UV spectrum acquired is not too much distorted by the presence of humic acids. Any compound detected by a SAMOS system, applied in the field, must be identified by MS, often by LC–MS. Another important application area of LC–PDA is in the field of natural-product analysis. Typical examples are the identification of flavonoids in plants [96], the analysis of carotenoids [97], and the determination of taxol in *Taxus* species [98].

10.3.5 Liquid chromatography–inductively coupled plasma mass spectrometry

While the on-line combination of inductively coupled-plasma (ICP) atomic-emission spectrometry is an important method in the qualitative (and quantitative) analysis of many metals as well as nonmetals, enabling sensitive detection [99,100], the combination with ICP-MS has made it even more useful. ICP-MS is currently the method of choice in element speciation. It has been coupled to a variety of separation techniques, including GC, LC, SFC, and capillary electrophoresis [101,102]. The coupling for liquid introduction consists of a pneumatic or ultrasonic nebulizer and a spray chamber.

From there, the aerosol is fed into the ICP torch, where atomization and ionization of the analytes takes place. Ions emerging from the ICP are sampled by an atmospheric-pressure ion source, similar to the ones applied in LC–MS coupling *via* electrospray or atmospheric-pressure chemical ionization (Sec. 4.1.1). A wide variety of applications of LC–ICP-MS in selective or multi-residue determination of elements have been described. Studies frequently focus on the determination of elements like As, Hg, Pb, and Se, but applications to other elements, like Pt, Cd, and Cr, have also been described. Such elements are determined in a great variety of matrices, like surface water, urine, seafood, and other biological samples. Some applications in the area of element speciation have recently been reviewed [101,102].

10.3.6 Liquid chromatography–Fourier-transform infrared spectroscopy

The on-line combination of LC and FT-IR is less popular than GC–FT-IR, but with the introduction of newer interfaces, the technique has reached analytical utility [103]. There are basically two approaches to LC–FT-IR, *i.e.*, *via* a flow-cell or *via* a solvent-elimination interface. The flow-cell approach is very similar to the LC–UV systems and generally suffers from poor detection limits, partly due to limited optical path length and partly due to interfering solvent peaks in the IR spectrum. In solvent-elimination LC–FT-IR, the mobile phase is removed prior to the IR assay of the analytes. They are deposited on a suitable substrate, enabling a solvent-interference-free infrared spectrum. Moreover, the deposition step may yield some concentration of the analyte zones and permit signal averaging. In the currently most successful solvent-elimination interfaces, the mobile phase is nebulized prior to deposition, and this enhances the evaporation of the solvent. The interfaces developed are somewhat similar to devices developed for LC–MS coupling, *e.g.*, thermospray nebulization and deposition on a moving-belt interface [104], pneumatic nebulization in combination with a particle-beam interface and deposition of a selected peak on a solid substrate [105], pneumatic nebulization on a rotating IR-reflective disk [106] or on a moving solid substrate [107], electrospray nebulization, and ultrasonic nebulization.

10.3.7 Liquid chromatography–nuclear magnetic resonance spectroscopy

Perhaps the most attractive on-line combination of chromatographic separation with spectroscopy is with nuclear magnetic resonance (NMR) spectroscopy, especially LC–NMR. After some initial studies in the late 1970s, LC–NMR developed in the 1980s to become the highly useful hyphenated technique it is at present. LC–NMR is beginning to enter routine laboratories involved in structure elucidation in pharmaceutical industries, sometimes even in an on-line combination with LC–MS [108,109]. The combination is applied in solving structure elucidation problems related to natural products [110–113] and drug impurities and metabolites [114–116].

10.4 LIQUID CHROMATOGRAPHY–MASS SPECTROMETRY

Combined liquid chromatography–mass spectrometry (LC–MS) [117], may be considered as one of the most important analytical techniques of the past decade: development occurred at a rapid rate, and LC–MS is now widely accepted and implemented. It has become the method of choice for analytical laboratories at various stages of drug development in the pharmaceutical industry and it plays an important role in environmental analysis, in biochemistry, and in biotechnology, *e.g.*, in the field of proteomics. Most of the current instrumental developments take place in the instrument manufacturers' laboratories and workshops, although such developments are, to a significant extent, steered and initiated by the demands of users, especially the pharmaceutical industries.

10.4.1 Principles and instrumentation

Investigations into the coupling of LC and MS started in the early 1970s. During the first 20 years, most of the attention had to be given to solving interface problems, and a varied collection of interfaces was developed, tested, applied, and rejected: the moving-belt interface, direct-liquid introduction, thermospray, continuous-flow fast-atom bombardment, and particle-beam interfacing [117]. Technological problems in interfacing appear to be solved with the introduction of interfaces based on the principles of atmospheric-pressure ionization, *i.e.*, electrospray (ESI) and atmospheric-pressure chemical ionization (APCI).

10.4.1.1 Atmospheric-pressure ionization

An atmospheric-pressure ionization (API) interface/source consists of five parts [117]: (1) the liquid introduction device or spray probe, (2) the actual atmospheric-pressure ion-source region, where the ions are generated by means of ESI, APCI, or by other means, (3) an ion-sampling aperture, (4) an atmospheric-pressure-to-vacuum interface, and (5) an ion optical system, where the ions are subsequently transported into the mass analyzer. A schematic diagram of an API source is shown in Fig. 10.7.

The column effluent from LC is nebulized and passed into the atmospheric-pressure ion-source region. Nebulization is either performed pneumatically (*i.e.*, in a heated nebulizer APCI), by the action of a strong electrical field (*i.e.*, in ESI), or by a combination of both (*i.e.*, in pneumatically assisted ESI). From the aerosol, gas-phase ions are generated by either a liquid-phase mechanism, related to charging of the droplets during ESI, or by a gas-phase ionization mechanism, initiated by the corona discharge during APCI. These ions, together with solvent vapor and nitrogen bath gas, are sampled by an ion sampling device into a first pumping stage. In most systems, the spray probe is positioned orthogonally to the ion-sampling device to avoid contamination of the latter by nonvolatile sample constituents. The mixture of gas, solvent vapor, and ions is supersonically expanded into this low-pressure region (10–100 Pa). The core of the

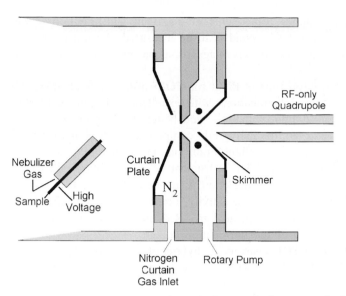

Fig. 10.7. Schematic diagram of a typical atmospheric-pressure ionization source for LC/MS.

expansion, containing the ions and other (neutral) material of higher molecular mass, is sampled by a skimmer into a second pumping stage (pressure 0.1–1 Pa), containing an ion-focussing and transfer device (RF-only multipole) to optimally transport and focus the ions in a suitable manner to the mass-analyzer region (pressure $< 10^{-3}$ Pa).

10.4.1.2 Developments in mass spectrometers

Almost as important to the optimization of API interfaces is the progress in MS instrumentation. The computer-controlled operation of the modern mass spectrometer helps it to achieve optimum performance. In addition, new developments in the instrumentation help to improve the overall performance of LC–MS. Most of the LC–MS applications are still run on single- and triple-quadrupole instruments. In recent years, the performance of triple-quadrupole (QqQ) instruments for MS–MS has been improved by replacing the RF-only quadrupole collision cell by various other designs, including the RF-only hexapole and octapole collision cells, as well as the linear accelerating high-pressure collision cell (LINAC). The latter allows shorter dwell-times in selective reaction monitoring (SRM), speed-up of the analysis, and/or monitoring of more SRM transitions during one (fast) chromatographic run.

API on an ion-trap mass spectrometer appears to be a very successful combination. While SRM on a QqQ instrument has to be preferred in low-level quantitative analysis [118], the API ion-trap instrument especially proves its power in qualitative analysis, where the multiple stages of MS–MS can be applied to achieve structure elucidation of the unknown compounds. Data-dependent acquisition further enhances the performance in the structure elucidation of minor components in a mixture [119,120]. If a low analyte signal in a relatively clean background is to be measured, the use of the ion accumulation features of an ion trap can be advantageous, as has been demonstrated for peptides [121,122].

Orthogonal-acceleration reflectron time-of-flight (oa-TOF) instruments combine the ability to perform accurate mass determinations with excellent full-scan sensitivity. Mass accuracies of better than 20 ppm without – and better than 5 ppm with – an internal lock mass can be achieved routinely. For compounds with a mass below 1000 Da, this accuracy provides an excellent confirmation of identity, based on calculated elemental compositions. An oa-TOF instrument is an integrating rather than a scanning system. In practice, the "all-ion-detection" capability of the oa-TOF system provides a 20- to 100-fold improvement in sensitivity, compared to a scanning QqQ system. Obviously, the QqQ instrument, operated in SRM mode, will provide better sensitivity, but only at the expense of the information content.

Given these features of an oa-TOF analyzer, a hybrid of a quadrupole front-end and an oa-TOF back-end for MS–MS (Q-TOF) would be an attractive system. The most interesting feature of the Q-TOF hybrid is its ability to perform accurate mass determinations at excellent sensitivity after conventional low-energy collision-induced dissociation (CID) in a collision cell. This greatly facilitates identification of unknown compounds, not only in the field of protein chemistry, for which the instrument was originally built, but also in studies related to impurities, degradation products, and metabolites of drugs.

Finally, multiply charging of proteins by ESI has also stimulated the use of Fourier-transform ion-cyclotron resonance mass spectrometry (FT-ICR-MS) [123,124]. Again, the enhanced resolution is an important feature. The high-resolution operation in combination with various dissociation techniques, such as CID, sustained off-resonance irradiation, infrared multiphoton dissociation, and surface-induced dissociation, enables the use of ESI FT-ICR-MS for advanced structure elucidation of proteins. In addition, reaction chemistry and gas-phase confirmation studies can be performed. Among these new and exciting developments, the role of ESI on magnetic sector instruments has diminished. The main reason for the use of magnetic sector instrument appears to be the use of an array-type of detector to enhance sensitivity. For many other potential applications of sector instruments, a TOF instrument is a viable alternative, being easier to operate and having a better price–performance ratio. The need for high-energy CID in peptide sequencing appears to be overcome to a large extent by the use of protein databases [125,126].

10.4.2 Applications

At present, the three major application areas of LC–MS technology are in the pharmaceutical, environmental, and biochemical fields. The applications in the field of drug development and testing as well as proteomics can be considered as the important driving forces in the current developments of LC–MS technology and applications.

10.4.2.1 Pharmaceutical applications

LC–MS technology is applied at virtually every stage of drug development. Sensitivity, selectivity, and speed are the key issues in this respect. Compared to LC–PDA, LC–MS provides enhanced confirmation of identity and often enhanced selectivity. In addition, the ease of operation and the achieved level of automation make LC–MS an attractive tool in drug development, especially because it can be applied to analysis and troubleshooting in many areas of drug research. The role of LC–MS commences in the drug discovery stage, no matter whether this is performed by conventional "intelligent" synthesis or by combinatorial chemistry. Powerful software tools have been developed for LC–MS to support its drug discovery tasks. The most generally applicable of these tools is Open Access [127,128]. This system transforms an API-MS instrument into a walk-up "black box" for synthetic chemists in need of rapid confirmation of the progress of their synthesis by molecular-mass determinations of their (intermediate) products. A remote computer serves as a log-in to the system, where the type of LC–MS experiments to be performed are selected. The sample is run automatically, and the resulting spectra are sent to the chemist by electronic mail or placed onto the laboratory information management system (LIMS network).

In drug discovery based on combinatorial synthesis, the combinatorial libraries must be screened for biological activity, while a rapid characterization of identity is also required. LC–MS technology is frequently applied in the latter step [129]. The analytical system consists of an x–y autosampler, enabling the use of a 96-well plate, connected to a column-bypass API-MS [130] or fast LC–MS system [131]. Single-quadrupole systems are often used for this purpose, although the use of oa-TOF instruments is of growing importance; the accurate-mass determination allows a better confirmation of identity. The software allows the rapid analysis of large series of samples at a high sample throughput (*i.e.*, up to 60 samples/h) in column-bypass mode [132]. On-line UV-PDA as well as other LC detectors, especially evaporative light-scattering detection, may be applied to establish compound purity [133]. The data-processing software not only automatically processes the data, but also presents them in a graphical representation of the 96-well plate, where the confirmed sample spots are colored green, while the others are colored red. More advanced color schemes have been described as well [134]. In some applications, the screening for biological activity and the mass spectral characterization are combined into one system [135], *e.g.*, by the application of on-line bioaffinity columns [136] or integrated biodetection systems, based on antigen–antibody or ligand–receptor interaction [137].

Another software tool combines the rapid screening of combinatorial libraries or of series of extracts from natural products with preparative-scale purification of the relevant biologically active compounds. The fractionation is controlled by the response of the compound of interest in LC–MS. Initially, these systems were developed with conventional LC columns, but preparative-scale LC columns are now applied as well [138–141]. The structure confirmation in Open-Access and combinatorial-chemistry strategies is primarily based on determining the molecular mass of the intact compound. In subsequent stages of drug development, *i.e.*, in impurity screening, identification of drug metabolites, and the search for degradation products in neat drugs and drug formulations, more elaborate structure elucidation is required. Screening strategies based on precursor-ion and neutral-loss scans in MS−MS are frequently applied to the search for structurally related compounds [142]. Based on the product-ion mass spectrum of the parent compound and educated guesses of possible products, selective neutral-losses and/or possible common-product ions are selected. Dedicated software tools, providing advanced automated operation and data processing for metabolite identification, have become available. The accurate-mass determination of intact metabolites as well as product ions in MS−MS, by means of FT-ICR-MS [143] or Q-TOF hybrid [144] instruments constitutes important progress in this field. The applications of multiple MS−MS stages, as available on ion-trap MS systems, can be of great help in structure elucidation [145]. In addition, the on-line combination of LC−NMR and LC−MS are applied in metabolite studies, with the aid of either (triple) quadrupole or ion-trap MS systems [108,109].

10.4.2.2 Quantitative bioanalysis

Quantitative bioanalysis is the most important application area of LC−MS, in terms of the number of instruments applied and the number of analyses performed. Quantitative bioanalysis is required to support pre-clinical and clinical drug testing and to provide pharmacokinetic and pharmacodynamic data. In addition, rapid quantitative bioanalysis is becoming more and more important as part of early studies of adsorption, distribution, metabolism, and excretion (ADME). Automated, unattended operation of the LC−MS instrumentation is required. Fast and routine LC−MS analysis also demands fast and automated sample pre-treatment strategies and advanced data-processing software. In order to speed up sample throughput, the use of generic methods is frequently explored, especially in early-ADME studies [146]. The keys to the success of LC−MS in quantitative bioanalysis are the excellent selectivity against possibly interfering compounds in the biological matrix, especially when operated in SRM mode, enabling typical detection limits in the pg- and, in favorable cases, even sub-pg-range, and the enhanced confidence in the identity of the compound(s) analyzed. LC−MS−MS in SRM mode on triple-quadrupole instruments is the method of choice in quantitative analysis in pharmaceutical industries. Unfortunately, for proprietary reasons, only a limited number of reports on successful LC−MS applications in quantitative bioanalysis is available in the open literature.

The higher selectivity achievable due to the use of SRM procedures is often immediately vitiated by a decrease in the quality of the sample pre-treatment and/or chromatographic separation. The reason for this is an increase in the sample throughput, which often puts serious demands on the sample pre-treatment methods and indicates the clear need for accelerating pre-treatment. Widely applied approaches are the use of parallel SPE procedures on short SPE columns or Empore disks, mounted in a 96-well plate format [147] and turbulent-flow LC−MS, enabling automated high-speed two- or four-channel parallel sample pre-treatment and chromatography [148]. While off-line sample pre-treatment seems to be preferred by most researchers, on-line strategies, *e.g.*, based on Prospekt SPE instrumentation, are described as well. An example is the rapid determination of pranlukast and metabolites in human plasma [149]. Serious matrix problems may be experienced in quantitative bioanalysis, especially in ESI. Signal suppression due to unknown matrix interferences is often observed, as described in a well-documented example by Matuszewski *et al.* [150]. Changes in the sample pretreatment procedures may be successful in solving the problem, but in some cases changing over to APCI, when applicable, appears to be the only feasible solution of signal suppression.

10.4.2.3 Environmental applications

Another important application area of LC–MS is environmental analysis [151]. Strategies different from those used in pharmaceutical applications are often required. While in pharmaceutical applications the analysis is directed to particularly one or only a few target compounds, screening is often required in environmental analysis, covering a multitude of compounds at often drastically different concentration levels. Although restriction to a more limited number of compounds or compound classes is applied for practical reasons, for regulatory purposes, screening must be broad. A clear example of the dilemma in developing analytical strategies in environmental analysis is the screening of surface and ground waters prior to their use in the production of drinking water. European regulations demand the qualitative and quantitative determination of individual pesticides at levels of 0.1 μg/L or above. This can only be achieved by pre-concentrating samples by, *e.g.*, off-line or on-line liquid–liquid extraction or SPE, in combination with SIM or SRM, *i.e.*, target-compound-directed approaches. The sample pre-treatment often results in the concentration of interfering material that may compromise the determination of the analytes of interest. These materials can hamper full-scan analysis and/or may produce matrix-related ion-suppression effects. Moreover, various compound classes give quite different responses in ESI or APCI. Some compounds are best analyzed in positive-ion mode, while others only provide sufficient response in negative-ion mode [152]. For optimum SRM performance, the tuning parameters of ion source and collision cell must be optimized, basically for each individual compound or compound class. The problems increase even further when not only pesticides but other environmental contaminants like surfactants, (azo) dyes, and drug residues, must be taken into account. In this respect, compounds that can act as endocrine disruptors, have received considerable attention lately [153]. A typical approach in environmental analysis is an integrated system called SAMOS (Sec. 3.4). Any compound detected by a SAMOS system, applied in the field, must be identified by MS, often by LC–MS. For this, a more target-compound-oriented approach can be applied, because often the compound class can be determined by UV-PDA. SAMOS-like systems can be on-line, coupled to LC–MS, as demonstrated by the research groups of Brinkman [154] and Barceló [155]. In recent years, the sample volume needed for pre-treatment could be reduced (from 100 to 10 mL), as a result of the improved performance of the LC–MS instrumentation.

10.4.2.4 Biochemical applications

ESI is frequently applied at various stages of the characterization of peptides and proteins: molecular-mass determination, amino acid sequencing, determination of the nature and position of chemical and post-translational modifications of proteins, investigations of protein tertiary and quaternary conformation, and the study of noncovalent association. In most cases, no on-line separation is needed, and the sample solution can be introduced directly *via* the ESI or nano-ESI interface. Impressive results have been achieved in this area [156–158]. On-line LC–MS for peptide and protein characterization has also been described, especially in relation to peptide sequencing and the characterization of secondary protein structure, *e.g.*, post-translational modifications. The determination of the *N*-glycosylation sites in recombinant Human Factor VIII protein by reversed-phase LC–ESI-MS [159] may serve as an example in this area. A microcapillary column-switching system, applied in combination with LC–MS has been described for the direct identification of peptides in major histocompatibility complex class I molecules[160].

10.4.3 Perspective

In addition to applications in these three main areas, LC–MS has frequently been used in many other fields, such as in the study of natural products, many of which are covered in Part B of this book. LC–MS is appreciated for its sensitivity and selectivity, its specificity and the information furnished, *e.g.*, on the molecular mass of the analyte. While it is easy to appreciate the benefits and advantages of LC–MS, its limitations appear to be somewhat neglected. The value of MS in LC–MS must be distinguished from its role in GC–MS.

Unlike in GC–MS, the analyte peaks must be searched for against a relatively high background of the solvent-related ion current in LC–MS. While a total-ion chromatogram (TIC) in GC–MS often reveals the presence of a number of compounds, even with small amounts of sample, this is not necessarily true in LC–MS. Low-level multi-residue screening and/or the search for unknown compounds at low concentrations is still difficult in LC–MS. The use of reconstructed-mass and/or base-peak chromatograms can be helpful in this respect, but the implementation of base-peak chromatograms in most commercial MS software packages is rather poor: in most cases the *m/z* range to be searched for base peaks cannot be specified. Whereas in GC–MS the identification of unknown analytes, based on excessive fragmentation in EI and helped by computer library searching, is clearly successful, the identification of unknowns by LC–MS is more difficult. The fragmentation of protonated molecules in low-energy CID often leads to only a limited number of fragments. This will often be insufficient for unambiguous identification, because the information in the MS–MS spectrum is insufficient, and the interpretation is hampered by the lack of insight into the fragmentation rules. Because of the influence of experimental conditions on the appearance of MS–MS or in-source CID spectra, generating generally applicable MS–MS libraries has proved to be difficult.

The softer and step-wise fragmentation achieved in CID in an ion trap is often very helpful in structure elucidation. The great number of ion-trap systems currently installed will perhaps allow the creation of dedicated spectral libraries. Current interest in the use of oa-TOF, Q-TOF, and FT-ICR-MS instruments for structure elucidation is based on the assumption that accurate mass determination of the (product) ions facilitates the interpretation of the fragments. Fortunately, identification of a completely unknown compounds is not always required. Generally, target-compound-like strategies can be applied, improvements in software facilitates the rapid optimization of experimental parameters, and the automated, unattended data acquisition as well as data processing are helpful. Excellent integrated and user-friendly software is available for data evaluation of quantitative analysis by LC–MS, and dedicated software packages are available for operating LC–MS in specific application areas, *e.g.*, open-access, screening of combinatorial libraries, peptide sequencing *via* (Internet) database searching, neonatal screening of blood for metabolic disorders, and MS-controlled fractionation in preparative LC. These types of applications will dominate the use of LC–MS in the coming decade. Further improvements in the instrumentation and, especially, in the software are to be expected, in order to adapt LC–MS to specific tasks within particular applications, and for some time to come, applications in the pharmaceutical industry and proteomics will continue to be the most important application areas for LC–MS.

REFERENCES

1   J.C. Giddings, *Anal. Chem.*, 56 (1984) 1259A.
2   R.W. Frei, *J. Chromatogr.*, 251 (1982) 91.
3   W.H. McFadden, *J. Chromatogr. Sci.*, 18 (1980) 97.
4   J.C. Giddings, in H.J. Cortes (Ed.), *Multidimensional Chromatography: Techniques and Applications*, Dekker, New York, 1990, Ch. 1.
5   J.C. Giddings, *J. Chromatogr. A*, 703 (1995) 3.
6   D.R. Deans, *Chromatographia*, 1 (1968) 18.
7   G. Schomburg and F. Weeke, in S.G. Perry and E.R. Adlard (Eds.), *Gas Chromatography 1972*, The Institute of Petroleum, United Kingdom, 1972, p. 285.
8   G. Schomburg, *LC-GC*, 5 (1987) 304.
9   G. Schomburg, *J. Chromatogr. A*, 703 (1995) 309.
10   J.B. Phillips and C.J. Venkatramani, *J. Microcol. Sep.*, 5 (1993) 511.
11   J.B. Phillips and J. Xu, *J. Chromatogr. A*, 703 (1995) 327.
12   J.B. Phillips and J. Beens, *J. Chromatogr. A*, 856 (1999) 331.
13   H.-J. de Geus, J. de Boer and U.A.Th. Brinkman, *J. Chromatogr. A*, 767 (1997) 137.
14   J.B. Phillips, R.B. Gaines, J. Blomberg, F.W.M. van der Wielen, J.M. Dimandja, V. Green, J. Granger, D. Patterson, L. Racovalis, H.-J. de Geus, J. de Boer, P. Haglund, J. Lipsky, V. Sinha and E.B. Ledford, Jr., *J. High Resolut. Chromatogr.*, 22 (1999) 3.
15   R.M. Kinghorn and P.J. Marriott, *Anal. Chem.*, 69 (1997) 2582.
16   J. Dallüge, L.L.P. van Stee, X. Xu, J. Williams, J. Beens, J.J. Vreuls and U.A.Th. Brinkman, *J. Chromatogr. A*, 974 (2002) 169.
17   G.S. Frysinger, R.B. Gaines and E.B. Lydford, Jr., *J. High Resolut. Chromatogr.*, 22 (1999) 195.
18   G. Schomburg, H. Husmann and E. Hübinger, *J. High Resolut. Chromatogr. Chromatogr. Commun.*, 8 (1995) 395.
19   J. Dallüge, M. van Rijn, J. Beens, J.J. Vreuls and U.A.Th. Brinkman, *J. Chromatogr. A*, 965 (2002) 207.
20   W.A. König, B. Gehrke, D. Icheln, P. Evers, J. Dönnecke and W. Wang, *J. High Resolut. Chromatogr.*, 15 (1992) 367.

21 R.E. Majors, *J. Chromatogr. Sci.*, 18 (1980) 571.

22 J.J. Vreuls, G.J. de Jong, R.T. Ghijsen and U.A.Th. Brinkman, *J. Assoc. Off. Anal. Chem.*, 77 (1994) 306.

23 K. Grob, *J. Chromatogr. A*, 703 (1995) 265.

24 K. Grob, G. Karrer and M.-L. Riekkola, *J. Chromatogr.*, 333 (1985) 129.

25 E.C. Goosens, D. de Jong, G.J. de Jong and U.A.Th. Brinkman, *Chromatographia*, 47 (1998) 313.

26 K. Grob and J.-M. Stoll, *J. High Resolut. Chromatogr. Chromatogr. Commun.*, 9 (1986) 518.

27 F. Munari, A. Trisciani, G. Mapelli, S. Trestianu, K. Grob and J.M. Colin, *J. High Resolut. Chromatogr.*, 8 (1985) 601.

28 K. Grob and E. Müller, *J. High Resolut. Chromatogr.*, 11 (1988) 388 and 560.

29 H.G.J. Mol, H.-G. Janssen and C.A. Cramers, *J. High Resolut. Chromatogr.*, 18 (1995) 19.

30 K. Grob, A. Artho and C. Mariani, *Fat Sci. Technol.*, 94 (1992) 394.

31 J. Beens and R. Tijssen, *J. Microcol. Sep.*, 7 (1995) 345.

32 T. Hyötyläinen, K. Grob and M.-L. Riekkola, *J. High Resolut. Chromatogr.*, 20 (1997) 657.

33 T. Hyötyläinen, K. Janho and M.-L. Riekkola, *J. Chromatogr. A*, 813 (1998) 589.

34 J. Roeraade, *J. Chromatogr.*, 330 (1985) 263.

35 E.C. Goosens, R.G. Bunschoten, V. Engelen, D. de Jong and J.H.M. van den Berg, *J. High Resolut. Chromatogr.*, 13 (1990) 438.

36 E. Noroozian, F.A. Maris, M.W.F. Nielen, R.W. Frei, G.J. de Jong and U.A.Th. Brinkman, *J. High Resolut. Chromatogr. Chromatogr. Commun.*, 10 (1987) 17.

37 T.H.M. Noij, E. Weiss, T. Herps, H. van Cruchten and J. Rijks, *J. High Resolut. Chromatogr. Chromatogr. Commun.*, 11 (1988) 181.

38 J.J. Vreuls, A.J.H. Louter and U.A.Th. Brinkman, *J. Chromatogr. A*, 856 (1999) 279.

39 Th. Hankemeier, S.P.J. van Leeuwen, J.J. Vreuls and U.A.Th. Brinkman, *J. Chromatogr. A*, 811 (1998) 117.

40 L.E. Edholm and L. Ögren, in I. Wainer (Ed.), *Liquid Chromatography in Pharmaceutical Development*, Aster, Springfield, OR, 1986, p. 345.

41 P. van Zoonen, E.A. Hogendoorn, G.R. van der Hoff and R.A. Baumann, *Trends Anal. Chem.*, 11 (1992) 11.

42 E.A. Hogendoorn and P. van Zoonen, *J. Chromatogr. A*, 703 (1995) 149.

43 L.A. Holland and J.W. Jorgenson, *Anal. Chem.*, 67 (1995) 3275.

44 E.A. Hogendoorn, P. van Zoonen and U.A.Th. Brinkman, *Chromatographia*, 31 (1991) 285.

45 A. Farjam, J. de Jong, U.A.Th. Brinkman, W. Haasnoot, A.R.M. Hamers, R. Schilt and F.A. Huf, *Chromatographia*, 31 (1991) 469.

46 A. Farjam, N.C. van der Merbel, A.A. Nieman, H. Lingeman and U.A.Th. Brinkman, *J. Chromatogr.*, 589 (1992) 141.

47 G.R. Rule, A.V. Mordehai and J.D. Henion, *Anal. Chem.*, 66 (1994) 230.

48 K.A. Bean and J.D. Henion, *J. Chromatogr. A*, 791 (1997) 119.

49 P.L. Ferguson, C.R. Iden, A.E. McElroy and B.J. Brownawell, *Anal. Chem.*, 73 (2001) 3890.

50 D.W. Roberts, M.I. Churchwell, F.A. Beland, J.-L. Feng and D.R. Doerge, *Anal. Chem.*, 73 (2001) 203.

51 A. Walhagen and L.-R. Edholm, *J. Chromatogr.*, 473 (1989) 371.

52 A. Walhagen and L.-E. Edholm, *Chromatographia*, 32 (1991) 215.

53 H.P.M. van Vliet, Th.C. Bootsman, R.W. Frei and U.A.Th. Brinkman, *J. Chromatogr.*, 185 (1979) 483.

54 M.-C. Hennion, *J. Chromatogr. A*, 856 (1999) 3.

55 A. Rudolphi, K.-S. Boos and D. Seidel, *Chromatographia*, 41 (1995) 645.

56 K. Racaityte, E.S.M. Lutz, K.K. Unger, D. Lubda and K.S. Boos, *J. Chromatogr. A*, 890 (2000) 135.

57 A.G. Mayes and K. Mosbach, *Trends Anal. Chem.*, 16 (1997) 321.

58 R. Koeber, C. Fleischer, F. Lanza, K.-S. Boos, B. Sellergren and D. Barceló, *Anal. Chem.*, 73 (2001) 2437.

59 A.C. Hogenboom, P. Speksnijder, R.J. Vreeken, W.M.A. Niessen and U.A.Th. Brinkman, *J. Chromatogr. A*, 777 (1997) 81.

60 J.L. Herman, *Rapid Commun. Mass Spectrom.*, 16 (2002) 421.

61 F.W. Karasek and R.E. Clement, *Basic Gas Chromatography–Mass Spectrometry*, Elsevier, Amsterdam, 1988.

62 W.M.A. Niessen (Ed.), *Current Practice of Gas Chromatography–Mass Spectrometry*, Dekker, New York, 2001.

63 F.W. McLafferty and F. Turecek, *Interpretation of Mass Spectra*, 4th Edn. University Science Books, Mill Valley, 1993.

64 R.M. Smith, *Understanding Mass Spectra, A Basic Approach*, Wiley, New York, 1999.

65 N. Ragunathan, K.A. Krock, C. Klawun, T.A. Sasaki and C.L. Wilkins, *J. Chromatogr. A*, 856 (1999) 349.

66 A.J. McCormack, S.C. Tong and W.D. Cooke, *Anal. Chem.*, 37 (1965) 1470.

67 C.A. Bache and D.J. Lisk, *Anal. Chem.*, 37 (1965) 1477.

68 P.L. Wylie and B.D. Quimby, *J. High Resolut. Chromatogr.*, 12 (1989) 813.

69 T.G. Albro, P.A. Dreifuss and R.F. Wormsbecher, *J. High Resolut. Chromatogr.*, 16 (1993) 13.

70 B.D. Quimby and J.J. Sullivan, *Anal. Chem.*, 62 (1990) 1027 and 1034.

71 Th. Hankemeier, J. Rozenbrand, M. Abhadur, J.J. Vreuls and U.A.Th. Brinkman, *Chromatographia*, 48 (1998) 273.

72 H. Tao, R.B. Rajendran, C.R. Quetel, T. Nakazato, M. Tominaga and A. Miyazaki, *Anal. Chem.*, 71 (1999) 4208.

73 L.V. Azzaraga, *Appl. Spectrosc.*, 34 (1980) 224.

74 G.T. Reedy, D.G. Ettinger, J.F. Schneider and S. Bourbe, *Anal. Chem.*, 57 (1985) 1602.

75 S. Bourne, A.M. Haefner, K.L. Norton and P.R. Griffiths, *Anal. Chem.*, 62 (1990) 2448.

76 P. Jackson, G. Dent, D. Carter, D.J. Schofield, J.M. Chalmers, T. Visser and M. Vredenbregt, *J. High Resolut. Chromatogr.*, 16 (1993) 515.

77 H. Budzinski, Y. Hermange, C. Pierard, P. Garrigues and J. Bellocq, *Analusis*, 20 (1992) 155.

78 J.W. Diehl, J.W. Finkbeiner and F.P. Disanzo, *Anal. Chem.*, 65 (1993) 2493.

79 K.S. Kalasinsky, B. Levine, M.L. Smith, J. Magluilo and T. Schaefer, *J. Anal. Toxicol.*, 17 (1993) 359.

80 J.C. Demirgian, *Trends Anal. Chem.*, 6 (1987) 58.

81 Th. Hankemeier, E. Hooijschuur, J.J. Vreuls, U.A.Th. Brinkman and T. Visser, *J. High Resolut. Chromatogr.*, 21 (1998) 341.

82 M.V. Pickering, *LC-GC*, 8 (1990) 846.

83 H.K. Chan and G.P. Carr, *J. Pharm. Biomed. Anal.*, 8 (1990) 271.

84 J.B. Castledine, A.F. Fell, R. Modin and B. Sellberg, *Anal. Proc.*, 29 (1992) 100.

85 F. Cuesta Sánchez and D.L. Massart, *Anal. Chim. Acta*, 298 (1994) 331.

86   M.D.G. Garcia, A.G. Frenich, J.L.M. Vidal, M.M. Galera, A.M. de la Peña and F. Salinas, *Anal. Chim. Acta*, 348 (1997) 177.

87   S. Gorog, M. Bihari, E. Csizer, F. Dravetz, M. Gazdag and B. Herenyi, *J. Pharm. Biomed. Anal.*, 14 (1995) 85.

88   T.W. Ryan, *Anal. Lett.*, 31 (1998) 651.

89   H. Engelhardt and T. König, *Chromatographia*, 28 (1989) 341.

90   B.K. Logan, D.T. Stafford, I.R. Tebbett and C.M. Moore, *J. Anal. Toxicol.*, 14 (1990) 154.

91   P.R. Puopolo, S.A. Volpicelli, D. MacKeen Johnson and J.G. Flood, *Clin. Chem.*, 37 (1991) 2124.

92   B. Dathe and M. Otto, *Chromatographia*, 37 (1993) 31.

93   Y. Gaillard and G. Pépin, *J. Chromatogr. A*, 763 (1997) 149.

94   J. Slobodník, E.R. Brouwer, R.G. Geerdink, W.H. Mulder, H. Lingeman and U.A.Th. Brinkman, *Anal. Chim. Acta*, 268 (1992) 55.

95   J. Slobodník, M.G.M. Groenewegen, E.R. Brouwer, H. Lingeman and U.A.Th. Brinkman, *J. Chromatogr. A*, 642 (1993) 259.

96   P. Pietta, P. Mauri, A. Bruno, A. Rava, E. Manera and P. Ceva, *J. Chromatogr.*, 553 (1991) 223.

97   P.M. Bramley, *Phytochem. Anal.*, 3 (1992) 97.

98   V. Nemeth-Kiss, E. Forgács, T. Cserháti and G. Schmidt, *J. Chromatogr. A*, 750 (1996) 253.

99   F. Laborda, M.T.C. De Loos-Vollebregt and L. de Galan, *Spectrochim. Acta B*, 46 (1991) 1089.

100 H. Sawatari, T. Asano, X. Hu, T. Saizuka and H. Haraguchi, *Anal. Sci.*, 7 (1991) 477.

101 K.L. Sutton and J.A. Caruso, *J. Chromatogr. A*, 856 (1999) 243.

102 C.A. Ponce de León, M. Montes-Bayón and J.A. Caruso, *J. Chromatogr.*, 974 (2002) 1.

103 G.W. Somsen, C. Gooijer and U.A.Th. Brinkman, *J. Chromatogr. A*, 856 (1999) 213.

104 A.M. Robertson, D. Littlejohn, M. Brown and C.J. Dowle, *J. Chromatogr.*, 588 (1991) 15.

105 R.M. Robertson, J.A. de Haseth and R.F. Browner, *Appl. Spectrosc.*, 44 (1990) 8.

106 J.J. Gagel and K. Biemann, *Anal. Chem.*, 58 (1996) 2184.

107 G.W. Somsen, R.J. van der Nesse, C. Gooijer, U.A.Th. Brinkman, N.H. Velthorst, T. Visser, P.R. Kootstra and A.P.J.M. de Jong, *J. Chromatogr.*, 552 (1991) 635.

108 K. Albert, *J. Chromatogr. A*, 856 (1999) 199.

109 I.D. Wilson, *J. Chromatogr. A*, 892 (2000) 315.

110 J.-L. Wolfender, S. Rodriguez and K. Hostettmann, *J. Chromatogr. A*, 794 (1998) 299.

111 B. Schneider, Y. Zhao, T. Blitzke, B. Schmitt, A. Nookandeh, X. Sun and J. Stockigt, *Phytochem. Anal.*, 9 (1998) 237.

112 I.D. Wilson, E.D. Morgan, R. Lafont, J.P. Shockcor, J.C. London, J.K. Nicholson and B. Wright, *Chromatographia*, 49 (1999) 374.

113 K. Pusecker, K. Albert and E. Bayer, *J. Chromatogr. A*, 836 (1999) 245.

114 A.E. Mutlib, J.T. Strupczewski and S.M. Chesson, *Drug Metab. Disp.*, 23 (1995) 951.

115 E. Clayton, S. Taylor, B. Wright and I.D. Wilson, *Chromatographia*, 47 (1998) 264.

116 W.J. Ehlhardt, J.M. Woodland, T.M. Baughman, M. Vandenbranden, S.A. Wrighton, J.S. Kroin, B.H. Norman and S.R. Maple, *Drug Metab. Disp.*, 26 (1998) 42.

117 W.M.A. Niessen, *Liquid Chromatography–Mass Spectrometry*, 2nd Edn. Dekker, New York, 1999.

118 G. Bartolucci, G. Pieraccini, F. Villanelli, G. Moneti and A. Triolo, *Rapid Commun. Mass Spectrom.*, 14 (2000) 967.

119 P.R. Tiller, Z. El Fallah, V. Wilson, J. Huysman and D. Patel, *Rapid Commun. Mass Spectrom.*, 11 (1997) 1570.

120   D.M. Drexler, P.R. Tiller, S.M. Wilbert, F.Q. Bramble and J.C. Schwartz, *Rapid Commun. Mass Spectrom.*, 12 (1998) 1501.

121   D. Figeys and R. Aebersold, *Electrophoresis*, 18 (1997) 360.

122   A. Marina, M.A. García, J.P. Albar, J. Yagüe, J.A. López de Castro and J. Vázquez, *J. Mass Spectrom.*, 34 (1999) 17.

123   D.A. Laude, E. Stevenson and J.M. Robinson, in R.B. Cole (Ed.), *Electrospray Ionization Mass Spectrometry*, Wiley, New York, 1997, p. 291.

124   E.R. Williams, *Anal. Chem.*, 70 (1998) 179A.

125   M. Mann, P. Hojrup and P. Roepstorff, *Biol. Mass Spectrom.*, 22 (1993) 338.

126   A.L. McCormack, D.M. Shieltz, B. Goode, S. Yang, G. Barnes, D. Drubin and J.R. Yates, III, *Anal. Chem.*, 69 (1997) 767.

127   F.S. Pullen, G.L. Perkins, K.I. Burton, R.S. Ware, M.S. Taegue and J.P. Kiplinger, *J. Am. Soc. Mass Spectrom.*, 6 (1995) 394.

128   R.C. Spreen and L.M. Schaffter, *Anal. Chem.*, 68 (1996) 414A.

129   J.N. Kyranos and J.C. Hogan, Jr., *Anal. Chem.*, 70 (1998) 389A.

130   G. Hegy, E. Görlach, R. Richmond and F. Bitsch, *Rapid Commun. Mass Spectrom.*, 10 (1996) 1894.

131   D.B. Kassel, M.D. Green, R. Wehbie, R. Swanstrom and J. Berman, *Anal. Biochem.*, 228 (1995) 259.

132   T. Wang, L. Zeng, T. Strader, L. Burton and D.B. Kassel, *Rapid Commun. Mass Spectrom.*, 12 (1998) 1123.

133   E.W. Taylor, M.G. Qian and G.D. Dollinger, *Anal. Chem.*, 70 (1998) 3339.

134   E. Görlach, R. Richmond and I. Lewis, *Anal. Chem.*, 70 (1998) 3227.

135   J.R. Krone, R.W. Nelson, D. Dogruel, P. Williams and R. Granzow, *Anal. Biochem.*, 244 (1997) 124.

136   M.L. Nedved, S. Habibi-Goudarzi, B. Ganem and J.D. Henion, *Anal. Chem.*, 68 (1996) 4228.

137   A.C. Hogenboom, A.R. de Boer, R.J.E. Derks and H. Irth, *Anal. Chem.*, 73 (2001) 3816.

138   L. Zeng, L. Burton, K. Yung, B. Shushan and D.B. Kassel, *J. Chromatogr. A*, 794 (1998) 3.

139   L. Zeng and D.B. Kassel, *Anal. Chem.*, 70 (1998) 4380.

140   D.M. Drexler and P.R. Tiller, *Rapid Commun. Mass Spectrom.*, 12 (1998) 895.

141   W.M.A. Niessen, J. Lin and G.C. Bondoux, *J. Chromatogr. A*, 970 (2002) 131.

142   N.J. Clarke, D. Rindgen, W.A. Korfmacher and K.A. Cox, *Anal. Chem.*, 73 (2001) 430A.

143   N.J. Haskins, C. Eckers, A.J. Organ, M.F. Dunk and B.E. Winder, *Rapid Commun. Mass Spectrom.*, 9 (1995) 1027.

144   G. Hopfgartner and F. Vilbois, *Analusis*, 28 (2000) 906.

145   P.R. Tiller, A.P. Land, I. Jardine, D.M. Murphy, R. Sozio, A. Ayrton and W.H. Schaefer, *J. Chromatogr.*, 794 (1998) 15.

146   L.F. Colwell, C.S. Tamvakopoulos, P.R. Wang, J.V. Pivnichny and T.L. Shih, *J. Chromatogr. B*, 772 (2002) 89.

147   B. Kaye, W.J. Heron, P.V. McRae, S. Robinson, D.A. Stopher, R.F. Venn and W. Wild, *Anal. Chem.*, 68 (1996) 1658.

148   J.L. Herman, *Rapid Commun. Mass Spectrom.*, 16 (2002) 421.

149   A. Marchese, C. McHugh, J. Kehler and H. Bi, *J. Mass Spectrom.*, 33 (1998) 1071.

150   B.K. Matuszewski, M.L. Constanzer and C.M. Chavez-Eng, *Anal. Chem.*, 70 (1998) 882.

151   D. Barceló (Ed.), *Applications of LC–MS in Environmental Chemistry*, Elsevier, Amsterdam, 1996.

152   E.M. Thurman, I. Ferrer and D. Barceló, *Anal. Chem.*, 73 (2001) 5441.

153   M. Petrovic, E. Eljarrat, M.J. Lopez de Alda and D. Barceló, *J. Chromatogr. A*, 974 (2002) 23.

154   J. Slobodník, A.C. Hogenboom, J.J. Vreuls, J.A. Rontree, B.L.M. van Baar, W.M.A. Niessen and U.A.Th. Brinkman, *J. Chromatogr. A*, 741 (1996) 59.

155   C. Aguilar, I. Ferrer, F. Borrull, R.M. Marcé and D. Barceló, *J. Chromatogr. A*, 794 (1998) 147.

156   S. Akashi, M. Shirouzu, T. Terada, Y. Ito, S. Yokoyama and K. Takio, *Anal. Biochem.*, 248 (1997) 15.

157   H. Deissler, M.S. Wilm, B. Genc, B. Schmitz, T. Ternes, F. Naumann, M. Mann and W. Doerfler, *J. Biol. Chem.*, 272 (1997) 16761.

158   G. Neubauer, A. Gottschalk, P. Fabrizio, B. Seraphin, R. Luhrmann and M. Mann, *Proc. Natl Acad. Sci. USA*, 94 (1997) 385.

159   K.F. Medzihradszky, M.J. Besman and A.L. Burlingame, *Anal. Chem.*, 69 (1997) 3986.

160   E. van der Heeft, G.J. ten Hove, C.A. Herberts, H.D. Meiring, C.A.C.M. van Els and A.P.J.M. de Jong, *Anal. Chem.*, 70 (1998) 3742.

Erich Heftmann (Editor)
Chromatography, 6th edition
Journal of Chromatography Library, Vol. 69A
© 2004 Elsevier B.V. All rights reserved

Chapter 11

Microfabricated analytical devices

ANDRAS GUTTMAN and JULIA KHANDURINA

CONTENTS

11.1 Introduction 431
11.2 Capillary electrophoresis on microchips 432
    11.2.1 Microfabrication methods 432
    11.2.2 Fluid manipulation and injection techniques 434
    11.2.3 Detection schemes 436
11.3 Applications 437
    11.3.1 DNA fragment analysis and sequencing 437
    11.3.2 Rapid protein analysis 442
    11.3.3 Separation of other biologically important molecules 446
    11.3.4 High-throughput screening 447
11.4 System integration 451
    11.4.1 Multi-functional micro-devices for sample processing and analysis 451
    11.4.2 Integration of monolithic stationary phases 454
    11.4.3 Micro-reactors 456
    11.4.4 Micro-preparative methods 458
11.5 Modeling by computational fluid dynamics 461
Acknowledgment 462
References 462

11.1 INTRODUCTION

The applicability of contemporary microfabrication technology to analytical devices initiated a rapidly advancing interdisciplinary field in separation science. Microfluidic systems are used for transporting and manipulating minute amount of fluids and/or biological entities through micro-channel manifolds and allow integration of various chemical processes into fast, automated, and monolithic structures. Microfabricated analytical devices, often referred to as lab-on-a-chip systems, include micro-separation units, miniaturized reactors, micro-arrays, and most possible combinations of the above. The latest achievements in this rapidly progressing field are summarized in a few recently published books [1–3], describing in detail the broad multidisciplinary subject of microfluidics, originating in different areas of physics, chemistry, biology, and engineering. Materials and micro-technologies, currently used to fabricate microfluidic devices, as well as the main principles and formats of typical electrokinetic manipulations, separations, and detection methods, have been detailed in several reviews [4,5]. Most of the progress in this area has been in the rapidly developing fields of the global analysis of genomes, proteomes, and metabolomes. The topics covered in this chapter are limited to the most recent developments and trends in microfluidic analyses of biological interest, mainly DNA, proteins, and complex-carbohydrate analysis and high-throughput screening. It has become evident that there is a tremendous market potential for micro-devices supporting diagnostics, drug discovery, and evaluation of new pharmaceuticals. Some of the main advantages of miniaturization include improved performance, separation speed and throughput, reduced costs and reagent consumption, and the promise of integrated and parallel analysis [4,6–9]. Microfabricated analytical devices are also expected to satisfy the urgent demand of large-scale applications.

11.2 CAPILLARY ELECTROPHORESIS ON MICROCHIPS

11.2.1 Microfabrication methods

A number of microfabrication techniques, initially developed for micro-electronics and micro-electro-mechanical systems (MEMS), are now penetrating the biotechnology field and are being used for building analytical devices. Compared to silicon-based micro-electronic devices, biochips are much more diverse, due to the large variety of materials, chemicals and fluids used. The microfabrication methods comprise photolithography in rigid materials (glass, quartz, silica) as well as fabrication in plastics and elastomers. Glass substrates are the most common because of good optical properties and well-developed microfabrication technology and surface chemistry. On the other hand, various polymeric materials are becoming more and more attractive for certain types of applications due to their potentially low manufacturing costs and concomitant disposability. Informative overviews as well as detailed descriptions of microfabrication methods used for patterning fine structures and assembling analytical micro-devices can be found in a number of recent publications [1–5,10].

A typical lithographic process consists of three successive steps, schematically shown in Fig. 11.1 [10]:

(1) coating a substrate with an irradiation-sensitive polymer layer ("resist");

(2) exposing the resist to light, electron or ion beams; and

(3) developing the resist image with a suitable chemical. Exposure can be effected either through a mask for parallel replication or by scanning a focused beam, pixel by pixel, to form a designated pattern. Conventional photolithography involves the interaction of an incident beam with a solid substrate. Absorption of light or inelastic scattering of particles affects the chemical structure of the resist by changing its solubility. The resist response can be positive or negative, depending on whether the exposed on unexposed portions are removed from the substrate after development. The development

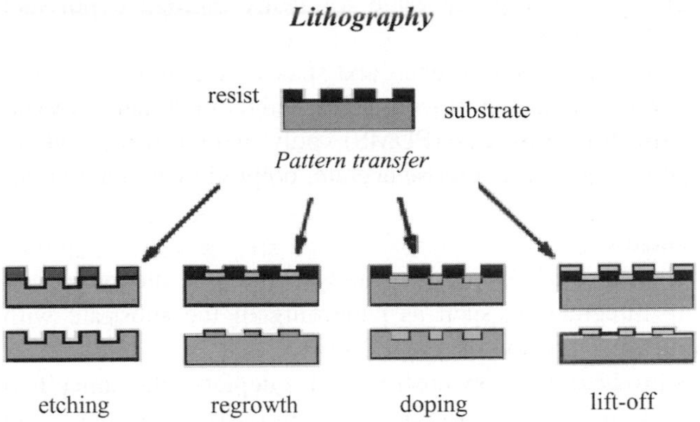

Lithography

Fig. 11.1. Schematic representation of lithography and pattern transfer techniques. (Reprinted from Ref. 10 with permission.)

step is followed by pattern transfer from the resist to the substrate. Different pattern transfer techniques are available: etching of unprotected areas (both wet chemical and dry plasma etching can be used), selective growth of materials in the grooves of the resist, and doping through the open areas by diffusion or implantation [10]. Multilevel fabrication is achieved by performing lithography and pattern transfer for each level. Conventional lithographic methods comprise electron-beam lithography for primary patterning directly from a computer-designed pattern; focused ion-beam lithography for highly localized micro-patterning; optical projection lithography used for mass production of integrated circuits; extreme UV lithography with laser-induced plasmas or synchrotron radiation; X-ray lithography and electron- and ion-projection lithography [11]. Among the emerging nonconventional microfabrication methods it is worthwhile to mention nano-imprint lithography for creating three-dimensional structures in heated polymers with a rigid mold; soft lithography or micro-contact printing, based on the use of an elastomer stamp to ink a solid substrate with self-assembling mono-layers, which then is used as a mask for wet etching (fine features, as small as 200 μm, can be obtained); near-field optical lithography; proximity-probe lithography, based on surface modification by means of scanning tunneling microscopy, atomic-force or scanning near-field optical microscopy; and some other methods [10,12]. Replication technologies widely employed for micro-device fabrication in plastics are very well summarized in recent reviews [4,6,13,14], which describe master fabrication, hot embossing, injection molding, casting and soft lithography, as well as some direct techniques, such as laser ablation, milling, optical lithography in deep resists, stereo-lithography, multi-layer approaches with sacrificial layers and thin-film growth. Other microfabrication aspects include sealing steps to form complete devices with enclosed micro-channels by direct bonding, lamination, gluing, and possibly welding (using a laser or ultrasonic-wave energy) [13].

Some important advantages of polymeric materials over more "traditional" glass and fused silica include:

(a) inert surface characteristics, especially favorable for many biological polymer separations and enzymatic reactions;

(b) feasibility of fabrication of high-resolution deep and shallow features with well-defined vertical walls, for which a number of well-developed methods and a variety of materials [*e.g.*, poly(dimethylsiloxane) (PDMS), poly(methyl methacrylate) (PMMA), polystyrene, polycarbonate, cellulose acetate, poly(ethylene terephthalate)] are readily available;

(c) a certain chemical resistance;

(d) low fabrication temperatures, suitable for performing some delicate biological assays and other on-chip integrations, such as patterning of the substrate with different chemicals, coatings, deposition of microelectrodes, etc.;

(e) applicability of fast mass production at low cost. Fig. 11.2 depicts schematically a typical fabrication process of the elastomer PDMS by means of soft lithography and replica molding – an excellent approach for rapid prototyping of cheap disposable microfluidic devices [14].

The first application of biodegradable polymer micro-structures and micro-devices was reported by Armani *et al.* [15]. The degradability in tissues holds an immense promise for such materials in designing implantable medical micro-devices. The authors described novel techniques for microfabrication, using biodegradable polycaprolactone, which enabled formation of 3-D microstructures *via* silicon micro-molding, transferring metal patterns to the plastic substrates, and sealing both dry and liquid-filled micro-cavities with a thin gold film [16,29]. Microfabrication of PMMA with the use of deep X-ray etching is being actively developed in Soper's group [16], and their new complex PMMA-based device, integrating nano-reactors and micro-separation devices is described in Sec. 11.4.

The applicability of a novel and promising polymeric material, Zeonor 1020, to the fabrication of microfluidic chips has been investigated at Cornell University [17]. The authors employed conventional embossing techniques to fabricate electrophoresis micro-channel structures, coupled with a micro-sprayer *via* a micro-liquid junction. The device provided efficient electrospray MS detection of CE-separated components. Zeonor polymer, commonly used in manufacturing compact disks, is inexpensive and possesses favorable physical characteristics, such as good chemical resistance to organic solvents, applicability of standard plastic microfabrication techniques, as well as good bonding and optical properties. All of the above could make this new polymer a desirable microchip manufacturing candidate. Some other recent plastic-based implementations of micro-fluidic devices for integrated processing and analysis of biological samples will be described in the following sections.

11.2.2 Fluid manipulation and injection techniques

For miniaturized analysis systems, sample handling and manipulation are of great importance. Specific problems resulting from the shrinking of macroscopic systems include failure of samples to be representative and of manipulations to be reproducible. This makes it necessary to incorporate specific micro-scale techniques and components in such devices. In particular, analytical performance of microfluidic devices is drastically affected by dead-volumes in the system and also by surface properties of the fluidic

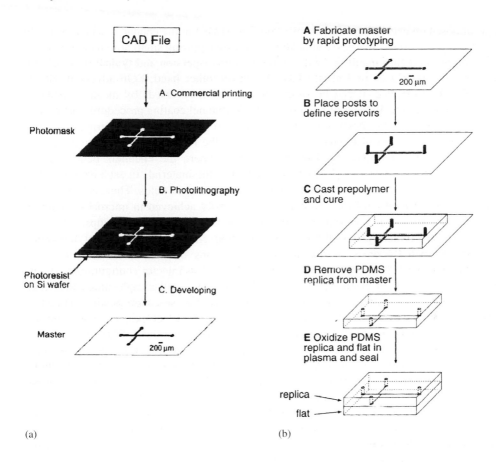

(a)   (b)

Fig. 11.2. Rapid prototyping and replica molding of microfluidic devices in polydimethylsiloxane (PDMS). (a) The channel design is created by a computer-assisted program and the file, printed on a high-resolution transparency, also serves as a photomask, in contact photolithography, to produce a positive relief master in a photoresist (SU-8). The exposed areas are polymerized and the rest is dissolved. (Reprinted from Ref. 14 with permission.) (b) Replica molding by casting a pre-polymer on a master, curing, surface treatment with oxygen/air plasma, and sealing a microdevice. (Reprinted from Ref. 182 with permission.)

channels. Microfluidic devices for separation-based analysis are typically operated by means of electrokinetic forces and/or pressure. Microchip-based, integrated CE systems were pioneered by Widmer's group in the early 1990s [18,19] and have been developing fast over the last decade. Transferring separation concepts of conventional CE, capillary electrochromatography (CEC), micellar electrochromatography, sample stacking and derivatization techniques to the planar microchip format, in conjunction with the integration of the various analysis steps, into a monolith system, resulted in dramatic improvements of analytical performance, speed, and throughput. Electrokinetic forces include both electrophoretic and electro-osmotic phenomena, which are often superimposed on each other. Electro-osmotic flow (EOF) in the channels, which is due to the presence of surface charges, can be utilized for transporting small amounts of material through the microchip manifolds with minimal zone dispersion and typical flow-rates in the order of 100 pL/sec to 1 nL/sec [20]. If, on the other hand, EOF adversely affects separation performance, it can be suppressed (or even reversed) by means of surface modification or passivation. A number of capillary/channel coating procedures, developed for CE or CEC [21–23], have been adapted for microfluidic chips. If external potentials are applied to a system of interconnected channels, the respective field strength in each channel will be determined by Kirchhoff's laws, where the channels behave as an electrical network of resistors [24]. Mass transport of the materials in each channel takes place according to the direction and magnitude of the electric fields. Thus, complex fluid manipulations at femtoliter to nanoliter scale are easily achieved in microchips without any active control devices or moving parts, such as valves or external pumps.

Samples are typically introduced into separation channels by means of integrated injectors, formed by cross or double-tee intersections of sample/sample waste and separation channels [18,20,25]. Such miniaturized valveless injector configurations enable precise volume-defined electrokinetic injections of very small sample plugs with high reproducibility [24]. Different injection configurations, *e.g.*, so-called pinched and gated modes, allow time-based analyte volume control [18,25]. Electrokinetic manipulations can be used on-chip for stack injection [25], zone manipulations [20,24–26], fraction collection, on-line sample derivatization, and other applications [4,26]. Hydrodynamic forces can be employed in microfabricated devices for reagent delivery [27,28], direct coupling to external flow-through pressure-driven analyzers for uninterrupted sampling [29], and precise on-chip pressure-driven mixing of reagents and liquids to carry out various reactions and assays [30]. A multi-port flow-control system for microfluidic devices has been reported [30] for dye mixing and enzyme titration experiments with pressure-driven flow only.

11.2.3 Detection schemes

Extremely small amounts of analyte require sensitive detection methods. Laser-induced fluorescence (LIF) has been the most popular detection scheme so far, due to its inherent selectivity and sensitivity [4]. Typically, a laser beam is focused inside the micro-channel as a small spot to excite flourescently labeled analyte molecules passing the detection window. This is achieved by either angular [31] or confocal illumination [32] and the fluorescent light is collected and collimated with a microscope objective, passed through a set of optical filters and dichroic mirrors, if necessary, to a detector, such as a photomultiplier tube (PMT), avalanche photodiode (APD), or charge-coupled device (CCD). Ultraviolet or visible-light absorption-based detectors have been also tested for microchips [33,34] with a specially fabricated detection cell and additional channels to integrate optical fiber guides to overcome sensitivity limitations.

Electrochemical methods offer an alternative approach to on-chip detection by electrode microfabrication [35–37], benefiting from the possibility of complete integration of the analysis and miniaturized detection on a single device. Mass spectrometry, as one of the most powerful detection and identification techniques, has been successfully interfaced with micro-devices and has significantly enhanced microchip applicability. Specific designs and applications of microchip/MS systems will be outlined in Sec. 11.3.2. Among other detection schemes are Raman spectroscopy [38] and holographic refractive-index detection [39].

11.3 APPLICATIONS

11.3.1 DNA fragment analysis and sequencing

Analysis of nucleic acids is one of the leading applications of microchip-based analysis today. Because deleterious effects of Joule heating during electric field-mediated separations in micro-channels are negligible and very small, well-defined sample plugs can be injected, the separation efficiency of microfluidic analytical devices is mainly diffusion-limited. This results in a performance that is superior to slab-gel and capillary electrophoresis. Rapid, high quality separations on-chip have been demonstrated for the analysis of oligonucleotides, RNA and DNA fragments, as well as in genotyping and sequencing applications. Typically, LIF is employed for detection, but other methods such as MS and electrochemical detection have been applied as well.

Shorter separation distances than in conventional CE present new challenges to the optimization of such separating conditions as electrokinetic manipulations, channel geometry, and sieving media. The geometrical effects of folded (serpentine) micro-channel structures on band-broadening have been extensively studied [40,41]. A variety of separation matrices, developed for CE instruments, are based on viscous solutions of entangled water-soluble polymers, and have been successfully applied to microchip electrophoresis of DNA. Linear polyacrylamide (LPA) and its derivatives, *e.g.*, poly(dimethylacrylamide), as well as poly(ethylene oxide) (PEO), poly(vinylpyrrolidone) (PVP), poly(ethylene glycols), hydroxyethylcellulose, and other polysaccharide and cellulose derivatives have been utilized for size separations of nucleic acid molecules in micro-channels [42]. Recently introduced novel thermoresponsive co-polymers, comprising hydrophobic and hydrophilic blocks, such as pluronics [43–45] and poly(isopropylacrylamide), grafted with poly(ethylene oxide) chains [46], have shown promising results. These matrices have pronounced temperature-dependent viscosity transition points, and this suggests promising implementations. In particular, thermoresponsive polymers can offer some practical advantages in micro-channel electrophoresis, enabling easier handling and loading of viscous polymer solutions without the requirement of a high-pressure manifold. Barron's group [47] has constructed interesting "viscosity-switch" materials, which respond to changes in temperature, pH, or ionic strength. These matrices are based on co-polymers of acrylamide derivatives with variable hydrophobicity and possess reversible temperature-controlled viscosity-switch characteristics (change of high-viscosity solutions at room temperature to low-viscosity colloid dispersions at elevated temperatures). Also, high resolving power and good DNA sequencing performance (463 bases in 78 min) was observed for these copolymers.

An example of a matrix-free approach to DNA separations has been presented by Han *et al.* [48]. A nano-fluidic channel was designed and fabricated to separate long DNA molecules, based on the so-called "entropic-trap" principle. The separation channel had relatively wide (1.5–3 μm) and narrow (75–100 nm) regions, causing size-dependent trapping of DNA at the start of each constriction (Fig. 11.3). The mobility differences created by this means enabled DNA separation without the use of any sieving polymer matrix or pulsed electric fields. Fragments in the 5–160 kbp range were successfully separated in a 1.5-cm channel. A device with parallel channels was also demonstrated. Simplicity of the microfabrication approach, its efficiency, and the possibility of

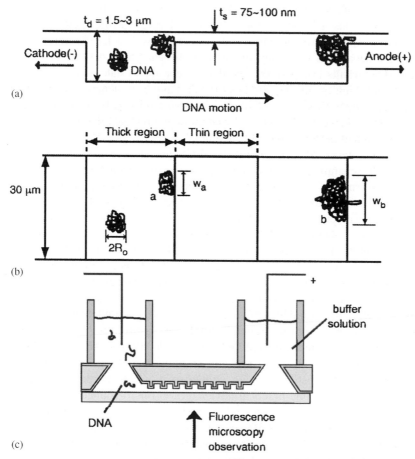

Fig. 11.3. Nano-fluidic separation device with entropic traps. (a) Schematic cross-sectional diagram of the device. DNA molecules are trapped during electrophoresis whenever they meet a thin region with the depth (t_s) much smaller than the radius of gyration (R_o). t_d, depth of the larger channel regions. (b) Top view of the device in operation. Trapped DNA molecules eventually escape, with an escape probability proportional to the length of the slit covered by the DNA molecule (w_a and w_b). Larger molecules have higher escape probability because they cover wider regions of the slit ($w_b > w_a$). (c) Experimental setup. (Reprinted from Ref. 48 with permission.)

theoretical modeling suggested a potential practical combination of these results into an integrated DNA analytical system.

The major applications of electrophoresis on microchips include sizing of double-stranded DNA fragments [5,49–53], short single-stranded oligonucleotides [54], and ribosomal RNA fragments [55]. High separation performance and speed (ranging from a few seconds to a few minutes) have been achieved. DNA genotyping on microchips enables quick identification of genes and can substantially enhance the capabilities of genomic, diagnostic, pharmacogenetic, and forensics tests. The identification of genes related to hereditary diseases, such as muscular dystrophy [56] and hemachromatosis [57], has been accomplished on-chip. An ultra-fast allelic profiling assay for the analysis of short tandem repeats (STR) has also been demonstrated [58]. Separation of the CTTv quadruplex allelic ladder system was accomplished in less than 2 min, using a 26-mm separation distance and resulting in an analysis time which was 10 and 100 times shorter than that for capillary or slab-gel electrophoresis, respectively [59]. Two-color multiplex analysis of eight STR loci has also been performed by the same group.

A number of publications have demonstrated the feasibility of DNA sequencing in microfluidic chips [30,63,64]. This is one of the most challenging tasks in DNA separation, since very high resolving power is required for accurate base calling. High-speed, high-throughput four-color sequencing analysis has been performed on-chip in 20 min with a read-length of over 500 bases [60]. Microfabricated devices can perform high-throughput DNA sequencing in a 96-channel array format [27]. Based on the best results reported so far, roughly 1200 bases per channel per hour can be identified, yielding 2.7 Mbases per day in a 24-h cycle with 96-lane chips [61]. Combining the capabilities of microfluidics with the sensitivity of fluorescence-based detection methods, a single DNA molecule can be detected on electrophoresis microchips. The single-molecule fluorescence-burst counting technique has been used to monitor DNA separations in microfluidic channels with the aid of an avalanche photodiode detector and sample-stream focusing, either electrodynamically or by physical tapering of the separation column in the detection region [62].

DNA field-inversion electrophoresis has been implemented on a microchip device to accommodate a shorter effective separation length [63]. The method employed a pulsed field, periodically switched at 180° angle with a certain optimized frequency at which DNA molecules exhibit a mobility minimum and band-inversion phenomena. Short oligonucleotides, 20-, 40-, and 60-mers, were effectively separated, using only a 6-mm separation distance. The authors suggested that this technique might have a great potential for single-strand conformation polymorphism (SSCP) analysis. An interesting approach to reducing the applied separation voltage, while maintaining high electric field strength, has been proposed [64]. An array of microfabricated electrodes divided a channel into smaller separation zones. A small DC voltage was sequentially applied to the proper pairs of electrodes to generate a moving electric field. Different electrode layouts and their effects on the field uniformity were investigated. This technique contributes to the versatility of tools for microfluidic manipulations. A number of studies have dealt with the optimization of micro-scale DNA amplification by polymerase chain reaction (PCR). Two possible embodiments can be envisioned, an enclosed micro-chamber and an open well format, both of which can be multiplexed to an array. Fast, real-time PCR analysis was demonstrated in silicon micro-chambers by several groups [65–68]. DNA amplification can be coupled with special microfluidic cartridges designed to carry out several sequential steps of DNA extraction from different sample types prior to the PCR [69]. DNA amplification in open format has not yet been reported, but other chemical and enzymatic reactions, conducted in nano- and pico-vials, have shown promising results [70–73]. There have been several reports on integration of DNA amplification and electrophoretic separation on a single microfabricated chip [35,74–78]. These devices contain small reaction wells, which are thermocycled to generate the amplicons, followed by the injection/separation/detection steps in the interconnected micro-channel network. An integrated system, combining fast on-chip DNA amplification by local thermocycling and pre-concentration techniques, followed by microchip electrophoretic sizing, was reported by Khandurina *et al.* [76]. DNA fragment pre-concentration was accomplished prior to injection by means of a microfabricated porous membrane structure [79] (Fig. 11.4). This method enabled reduction of the PCR cycle number and speeded up the total analysis time (<20 min) [76]. A microchip device, developed by Mathies' group, was coupled to a special micro-chamber with fast heating/cooling capabilities for effective DNA amplification [35]. Recently, the same group developed an integrated monolithic system incorporating several 280-nL PCR chambers etched into a glass structure and connected to microfluidic valves and hydrophobic vents for sample introduction and immobilization during the cycling (Fig. 11.5) [77,78]. The low thermal mass and use of thin-film heaters enabled cycle times as short as 30 sec. The amplified products were directly injected into the corresponding gel filled microchannels and detected by laser-induced fluorescence. The excellent detection sensitivity attained (20 copies/μL) suggested that it might be possible to perform stochastic single-molecule PCR amplification.

It is worth noting that there are similar trends of system integration in capillary electrophoresis. An automated capillary array machine for DNA typing directly from blood samples [80], as well as nano-reactor-based cycle sequencing instrument [81], have been developed by Yeung's group. Microfluidic elements and a micro-thermocycler were incorporated to co-amplify STR loci or carry out sequencing reaction, as well as on-line CZE purification of the products, loading/injection, and regeneration/cleaning. An interesting alternative to a micro-chamber reaction format has been proposed by Manz's group, who demonstrated a continuous-flow PCR system [82,83]. A PCR cocktail was pumped continuously through a serpentine glass channel, periodically passing the three different temperature zones to perform denaturation, annealing and amplification steps. Although, total reaction time was relatively long (50 min for 20 cycles), multiple simultaneous reactions can be carried out by sequential introduction of separate reactions in each loop.

Ultra-short thermal cycling times of 17 sec per cycle have been achieved by employing IR radiation-mediated heating and compressed air cooling to amplify DNA in a capillary or on-chip microchambers by Landers' group [84–86]. A tungsten lamp, thermocouple feedback, and computer interface were used to accurately control the temperature by a combination of light intensity and air-flow. A novel thermocycling system, based on a capillary, equipped with bi-directional pressure-driven flow and *in situ* optical position sensors, has been presented by Ehrlich's group [87]. A 1-μL PCR mix droplet was

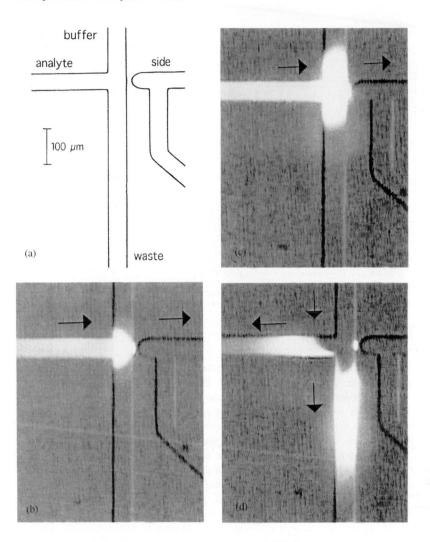

Fig. 11.4. Overlay of injection tee and porous membrane. (a) Charge-coupled device images of analyte concentrated for (b) 2 and (c) 3 min. (d) Injection of concentrated analyte plug. Porous membrane region 7 μm width. All channels filled with 3% linear polyacrylamide in 1 × TBE buffer. DNA sample, ΦX-174-Hae III digest with intercalating dye, TO PRO, added. In pre-concentration mode (b, c) 1 kV was applied between the analyte and side reservoirs with no potential at the buffer and waste ports. During injection (d), relative potentials applied to buffer, analyte, and waste ports were 0, 0.4 and 1.0 kV, respectively (with no potential at the side). (Reprinted from Ref. 79 with permission.)

controllably moved between three different heated zones in a 1-mm-ID oil-filled capillary. A light-scattering detector was used to control the droplet position. Thirty PCR cycles were completed in 23 min with good amplification efficiency. The maximum possible speed with this arrangement was estimated to be as high as 2.5 min. A hybrid PDMS-glass

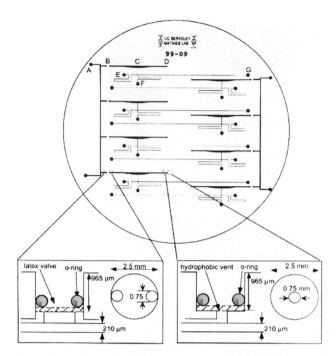

Fig. 11.5. Monolithic, integrated, microfluidic PCR amplification and CE analysis system. Schematic drawing of 100 mm mask for PCR-CE device. The sample is loaded in the fluidic bus reservoir, A. It then travels through the valve port, B, into the 280-nL PCR chamber, C, and stops at the vent port, D. (Hydrophobic vents are used to locate the sample and to eliminate gas.) The PCR chamber is directly connected to the injection cross and separation channel with its cathode, E, and anode, G, reservoirs. The insets show the side and top views of the valve and vent structures. This multireactor system permits the introduction of a single sample into four different reactors for multiple analyses, if desired. Two aluminum manifolds, for the vents and valves, are placed into the respective ports and clamped in place by applying vacuum. The manifolds are connected to the external solenoid valve for pressure and vacuum actuation. Thermocycling is accomplished with a resistive heater and a miniature thermocouple below the PCR chamber. (Reprinted from Ref. 77 with permission.)

microchip for functional integration of DNA amplification and gel electrophoresis was recently reported [88]. Thermoelectric heating/cooling control was used in this case. Such devices can be disposable, because they are inexpensive and relatively simple to fabricate.

11.3.2 Rapid protein analysis

In conventional proteomics proteins are extracted from cells, separated and detected by 2-D gel electrophoresis, bands are cut out and digested, and then the resulting peptide mixture is analyzed by MS. These techniques are rather slow and labor-intensive.

Integrated, fast, and high-throughput multi-sample systems are in great demand. Microfluidic devices offer new opportunities, unavailable in traditional protein analysis technologies, for control and automation of multiple sample processing steps. The majority of the recent developments in microfluidics for protein applications is related to the combination of single- or multi-dimensional micro-channel separations and MS detection [89,90]. Here, we will mention only a few interesting examples of the most recent microfluidics-based solutions in the rapidly developing world of proteomics.

A number of research groups focused their efforts on performing traditional 2-D gel electrophoresis in microchip format. The problems of protein purification and desalting prior to the analysis were addressed in microfabricated structures by the use of semi-permeable membranes, sandwiched between microfluidic manifold layers, and connecting the buffer and analyte counter-flows [91,92]. This micro-dialysis enabled effective clean-up and separation of analytes from low-molecular-weight compounds, and enhanced subsequent MS characterization. Similar fluidic constructions, microfabricated in plastic [92,93], are discussed in more detail later in this chapter. Examples of isoelectric focusing (IEF) of proteins in microchips demonstrate the potential of micro-scale techniques [89]. Chip-based IEF has been accomplished in 30 sec in a 7-cm channel with a capacity of *ca.* 30–40 peaks [94]. Electro-osmotically driven mobilization of the focused zones was found to be the most suitable technique for a microchip, due to easy implementation and high speed. Nevertheless, IEF analysis has been performed so far on simple samples only, and the resolving power achieved was low in comparison with conventional immobilized pH gradient IEF gels. An integrated IEF/ESI-MS plastic microfluidic device, coupling the electrospray tip to the microchip, has also been reported [95].

Chromatography and electrophoresis represent the two major separation techniques for proteins. The latter is by far more popular in microchip applications, because it is easier to apply in miniaturized formats. Nevertheless, on-chip chromatographic separations have recently been attempted by trapping coated beads in a micro-cavity within a micro-channel network [96], or by using *in situ* microfabricated structures [97] or polymerized [98] beds. Numerous examples of microchip electrophoretic separations of peptides and proteins in free-zone and micellar electrokinetic chromatography (MEKC) format have been demonstrated and reviewed [4,6,8,89]. SDS gel electrophoresis has also been performed in micro-channels [99]. Six proteins in the 9- to 116-kDa range were separated in 35 sec, which is five times as fast as the fastest SDS-protein separation in CE [100].

The development of more sophisticated microfluidics systems with both "horizontal" integration – by building parallel analysis lanes for high-throughput applications – and "vertical" integration – by implementing several functions on a single device – is perhaps the most exciting trend in microchip technology. Recently, a new microfluidic glass device was developed for integrated protein-sizing assay. It performs separation, staining, virtual de-staining and detection in less than 30 min and can sequentially analyze 11 different samples [101]. Universal noncovalent fluorescent labeling on-chip was employed in combination with post-separation dilution to reduce fluorescent background, associated with SDS micelle-bound dye, to increase the signal-to-noise ratio. Hydrophobic regions of the SDS/protein complexes bind noncovalent fluorogenic dyes, such as Sypro dyes, Nile Red, and ethidium bromide. This results in fluorescence enhancement, similar to that of intercalating dyes in DNA analysis, and offers certain advantages for protein labeling over the traditional chemical attachment [102]. The applicability of this technique to micro-scale protein separations has been investigated by several research groups [103–105]. High speed and sensitivity of microseparations, and the option of on-chip pre- or post-separation labeling, mixing, destaining, etc., proved to be clearly advantageous in a combination with noncovalent protein-staining dyes. High separation efficiency and low detection limits (in some cases as low as pg- and attomol level [104]) were demonstrated.

An ultimate microfabricated analog to 2-D IEF/SDS-PAGE has not yet been reported, but examples of two-dimensional analysis in microchips and micro-gels [106] have been published recently. Multi-dimensional separations are superior to one-dimensional methods, due to the multiplicative increase in peak capacity. Open-channel electro-chromatography (OCEC) and capillary electrophoresis, as first and second dimensions, respectively, have been implemented on a single glass chip [107] (Fig. 11.6). A 25-cm spiral OCEC channel, coated with octadecyl groups, was connected with a 1.2-cm straight CE channel through a cross-injection junction. The effluent from the first dimension was repeatedly injected into the CE channel every few seconds. The sampling (injection) rate of the effluent of the first dimension into the second one was optimized to perform representative CE analysis of the OCEC separation products. Fluorescently labeled tryptic digests of β-casein were analyzed in 13 min in this device. Another, similar 2-D microfabricated device [108] combined micellar electrochromatography (MEKC) and CE for the analysis of peptide mixtures. The peak capacity of the 2-D device was estimated to be in the 500–1000 range, greatly improving the separation efficiency of each dimension alone. The approach can be useful for rapid automated fingerprinting of proteins and peptides with possible coupling to MS detection.

Immunoassays represent a very important tool in clinical diagnostics and medical research. Various miniaturized immuno-assays mostly based on competitive antigen/anti-body interactions, have been performed successfully in electrophoresis microchips, demonstrating greater speed and the feasibility of automated analysis in a portable format. Earlier progress in this field has been described in several reviews [8,89]. Among the most recent developments, Gottschlich *et al.* [109] reported an integrated microchip device that sequentially performed enzymatic reactions, electrophoretic separation of the products, post-separation on-line labeling of the peptides and proteins, and fluorescent detection. Tryptic digestion of oxidized insulin B-chain was performed in 15 min under "stopped flow" conditions in a locally thermostated channel. The electrophoretic analysis was completed in 1 min, and the separated peptides were labeled with naphthalene-2,3-dicarboxaldehyde (NDA) by on-chip mixing. The same device was also used for on-chip reduction of disulfide bridges in insulin.

A new class of microfluidic immuno-assays, based on solid-supported lipid bilayers, has been described recently [110]. The bilayers, created on PDMS surfaces of an array of parallel microchannels, contained dinitrophenyl (DNP)-conjugated lipids for binding with fluorescently labeled anti-DNP antibodies of different concentration in each channel. The methodology can be used for performing rapid and accurate heterogeneous assays in a single experiment, and the amount of proteins required is significantly less than for conventional methods. Nano-vials are another example of the use of microfabricated devices for protein analysis [70,71]. These very small wells, fabricated in different

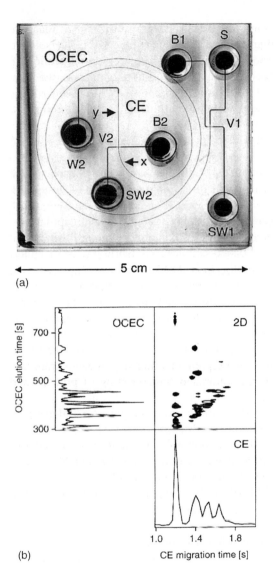

(a)

(b)

Fig. 11.6. (a) Microchip used for 2-D separations. The separation channel for the first dimension, open-channel electrochromatography (OCEC), extends from the first valve, V1, to the second valve, V2. The second dimension, capillary electrophoresis (CE), extends from V2 to the detection point, y (0.8 cm from V2). Sample (S), Buffers 1 and 2 (B1, B2), sample waste 1 and 2 (SW1, SW2), and waste (W) reservoirs are positioned at the terminals of each channel. The arrows indicate the detection points in the OCEC channel (x) and CE channel (y). (b) 2-D separation of tetramethylrhodamine isothiocyanate (TRITC)-labeled tryptic digest of β-casein. The projections of the 2-D separation into the first dimension (OCEC) and the second dimension (CE) are shown to the left and below the contour plot, respectively. The field strengths were 220 V/cm in the OCEC and 1890 V/cm in the CE channels. The buffer was 10 mM sodium borate with 30% (v/v) acetonitrile. (Reprinted from Ref. 107 with permission.)

substrates, feature a high surface-to-volume ratio and have shown higher reaction efficiency (*e.g.*, in enzymatic digestion) than conventional micro-fuge vials. The solvent evaporation problem was addressed by periodic solvent addition and/or humidification of the surrounding environment, as well as by continuous compensation for evaporation with an array of solvent-delivery capillaries [71]. Performing chemical reactions on a sub-nanoliter scale in a parallel fashion opens new opportunities in high-throughput proteomics, particularly when combined with MALDI-TOF-MS.

Since MS is the favored method of protein detection and characterization, a number of reports on microfluidic devices coupled to mass spectrometers have appeared [89,95, 111–117]. Such a combination enables automated sample delivery and enhanced MS analysis efficiency by, for instance, integrated sample processing/enrichment/clean-up and fractionation prior to detection. These devices can transport the analyte fluid electrokinetically or by pressure and generate electrospray *via* an attached capillary or more complex emitter couplings. Enzymatic digestion was monitored in real time with high detection sensitivity (0.1–2 pmol/μL of loaded sample) on a hybrid microchip nano-electrospray device by peptide mass fingerprinting [117]. The same research group used a microfabricated, electrically permeable, thin glass septum to generate electrospray by electroosmotically pumping the solutions through field-free channels, past the point where the CE electric potential was applied [116]. Automated MS analysis was reported by Ekstrom *et al.* [118], who utilized a porous, microfabricated digestion chip, integrated with a sample pre-treatment robot and micro-dispenser for transferring reaction products to a MALDI target plate. A new design for high-throughput microfabricated CE/ESI-MS with automated sampling from a micro-well plate has been reported by Karger's group [119,120]. The assembly combined a sample loading port, separation channel, and a liquid junction for coupling the device to the MS with a miniaturized sub-atmospheric electrospray interface (Fig. 11.7). The micro-device was attached to a polycarbonate manifold with external reservoirs, equipped for electrokinetic and fluid-pressure control, which simplified the micro-device design and enabled extended automatic operation of the system. A computer-activated electro-pneumatic distributor was used for sample loading and washing the channels. Automated CE/ESI-MS analysis of peptides and protein digests was successfully demonstrated.

11.3.3 Separation of other biologically important molecules

Applications of microfluidic devices to the analysis of complex carbohydrates and other biologically important molecules, such as lipids and fatty acids, have been very sparse so far, probably due to the limited development of CE techniques for these analytes, in contrast to polynucleotides and proteins [121]. Fractionation of lipids and fatty acids is complicated mainly because of their poor solubility in aqueous solutions. They therefore require nonaqueous organic media or surfactant micellar systems. Carbohydrate analysis, on the other hand, is complex due to wide diversity of carbohydrate structures [122]. Also, derivatization is often necessary to introduce charges and UV or fluorescent labels into the neutral sugar molecules. In spite of the lack of examples of microchip applications for the analysis of these biopolymers, it is reasonable to anticipate its rapid emergence in the future.

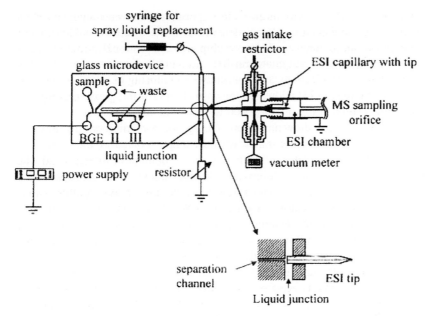

Fig. 11.7. Diagram of the micro-device with a subatmospheric electrospray interface. The expanded view shows the coupling of the ESI tip with the separation channel in the liquid junction. (Reprinted from Ref. 119 with permission.)

A mixture of sucrose, *N*-acetylglucosamine, and raffinose was separated in 17 sec on a synchronized cyclic capillary electrophoresis chip with a square channel, using a holography-based refractive index detection method [39]. Although the initial detection limits were poor compared with alternative techniques, the potential of a small-volume universal detection for chip-based separations was demonstrated. Wang *et al.* [123] achieved simultaneous microchip CE-based bioassays of a mixture of compounds (glucose, uric acid, ascorbic acid, and acetaminophen). Enzymatic oxidation of glucose was carried out in a micro-channel, followed by the separation and electrochemical detection of the neutral peroxides and anionic uric and ascorbic acids. The chip-based procedure allowed integration of miniaturized bioassays, separations and amperometric detection on a single device, competitive with benchtop analyzers and conventional biosensors in terms of performance, speed, sample volume, and size.

11.3.4 High-throughput screening

In research-based industrial projects, such as genomics, proteomics, metabolomics, and especially in drug discovery, the need to carry out a large number of complex experiments poses a real challenge in terms of efficiency, data quality, and cost. For example, in primary drug screening alone, major pharmaceutical companies test-score their new biological targets against compound libraries of hundreds or thousands of compounds per year, generating millions of data points in search of new targets. Therefore, high-throughput systems that perform large-scale experimentation and analysis while using minute amounts of reagents are in great demand. In spite of all the advantages of micro-scale instrumentation, an interface between chip and conventional liquid handling and analysis is often required to automate further the sample introduction/collection process for high-throughput applications and continuous monitoring of analytes. The so-called Sipper Chip Technology [124] addresses this issue by performing experiments in a serial, continuous-flow fashion at the rate of 5,000 to 10,000 experiments per channel per day. The system uses capillaries to draw nanoliters of reagents from micro-well plates into the channels of the chip, where they are mixed with the target molecules and a series of processing steps are carried out to determine whether the compound of interest binds the target. A range of assays, including fluorogenic kinase assays with electrophoretic separation of the reaction products from substrates, mobility shift assays, cellular assays with Ca^{2+}-sensitive fluorescent indicators yielding $50-100$ cells per data point, have been demonstrated, using various chip designs [125]. An example of a single-sipper microfluidic device for continuous-flow enzyme inhibition assays with direct sampling of the compounds from micro-titer plates is depicted in Fig. 11.8.

Another possible solution to this problem has been implemented recently by Attiya *et al.* [28]. The device contained a large sample-introduction channel with a volume flow-resistance $>10^5$ times lower than that in the analysis micro-channels. This approach enabled interfacing the large sample-introduction channel with an external pump (flow-rate up to 1 mL/min) for pressure-driven sample delivery without perturbing the solutions and electrokinetic manipulations within the rest of the micro-channel network. On-chip mixing, reaction, and separation of ovalbumin and anti-ovalbumin were demonstrated with good performance and reproducibility. Such a strategy for decoupling the electrokinetic flow in the micro-channels from the external environment extends the applicability of microchip analysis and provides a useful alternative to mechanical valve flow control. Shi *et al.* [27] introduced a pressurized capillary-array system for simultaneously loading 96 samples into 96-sample wells of a radial micro-channel-array electrophoresis micro-plate for high-throughput DNA sizing (Fig. 11.9) (*cf.* Fig. 9.1). As a result, 96 samples were analyzed in less than 90 sec per micro-plate, demonstrating the power of microfabricated devices for large-scale and high-performance nucleic acid characterization. An interesting approach to parallel micro-scale separations has been developed by Ewing's group [126], who used ultra-thin (57-μm) slab-gel electrophoresis with its parallel processing capabilities in a combination with a capillary sample-delivery system. The technique allowed accurate and rapid automated injection (deposition) of multiple samples (PCR products, STR fragments, oligonucleotides, or amino acids) at the edge of an ultra-thin slab gel for subsequent separation and detection. This simple system eliminates cross-channels and additional buffer reservoirs, required for analyte injection in microchannel devices, and yields high analysis throughput, and requires minimal operator intervention. Recently, the same group demonstrated continuous parallel separations in an array of separation lanes ($500-700$ μm wide and $250-350$ μm deep) fabricated in glass chips, using the same capillary sample-introduction method [127].

Membrane-mediated sample loading was introduced by Guttman and co-workers [128] in conjunction with micro-scale electrophoresis based on ultra-thin (≤ 190-μm) gel slabs. Minute amounts of DNA samples (0.2 μL) were spotted manually or robotically onto the

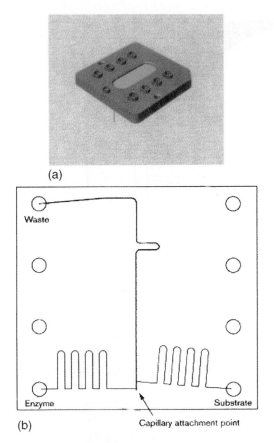

(a)

(b)

Waste

Enzyme

Substrate

Capillary attachment point

Fig. 11.8. A microfluidic device for continuous-flow enzyme-inhibition assay. (a) Test compounds are injected *via* a fused-silica capillary, perpendicularly attached to the microchip. (b) Schematic layout of the microfluidic channel structure of the chip. (Reprinted from Ref. 125 with permission.)

surface of the tabs of a membrane loader, which was then placed into the cathode side of the micro-cassette, in intimate contact with the straight edge of the ultra-thin layer of agarose gel. Under the applied electric field strength, the sample DNA molecules migrated into the gel, forming finely defined, sharp bands. This procedure facilitated the introduction of a large number of samples (96 parallel analysis lanes) into microgels, eliminating the need for forming individual wells and enabling convenient sample loading on the bench-top, outside the separation platform. The membrane-mediated injection method is readily automated (robotic spotting) and can be applied to most high-throughput applications, such as automated DNA sequencing [129].

Researchers in the drug discovery field also need microfluidics to cope with the drastically growing scale of combinatorial chemistry libraries and screening processes [125]. The goal is the fabrication of a chemical microprocessor that combines the reactions and screening on a single microdevice. One of the latest prominent industrial

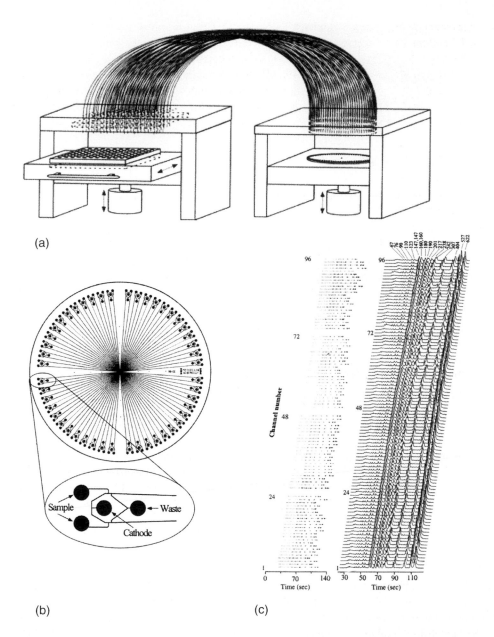

(a)

(b)   (c)

Fig. 11.9. (a) A 96-sample capillary-array loader. Pressurization of the micro-titer dish chamber to 21 kPa (*ca.* 3 psi) is used to transfer 96 samples (transfer volume *ca.* 1 μL) to the sample reservoirs of the radial micro-plate. (b) Mask pattern for the 96-channel radial capillary-array electrophoresis micro-plate. The substrate is 10 cm in diameter. Channels are 110 μm wide and 50 μm deep. The distance from the injector (double-T) to the detector is 3.3 cm. (c) Image and electropherograms of a simultaneous separation of 96 bBR322 *Msp*I/TOTO complex samples. Separation conditions: 200 V/cm field strength; 1% (w/v) hydroxyethylcellulose in TPA {[tetrapentylammonium 3-*tris* developments of the kind is a device which combines multiple reaction units on one chip [130]. These units are enclosed but connected to the outside by fluidic channels [131]. Two different reagents are fed into each reaction well, where the reaction takes place, followed by spectroscopic detection/identification. This biochip requires only nanoliters of reaction volumes and can be fabricated using materials appropriate for various chemical conditions. Three-dimensional microfluidic structures for complex integrated processing are currently being developed to serve in both chemical synthesis and genomic screening.

11.4 SYSTEM INTEGRATION

11.4.1 Multi-functional micro-devices for sample processing and analysis

Downscaling of conventional analytical methods and instrumentation has been one of the central but not the only trend driving the development of the μ-TAS (micro total analysis systems) concept. Research has therefore focused on exploring ways of building integrated devices that combine separations with pre- and post-processing steps. Among the latter are PCR-based DNA amplification, restriction digestion, other enzymatic assays, analyte pre-concentration, filtering, dialysis, and various post-derivatization and detection methods. Some of these complex devices, reported recently, have already been mentioned in Sec. 11.3. The ultimate integration of all of these steps into a single microchip has not been realized so far, although higher-level system integration and complex microdevices are rapidly emerging. Very recently, the first commercial products (Agilent and Shimadzu), for chip-based electrophoretic analysis of DNA, RNA, and proteins, have been introduced [132,133]. Another commercial product being developed intends to provide 96-channel microfluidic plastic LabCards [134] for high-throughput miniaturized biochemical assays, addressing the whole range of common laboratory procedures, such as mixing, incubation, metering, dilution, purification, capture, concentration, injection, separation, and detection. Other integrated microchips for clinical applications have been designed and prototyped [68,135,136]. A dual-function microchip [68] has integrated two key steps in the analytical procedure – cell isolation and PCR. White blood cells were isolated from whole blood on 3.5-μm elements of weir-type filters, etched in silicon-glass chips, followed by direct PCR amplification of isolated genomic DNA. Modification of filter size and shape and/or specific capture agents can be used to increase the selectivity of the devices.

A simple, elegant method of thermocapillary pumping, based on surface tension changes under localized heating, was incorporated into a microchip structure for guiding nanoliter solution droplets in a controllable manner through the channels, mixing and incubating them to perform enzymatic reactions [135,136]. This integrated system also incorporates thermocycling chambers, microfabricated gel electrophoresis channels, and a set of

---

(hydroxymethyl) methylamino]-1-propanesulfate}–TAPS {3-*tris*[(hydroxymethyl)methylamino]-1-propanesulfate}–EDTA (ethylenediaminetetraacetic acid) buffer (pH 8.4) as a sieving matrix; DNA concentration, 1 ng/μL; intercalating dye TOTO/DNA ratio 1:25. The numbers at the top indicate fragment sizes in base pairs. (Reprinted from Ref. 27 with permission.)

doped-diffusion diode elements, fabricated in silica, for the detection of β-radiation (^{32}P-labeled DNA). The thermocapillary pumping idea was recently also developed by Troian *et al.* [137], who proposed configurable thermofluidic arrays to pattern flow at the microscale level. The development of this approach will enable "pixelating" the chip surface for precise temperature control, pixel by pixel, and, consequently, surface tension and viscosity of the liquid droplets. By this means, it is possible to move, mix, or otherwise manipulate small volumes. Another integrated device, developed recently in Burns' group [138], employed photo-definable polyacrylamide gels as a sieving medium for DNA electrophoresis with a locally controlled gel interface and electrode-defined sample compaction and injection technique (Fig. 11.10). The latter approach was based on integrated microfabricated electrodes and helped achieve sample compaction without migration into the gel, enabling control over the size and application of the sample plug.

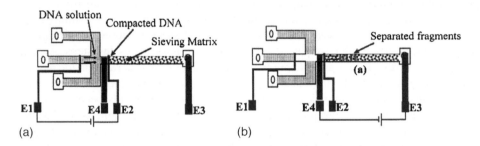

Fig. 11.10. Electrode arrangements in a microfabricated sample injection and separation system. Fifty-μm thin electrodes (E1, E2) are used for sample compaction and separation. A thick electrode (E4) is introduced to allow the use of higher voltages during the sample release, and separation phase. (a), (b) Schematic drawings of the operation of electrode-defined sample compaction, release and subsequent separation for the electrode system. (Reprinted from Ref. 138 with permission.)

An interesting alternative approach to combining informatics, bioassays, and miniaturization is a "laboratory on a disc", which recently has been implemented [139]. The so-called LabCD is a consumable compact disc with micro-scale fluid paths, reaction chambers, and valves. Fluid is moved along these pathways by capillary action and centrifugal forces, generated by the rotation of the disk, enabling it to perform various assay types. Digital information (software) is encoded on the other side on the same disc, automatically controls the disc motion and operation, and allows recording and analyzing of bioassay data from the microfluidic layer. Microfluidic processes, such as liquid transport, valving, mixing, and incubation are accomplished by geometrically designing chambers and interconnecting channels and programming the sequence of disc rotation. Integrated DNA amplification of blood and bacteria samples was performed by mixing individual components in the PCR chamber and then thermocycling [140]. LabCDs are intended to operate with "large" sub-μL to sub-mL volumes of solutions or biological liquids, and this would make them excellent practical diagnostic devices for point-of-care applications.

More recently, a flexible elastomer chip has been introduced, fabricated by multi-layer soft lithography [141] and incorporation of arrays of micro-scale valves and pumps for precise mechanical pumping and switching of minute quantities of fluids. This technology is based on the studies of Quake and co-workers [142], who also developed interesting miniaturized devices for cellular and particle manipulation, such as a fluorescence-activated flow cytometer. This represents the first demonstration of cell sorting in microfabricated fluidic channels, based on their fluorescent properties. These disposable sorters were made of flexible elastomers, using soft lithography techniques. The cells or beads of different colors were spatially confined in the narrow (3-µm) T-shaped junction of the manifold, eliminating the need for hydrodynamic focusing, which is used in conventional flow cytometers. The cells were passing through the detection point (probing laser beam), one by one, and sorted by simply redirecting the electrokinetic flow into one of the two output channels. Similar microfabricated silicone structures have been used by the same research group for sizing and sorting single DNA molecules in the 2–200 kbp range [143]. Such devices directly measured DNA length, based on fluorescence, and allowed absolute molecule length measurements without the need for a sizing standard. The resolution of single-molecule DNA sizing devices, unlike that in conventional gel electrophoresis, increases as DNA molecules become longer, due to the improved signal-to-noise ratio. In addition, the method does not depend on analyte mobility, it is much faster than pulse-field gel electrophoresis, and it requires but a millionth of the amount of sample. Combining multi-layer soft lithography technology and the unique properties of elastomer materials, mechanical valves and pumps can be microfabricated, resulting in precise fluid control over a wide range of conditions [144]. These valves are pneumatically actuated, very fast (msec response levels), and compact (as small as 20×20 µm). Peristaltic pumps are implemented by arranging three consecutive valves in a row. These valves and pumps have negligible dead-volume, can transport fluids up to a few nL/sec, and overall, represent an approach to microfluidics combined with electrokinetics. The technology is currently being commercially developed and is expected to find application in a variety of life sciences areas, including high-throughput screening and drug discovery.

A complex PMMA-based fluidic device, built by Soper's research group [16], coupled capillary nano-reactors to micro-separation platforms (electrophoresis chips) for the generation of sequencing ladders and PCR products. The nano-reactors consisted of fused-silica capillary tubes with a few tens of nL reaction volume, which can be interfaced with the chips *via* connectors, micro-machined in PMMA, by using deep X-ray etching. A DNA template was immobilized on the nano-reactor walls *via* a biotin/streptavidin linkage. Following fast thermocycling in a special chamber oven, the DNA fragments were directly (electrokinetically) injected into the microchip device for fractionation. A dual fiber-optic component was micro-machined into the plastic chip for integrated laser-induced fluorescence detection, and consisted of two channels, accommodating laser delivery and emission collection fibers. These results look promising for the fabrication of automated devices which consume minute amounts of costly reagents.

A recent series of publications by Locascio and others [92,93,145] had demonstrated successful applications of PDMS, PMMA, copolyester, and their combinations for assembling integrated fluidic structures that perform affinity dialysis and concentration for fast and sensitive ESI-MS analysis of various compounds. Multi-layer devices were assembled from separate plastic pieces, imprinted with micro-channels by silicon template imprinting and capillary molding techniques. 3-D fluidic devices were constructed by sandwiching a polyvinylidene fluoride membrane between the substrate layers with appropriate channels and simply clamping the assembly together, taking advantage of the good adhesive properties of the polymeric materials. Access holes were drilled in the plastic to interconnect fluidic layers and interface them with external inlet/outlet lines. These devices performed affinity capture, concentration, and direct identification of a targeted compound by ESI-MS (coupled to the chip through the capillary and micro-dialysis junction), as well as miniaturized ultrafiltration of affinity complexes of antibodies [92]. The analyte solutions were pumped through the channel counter-currently to the buffer flow in the adjacent channel (*via* the semi-permeable membrane). By this means, it was possible to perform dialysis and separation of aflatoxin affinity complexes with their antibodies from unbound compounds, followed by passing the solution of affinity complexes against an air counter-flow into another fluidic layer for water evaporation and analyte concentration (Fig. 11.11). Another, similar device performed ultrafiltration of affinity complexes of a phenobarbital antibody and barbiturates, including sequential loading, washing, and dissociation steps. These microfluidic devices significantly reduced dead-volume and sample consumption, and increased MS detection sensitivity by 1–2 orders of magnitude. An integrated platform, based on similar PDMS/membrane assemblies, was presented [93] for rapid and sensitive protein identification by on-line digestion and consequent analysis by ESI-MS or transient capillary isotachophoresis/capillary zone electrophoresis with MS detection. A miniaturized membrane reactor was fabricated, using a porous membrane with adsorbed trypsin separating two adjacent channels. Due to the large surface-to-volume ratio of the porous membrane structure, extremely high catalytic turnover was achieved. It is worthwhile to mention the most recent publication of the same group [146], where a microfluidic chip has been adopted for a sandwich hybridization assay to detect pathogenic bacteria. A PDMS channel was sealed to gold-coated glass, which was used for immobilization of thiolated capture probes and self-assembling layers of blocking liposomes. The reporter probes were tagged with carboxyfluorescein-filled liposomes, enabling detection of very low concentration of targets. The small dimension of the channel and constant flow made it possible to reduce the amounts of reagents and accelerated the assay by eliminating the diffusion limitations.

11.4.2 Integration of monolithic stationary phases

A number of research groups, working in the rapidly growing area of monolithic columns for capillary electrochromatography (CEC), have adopted and developed these approaches for microfabricated devices [147–154]. Such monolithic stationary phases offer interesting possibilities in microanalysis, since they provide a simple way to create spatially definable cast-to-shape, UV-cured polymer packings within micro-channels. Conventional open-channel geometric configurations in microfluidic chips feature relatively small surface-to-volume ratios, limiting certain applications, such as chromatographic separations, heterogeneous catalysis, and solid-phase extraction, which are based on interaction with a solid surface. The developed photo-polymerization techniques enable rather elegant preparation of polymer structures, precisely positioned in the micro-channels, with good control over porosity and introduction of specific chemical

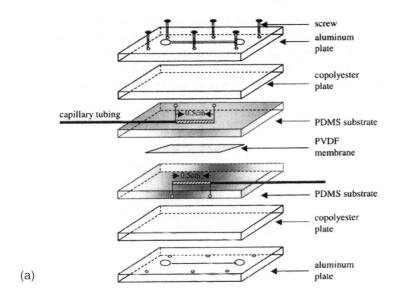

(a)

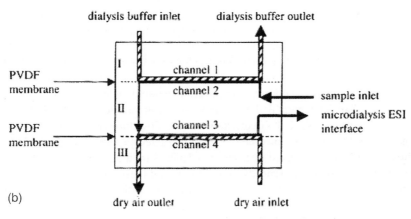

(b)

Fig. 11.11. (a) Assembly of a complete miniaturized affinity ultrafiltration system, containing preformed PDMS micro-channels. The co-polyester pieces provide structural support to the soft PDMS substrates. (b) Schematic side view of the miniaturized affinity dialysis and concentration system. I–III indicate top, middle, and bottom imprinted co-polyester pieces, respectively. Piece II is imprinted on both sides. Two poly(vinylidene fluoride) (PVDF) membranes separate the co-polyester channels. (Reprinted from Ref. 92 with permission.)

properties. Separations of polycyclic aromatic hydrocarbons [151], other small molecules [152], as well as peptides and amino acids [154] have been performed in micro-channels, containing acrylate-based porous polymer monoliths. In addition, de-polymerization methods are being developed, based on short-wave UV light and thermal incineration [151,154], to enable removal of polymer phases either in the detection zone or the entire channel for re-use of microchips. Svec's group reported an enzymatic on-chip reactor, consisting of trypsin, immobilized on macroporous polymer monoliths, molded in microfluidic channels [148], as well as an on-chip solid-phase extractor and pre-concentrator [149], using photo-co-polymerization of acrylate monomers in the presence of porogenic compounds. These examples are promising and broaden the applicability of microfluidics-based analysis.

11.4.3 Micro-reactors

Miniaturization has significant advantages with respect to cost, safety, throughput, kinetics, and scale-up. Micro-reactors are in great demand in biochemical processing and in the pharmaceutical industry for accelerating and increasing the efficiency of pre-clinical drug discovery, chemical development and manufacturing, and mass screening. Typically, chemical synthesis devices handling sub-microliter volumes are considered to be micro-reactors, and they require integration of specially designed components, such as micro-mixers, micro-pumps and -valves, micro-reaction chambers, miniaturized heat exchangers, micro-extractors, and micro-separators, depending on the applications. Informative overviews of the most recent progress in this area can be found in a number of publications [9,155,156]. In this chapter, we will restrict ourselves to a brief outline and some interesting developments, based on microfluidic analytical devices. One of the most important features of a micro-reactor is its remarkably large surface-to-volume ratio. Thus, extremely rapid and highly exothermic reactions can be carried out under isothermal conditions, and higher selectivity and more precise kinetic information, compared to conventional-scale methods, are available, owing to very low mass transfer distances [9]. Micro-reactors enable on-site production at the point of demand. Micro-reactors can be gas- or liquid-phase. The fundamental design and operation problem of the latter is adequate mixing and fluidic control. To overcome laminar-flow limitations and increase diffusive mixing, different configurations have been constructed [9,155]. Pumping of the fluids within the chips has been achieved by a variety of means including pressure, electro-osmotic, electrophoretic, and electrohydrodynamic flows, and their combinations.

Orchid Biocomputer has developed a microfabricated valve within a multilayer device that enables precise fluid distribution to 24 independent reactors simultaneously by means of pressure pumping [156]. A capillary hole prevents the fluid from passing into the reaction well or adjacent wells. The capillary valve opens when pressure is applied, letting fluid into the reaction well, thus enabling precise control of liquid delivery to several reactors. Application of electro-osmotic flow has certain advantages over pressure-driven flow, namely, a flat flow profile, experimental simplicity, absence of moving parts and valves, minimal back-pressure effects, increased separation efficiency increase and the ability to control multiple channels with just a few electrodes. Other reported methods of fluid control have been electrohydrodynamic pumping, creating gradients in surface pressure with small voltages and surfactant molecules with redox groups, use of hydrophobic/hydrophilic barriers on the surface [156], and thermocapillary pumping [135–137]. Chemical synthesis in micro-reactors can be conducted in both continuous-flow and batch modes, the first being typically utilized for synthesis of one product and the latter for parallel processing. Illustrative examples can be found in the literature [9,155,156].

Electrophoretic microfluidic chips feature a number of micro-reactor characteristics and have been used for conducting chemical reactions in the channels and microfabricated chambers, mixing reagents, micro-extraction and micro-dialysis, post- and pre-separation derivatizations, etc. [26]. Ultra-fast micro-PCR, carried out in a micro-chamber or micro-channel, as well as in an open well, is another example of the micro-reactor approach, and such integrated microfabricated devices are described in Sec. 11.3.1. Other integrated micro-devices with micro-reactor features for cell sorting, enzymatic assays, protein digestion, affinity-based assays, etc., are mentioned throughout this chapter. A new, versatile architecture of microfluidic glass devices, consisting of three wafer layers bonded together, has recently been reported for flow-injection analysis of ammonia and for the Wittig synthesis [157]. This three-dimensional glass structure has greater chemical stability than silicon-based systems and provides more design freedom.

A filter-chamber array, microfabricated in silicon and sealed with glass, enabled real-time parallel analysis of three different samples on beads in a volume of 3 nL on a 1-cm^2 chip [158]. Each filter chamber contained microfabricated pillars to trap and localize reacting particles. Single-nucleotide polymorphism (SNP) analysis by solid-phase pyro-sequencing has been performed in this chamber, using biotinylated primers, attached to the streptavidin-coated beads. Allele-specific pyro-sequencing is based on the difference in DNA polymerase extension reaction between match and mismatch primers, hybridized to the target DNA. The nucleotide incorporation results in the release of pyrophosphate, which is enzymatically converted to ATP, and detectable by luciferase-generated light. Passive fluidic valves, consisting of hydrophobic patches of plasma-deposited octafluor-ocyclobutane, were incorporated in the device and served as physical barriers between liquids, to allow controllable sequential loading of different samples without mutual interference. The device is re-usable, enables parallel sample handling, and is adaptable to the implementation of complex biochemical assays on beads (Fig. 11.12).

Microfluidic manipulations at the level of a single cell can be considered as a further development of the micro-reactor approach and offer new and exciting possibilities in biotechnology and analysis. Single cells represent an extremely complex natural micro-reactor, and studies at this level help investigate important processes *in vivo*. A number of studies have focused on using optical forces for the manipulation of small objects (cells, bacteria, beads, micro-droplets, etc.) in liquids. This technique, known as optical tweezers or laser trap, makes it possible to select, capture, and guide single cells or microorganisms. A new micro-system, presented by Reichle *et al.* [159], is based on the combination of optical and chip-based di-electrophoretic trapping of cells and beads with micrometer precision. Another microfluidic system [160] has been designed for high-speed separation/isolation of a single living cell or microorganism in the presence of a large number of other cells in the solution. This was achieved by integrating laser trapping and di-electrophoretic forces. The flow-velocity within the micro-channel was adjusted to balance the optical force on the cell. A target entrapped in the laser focal point can be transported through the micro-channel system and isolated. A micro-reactor system, based on precise optical manipulation of water droplets in oil emulsions, has also been described [161]. Extremely small quantities of samples (down to the single DNA molecule level) can be brought into contact by the fusion of two microdroplets induced by the optical pressure of the laser beam (Fig. 11.13).

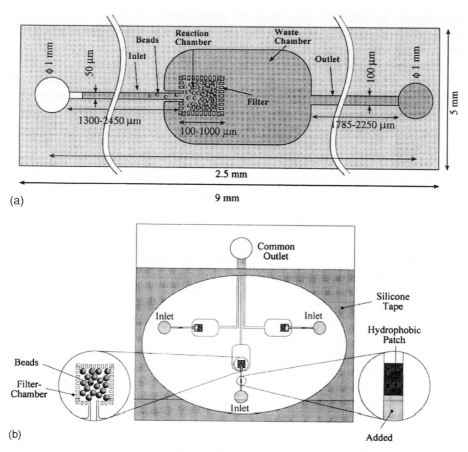

Fig. 11.12. (a) Schematic drawing of a micro-machined, single-filter chamber device. (b) Conceptual drawing of the filter-chamber array. The two key elements, the filter chamber and the passive valve, are magnified. (Reprinted from Ref. 158 with permission.)

Scientists at Göteborg University have described a microfluidic device for combinatorial electrofusion of liposomes and cells [162]. Optical trapping was used to transport individual liposomes and cells through micro-channels into the fusion container, where selected liposomes were fused by means of microelectrodes. This device can also facilitate single cell/cell electrofusions (hybridoma production). Exploitation of these micro-scale tools, based on micro-reactor chips, will eventually reap great rewards and provide new opportunities in drug discovery and development, biotechnology, as well as fundamental studies.

11.4.4 Micro-preparative methods

Preparation of minute samples is quite challenging, due to such problems as sample loss by adsorption and evaporation, the statistical significance of fluctuations at low concentrations, obtaining a representative sample from nonhomogeneous specimen, and detection sensitivity. Progress in the development of on-chip sample preparation has been substantial, and there are already devices that can sort and isolate cells, *e.g.*, white cells from whole blood, and isolate specific DNA sequences [163]. Integration of sample preparation with other analytical procedures, such as PCR, has also been described. Micro-filtration systems with different micro-filter designs for preparative blood cell separation and manipulations were fabricated in silicon by wet and reactive ion etching [164,165]. These structures contain channels and microfabricated filter elements and are capped with glass. Samples are introduced by mechanical pumping. Microchips with an integrated filter for specific cell capture and PCR chambers have also been designed for effective isolation and direct analysis of genomic DNA targets [166]. Microfabricated arrays of posts in silicon chips have been used to separate preparatively nucleic acid sequences from complex biological specimens, based on differences in charge, size, and sequence [163]. Such complexity reduction captured specific sequences using an electrophoretically driven hybridization process, and provided a rapid enrichment method, potentially quantitative and suitable for integration with other processing and analysis steps.

While electric-field-mediated separations in micro-channel and capillary dimensions are primarily considered to be analytical tools, they can also be used in micro-preparative applications. The feasibility of capillary-based micro-preparation and fraction collection following electrophoretic separation was demonstrated for DNA fragments, peptides, proteins, and oligosaccharides, and automated multi-capillary systems have been developed, capable of collecting hundreds of samples. Precise timing is essential for efficient fraction collection to avoid cross-contamination of closely migrating peaks. Microfabricated fluidic devices allow electrokinetic manipulation of analyte zones during separation, as selected zones can be redirected from a separation channel to a side-channel for collection or other assays [167–169]. Microchip electrophoresis can be adapted for rapid and high-resolution fraction collection by using a simple monolithic cross-channel device [167]. DNA fragments of various sizes were separated and collected by simply redirecting the desired portions of the detected sample zones to corresponding collection wells by appropriate voltage manipulations. The efficiency of sampling and collection of the fraction of interest was enhanced by placing a cross-channel configuration at or immediately after the detection point. Upon detection of the band of interest, the potentials were switched to the collection mode in such a way that the selected sample zone migrated to the collection well and the rest of the analyte components in the separation channel were stopped or slightly reversed, thus increasing the spacing between the zone being collected and the one immediately following it. The separation/collection cycle was repeated until all required fractions were physically collected. The amounts of collected DNA fractions were sufficient for further processing, *e.g.*, for conventional PCR-based analysis [167]. An integrated fraction collection system, based on a micro-machined microfluidic cross-connector module coupled to fused-silica capillaries, has also been described by the same authors [183] (Fig. 11.13). A microfluidic cross-connector was micro-machined in an acrylic substrate by drilling precisely centered, flat end- and through-channels, producing capillary connections with essentially zero dead-volume joints. The assembly featured integration of microfluidic functionality, facilitating precise and fast electrokinetic manipulations for fraction

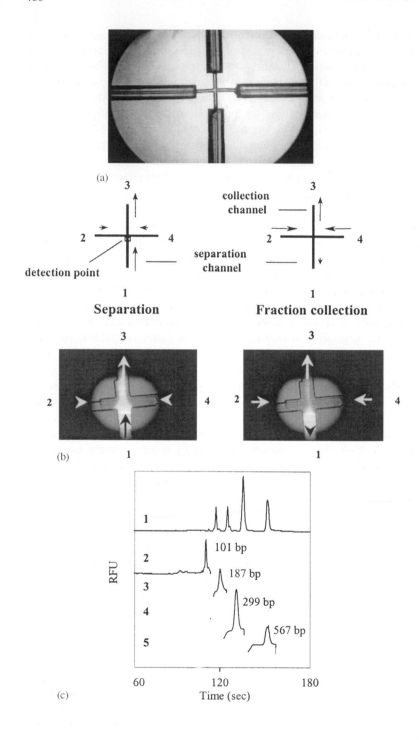

(a)

3

2 ── 4

detection point

1

Separation

collection channel ──

3

2 ──── 4

separation channel ──

1

Fraction collection

(b)

3   3

2   4   2   4

1   1

(c)

collection, and long separation pathways, while minimizing necessary fluidic reservoirs and electrical connections. Interfacing the system with sample and collection micro-titer plates by robotic positioning is straightforward and will provide an automated large-scale micro-preparative fractionation method.

11.5 MODELING BY COMPUTATIONAL FLUID DYNAMICS

Modeling and simulation by computational fluid dynamics (CFD) can provide valuable insight into a plethora of liquid-phase processes, such as mixing, fluid transport, and separation [170]. Computer simulation is primarily considered to be a design tool, but it is also used to support the interpretation of experimental data. A general, steady-state, finite element approach to simulate two- and three-dimensional fluid flows, thermal fields, and chemical concentrations in microfabricated devices is based on the Galerkin finite element method [171]. The model system consisted of two coupled domains, the flow channel and the surrounding solid material exchanging heat with the flow channel. This simulation method facilitated the exploration of various micro-device designs and demonstrates the value of computer simulation in the development of membrane-mediated systems [172]. The macroscopic description of electrokinetic transport in microfabricated channel networks is based on the principles of multi-component continuous media mechanics, coupled with equations describing the electric field [173]. If the frequently quoted "lab-on-a-chip" approach is to solve ever more complicated chemical problems, it will have to incorporate correspondingly sophisticated fluidic designs and complex electrokinetic control methods. In addition to solely relying on experimentation to develop microfluidic structures, computer-aided design and analysis are being enlisted more and more in the prototyping and refining of fluidic arrangements. A significant reduction is expected in both time and expense when such tools are implemented. The principal transport mechanisms and developed theory of electric-field-mediated migration (electrophoresis and electro-osmosis) and diffusion phenomena should be utilized in the design and optimization of microfabricated devices.

Fig. 11.13. High-precision micro-preparative separation system. (a) Micro-machined acrylic cross-connector, coupled with 4 fused-silica capillaries. (b) Schematic drawings and corresponding images of electrokinetic sample manipulations during separation (left panel) and collection (right panel) modes with the plastic cross-connector/capillary assembly. Arrows indicate direction and relative migration velocities of buffer/sample flow (not to scale). For imaging, the channels and Reservoirs 2, 3, and 4 were filled with $0.5 \times$ TBE buffer (45 mM Tris, 45 mM boric acid, 1 mM EDTA), containing 0.2% poly(vinylpyrrolidone). Reservoir 1 contained 50 µM sodium fluorescein dye in TBE buffer. (c) Collection of multiple fractions with cross-connector/capillary assembly. Separation and collection distances were 5 cm. Electrophoretic patterns of PCR product mixture (1). Detection of the first (2), second (3), third (4), and fourth (5) peak, each followed by voltage reconfiguration to collection mode, during which larger DNA fractions were retained in the separation channel. For better visualization, patterns 3, 4, and 5 were offset along the x-axis by the values corresponding to the times of potential reconfiguration for the collection of the preceding bands. (Reprinted from Ref. 183 with permission.)

A comprehensive computer program, simulating the mass transport in any two-dimensional channel structure was developed by Ermakov *et al.* [174]. Their first approximations considered uniform electric conductivity, insignificant Joule heating, as well as, constant thermo-physical and surface properties. Computer simulation assisted in the examination of different aspects of operations and helped to reveal optimal operation parameters for various cross-channel-based injection methods (gated and pinched type) [175,176]. It was also demonstrated that such basic micro-device elements as the simple T-structure and channel-cross can readily accommodate sample mixing and electrokinetic focusing, respectively [177]. Effectiveness of cell separation from various mixtures was modeled prior to microfabrication by using the MATLAB software package [178]. Accurate flow and thermal predictions were also obtained by the lattice-Boltzman simulation method for microfabricated devices, even in case of complex geometric designs. Additional terms, such as surface forces, can be included to simulate discrete liquid/vapor- and liquid/liquid-phase interfaces, as well as surface adsorption properties [179]. Analyses of electrokinetic injection phenomena [180] and predictions of electrophoretic motion in microfabricated channels were obtained with a finite-element commercial simulator, FlumeCAD (Coventor, Inc.) [181]. This highly sophisticated software package enables the design of integrated chemical processes and/or analysis systems and also supports the three-dimensional simulation of chemical transport in electrophoretic, electro-osmotic, and pressure-mediated systems.

ACKNOWLEDGMENT

The authors acknowledge the kind support of Syngenta Research and Technology.

REFERENCES

1   M. Koch, A. Evans and A. Brunnschweiler, *Microfluidic Technology and Applications*, Research Studies Press, Badlock, Hertfordshire, 2000.
2   A. Manz and H. Becker, *Microsystem Technology in Chemistry and Life Sciences*, Springer, Berlin, 1999.
3   M. Heller and A. Guttman, *Integrated Microfabricated Biodevices*, Dekker, New York, 2001.
4   V. Dolnik, S. Liu and S. Jovanovich, *Electrophoresis*, 21 (2000) 41.
5   C.S. Effenhauser, G.J.M. Bruin and A. Paulus, *Electrophoresis*, 18 (1997) 2203.
6   G.J.M. Bruin, *Electrophoresis*, 21 (2000) 3931.
7   J.P. Kutter, *Trends Anal. Chem.*, 19 (2000) 352.
8   G.H.W. Sanders and A. Manz, *Trends Anal. Chem.*, 19 (2000) 364.
9   T. Chovan and A. Guttman, *Trends Biotechnol.*, 20 (2002) 116.
10 Y. Chen and A. Pepin, *Electrophoresis*, 22 (2001) 187.
11 H.I. Smith and H.G. Craighead, *Phys. Today*, (1990) 24.
12 J.J. Wilbur and G.M. Whitesides, in G. Timp (Ed.), *Nanotechnology*, Springer, New York, 1998, p. 331.
13 H. Becker and C. Gartner, *Electrophoresis*, 21 (2000) 12.
14 J.C. McDonald, D.C. Duffy, J.R. Anderson, D.T. Chiu, H. Wu, O.J.A. Schuller and G.M. Whitesides, *Electrophoresis*, 21 (2000) 27.

15 D.K. Armani and C. Liu, *J. Micromech. Microeng.*, 10 (2000) 80.
16 S.A. Soper, S.M. Ford, Y. Xu, S. Qi, S. McWhorter, S. Lassiter, D. Patterson and R.C. Bruch, *J. Chromatogr. A*, 853 (1999) 107.
17 J. Kameoka, H.G. Craighead, H. Zhang and J. Henion, *Anal. Chem.*, 73 (2001) 1935.
18 D.J. Harrison, A. Manz, Z. Fan, H. Ludi and H.M. Widmer, *Anal. Chem.*, 64 (1992) 1908.
19 A. Manz, D.J. Harrison, E. Verpoorte, J.C. Fettinger, A. Paulus, H. Ludi and H.M. Widmer, *J. Chromatogr.*, 593 (1992) 253.
20 C.S. Effenhauser, in A. Manz and H. Becker (Eds.), *Microsystem Technology in Chemistry and Life Sciences*, Springer, Berlin, 1999, p. 51.
21 P.G. Righetti, C. Gelfi, B. Verzola and L. Castelletti, *Electrophoresis*, 44 (2001) 603.
22 C.Y. Liu, *Electrophoresis*, 22 (2001) 612.
23 J. Horvath and V. Dolnik, *Electrophoresis*, 22 (2001) 644.
24 K. Seiler, D.J. Harrison and A. Manz, *Anal. Chem.*, 65 (1993) 1481.
25 S.C. Jacobson, R. Herdenroder, L.B. Koutny and J.M. Ramsey, *Anal. Chem.*, 66 (1994) 1107.
26 J. Khandurina and A. Guttman, *J. Chromatogr.*, 943 (2002) 159.
27 Y. Shi, P.C. Simpson, J.R. Schere, D. Wexler, C. Skibola, M.T. Smith and R.A. Mathies, *Anal. Chem.*, 71 (1999) 5354.
28 S. Attiya, A.B. Jemere, T. Tang, G. Fitzpatrick, K. Seiler, N. Chiem and D.J. Harrison, *Electrophoresis*, 22 (2001) 318.
29 Y.H. Lin, G.B. Lee, C.W. Li, G.R. Huang and S.H. Chen, *J. Chromatogr.*, 937 (2001) 115.
30 R.L. Chien and J.W. Parce, *Fresenius J. Anal. Chem.*, 371 (2001) 106.
31 S.C. Jacobson, R. Herdenroder, A.W. Moore and J.M. Ramsey, *Anal. Chem.*, 66 (1994) 4127.
32 R.A. Mathies and X.C. Huang, *Nature*, 359 (1992) 167.
33 Z.H. Liang, N. Chiem, G. Osvirk, T. Tang, K. Fluri and D.J. Harrison, *Anal. Chem.*, 68 (1996) 1040.
34 K.B. Mogensen, N.J. Petersen, J. Hubner and J.R. Kutter, *Electrophoresis*, 22 (2001) 3930.
35 A.T. Woolley, D. Hadley, P. Landre, A.J. deMello, R.A. Mathies and M.A. Northrup, *Anal. Chem.*, 68 (1996) 4081.
36 P.F. Gavin and A.G. Ewing, *Anal. Chem.*, 69 (1997) 3838.
37 A.J. Gawron, R.S. Martin and S.M. Lunte, *Electrophoresis*, 22 (2001) 243.
38 P.A. Walker, M.D. Morris and M.A. Burns, *Anal. Chem.*, 70 (1998) 3766.
39 N. Burrgraf, B. Krattiger, A.J. deMello, N.F. deRooij and A. Manz, *Analyst*, 123 (1998) 1443.
40 C.T. Culbertson, S.C. Jacobson and J.M. Ramsey, *Anal. Chem.*, 70 (1998) 3781.
41 B.M. Paegel, L.D. Hutt, P.C. Simpson and R.A. Mathies, *Anal. Chem.*, 72 (2000) 3030.
42 M.N. Albarghouthi and A.E. Barron, *Electrophoresis*, 21 (2000) 4096.
43 R.L. Rill, B.R. Locke, Y. Liu and D.H. Van Winkle, *Proc. Natl Acad. Sci. USA*, 95 (1998) 1534.
44 C. Wu, T. Liu and B. Chu, *Electrophoresis*, 19 (1998) 231.
45 D. Liang and B. Chu, *Electrophoresis*, 19 (1998) 2447.
46 D.H. Liang, L.G. Song, S.Q. Zhou, V.S. Zaitsev and B. Chu, *Electrophoresis*, 20 (1999) 2856.
47 B.A. Buchholz, E.A.S. Doherty, M.N. Albareghouthi, F.M. Bogdan, J.M. Zahn and A.E. Barron, *Anal. Chem.*, 73 (2001) 157.
48 J. Han and H.G. Craighead, *Science*, 288 (2000) 1026.
49 A.T. Woolley and R.A. Mathies, *Proc. Natl Acad. Sci. USA*, 91 (1994) 11348.
50 R.M. McCormick, R.J. Nelson, M.G. Alonso-Amigo, J. Benvengu and H.H. Hoopwer, *Anal. Chem.*, 69 (1997) 2626.
51 C.S. Effenhauser, G.J.M. Bruin, A. Paulus and M. Ehrat, *Anal. Chem.*, 69 (1997) 3451.
52 D.C. Duffy, J.C. McDonald, O.J.A. Schueller and G.M. Whitesides, *Anal. Chem.*, 70 (1998) 4974.

53   Z. Ronai, C. Barta, M. Sasvari-Szekely and A. Guttman, *Electrophoresis*, 22 (2001) 294.
54   C.S. Effenhauser, A. Paulus, A. Manz and H.M. Widmer, *Anal. Chem.*, 66 (1994) 2949.
55   M. Ogura, Y. Agata, K. Watanabe, R.M. McCormick, Y. Hamaguchi, Y. Aso and M. Mitsuhashi, *Clin. Chem.*, 44 (1998) 2249.
56   L. Cheng, L.C. Waters, P. Fortina, J. Hvichia, S.C. Jacobson, J.M. Ramsey, L.J. Kricka and P. Wilding, *Anal. Biohem.*, 257 (1998) 101.
57   A.T. Woolley, G.F. Sensabaugh and R.A. Mathies, *Anal. Chem.*, 69 (1997) 2181.
58   D. Schmalzing, L. Koutny, D. Chisholm, A. Adourian, P. Matsudaira and D. Ehrich, *Anal. Biochem.*, 270 (1999) 148.
59   D. Schmalzing, L. Koutny, A. Adourian, P. Belgrader, P. Matsudaira and D. Ehrich, *Proc. Natl Acad. Sci. USA*, 94 (1997) 10273.
60   S.R. Liu, Y.N. Shi, W.W. Ja and R.A. Mathies, *Anal. Chem.*, 71 (1999) 566.
61   E. Carrilho, *Electrophoresis*, 21 (2000) 55.
62   B.B. Haab and R.A. Mathies, *Anal. Chem.*, 71 (1999) 5137.
63   M. Ueda, Y. Endo, H. Abe, H. Kuyama, H. Nakanishi, A. Arai and Y. Baba, *Electrophoresis*, 22 (2001) 217.
64   Y.-C. Lin and W.-D. Wu, *Sensors Actuat. B*, 73 (2001) 54.
65   P. Belgrader, W. Benett, D. Hadley, J. Richards, P. Stratton, R. Mariella and F. Milanovich, *Science*, 284 (1999) 449.
66   M.A. Northrup, B. Benett, D. Hadley, P. Landre, S. Lehew, J. Richards and P. Stratton, *Anal. Chem.*, 70 (1998) 918.
67   T.B. Taylor, E.S. Wimm-Deen, E. Picozza, T.M. Woudenberg and M. Albin, *Nucleic Acids Res.*, 25 (1997) 3164.
68   P. Wilding, L.J. Kricka, J. Cheng, G. Hvichia, M. Shoffner and P. Fortina, *Anal. Biochem.*, 257 (1998) 95.
69   M.T. Taylor, P. Belgrader, R. Joshi, G.A. Kintz and M.A. Northrup, in A. Van den Berg (Ed.), *Micro Total Analysis Systems*, Kluwer, Dordrecht, 2001, p. 670.
70   E. Litborn, A. Emmer and J. Roeraade, *Anal. Chim. Acta*, 401 (1999) 11.
71   E. Litborn, A. Emmer and J. Roeraade, *Electrophoresis*, 21 (2000) 91.
72   R.A. Clark and A.G. Ewing, *Anal. Chem.*, 70 (1998) 1119.
73   D.D. Bernhard, S. Mall and P. Pantano, *Anal. Chem.*, 73 (2001) 2484.
74   L.C. Waters, S.C. Jacobson, N. Kroutchinina, J. Khandurina, R.S. Foote and J.M. Ramsey, *Anal. Chem.*, 70 (1998) 5172.
75   L.C. Waters, S.C. Jacobson, N. Kroutchinina, J. Khandurina, R.S. Foote and J.M. Ramsey, *Anal. Chem.*, 70 (1998) 158.
76   J. Khandurina, T.E. McKnight, S.C. Jacobson, L.C. Waters, R.S. Foote and J.M. Ramsey, *Anal. Chem.*, 72 (2000) 2995.
77   E.T. Lagally, P.C. Simpson and R.A. Mathies, *Sensors Actuat. B*, 63 (2000) 138.
78   E.T. Lagally, I. Medintz and R.A. Mathies, *Anal. Chem.*, 73 (2001) 565.
79   J. Khandurina, S.C. Jacobson, L.C. Waters, R.S. Foote and J.M. Ramsey, *Anal. Chem.*, 71 (1999) 1815.
80   N. Zhang, H. Tan and E.S. Yeung, *Anal. Chem.*, (1999) 71.
81   G. Xue, H.M. Pang and E.S. Yeung, *J. Chromatogr. A*, 914 (2001) 245.
82   M.U. Kopp, M.B. Luechinger and A. Manz, in D.J. Harrison and A. van den Berg (Eds.), *Micro Total Analysis Systems '98*, Kluwer, Dordrecht, 1998, p. 7.
83   M.U. Kopp, A.J. deMello and A. Manz, *Science*, 280 (1998) 1046.
84   B.C. Giordano, E.R. Copeland and J.P. Landers, *Electrophoresis*, 22 (2001) 334.
85   A.F.R. Huhmer and J.P. Landers, *Anal. Chem.*, 72 (2000) 5507.

86 R.P. Oda, M.A. Strausbauch, A.F.R. Huhmer, N. Borson, S.R. Jurrens, J. Craighead, P.J. Wettstein, B. Eckloff, B. Kline and J.P. Landers, *Anal. Chem.*, 70 (1998) 4361.

87 J. Chiou, P. Matsudaira, A. Sonin and D. Ehrlich, *Anal. Chem.*, 73 (2001) 2018.

88 J.W. Hong, T. Fujii, M. Seki, T. Yamamoto and I. Endo, *Electrophoresis*, 22 (2001) 328.

89 D. Figeys and D. Pinto, *Electrophoresis*, 22 (2001) 208.

90 S. Mouradian, *Curr. Opin. Chem. Biol.*, 6 (2002) 51.

91 N. Xu, Y. Lin, S. Hosttadler, D. Matson, C. Call and R. Smith, *Anal. Chem.*, 70 (1998) 3553.

92 Y. Jiang, P.-C. Wang, L.E. Locascio and C.S. Lee, *Anal. Chem.*, 73 (2001) 2048.

93 J. Gao, J. Xu, L.E. Locascio and C.S. Lee, *Anal. Chem.*, 73 (2001) 2648.

94 O. Hofmann, D. Chi, K.A. Cruickshank and U.R. Muller, *Anal. Chem.*, 71 (1999) 678.

95 J. Wen, Y. Lin, F. Xiang, D.W. Matson, H.R. Udseth and R.D. Smith, *Electrophoresis*, 21 (2000) 191.

96 R.D. Oleschuk, L.L. Shultz-Lockyear, Y. Ning and D.J. Harrison, *Anal. Chem.*, 72 (2000) 585.

97 B. He, J. Li and F. Regnier, *J. Chromatogr. A*, 853 (1999) 257.

98 C. Ericson, J. Holm, T. Ericson and S. Hjerten, *Anal. Chem.*, 72 (2000) 81.

99 S. Yao, D.S. Anex, W.B. Caldwell, D.W. Arnold, K.B. Smith and P.G. Schultz, *Proc. Natl Acad. Sci. USA*, 96 (1999) 5372.

100 K. Benedek and A. Guttman, *J. Chromatogr. A*, 680 (1994) 375.

101 L. Bousse, S. Mouradian, A. Minalla, H. Yee, K. Williams and R. Dubrow, *Anal. Chem.*, 73 (2001) 1207.

102 A. Guttman, Z. Ronai, Z. Csapo, A. Gerstner and M. Sasvari-Szekely, *Electrophoresis*, 894 (2000) 329.

103 C.L. Colyer, S.D. Mangru and D.J. Harrison, *J. Chromatogr. A*, 781 (1997) 271.

104 Y. Liu, R.S. Foote, S.C. Jacobson, R.S. Ramsey and J.M. Ramsey, *Anal. Chem.*, 72 (2000) 4608.

105 Z. Csapo, A. Gerstner, M. Sasvari-Szekely and A. Guttman, *Anal. Chem.*, 72 (2000) 2519.

106 A. Guttman, Z. Csapo and D. Robbins, *Proteomics*, 2 (2002) 469.

107 N. Gottschlich, S.C. Jacobson, C.T. Culbertson and J.M. Ramsey, *Anal. Chem.*, 73 (2001) 2669.

108 R.D. Rocklin, R.S. Ramsey and J.M. Ramsey, *Anal. Chem.*, 72 (2000) 5244.

109 N. Gottschlich, C.T. Culbertson, T.E. McKnight, S.C. Jacobson and J.M. Ramsey, *J. Chromatogr. B*, 745 (2000) 243.

110 T. Yang, J.-Y. Jung, H. Mao and P.S. Cremer, *Anal. Chem.*, 73 (2001) 165.

111 L. Licklider, X. Wang, A. Desai, Y.-C. Tai and T. Lee, *Anal. Chem.*, 72 (2000) 367.

112 F. Xiang, Y. Lin, J. Wen, D.W. Matson and R.D. Smith, *Anal. Chem.*, 71 (1999) 1485.

113 J. Li, P. Thibault, N.H. Bings, C.D. Skinner, C. Wang, C. Colyer and D.J. Harrison, *Anal. Chem.*, 71 (1999) 3036.

114 N.H. Bings, C. Wang, C.D. Skinner, C.L. Colyer, P. Thibault and D.J. Harrison, *Anal. Chem.*, 71 (1999) 3292.

115 B. Zhang, H. Liu, B.L. Karger and F. Foret, *Anal. Chem.*, 71 (1999) 3258.

116 I.M. Lazar, R.S. Ramsey, S.C. Jacobson, R.S. Foote and J.M. Ramsey, *J. Chromatogr. A*, 892 (2000) 195.

117 I.M. Lazar, R.S. Ramsey and J.M. Ramsey, *Anal. Chem.*, 73 (2001) 1733.

118 S. Ekstrom, P. Opperfjord, J. Nilsson, M. Bengsson, T. Laurell and G. Marko-Varga, *Anal. Chem.*, 72 (2000) 266.

119 B. Zhang, F. Foret and B.L. Karger, *Anal. Chem.*, 72 (2000) 1015.

120 B. Zhang, F. Foret and B.L. Karger, *Anal. Chem.*, 73 (2001) 2675.

121 S.N. Krylov and N.J. Dovichi, *Anal. Chem.*, 72 (2000) 11R.

122 A. Guttman, *Nature*, 380 (1996) 461.

123 J. Wang, M.P. Chatrathi, B. Tian and R. Polsky, *Anal. Chem.*, 72 (2000) 2514.
124 http://www.calipertech.com/products/throughput_sipper.html.
125 J. Khandurina and A. Guttman, *Curr. Opin. Chem. Biol.*, (2002) 359.
126 P.B. Hietpas, K.M. Bullard, D.A. Guttman and A.G. Ewing, *Anal. Chem.*, 69 (1997) 2292.
127 E.M. Smith, H. Xu and A.G. Ewing, *Electrophoresis*, 22 (2001) 363.
128 A. Guttman and Z. Ronai, *Electrophoresis*, 21 (2000) 3952.
129 A. Gerstner, M. Sasvari-Szekely, H. Kalas and A. Guttman, *BioTechniques*, 28 (2000) 628.
130 http://www.orchidbio.com
131 M. Leach, *Drug Discov. Today*, 2 (1997) 253.
132 www.chem.agilent.com
133 S.M. Wishnies, A. Arai, H. Tanaka, Y. Aso, H. Abe, S. Maruyama and M. Ueda, in Pittcon 2001, New Orleans, 2001, p. 1722P.
134 http://www.aclara.com
135 M.A. Burns, B.N. Johnson, S.N. Brahmasandra, K. Handique, J.R. Webster, M. Krishnan, T.S. Sammarco, F.P. Man, D. Jones, D. Heldsinger, C.H. Mastrangelo and D.T. Burke, *Science*, 282 (1998) 484.
136 M.A. Burns, C.H. Mastrangelo, T.S. Sammarco, F.P. Man, J.R. Webster, B.N. Johnson, B. Foerster, D. Jones, Y. Fields, A.R. Kaiser and D.T. Burke, *Proc. Natl Acad. Sci. USA*, 93 (1996) 5556.
137 A.A. Darhuber, S.M. Troian and W.W. Reisner, *Phys. Rev. E*, 64(3-1) (2001) 031603.
138 S.N. Brahmasandra, V.M. Ugaz, D.T. Burke, C.H. Mastrangelo and M.A. Burns, *Electrophoresis*, 22 (2001) 300.
139 http://www.tecan.com/index_tecan.htm
140 B.L. Carvalho, T.E. Arnold, D.C. Duffy, G.J. Kellogg, N.F. Sheppard, in ACS National Meeting, San Diego, April 1–5, 2001, Paper 218.
141 M.A. Unger, H.-P. Chou, T. Thorsen, A. Schere and S.R. Quake, *Science*, 288 (2000) 113.
142 A.Y. Fu, C. Spence, A. Schere, F.H. Arnold and S.R. Quake, *Nat. Biotechnol.*, 17 (1999) 1109.
143 H.-P. Chou, C. Spence, A. Scherer and S. Quake, *Proc. Natl Acad. Sci. USA*, 96 (1999) 11.
144 S.R. Quake and A. Schere, *Science*, 290 (2000) 1536.
145 D. Ross, T.J. Johnson and L.E. Locascio, *Anal. Chem.*, 73 (2001) 2509.
146 M.B. Erch, L.E. Locascio, M.J. Tarlov and R.A. Durst, *Anal. Chem.*, 73 (2001) 2952.
147 E.F. Hilder, F. Svec and J.M. Frechet, *Electrophoresis*, 23 (2002) 3934.
148 D.S. Peterson, T. Rohr, F. Svec and J.M. Frechet, *Anal. Chem.*, 74 (2002) 4081.
149 C. Yu, M.N. Davey, F. Svec and J.M. Frechet, *Anal. Chem.*, 73 (2001) 5088.
150 T. Rohr, C. Yu, M.N. Davey, F. Svec and J.M. Frechet, *Electrophoresis*, 22 (2001) 3959.
151 Y. Fintschenko, W.Y. Choi, S.M. Ngola and T.J. Shepodd, *Fresenius J. Anal. Chem.*, 371 (2001) 174.
152 I.S. Lurie, D.S. Anex, Y. Fintschenko, W.Y. Choi and T.J. Shepodd, *J. Chromatogr. A*, 924 (2001) 421.
153 S.M. Ngola, Y. Fintschenko, W.Y. Choi and T.J. Shepodd, *Anal. Chem.*, 73 (2001) 849.
154 D.J. Throckmorton, T.J. Shepodd and A.K. Singh, *Anal. Chem.*, 74 (2002) 784.
155 W. Ehrfeld, V. Hessel and H. Lehr, in A. Manz and H. Becker (Eds.), *Microsystem Technology in Chemistry and Life Sciences*, Springer, Berlin, 1999, pp. 233–252.
156 S.H. DeWitt, *Curr. Opin. Chem. Biol.*, 3 (1999) 350.
157 A. Daridon, V. Facsio, J. Lichtenberg, R. Wutrich, H. Langen, E. Verpoorte and N.F. de Rooij, *Fresenius J. Anal. Chem.*, 371 (2001) 261.
158 H. Anderson, W. van der Wijngaart and S. Stemme, *Electrophoresis*, 22 (2001) 249.
159 C. Reichle, K. Sparbier, T. Muller, T. Schnelle, P. Walden and G. Fuhr, *Electrophoresis*, 22 (2001) 272.

160 F. Arai, A. Ichikawa, M. Ogawa, T. Fukuda, K. Horio and K. Itoigawa, *Electrophoresis*, 22 (2001) 283.

161 S. Katsura, A. Yamaguchi, H. Inami, S.-I. Matsuura, K. Hirano and A. Mizuno, *Electrophoresis*, 22 (2001) 289.

162 A. Stromberg, A. Karlsson, F. Ryttsen, M. Davidson, D.T. Chiu and O. Orwar, *Anal. Chem.*, 73 (2001) 126.

163 J. Cheng, L.J. Kricka, E.L. Sheldon and P. Wilding, in A. Manz and H. Becker (Eds.), *Microsystem Technology in Chemistry and Life Sciences*, Springer, Berlin, 1999, p. 215.

164 J. Cheng, P. Fortina, S. Sorrey, L.J. Kricka and P. Wilding, *Mol. Diagn.*, 1 (1996) 183.

165 P. Wilding, J. Pfahler, H.H. Bau, J.N. Zemel and L.J. Kricka, *Clin. Chem.*, 40 (1994) 43.

166 P.K. Yuen, L.J. Kricka, P. Fortina, N.J. Panaro, T. Sakazume and P. Wilding, *Genome Res.*, 11 (2001) 402.

167 J. Khandurina, T. Chovan and A. Guttman, *Anal. Chem.*, 74 (2002) 1737.

168 C.S. Effenhauser, A. Manz and H.M. Widmer, *Anal. Chem.*, 67 (1995) 2284.

169 J.W. Hong, H. Hagiwara, T. Fuji, H. Machida, M. Inoue, M. Seki and I. Endo, in J.M. Ramsey and A. van den Berg (Eds.), *Micro Total Analysis Systems 2001*, Kluwer, Dordrecht, 2001, p. 113.

170 K.F. Jensen, *Chem. Eng. Sci.*, 56 (2000) 293.

171 I.M. Hsing, R. Srinivasan, M.P. Harold, K.F. Jensen and M.A. Schmidt, *Chem. Eng. Sci.*, 55 (2000) 3.

172 D.J. Quiram, I.M. Hsing, A.J. Franz, K.F. Jensen and M.A. Schmidt, *Chem. Eng. Sci.*, 55 (2000) 3065.

173 L.A. Mosher, D.A. Saville and W. Thormann, in B.J. Radola (Ed.), *Electrophoresis Library*, VCH, Weinheim, 1992.

174 S.V. Ermakov, S.C. Jacobson and J.M. Ramsey, in D.J. Harrison and A. van den Berg (Eds.), *Micro Total Analysis Systems '98*, Kluwer, Dordrecht, 1998, p. 149.

175 S.V. Ermakov, S.C. Jacobson and J.M. Ramsey, *Anal. Chem.*, 72 (2000) 3512.

176 S.V. Ermakov, S.C. Jacobson and J.M. Ramsey, in A. van den Berg, W. Olthuis and P. Bergveld (Eds.), *Micro Total Analysis Systems 2000*, Kluwer, Dordrecht, 2000, p. 291.

177 S.V. Ermakov, S.C. Jacobson and J.M. Ramsey, *Anal. Chem.*, 70 (1998) 4494.

178 D. Holmes, M. Thomas and H. Morgan, in A. van den Berg, W. Olthuis and P. Bergveld (Eds.), *Micro Total Analysis Systems 2000*, Kluwer, Dordrecht, 2000, p. 115.

179 D.R. Rector and B.J. Palmer, in IMRET 3, Frankfurt, Germany, 1999

180 M. Deshpande, K.B. Greiner, J. West and J.R. Gilbert, in A. van den Berg, W. Olthuis and P. Bergveld (Eds.), *Micro Total Analysis Systems 2000*, Kluwer, Dordrecht, 2000, p. 339.

181 T. Roussel, D. Jackson, M. Crain, R. Baldwin, J. Naber, K. Walsh and R. Keynton, in PITTCON 2001: The Pittsburgh Conference on Analytical Chemistry and Applied Spectroscopy, New Orleans, 2001

182 D.C. Duffy, J.C. McDonald, O.J.A. Schuller and G.M. Whitesides, *Anal. Chem.*, 70 (1998) 4974.

183 J. Khandurina and A. Guttman, *J. Chromatogr.*, 979 (2002) 105.

Erich Heftmann (Editor)
Chromatography, 6th edition
Journal of Chromatography Library, Vol. 69A

Chapter 12

Instrumentation

ROBERT STEVENSON

CONTENTS

12.1 High-performance liquid chromatography 470
    12.1.1 Control and co-ordination 474
    12.1.2 Pumps 475
        12.1.2.1 Pressure 475
        12.1.2.2 Flow-rate 476
        12.1.2.3 Pulse dampers and gradient elution 477
    12.1.3 Sample injection 478
    12.1.4 Transfer lines and plumbing 480
    12.1.5 Columns 480
        12.1.5.1 Monoliths 495
        12.1.5.2 Capillary columns 496
        12.1.5.3 Column temperature 496
    12.1.6 Detectors 497
        12.1.6.1 Specifications and properties 497
        12.1.6.2 Universal detectors and detection techniques 497
        12.1.6.3 Mass spectrometry 498
            12.1.6.3.1 Interfacing with liquid chromatography 498
            12.1.6.3.2 Ionization modes 500
            12.1.6.3.3 Ion analyzers 500
        12.1.6.4 Solute-specific detectors 502
        12.1.6.5 Multiple detectors 504
    12.1.7 Data stations 504
    12.1.8 Fraction collectors 505
    12.1.9 Instrument design 505
        12.1.9.1 Design Standards 505
        12.1.9.2 Modular *vs.* integrated design 506
        12.1.9.3 System dead-volume 506
        12.1.9.4 Materials of construction 507

12.2   Gas chromatography 507
        12.2.1   Instrumentation   510
        12.2.2   Sample preparation   511
12.3   Thin-layer chromatography   511
12.4   Supercritical-fluid chromatography 512
12.5   Flash chromatography 512
12.6   Electrophoresis   513
12.7   Electrochromatography   513
12.8   Future developments 515
Acknowledgments   515
References 516

12.1 HIGH-PERFORMANCE LIQUID CHROMATOGRAPHY

Space limitations make it necessary to cover only one group of instruments in detail, namely those for high-performance liquid chromatography (HPLC), but there will also be some links to equipment for other techniques. Current information can be found in the reviews that appear annually in Analytical Chemistry [1,2] and the yearly reviews of columns in *LC/GC* [3–5], *American Laboratory News* [6,7], and *American Laboratory* [8,9].

In 2002, HPLC was by far the largest segment of the analytical instrument business, with global sales of $2.5 billion out of a total of $21 billion dollars. The size of the market, and the ingenuity of the estimated 120,000 users has led to a proliferation of products. Experienced chromatographers are expected to know the idiosyncrasies of the instrumentation and columns. Clearly, to select those most suitable for the task at hand takes experience and competency. Of course, the scientific literature and, now, the internet (Chap. 24) are rich sources of information, and computer-aided optimization is also helpful [10,11] (Chap. 1). The earliest chromatographic separations were performed under gravity flow in open glass tubes, filled with some sorbent [12]. This is a simple technique, limited to rather coarse sorbent particles, because fine particles increase the resistance to flow. In contrast to gravity-flow liquid chromatography (LC), HPLC requires complex electromechanical systems, consisting of many sub-systems. This chapter will survey the presently available systems. Although the performance and design of most components has been relatively stable for two decades, there is reason to expect some design changes as a result of new demands for capillary and monolith columns. A typical HPLC system consists of a computer, which controls and coordinates the various sub-systems, solvent reservoirs, pumps, injector, column, detector, data station, printer, and fraction collector. This is the sequence in which they will be treated in this section.

Table 12.1 provides a list of commercial sources of instruments and modules for HPLC. The HPLC business is more than twice as large as any other segment in analytical instrumentation. This has attracted a huge number of commercial firms over the years.

TABLE 12.1

VENDORS OF HPLC INSTRUMENTS (ABSTRACTED FROM LABORATORY EQUIPMENT 34, NO. 12 (1998) 106 AND VENDOR-SUPPLIED DOCUMENTS) [67]

Company Name	A	B	C	D	E	F	G	H	I	J	K	L	M	N	O	P	Q	R	S	T	U	V	W	X	Y	Z	AA	BB	CC	DD	EE	FF
Alcott Chromatography																														X	X	
Alltech Associates Inc.				X		X	X					X			X	X			X	X	X				X	X	X			X	X	X
Amicon Inc.			X	X			X		X																							
Analytical Scientific Instruments																													X			X
Beckman Instruments	X									X				X								X	X									
Bioanalytical Systems	X														X		X			X										X		
Cecil Instruments Limited	X															X					X	X								X		
Chiral Technologies Inc.			X					X																								
Cohesive Technologies, Inc.	X	X	X																													
Dionex Corporation	X									X		X	X	X		X									X			X	X			

(Continued on next page)

TABLE 12.1 (continued)

Company Name	A	B	C	D	E	F	G	H	I	J	K	L	M	N	O	P	Q	R	S	T	U	V	W	X	Y	Z	AA	BB	CC	DD	EE	FF
D-Star Instruments	X																X															
Eldex Laboratories, Inc.																									X	X			X			
EM Science																																
ESA Inc.	X								X					X								X										
Thermo Finnigan			X															R														
Gilson, Inc.	X	X											X				X															
Groton Technology Inc	X							X					X	X	X		X	X														
Hewlett Packard Company	X	X		X				X					X				X	X	X	X	X	X			X		X	X		X		X
Isco, Inc.	X	X																X	X						X				X	X		
JASCO, Inc.	X	X				X				X			X		X		X	X	X										X	X		
Lab Alliance	X								X								X	X	X	X	X	X										
Micro Tech Scientific, Inc	X			X									X						X	X	X				X					X		
PDR, Inc		X					X																									
Perkin-Elmer Corp.	X			X				X					X				X	X	X	X	X	X			X	X	X	X		X	X	X
Pickering Laboratories, Inc.																								X					X			
Polymer Laboratories, Inc.	X							X							X			X	X	X												
Rheodyne L.P.									X																							
Shimadzu Scientific Instruments	X	X		X				X			X	X	X	X			X	X	X	X	X	X	X		X	X	X	X	X	X	X	X
Sievers Instruments										X																						

St. John Associates										X					X																
Thermo Separation Products	X				X				X			X	X			X				X	X		X	X	X							
Unimicro Technologies									X												X											
Valco Instruments., Inc																																X
Varian Instruments	X												X	X		X				X	X	X	X			X				X		X
Viscotek Corp.								X									X			X												
Waters	X	X		X				X	X			X	X	X		X				X	X	X	X	X			X	X	X	X	X	
Wyatt Technology Corp.	X							X									X			X												

Columns: A, Analytical; B, Preparative; C, Process; D, HPLC/MS; E, Ion chromatography; F, Affinity chromatography; G, Chiral; H, Size-exclusion; I, Capillary. *Detectors*: J, Amperometric; K, Chemiluminescent; L, Conductivity; M, Diode-array; N, Electrochemical; O, Evaporative light-scattering; P, Fluorescence; Q, Laser light-scattering; R, Mass spectrometry; S, Radioactivity; T, Refractive-index; U, UV/VIS. *Data Handling*: V, Dedicated workstations; W, Software; X, Integrators; Y, Calibration standards; Z, Column heaters; AA, Fraction collectors; BB, Gradient controllers; CC, Post-column reactors; DD, Pumps; EE, Sample injectors; FF, Valves.

Many of the smaller ones try to enter with niche or specialty products. It is important to note that products are introduced and withdrawn from time to time and that there are also mergers. The internet is the best source of current addresses.

12.1.1 Control and co-ordination

The HPLC revolution occurred simultaneously with the microprocessor revolution in electronics. The first instruments (before *ca.* 1975) had no resident software. Operating parameters were entered with dials, thumb wheels, or even mechanically. Today, almost every sub-system in the chromatograph has a software component. Some fault-detection software will even check for the correct selection and operation of the mixer and column. System software comes at a high price: About 60% of the product development cost of HPLC instrumentation is attributed to software. Changes in computer operating systems continually drive revision of the control and evaluation software. Updates last for only a few years. Thus, one should expect to budget for a new version every three to five years.

The first level of control is the front of the instrument. Some instruments and modules will have a limited set of controls that can be used to start and stop, or perhaps select detector range or wavelength. Modern design is moving away from putting anything but the basics on the front panel. Some modules have only a power switch. The sophisticated or less frequently accessed controls are usually implemented in software. This software is accessed with key pads (usually with difficulty) or *via* a personal computer (PC) with a keyboard and mouse. The computer, usually a PC operating in a Windows environment, can set the parameters of each module from a method file, downloaded from the computer. The run can be initiated upon completion of the prior run and completion of a limited operational qualification (OQ) sub-routine. The OQ subroutine can check for the presence of a vial in the autosampler, sufficient solvent, column temperature within set tolerance, or even download the expected results for the next run. The latter sub-routine is used in some purification procedures, where only compounds with the expected mass will be collected. Fractions containing products not meeting the chosen criteria are sent to waste. Individual modules will also be controllable from a touch-panel interface. The overall status is often displayed on the modules as well as the monitor. Behind the operating panel, fault-detection sub-routines, coupled with light-emitting diode displays will aid the service technician in locating a malfunctioning circuit board or, occasionally, a component.

When under control of the computer, each module provides an electronic "hand shake" with the system-control software. When unusual events are detected, such as a loss of pressure or erratic temperature, the software will note the event, and may try to rectify it. In all cases, it will be capable of sending an audible or electronic signal to notify the operator that attention is required. The software command and communication protocols are also important. Recognizing this, a working committee of the Analytical Instrument Association (AIA) [13] developed the analytical-data interchange protocols (ANDI Protocols) to increase laboratory efficiency and productivity by setting standards for the integration and use of data from multiple vendors' instruments. In the early 1990s the AIA released ANDI Protocols for chromatographic data and for mass spectral data. The ANDI Protocols provide a standardized format for the creation of raw data files or results files. This standard format has the extension ".cdf" (derived from Net CDF), regardless of whether the files originate from chromatographic or mass spectral data. Most vendors of HPLC and LC/MS have adopted these protocols, but some are proprietary. This may limit the selection of hardware components. Many instruments are designed for remote operation, enabling service personnel to examine it from a service center. Key pads are less expensive but less versatile than a computer. On the other hand, the latter can also be used for data processing and report generation.

12.1.2 Pumps

Pumps are used to force the mobile phase through the column. A uniform flow-rate is essential for analytical HPLC, and metering pumps deliver mobile phase at a carefully controlled rate [14]. They usually allow the user to key in the flow-rate. This is so convenient that they are almost universally adopted, although pump designs that require calibration can be purchased for 10% of the price.

12.1.2.1 Pressure

In the early days, when HPLC was the abbreviation for high-pressure liquid chromatography, chromatographs were designed to operate at a pressure of 6000 psi (400 bar) and occasionally as high as 10,000 psi (650 bar). Use of higher pressure was too expensive, and the columns were unable to handle much more than a few thousand psi. The second- and third-generation chromatographs had a maximum pressure of *ca.* 5000 to 6000 psi; most were operated at no more than 2000 psi (130 bar). More recently, pressures as high as 100,000 psi (*ca.* 7000 bar) have been explored. They allow the use of longer columns with particles less than 1.5 μm in diameter. Of course, longer columns provide greater resolution and higher peak count (Chap. 2). A pressure of 6000 psi is probably sufficient for nearly all analytical work. One can anticipate that new instruments may enter into a psi race, probably up to 25,000 psi (1500 bar), but this will probably entail a much higher price. For preparative pumps, the flow-rate range is directly proportional to the square of column diameter. Since preparative columns are packed with larger particles, the pressure requirement is much less. Many preparative chromatography pumps are designed for operation at a maximum pressure of 1000 psi, and routine operation is seldom above 500 psi.

Most malfunctions in the fluid part of the chromatograph will show up as a change in the nominal pressure. The pressure of the system is usually measured in the pump or just after the pump, in case there is a pressure gauge. Gauges, however, are seldom used today, because of their dead-volume and difficulties in integration into a digital system controller. Electronic pressure monitors utilize strain gauges, which change in resistance with pressure. Micro-electronics has reduced their size to just a few square mm. The signal is amplified, digitized, and sent to the overall system controller. It may also be sent to the pump controller for compressibility correction. Shimadzu and Dionex have incorporated a pressure transducer into the end of the piston chamber. This enables a direct comparison of the pressure on both sides of the check valve. Thus, the pressure in the piston can be rapidly increased until it is equal to the pressure at the head of the column, and then it switches over to metered delivery. In addition to providing overall status indication, electronic pressure monitors can be invaluable in fault detection. For example, a leak or an empty solvent reservoir can be detected by a loss in pressure. This event can be programmed to shut down the instrument and send an alarm to a designated station. Pressure above a user-selected threshold indicates a blockage of the flow-path, due to, *e.g.*, an accumulation of contaminants at the top of the column. Thus, the maximum and minimum pressure are usually set as part of the operational qualifications.

12.1.2.2 Flow-rate

Flow-rate reproducibility is the most important specification of a metering pump, since it governs the run-to-run reproducibility upon which chromatographic identification is mainly based. Typical values are $+/-0.1\%$, but actual performance is usually much better. Irreproducible performance is usually attributed to leaks in check valves or piston seals. However, there can be other causes, such as uncontrolled temperature or cavitation, caused by dissolved gas in the mobile phase. Cavitation can be eliminated with degasser modules or helium sparging. Since helium has a very low solubility in most solvents, a gentle stream of helium, bubbled through the mobile phase, can drive off other gases with greater solubility. Accuracy of the flow-rate is a much less important specification in practice. It is mainly needed to reproduce methods or to transfer methods from one instrument to another. The accuracy is easily checked by collecting the eluent. Troubleshooting of pump performance has been reviewed repeatedly [15–18].

Metering pumps are designed for particular ranges of flow-rate, with a useful operating range of *ca.* 1000. While specifications indicate a very wide range, the best performance is obtained at a specific flow-rate (design point) (Table 12.2). Gilson and several other

TABLE 12.2

FLOW-RATES OF METERING PUMPS

Purpose	Flow-rate range	Design point	Best pump design
Preparative	1 to 1000 mL/min	100 mL/min	Dual-piston Metering
Analytical	10 μL to 10 mL/min	1 mL/min	Dual-piston Metering
Capillary	100 nL to 1 mL/min	20 μL/min	Syringe

manufacturers produce designs where the same motor drive can activate piston assemblies of different diameter. This allows the user to change from a capillary to a preparative range by switching pump heads. In capillary chromatography, the flow-rate range is so small, that one can use an analytical pumping system with a splitter to divert a small fraction of the flow to the capillary column. Splitters divide the flow between two paths by adjusting the resistance to flow in the high-flow path to provide the desired flow in the low-flow path [19]. Split flow works well, but it may be difficult to keep the flow-rate constant and reproducible from run to run. A small change in the resistance at the capillary side can occur, due to temperature-induced changes in viscosity or partial plugging of the column with particles. Agilent has developed an electronic flow controller than uses heated pulses to measure flow in the capillary leg. It automatically adjusts the pressure on the capillary leg to compensate for any change in flow resistance. Until electronic flow control was introduced, syringe pumps were preferred for capillary chromatography, since they are compatible with the low flow-rates that capillaries require and do not waste solvents in gradient elution.

As water is much less compressible than organic solvents, pumps for HPLC usually have compressibility compensation circuits built into the motor drive to compensate for this. For example, during the compression stroke of a metering pump, no liquid passes from the check valve into the column until the pressure in the piston chamber exceeds the opening pressure of the check valve. Without correction, organic solvents would not be metered at the set flow-rate, making it difficult to correlate the solvent composition blended by the pumps with that obtained with a pre-mixed solvent of the same nominal composition. Pumping water in a pump that was optimized for organic solvents, the motor pushes the piston too fast during the compression stroke, adding extra water to the flow-stream. The extra pressure can also show up as a small pressure pulse, seen in the baseline of flow-sensitive detectors. This flow-induced noise can degrade detection limits. There are several engineering solutions to this problem. For instance, pressure transducers in the piston chambers can be used to control the motor during the refill, compression, and delivery segments of the pumping cycle. During the compression segment, the pump motor runs quickly to compress the solvent in the working cylinder and then, as it approaches the working pressure, it decelerates to the delivery rate. This produces a very uniform flow-rate, even when the back-pressure of the system changes. Today, all but the lowest-cost HPLC pumping systems have effective compressibility compensation.

Syringe pumps also can suffer from a problem related to solvent compression. During gradient elution, the differential compressibility of water and most organic solvents can produce inaccurate or even unstable flow-rates, if the piston volume is large. For example, when one syringe is filled with methanol and another filled with water, the gradient will produce a two-fold increase in back-pressure at about 50% methanol. Then, while pumping 70–100% methanol, the viscosity will decrease to about one-third of that of water. If the volume in the methanol piston is large, the lower viscosity will allow more methanol to enter the mixer, since it is more compressible or, in this case, expandable. This produces an overshoot in the solvent composition and causes the peaks to move ahead [20]. Today, the problem is avoided by the use of small-volume pistons. These require frequent refilling, but this can be fast and built into the program.

12.1.2.3 Pulse dampers and gradient elution

Early reciprocating pumps for HPLC produced a pulsed flow during the refill and compression parts of the cycle. Since most detectors are flow-sensitive, the pulses show up on the baseline as a periodic noise. Pulse dampers have been designed to smooth out the flow and improve the detection limit. The major problem with pulse dampers is that they add to the dead-volume of the system, and the actual volume changes with the back-pressure. As the column ages, the pulse damper expands and increases the system volume. This may cause a subtle change in the solvent gradient, adversely affecting the reproducibly of retention time, especially in chromatographs with a single pump, fitted with proportioning valves. Today, most pulse dampers are small and can be ignored. The flow-sensitivity of the detectors has been decreased by careful engineering, usually involving thermostating. The index of refraction of solvents and of quartz in detector cells varies slightly with temperature. Heat is transferred to and from the surroundings, and the transfer depends upon the flow-rate. Thus, before temperature control was introduced, the pulses in the flow also had a non-uniform temperature profile. Pulsations changed the focal point of the lamp, registering as flow-induced noise. This noise has been nearly eliminated by careful design of the flow-cell, including heat exchangers in the detector plumbing. Also, the electronics controlling the pump speed are much better now and even more effective in eliminating pulses.

Over the years, many designs have been used for forming solvent gradients, including mixing dilution flasks and segmented reservoirs. Today, these are all but forgotten, having long been replaced by the multiple pump and the lower-priced pre-pump design (Chap. 2). Proportioning metering pumps utilize valves prior to the compression piston to control the feed during the suction part of the pump cycle. These valves can select up to four solvents. Usually, the amount of each mobile-phase component is calculated by the computer, based upon the piston displacement. The proportioning valves are very fast, with opening and closing times of a few milliseconds. This is adequate for columns larger than about 2 mm ID. For smaller-diameter columns, the delay volume is so large that the gradients are too long. Multi-pump gradient systems, where each pump is dedicated to one component of the mobile phase and delivers the solvent at the rate specified by the gradient controller, are best.

12.1.3 Sample injection

In the early days of HPLC, a variety of devices were designed for sample injection [21]. Many were difficult to operate manually, automated operation was an engineering nightmare, and some caused excessive band-broadening. Around 1985, multi-port injection valves were introduced by Rheodyne and Valco. They are most convenient and reliable, and the six-port valve is easy to automate. Today, injection valves account for over 90% of the injection devices, but most are incorporated into automated samplers. Fig. 12.1 shows two designs of six-port valves for sample injection. The front-loading valve (Rheodyne 7125) is used for manual injection with a syringe. The sample is loaded with a syringe into the port on the front of the valve, which is connected to the sample loop. Once the loop is partially loaded or completely filled, the crank is turned to place the loop in the flow-path. The six-port 7010 valve has a sample loop that is filled and then rotated into the sample stream in a similar manner. The major difference is the location of the fill port, the 7010 design being usually preferred in autosamplers. With both valves, it is essential to leave the valve in the flow-path long enough for the sample to be swept into the column.

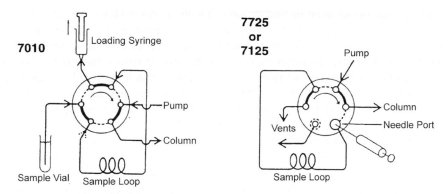

Fig. 12.1. Schematic drawing of two six-port rotary injection valves for HPLC. Model 7010 is designed for automated loop filling, Model 7125 for manual injection from the front through the needle port. (Reproduced by courtesy of Rheodyne Corp.)

The most critical figure of merit for an injector is band-broadening. In general, band-broadening due to the injector should be less than 5% of the band-broadening due to the column. Since the band-broadening is usually due to the sample volume itself, most users select sample loops that meet the 5% criteria. However, using a small sample volume does not assure narrow peaks. It is best to inject the sample in a solvent that is weaker than the eluent. This will cause the sample to be focused at the top of the column when the chromatographic process starts. The diameter along the flow-paths should be kept constant to minimize the "fingering" effect, a finger-like distortion of peaks. Fingering can be a particular problem in capillary chromatography, because the sample size is very small and the surface-to-volume ratio is larger than in 4-mm-ID columns. Manual injection is rarely used today. Technically, manual injection is justified only when the sample is very limited, or the instrument is used infrequently. Sample loss can be minimized by careful manipulation of the syringe. In most cases, the added cost of an autosampler is justified by improved precision and the ability to track and process samples automatically.

Autosamplers not only automate the sampling and injection process, but some will also process the samples by filtration and solid-phase extraction. Since most samples are best processed in 96- or 384-well plates, efficient automation has been developed around this format. Older designs used carousels, but this is bulky, and manual loading of the samples is an error-prone process. Autosamplers differ in the filling methods. Depending upon the application, there can be significant advantages to one mode or another. Gas displacement to fill the injection loop is a proven design that is suitable for applications where the amount of sample is not limited. It is necessary to flush sufficient sample through the loop to minimize carry-over and discrimination effects. The interval between injections also is long, due to the long wash-and-load cycles, but if the run times are longer than 3 min, this causes no problem. Some autosamplers sip the sample from a sample tube or well and transfer it directly to the injection valve. In some cases, the piston movement can be programmed to provide variable-volume injection. This type of autosampler is appropriate for small samples and fast cycle times. Some samplers are capable of making two or more injections per minute. Autosamplers that incorporate sample preparation are very useful in improving the precision of the method, since about 80% of the experimental error in HPLC is due to steps prior to column chromatography. All injectors suffer from carry-over to some extent. The best way to evaluate carry-over is to inject a concentrated sample and follow this with a second injection of eluent blank. If necessary, longer or repeated wash cycles or interspersed blanks should be used, and this can be incorporated in the program.

12.1.4 Transfer lines and plumbing

Up to the injector, one can use tubing with an ID as large as 1 mm for analytical pumps and 2–4 mm for preparative pumps. However, from the injector to the last detector, details of the plumbing need careful attention. The dead- or dwell-volume should be held to a minimum, and lines should be as narrow as possible without the risk of clogging. Stainless-steel tubing with internal diameters as small as 0.005 in. (0.12 mm) or even smaller is available, but it is rarely used for routine work with columns larger than 2 mm ID, since plugging occurs frequently. Most analytical work is done with tubing of 0.10 in. (0.25 mm) diameter. The volume per inch of 0.005-in.-ID tubing is only 0.32 μL. For 0.010-in. tubing, the volume is 1.29 μL per inch. Thus, a few inches of 0.010-in. tubing will not reduce the quality of a separation significantly. For capillary LC, fused-silica tubing, ranging from 530 μm ID to as narrow as 20 μm can be used. Although plugging often occurs, fused silica is inexpensive, and there is no practical alternative. Another, more subtle effect arises from changes in flow-path diameter. The profile of a liquid pumped through a line is parabolic, the center moving faster than the liquid at the wall, and there is more liquid at the wall than in the center. If one connects a larger-diameter capillary to a smaller one, the center of the larger one will enter the center of the smaller one first. There is a streamline of flow, where the liquid at the walls in the larger diameter is effectively separated and moves much more slowly. When the material along the wall diffuses into the center stream, tailing occurs, and this adds to the variance of the peak. It is therefore best to use transfer lines with a small and constant diameter.

12.1.5 Columns

A partial list of the more common brands of HPLC columns is presented in Table 12.3. These are selected from a list of more than 1000 items, which is growing at the rate of more than 50 entries per year. Each brand may include literally hundreds of combinations of particle size, column diameter and length. The newcomer to HPLC may feel overwhelmed, especially if the only reference in the literature is to one of the obscure band names. However, help is just a phone call away. The larger column vendors, such as Phenomenex, Waters, Thermo Keystone, Supelco, and Agilent, maintain applications laboratories staffed with technicians having considerable experience in recommending columns that offer performance similar to that of columns from other current or defunct vendors. Although there is an obvious commercial bias, one can usually buy a column that will be close enough without much trouble. The web pages of suppliers are also a good

TABLE 12.3

PROPERTIES OF SELECTED HPLC COLUMNS

Brand or column name	Vendor	Core Material	dp, µm	Pore diam Å	Surface area m²/g	Surface material	% Carbon	Endcaped
(3S, 4R)-Pirkle 1-J	Regis	Silica	5	100	350	DBPL	N/A	N/A
(R,)Alpha-Burke II	Regis	Silica	5	100	350	DDBA	N/A	N/A
(R,R) ULMO	Regis	Silica	5	100	350	DDDD	N/A	N/A
(R,R)Beta-Gem	Regis	Silica	5	100	350	DAPP	N/A	N/A
(R,R)Whelk-O1	Regis	Silica	5	100	350	DTM	N/A	N/A
(R,R)Whelk-O2	Regis	Silica	5	100	350	DTT	N/A	N/A
(S)Alpha-Burke II	Regis	Silica	5	100	350	DBAD	N/A	N/A
(S,S) DACH-DNB	Regis	Silica	5	100	350	DDDX	N/A	N/A
(S,S) ULMO	Regis	Silica	5	100	350	DDDD	N/A	N/A
(S,S)Beta-Gem	Regis	Silica	5	100	350	DAPP	N/A	N/A
Abzelute ODS-DB	Varian	Silica		80	220	C18	16	Yes
Accubond C18,	ChromRes	Silica		120		C18		N/A
Accusphere C18,	Agilent (NLA)	Silica		120	170	C18	10	Yes
AcquaSil™ C18	Thermo Hypersil Keystone	Silica	3 & 5	100	310	C18	12	–
Adsorbosil C18	Alltech	Silica	3 & 5	60	450	C18	15	Yes
Adsorbosphere C18	Alltech	Silica	3 & 5	80	200	C18	12	Yes
Allsphere ODS-1	Alltech	Silica	3 & 5	80	220	C18	7	Partial
Allsphere ODS-2	Alltech	Silica	3 & 5	80	220	C18	12	Yes
Alltima C18	Alltech	Silica	3 & 5	100	350	C18	16	Yes

(Continued on next page)

TABLE 12.3 (continued)

Brand or column name	Vendor	Core Material	dp, μm	Pore diam Å	Surface area m²/g	Surface material	% Carbon	Endcaped
Allure C18	Restek	Silica	3,5,10	60	520	C18	27	Yes
Allure PFPP	Restek	Silica	3,5,10	60	520	PP	17	Yes
Alox RP	Gyncothek	Alumina						
AlphaBond C18	Alltech	Silica	3 & 5	125	300	C18	10	Yes
Alphasil C18	hplc	Silica	3 & 5	100		C18		
Aminex HP-87 C	BioRad	Polymer	8			Cation Ex.		
Apex I C18	Jones	Silica	3 & 5	100	170	C18	10	
Aquapore OD300	Perkin-Elmer	Silica	5	300		C18		
Asahipak ODP-50	Showa Denko	Polymer	3 & 5			C18	17	
ASTEC C18	ASTEC	Silica		80		C18	12	Yes
Bakerbond C18	JT Baker	Silica	5 & 10	120	170	C18	12	Yes
BetaBasic™ C18	Thermo Hypersil Keystone	Silica	3, 5 & 10	150	200	C18	13	Yes
BETASIL® C18	Thermo Hypersil Keystone	Silica	3, 5 & 10	100	310	C18	20	Yes
BioBasic® C18	Thermo Hypersil Keystone	Silica	3, 5 & 10	300	100	C18	9	Yes
BioSil C18 A/B	BioRad	Silica	3 & 5	90		C18 DB		

Bondclone C18	Phenomenex	Silica	10	150	300	C18		
Bondesil C18	Analych/Varian	Silica		50	300	C18	18	
Bondex C18	Phenomenex	Silica		120	300	C18	10	Yes
Capcell PAK UG C18	Shiseido	High-purity silica	3, 5 & 10	120	300	C18	15	Yes
CarboPac	Dionex	Polymer						
Carbosorb ODS-2	Varian	Silica	5	100	360	C18	20	
Chromegabond C18	ESIndustries	Silica	3, 5 & 10	60		C18	18	Yes
Chromolith RP18e	Merck KGaA	Silica	monolith	130	300	C18	17	Yes
Chromolith Si	Merck KGaA	Silica	monolith	130	300	-	-	-
Chromosorb LC-7	Manville	Silica	10 & 15	100	250 +	C18	-	
Chrompack Carboh. Ca	Varian	Polymer PSDB 10μm	10	120	165	Ca	-	
ChromSpher B	Varian	Shielded silica (poly)	3 & 5	110	165	C18	8.3	
ChromSpher UOP C18	Varian	Non-porous silica	3 & 5	-	-	C18	-	NO
CoraSil	Waters	Pellicular silica	30		8	C-18	30	
Cosmosil C18	Nakalai	Low-metal-content silica		110	330	C18	20	Yes
CP-EcoSpher PAH	Varian	Silica	4	100	320	Reversed phase	-	NO
CP-Micorspher C18	Varian	Silica	3 & 5			C18		
Cyclobond I	ASTEC		5					
D-Leucine	Regis	Silica	5	100	350	DNBL	N/A	N/A
D-Phenylglycine	Regis	Silica	5	100	350	DNBP	N/A	N/A

(Continued on next page)

TABLE 12.3 (continued)

Brand or column name	Vendor	Core Material	dp, μm	Pore diam Å	Surface area m²/g	Surface material	% Carbon	Endcaped
Delta-Pak C18	Waters	Silica	5	100	300	C18	17	Yes
Deltabond™ ODS	Thermo Hypersil Keystone	Silica	3 & 5	300	100	ODS	12	No
Deltapak 300 C18	Waters	Silica	5	300	160	C18	6.9	Yes
Develosil 300ODS-HG	Nomura	Silica	3 & 5	250		C18	11	Yes
Discovery BIO Wide Pore C18	Supelco	High-purity inert silica	5	300	100	C18		
Dynamax C18	Varian	Silica	3 & 5	60		C18		
Econosil C18	Alltech	Silica	5 & 10	60	450	C18	15	Yes
EnviroSep-PP	Phenomenex	Silica				–		
Exmer Series	SGE						–	–
Fluophase™ PFP	Thermo Hypersil Keystone	Silica	5	100	310	PFP	12	Yes
GammaBond RP-18	ESIndus					Polyb-C18		
Great 8 ODS	Regis	Silica	5	80	220	C18	7	
GromSIL Octyl-1 B	Grom	Base-deactivated silica	3, 5 & 10	100	200	C8	6.5	No

Name	Manufacturer	Material				Cation Exch.		
Hamilton HC75 Calcium	Hamilton	Polymer				Cation Exch.	–	–
Helix DNA	Varian	–		–	–	–		–
HiChrom ODS	Regis	Silica	5	80	220	C18	7	No
HiPak C8 AB	Bischoff	silica		120	200	C8	8	yes
Hisep	Supelco	Silica		120	170	SHP	9	No
Hitachi Gel 3053	Hitachi	Polymer		75	300	C18		
Hy-Tach Pro/Peptide	Glycotech					C18		
Hyperbond™ C18	Thermo Hypersil Keystone	Silica		150	300	C18	10	Yes
Hypercarb®	Thermo Hypersil Keystone	Carbon		250	120	Graphitized carbon	100	–
HyperREZ™ XP Carbohydrate Ag$^+$	Thermo Hypersil Keystone	Polymer		–	–	Carbohydrate Ag$^+$	–	–
Hypersil® 100 C18	Thermo Hypersil Keystone	silica	3, 5 & 10	100	300	C18	16	Yes
HyPURITY™ C18	Thermo Hypersil Keystone	silica	3, 5 & 10	190	200	C18	13	Yes
IB-Sil C18	Phenomenex	Silica	3, 5	125	165	C18	11	Yes
IC-Allsep Anion	Alltech	Polymer, 7µ	7			Quat Amine	–	–
IC-Pak Anion	Waters							

(Continued on next page)

TABLE 12.3 (continued)

Brand or column name	Vendor	Core Material	dp, μm	Pore diam Å	Surface area m²/g	Surface material	% Carbon	Endcaped
IC-Pak Kation	Waters							
IC-Unversal Cation	Alltech	Silica, 7μ	7			Poly-butadiene	–	–
ICN Silica RP18	ICN	Silica	7, 10, 20	60		C18	16.5	
Inertsil ODS-80A	GL Science	Low-metal-content silica	3, 5 & 10	80	450	C18	18	
Interaction ION-300	Interaction	Polymer						
Ion Pac CS5	Dionex	Polymer				Cation exch.		
Ion Pac ICE-AS 1	Dionex	Polymer						
IonoSpher A	Varian	Silica	5	120	165	Anion	3.8	NO
IonoSpher C	Varian	Silica	5	120	165	Cation	6.3	NO
Jupiter C18	Phenomenex	Ultra-pure silica	3, 5 & 10	300	170	C18	13.34	Yes
Keystone® ODS1	Thermo Hypersil Keystone	silica	3, 5 & 10	80	250	ODS1	7	Yes
Keystone® ODS2	Thermo Hypersil Keystone	silica	3, 5 & 10	80	250	ODS2	12	Yes

Kingsorb C18	Phenomenex	Silica			C18		18	Yes
Kovasil MS-C14	Chemie Ueticon	Non-porous silica 1.5μm	1.5	427	C-14	90		Yes
Kromasil ODS (C18)	E. Nobel	Silica	3, 5 10	100	C18	340	19	Yes
Kromasil Silica	E. Nobel	Silica	3, 5 10	100	C18	340	0	N/A
L-leucine	Regis	Silica	5	100	DNBL	350	N/A	N/A
L-Phenylglycine	Regis	Silica	5	100	DNBP	350	N/A	
LiChrosorb RP18	Merck KGaA	Silica	5 & 10	100	C18	300	16.5	no
LiChrospher 100 RP18	Merck KGaA	Silica	5 & 10	100	C18	350	21	no
LiChrospher 100 RP18e	Merck KGaA	Silica	5 & 10	100	C18	350	21.6	yes
Little Champ ODS	Regis	Silica	5	100	C18	220	7	Yes
Luna C18	Phenomenex	Ultra-pure silica	3 & 5	100	C18	440	19	
Matrex MC 250	Amicon	Silica	3, 50, 10 15	250	C8	130	5	Yes
Maxsil C18	Phenomenex	Silica	3 & 5	65	C18	500	12.5	Yes
MetaSil AQ C18	Varian	Silica	3, 5 & 10	80	C18	220	19	YES
Micra C30 NPS	Eichrom	Silica	1.5 & 3	Non-porous	C-30	<3	1.5	Yes
Micra ODS	Eichrom	Silica	1.5 & 3	Non-porous	C18	<3	1.5	Yes
MicroPak C18/CN	Varian	Silica	5 & 10	150	C18 & CN		5	
Microsorb C18	Varian	Silica	3, 5 & 10	60/100/300	C18	-/190/80	–	YES
MicroSpherogel Carbohydrate	Beckman	Polymer			Ca++			

(Continued on next page)

TABLE 12.3 (continued)

Brand or column name	Vendor	Core Material	dp, μm	Pore diam Å	Surface area m²/g	Surface material	% Carbon	Endcaped
MicroSpherogel TSK GPC	Beckman	Polymer	3, 7 10	Range		NA		
MonoChrom C18	Varian	Silica	3, 5 & 10	110	300	C18	19	YES
Multospher PAH III	CS	Silica						
Naphthylleucine	Regis	Silica	5	100	350	N1NL	N/A	N/A
Nova-Pak C18	Waters	Silica	5 & 10	60	120	C18	7.3	Yes
Nucleosil 100 C18	MachereyNagel	Silica	3, 5 & 10	100	350	C18	14	Yes
ODSpak F series	Showa Denko					C18		
OligoPore	Polymer Laboratories	PS/DVB	6	GPC 100Å		Organic size exclusion		
OmniSpher	Varian	Silica	3 & 5	110	320	C18	20	NO
Optisil C18	Phenomenex	Silica	3 & 5	90	330	C18	11	Yes
Partisil ODS	Whatman	Silica	10	85	350	C18	5	No
PartiSphere 300 C18	Whatman	Silica	10	300		C18		Yes
Pecosil C18	PerkinElmer	Silica	5 & 10	60		C18		
Perkex C18 HS	Phenomenex					C18		
PetroSpher B	Varian	Silica	5	120	165	–	4.2	YES
Pico-Tag AA	Waters	Silica		60	120	C18	7	Yes
Pinkerton ISRP	Regis	Silica	5	80	120		8	No
Pinnacle ODS	Restek	Silica	3,5,10	120	170	C18	12	Yes
PL-GFC 4000	Polymer Laboratories	Hydrophilic polymer	8 μm	4000		Gel filtration		

Name	Manufacturer	Material	Particle size	Pore size	Chemistry			
PL-Hi-Plex Pb	Polymer Laboratories	PS/DVB	8 µm		Ligand exchange			
Platinum C18	Alltech	Silica	3 & 5	100	C18	200	6	Yes
PLgel	Polymer Laboratories	PS/DVB	3,5,10,20	GPC 50 Å-10E6A plus mixed gel	Organic size exclusion			
Polaris C18-A	Varian	Silica	3 & 5		C18	–	–	YES
PolyCAT A	PolyLC	Silica		300, 1000, 1500	Polyaspartic acid (WCX)			
PolyEncap A	Bischoff	silica	3, 5 & 10	100	C4 polymeric encapsulated	200	8	yes
PolyEncap B	Bischoff	silica	3, 5 & 10	100	C18 polymeric encapsulated	200	12	yes
PolyEncap WCX	Bischoff	silica	3, 5 & 10	100	COOH polymeric encapsulated	200	5	no
PolyETHYL A	PolyLC	Silica	3, 5 & 10	300, 1000	Polyethyl aspartamide (HIC)			
PolyGLYCOPLEX	PolyLC	Silica	3, 5 & 10		HILIC			
Polygosil 60	Macherey-Nagel	Silica	5, 7 & 10					
Polygosil C18	Macherey-Nagel	Silica	5, 7 & 10	60	C18	500	12	Yes
PolyHYDROXYETHYL A	PolyLC	Silica		60, 100, 200, 300, 500, 1000	Poly-hydroxyethyl aspartamide (HILIC)			

(Continued on next page)

TABLE 12.3 (continued)

Brand or column name	Vendor	Core Material	dp, μm	Pore diam Å	Surface area m²/g	Surface material	% Carbon	Endcaped
PolyMETHYL A	PolyLC	Silica		300, 1000		Polymethyl aspartamide (HIC)		
PolyPROPYL A	PolyLC	Silica		300, 1000, 1500		Polypropyl aspartamide (HIC)		
PolySULFOETHYL A	PolyLC	Silica		200, 300, 1000		Polysulfoethyl aspartamide (SCX)		
PolyWAX LP	PolyLC	Silica		100, 300, 1000, 1500		Linear polyethyleneimine (WAX)		
Porasil	Waters	Silica, 15-20 μm	125	330			N/A	n/a
Prism™ RP	Thermo Hypersil Keystone	silica		100	–	RP	12	Yes
Prodigy C18 (ODS 2)	Phenomenex	High-purity silica	3 & 5	150	310	C18	18.4	Yes
ProntoSIL 300 C18 H	Bischoff	High-purity silica	3 & 5	300	100	C18	9	yes
Prosphere 300 C18	Alltech	Silica	3 & 5	300	100	C18	9	Yes
ProTec - C18	ES Indust	Silica	3 & 5	60	100	C18	9	Yes

Protect I	MachereyNagel	Base-deactivated silica				C18		
Protein-pak +	Waters	Silica	5					
Protesil Diphenyl	Whatman	Silica		300	250	Diol + Phenyl	8	Yes
PRP-1	Hamilton	Polymer						
Purospher RP18	Merck KGaA	Silica	na	80	500	C18	18.5	no
Regis SPS C18	Regis	Silica	5	100		C18	12	No
Reliance (CartCol)	duPont					Zor +		
Res Elut C18	Varian	Silica		90	200	C18	10.2	No
Resolve C18	Waters	Silica		90	100	C18	14	Yes
Rexchrom C18	Regis	Silica	5	300	210	C18	16	Yes
Rocket Alltima C18	Alltech	Silica	1.5 & 3	100	400	C18	17	Yes
RoSil C18 HL DA	Alltech	Silica	5	80	330	C18	13	Yes
Selectosil C18	Phenomenex	Silica	5	100		C18		Yes
Shim-pack CLC ODS	Shimadzu	Silica		120		C18		
Shodex GPC	Showa Denko	Polymer	7, 10 & 14					
Sperogel TSK Phenyl PW	Beckman	Polymer	7			Phenyl		
Sperogel TSK SW CM	Beckman	Silica	7			Carboxy-methyl		
SphereClone ODS 1	Phenomenex	Silica	5 & 10	80	220	C18	7	
SphereClone ODS 2	Phenomenex	Silica	5 & 10	80	220	C18	12	
Spherex C18	Phenomenex	Silica	5 & 10	100	180	C18	11	Yes
Spheri ODS poly	PerkinElmer	Silica	5 & 10	80		C18		
Spherisorb ODS	Waters	Silica	3, 5 & 10	80	220	C18	7	No
Spherisorb ODS2	Waters	Silica	3, 5 & 10	80	220	C18	12	Yes
Spherogel HIC	Beckman	Silica				Ethoxy propyl		
Star-ION A 300	Phenomenex	Polymer PSDVB						

(Continued on next page)

TABLE 12.3 (continued)

Brand or column name	Vendor	Core Material	dp, μm	Pore diam Å	Surface area m²/g	Surface material	% Carbon	Endcaped
Supelcogel Ca	Supelco	Polymer PSDVB (9μm)		N/A	N/A	Ca		
Supelcosil LC-18	Supelco	Silica	3, 5 & 10	120	170	C18		
Superspher 100 RP18	Merck KGaA	Silica	na	100	350	C18	n/a	no
Suplex pKb-100	Supelco	Silica		120	170	Alkylamide		
Swift	ISCO	Polymer monolith	no particle	120	170	C18		
Symmetry 300	Waters	Silica	3 & 5	300	110	C18	8.5	Yes
SymmetryPrep C18	Waters	Silica	3 & 5	100	335	C18	19.1	Yes
SymmetryShield C18	Waters	Silica	3 & 5	100	335	C18	17	Yes
Synchropak RP-8	Eichrom	Silica			300	C8		
Taxsil	Varian	Silica		80	220		7	
Techopak C18	hplc	Silica				C18		Yes
Techsil C18	hplc	Silica		60	500	C18	11	Yes
TSKgel Butyl-NPR	Tosoh Biosep	Polymer	2.5	N/A	2–3	Butyl	unavailable	N/A
TSKgel Ether-5PW	Tosoh Biosep	Polymer	10 & 13 & 20	1000	80–100	Oligo ethylene-glycol	unavailable	N/A
Ultra Aqueous C18	Restek	Silica	3,5,10	100	310	C18	17	no
Ultra C18	Restek	Silica	3,5,10	100	310	C18	20	Yes
Ultra-Techsphere C8	hplc	Silica		100	200	C8		Yes
Ultracarb ODS 20	Phenomenex	Silica	3 & 5	90	370	C18	22	Yes
Ultraffinity EP	Beckman	Silica	3 & 5	300		Epoxide		
Ultrasep C18	Bischoff	silica	3 & 5	100	200	C18	13	yes

Ultrasil ODS	Beckman	Silica	3 & 5			C18		
Ultrasphere C18	Beckman	Silica	3 & 5	80	200	C18	14	Yes
UltraSpherogel (SEC)	Beckman	Silica	3 & 5	Range		NA		
Ultrastyragel	Waters	Polystyrene/DVB	7, 10 & 13	Range		GPC/SEC		
Ultremex C18	Phenomenex	Silica	5	80	200	C18	13	Yes
Uniphase™ C18	Thermo Hypersil Keystone	silica	3 & 5	–	–	C18		Yes
Unisphere C8	Biotage	Alumina				C8/PBD		
Val-u-pak ODS	Regis	Silica	5	100		C18	7	Yes
VeloSep C18	PerkinElmer	Silica				C18		
Versapak C18	Alltech	Silica		80	200	C18	10	Yes
Vydac 101SC	Grace Vydac	Silica	30	Non-porous	N/A	N/A	N/A	No
Vydac 201HS	Grace Vydac	Silica	3, 5, 10, 10–15, 15–20, 20–30	90	270	C18	13	Yes
Wakosil C18 200	Wako Chem.	Silica		200		C18	12	Yes
Wakosil II C18HG	Wako Chem.	Silica		120		C18	17	Yes
Waters Spherisorb ODS2	Waters	Silica	3, 5 & 10	80	220	C18	11.5	Yes
Workhorse ODS	Regis	Silica	5	100	200	C18	7	Yes
XTerra C18	Waters	Silica	3, 5, 7 & 10	125	175	C18	15.5	Yes
YMC ODS A	YMC	Silica	5 & 10	125	335	C18	17	Yes
Zipax (obs.)	Agilent	Silica	35	<1	0	Many	1	no
ZirChrom-CARB	ZirChrom	Zirconia	3 & 5	300	23	Carbon	1	No
ZirChrom-PBD	ZirChrom	Zirconia	3 & 5	250	23	Poly-butadiene	2	No

(*Continued on next page*)

TABLE 12.3 (continued)

Brand or column name	Vendor	Core Material	dp, μm	Pore diam Å	Surface area m²/g	Surface material	% Carbon	Endcaped
Zorbax Extend-C18	Agilent	Silica 3.5, 5 μm	3 & 5	80	180	C18	13	Yes
Zorbax ODS	Agilent	Silica	3 & 5	70	330	C18	20	Yes
μBondapak C18	Waters	Silica	10	125	330	C18	9.8	Yes
μPorasil	Waters	Silica	10 & 22	125	330		N/A	n/a
μStyragel	Waters	Polymer	8, 10 & 15			GPC/SEC		

DBPL, 3-(3,5-dinitrobenzamido)-4-pheny-β-lactam; DDBA, dimethyl N-3,5-dinitro-benzoyl-α-amino-2,2-dimethyl-4-pentenyl phosphonate covalently bound to 5 μm mercaptopropyl silica; DDDX, 3,5-dinitrobenzoyl derivative of 1,2-diaminocyclohexane; DDDD, 3,5-dinitrobenzoyl derivative of diphenylethylene-diamine; DAPP, N-3,5-dinitrobenzoyl-3-amino-3-phenyl-2-(1,1-dimethylethyl)-propanoate; DTM, 1-(3,5-dinitro-benzamido)-1,2,3,4,-tetrahydrophenanthrene–mono functional linkage; DTT, 1-(3,5-dinitrobenzamido)-1,2,3,4,-tetrahydrophenanthrene–trifunc-tional linkage; DBAD, dimethyl N-3,5-dinitro-benzoyl-α-amino-2,2-dimethyl-4-pentenyl phosphonate covalently bound to 5-μm mercaptopropyl silica; PP, Pentafluorophenyl - propyl; DNBL, 3,5-dinitrobenzoyl leucine, covalently bonded to 5-μm aminopropyl silica; DNBP, 3,5-dinitrobenzoyl phenylglycine, covalently bonded to 5-μmm aminopropyl silica; N1NL, N-(1-naphthyl) leucine; N/A not applicable source of information (Chap. 24). SciQuest maintains a database of HPLC columns, available on a subscription basis.

Various scientists have attempted to classify the columns from various vendors. They have been unsuccessful, since the batch-to-batch and column-to-column variability is significant. Also, it is now clear that lab-to-lab variance is potentially very large. This makes it even more difficult to classify columns in a meaningful manner. The US Pharmacopoeia has tried a different approach of classifying columns by general specifications, but this classification has not been adopted outside the pharmaceutical industry for at least two reasons: advances in column technology have been too rapid, and descriptions of the manufacturing process has been found inadequate for defining the product and performance, particularly the selectivity. The European Commission has tried still a different approach: It commissioned the development of a European Standard Column that is supposed to serve as a reference column, in the same manner as a standard or certified reference material. After more than six years in development, the column has not been released. Round-robin testing first gave unacceptably large variances in response to many tested parameters.

Columns for ion chromatography are a special case, because there are only a few vendors. Dionex has developed many proprietary column packings that offer unique performance. All of its brand names begin with "IonPak," so that at least the vendor's identity is not in doubt. Metrohm, Showa Denko, Alltech, and others also offer columns for ion chromatography.

If all else fails, one can resort to comparing the performance of different columns. This can be done with automated column-switching valves, which are available from Agilent, Rheodyne, or Valco. The valves can be set up to inject the same sample into 6 to 10 columns in succession in a technique called column scouting. Thus, an experimenter can run several gradients per column in unattended operation (usually overnight) and quickly scan the results. A factorial design or optimization program, such as DryLab (LC Resources) [22] or Chromsword (Merck) [23] can be used to assist in method development.

12.1.5.1 Monoliths

Monoliths are characterized by a continuous bed rather than packed particles. The original name, "continuous-bed columns", was later changed to "monolithic phases." In 2002, the leading monolith product was the Chromolith from Merck. The silica is precipitated as a rod and then inserted into a plastic tube [24]. While these columns provide an efficiency equivalent to that of columns packed with 5- or 3-μm particles, the pressure drop is only *ca.* 5% to 10% of the corresponding packed-bed columns. This allows the use of higher flow-rates than with packed beds. Columns can be connected in series to provide separation efficiencies of 200,000 plates, which is *ca.* 5 to 10 times the practical limit of packed-bed columns. ISCO has introduced monolithic columns (called "Swift") made from organic monomers that are polymerized *in situ*. Stationary phases are suitable for ion-exchange and RP chromatography [25] as well as electrochromatography.

12.1.5.2 Capillary columns

   Capillary columns are usually defined as having a diameter of less than 1 mm. They can be divided into two general classes.
   (a) Open-tubular columns are analogous to open-tubular columns in gas chromatography. For HPLC, the inside diameter needs to be less than 20 μm, and the narrower the better. At these dimensions, capillaries have a high resistance to flow and low load capacity. Peak volumes are so small that they are beyond the detection capabilities of current commercial instrumentation.
   (b) Packed capillaries for HPLC are increasingly popular, especially for LC/MS, since the low flow-rates are compatible with the vacuum systems. In practice, narrow capillaries (100 μm ID) provide greater efficiency, but with larger diameters (typically, 500 μm) the dynamic range is better, and imperfections in plumbing are less troublesome.
Monolithic silica columns in capillary dimensions were developed by Kyoto Monoliths [26] and introduced in the USA by GL Sciences. Svec's group [27–30] has published directions for preparing polymeric monolithic capillary columns. A mixture of monomers, cross-linkers, poragen, and polymerization initiator is pumped into the capillary. The reaction is initiated by heat or light. After washing the poragen from the resulting column, it is ready for use.

12.1.5.3 Column temperature

   The effect of column temperature is receiving increasing attention [31,32]. Before temperature control was instituted, RSDs in the 1 to 3% range were accepted, but precision can be improved by a factor of *ca.* 10 by maintaining the temperature at *ca.* 5°C to 10°C above room temperature. It is typically set at 30°C or 35°C. Failure to control temperature contributed to excessive variations in the results of evaluating the European Reference Column [33] (see above). Ion-exchange or steric-exclusion chromatography of polymers requires the use of higher temperature. The accompanying increase in selectivity has received scant attention. Advances in column technology, particularly the use of polymers beads and, later on, zirconia particles, have led to the use of elevated temperature to modify elution, particularly in RP-LC. Selerity Technologies introduced the Series 8000 HT/HPLC, where HT stands for high temperature, *i.e.*, a temperature up to 200°C. Temperature programming is possible at heating rates as high as 50°C/min, and isothermal operation at 100°C provides a substantial time saving. Water at 140 °C has a polarity similar to 60% aq. methanol [34]. This is of particular advantage in LC/NMR, since D_2O is much less expensive than perdeuterated organic solvents. Separation is achieved with D_2O by temperature programming. The transfer line from the column to the NMR coil is cooled to improve detection. Programming the column temperature in HPLC is much more difficult than in GC. The heat capacity of the eluent is high, and the heat conduction inside the column is severely restricted by the column packing. This becomes a major problem as the column diameter increases above 1 mm. Selerity has solved this problem by heating the connecting tube between the injector and the column to supplement column heating. One of the more interesting uses of temperature control is called denaturing HPLC (DHPLC) (Chap. 19). It is used by molecular biologists to locate polymorphisms in double-stranded DNA. DHPLC is a specialized techniques that requires good control of the column temperature for success [35–37].

12.1.6 Detectors

Detectors are said to be either bulk-property or solute-specific detectors. Bulk-property detectors measure some property of the solution, such as refractive index, whereas solute-specific detectors focus on some property of the solute, such as light absorbance. Detectors can also be classified as concentration- or mass-sensitive detectors. Most detectors for HPLC are concentration-sensitive, but in GC, many of the detectors are mass-sensitivity, *e.g.*, the FID. Two books present an in-depth review of HPLC detectors, although some of the specific examples are out of date [38,39]. More recently, Scott has reviewed LC detectors [40] and tandem LC detectors, including MS, MS^n and NMR [41].

12.1.6.1 Specifications and properties

Detectors are characterized by several specifications [42]:
(a) *Linearity* refers to a response of the detector which is linear with concentration.
(b) *Linear dynamic range* is the concentration range over which the response of the detector is linear. Most detectors will continue to respond to changes in concentration up to very high levels, but not linearly, due to saturation of the detection elements.
(c) *Noise level* is the signal recorded without sample in the cell. In HPLC, it is usually measured with water flowing through the cell. It is important to know what contributes to the noise level. Changes in the noise level are usually significant; an increase in noise is often an early indicator of malfunction.
(d) *Drift* is the change in signal over a long period of time, typically 1 h. Usually, drift is *ca.* 10 times the noise level.
(e) *Detection sensitivity*. The minimum detectable concentration or quantity is the minimum amount of sample that can be detected. Usually, this is the amount that produces a response of 2–3 times the noise level.
(f) *Flow sensitivity* is the change in detector response due to a change in flow-rate, and it is usually related to temperature sensitivity.
(g) *Temperature sensitivity* is the change in detector signal arising from a change in temperature. The mechanism responsible for the temperature sensitivity varies with the detector design. For optical detectors, it is usually due to a change in the image or focal point. The spectrum can also change significantly with temperature.

12.1.6.2 Universal detectors and detection techniques

In the early days of HPLC, analytes that were not suitable for detection by UV absorbance were measured with differential refractometers. Most solvents used in HPLC have a low refractive index (RI), whereas analytes generally have higher values. Thus, the RI changes when almost any solute passes through the flow-cell. Detection limits were then in the low-microgram range, but in the last 30 years, the design has improved so that detection limits are now in the low-nanogram range. Typically, the linear dynamic range is 10^4. Since the refractive index is strongly influenced by temperature, some detectors provide temperature control of the column effluent. Many designs have been proposed and later dropped. Today, the major vendors (Waters and Agilent) use a beam-deflection design. Wyatt Technology uses an interferometer, which improves the detection limit ten-fold. Presently, the major use of the RI detector is for measuring the concentration of polymers, often as part of a system including low-angle laser-light scattering (LALLS). For other analytes, the RI detector has largely been replaced by the evaporative light-scattering detector (ELSD), which is much more robust, gradient-compatible, easier to use, and often more sensitive.

Evaporative light scattering entails nebulization of the column effluent, followed by sweeping the nebulization cell with a warm stream of dry gas. This causes the solvent to evaporate but nonvolatile solids to condense. They are swept along to the flow-cell, which consists of a light source and a photodetector. Nonvolatile analytes scatter the light, which produces an increase in the signal. The ELSD is popular in combinatorial chemistry, where it provides a quick check on the progress of a reaction. It is also useful as a gradient-elution-compatible alternative to RI detection. Unfortunately, volatile analytes are often not detected, but Shimadzu's ELSD flow-cell is at a lower temperature to improve their detection.

When universal detectors are not suitable, it is sometimes possible to make a selective detector function as a nearly universal detector by using the vacancy effect to provide indirect detection. The indirect detection technique makes use of a reporting molecule, dissolved in the mobile phase. Most commonly, this is a compound with strong UV absorbance. With a UV detector, this produces a constant signal of high absorbance in the absence of any sample. Compounds invisible to the UV detector will displace and, hence, dilute the reporting compound, thus producing a loss of signal, which can then be measured. This technique was earlier called vacancy detection. It requires enough of the compound to provide a signal, but not so much that it absorbs the signal of the reporting compound, which is usually light.

12.1.6.3 Mass spectrometry

Mass spectrometry can be universal, as when the total ion current is collected, But the mass is specific to the solute, as used in single-ion monitoring, or solute-specific, as in single-ion monitoring.

12.1.6.3.1 Interfacing with liquid chromatography

During the past decade, MS, which is an established detection method in GC, has emerged as the most powerful detection technique for HPLC. It is also very useful for CE and should find application in CEC as well. Initial attempts at combining LC with MS were unsuccessful due to difficulties in introducing the column effluent into the mass spectrometer. Since MS requires a high vacuum, even a modest flow-rate of eluent was sufficient to overcome the pumps and ruin the spectrometer. Initial attempts centered around some variety of a detector where the column effluent was deposited on a moving wire and then evaporated in a heated zone [43,44]. The wire was next passed into a pyrolysis zone, which was connected to the inlet of the mass spectrometer. However, the heat for evaporating the solvent also evaporated some of the more volatile target compounds, thus degrading detection limits and reproducibility. Moreover, compounds tended to form a puddle during drying, producing very high concentrations when they reached the pyrolysis zone.

In present-day instruments the effluent enters the spectrometer either from a capillary [45] or from a splitter [46] (Fig. 12.2). The splitter takes the effluent from a larger-diameter column and reduces it to a level compatible with the vacuum pumps. Alternatively, one can use a splitter upstream of the column to facilitate the use of capillary columns. Some splitters are simply "T"-pieces with outlets having different resistances to flow. The balance between its two legs is adjusted empirically, often by cutting the length. Thermo Finnigan and other instrument manufacturers have added multi-dimensional chromatographic inlets to their MS/MS systems, particularly the ion traps. This is especially useful for structure determination of peptides. Recently, Rheodyne developed a mass-rate attenuator (MRA), which is a rapidly actuated internal loop valve that can be programmed to divert small aliquots of liquid to the MS line, while passing most to the line leading to the fraction collector [47]. This provides split ratios of from 100:1 to 100,000:1.

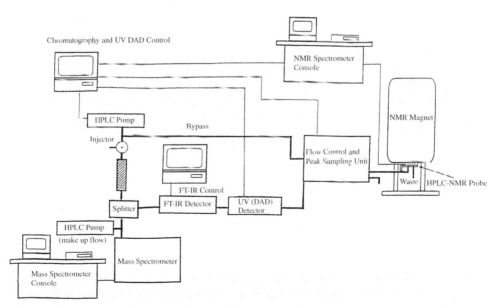

Fig. 12.2. Schematic drawing of LC/FTIR/UV/NMR systems for qualitative and quantitative analysis of drug metabolites. The stream for MS is split off from the stream for the other detectors. (Reproduced by courtesy of I. D. Wilson and American Laboratory.)

12.1.6.3.2 Ionization modes

The first really useful ion source for LC/MS was the thermospray [48]. The effluent from the column was passed to a stainless-steel needle, heated with a small cartridge heater. This would heat the liquid and cause it to evaporate upon exiting the capillary. When the column effluent contained ammonium ions, the evaporation process produced charged clusters during the evaporation, which were attracted electrostatically to the skimmer and onto the analyzer. Another option was to irradiate the spray with electrons from a hot filament or from a glow discharge. Thermospray is a very soft mode of ionization, which provides strong molecular-ion peaks with little fragmentation. Thermospray has been largely replaced by atmospheric-pressure chemical ionization (APCI) and electrospray ionization (ESI).

Since heating seemed to work, there was a strong effort to increase the range of analytes that could be ionized. APCI substituted the ammonium ions in thermospray with a highly charged (*ca.* 2000 V) needle, placed in the post-capillary nebulization region. Another modification makes use of a sheath of dry nitrogen as a nebulizing gas. The corona discharge of the needle generates hydrogen ions (from traces of water vapor in the eluent and curtain gas), which collide with the analyte to impart a charge; the process occurs at atmospheric pressure.

APCI was quickly followed by ESI, where the effluent from the capillary passes through a conducting needle, which is at a high voltage (*ca.* 4000 V). This imparts a charge to the liquid and thus to the droplets that form during the nebulization process. As the drops contract due to evaporation, the charge density builds, eventually exploding the micro-drop. Many of the products are charged and thus attracted to the ion optics of the spectrometer. Neutral molecules are swept away. ESI usually produces molecules with multiple charges. Several software programs have been created to calculate the mass of a singly charged ion from the ESI patterns.

Matrix-assisted laser desorption ionization (MALDI) can be used to introduce large (*ca.* 100,000 Da) molecules or some that are difficult to ionize into the ion source. By now, MALDI is an accepted ionization mode for qualitative analysis. Improvements in the design, especially the vacuum deposition interface, indicate that a RSD of $+/- 10\%$ can be attained with a MALDI interface.

In photo-ionization the column effluent is vaporized in a nebulizer and then irradiated with light. The ionization potential of suitable analytes must be lower than the ionization potential of common solvents, such as water, methanol, and acetonitrile. In favorable cases, photo-ionization produces a low-background signal, which can improve detection limits.

12.1.6.3.3 Ion analyzers

Presently, there are four main choices for the ion-analyzer stage of LC/MS. These can be combined to provide tandem MS (MS/MS) or multiple MS (MS^n). In MS/MS, the ion from the first stage is caused to fragment, usually by adding a collision gas, and then the pieces are analyzed. MS/MS is particularly useful in determining the structure of complex molecules, such as peptides, proteins, and carbohydrates.

Quadrupole analyzers are the least expensive and probably the most rugged. The analyzer consists of four rods that are connected to a radiofrequency (RF) oscillator. At a particular RF setting, only ions of a particular mass-to-charge ratio (m/z) will have the proper frequency to pass through the analyzer onto the detector. The frequency of the field is scanned over a few seconds to produce the mass spectrum. The principal problem with quadrupole mass analyzers is that they are relatively slow. Chromatographic peaks may be only a few seconds wide. Thus, the rise and fall of the concentration profile is superimposed on the MS scan. This makes comparison of results very difficult, since a low peak in the MS can arise from a low concentration or just poor timing. Thus, quadrupole LC/MS detectors are used for peaks that are much wider than the scan time of the mass analyzer. When this is possible, the quadrupole analyzer generally provides the most reproducible qualitative and quantitative analytical results.

Ion traps (IT), which hold ions in a RF trap, are much faster in response than quadrupoles, but they have a lower dynamic range, typically less than 1000. The ions enter the ion trap, which is a cavity formed by electrodes on which a RF field is imposed. The ions are held in the trap, and expelled to the detector when the frequency is correct. This process occurs much faster, typically 10 to 50 times per second, and facilitates following the fastest chromatographic peaks. The IT can be pulsed with a gas, the so-called collision gas, to induce fragmentation. This is similar to the process used in quadrupoles, but since scanning the mass with the IT is so much faster, one can perform multiple fragmentations. This facilitates focusing on smaller and smaller fragments. Then the spectrometer returns to the main or parent peak and repeats the process. Since it is fast, structure determinations are also fast. Ion traps have been particularly useful in the determination of the structure of peptides by inducing stepwise fragmentation, corresponding to the stepwise loss of the constituent amino acids.

Time-of-flight (TOF) mass analyzers are based on the time required to traverse the distance from the ion source to the detector element. TOF mass analyzers were not widely accepted until *ca.* 1990, when MALDI-TOF instruments were introduced, primarily for qualitative analysis. TOF facilitates the analysis of very large ions, even as large as half a million daltons, and the repetition rate can be extremely fast. As many as 1000 spectra per second can be collected for recording or for averaging. TOF and TOF/TOF analyzers are particularly useful for protein analysis.

In the *ion cyclotron resonance* (ICR) and *Fourier-transform* (FT)-ICR mass analyzer ions are held in a magnetic field that is very stable. Hence, it provides very high mass accuracy, often better than one part per million [49]. Frequency analysis by FT further improves the signal quality. Generally, ICR is used with an ESI interface. The instruments are large and expensive and they require a continuous supply of liquid helium. Recently, Bruker introduced an instrument for about half a million dollars, a significant price reduction compared to the prior model.

Accelerator mass spectrometers (AMS) use a linear accelerator to propel the ions into a detection foil for counting. The complex ion optics are capable of counting one isotope in the presence of 10^{15} isobaric atoms. In contrast to MALDI and ICR, the AMS provides very accurate signals, suitable for quantitative analysis. It is about 1000 times more sensitive than radio-labeling and scintillation counting. Cost and size are the two main drawbacks. One instrument is housed in an obsolete mid-sized airplane hangar. Operating costs are proportional to the size. The advent of stable-isotope labeling has increased the number of core facilities providing AMS on a service basis.

12.1.6.4 Solute-specific detectors

The solute-specific detectors measure some property of the analyte that is absent from the solvent, such as light absorption, fluorescence, or chemiluminescence. Flow-cells are important design feature in absorption spectrometers, which must be matched to the rest of the system, especially the injector and column. The response of absorbance detectors increases with light-path length in the flow-cell. However, cell design is a compromise between increasing the volume to increase detector sensitivity and decreasing the volume to maximize chromatographic resolution. For analytical work, a long path-length (typically, 10 mm) will be used, while preparative chromatographs will have flow-cells with a short path-length (1 mm or less). Not only path-length, but cell diameter must be considered. A larger diameter allows more light to be transmitted and improves the signal-to-noise ratio. The standard flow-cell has an internal diameter of 1 mm and a cell volume of 8 µl. For capillary columns, a narrower flow-cell can be used, but that limits the amount of light that can reach the detector. That, together with schlieren effects, increases the noise. Moreover, the flow resistance of the cell can be so high as to cause leaks.

Noise levels can be lowered by thermostating the flow-cell (besides using pumps with low pulsation). Noise may be due to the temperature dependence of the index of refraction, which affects the focal point in the optical system. In both absorbance and fluorescence detectors, the response of some compounds can change by 1% or more per degree centigrade. Failure to control the temperature can show up as a drift in the output, requiring frequent calibration.

Photometers are the least expensive and most sensitive and stable absorbance detectors. They are suitable, if the wavelength of the light source coincides with the absorbance of the compounds of interest. Most photometers are equipped with low-pressure mercury lamps, which provide radiation at 254 nm. Other wavelengths can be supplied by phosphors or lamps with other elements, but none of them rival the 254 nm line of mercury.

Spectrophotometers allow the selection of wavelengths other than those offered by ordinary photometers. This is important, *e.g.*, for determining compounds, such as peptides, which have a stronger absorption below 200 nm. Since solvents start to absorb at 190 nm, there is a very narrow, but useful window at *ca.* 195 nm. Generally, the signal-to-noise ratio of spectrophotometers is *ca.* 10 times poorer and the cost 2−3 times higher than that of a photometer of comparable quality. Some spectrophotometers can be programmed to provide measurements at different wavelengths as a function of time. This is a very useful feature, especially for complex samples, since detection can move to wavelengths of greater sensitivity for certain analytes and avoid wavelengths of interfering material. Usually, spectrophotometers are equipped with deuterium lamps for measurements at wavelengths from 190 to 380 nm and with tungsten lamps for 380 nm to 800 nm. The deuterium lamp is expensive and has a lifetime of only a few thousand hours.

The next-higher step in the cost of absorbancy detection is the *diode-array detector* (DAD). It can measure absorbance over a wide range of wavelengths, typically from *ca.* 190 nm to 380 nm, or even to 800 nm, if a tungsten lamp is used. A diffraction grating after the flow-cell spreads the light onto the elements of a photo-diode array. Typical arrays comprise 512 or 1024 elements. Thus, spectral resolution is typically *ca.* 2 nm/diode for the 512-, and 1 nm/diode for the 1024-element array. Since absorbance is measured over a wide spectral range, one can obtain the absorption spectrum of the analytes as they are eluted from the column. This may be sufficient for identification and for detection of overlapping chromatographic peaks. DADs are only now approaching the signal-to-noise performance of spectrophotometers. Noise specifications should be examined carefully to make sure that the low noise claimed for a diode array is not obtained by excessive electrical filtering or noise averaging. A good DAD costs about twice as much as a spectrophotometer of comparable quality and 7 times as much as a photometer.

Fluorescence can be selected as a detection mode when the analytes fluoresce and when higher sensitivity is required than UV absorbance provides. However, fluorescence detection is sensitive to incidental effects, such as quenching and hypsochromic and bathochromic solvent shifts, which can degrade reproducibility. Quenching by certain solvents, such as acetonitrile, is also a factor to be considered. The flow-cell is designed to be as large as possible to increase sensitivity without contributing more than 10% to band-broadening, or peak variance, of the column.

Spectrofluorometers substitute grating monochromators for both the excitation and emission optics. This quadruples the price, while the light level is often restricted, since so much is discarded in the excitation monochromator. Therefore, these detectors are not very popular, but they are useful in method development for rapidly selecting the appropriate excitation and emission wavelengths. For compounds lacking appreciable fluorescence, derivatization with fluorescent reagents may be used. This works well if only one reaction site is present. For larger molecules, such as proteins, derivatization is often incomplete and variable.

Compounds lacking appreciable fluorescence will fluoresce when irradiated with high-intensity photons, such as provided by a laser. *Laser-induced fluorescence* (LIF) is effected by focusing the intense light from the laser on a small volume. Thus, it works especially well with capillaries columns in CE, CEC, and HPLC. There are two major drawbacks: acquisition and operating costs are often high, and the reproducibility of the laser from pulse to pulse is variable. This makes it difficult to achieve good precision (better than $+/- 5\%$) with LIF.

Electrochemical reactions can be used to detect analytes that contain oxidizable or reducible groups. Electrochemical techniques include potentiometry, amperometry, and coulometry. All of them measure the current generated by an applied potential. Detection limits are typically at the nanomole and occasionally picomole level. Originally, the detectors were difficult to use, since the electrode surfaces tended to foul easily and irreversibly. Today, pulse programs are designed to regenerate the surface prior to the next measurement sweep. Generally, potentiometric detectors are used in well-engineered analyzers, where the vendor has developed the application and even validated it.

12.1.6.5 Multiple detectors

The specificity of the mass spectrometer can be used to confirm the identity of a chromatographic peak, based upon retention times and measured by UV absorbance. MS/MS, where fragment ions are further examined, can serve for further identification. However, MS is seldom useful for discriminating between optical isomers. In contrast, NMR is very useful in the structural elucidation of material that is isomerized during the ionization process in MS. For example, the tocopherols are a class of natural products with numerous cis/trans as well as positional isomers. Their mass spectra are similar, due to isomerization to a common molecular ion, but NMR enables the qualitative analysis of the isomers [50].

LC/MS/NMR/FTIR is an exceptional example of multiple detectors, which has been used for the qualitative and quantitative on-line analysis of complex mixtures from, *e.g.*, metabolism studies of drug candidates [51]. By performing the analysis with all detectors on-line, the problems of correlating data taken in different labs at different times can be avoided [52]. The cost of such a system can be over three million dollars, but if it saves a day in development of a blockbuster drug, it is worth it [53]. As shown in Fig. 12.2, the sample stream is split prior to the MS detector. Typically, about 90% of the flow passes to the optical detectors, first to the FT-IR and then to the diode-array UV detector. The NMR detector is last, since the strong magnetic field requires that the rest of the instruments remain several feet away, even with active shielding. Proton NMR works best with deuterated solvents, which lower the background, but deuterated solvents are expensive. Solvent exchange from proton-containing solvents to deuterated solvents can improve the signal/noise at the expense of going off-line. Bruker has developed an automated solvent exchange unit, in which the column effluent is diluted with water and passes through a C_{18} SPE cartridge, where it is adsorbed and concentrated. Adsorption can be repeated several times in succession to recover more sample and thus improve detection [54]. The cartridge is eluted with trideutero-acetonitrile, which is routed back to the NMR cell.

12.1.7 Data stations

Early detectors provided only analog output, usually with a signal up to 10 mV, that was then passed to a potentiometic recorder and, occasionally, a digital integrator. Beginning in 1975, analog outputs were replaced by digital signal processing, with direct transmission to data stations for further processing and report generation. As the cost of microprocessors dropped, it became possible to add features, such as event programming, gain control, and fault detection, to the individual modules. Today, most HPLC modules communicate with each other. While this adds features that would be difficult to implement in hardware, it increases development cost. Software development now typically accounts for more than half of development for a detector or pump. Strip-chart recorders and digital integrators have been almost completely replaced by computerized data stations. Nowadays, a modern computer costs less than a strip-chart recorder cost in 1975. However, the software may be 2–10 times more expensive than the computer.

For example, the early editions of Millennium software for chromatography were very difficult to learn and have been replaced with a new program, called Empower. Empower has four different user interfaces: a Millennium emulation for users familiar with Millennium software, a basic level for inexperienced operators of the instrument, and intermediate and advanced levels for specialists with increasing skill and responsibility. Software has a short product life, typically only 3 to, at most, 7 years. Development usually continues with frequent releases and updates that patch problems or add new capability found on other software. However, each improvement can have unintended consequences, such as decreasing the reliability of the original code. This is compounded by the evolving requirements of the Food & Drug Administration (FDA) for controlling software and auditing methods and results. This is called current good laboratory practice (cGLP) and 21CFR part 11. Following the lead of the FDA, the Environmental Protection Agency (EPA) is beginning to set up similar regulations for environmental laboratories. Being behind the FDA in deploying these requirements, the EPA is expected to avoid many of the difficult issues, such as what to do with legacy data systems. Considering the technical properties of the data station and detector, it is desirable for sampling to be fast enough to take at least 10 measurements across the narrowest peak, which is usually the first one. Faster sampling only degrades the signal quality due to noise and creates excessively large data files.

12.1.8 Fraction collectors

Eluate fractions can be collected in vials or multi-well plates. Fraction collectors can be controlled by a computer, by a timer, or by the detector output. Thus, Kassel *et al.* [55] devised a multiplexed mass spectrometer for chromatographic purification of combinatorial-chemistry products. The crude reaction products were placed in an autosampler, and the solvent delivery system was set to run a gradient at 40 mL/min, the flow being split equally between four matched columns. The UV detectors provided the primary detection of peaks by measuring the absorbance at 220 and 254 nm, and the splitters routed 99.8% of the liquid either to the fraction collection plate or to waste and only 0.2% to the multiplexed mass spectrometer.

12.1.9 Instrument design

12.1.9.1 Design Standards

Analytical instruments are subject to design codes for safety and quality assurance. Compliance with the International Standards Organization (ISO) is not very meaningful, but design codes for safety are important. In Europe, the CE mark is required for instruments, such as chromatographs, that are powered by electricity or contain gas under high pressure. A product bearing the CE mark meets the relevant European Union directives [56]. In the USA, most instruments are designed to comply with Underwriters Laboratories (UL) Code 1262L. This places stringent requirements, particularly, on electrical design. Certain local authorities have issued their own design codes. Some will accept UL certification, others will not. Canada once had unusually stringent design codes, called CSA, but has recently bowed to pressure for harmonization by recognizing standards of other organizations. Japan still has codes that appear to be non-tariff barriers to importers, particularly on products employing high-pressure gases and radioactivity. The former affects, in particular, the import of supercritical-fluid instruments. Buyers of instruments should ask about both ISO certification and compliance with safety codes. It is also incumbent upon the operator to make sure that instruments are operated within their design and safety limits. For example, protective barriers and circuits for lasers or high voltage should be used and not disabled.

12.1.9.2 Modular vs. integrated design

Instruments can be either modular or integrated, housed in a single cabinet. While an instrument assembled from modules provides flexibility, integrated instruments are designed for a specific use, such as portability for field monitoring or high-temperature operation, *e.g.*, a size-exclusion chromatograph for the analysis of polymers [57]. Modular systems may have redundant power supplies and communications links, which increase cost, but allow replacement of obsolete modules at a lower cost than replacement of integrated instruments. Waters has marketed a hybrid HPLC instrument, called "Alliance", which consists of a single cabinet, housing the solvent reservoirs, pumps, injector or autosampler, column thermostat, and column. Detector and the Empower data-and-control program are added as modules.

12.1.9.3 System dead-volume

The dead-volume of the instrument is a key parameter affecting chromatographic performance, especially in gradient elution (Chap. 2). After operator error, the system volume is responsible for the majority of problems found in method transfer, especially in trying to copy a method from the literature. Most of the current chromatographs are optimized to pump mobile phase at 1 mL/min. Fortuitously, this is near the minimum in the HETP *vs.* velocity curve (Chap. 1) for columns with an internal diameter of *ca.* 4 mm. Thus, it is a simple exercise to calculate the delay time associated with a change in conditions. For example, if a gradient chromatograph is constructed with high-pressure mixing, the delay time corresponds to the volume of the mixer, injector, and the transfer lines to the top of the column. For most chromatographs with high-pressure mixing, this is much less than 1 mL. High-pressure mixing produces a gradient with a fast response. Pre-pump gradients mix solvents before the pump. This lowers the cost but slows the response to changes in composition. Usually, the mixing chamber must be larger, since the solvent is introduced for controlled time intervals. Typical pre-pump gradient systems have a volume of *ca.* 4 mL. This is trivial in isocratic elution, but in gradient elution the effect can be dramatic.

12.1.9.4 Materials of construction

Most analytical chromatographs are made with stainless-steel plumbing in the high-pressure section, which extends from the outlet of the piston chamber to the outlet of the detector. Stainless steel is strong but susceptible to corrosion [58]. Corrosion can contaminate the eluent with metal ions, which may react with the sample, or the sample may react directly with the corrosion deposit. The fritted disks ("frits") in columns are particularly suspect. Irreversible adsorption of sample components on the column or frits can be detected as an intercept in a plot of sample added *vs.* response. Stainless steel can be made inert through passivation, which usually entails a preliminary washing of the plumbing with 50% aq. nitric acid. This treatment reestablishes the oxide coating on the steel. To test frits for passivation, they are soaked in water overnight. Corrosion is indicated by discoloration of the frit. In the early days, HPLC systems were made "biocompatible" by replacing stainless-steel parts with glass, plastic, or titanium. Titanium parts are unusually resistant to corrosion but expensive. Titanium is difficult to machine, and titanium tubes are difficult to bend. Today, instruments for chromatography of proteins, nucleic acids and carbohydrates are usually made with polyether ethyl ketone (PEEK) tubing. PEEK is also preferred for ion-exchange chromatography, including ion chromatography. This is a plastic material strong enough to withstand several hundred atmospheres of pressure when used with most aqueous eluents at room temperature. However, some organic solvents, particularly concentrated tetrahydrofuran and acetonitrile, will slowly soften the plastic, and should be avoided.

Piston seals are another source of contamination. The seals are often made of composite plastics containing lubricating agents, which are designed to wear slowly. Occasionally, this can cause clogging of the column inlet frit. A low dead-volume 2-μm-porosity filter ahead of the injector is recommended to catch wear particles from the pump. Check valves are usually sapphire balls housed in plastic sealing washers. The balls or seats can fail due to contaminants. They can usually be rescued by pumping with a dilute solution of laboratory detergent or soaking for a few hours in 50% nitric acid. Inlet lines, feeding the pump, are usually made of Teflon. Since Teflon is permeable to atmospheric gases, the outlet of the degasser should be connected to the pump with stainless-steel tubing.

12.2 GAS CHROMATOGRAPHY

In 2002, gas chromatography celebrated its 50th anniversary as an instrumental technique. In most countries of the world it is now considered a mature technology with all the factors associated with product maturity: Generally, the number of vendors has declined to one dominant vendor (Agilent) and a few secondary vendors (Table 12.4). There is very little technical innovation, limited primarily to sample preparation technology and improved injector design. China is an exception to the generalization of a mature market. The rapid development of China's infrastructure has created a demand for low-priced gas chromatographs. The majority of these instruments utilize designs that

TABLE 12.4

VENDORS OF GC INSTRUMENTS

Vendor	Instruments	MS	Other	Injector	Sample prepn	Portable	Comments
ABB	P						Process control
AC Analytical Controls	L		x	x			Petroleum analyzers
Aerograph			x	x			Obsolete trade name of Varian
Agilent	L	M	M	M	x	X	Leading vendor for lab. GC
Alltech	x		x	x	M		Low-priced basic chromatograph
Antek	L		U	x			Focus on industrial chemicals
ATAS (Div. of GL)	L		X	M			Innovative injection technology
BeifenRuili	L		x	x			Leading manufacturer in China
Bruker		M					OEM *via* Agilent
Buck Scientific	L		x	x			Low-priced basic chromatograph
Chromato-Sud	P		x				Several process analyzers
Chrompack							Acquired by Varian
Detector Technology			M				Leader in thermionic detection
F&M							Acquired by Hewlett-Packard, now Agilent
Fuli Analytical Instrument Co., Ltd.	L		M	M			Chinese Manufacturer of GC
Gerstel					M		Stir bar extraction
GL Sciences	L		M	M			Major vendor in Japan
GOW-MAC	L		x	x			Leader in thermal conductivity detection

Company					Remarks
Hewlett-Packard	L				Name changed to Agilent
Hitachi	L	M	M		Limited to Japan
JEOL			x		Popular in Japan
KC[1]	L	M		X	Chinese Manufacturer of GC
Konik-Tech	L	x	x		GC developer in Spain
Leap Technologies					Autosampler
LECO	L	U	x		Leader in GC X GC
OI Analytical	L & P	x		X	Strong position for GC of petrochemicals
Perkin-Elmer	L	x	x		Global vendor for lab. Inst.
RIPP[2]	L	x			Chinese Mfg. for petrochemical GCs
Rosemount Analytical	F	x			Focus on process control
Shandong Luan Ruihong Instruments	L	x			Chinese Manufacturer of GC
Shimadzu	L	M	M		Global vendor for lab. inst.
SICT[3]	L	x	x		Chinese Manufacturer of GC
Siemens Applied Automation	P	x			Process control
SRI Instruments	L	X	x		Leader in low-priced GC
Techcomp Ltd	L	M	M		Chinese Manufacturer of GC
Tekmar Dohrmann		M	M		Purge-and-trap injection
Thermo Electron	L	M	x	x	GC strong in Europe, leading in MS
Valco		U			Helium-ionization detector
Varian	L	M	M	X	Global vendor for lab. inst.

1, Shanghai Ke Chuang Chromatograph Instruments Co., Ltd.; 2, Research Institute of Petroleum Processing; 3, Shanghai Haixin Chromatographic Instruments Co., Ltd.; X, Significant products and market share; x, Small market share; M, Major vendor; U, Unique; L, Laboratory instrument; P, Instruments for on-line or process control were typical of research-grade instruments about 15 years ago, but performance is more than adequate for routine work in an environment where labor is not yet expensive and high precision or sensitivity is not required.

12.2.1 Instrumentation

Over the last 15 years, instrumentation of GC has continued to evolve, albeit slowly. Smart electronic flow-controllers for the carrier gases and exquisitely designed ovens combine to provide a retention time RSD better than 0.05%. Similar precision can be achieved for peak areas, especially with the use of internal standards. Computer control of all functions and parameters is now standard. Electronic flow-controllers have been the most significant advance in the pneumatic area. Algorithms have been developed by Agilent and PerkinElmer to tune the instrument automatically for optimum results. The error adjustments can be accessed for diagnostics or call for preventive maintenance. All this comes at a price, and most assays do not require the precision that is now routinely purchased. Less expensive instruments with precision ten times poorer and less than half the price are relegated to developing countries, even though these are designed and manufactured by some of the market leaders.

Detectors are an important and still evolving accessory. As the cost of mass spectrometers is declining, more than half of the gas chromatographs sold recently are equipped with MS detectors. However, the buyer of such detectors may be overlooking the significant savings in the use of element-specific detectors, such as flame photometers for sulfur-containing compounds and thermionic detectors for nitrogen- and phosphorus-containing compounds [59]. To save energy, column ovens can be replaced by embedding the capillary in a microwave cavity. The capillary must be coated with a polymer that absorbs the microwave radiation. This provides very fast heating and cooling with low power consumption. Other designs include a heater, wire wound in a spiral around the capillary, which is enclosed in an insulating tube. The result is not only lower power consumption, but also a faster heating rate than that provided by a column oven.

Injection of liquid samples into capillary GC columns can strip the stationary phase from the column, and solvents can disrupt or even extinguish the flame in flame-based detectors. Instead of passing all the solvent through the column, the injector is cooled to just above the boiling point of the solvent, and a side-vent is opened. The gas supply is sufficient to keep the mobile phase running through the column, while the solvent is vented through the open splitter. After the solvent is removed, the vent valve is closed, and the temperature of the injector is increased rapidly to produce the chromatogram in the usual manner. Similarly, on-column injection, split/splitless injection, programmed-temperature vaporization, and large-volume injection, can be performed on the same instrument. With a large-volume injector, samples as large as 3 mL can be injected to detect analytes in the part-per-trillion range.

Comprehensive GC [60,61] is the name given to separations implemented with two columns, connected in series with a trapping interface (Chap. 8). The longer column provides the basic separation, while the shorter column, which has a very different polarity, separates the analytes that were unresolved on the first column. A stimulus, usually heat, is applied to the trap to release a pulse of sample to the second column.

The output is generally a two-dimensional contour plot. In addition to the improved separation, the trapping interfaces can concentrate the analytes once again after passing through the long column. This usually improves the limit of detection 10 to 30 times.

12.2.2 Sample Preparation

The advances in instrumentation are minor compared to the novelty seen in sample preparation for GC. These advances not only reduce labor, but also improve precision, by as much as 80%. The change from packed to capillary columns has hastened the development of sampling technology for small samples. Solid-phase micro-extraction (SPME) is very convenient and widely used (Chap. 8). The SPME technique depends on a sorbent-coated fiber inside a syringe needle. The fiber is extended into the sample, where it extracts the analyte. After extraction, the loaded fiber is withdrawn into the needle, which serves as a sheath. During injection, the needle pierces the septum of the injector, while protecting the fiber from mechanical damage. The fiber is immediately extended into the hot injector, the analytes are thermally desorbed, and then enter the column. Generally, SPME works best with small amounts of nonpolar analytes.

Stir-bar extraction is performed in a closed vial with a polymer-coated magnetic micro stir-bar. Polydimethylsiloxane and other coatings are available. Injectors, such as the Gerstel TDS2, have been designed to pass the stir-bar into a heated zone for thermal desorption. The principal advantage of stir-bar extraction is that the capacity of the coating on the stir-bar can be much larger than that of the SPME fiber, and this improves the detection limit thousand-fold.

In the purge-and-trap method the analytes are trapped on a strong sorbent, often at low temperature. Rapid heating of the sorbent then drives off the analytes in a narrow pulse, and they are swept into the column. The GC Cryo-Trap from Scientific Instrument Services is designed to concentrate volatile analytes at the top of a capillary column by cooling with liquid CO_2. They are then flashed off in an controlled manner. An electronic temperature program controls the heating coil, wrapped around the megabore capillary section.

Autosamplers generally process racks of samples that have been prepared off-line. The key parameters in evaluating commercial autosamplers are: minimal sample volume, cycle time, %RSD in repetitive injection, and carry-over. Carry-over from a concentrated sample can be minimized by interspersing blank injections.

12.3 THIN-LAYER CHROMATOGRAPHY

Another mature technology, thin-layer chromatography (TLC), continues to evolve slowly (Chap. 6). The principal users are concentrated in Europe, particularly in the large pharmaceutical laboratories. In the United States, the preference is for HPLC, since it can be automated and operated by technicians with less formal education. In 2002, Merck introduced the Ultra Thin-Layer Chromatography (UTLC) plates [62] with a sorbent layer made of hybrid silica, similar to the Xterra phases for HPLC from Waters. The sorbent layer is a 10-μm-thick continuous phase rather than a layer of granules. With UTLC, 2- to

5-cm development takes less than 5 min. Densitometers with slit optics may be undesirable for such a small object, since the resolution is degraded by the slit width. It is probable that video densitometers will be more satisfactory, as was the case with the postage-stamp-size gels used so successfully in the 1990s by Amersham Pharmacia Biotech.

Instrumentation for TLC continues to evolve as engineers attempt to automate the manual steps and improve reproducibility. One example is the continuing struggle to improve OPLC, which stands for Optimum Performance Laminar Chromatography [63] or Overpressure Liquid Chromatography, depending upon the vendor. In OPLC (Chap. 6) a thin-layer plate is sandwiched between two plates. Gaskets close off the sides so that eluent can be forced through the sorbent layer of the TLC plate. This is similar to HPLC, but the column is a planar bed. The advantage is that many − up to 96 − separations can be run in parallel. OPLC can also be used for preparative purification by streaking the sample on the plate, but, of course, the amount of sample that can be processed is limited.

12.4 SUPERCRITICAL-FLUID CHROMATOGRAPHY

Supercritical Fluid Chromatography (SFC)(Chap. 2) is experiencing a revival after a period of quiescence in the early 1990s. The new growth is in response to needs in three applications areas:
  (a) Characterization of the unsaturated fractions of petroleum fuels and feedstock.
  (b) Analysis and purification of synthetic drugs.
  (c) Fractionation and purification of natural products.
SFC is well suited for the separation of compounds that are soluble in methanol or less polar solvents. Instrumentation for SFC resembles that for HPLC with some important differences: The flow-cells for optical detectors need to withstand much higher pressure than in HPLC. SFC requires a pressure controller at the outlet of the column, and this is accomplished with electronic pressure controllers. The pumps require more complex control of the fill- and compression-stroke than HPLC pumps. Carbon dioxide is the most commonly used eluent. Since it is much more compressible than the commonly used solvents for HPLC, pumps that are designed specifically for pumping CO_2 must include elaborate compressibility correction algorithms. Berger Instruments has developed solvent delivery systems that provide composition reproducibility and accuracy equal to those for HPLC. Next to Berger Instruments, the following vendors supply pumps and other accessories for SFC: Selerity Technologies, Thar Designs, and JASCO.

12.5 FLASH CHROMATOGRAPHY

Before leaving the topic of instrumentation for chromatography, a variant of the oldest form should be mentioned. Flash chromatography is a very useful technique for the rapid purification of reaction products in synthetic organic chemistry, particularly in the pharmaceutical industry, when the resolution required is low, but the capacity needed is high. Packed columns, containing *ca.* 100 g of irregularly shaped silica gel particles are eluted with organic solvents by gravity flow. To speed up the process, gas pressure (up to 100 psi) can be applied to the solvent reservoir. Highly automated instruments for flash chromatography are available from a variety of firms, including ISCO, Biotage, and Argonaut Technologies.

12.6 ELECTROPHORESIS

For the last ten years, the market has shifted from chemicals for preparing slab gels to precast gels for PAGE (polyacrylamide gel electrophoresis) and less so for agarose gels. They are not only convenient to use, but also of consistent quality. Capillary (CE) is used not only for proteins, nucleic acids, and their components, but also for the analysis of chiral compounds, carbohydrates, in the identification of drugs, and in clinical diagnostics. CE instrumentation (Chap. 9) is available from several dealers (Table 12.5). The leading vendors of CE equipment for general analytical and clinical diagnostic applications are Beckman-Coulter, Agilent, and CE Technologies. Unimicro Technologies and Agilent also sell instruments capable of capillary electrochromatography (CEC). Beckman offers CE instruments with eight capillaries, while Amersham Pharmacia Biotech and Applied Biosystems have developed analyzers that can process 96 samples at a time for DNA sequencing. CombiSep and Spectrumedix provide applications tuned to high-throughput screening and drug evaluation. Some models have capillary bundles of 96 and 384 capillaries that take samples from the multi-well plates. Detection is one of the areas in CE that needs improvement. UV detection is limited by the short path length of the cell. Fluorescence suffers from a similar problem, since the cell volume is also small. Laser-induced fluorescence (LIF) (Sec.12.1.6.4) is a partial solution, but the cost is high. Beckman, in cooperation with Thermo Finnigan, has coupled CE with ion-trap MS. This is a useful improvement, but again, at significant increase in price. Capillaries for high-performance (HP)CE are usually fabricated from fused-silica capillaries. The interior surface is coated to control or modify the surface charge and, thus, electro-osmotic flow [64]. The future belongs to HPCE on solid chips (Chap. 11), first commercially available from Agilent and Caliper.

12.7 ELECTROCHROMATOGRAPHY

Capillary electrochromatography (CEC) combines some of the advantages of HPLC and CE [65] (Chap. 7). This seems to be true with respect to the range of sample capacity. Chromatography (HPLC and CEC) offers a dynamic range of about 10^6 in concentration, while it is 10^3, at best, for CE. Many separations that require gradient elution in HPLC can be performed isocratically with CEC, but controlling the flow-rate is not as easy as in HPLC. Still, CEC will be a powerful analytical tool when the major problems are solved or controlled. A list of vendors for CEC instruments is presented in Table 12.5.

TABLE 12.5

VENDORS OF CAPILLARY ELECTROPHORESIS (CE) AND CAPILLARY ELECTROCHROMATOGRAPHY (CEC) PRODUCTS

Vendor or Brand Name	Capillary Electrophoresis			Capillary Electro-chromatography		
	Status	Instrument Type	Capillaries and kits	Instrument type	Capillaries	Comments
Agilent 2100	M	Chip	Kits			Lab-on-chip
Agilent 3D	M	G	C	G		Popular general purpose instrument
Amersham Biotech	W		DNA Arrays			Market collapsed after HUGO completed
Applied BioSystems	M	CAE	DNA Arrays			Leader in DNA sequencing
Beckman	M	G + DNA	C & K			Leader in clinical HPCE
BioCal	M	SNPS	K			Focus on SNPS
Caliper	M	G				Lab-on-chip
Capel						See Luminex
CE Technologies	M	G		G		Manufactured in Singapore
Combisep	M	CAE	HTS			Many applications for Pharmaceutical assays
Luminex	M	G				Manufactured in Russia
Megabase	W	CAE				See Amersham
Microsolve	M		C & K			Leading vendor of application kits for HPCE
MicroTech Scientific	M	G	C	G	C	Innovative vendor of HPCE, CHPLC and CEC
Prince Technologies	M	G				Manufactured in Netherlands
UniMicroTechnologies	M	G	C	G	C	CE, CEC and CHPLC with same instruments

C, Capillaries; CAE, Capillary-array electrophoresis; Chip, CE on a chip; DNA, Optimized for DNA sequencing; G, General-purpose instrument; HTS, High-throughput screening; K, Kits; M, In production at time of writing; SNP, Single-nucleotide polymorphisms; W, Withdrawn from market.

12.8 FUTURE DEVELOPMENTS

The next decade should see a steady evolution in design, mostly along conventional lines. Since the largest firms want to avoid marketing blunders, true innovation will probably come from the smaller firms and research institutions. Miniaturization and parallel processing are clearly in the immediate future. For example, a sixteen-channel HPLC array for denaturing HPLC has been devised for locating SNP (single-nucleotide polymorphism) [66] (Fig. 12.3). The parallel design increases throughput.

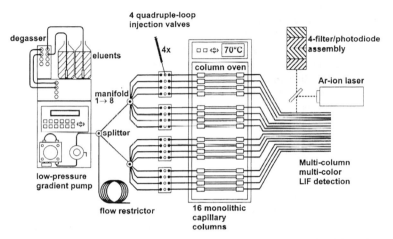

Fig. 12.3. Sixteen-channel HPLC for analysis of single nucleotide polymorphisms (SNP) with monolithic columns and multiplexed multichannel laser-induced fluorescence for detection. (Reproduced by courtesy of P. J. Oefner, Stanford University [68].)

An expected trend is the growth of HPCE and CEC in lab-on-a-chip (LOC) applications. The problem of performing HPLC on a chip is not yet solved. Detection limits of MS still do not meet the needs of some analysts. It is probable that improvements in ion source and interface design will enable a 100-fold improvement in detection limits within the next decade. Detectors for capillary separations, based upon transmission through a cell appear to be running into a brick wall. It is likely that detection principles of the future will be based upon surface phenomena or interference phenomena. Finally, data processing, including signal extraction, is ripe for further advances. The improvements in computational power will enable some of these. This will be combined with new search algorithms to quickly find, retrieve, and present relevant information from masses of data from around the globe.

ACKNOWLEDGMENTS

I want to thank many friends and colleagues who responded so quickly to my requests for help on columns and instruments. These include: Andy Alpert of PolyLC, Uwe Neue of

Waters, Bingwen Yan and Peter Carr of Zirchrom, Francis Mannerino of Regis, John Hobbs of Beckman, Howard Jordi of Jordi Assoc., Eric Stover and Dick Henry of Thermo Hypersil-Keystone, Tracy Ascah-Ross and Russell Gant of Supelco, Norikazu Nagae of Nomura Chemical Co. Ltd., Ron Majors of Agilent Technologies, Fred Rabel of E. Merck, John Dolan of LC Resources, and Peter Oefner of Stanford University.

REFERENCES

1   W.R. LaCourse, *Anal. Chem.*, 72 (2000) 37R.
2   W.R. LaCourse, *Anal. Chem.*, 74 (2002) 2813.
3   R.E. Majors, *LC/GC Europe*, 14 (2002) 248; 15 (2003) 278.
4   R.E. Majors, *LC/GC*, 19 (2001), 272 &ff.
5   R.E. Majors, *LC/GC*, 18 (2000) 262; 18 (2003) 356.
6   R.L. Stevenson, *Am. Lab. News*, 34 (2002) 6.
7   R.L. Stevenson, *Am. Lab.*, 34 (2002) 13.
8   R.L. Stevenson, *Am. Lab.*, 34 (2002) 12.
9   R.L. Stevenson, *Am. Lab.*, 33 (2001) 40.
10 DryLab from LC Resources, Walnut Creek, CA, http://www.lcresources.com/software/software.html.
11 Chromsword from E. Merck, Darmstadt, Germany, http://www.merck.de/english/services/chromatographie/sword/csa2web2.pdf.
12 R.L. Wixom, The beginnings of chromatography – the Pioneers (1900 to 1960), in C.W. Gehrke, R.L. Wixom and E. Bayer (Eds.), Chromatography, a Century of Discovery 1900–2000, Elsevier, Amsterdam, 2001.
13 For more information about the ANDI Protocols contact the AIA at 225 Reinekers Lane-Suite 625, Alexandria, VA 22314, voice: (703) 836-1360, fax: (703) 836-6644, e-mail: AInstA@aol.com.
14 R.L. Stevenson, Mobile phase delivery systems for HPLC, in E. Katz, R. Eksteen, P. Shoemakers and N. Miller (Eds.), Handbook of HPLC, Dekker, New York, 1998, pp. 499–530.
15 J.W. Dolan, Troubleshooting HPLC Systems, appearing monthly in LC/GC.
16 J.W. Dolan and L.R. Snyder, *Troubleshooting LC Systems*, Humana Press, Clifton, UK, 1989.
17 E. Neue, LC Troubleshooting, appearing monthly in American Laboratory.
18 R.L. Stevenson, Mobile phase delivery systems for HPLC, in E. Katz, R. Eksteen, P. Shoemakers and N. Miller (Eds.), Handbook of HPLC, Dekker, New York, 1998, pp. 518–530.
19 Anon. Dionex Corp. 2002.
20 J.C. Helmer, *Anal. Chem.*, 48 (1976) 1741.
21 R.A. Henry, Injection devices, in E. Katz, P. Eksteen, R. Shoemakers and N. Miller (Eds.), Handbook of HPLC, Dekker, New York, 1998, pp. 559–580.
22 J.W. Dolan and L.R. Snyder, *LC/GC*, 20 (2002) 1016.
23 T. Baczek, R. Kaliszan, H.A. Classens and M.A. Van Straten, *LC–GC Europe*, 14 (2001) 304.
24 R.L. Stevenson, *Am. Lab. News*, May (2001), p. 23 &ff.
25 R.L. Stevenson, *Am. Lab. News*, May (2002), p. 6 &ff.
26 N. Tanaka, H. Kobayashi, N. Ishizuka, H. Minakuchi, K. Nakanishi, K. Hosoya and T. Ikegami, *J. Chromatogr. A*, 965 (2002) 35.

27 E.C. Peters, M. Petro, F. Svec and J.M.J. Frechet, *Anal. Chem.*, 69 (1997) 3646.

28 E.C. Peters, M. Petro, F. Svec and J.M.J. Frechet, *Anal. Chem.*, 70 (1998) 2288.

29 E.C. Peters, M. Petro, F. Svec and J.M.J. Frechet, *Anal. Chem.*, 70 (1998) 2296.

30 S. Xie, R.W. Allington, J.M.J. Frechet and F. Svec, in T. Scheper (Ed.), *Advances in Biochemical Engineering/Biotechnology*, Vol. 76, Springer-Verlag, Heidelberg, 2002.

31 T. Greibrokk, *Anal. Chem.*, 74 (2002) 375A.

32 J.K. Swadesh, Temperature control in analytical high performance liquid chromatography, in E. Katz, R. Eksteen, P. Shoemakers and N. Miller (Eds.), Handbook of HPLC, Dekker, New York, 1998, pp. 607–615.

33 R.M. Smith, K.K. Unger, C. du Frense von Hohensche, H. Engelhardt, F. Steiner, C.A. Cramer, H. Classens, R. Arras, K. Bischoff, S. Lamotte, D. Sanchez and U. Berger, *Evaluation of HPLC Columns by Round Robin tests*, 24th International Symposium on Chromatography, Leipzig, Germany, September 15–20, 2002.

34 Abstract of Poster 2296 P, PittCon (2002)

35 W. Xiao and P.J. Oefner, *Human Mutation*, 17 (2001) 445.

36 P.J. Oefner and P.A. Underhill, *Am. J. Hum. Genet.*, 57(Suppl.) (1995) A266.

37 A. Premstaller and P.J. Oefner, *Am. Genomic/Proteomic Technol.*, 2 (2002) 27.

38 R.P.W. Scott, *Liquid Chromatography Detectors*, Elsevier, Amsterdam, 1977.

39 T.M. Vickery, *Liquid Chromatography Detectors*, Dekker, New York, 1983.

40 R.P.W. Scott, LC Detectors, in E. Katz, R. Eksteen, P. Shoemakers and N. Miller (Eds.), *Handbook of HPLC*, Dekker, New York, 1998, pp. 531–558.

41 R.P.W. Scott, Tandem liquid chromatography systems, in E. Katz, R. Eksteen, P. Shoemakers and N. Miller (Eds.), Handbook of HPLC, Dekker, New York, 1998, pp. 581–606.

42 R.P.W. Scott, LC detectors, in E. Katz, R. Eksteen, P. Shoemakers and N. Miller (Eds.), Handbook of HPLC, Dekker, New York, 1998, pp. 531–558.

43 W.T. Kok, Principles of Detection, in E. Katz, R. Eksteen, P. Shoemakers and N. Miller (Eds.), Handbook of HPLC, Dekker, New York, 1998, pp. 163–166.

44 W.F. McFadden, H.L. Schwartz and S. Evens, *J. Chromatogr.*, 122 (1976) 389.

45 J. Yates, *Am. Genomics/Proteomics Technol.*, 1(1) (2001) 42.

46 I.D. Wilson, *Spectroscopy Perspectives, supplement to American Laboratory*, March (2002) 14.

47 H. Cai, J.P. Kiplinger, W.K. Goetzinger, R.O. Cole, K.A. Laws, M. Foster and A. Schrock, *Rapid Commun. Mass Spectrom.*, 16 (2002) 544.

48 T. Covey and J. Henion, *Anal. Chem.*, 55 (1983) 2275.

49 A.G. Marshall, C.L. Hendrickson and S.D.H. Shi, *Anal. Chem.*, 74 (2002) 253A.

50 K. Albert, Paper HPLC 2002, in press

51 W. Schradr, J. Geiger, T. Hoffmann, B. Warscheid, U. Margraf, Abstr. Pap. American Chemical Society 2001, 221st meeting ENVR-129

52 N.J.C. Bailey, P.D. Stanley, S.T. Hadfield, J.C. Lindon and J.K. Nicholson, *Rapid Commun. Mass Spectrom.*, 14 (2000) 679.

53 I.D. Wilson, *American Laboratory, Spectroscopy Perspectives supplement*, (2002) 13–18.

54 O. Corcoran, P.S. Wilkinson, M. Godejohann, U. Braumann, M. Hofman and M. Spraul, *American Laboratory, Chromatography Perspectives supplement*, May (2002) 18–23.

55 R. Xu, T. Wang, J. Isbell, Z. Cai, C. Sykes, A. Brailsford and D.B. Kassel, *Anal. Chem.*, 74 (2002) 3055.

56 R.L. Stevenson and P.J. Jenks, *Marketing to the European Laboratory*, HNB Publishing, New York, 2001, pp. 108–110.

57 5.7.L. DeFrancesco, *Anal. Chem.*, 74 (2002) 275A.

58 K.E. Collins, C.E. Colins and C.A. Bertran, *LC/GC*, 18 (2000) 268.

59 P. Paterson, "Selective Thermionic Ionization Detection Coupled to Gasless Gas Chromato-
    graph" Poster 1387 Pitt Con 2002, March 17–22, New Orleans, LA
60 J.B. Phillips and J. Beens, *J. Chromatogr. A*, 856 (1999) 331.
61 P. Marriott and R. Shellie, *Trends Anal. Chem.*, 21 (2002) 573.
62 C. Schafer, Presented at the 24th International Symposium on Chromatography, Leipzig,
    Germany, September 15–20, 2002
63 Anonymous OPLC™ prepared by Bionisis SA of Paris France, 2002
64 R. Weinberger, *Practical Capillary Electrophoresis*, Academic Press, 2nd Edn., 2000, pp. 38,
    59–61, 89–91.
65 I.S. Krull, R.L. Stevenson, K. Mistry and M.E. Schwartz, *Capillary Electrochromatography
    and Pressurized Flow Capillary Electrochromatography, An Introduction*, HNB, New York,
    2000, p. 1.
66 A. Permstaller, H. Oberacher, A. Rickert, C.G. Huber and P.J. Oefner, *Genomics*, 79 (2002)
    793.
67 Abstracted from *Laboratory Equipment*, Vol. 34 #12, April 1998 (p. 106 to 111) and vendor
    supplied documents
68 W. Xiao and P.J. Oefner, *Human Mutation*, 17 (2001) 442.

Subject Index

A

Absolute mobility B655–657
Absorption A47–54
Accelerator mass spectrometery A501–502
Acceptor forces A325
Acceptor interactions A60
Acceptor phases A112
Aceclofenac B954
Acetaminophen B957, B966
Acetate B539, B563
Acetic acid B545
Acetylcodeine B1076–1078
Acetyldigitoxin B1105
Acetylene B524
N-Acetylneuraminic acid B869
cis-, trans-Aconitate B539
Aconitine B1057, B1105
Actinide B523, B526
Acyclovir B973
Acyl phosphatase B658, B659
Adhesion A232–233
Adsorption
    hydrophobic interactions B705
    solute-zone separation A54–60
    stoichiometry B703
Affinity capture A454, A455
Affinity chromatography A139–161
    adsorption A141, A151
    kinetics A150–151
    ligands for A142–143, A149–150,
        A153–160
    purification schemes A160
    thermodynamics of A150–151
    tags A145–148
Aflatoxins A411
Agarose A217, B690–691, B879
Aggregate separations B721–723, B724
Aggregation stability A238

Alachlor B1014
Alcetochlor B1014
Aldehydes B799
Alkaline earths B569, B570–571
Alkaline phosphatases B822
Alkaloids B1057, B1058, B1059
Alkanes B524
Alkali metals B522, B535–536, B559
Alprenolol B959
Aluminum B526, B527, B551, B570
Amanitine B1105
Americium B523, B525
Amino acids B588–610
    analyzers for B846
    aromatic B605–607
    chiral separations of B598, B600–606
    derivatization reagents for B590, B595–598
    electrophoresis of B588, B591–599,
        B601–610
    underivatized B597–598
p-Aminobenzenearsonate B574
Aminobutyric acid B1100
Amino-cyano phase sorbents B854
7-Aminoflunitrazepam B1098
p-Aminophenylarsonic acid B575
Amino sugars B869
Amlodipine B958
Ammonium B552
Amperometric detectors A118–119,
        A183–184, B598
Amphetamines B951, B1085–1092
Anabolic agents B1106
Androstanediols B1107
Androstenedione B1107
Anesthetics B969–972
Angiotensin A309
Anhydroecgonine methyl ester B1083–1085
Anion-exchange chromatography B840–841,
        B863–864, B866–875, B915

Anions
acid anions B549
electroseparation of B559–665
gas chromatography of B530–531
liquid chromatography of B537–550
Anthocyanins B1051
Antibiotics B999, B1020–1021
Antibodies B675–677, B723
Anti-circular development modes A268, A269
Anti-diabetic drugs B1100–1101
Antimony B522, B532, B536, B557
Antimony hydrides B524
Antioxidants B681–682
Antipyrine B957, B965, B967
Antiseptics B1022
Arachidonate B781–782
Argentation chromatography A62
Argon B523
Arsenate B540, B557, B559
Arsenazo III B551
Arsenic B522, B525, B532, B536, B557, B559, B575
Arsenic hydrides B524
Arsenite B540, B559
Arsenobetaine B557, B574, B962
Arsenocholine B557, B574
Arsine B523
Arson accelerants B1112–1113
Atmospheric-pressure chemical ionization A226, A418–419, A500, B1004–1005
Atomic absorption spectrometry A121
Atomic emission A360–361, A414–415, B521–522
Atomic spectroscopy B540
Atovaquone B948
Atracurium B1102
Atracurium besylate B964–965
Atrazine B992, B1014
Atropine B958, B1058
Aurones B1051
Automated multiple development A271, A276–277
Autoradiography A281, B644
Autosamplers A102–103, A479–480
Azide B563, B1111
Azobactam B951

B

Back flushing A334
Background electrolytes B560–561, B563
Bacteria B890
Bacterial endospores B1109
Barium B523, B569, B570
Benazepril B958, B973
Benzene B1097–1098
Benzidine B992
Benzimidazoles B1108–1109
Benzoates B563, B789–790
Benzodiazepines B975, B1098
Benzodioxoazolylbutanamine B1090
Benzoylecgonine B1082–1085, B1104–1105
Benzphetamine B1089
Berberine B1058
Bergamot oil A347
Beryllium B526, B527, B564
Beta-receptor blocking drugs B1107
Bicarbonate B544
Biflavonols B1051
Binary compounds B523–526
Binding constants B713
Bioinformatics resources B1148–1149
Biological samples
cell separation/concentration B909–910
fluids B1086–1092
preparation of B909–912, B947–953
Biomimetic ligands A155–157
Biopolymers A446–447
Biosensors A394
Biperiden B951
Bismuth B522, B523, B524, B572
Bisphenol A B1023–1024
Blocked co-polymers A236–237
Blood B1093–1095, B1097
Blood coagulation factor B722–723, B724
Bonded amide sorbents B854
Bonded phases A104–107, A381, B853–855
Boranes B524
Borate B549
Boron B522
Boronic acid A157
Bosentan B954
Brassicaceae seed oils B795
Bromate B539, B543
Bromide B539, B563, B564
Brominated compounds B1010–1011

Bromine B531
Buffered eluents A125–127
Buffering capacity B718–719
Buffers
   amphoteric B634–636, B637
   acrylamido B639
   additives A377–379
   multiphase B642–643
   poly-ampholyte B719
   acidic B647–649
   types of A382–384
Bulk-property detectors A358–364, A497
Bupivacaine B952
Buprenorphine (BP) B1080–1081
1,4-Butanediol B1100
Butane B1098
Butanesulfonate B563
Butanol B1097–1098
Butyl acetate B1097–1098
Butyltin compounds B996
Butyrate B563
Butyrolactone B1100

C

Cadmium, B525, B551, B553, B569, B570,
   B572
Caffeic acid B1052–1053
Caffeine B957, B967, B975, B1076, B1107
Calcium B550–553, B569–570
Calibration A228–230, A234, B877–878,
   B879–880
Californium B523
Calmus oil A23–24
Camag sample applicator A287, A289
Cannabinoids B1092–1096, B1097
Capacity factors B700–701, B856
Capillary arrays A370–374, A448–449, A450,
   B934
Capillary chromatography
   of amino acids B588, B591–593, B603–610
   of peptides B618–620
Capillary columns A219, A496
Capillary electrochromatography B929–930
   dispersion A305, A307–310, A312–313
   instrumentation for A513–514
   stationary phases for A304–315
Capillary electrophoresis
   instrumentation for A513, A514

on microchips A432–437
   microfabrication methods for A432–434
   PCR-based technologies for B929
Capillary flow-control A256, A257
Capillary gas chromatography A329–331,
   B521–522, B884
Capillary isoelectric focusing A385–386
Capillary zone electrophoresis A369–395
   additives A377–379
   capillary array A370–374
   capillary isoelectric focusing A385–386
   covalently-bonded phases A381
   electro-osmotic flow A375–381
   folding transitions B658–659
   four-color confocal detection A372–373
   instrumentation for A370–374
   open-tube A384–385
   operational modes A384–389
   two-dimensional A389
Carbamates B996, B1001, B1108–1109
Carbendazim B1001
Carbohydrates B839–893
   anionic B865
   affinity chromatography of B878, B881
   capillary electrophoresis of B893
   derivatization B844–847
   detection of B841–844, B876–877,
      B888–889
   electrospray ionization-mass spectrometry
      B875, B891
   gas chromatography of B841, B882–887
   high-performance anion-exchange chroma-
      tography of B863–875
   hydrophobic-interaction chromatography of
      B855–861
   liquid chromatography of B839–881
   microfabricated devices A446–447
   planar chromatography of B887–892
   size-exclusion chromatography of
      A239–240, B840, B875–878,
      B879–880
   thin-layer chromatography of B887–892
   two-dimensional liquid chromatography of
      B862
   underivatized B850–851, B858–859
Carbon B522
Carbonates B539, B544, B563
Carbon monophosphide B523
Carbon trioxide B564

Carboranes B524
Carbosilanes B524
Carboxylic acids A199, A200
Cardiolipin B745–747
Carotenes B1048, B1049
Carrageenans B879
Carrier ampholytes B634–636, B637
Carrier gases A323–325
β-Casein B650–651
Catechin B1055
Cathine B1107
Cathinone B1086
Cation-exchange chromatography B678–679
Cations B550–555
Cells B909–910
Cellulose B879
Cephalomannine B1046
Ceramides B787–800
Cerium B553, B569–570
Cesium B523, B570
Cevadine B1105
Chalcones B1051
Chambers A262–266, A457, A458
Chaotropes B645
Chaotropic salts B711
Charge/mass ratio B655–657
Charge-remote fragmentation B756
Chelate effect A61–62
Chemical cleavage A147–148
Chemical warfare agents B1109–1110
Chip technology B728, B1057–1058,
        B1060–1061
    see also Microchip technology
Chiral separations
    coupled-column LC A411
    drug analysis B957–959
    solute-zone complexation A63
Chitin B879
Chito-oligosaccharides B860
Chitosan B879
Chlofetezine B1001
Chlorate B539, B563, B1111
Chlordiazepoxide B975
Chlorinated compounds B1007–1010
Chlorine B523, B524, B542, B550
Chlorite B539
Chloroacetamide B1014
Chloroacetanilide herbicides B1001, B1013
Chlorodifluoromethane B1098

Chlorofluorocarbons B525
Chloroform B1097
Chlorothiazide B1108
Chlorpheniramine B948
Chlorporopamide B1101
Chlorpyrifos B1013
Chromarod A285
Chromate B539
Chromatofocusing B718–720
Chromatographic optimization function
        A73–74
Chromatographic resolution statistic A74
Chromatography
    coupling with spectrometry A413–425
    theory of A1–89
Chromium B526, B527, B530, B551, B572
Chromolith B694–695
Chromophore detection A116, A117
Chromosome-specific polymorphisms
        B1119–1120
Cimetidine B1108
Circular development modes A268–269
Cisplatin B527
Citrate B539, B563
Citric acid B545
Clarithromycin B962
Clausius–Clapeyron Equation A45–46
Clean-up methods B998–1001
Cleavage strategies A147–148
Clemastine B971
Clenbuterol B952, B1107, B1108
Clevidipine B969
Clinafloxacin B959
Clobenzorex B1089
Clonazepam B1100
Cobalt B526–528, B551–553, B558,
        B569–572
Cocaethylene B1082, B1084
Cocaine B1081–1085
Codeine B1058, B1075–1077, B1078
Collagens B616, B617
Colloidal dispersions B672–673
Column liquid chromatography A96–131
    capillary B884
    columns for A103–115
    detection in A115–121
    equipment for A98–104
    high-temperature separations in A129
    large-scale separations in A130–131

miniaturization in A129–130
modes of A97–98
optimization methods for A128
sample preparation in A128–129
stationary-phase materials for A104–115
Column rotation planar chromatography
A275–276
Columns
dispersion contributions A17–38
fused-silica A374–375
monolithic A109, A218–219, A306–310,
A495
material A193–194
switching B949–951
temperature A496–497, B919–920
Combinatorial libraries A159–160, A421,
A449, A450
Combined chromatographic techniques
A403–425
Combined identification methods B761–763
Complexation A60–63
Complexes B721–723, B724
Computational fluid dynamics A461–462
Concanavalin B1060, B1061
Concentration shift method B924
Concentrator columns B543
Condensation B846
Conductivity detection
in column liquid chromatography A119
in gas chromatography A358
in ion chromatography B539, B543
theory of A180–183
Confocal systems A371–374
Coniine B1058
Conjoint liquid chromatography B695–696
Conotoxin B1110
Continuous-bed columns see Monolithic
columns
Contraceptives B1018–1020
Control equipment A474–475
Convallatoxin B1105
Convective mixing A305
Co-ordination compounds
electroseparation of B571–572, B573–574
gas chromatography of B526–530
liquid chromatography of B555–558
Co-polymers A236–237
Copper B523, B526–527, B551–553,
B569–572

Coulombic attraction A42–43
Coulometric detectors A118–119
Coumaric acid B545, B1052–1053
Coumarins B955, B1052–1053
Counter-ions A126–127
Coupled chromatographic techniques
A403–425
Coupled column liquid chromatography
A410–412
Coupled columns A120–121, A332, A347,
A405–412, A413–414, B521–522,
B540
Crack cocaine B1082
Creatine B552
Creatinine B552
Cryptand columns A198
Curium B525
Cyanide B547, B548
Cyanogen bromide B526
Cyanogen chloride B526
Cyclic reducing agents B646–647
Cyclodextrin A326, B855, B872, B885–886
Cyclohexane B1097
Cyclosporin B949, B950
Cysteine B609–610
Cytisine B1058

D

Darifenacin B952
De-acylation B819
Dead-volume A506
De-amidation B680–681
Dean's switch A335
Debrisoquine B955
Debye forces A40–41
Debye–Hückel equation B698
Debye length A17, A301, A302, B715
Debye screening factor B656
De-glyceration B819–820
Dehydroepiandrosterone B1107
Denaturing gel electrophoresis A239
Denaturing gradient gel electrophoresis B924
Denaturing high-performance liquid
chromatography B919–921
3-Deoxy-D-manno-octulosonic acid B869
Deoxyribonucleic acid B906–908
capillary electrophoresis A387–388,
B926–928

column temperature B919–920
fluorescent dyes B933
fragment analysis A437–442
ion-pair chromatography B918–919
mutations B928
polymorphisms B1115–1116
sequencing A439–442, B928
typing A440–442, B1113–1121
De-phosphorylation B819–820, B821
Deramciclane B974
Desalting B723–725
Desflurane B971
Designer drugs B1085–1087, B1089–1091
Design standards A505–507
Desipramine B948
Deslanoside B1105
Desmethylflunitrazepam B1098, B1100
Desorption phases A141, A151
Detectors
    amperometric A118–119, A183–184, B598
        pulsed-amperometric B842–843,
        B867–868
    atomic emission A360–361, A414–415
    atomic spectroscopy A121, B540
    bulk-property A358–364, A497
    chemiluminescent A361–362, B551, B597
    chiral A119
    electrochemical A118–119, B540,
        B842–843
    electron-capture A353, A355–356
    flame A285, A352–355, A359–360
    fluorescence A116, A118 A233, A503,
        B551, B597
    four-color confocal A372–373
    indirect B567, B598, B599
    infrared A362
    ionization A352–358
    light-scattering A113
        evaporative A421, A498, B597, B802,
        B841
        multi-angle laser A222–223, B876
    microfabricated devices A436–437
    nitrogen B597
    olfactory A363
    optical A358–362
    photoionization A353, A356–357
    photometric A358–362, A502, B567
    pulse-discharge A353, A357
    spectroscopic A115–118, A121, B540

ultraviolet B932–933
thermal-conductivity A358
thermionic ionization A353, A355
universal A497–498
variance A37
Development modes A267–272
Dexamethasone B973, B1107
Dextran-methylprednisolone succinate
    A221–222
Diacylglycerols B746–747, B751–752,
    B787–793
Diagnostic contrast media B1022–1023
Diazepam B974–975, B1104–1105
Diborane B525
Dichloroacetate B544
Dichloromethane B1097
Dichlorprop B992
Diclofenac B954
Diesel oil A347
Diethylpropione B1085–1086
Diffusion
    coefficients A11–14, A18–19, A20, A31
    in electrokinetic chromatography A312
    in gas chromatography A323
Diflubenzuron B1001
Digital autoradiography A281
Digitalis B1105
Digitoxigenin B1105
Digitoxin B1105
Digoxin B1105
Dihydrocodeine B1080–1081
Dihydromorphine B1080–1081
o-Dihydroxybenzoate B545
Dihydroxymetabolites B785–786
Dihydroxymethamphetamine B1090
β-Diketones B526–529, B530
Dimethylamphetamine B1089
Dimethylantipyrine B957
Dimethylarsinate B559
Dimethylarsinic acid B557, B574–575
Dinitrobenzene B1111
Dinitrobenzoyl derivatives B790–791
Dinitrotoluene B1111
cis-Diol cyclic complexes A157
cis-Diol titration B719–720
Dioxins B991, B1009
Diphenylamine B1111
Diphenylarsonic acid B574
Diphenylhydramine B948

Diphenyllead chloride B574
Dipole induction forces A40–41
Dipole moments A41
Dipole orientation forces A41–42
Diquat B1014
Diradylglycerols B787–793
Direct detection B567, B575, B597,
    B613–614
Disc electrophoresis A388–389
Dispersion
    extra-column contributions A17–38
    forces A39, A325–326
    solute-zone A9–38
    theory A9–38
Dissolved-organic-carbon detectors A223
Distribution coefficient A199–200, A299
Distribution constant A322–323
Diterpenes B1044, B1046–1047
Dithioalkylates B528
Ditrate B539
Diuretics B1106, B1108
Diuron B991
Donnan dialysis B540–541
Donor/acceptor forces A325
Donor/acceptor interactions A60
Donor/acceptor phases A112
Doping agents B1106–1108
Doxazosin B958, B973
Doxepin B957, B1108
Drift A497
Drug analysis B945–975
    column/column liquid chromatography in
        A412
    gas chromatography in B969–972
    high-performance liquid chromatography in
        A416, B953–968
    ion chromatography in B971–972
    multiple detectors for B966–968
    overpressured-layer chromatography in
        B973–974
    photodiode-array spectrometry A416
    supercritical-fluid chromatography in
        B968–969, B970
    solid-phase extraction for B948–995
    thin-layer chromatography in B972–975
    turbulent-flow chromatography in B953
Drugs of abuse B1074–1098
Dual detection A364
Dye ligands A155–157

Dynamic coating ion-pair chromatography
    B546
Dysprosium B569

E

Ecgonidine B1082–1085
Ecgonine B1082–1085
Ecgonine methyl ester B1082–1085
Ecstasy B1085–1087, B1089–1091
Edible oils A409
Egg-shell proteins B619, B620
Egg yolks B810–811
Elastomer chips A452–453
Electrically driven electrokinetic chromato-
        graphy A315–316
Electric-field-mediation separations A459
Electro-antennograph A363–364
Electrochemical detection A118–119, B540
Electrochemically regenerated ion neutralizer
        B542
Electrochemical reactions A503
Electrochemical regeneration of suppressors
        A178–180
Electrochromatography
    see Capillary electrochromatography
Electrodes B842–843
Electrodialysis B541
Electrokinetic chromatography A297–317
    capillaries A301–302, A303–304,
        A306–310, B561
    electrically driven A315–316
    electro-osmotic flow A301–304
    enantiomer separation A314–315
    modes A298–301
    open capillaries A301–302, A306
    packed capillaries A301–302, A303–304,
        A306–310
    peak dispersion A305, A307–310,
        A312–313
    pressure-driven A315–316
    pseudo-stationary phases A310–315
    selectivity A307
    stationary phases A304–315
Electrolytic membrane suppressor A178–180
Electron-capture detector A353, A355–356
Electron donor/acceptor interactions A60
Electronic variance A37

Electron ionization mass spectrometry B780–781

Electro-osmotic flow
in capillary zone electrophoresis A375–381
electrochromatography A298, A301–304
open capillaries A301–302
porous beds A303–304

Electrophoresis
instrumentation for A513, A514

Electrophoretic migration B641

Electrophoretic mobility A311, B560–562, B566, B568, B656–657

Electro planar chromatography A258, A276

Electrospray ionization A418, A446, A447, A500, B846

Electrospray ionization mass spectrometry B755–758, B763, B792, B802–806, B875, B891

Electrostatic interaction B715–717

Electrostatic ion chromatography A201–203

Elemental analysis A417, B521–536

Eluents
generation of A175–176
ion chromatography A174–176, B546–547, B548
inorganic species B538–539
selectivity A187, A188–190
ion-exclusion chromatography B548
ionic strength A221
ion-interaction chromatography B546–547
ion-pair chromatography B546–547
peptide chromatography B613–614

Elution A141, A123–125, A282

Enantiomers A314–315, A411, B598, B600–606, B885–886

End-capillary conductometric detection B567–568

Endocrine disruptors A423, B1018

Energy-transfer primers B933

Enflurane B1097

Enhanced separations A332–340

Enhancement columns B549

Entacapone B951

Environmental analysis B987–1036
by gas chromatography B532
by ion chromatography B543
by liquid chromatography A416, A423–424, B543
detection in B1006–1007

mass spectrometry in A423–424
organometallics B532–536
persistent organic pollutants B1007–1011
personal care products B1018–1022
pharmaceuticals B1018–1022
photodiode-array spectrometry in A416
plasticizers B1023–1024
quantitative B1004–1024
sampling B989–1003
solid samples B994–1002
surfactants B1014–1018
toxins B1108–1109
turbulent-flow chromatography in B1004
water samples B989–994

Enzymatic characterization B820–822

Enzymatic cleavage A147–148

Enzymatic reactions A455

Enzyme-inhibition assay A448, A449

Enzyme-linked immunosorbent assay A143, A144, B841

Ephedrine B949, B1086, B1107

Epitestosterone B1106, B1107

Epoxypolyunsaturated fatty acids B785–786

Equilibrium constant A46–47, A49–50, A68–70, A81

Equivalent chain-lengths B749

Erbium B569

Erucic acid B795

Erythropoietin B675–677

Escherichia coli B682–683

Essential oils A334, B1040–1042, B1043

Estradiol fatty acid esters B747

Estrogens A411, B999, B1019

Ethanesulfonate B563

Ethanol B1098

Ether lipids B792

Ethinylestradiol B965

Ethylamine B552

Ethyleneglycol dinitrate B1111

Ethylenethiourea in water A411

Ethylmercury B521, B534–535, B557

Ethynodiol B959

Europium B523, B530

Explosives B1110–1112

Expression systems B682–683

Extrachrom A274–275

Extra-column contributions, variance A33–38

Extraction
see also Solid-phase...; Solid-phase micro...
immuno-affinity B952, B992
lipids B741–747
organic pollutants B998–1001
solid samples B997–1002

F

Famotidine B1108
Fast-atom bombardment B754–755,
B809–810, B846, B891
Fast gas chromatography A336–337
Fast separations A332–340
Fatty acid esters B746–747
Fatty acid methyl esters B779
Fatty acids B765–769, B779–782
Fenoprop B992
Fenspiride B971
Fentanyl B971
Fermium B523
Ferulic acid B545, B1052–1053
Fick's laws A10
Field analysis A208
Field-inversion electrophoresis A439–440,
B931–932
Film thickness A27, A28–29
Filter-chamber arrays A457, A458
Flame retardants B1002
Flash chromatography A512–513
Flavonoids B1050–1052, B1054, B1055,
B1061
Flavonol glycosides B1055
Flobufen B958
Flow rate A84–85, A266–267, A476–477
Flubiprofen B967
Fluid manipulation A434–436
Flunitrazepam B951, B1098–1099
Fluorescent dyes B933
Fluoride B539, B544, B563
Fluorine B531, B564
Fluorogenic reagent reactions B846
Fluoroquinolone B959
Fluoxetine B958, B971, B973
Fluprostenal B971
Flurazepam B975
Flurbiprofen B965
Fluticasone B964–965
Folch extraction B742, B743

Folding transitions B658, B659, B725
Forced-flow chambers A266
Forced-flow planar chromatography
A255–258, A269, A287
method development A289–290
preparative A284–285
Forensic analysis B1073–1121
amphetamines B1085–1092
arson accelerants B1112–1113
biological warfare agents B1109–1110
cannabinoids B1092–1096, B1097
chemical warfare agents B1109–1110
chromatographic screening B1102–1105
chromosome-specific polymorphisms
B1119–1120
cocaine B1081–1085
DNA typing B1113–1121
doping in sport B1106–1108
drugs of abuse B1074–1098
environmental toxins B1108–1109
explosives B1110–1112
hallucinogens B1085–1092
inks B1113, B1114
mitochondrial DNA B1119–1120
natural compounds B1105
nucleic acids B1113–1121
occupational toxins B1108–1109
opioids B1075–1081
printing media B1113, B1114
sex chromosomes B1119–1120
therapeutic drugs B1098–1102
volatile narcotics B1096–1098
writing media B1113, B1114
Formate B539, B563
Formic acid B545
Forward recoil spectrometry A233
Four-color confocal detection A372–373
Fourier-transform infrared spectrometry A362,
A415–416, A417
Fourier-transform ion-cyclotron resonance
spectrometry A420
Fraction collection A459, A461, A505
Fragrances B996, B1000, B1021–1022
Free-zone electrophoresis B647–660
Freundlich equation A59
Fructans B872
Full adsorption/desorption A235
Fuller–Schettler–Giddings Equation A12
Fumaric acid B545

Fungicides B993
Furans B991
Furfurals B845
Furosemide B1107
Fusarenone B1044, B1045

G

Gadiodiamide B962
Gadolinium B569
Galactarate B563
Galactoglycerolipids B768
Gallamine B1102
Gallic acid B545
Gallium B551
Gallopamil B958
Gangliosides B767, B890–891
Gas chromatography A319–364
    atomic emission detection in A414–415
    columns for A321–323, A332
    combined techniques A405–410,
        A413–416, B1012
    detection in A348–364
    enhanced separations in A332–340
    extra-column variance A35
    Fourier-transform infrared spectrometry in
        A415–416
    headspace A346–347
    injection techniques in A340–347
    instrumentation for A507–511
    mass spectrometry in A413–414, B761,
        B786, B1012
    molecular species resolution B784–785,
        B793
    multi-dimensional GC A333–336,
        A337–340
    negative ion-chemical ionization mass
        spectrometry in B787–788
    operating variables A321–331
    pressure tuning A332–333
    programmed-temperature vaporizer
        A345–346
    sample introduction in A340–347
    solute-zone separation A50, A51
    stationary phases for A325–327
    tandem A81
    total profiling B813–814
    two-dimensional A337–340, B1005–1006
    univariate optimization A80–81

Gas-phase diffusion coefficients A12–13
Gastric juice B806
Gel electrophoresis A239, B924
Gel filtration A70
Gel permeation A70
    see also Size-exclusion chromatography
Gel slab electrophoresis B925
Genomics
    computer/software resources B1146–1147,
        B1149, B1153–1154
    sequencing B726–727, B928, B929, B935
Gentisic acid B545
Geometrical isomerism B530
Germanes B523, B524
Germanium B522, B524, B536
Gibberellins B1046–1047
Gibbs free energy A44
Gitaloxin B1105
Gitoxin B1105
Gleevec B962
Glibenclamide B1100–1101
Glibornuride B1101
Gliclazide B973, B1101
Glimepiride B1101
Glipizide B1101
Gliquidone B1101
Glisoxepide B1101
Globin peptids B652
Glucans B872, B875
Glucose monophosphates B870
Glutamate B563
Glutathione B609–610
Glycerophospholipids B757, B769–773,
        B800–811
    acidic B807–809
    neutral B800–807
Glycoconjugates B873–874, B876, B892
Glycolipids B849
Glycopeptides B874
Glycoproteins A154–155, B861
Glycosaminoglycans B873
Glycosidases B677–680
Glycosphingolipids B778, B892
Glycosyl phosphatidylinositols B778
Glycosyltransferases B677–680
Glyphosite B992
Golay Equation A20–21
Gold B523, B547, B548
Gradient elution A315, A477–478

Gradient formation B719–720
Gradient gel electrophoresis B924
Gradient polymer elution chromatography
    A231–232
Graphitized carbon A110, B859–860
Group-specific ligands A145

H

Hagen–Poiseuille Equation A15
Halides B525, B531, B535
Hallucinogens B1085–1092
Halogens B522
Haloperidol B1104
Halothane B971, B1097, B1098
Hashish B1092–1096, B1097
Heart-cutting A333
Helmholtz–Smoluchowski Equation A17, A302
Hemoglobin B654, B655, B720
Henry equation A45, A301, B656
Henry's correction factor A301
Heparin A154, A229, B879
Heptafluorobutyrate B884–885
Herbicides B992–993, B1001, B1002, B1013
Heroin B1075–1076, B1078
Hexane B1097
High-performance affinity chromatography
    B881
High-performance anion-exchange chromato-
    graphy B840–841, B863, B866–875
High-performance capillary electrophoresis
    B893
High-performance liquid chromatography B909
    see also Column liquid chromatography
    with inductively coupled-plasma mass
        spectrometry B962–964
    with mass spectrometry B845, B846–847,
        B959–964, B966–968
    with nuclear magnetic resonance B964–968
High-performance molecular-sieve chromato-
    graphy A217
High-performance size-exclusion chromato-
    graphy A223–224, B875–876
High-performance thin-layer chromatography
    A255, A259–260, A280, A288–291,
    B887, B748–749, B753
High-pH chromatography B840–841, B863,
    B866–875
High-speed analysis A207–208

High-temperature separations A129
High-temperature size-exclusion chromato-
    graphy A232
High-volume direct injection A203–204
Hildebrand solubility parameter A47–48
Hirschfelder-Bird-Spotz Equation A11
Holmium B569
Homocysteine B609–610
Hormones A238, B1018–1020
Host cell expressions B682–683
Hückel–Onsager Equation A301–302
Human growth hormone A238, B674, B676,
    B680–681
Humic substances A240
Hyaluronan oligosaccharides A239–240
Hybrid silica packing B956–957
Hydrides B523, B524–525
Hydrocarbons B542
    aromatic B1096–1097
Hydrochlorothiazide B973, B1108
Hydrodynamic chromatography A67–68
Hydrodynamic permeability A216–217
Hydrogen chloride B524
Hydrogen selenide B523
Hydrogen sulfide B523, B524
Hydrolysis B588–589, B816–819
Hydrophilic-interaction chromatography
    B847–855, B861–862
Hydrophobic-interaction chromatography
    adsorption B705
    purification B707, B708
    retention mechanisms B697–703
Hydrophobic interactions A301
Hydrophobic polymers A379
Hydroxide generators B538–539
Hydroxyapatite B692–693
p-Hydroxybenzoate B545
Hydroxybutrate B1098
Hydroxycinnamic acids B1052–1053
Hydroxyl radical B564
Hygiene B543, B1018–1022
Hypericin B967
Hyphenated chromatographic techniques, see
        Coupled chromatographic techniques

I

Iatroscan A285
Ibuprofen B952, B965, B966, B967

Ilperidone B966
Imipramine B957
Immobilized affinity ligands A142–143
Immuno-affinity A155, A412, B952, B992
Immuno-assays A143–147, B841, B888
Immunoglobulin B675–677, B721–722, B723
Immunosorbents A143, A144, B841, B991
Incapacitating drugs B1098–1110
Indirect chiral separations B605–606
Indirect detection B598, B599
Indium B551
Indole B1059
Indomethacin B967
Induction forces A40–41, A325
Inductively coupled-plasma mass spectrometry
    A121, A225–226, A416–417,
    B962–964
Injection
    column liquid chromatography A101–103,
      A478–480
    cool on-column injection A343–346
    gas chromatography A340–347, A510
    microfabricated devices A434–436, A452
    variance A35
Inks B1113–1114
Inorganic anions
    detection of B538–540
    electroseparation of B559–665
    gas chromatography of B530–531
    ion chromatography of B537–550
    liquid chromatography of B536–550
Inorganic cations
    detection of B561–562
    co-ordination compounds B555–556,
      B557–558
    electroseparation of B566–571
    ion chromatography of B550–551
    ion-pair chromatography of B551–554
    liquid chromatography of B550–559
    organometallic compounds B556–559
Inorganic sorbents B692–693
Inorganic species B519–578
    anions B530–531
    electroseparation of B559–577
    gas chromatography of B520–536
    ion-exclusion chromatography of B548–550
    liquid chromatography of B536–559
    organometallics B532–536
Insecticides B996

Instrumentation A469–516
Insulin B1101
Integrated design A506
Interaction energy A55–57
Interferon B678
Interhalogens B525
Internal energy A42
Internet B1135–1156
Interpretive methods A77–79
Intrinsic viscosities B673
Inverse sum of resolution A73
Iodate B539
Iodide B546
Iodine B522–523, B531, B547, B564
Ion analyzers A500–502
Ion-beam lithography A433
Ion chromatography A171–209
    combustion B542
    conductivity detection theory A180–183
    chelation B554–555
    detection in A180–186, B539–540
    instrumentation for A173–186
    sample handling for B540–543
    sample preparation for A203–207
    selectivity A186–203
    solute-zone separation A64–67
    suppressors A176–180
Ion-exchange chromatography A191–197
    chromatofocusing B718–720
    retention mechanisms B712–717
    solute-zone separation A64–67
    sorbents B717–718
    types A194–197
Ion-exclusion chromatography
    detection B549
    eluents B548
    selectivity A198–201
    solute-zone separation A64–67
    stationary phases for B548
Ionic strength B700
Ion-induced dipole forces A42
Ion-interaction chromatography *see* Ion-pair
    chromatography
Ion/ion forces A42
Ionization modes A500
Ion-moderation partitioning B863–866
Ion-pair chromatography
    detection B547
    eluents B546–547

liquid chromatography B546–547
   stationary phases for B546–547
Ion-pairing reagents A127, B546–547,
   B613–614
Ion-pair reversed-phase chromatography
   B917–921
Ion-projection lithography A433
Ion-traps A419–420, A501, B1012
Irgarol B991
Iridium B525, B527
Iron B552, B553, B558, B570
Isoamyl acetate B1097–1098
Isobutane B1098
Isoelectric buffers B647–648, B649
Isoelectric focusing A385–386, B634–640,
   B641
Isoeluotropic A53
Isoflavonols B1051
Isoflurane B1097
Isolectin B1060, B1061
Isomerism B530
Isoprenoids B1040–1050
Isoprostaglandin B784 785
Isoprostanes B766
Isoproturon B1014
Isosorbide B969
Isotachophoresis A388–389
Isotherms A46–47, A59–60
Isotopes B526

J

Joule heating A305, A312

K

Kaempferol B1055, B1061
Kessom forces A41–42
Ketoconazole B973
Ketomalonate B539
Ketoprofen B958
Kinetic Langmuir equation A59–60
Knudsen flow A14–15
Kozeny-Carman Equation A15

L

Labile proteins B709, B710
α-Lactalbumin B711
Lactic acid B545

Lactoferrin B701–702
Laminar flow A15–16
Langmuir equation A59–60
Lansoprazole B1108
Lanthanides
   electroseparation of B570
   gas chromatography of B525, B526, B530
   ion chromatography of B555
   ion-pair chromatography of B551
Lanthanum B553, B569, B570
Large-scale separations A130–131
Large-volume injection procedures A346
Laser-induced fluorescence B594–595,
   B605–607, B933–935
Latanoside B1105
Layer-thickness gradients A286–287
Lead B521–523, B527, B532, B552–553,
   B569–572
Leaded gasoline B534
Lectins A154–155, B889–890, B1060, B1061
Leukotriene B783
Levonantradol B1092
LiChrosorb B854–858, B878, B881
Lidocaine B948
Ligands A113, A142–145, A155–159, B1052
Limit of detection B593–595, B598
Linear alkylbenzene sulfonates B1014–1015
Linear development modes A267–268
Linear velocity A21–27, A31–33
Linolenic acid B759, B795–796
Lipids B739–822
   chemical characterization of B816–820
   combined identification methods for
      B761–763
   extraction of B741–747
   gas chromatography of B749–750, B753
   liquid chromatography of B750–752
   infrared spectrometry B763–765
   ion-exchange chromatography of B752
   isolation, individual classes B765–778
   mass spectrometry B754–763
   microfabricated devices A446–447
   molecular species resolution B779–811
   neutral glycerolipids B787–800
   nuclear magnetic resonance B763–765
   peak identification B753–765
   profiling B811–816
   quantification of B753–765
   separation methods for B748–753

supercritical-fluid chromatography of
    B752–753
size-exclusion chromatography of A240
solid-phase extraction of B745–747
thin-layer chromatography of B748–749,
    B753–754
Lipolysis B820–821
Liquid chromatography
    see also Column liquid chromatography
    with Fourier-transform infrared spec-
        trometry A417
    with gas chromatography A223, A347,
        A407–410
    with inductively coupled plasma mass
        spectrometry A416–417
    with mass spectrometry A416–25,
        B845–847, B959–964, B966–8,
        B1004, B1017–1018
    with nuclear magnetic resonance spec-
        trometry A417
    with photodiode-array spectrometry A416
    with tandem mass spectroscopy B1004,
        B1017–1018
Liquid fuels B542
Liquid/liquid extraction B948
Liquid/liquid partitioning A316
Liquid-phase diffusion coefficient A13–14
Liquid polymers A387–388
Liquid secondary-ion mass spectrometry B891
Lithiated adducts B803–804
Lithium B523, B552, B569–570
Lithography A432–433
Loading phase A141
London forces A39
Longitudinal diffusion A18–19, A312, A323
Lutetium B569
Lysergic acid diethylamide B1091–1092

M

McReynolds constant A327–328
Macrocyclic effects A61–62
Macromolecular porosimetry A228–229
Magnesium B550–553, B569, B570
Magnetic-sector mass spectrometers B1006
Maleate B539
Maleic acid B545
Malonate B539
Manganese B552, B553, B569, B570

Mannitol B719–720
Manual injection A479
Marijuana B1092–1096, B1097
Marinol B1092
Mass spectrometry
    atmospheric-pressure A226
    chemical ionization A226
    column liquid chromatography A120
    combined techniques A413–414,
        A416–417, A418–425
    computer resources B1144–1145, B1149,
        B1151–1153
    electron ionization B780–781
    fast-atom bombardment B754–755,
        B809–810
    in gas chromatography A413–414, B761,
        B786–788, B1012
    in high-performance anion-exchange
        chromatography B875
    in high-performance liquid chromatography
        A498–502, B845–847, B959–968
    inductively coupled plasma A121,
        A225–226, A416–417, B962–964
    in ion chromatography A186
    liquid secondary-ion B891
    magnetic-sector B1006
    matrix-assisted laser desorption ionization
        A226–227, A229, A500, B761,
        B846–847
    matrix-assisted laser desorption ionization
        time-of-flight A226–227, B759–760,
        B806–809
    microfabricated devices A446–447
    negative-ion chemical ionization
        B761–762, B787–788
    negative-ion liquid secondary-ionization
        B806
    in planar chromatography A281
    in size-exclusion chromatography
        A224–227
    tandem B815–816, B959–968
    time-of-flight A501, B812–813, B846–847,
        B1006–1007, B1012
Mass transfer A19, A20, A313, A323
Matrix effects A152–153
Matrix removal A205–206
Maximum permissible time constant A37–38
Mean free path A15
Mefenorex B1087, B1089

Membrane suppressor A178, A179
Mendelevium B523
Mercury
    electroseparation of B570, B572
    gas chromatography of B521–523, B525,
        B527
Metal chelate affinity A157–159
Metal-chelate interaction chromatography
        A157–159, B613
Metal chelates B556
Metal ion derivatizing reagents B527
Metallic oxides B525
Metallo-cyanides
    ion-interaction chromatography of B547
    ion-pair chromatography of B547, B548,
        B551, B554
Metalloid hydrides B524–525
Metalloporphyrins B529
Metals
    inorganic cations B569, B570–571
    ion chromatography B554
    ion-pair chromatography B551, B554
    trace determination B526–529
Metamizole B1104
Methacatinone B1086
Methadone B1077–1080, B1104–1105
Methamphetamines B1085, B1090–1091
Methandrosterone B971
Methane B524
Methaqualone B1076
Methenolone B1106
Methoxycinnamic acid B1052–1053
Methscopolamine B973
3-Methyl-L-histidine B552
Methyl-DL-selenocystine B558
Methylamine B552
N-Methylbenzoxodiazolylbutanamine B1088
Methyldigoxin B1105
Methylecgondine B1083–1085
Methylenedioxyamphetamine B1085–1091
Methylenedioxyethylamphetamine
        B1085–1091
Methylenedioxymethamphetamine
        B1085–1091
Methylephedrine B1107
Methyl glycosides B884
Methylmercury B521, B534–535, B557
Methylphenidate B1085–1086
Methylsulfonate B539

Methyltestosterone B1106
Methyltetrabutyl ether B1005
Metolachlor B1014
Metoprolol B959, B969
Metronidazole B1104
Micacurium B1102
Micellar electrokinetic chromatography
        A390–393
Micellar overload A312–313
Micro-bore separations B953–955
Micro-emulsion electrokinetic chromatography
        A301, A310–313
Microfabricated analytical devices A431–462,
        B911–912
    capillary-array systems A448–450
    capillary electrophoresis A432–437
    computational fluid dynamics A461–462
    field-inversion electrophoresis A439–440
    high-performance liquid chromatography
        B602
    immuno-assays A444–446, A447
    liquid chromatography B1057–1058,
        B1060–1061
    mass spectrometry A446–447
    micro-preparative methods A458–459,
        A461
    micro-reactors A456–458, A460
    multi-dimensional separations A444
    preparative methods A458–459, A461
    sample processing A451–454, A455
    screening A447–451
    system integration A451–461
Microfluidic immuno-assays A444–446
Microfluidic systems A431
Microheterogeneity B674–684
Micro-particulate resins and gels B864–866
Micro-preparative methods A458–459, A461
Micro-reactors A456–458, A460
Micro-satellites B1115–1120
Micro-separation B1057–1058, B1060–1061
Microwave-induced plasmas A414–415
Midazolam B1100
Migration A255, A304–305, A311, B641
Miniaturization A129–130
Minisatellites B1115–1120
Mitochondrial DNA B1119–1120
Mobile phases
    gradients A286
    selectivity A187–190

multivariate optimization A86–88
resistance to mass transfer A19
role in adsorption A57–58
univariate optimization A82–83
Mobility determination B655–657
Modular design A506
Molal surface tension B699
Molar enthalpy A43–44
Molar entropy A44
Molar Gibbs free energy A44
Molar mass distribution A235
Molecular flow A14–15
Molecularly imprinted polymers A114–115, B615, B952, B991–992
Molecular interactions A39–43
Molecular-mass distribution A230–237
Molecular-sieve chromatography A70
Molecular species resolution B779–811
Molecular-weight distribution A69
Molybdate B563
Molybdenum B522, B525, B530, B564
Monoacetylmorphine B1078
Monoacylglycerols B746–747, B788–789
Monochloroacetate B539
Monofluorophosphate B544, B563
Monomeric additives A301
Monomethylarsonate B559
Monomethylarsonic acid B557, B574, B575
Mono-specific ligands A144–145
Monoterpenes B1040–1042, B1043
Morphine B952, B975, B1057, B1058, B1075–1076, B1078
Moving-bed separations A130–131
Multi-angle laser light-scattering A222–223, B876
Multi-capillary systems A459
Multi-dentate bonded silica-based stationary phases A106–107
Multi-dimensional development modes A268, A269–272, A276–277
Multi-dimensional gas chromatography A333–336, A337–340, A405–407
Multi-dimensional planar chromatography A290
Multi-dimensional separations A389, A444
Multi-locus probes B1116–1117
Multi-modal separations A290–291
Multi-path dispersion A17–18
Multiple detectors A504, B966–968

Multiple reaction monitoring B886–887
Multiplexing B960
Multi-variate methods A84–88
Muramic acid B869
Muscle relaxants B1101–1102
Musks B996, B1000, B1021–1022
Mutations B674, B920
Myricetin B1061

N

Nabilone B1092
Naftopidil B965
Nandrolone B1106
Naproxen B952, B965, B967
Narcotics B1096–1098
Narrow-bore separations B953–955
N-chamber A264–266
Neodymium B553, B569
Neomycin B973
Neptunium B525
Neritaloside B1105
Nernst distribution coefficient A299
Neurodegenerative disease markers A237
Neuropeptides A412
Neutral lipids B765–769, B779–782
Nickel B522–523, B526–527, B552–553, B569, B572
Nicotinamide B1076
Nicotine B1058
Niobium B558
Nitrate B539, B544, B547, B549, B563, B1111
Nitrazepam B974–975
Nitric oxide B609–610
Nitrite B544, B547, B563, B1111
Nitroaniline B992
Nitrogen B523
Nitrogen dioxide B531, B564
Nitrogen tetrahydride B550, B570
Nitrogen trioxide B531, B550, B564
Nitroglycerin B1111
Nitrophenylamines B1111
4-Nitrophenylarsonic acid B575
Nivalenol B1044, B1045
Nizatidine B1108
Non-cholesterol plasma sterols B799–800
Non-enzymatic glycation B684
Non-esterified fatty acids B746–747
Non-glycosylated proteins B680–681

Non-ionic surfactants B1015−1018
Non-methylene-interrupted fatty acids B781
Non-sialylated sphingolipids B767
Non-specific interactions A143−144
Norandrostendiol B1106
Norandrostendione B1106
Norbuprenorphine B1080−1081
Norcocaine B1084
Norcodeine B1078
Nordiazepam B975
Norephedrine B1086, B1107
Normalized product of resolution A73
Normal-phase high-performance chromato-
graphy A98, A104−105, A121−124,
B686−697, B751, B791−792,
B802−805
Normorphine B1078
9-Nortestosterone B1108
Nortriptyline B948, B957
Noscapine B1075
Nuclear magnetic resonance A120, A227,
A281, A417, B763−765, B964−967
Nucleic acids
anion-exchange chromatography of B915
capillary electrophoresis of B921−932
column liquid chromatography of B909,
B912−913, B916−917
detection of B932−935
ion-exchange chromatography of B913−915
ion-pair chromatography of B917−921
quantitation methods B912−921
preparation of B908−912
sequencing A439−442
temperature-gradient electrophoresis of
B922−923
Nucleosides B907, B916−917, B918, B930
Nucleotides B907, B913−914

O

Occupational hygiene B543
Occupational toxins B1108−1109
Ochratoxin A B1110
Octadecylsilica sorbents B857−859
Odoroside B1105
Off-line separations A283−284
Oleandrin B1105
Oligo-amides A235
Oligomeric additives A301

Oligonucleotides B908−909, B1117−1118
Oligosaccharides
high-performance anion-exchange
chromatography of B870−872,
B873−874
reversed-phase liquid chromatography of
B860
size-exclusion chromatography of
A239−240
Omeprazole B1108
On-capillary complexing B567
On-column derivatization B595, B597
On-column injection A343−346
On-line sample pre-treatment A206−207
On-line solid-phase extraction A409−412,
B949−951, B993
Open capillaries A301−302, A306
Open-tube capillary electrophoresis
A384−385
Opiates B1075−1077, B1078−1079
Opioids B1075−1081
Optical isomerism B530
Optical spectroscopy A279−280
Optimization A72−88, A128
Optimum velocity A21−27, A31−33
Organic pollutants B998−1001
persistent B1007−1011
Organoarsenic B559
Organochlorine pesticides B991, B996
Organolead B534−535
Organomercuric halides B535
Organomercury B534−535
Organometallics B532−536, B556−559,
B572−575
Organophosphate B1109−1110
Organophosphorus B996
Organotin B522, B535−536
Orientation forces A325
Orthogonal-acceleration reflectron time-of-
flight A420
Orthophosphate B544
Osmium B523, B525
Overlay binding B888−891
Overpressured-layer chromatography
A256−257, A273−274, A288
Oxalate B539, B544
Oxazepam B955
Oxidized acyl chains B810−811
Oxo-acylglycerols B797−799

Oxo-fatty acids B782–787
Oxo-halides B540
Oxo-phospholipids B809–811
Oxo-triacylglycerols B768–769
Oxprenolol B959
Oxyphenbutazone B1107

P

Packed-bed suppressors A177–178
Packed capillaries A301–302, A303–304,
        A306–310
Packed columns A327
Packing materials B612–613
Paclitaxel B955, B969
Palladium B523, B526, B527
Pancuronium B1101–1102
Pantopraxole B1108
Papaverine B1075
Paracetamol B957, B965–966, B975,
        B1104
Paraquat B1014
Passive sampling B990–991
Poly(chlorinated biphenyls) B1009
Polymerase chain reaction-based technologies
        B929
Peak capacity A215
Peak dispersion A305, A307–310,
        A312–313
Peak identification B753–765
Pectins B872–873, B879
Pentaerythritol tetranitrate B1111
Pentamidine B952
Pentanesulfonate B563
Pentose B906
Pentylated α-cyclodextrin B885–886
Peptides
    absolute mobility B655–657
    affinity chromatography of A159–160
    capillary electrophoresis of B618–620,
        B647–660
    capillary electrochromatography of
        B615–620
    gel electrophoresis of B634–647
    hydrolysis B588–589
    isoelectric focusing B636, B637
    liquid chromatography of A423–424,
        B611–619
    mapping B709

    microfabricated devices for A455
    tetrameric A159
Perbenzoylation B857–858
Perchlorate B543–544, B1111
Performance criteria A72–74
Perfusion chromatography B697
Periodate oxidation B845–846
Permanent-coating ion-pair chromatography
        B546
Permeability A216–217, A322
Personal-care products B1018–1022
Pesticides
    column/column liquid chromatography of
        A411–412
    gas chromatography of A407, B1011–1012
    liquid chromatography/mass spectrometry of
        A423, B1012–1013
    quantitative analysis of B1011–1014
    size-exclusion chromatography of
        A240–241
    thermal lability B1012
    in water B992
    in wine A409
Petroleum A334
Petroleum fractions A409
pH gradients B636–640, B641, B719–720
Pharmaceuticals A421–422, B1018–1022
Phase ratio A27–29, A321–322
Phase types A327–331
Phenacetin B955, B957, B966–968
Phenazone B975
Phenethylamines B1085–1087, B1089–1091
Phenobarbital B1076
Phenolic compounds B1050–1055
Phenolphthalin B1076
Phenylalanine B552, B605–607
Phenylarsenic oxide B575
Phenylarsonate B559, B574–575
Phenylboronic acid A157
Phenylbutazone B1107
Phenylmercuric acetate B574
Phenylpropanes B1052–1053
Phenylpropanolamine B1086
Phenyl Sepharose B709–710
Phenylurea B992, B1013
Phenytoin B952
Phosfamide B966
Phosphate B539, B544, B563, B564
Phosphatidylcholine B746–747

Phosphatidylethanolamine B745–747
Phosphatidylglycerol B745–747
Phosphatidylinositols B745–747, B758–759, B773–778
Phosphatidylserine B745–747
Phosphine B523–524
Phosphite B563
Phospho-amino acids B608
Phospholipase C B751–752
Phospholipase D B822
Phospholipids B742–743, B769–777
Phosphorus B522–523, B525
Photodiode array A223–224, A416, B844
Photolithography A432–433
Photon correlation spectrometry A233
Phthalates A409, B539, B1023, B1111
Phthalic acid B545
Phytic acid B747
Phytochemical analysis B1037–1061
   alkaloids B1057–1059
   micro-separation B1057–1058, B1060–1061
   phenolic compounds B1050–1055
   sample preparation B1038–1040
   terpenoids B1040–1050
   tocopherols B1054–1056
Piperacillin B951
Piperidine alkaloids B1059
Planar chromatography A253–291
   chambers A262–266
   development modes for A267–272
   flame-ionization detection A285
   flow velocity in A266–267
   instrumentation for A273–278, A287, A289
   layer-thickness gradients A286–287
   mobile-phase in A255, A286
   preparative methods in A282–285
   principles of A257–273
   qualitative analysis A277–281
   quantitative analysis A282
   sample application for A260, A287, A289
   stationary phases for A255, A258–260
   vapor phase in A262–266
Plants B1037–1044, B1075–1076
Plasma lipoprotein phospholipids B789
Plasticizers B1023–1024
Plate heights A3–6, A20–33, A322
Plate number A7–8, A20–33, A88
Platinum B523, B527, B572

Plutonium B525
Poiseuille flow A15–16
Poisson-Boltzmann equation B715
Polarizability A39–40
Polyacrylamide gel electrophoresis B640–647
Polyacrylamide polymer sorbents B691–692
Polybrominated compounds B1000
Poly(brominated diphenyl ethers) B1010–1011
Poly(chlorinated biphenyls) A339, B991, B996, B1005, B1007–1010
Polychlorinated compounds B1000
Polycyclic aromatic hydrocarbons A455, B991, B992, B996, B1005
Poly(dimethylsiloxane) A434–435, A453–454
Polyesters A236
Polygalacturonic acids B872
Polymer additives A313–314
Polymerase chain reaction A439–440, B929, B1116–1117
Polymeric packing columns B956–957
Polymers A230–237, A379, A387–388, A434
Poly(methyl methacrylate) A453–454
Polymorphisms B920, B929, B1115–1120
Polypeptides A237–238
Polyphosphates B544
Polysaccharides B876–880
Polystyrene sorbents B691–692
Polystyrene standards A69–70
Polystyryllithium A234–235
Polyterpenes B1049–1080
Polythionates B547–548
Polyunsaturated fatty acids B765–766
Polyvinylstyrene sorbents B691–692
Pore size A69–70, B688–689
Porous beds A303–304
Porous graphitic carbon A110
Position- and time-resolved ion counting B875
Post-column derivatization A279, B595–596
Post-column reaction detection A185
Post-translational modifications B675–677
Potassium B523, B550, B552, B569–570
Praseodymium B553, B569
Pre-capillary complexing B567
Pre-column derivatization A279, B595–596, B605–606
Pre-concentrator columns A204–205
Preferential-interaction parameters B702–703
Preparative chromatography A282–285, B1040
Pressure A84, A315–316, A332–333

Pressurized capillary electrochromatography B930

Progestogens B1019

Programmed-temperature vaporizer A345–346

Promethazine B949, B958

Propane B1098

Propanesulfonate B563

Propionate B539, B563

Propranolol B952, B958, B959

Proscilardin B1105

Prostaglandins B782–783

Prostanoids B782–787

Proteases B678–679

Proteins
    capillary electrophoresis of B647–655, B658–659
    chromatofocusing of B718–720
    chromatography of B669–728
    folding/unfolding B658–659
    free-zone electrophoresis of B647–660
    gel electrophoresis of B634–647
    hydrophobic-interaction chromatography of B697–709
    intrinsic viscosities B673
    ion-exchange chromatography of B712–720
    liquid chromatography/mass spectrometry of A420, A423–424
    microfabricated devices A442–446, A447
    microheterogeneity of B674–684
    normal-phase liquid chromatography of B686–697
    precipitation of B947–948
    rapid analysis of A442–447
    retention factor B720–721, B722
    reversed-phase liquid chromatography of B697, B703–705, B707, B708–712
    salting-out chromatography of B697–703
    size-exclusion chromatography of A237–238, B720–726
    self-interaction chromatography of B673–674
    stability B685–686
    structure B670–674

Protein/sodium dodecylsulfate complexes B641

Proteomics B726–727
    microfabricated devices A442–446, A447
    size-exclusion chromatography A238
    software resources B1153–1154

Protocatechuic acid B545

Pseudo-cell size-exclusion chromatography/ mass spectrometry A225

Pseudoephedrine B1086, B1107

Psilocin B1092

Psilocybin B1092

Pulsed-field gel electrophoresis B922

Pulse dampers A477–478

Pumps
    for column liquid chromatography A100–101, A102, A475–478
    for microfabricated devices A451–453, A456

Purine B932

Purnell Equation A6

Pyrethroid pesticides B996

4-(2-Pyridylazo)resorcinol B551

Pyrimidine B932

Pyruvate B539

Pyruvic acid B545

Q

Quadrupole mass spectrometry A419, A501, B1012

Qualitative analysis A277–281

Quantitative analysis A282, A422–423, B753–765, B912–921, B1004–1024

Quantitative structure/activity relationships A127–128

Quercetin B967, B1054, B1055, B1061

Quinate B539

Quinine B1058

R

R_F value A269, A277–278

hR_F value A278

R_M value A278

Rabeprazole B1108

Radio-immuno-assay A143, A144

Raman spectroscopy A227, A280

Ranitidine B1108

Raoult's Law A45

Rare earths B554, B569–571

Rate theory A9, A17

Reboxetine B949

Recombination proteins B711

Reducing agents B646–647

Refolding B725

Refractive-index detector A118
Regeneration phase A141–142
Regression methods A77–79
Resistance-to-mass transfer A19–20, A26, A28–30
Resolution A2–9, A38, A73–74, A88
Response factor A349–351
Restricted access materials A411–412, B951, B992–993
Resveratrol B1052, B1054–1055, B1061
Retention factor
    electrokinetic chromatography A299, A304–305, A311
    hydrophobic-interaction chromatography B697–703
    ion chromatography A197
    ion-exchange chromatography B712–717
    multivariate optimization A85
    resolution A7–9
    reversed-phase liquid chromatography B703–705
    size-exclusion chromatography A220, B720–722
    solute-zone dispersion A28–31
    theory of A3–9, A88
    univariate optimization A82–83
Retention time A3–9
Retention volume B714
Reversed-phase liquid chromatography A98, A105–107, A124–128, B751, B792–798, B855–862
    purification by B709–712
    retention in B703–705
    sorbents for B707–708
Reynolds number A14
Rhenium B525
Rhodium B523, B526–527, B530
Ribonucleic acid A439, B906–907, B924
Rice-Whitehead Equation A17
Rocuronium B1101–1102
Rohypnol B1098
Roridin B1044–1045
Rotation planar chromatography
    classification A255
    forced-flow A257–258, A266–268
    instrumentation for A274–276, A288
Roxifiban B966
Rubber B1049–1080
Rutin B1055

S

Salbutamol B948
Salicylamide B967
Salicylic acid B967
Salting-out chromatography B697–703
Samarium B569
Sameridine B971
Sample handling B540–543
Sample injection A448–450
    in gas chromatography A340–347, A510
    in liquid chromatography A478–480
    in microfabricated devices A452
Sample preparation
    amplification B911
    integrated systems B911–912
    pre-treatments A206–207
    purification B910–911, B1038–1040
Sample Concentrator and Neutralizer B542
Sapogenins B1047–1048
Sarin B1109–1110
Screening A447–451, B1102–1105
Selected-ion monitoring B886
Selectivity
    in electrokinetic chromatography A307
    factor A5–9, A88
    in gas chromatography A348
    in ion chromatography A186–203
Selegiline B1089
Selenate B557
Selenides B526
Selenite B539, B557
Selenium B522, B524, B532, B536
Selenium trioxide B531, B564
Seleno-amino acids B610
Selenocystamine B574
Selenocysteine B588, B610
Selenocystine B557, B558, B574
Selenoethionine B558
Selenomethionine B557, B558, B574, B610
Self-interaction chromatography B673–674
Self-nucleation A232–233
Sensitivity
    of detectors A497
    in electrokinetic chromatography A315–316
    in gas chromatography A348
Separation A272–273
    chromatographic theory A72–88
    multivariate methods A84–88

parameters A74
performance criteria A72–74
regression methods A77–79
sequential methods A75–77
simultaneous methods A75
theoretical methods A78
univariate methods A79–84
Sequence determination B588–589
Sequence-specific oligonucleotide probes
        B1117–1118
Sequential methods A75–77, A286
Sesquiterpenes B1042, B1044–1045
Sevoflurane B1097
Short tandem repeats A439, B929,
        B1115–1120
Sialic acids B869–870, B873
Silanes B523–524
Silanization A105–106
Sildenafil B954, B1108
Silica-based anion exchangers B863–864
Silica-coated packings A217–218
Silicon B522
Silver B523
Simazine B1014
Simendan separations B959
Simulated moving-bed separations A130–131
Sinapic acid B1052–1053
Single-cell analysis A457–458, A460
Single-locus probes B1116–1117
Single-nucleotide polymorphism A457, A515,
        B925, B1120–1121
Size-exclusion chromatography A213–241
    aggregate separations B721–724
    applications of A230–241
    calibration of A228–230
    columns for A217–219
    desalting B723–725
    detectors for A222–227, B725–726
    inverse B688–689
    mobile phases for A219–222
    retention mechanisms in B720–722
    solute-zone separation A68–70
    sorbents for B720–722
    theory of A215–217
Size-exclusion electrochromatography A233
Sodium B552
Sodium chloride B523, B550, B569–570
Sodium dodecylsulfate/polyacrylamide gel
        electrophoresis B640–643

Solanine B1058
Solid-phase extraction
    for biological sample preparation B948–954
    for gas chromatography A409–410
    for ion chromatography A205–206,
        B541–542
    for liquid chromatography B541–542
    on-line A409–412, B949–951, B993
Solid-phase micro-extraction B993–996
    for biological sample preparation B951
    for gas chromatography A346–347
    for liquid chromatography B995
    instrumentation A511
Solubility parameters A47–54
Solute polarity A50
Solute-specific detectors A497, A502–504
Solute-zone dispersion A9–38
Solute-zone separation
    adsorption A54–60
    chromatographic theory A38–72
    complexation A60–63
    hydrodynamic chromatography A67–68
    ion exchange A64–67
    ion-exclusion chromatography A64–67
    ion pairing A66–67
    molecular interactions A39–43
    partition A47–54
    size-exclusion chromatography A68–70
    thermodynamics A43–47
Solvating gas chromatography A324–325
Solvents
    for column liquid chromatography
        A100–101
    delivery A100–101
    exchange B883
    extraction B742–744
    migration A255
    polarity parameters A123
    strength A57–58, A327
    system optimization A261–262, A263
Sorbents
    composite sorbents B693
    for hydrophobic-interaction chromatography
        B706–707
    for ion-exchange chromatography
        B717–718
    for normal-phase chromatography
        B686–695

for reversed-phase liquid chromatography
  B707–708
for size-exclusion chromatography
  B720–721, B722
polymer-based B690–692,
  B861–862
silica-based B847–855
unmodified silica B848–850
vinyl polymer B691–692
Southern blotting B1116–1117
Soybean oil B797–798
Spectrofluorometers A503
Spectroscopy
  atomic A121, B540
  atomic emission B521–522
  infrared A120, A280, A415–417,
    B763–765
  nuclear magnetic resonance A417
  photodiode-arrays A416
  photon correlation A233
  optical A279–280
  Raman A227, A280
  stopped-flow B964
Sphingolipids B767
Sphingomyelins B778
Split/splitless injection A341–343, A344
Squalene B1047, B1048
Starch B879
Static light scattering B725–726
Stationary phases
  alumina-based A109–110
  chiral A111–114
  cellulose-based A112–113
  composition of A80–82, A85–88
  co-ordination complexes B529–530
  covalently-bonded A381
  hybrid silica A107–109
  for hydrophobic-interactions B706–707
  molecularly imprinted B615
  macrocyclic A114
  materials A104–115
  for normal-phased liquid chromatography
    B688–689, B694–695
  polymeric A110
  pseudo-stationary phases A310–315
  resistance to mass transfer A20
  silica-based A104–109, B853–855
  titania based A109–110
  zirconia-based A109–110

Steric mass action B712
Sterol esters B799–800
Sterols B799–800, B1047–1048
Stilbenes B1052, B1054–1055, B1061
Stimulants B1106
Stoichiometric displacement models B712
Stokes-Einstein Equation A13
Stokes' Law B657
Stokes radius A229–230
Strontium B569–570
Strophantidin B1105
Strychnine B1058
Succinic acid B545
Succinylcholine B1102
Sugars B845, B883, B885
Sulfate B539, B544, B549, B563, B1111
Sulfides B526, B549
Sulfite B549
Sulfonamides B967
Sulfonylurea B1100–1101
Sulfur A362, B522, B523, B525, B527–529,
    B542, B564, B609–610
Sulfur dioxide B524
Sulfur hexafluoride B525
Sulfur hydrides B524
Sulfur oxides B524, B550, B564
Sunscreen agents B1022
Supercritical-fluid chromatography
  instrumentation for A512
  solute-zone separation A51, A53–54
Supercritical-fluid extraction B743–744
Suppressed-conductivity detectors A119
Suppressors A176–180
Surface-enhanced Raman spectroscopy (SERS)
    A227
Surfactants
  anionic B1014–1015
  non-ionic B1015–1018
  solid samples B998–999
Suxamethonium B1102
System integration A451–461

T

Tacrolimus B949
Tamoxifen B952
Tantalum B558
Tartaric acid B545
Taurine B609–610

Taxol B1044, B1046, B1105
Taylor dispersion A306
Technicium, B525
Tellurium B522, B524
Temperature-gradient gel electrophoresis B922–923
Temperature-gradient interaction chromatography A234–235
Terbium B569
Terpenoids B1040–1050
Testosterone B1106
Tetrachlorodibenzodioxins A413
Tetrachloroethylene B1098
Tetradentate ligands B527
Tetraethyllead chloride B574
Tetrahydrocannabinol 9-carboxylic acid B1086–1087, B1093–1097
Tetramethylarsonium B557
Tetramethyllead chloride B574
Tetraterpenes B1048, B1049
Thalidomide B958
Thebaine B1058, B1075–1076
Theobromine B1107
Theophylline B952
Therapeutic drugs B1098–1102
Therapeutic proteins A148, B685
Thermal energy analyzer A362
Thermal lability B1012
Thermocapillary pumping A451–452, A456
Thermocycling A440–442
Thermodynamics A43–47, A150–151
Thermospray A500
Thiabendazole B1001
Thiazide diuretics B1108
Thin-layer chromatography A255–291, A511–512
Thiocarbamide B545
Thiocyanate B544, B564, B1111
Thiols B646–647
Thionyl chloride B524
Thiosulfate B539, B547, B548, B563
Thioxanthates B528
Thorium B557
Thulium B569
Thyroid-stimulating hormone releasing factor B594
Time constants A37–38
Time-resolved ion counting B875
Time slicing B964

Tin B521, B522, B524, B532
Tissue B1086–1089
Titanium B523, B525, B532
Tocopherols B766, B1054–1056
Tolbutamide B1101
Tolfenamic acid B965
Toluene B1097–1098
Tomatine B1058
Toxins B890, B1105, B1108–1109
Tramadol B952, B972
Transition metals B522
   as catalysts A232
   gas chromatography of B525
   ion-pair chromatography of B551, B555
   solute-zone separation A62–63
Triacylglycerols B746–747, B755–760, B788–789, B793–799
Triazines B992, B993, B1001–1002, B1013
Triazolam B1100
Tributyltin B558, B574
Tricarballylate B539
Trichlormethiazide B1108
Trichloroacetate B544
1,1,1-Trichloroethane B1098
Trichothecenes B1042, B1044–1045
Triclosan B999
Trifluoroacetate B539
Trifluorperazine B966
Trifluralin B1013
Trihalomethanes B996
Trihexyphenidyl B951
Trimethoxycinnamic acids B1052–1053
Trimethylamine B552
Trimethylarsine oxide B557
Trimethylsilylated oxime derivatives B882–884
Trimethylstilboxide B557
Trimethyltin B574
Trimipramine B948
Trinitrobenzene B1111
Trinitrotoluene B1111
Triphenyltin B558
Tripolidine B973
Triterpenes B1047–1048
Tropane B1059
Tryptic protein fragments B648–653
Tryptophan B607–608
TSK gel Amide-80 sorbent B853
Tubocurarine B1102

Tungstate B539, B563
Tungsten B522, B525
Turbulent flow A16–17
Turbulent-flow chromatography A412, B953,
        B1004
Two-dimensional gas chromatography
        A337–340, A405–407, B1005–1006
Two-dimensional high-performance liquid
        chromatography B727–728, B862
Two-dimensional polyacrylamide gel electro-
        phoresis B643–647
Tyrosine B552
Tyrosinemia B605–608

U

Ultra-thin-layer chromatography A511–512
Ultra-thin-layer gel electrophoresis B925
Ultraviolet detectors B932–933
    for column liquid chromatography A116,
        A117
    for ion chromatography A184–185
    for size-exclusion chromatography A223
Ultraviolet filters B1022
Univariate methods A79–84
Universal calibration curves A234
Uranium B525, B570
Uranyl B557
Urea shift method B924

V

Valerate B539, B563
Validation A315
Vanadium B525–528, B558
Van Deemter Equation A324, A336–337
Van der Waals interactions B716
Van't Hoff Equation A44–45
Van't Hoff plots A307
Vaporizers A345–346
Variance A17, A18, A20, A33–38
Vario-KS chamber A265–266
Vecuronium B1102
Vegetable oils B798–799
Verapamil B958, B959
Veratridine B1105
Verrucarin B1044, B1045
Viagra B1108
Vincamine B958

Viruses B890
Viscosity A323–324, B673, B877

W

Wall-adsorption-related dispersion A313
Warfare agents B1109–1110
Warfarin B958–959
Washing phase A141
Water
    contaminants B991
    drinking B543
    environmental analysis B543,
        B989–994
    fungicides in B993
    gas chromatography B524
    herbicides in B993
    passive sampling of B990–991
    pesticides in B992
    phthalates in A409
    triazines in B992
Weak-acid analytes A200
Weak-acid anions B549
Whey proteins B709, B710
Wilke–Chang Equation A13
Window diagrams A77–78
Wines A409

X

Xanthates B528
Xanthophylls B1048–1049
X-ray lithography A433
Xylenes B1097

Y

Y-chromosome-specific polymorphism
        B1119–1120
Yeast B683
Ytterbium B523, B569

Z

Zeta potential A302, A306, A375
Zinc B525, B527, B551–552, B569–570
Zinc phosphide B524
Zirconium B525
Zone compression A203–204

JOURNAL OF CHROMATOGRAPHY LIBRARY

A Series of Books Devoted to Chromatographic and Electrophoretic Techniques and their Applications

Although complementary to the Journal of Chromatography, each volume in the library series is an important and independent contribution in the field of chromatography and electrophoresis. The library contains no material reprinted from the journal itself.

Recent volumes in this series

Volume 60   **Advanced Chromatographic and Electromigration Methods in BioSciences**
edited by Z. Deyl, I. Mikšík, F. Tagliaro and E. Tesařová

Volume 61   **Protein Liquid Chromatography**
edited by M. Kastner

Volume 62   **Capillary Electrochromatography**
edited by Z. Deyl and F. Svec

Volume 63   **The Proteome Revisited: theory and practice of all relevant electrophoretic steps**
by P.G. Righetti, A. Stoyanov and M.Y. Zhukov

Volume 64   **Chromatography - a Century of Discovery 1900-2000: the bridge to the sciences/technology**
edited by C.W. Gehrke, R.L. Wixom and E. Bayer

Volume 65   **Sample Preparation in Chromatography**
by S.C. Moldoveanu and V. David

Volume 66   **Carbohydrate Analysis by Modern Chromatography and Electrophoresis**
edited by Z. El Rassi

Volume 67   **Monolithic Materials: Preparation, Properties and Applications**
by F. Svec, T. Tennikova and Z. Deyl

Volume 68   **Emerging Technologies in Protein and Genomic Material Analysis**
by G. Marko-Varga and P. Oroszlan